PRINCIPLES OF WIRELESS ACCESS AND LOCALIZATION

PRINCIPLES OF WIRELESS ACCESS AND LOCALIZATION

Kaveh Pahlavan

Worcester Polytechnic Institute, Worcester, Massachusetts, USA

Prashant Krishnamurthy

University of Pittsburgh, Pittsburgh, Pennsylvania, USA

© 2013, John Wiley & Sons, Ltd

Registered office
John Wiley & Sons Ltd, The Atrium, Southern Gate, Chichester, West Sussex, PO19 8SQ, United Kingdom

For details of our global editorial offices, for customer services and for information about how to apply for permission to reuse the copyright material in this book please see our website at www.wiley.com.

Library of Congress Cataloging-in-Publication Data has been applied for.

A catalogue record for this book is available from the British Library.

ISBN 978-0-4706-9708-5 (hardback)

Set in 9.5/11.5pt Times by Aptara Inc., New Delhi, India
Printed and bound in Malaysia by Vivar Printing Sdn Bhd

1/2013

*To our wives Farzaneh and Deepika and
our children Nima, Nasim, Shriya and Rishabh*

Contents

Preface

Engineering disciplines are going through a "transformation" from their traditional focused curriculum to a "multi-disciplinary" curriculum and "inter-disciplinary" research directed toward innovation and entrepreneurship. This situation demands more frequent updates and adjustments in the curriculum, project-oriented delivery of educational content, and the ability to form inter-disciplinary cooperation in research programs. A successful transformation of this form demands entrepreneurship and visionary talents to adapt to these frequent changes and industrial experiences to direct the transformation toward emerging inter-disciplinary industries. Wireless access and localization is an excellent example and one of the flagships of a multi-disciplinary area of research and scholarship, which has emerged in the past few decades. Material needed for teaching wireless access and localization includes several disciplines such as signal processing, digital communications, queueing theory, detection and estimation theory, and navigation. The content of courses on wireless access and localization are useful for traditional Electrical and Computer Engineering (ECE) and Computer Science (CS) students as well as students in emerging multi-disciplinary programs such as Robotics and Biomedical Engineering and traditional Mechanical and Civil Engineering programs, which are similar to ECE, shifting toward inter-disciplinary curriculums. Cyber physical systems play an important role in the future of these multi- and inter-disciplinary engineering programs and wireless access and localization is essential in the integration of all of these systems. Therefore, there is a need for academic courses and a comprehensive textbook to address the principles of wireless access and localization to be taught in these multi-disciplinary programs.

To prepare a textbook to be taught in academic courses in a multi-disciplinary area of technology, we need to provide selected details of practical aspects of a number of disciplines to give to the readers an intuitive feeling of how these disciplines operate and interact with one another. To achieve this goal in our book, we describe important wireless networking standards and localization technologies, classify their underlying science and engineering in a logical manner, and give detailed examples of successful science and engineering that has turned into popular applications. Selection of detailed technical material for teaching courses in a multi-disciplinary area with a large and diversified set of technical disciplinary is very challenging and these challenges become more defying in teaching wireless access and localization because in this area the emphasis of the skills needed to be taught in the course shifts in time.

The success of wireless information networks in 1990's was a motivation behind a series of textbooks describing wide and local area wireless networks [Pah95, Goo97, Wal99, Rap03, Pah02]. The technical focus of these books was on describing wide area cellular telephone networks and local wireless data networks. These books were written by professors of Electrical and Computer Engineering with different levels of emphasis on detailed description of the lower layers issues and system engineering aspects describing details of implementation of wireless networks. Wireless localization has gained significant importance in the past decade and these books do not lay emphasis on the details of wireless localization techniques. As a result, currently, there is no single textbook that integrates wireless access and localization. Wireless access and localization are extremely interrelated in applications and fundamentals of design and operation. Understanding of these technologies have tremendous amount of similarities in the implementation of the physical layer and in the understanding of fundamentals of the radio propagation in the environment.

This book provides a comprehensive treatment of the wireless access and localization technologies. The novelty of the book is that it places emphasis on radio propagation and physical layer issues related to the formation and transmission of packets as well as how the received signals can be used for RF localization in a variety of networks. The structure and sequence of material for this book was first formed in a lecture series by the principal author at the graduate school of the Worcester Polytechnic Institute (WPI), Worcester, MA entitled "Wireless Access and Localization". The principal author also taught shorter versions of the course focused on either of the two topics at different conferences and universities. The co-author of the book has taught material from this book at the University of Pittsburgh for first-year graduate and junior/senior undergraduate students in information science and telecommunications.

We have organized the book as follows: we begin with an overview of the evolution of wireless access to public switched telephone network (PSTN) and the Internet for voice-oriented and data-oriented information and an overview of wireless localization techniques followed by four parts each including several chapters. Part I contains chapters 2 to 4 and explains the principles of design and analysis of physical layers of wireless networks. In chapter 2, we begin this part by describing multipath characteristics of radio channel in indoor and urban areas, where all wireless access and localization techniques used in emerging smart wireless devices are applied. Then we explain how multipath arrival of the signal affects waveform transmission for wireless access and localization. In chapters 3 and 4, we discuss how bits are transmitted and how packets of information are formed for transmission, respectively. Part II of the book is devoted to principles for design of wireless network infrastructure. Three chapters of this part, chapters 5–7, cover deployment, operation, and security of these networks, respectively.

Part III is devoted to wireless local access technologies. Three chapters in this part cover traditional wireless local area networks (chapter 8) as well as low-power sensor technologies (chapter 9) and technologies striving for gigabit wireless access (chapter 10). Part IV of the book describes technologies used for wide area wireless cellular networks with three chapters addressing TDMA technology (chapter 11), CDMA technology (chapter 12), and OFDM/MIMO technologies (Chapter 13) employed in 2G, 3G, and 4G cellular networks, respectively. Part V covers wireless localization techniques with three chapters describing systems aspects (chapter 14), principles of wireless localizations (chapter 15), and practical aspects (chapter 16) of these technologies.

The partitioned structure of the book allows flexibility in teaching the material that is essential when it is used in different disciplines. We believe that the most difficult part of the book for the students is chapters 2–5 and chapters 15 and 16, which provide a summary through mathematical description of numerous technologies and algorithms. The rest of the chapters of the book appear mathematically simpler but carry more details of how systems work. To make the difficult parts simpler for the students, an instructor can mix these topics as appropriate. For example, the lead author teaches similar material in one of his undergraduate courses in wireless networking by first introducing the channel behavior (chapter 2), then describing assigned access methods (chapter 4) before describing TDMA cellular networks (chapter 11). Then he introduces spread spectrum modulation and coding techniques (chapters 3) and CDMA cellular networks (parts of chapters 4 and 12), and at last he covers multi-dimensional constellations (chapter 3) before he discusses wireless LANs (chapter 8). His new graduate-level course on wireless access and localization mostly covers chapters 1–5 and chapters 14–16 in depth.

In fact, we believe that this is an effective approach for enabling the understanding of the fundamental concepts of wireless access and localization in students. Therefore, depending on the selection of the material, depth of the coverage, and background of the students and the instructor, this book can be used for senior undergraduate or first- or second-year graduate courses in CS, ECE or Robotics, Biomedical, Mechanical or Civil Engineering as one course or a sequence of two courses.

The idea of writing this book first came to the authors in 2007 because of the need for a revision for the authors' previous book, Principles of Wireless Network – A unified Approach, and expanded that to include emerging wireless localization techniques. When the book was completed just before 2013, it was substantially different from the previous book and we decided to publish it as an independent book with a more relevant title: Principles of Wireless Access and Localization.

Much of the writing of the lead author in this book was accomplished during his sabbatical leave from Worcester Polytechnic Institute, Worcester, MA at School of Engineering and Applied Science of the Harvard University, Cambridge, MA during the spring semester of 2011. He would like to express his deep appreciation to the Worcester Polytechnic Institute and the Harvard University for providing him this opportunity. In particular, he thanks Prof. Vahid Tarokh of the Harvard University for his timely arrangement of the visit and Dean Cherry A. Murray of the Harvard School of Engineering and Applied Sciences for granting the visit. Also, he thanks Prof. Fred Looft, Head of the WPI ECE Department, and Provost John A. Orr of WPI at that time for their support of his sabbatical leave for the work on this project.

Much of the new material in localization and body area networking are extracted from the research work of the students at the Center for Wireless Information Network Studies (CWINS), WPI. We are pleased to acknowledge the students' and colleagues' contributions to advancing the understanding of wireless channels and its application in wireless access and localization techniques. In particular, the authors would like to thank Dr. Xinrong Li, Dr. Bardia Alavi, Dr. Nayef Alsindi, Dr. Mohammad Heidari, Dr. Ferit Akgul, Dr. Muzzafer Kanaan, Dr. Yunxing Ye, and Umair Khan of the CWINS, Prof. Sergey Makarov of WPI, Prof. Pratap Misra of Tufts University, and Mr. Ted Morgan and Dr. Farshid Alizadeh of Skyhook Wireless, who have directly or indirectly helped the authors to extend their knowledge in this field and shape their thoughts for the preparation of the new material in this book. We owe special thanks to the National Science Foundation (NSF), Defense Advanced Research Projects Agency (DARPA), National Institute of Standards and Technology (NIST), Department of Defense (DoD), and Skyhook in the United States as well as Finnish Founding Agency for Technology and Research (TEKES) and Nokia in Finland, whose support of the CWINS program at WPI enabled graduate students and the staff of CWINS to pursue continuing research in this important field. A substantial part of the new material in this book has flowed out of these sponsored research efforts.

The authors also would like to express their appreciation to Dr. Allen Levesque, for his contributions in other books with the lead author, which has indirectly impacted the formation of thoughts and the details of material presented in this book. The authors also acknowledge the indirect help of Prof. Jacques Beneat of Norwich University, VT, who prepared the solution manual of our other book, Principles of Wireless Networks – A Unified Approach. A significant number of those problems, and hence their solutions, are used in this book. They also thank Drs. Mohammad Heidari and Yunxing Ye and Bader Alkandari who are preparing the solution for this book based on the solutions in the previous book and Guanqun Bao and Bader Alkandari for their careful review of several chapters. The second author expresses his gratitude to Drs. Richard Thompson,

David Tipper, Martin Weiss, and Taieb Znati of the Graduate Program in Telecommunications and Networking at Pitt. He has learnt a lot and obtained different perspectives on networking through his interaction and association with them. Like the lead author, he would like to thank his current and former students who have directly or indirectly helped him to extend his knowledge in this field and shape his thoughts for the preparation of the new material in this book. Similarly, we would like to express our appreciation to all graduates and affiliates of CWINS laboratory at WPI and many graduates from the Telecommunications Program at Pitt whose work and interaction with the authors have directly or indirectly impacted the material presented in this book.

We have not directly referenced our referral to several resources on the Internet, notably Wikipedia. While there are people who question the accuracy of online resources, they have provided us with quick pointers to information, parameters, acronyms, and other useful references, which helped us to build up a more comprehensive and up-to-date coverage of standards and technologies. We do acknowledge the benefits of these resources.

The authors also would like to thank Mark Hammond, Sarah Tilley, and Sandra Grayson of John Wiley & Sons for their assistance and useful comments during various stages of the production of the book and Shikha Jain of Aptaracorp for her help during the manuscript proofs. Finally, we would like to thank John Wiley & Sons for hosting the book's website at: http://www.wiley.com/go/pahlavan/principles.

1

Introduction

1.1 Introduction

Technological innovations by engineers during the past century have brought a deep change in our life style. Today, when we fly over a modern city at nighttime, we see a planet full of the footprints of the modern civilization made by engineers. The glowing lights below remind us of the impact made by electrical engineers, the planes we fly in and the moving cars under them remind us of the contributions of mechanical engineers, and high rise buildings and complex road systems remind us of what civil engineers have done. Through the eyes of an engineer, the glow of light, the movement of cars, and the complexity of civil infrastructure display the challenges in implementation and the size of the market for this industry and demonstrate the impact of this technology on human life. There is one industry, whose infrastructure is not seen from an airplane because it is mostly buried under the ground, but it is the most complex, it owns the largest market size, and it has enabled us to change our life style by entering the age of information technology. This industry is the *information networking industry*.

Perhaps the most prominent feature of the human species over other living species on the earth is the ability to create a sophisticated linguistic that allows us to generate information based on our experiences in life and to communicate that with others, store them in writing, and retrieve them by reading. As a result, while other species have little knowledge of their peers' experiences in other places or even living close to them, our lives are based on the retrieval of cumulative information that has been collected and stored over several thousands of years around the world. The availability of this vast treasure of information has allowed us to create an advanced civilization that is by far above the other species living on planet earth. Therefore, the availability of information has been the most important factor in the growth of our civilization. Information networks facilitate the transfer of information across the world. In the same way that highway systems facilitate the physical transfer of merchandise and people across the continents to nurture economic growth, information networks facilitate the transfer of merchandise descriptions and human thoughts to stimulate the economy. Highway systems facilitate their physical presence in diversified locations and information networks facilitate the close to instantaneous virtual presence of information about them in diversified locations. The importance of existent of information in diversified locations in the growth of our economies has resulted in huge investments in the infrastructure for information networking and the emergence of this industry as the largest industry made by engineers.

Principles of Wireless Access and Localization, First Edition. Kaveh Pahlavan and Prashant Krishnamurthy.
© 2013 John Wiley & Sons, Ltd. Published 2013 by John Wiley & Sons, Ltd.

To have an intuitive understanding of the size of the information industry, it is illustrative to notice that the size of the budget of American Telephone and Telegraph (AT&T) Corporation in the early 1980s, before its divestiture, was close to the budget of the *fifth largest economy* of the world at that time. AT&T was the largest telecommunication company in the world and its core revenue at that time was generated mainly from wired connections to the public switched telephone network (PSTN) just for the basic telephone call application that was first patented in 1876. During the past three decades, the cellular telephone industry augmented the income of the prosperous circuit-switched telephone services with subscriber fees from approximately seven billion cellular telephone users worldwide. Today the income of the wireless industry has already surpassed the income of the wired telephone industry and this income is still dominated by the revenue from cellular telephone calls for wireless access to the PSTN and their recurring subscriber fees.

In the mid-1990s the Internet brought the data-oriented packet switched computer communication industry from a business-oriented office industry to an "everyone-use" home-oriented industry that soon generated an income comparable to that of the wired telephone and wireless access industries. At the time of writing, the *information networking industry* (including fixed and wireless telephones as well as Internet access industries) has annual revenues of a few trillion dollars and by far is the largest engineering industry in the world. The largest portion of earnings of the wireless industry is made from the revenue generated by cellular telephone calls. However, this trend is rapidly changing and the future of this industry relies on broadband wireless Internet access that has shown a rapid and continual growth to support the emerging multimedia communication networking industry and ad hoc wireless sensor networking. Sensor networks are becoming important for emerging cyber physical systems in different areas such as medicine and transportation.

The main forces behind the growth of the necessity for packet switched wireless data networks in the past few years were the sudden success of the smart phones that became an epidemic after their introduction and the unprecedented popularity of the iPhone in 2007. Smart phones, and in particular the iPhone, opened a new paradigm for a variety of data applications and nurtured the growth of social networking that was another revolution in networking applications. The exponential growth of the volume of information transfer using wireless data for multimedia and Internet browsing applications in the late 2000s caused an exponential growth in the wireless local area networking industry and forced the cellular telephone industry to shift its focus from the traditional telephone application and its quality of service to the emerging multimedia data applications which demand higher data rates but are more tolerant of delay.

The amount of information produced by these emerging devices is so vast that we need a method to filter them and capture the most useful parts for useful applications. The most popular filtering is through the association of information to the time and location (space). As a result, measuring time and location is an essential part of information processing, and engineers have tried to measure them ever more accurately throughout the centuries. In the past few centuries, we have found technologies for the precise measurement of time and the ways to make them available to a variety of applications. The localization industry for day-by-day use started in the past few decades by using radio frequency (RF) signals to measure the distance between a landmark and a mobile electronic device. First, Global Positioning System (GPS) was introduced for outdoor environments [Mis10], then the cell tower and Wi-Fi localization complemented that to extend the coverage to indoor areas [Pah02] and more recently localization is under research for inside the human body [Pah12a].

The iPhone, followed by other smart phones, also introduced the first popular and inexpensive wireless localization techniques on a massivescale. The availability of localization and the popularity of mobile computing initiated another round of growth in application development on smart devices using wireless localization. In early 2007, the localization for smart devices was built on a few popular applications such as turn by turn direction finding. By the year 2010 around 15%

of over 100 000 applications developed for the iPhone were using wireless localization [Mor10]. The popularity of multimedia and location-enriched applications on mobile smart devices has radically shifted the habits of humans in their communications and information processing and it has profoundly affected the way that we live and relate to others.

The purpose of this book is to provide the reader with a textbook for understanding the principles of wireless access and localization. Wireless access and localization is a multidisciplinary technology; to understand this industry we need to learn about a number of disciplines to develop an intuitive feeling of how these disciplines interact with one another. To achieve this goal we provide an overview of the important wireless access and localization applications and technologies, describe and classify their underlying science and engineering principles in a logical manner, give detailed examples of successful standards and products, and provide a vision of the evolving technologies. In this first chapter, we provide an overview of the wireless industry and its path of evolution. The next three chapters describe the fundamental principles of the radio propagation, transmission schemes, and medium access control techniques in wireless networks. The succeeding three chapters examine principles of wireless network infrastructure deployment, operation and security. The following three chapters describe the popular wireless local area networks and personal area networks that have evolved to complement them by supporting low-power sensor networking and high-speed gigabit wireless multimedia applications. The next three chapters provide the details of different generations of wireless wide-area cellular networks. The last three chapters of the book are devoted to wireless localization techniques.

In the remainder of this chapter, we first provide the elements of a wireless network and then we give a summary of the evolution of important standards and technologies for wireless networking as well as evolution of technologies for wireless localization. Finally, we give an outline of the chapters of this book and how they relate to one another.

1.2   Elements of Information Networks

Information networks have evolved to interconnect networking enabled devices over a geographical area to share information generated by an application in the device. Figure 1.1 illustrates the abstract of this basic concept. The information source could be the voice of a human being creating an electronic signal on a telephone device connected to a local public branch switch or the PSTN to transfer that information to another geographical location. The information source could be a video stream from a video camera or sensor data from a robot that is sent through a networking interface card to a local area network or the Internet to be delivered to another networking enabled device in a geographically separated location. The sensor data for example, could be used for remotely navigating the robot. The information could be a simple on–off signal generated by a light switch in one location to be transferred by a communication networking interface protocol to another location to turn a light bulb on. What is common among all of these examples is an *application* that needs the *transfer* of a certain amount of *information* from one location to another, a *network* that can carry the information and an *interface device* that shapes the information to a format or protocol suitable for a particular networking technology.

Figure 1.2 shows a diagram of the elements affecting information networks and the relationships among them. Information generated by an application is delivered to a communication device that uses the network and delivers that information to another location. When the network includes multiple service providers, the interface between the device and the network should be *standardized* to allow communication among different network providers and various user devices. Standardization also allows multivendor operation so that different manufacturers can design different parts of the

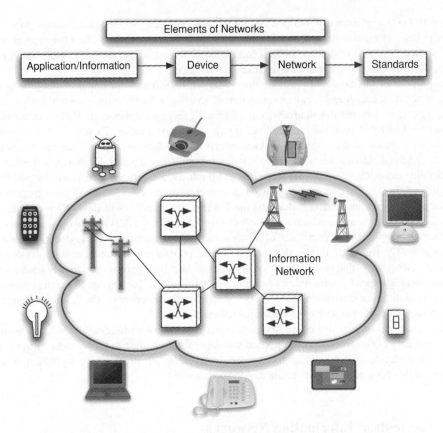

Figure 1.1   Abstract of the general concept of information networking.

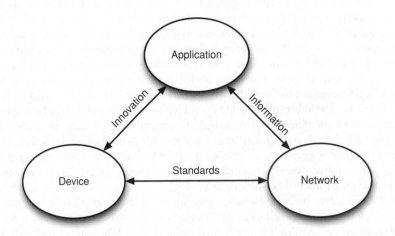

Figure 1.2   Elements of information networking.

network. Applications, telecommunication devices, and communication networks evolve in time to support innovations that enable new applications. These are the new applications that fuel the economy and the progress in the quality of life over time. For example, the introduction of iPhones and iPads opened a new horizon for hundreds of thousands of new applications in the past few years. The evolution of these devices was enabled by the availability of reliable wireless mobile data cellular services, Wi-Fi and Bluetooth technologies for wireless access to the PSTN and Internet, as well as GPS chipsets, Wi-Fi, and cell tower wireless localization technologies for localization using radio frequency (RF) signals. These applications are changing how we work, eat, and socialize; so in fact they are instrumental in the evolution of our habits.

1.2.1 Evolution of Applications, Devices, and Networks

Figure 1.3 illustrates the evolution of applications, devices, and networks. The first communication device that enabled a popular application was the Morse pad for the telegraph application that was invented in the 1837. The telegraph was the very first short messaging system (SMS). It needed two operators familiar with the Morse code to transfer a message between two nodes of the telecommunication network. The operator at one node would read the message and re-route it to another location in the network that was closer to the destination. The message would go along the network from node to node until it reached the destination. These operators were like "human routers" for the first telecommunication network. The operators could have a coffee between the time they received a message and the time they transmitted it to the next node because data

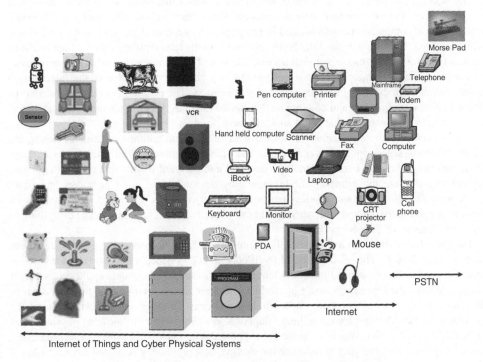

Figure 1.3 Evolution of applications, devices, and networks.

applications can tolerate such delay to a certain extent. The transmission technique for the device was digital communication. Therefore, the telegraph could be considered the first packet switched digital network with human routers designed for data burst SMS applications.

The more popular telephone network, which was invented in 1876, operated using analog telephone devices. The user of the device would connect to the operator and the operator would communicate with other operators to establish a line between the source and the destination before conversation starts and information gets transferred along the network. The operator in this application had to work hard to establish the connection fast enough and to maintain that connection during the period of information transmission or streaming of the conversation in both directions. The operator in this case was a human switch that was expected to establish the connection quickly and to maintain that connection during the communication period. Therefore, the telephone network was an analog connection-based circuit-switched network originally designed for voice applications. The Morse pad that was the device used for the telegraph network needed a specialized operator capable of using the code for data communications; as a result the telegraph industry evolved as an office-based application with certain limitations on its size. The telephone devices, however, could be used by anyone and they penetrated the home market; thus orders of magnitude higher numbers of telephone devices were sold and the telephone network became much larger than telegraph network generating tremendously larger revenue for the company. By considering the telephone and telegraph networks, we observe that at the beginning of the twentieth century, the telecommunications industry had already been exposed to a number of important issues, which played similar roles during the entire course of the past century and culminated in the emergence of modern wireless networks. Among these important issues were analog versus digital, voice versus data, packet-switched versus circuit-switched networking, and home versus office networking.

The next popular telecommunication devices related to information networks were voice-band modems. These devices emerged after the Second World War to allow communication between computers and computer terminals located in geographically separated areas. Computer networks, which evolved that way, extended the SMS supported by the telegraph to other data applications such as file transfer and remote terminal access. The size of the computer communication industry was still very small compared to the telephone industry until the penetration of the Internet into homes and through the use of desktop and laptop computers. The evolution of computer networks opened up new applications and communication devices such as printers, scanners, fax machines, video cameras, and monitors that could attach to them.

The popularity of wireless networks started with cellular and cordless telephones during the 1980s, extending voice applications across local and wide area networks. During the 1990s, wireless local area networking (WLAN) technology emerged and nurtured mobile computing to connect laptops (which were the primary mobile computing devices at the time) in homes and small office networks. In the 2000s, wireless personal area networking (WPAN) technology allowed communications between and with sensors that can virtually connect the Internet to everything to create the Internet of Things.

The latest devices that heavily impacted the evolution of information networking technology were mobile smart devices. The introduction of the iPhone in 2007 opened a new horizon for wireless data applications that demanded more efficient networks to support these data applications. Smart phones, lead by the iPhone, created a platform for running data-consuming applications such as YouTube access and web browsing on a wireless platform. This demand further increased the popularity of WLANs and forced cellular telephone service providers to move to physical layer technologies used in WLANs to increase the supported data rates. At the time of writing cyber physical systems are emerging to facilitate the massive data processing collected from distributed sensors for medical, transportation, power distribution, and other applications.

1.2.2   Information Network Infrastructures and Wireless Access

To support the transmission of voice, data, and video, several wired information network infrastructures have evolved throughout the past century. Wireless networks allow a mobile wireless
device to access these wired information network infrastructures. At first glance, it may appear that
a wireless network is only an antenna site or a base station connected to one of the switches or
routers in the wired information infrastructure that enables a mobile terminal to be connected to the
backbone network. In reality, in addition to the antenna site, a wireless network also needs to add
its own mobility-aware switches, databases, and base station control devices to be able to support
mobility and manage scarce radio resources when a mobile terminal changes its connection point
to the network. Therefore, a wireless network has its own fixed infrastructure with mobility-aware
switches and networked connections, similar to other wired infrastructures, as well as antenna sites
and mobile terminals.

When the geographical coverage area of a network is very large, the cost of deployment and
maintenance of the infrastructure is very high and a service provider makes the investment to build
the network infrastructure. To compensate for that large investment, the service provider leases the
infrastructure access to subscribers. We refer to these large infrastructures as backbone or wide-area
wired backbone networks. The two major examples of these backbone networks are the PSTN and
the Internet, each having a number of service providers in different countries. Wireless access to
these networks is either through wide-area wireless cellular networks, which allow for wireless
access over a large area of coverage through a service provider, or smaller networks, owned by
private enterprise or individuals. These smaller networks form the so-called local, personal, and
body area networks. Local area networks are either wired or wireless and the backbone networks
are mostly wired networks. In this book we address wireless networking technologies while details
of wired wide and local area networks are addressed in [Pah09].

Figure 1.4 shows the overall picture for wired and wireless telephone services using PSTN.
The PSTN, which was designed to provide wired telephone services, is augmented by a wireless
fixed infrastructure to support the mobility of a mobile device that communicates with several base
stations mounted over antenna posts. The PSTN infrastructure consists of switches, point-to-point
connections, and computers used for the operation and maintenance of the network. The fixed
infrastructure of the cellular telephone service has its own mobility-aware switches, point-to-point
connections, and other hardware and software elements that are needed for the mobile network

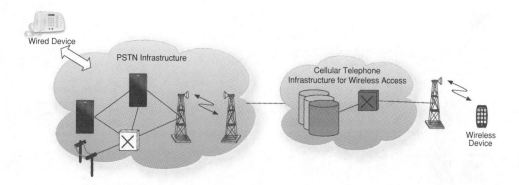

Figure 1.4   The PSTN and its extension to cellular telephone services.

operation and maintenance. A wireless telecommunication device, for example a smart phone, can connect to the PSTN infrastructure by replacing the wire attachment with radio transceivers. But, for the wireless device to change its point of contact, switches in the PSTN must be able to support mobility. Switches in the PSTN infrastructure were not originally designed to support mobility. To solve this problem, cellular telephone service providers have added their own fixed infrastructure with mobility-aware switches. The fixed infrastructure of the cellular telephone service provider is an interface between the base stations and the PSTN infrastructure that implements the environment to support mobility. The simplest wireless access to the PSTN is though a cordless telephone. This does not have any switch in the infrastructure and basically operates as a wireless connection between a handset and a telephone connected by wire to the PSTN and mostly through a standard or a proprietary protocol.

In the same way that a telephone service provider needs to add its own infrastructure to allow a mobile telephone to connect to the PSTN, a wireless data network provider needs its own infrastructure to support wireless Internet access. Figure 1.5 shows the traditional wireless data infrastructure and the additional wireless data infrastructure that allows wireless connection to the Internet. The traditional data network consists of routers, point-to-point connections, and computers for operation and maintenance. The elements of a wireless network include mobile devices, access points, mobility aware routers, and point-to-point connections. If the wireless data access intends to provide wide area coverage for the wireless data service, the new infrastructure has to support all the functionalities needed to support mobility. In simpler applications, such as a hot-spot or for home access, the wireless infrastructure does not necessarily need to be aware of mobility because connection to the Internet is through one access point only. However, to allow users with mobile devices to be able to connect to different access points, there is a need to support mobility through protocols and hardware.

The main difference between wireless access to the PSTN and the Internet is that wireless access to the PSTN, shown in Figure 1.4, is a connection-based voice-oriented network and wireless access to the Internet, shown in Figure 1.5, is a connectionless data-oriented network. A connection-based network needs a dialing process and, after dialing, a minimum quality of service is guaranteed to the user during the communication session. In connectionless networks, there is no dialing and the terminals are always connected to the network, but a uniform quality of service is not guaranteed. Figure 1.6 illustrates the basic difference between a packet-switched and a circuit-switched network in the handling and delivery of packets from a source to a destination terminal. In a connectionless

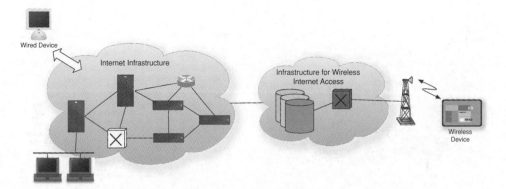

Figure 1.5   The Internet and its extension to cellular telephone services.

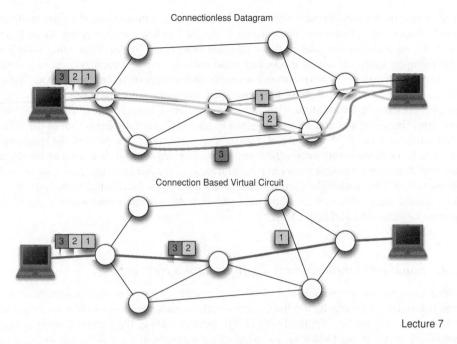

Figure 1.6 Connection-less packet-switched Internet versus connection-based circuit-switched PSTN.

data gram network, information packets takes routes that are determined by the routers, hub-by-hub, as based on the traffic and resources arriving and leaving the hub. As a result, consequent packets from a single information source may take different paths to arrive at the receiver. This approach provides a more efficient method to utilize the transmission line capabilities but has no guarantee for the delay of the arriving packets with respect to one another, which challenges support to maintain a prescribed quality of service for the user. In connection-based networks a virtual path is established between the source and destination, and the consecutive data packets take the same route. This formation allows more control on the delay and consequently the quality of service provided to the user.

1.2.3   Connection Between Wireless Access and Localization

Wireless localization is tied with wireless access through two connections. First, popular wireless localization techniques, such as Wi-Fi localization and cell tower localization, use the existing infrastructure and the transmitted signals originally established for wireless access and communications, to localize a mobile terminal. The data base of the location of the Wi-Fi access points or cell tower base stations is used as the landmark and the received signal strength or the time of flight of the signal between the landmark and the mobile terminal is used to estimate the distance of the terminal from the landmarks. The distances from several landmarks are used to estimate the location of the terminal. Using the existing infrastructure and the received signal strength is the most inexpensive and commercially popular method currently used for wireless localization of smart devices.

The second tie between wireless access and localization lies in understanding the multipath channel characteristics that cause deformation of the transmitted waveforms due to multipath effects. As we will describe later in this book, this deformation of the transmitted waveform by the multipath characteristics of indoor and urban areas for wireless communications imposes restrictions on the highest symbol transmission rate for communication applications. In localization application using the time of flight of the transmitted waveform, which provides a more precise measure for ranging the distance from a landmark, deformation of the waveform caused by multipath causes errors in estimations of the time of flight. The time of flight of the signal is often calculated from a reference location of a feature of a waveform, for example, the peak of the transmitted waveform. In the multipath environment the peak of the received signal is dislocated by the effects of the multipath causing an unwanted error in estimations of the distances using the time of flight estimation. Therefore, both high-speed wireless access and precise localization techniques need a careful understanding of the nature of the multipath arrivals in wireless media that is one of the important subjects addressed in this book.

1.2.4 Standards Organizations for Information Networking

The increasing number of portable and mobile applications on different communication devices demands a variety of standardized wireless access technologies operating on different frequency bands. Frequency bands are regulated by national agencies such as the Federal Communication Commission (FCC) in the United States. Wireless technologies that are discussed in this book include cellular telephone and personal communication systems that are operating within licensed bands and WLAN and WPAN technologies that are operating in unlicensed bands. Licensed bands are like a privately owned backyard. The owner of the band needs to invest a substantial amount of money and effort to obtain permission for using that band in a certain geographical area. These bands usually allow higher transmission power but they are more restricted in the size of the bandwidth. Unlicensed bands are similar to public gardens; users of these bands have access to a wider bandwidth but with restrictions on their transmission power. Figure 1.7 illustrates several licensed and unlicensed bands in the United States that are used both for different generations of cellular networks and for cordless telephones and several unlicensed bands used for WLAN and WPAN applications.

Standards define interface specifications between elements of a wireless network infrastructure allowing a global multivendor operation, which facilitates the growth of the industry. Figure 1.8 provides an overview of the standardization process in information networking. The standardization process starts in a special interest group of a standards developing body such as the Institute of Electrical and Electronics Engineers (IEEE 802.11) or Global System for Mobile (GSM) communications, which defines the technical details of a networking technology as a standard for operation. The defined standard for implementation of the desired network is then moved for approval by a regional organization such as the European Telecommunication Standards Institute (ETSI) or the American National Standards Institute (ANSI). The regional recommendation is finally submitted to world-level organizations, such as the International Telecommunications Union (ITU), International Standards Organization (ISO), or International Electrotechnical Commission (IEC), for final approval as an international standard. There are a number of standards organizations involved in information networking. Table 1.1 provides a summary of the important standards playing major roles in shaping the information networking industry, which are also mentioned in this book.

The most important standard developing organizations for technologies described in this book are the IEEE 802-series standards for personal, local, and metropolitan area networking. The IEEE

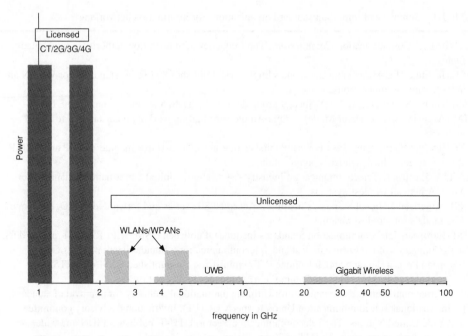

Figure 1.7   Samples of licensed and unlicensed band spectrums in the United States.

Figure 1.8   Standard development process.

Table 1.1   Summary of important standard organizations for information networking

---

FCC (Federal Communication Commission): The frequency administration authority in the United States.

IEEE (Institute of Electrical and Electronics Engineers): Publishes 802 series standards for WLAN and WPAN wireless applications.

GSM (Global System for Mobile): Special group defined 2G TDMA standard sponsored ETSI.

ATM (Asynchronous Transfer Mode) Forum: An industrial group working on a standard for ATM networks.

IETF (Internet Engineering Task Force): Publishes Internet standards that include TCP/IP and SNMP. It is not an accredited standards organization.

EIA/TIA (Electronic/Telecommunication Industry Association): United States national standard for North American wireless systems.

ANSI (American National Standards Institute): Accepted 802 series and forwarded to ISO. Developed JTC models for wireless channels.

ETSI (European Telecommunication Standards Institute): Published GSM, HIPERLAN-1, and UMTS.

CEPT (Committee of the European Post and Telecommunication): Standardization body of the European Posts Telegraph and Telephone (PTT) ministries. Co-published GSM with ETSI.

IEC (International Electrotechnical Commission): Publishes jointly with ISO.

ISO (International Standards Organization): Ultimate international authority for approval of standards.

ITU (International Telecommunication Union formerly CCITT): International advisory committee under the United Nations. The Telecommunication Sector, UTU-T, publishes ISDN and wide area ATM standards. Also works on IMT-2000.

---

is the largest engineering organization in the world, publishing a number of technical journals and magazines and organizing numerous conferences worldwide. The IEEE 802 community is involved in defining standard specifications for information networks. The number 802 was simply the next free number IEEE could assign to a committee at the inception of the group on February 1980, although "80-2" is sometimes associated with the date of the first meeting. Regardless of the ambiguity of the name, the IEEE 802 community has played a major role in the evolution of wireless information networks by introducing IEEE 802.11 WLANs, IEEE 802.15 WPANs, IEEE 802.16 WMAN, and other standards which are discussed in detail in this book.

Another important standard developing organization is the Internet Engineering Task Force (IETF) which was established in January 1986 to develop and promote Internet standard protocols around the Transmission Control Protocol/Internet Protocol (TCP/IP) suite for a variety of popular applications. In the 1990s, the Asynchronous Transfer Mode (ATM) Forum was an important standard developing group trying to develop standards for connection based fixed packet length communications for the integration of all services. This philosophy was in contrast with Internet/Ethernet networking using connection-less communications with variable and long length packets and it has lost its momentum.

The Telecommunication/Electronic Industry Association (TIA/EIA) is a United States national standards body defining a variety of wire specifications used in local, metropolitan and wide area networks. The TIA/EIA is a trade association in the United States representing several hundred telecommunications companies. The TIA/EIA has cooperated with the IEEE 802 community to define the media for most of the wired Local Area Networks (LANs) used in fast and gigabit Ethernet. TIA/EIA also defines cellular telephone standards such as Interim Standards (e.g., IS-95) or cdmaOne second generation (2G) cellular networks and the IS-2000 or CDMA-2000 third generation (3G) cellular telephone networks. ETSI and the Committee of the European Post and Telecommunications (CEPT) are the European standardization bodies publishing wireless networking standards, such as GSM for the 2G cellular networks and Universal Mobile Telephone Standard (UMTS) for 3G cellular, in the European Union.

The most important international standards organizations are the International Telecommunication Union (ITU), the International Organization for Standardization (ISO), and the International Electrotechnical Commission (IEC); and they are all based in Geneva, Switzerland. Established in 1865, ITU is an international advisory committee under United Nations and its main charter includes telecommunication standardization and allocation of the radio spectrum. The Telecommunication Sector, ITU-T, has published, for instance, the Integrated Service Data Network (ISDN) and wide-area ATM standards, as well as International Mobile Telephone-2000 (IMT-2000) for 3G cellular networks.

In 2009, the Radio communication sector of ITU (ITU-R) defined the IMT-Advanced requirements for fourth generation (4G) cellular networks. At the time of writing, the so-called Long Term Evolution (LTE) of UMTS is becoming the favorite choice of this standard. The World Administrative Radio Conference (WARC) was a technical conference of the ITU where delegates from member nations of the ITU met to revise or amend the entire international Radio Regulations pertaining to all telecommunication services throughout the world. ISO and IEC are composed of the national standards bodies, one per member economy. These two standards often work with one another as the ultimate world standard organization. Established in 1947, ISO nurtures worldwide proprietary industrial and commercial standards that often become law, either through treaties or national standards. The ISO seven-layer model for computer networking was one of the prominent examples of ISO standards. The IEC started in 1906 and it is a non-governmental international standards organization for "electrotechnology" which includes a vast number of standards from power generation, transmission and distribution to home appliances and office equipment, to telecommunication standards. The IEC publishes standards with the IEEE and develops standards jointly with the ISO as well as the ITU.

1.2.5   Four Markets in the Evolution of Wireless Networking Standards

The market for wireless networks has evolved in four different segments that can be logically divided in two classes: *voice-oriented* and *data-oriented*. The voice-oriented market evolved around wireless connections to the PSTN for wireless telephone applications. These services further evolved into local and wide-area markets. The local wireless access to the PSTN is based on low-power, low-mobility devices with a higher quality of voice that evolved around the cordless telephone application. The wide-area wireless access to the PSTN market evolved around cellular mobile telephone services that are using terminals with higher power consumption, comprehensive coverage, and lower quality of voice. Figure 1.9 (a) compares several features of these two sectors of the wireless access to the PSTN market.

The wireless data-oriented market evolved around wireless access to the Internet and computer communication network infrastructure. The data-oriented wireless access services are divided into broadband local, ad hoc, and wide-area mobile data markets. The wide-area wireless data market provides for wireless Internet access for mobile users with a comprehensive coverage similar to that of the cellular telephone. Local broadband and ad hoc networks include wireless local and personal area networks that provide for high-speed Internet access as well as evolving ad hoc wireless consumer product markets with a local spot coverage similar to the cordless telephone systems. Figure 1.9 (b) illustrates several differences among the local and wide-area wireless data networks.

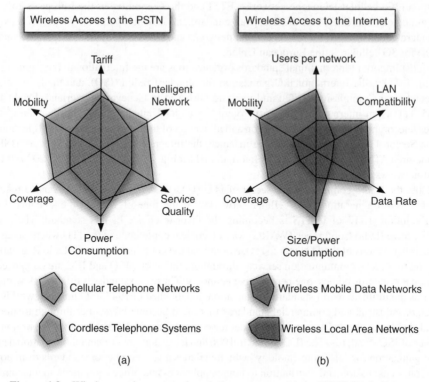

Figure 1.9   Wireless market sectors for wireless access to (a) the PSTN (b) the Internet.

Standards for wireless access to the PSTN and the Internet evolved around these four markets, described in Figure 1.9, for wireless telephone and wireless data applications. The evolutionary paths for these technologies were different because the market demand for data and voice applications was highly affected by the emergence of the Internet in the middle of the evolution of wireless networking technologies. The earlier standards were focused on voice and telephone applications using PSTN, which was the dominant source of revenue generation for the telecommunication network operating companies (and at the time of writing it still is). However, this trend started to change with the emergence of the popularity of data applications and the Internet in the late 1990s and the introduction of smart phones in the late 2000s. Smart phones increased the usage of the wireless data applications exponentially and that demand forced the cellular network providers to shift the focal point of the emerging wireless cellular networking systems toward data applications. As a result, as we will see in the following two sections, although the technologies for medium access control and physical layer in the four major markets differed substantially in the early days of this industry, they are evolving towards technologies that were originally explored for WLAN and wireless local data applications.

1.2.6   Trends in Wireless Data Applications

Wireless access and localization techniques are multidisciplinary systems engineering fields. During the evolution of these technologies the focal point of applications and the market as well as

Figure 1.10   Mobile and personal users of voice telephone applications.

research and development have shifted over time. During the 1990s and early 2000s the cellular telephone network industry was growing exponentially. The cellular telephone industry runs a connection-based telephone voice application with a bandwidth per user of around 10 Kbps but numerous simultaneous users each running a telephone call for a few minutes. Figure 1.10 shows the main two application environments for personal and mobile radio communications. The main applications are connected to the network for only a few minutes, they are moving around and they are sensitive to real-time delay. Thus, the wireless access network providers focus on increasing the number of simultaneous users and the steadiness of the quality of service during the connection time, as the users are moving around. As a result of that growth and these requirements the Time Division Multiple Access (TDMA) and Code Division Multiple Access (CDMA) medium access technologies emerged and frequency administration agencies released more bands for these applications to support the growth of the number of users. Service providers who had started their business by installing macro-cells to cover tens of kilometers at an approximate cost of around a million dollars per cell started deploying smaller micro-cells for dense urban areas that cover several hundred meters at a cost that was an order of magnitude lower than the cost of a macro-cell for dense urban areas; and then they shifted to pico-cells, for inside large buildings, which covers and costs about an order of magnitude less than the micro-cells. More recently the number of telephone voice users reached its steady large value of around seven billion users worldwide.

As shown in Figure 1.11, with the emergence and immediate popularity of smart phones in late 2000's, wireless data applications started to grow exponentially and this growth is expected to continue for the next few decades. At the time of writing, over 70% of this data is carried through Wi-Fi connections that are deployed randomly by individual users and private institutions to cover their work space area, not by the service providers. Service providers provide a complementary coverage in the areas that Wi-Fi does not cover and a more reliable connection for mobile data users. As shown in Figure 1.12 data applications are dominated by downloading from a cloud somewhere in the Internet into a stationary or semi-stationary device. The Internet is made of a fiber optics backbone and can provide the data pipe from the cloud up to the wireless access point; the bottleneck is the fast transmission from the access point to the user. Data applications are keen on speed of transmission because their applications are often bursts of data with a wide variety of sizes, from a short message to streaming videos. This demand for higher data rates resulted in the emergence of Orthogonal Frequency Division Multiplexing (OFDM), space-time coding, and

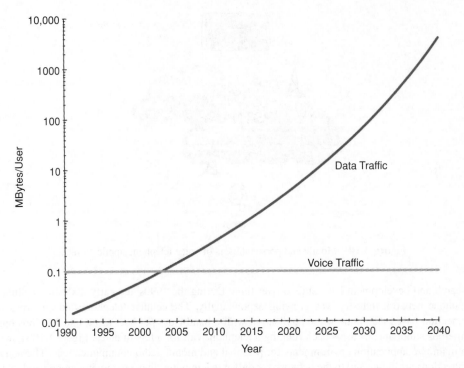

Figure 1.11 Trends in the growth of voice and data applications in the next few decades.

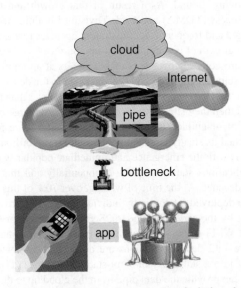

Figure 1.12 Stationary and semi-stationary users of wireless data applications.

Multiple-Input-Multiple-Output (MIMO) antenna system technologies first in Wi-Fi and then in LTE systems in 4G cellular networks. Wi-Fi supports local wireless data access and 4G is designed for the wide-area backup support of data applications. To support the growth of data more bandwidth efficient modulation techniques, release of wider bands from frequency administration agencies, and using smaller cells are the trends of the future. The wireless local industry is searching for gigabit wireless and the cellular networking industry is examining femto-cell technology.

1.3   Evolution of Wireless Access to the PSTN

Wireless access to the PSTN evolved around two applications, the cordless telephone for local access and the cellular telephone for wide-area wireless access. Table 1.2 shows a brief chronology of the evolution of wireless access networks to connect to the PSTN. The local area networks for wireless access to PSTN, mostly for home and small office applications, started with the introduction of the cordless telephone that appeared in the market in the late 1970s. A cordless telephone provides a wireless connection to replace the wire between the handset and the telephone set. The radio technology for implementation of a cordless telephone was similar to the technology used in walkie-talkies that existed since the Second World War. The important feature of the cordless telephone was that, as soon as it was introduced to the market, it became a major commercial success, selling on the order of tens of millions in numbers and generating a gross sales income exceeding several billions of dollars. The success of the cordless telephone encouraged further developments in this field. The first digital cordless telephone was cordless telephone (CT) and CT-2, a standard developed in the United Kingdom in the early 1980s. The next generation for cordless telephones was wireless multichannel local cordless telephone with a higher transmission rate to

Table 1.2   History of voice-oriented wireless networks

Exploration of first generation mobile radio at Bell Laboratories: early 1970s
First generation cordless phones: late 1970s
Exploration for second generation digital cordless CT-2: 1982
Deployment of first generation NORDIC analog NMT: 1982
Deployment of United States AMPS: 1983
Exploration of the second generation digital cellular GSM: 1983
Exploration of wireless PBX; DECT: 1985
Initiation for GSM development: 1988
Initiation for IS-54 TDMA digital cellular: 1988
Exploration of the QUALCOMM CDMA technology: 1988
Deployment of GSM: 1991
Deployment of PHS/PHP and DEC-1800: 1993
Initiation for IS-95 standard for CDMA: 1993
PCS band auction by FCC: 1995
PACS finalized: 1995
3G Standardization started: 1998 (add WiMax and LTE)
WiMax Forum formed: 2001
WiMax fixed: 2004
WiMax mobile: 2005
LTE products: 2006
LTE-advanced standardization: 2009

support wireless data using the digital European cordless telephone (DECT) standard. Both CT-2 and DECT had minimal network infrastructures to go beyond the simple cordless telephone and cover a larger area and multiple applications. However, in spite of the huge success of the traditional cordless telephone, neither CT-2 nor DECT became a commercially successful system immediately. These local systems soon evolved into the so-called Personal Communication Systems (PCS) that were complete systems with their own infrastructure, very similar to the cellular mobile telephone.

1.3.1   Cordless Telephone Systems

In the technical communities of the early 1990s, PCS systems were differentiated from the cellular systems in the way they are presented in Figure 1.9 (a). A PCS service was considered as the next generation cordless telephone designed for residential areas, providing a variety of services beyond the cordless telephone. The first real deployment of PCS systems was the Personal Handy Phone (PHP), later renamed the Personal Handy System (PHS), introduced in Japan in 1993. At that time, the technical difference between PCS services and cellular systems was perceived to be smaller cell size, better quality of speech, lower tariff, less power consumption, and lower mobility. However, from the user's point of view the terminals and services for PCS and cellular looked very similar and the only significant difference was marketing strategy and the way that they were introduced to the market. For instance, around the same time, in the United Kingdom, Digital European Cordless DEC-1800 services were introduced as a PCS service. The DEC-1800 service was using the second generation (2G) cellular GSM-like technology at a higher frequency of 1800 MHz but it employed a different marketing strategy. The last PCS standard was Personal Access Communications System (PACS) in the United States, finalized in 1995. Altogether, none of the PCS standards became a major commercial success and a competitor to cellular services.

In 1995 FCC in the United States auctioned the frequency bands around 2 GHz as PCS bands but PCS specific standards were not adopted for these frequencies. Eventually, the name PCS started to appear only as a marketing pitch by some service providers for digital cellular services, in some cases not even operating in the PCS bands. While the more advanced and complex PCS services evolving from the simple cordless telephone application did not succeed and eventually merged into the cellular telephone industry, the simple cordless telephone industry itself still remains active. In the early 2000s the frequency of operation of cordless telephone products was shifted into unlicensed industrial, scientific and medical (ISM) bands rather than the licensed PCS bands. Cordless telephones in the ISM bands could provide a more reliable link using spread spectrum technology. With the growing popularity of WLAN technology for home networking, interference between cordless telephones and wireless Internet access in the home and small offices captured the attention of manufacturers in this area. More recently, the DECT standard has attracted considerable renewed attention for the implementation of cordless telephones and DECT devices are flooding this market, replacing cordless telephones in the ISM bands. These devices are using 1.8 and 1.9 GHz PCS bands, which do not interfere with the ever-growing popular WLAN application for wireless networking inside residences and small offices. The DECT standard uses TDMA technology with a Time Division Duplex (TDD) option, in which a single carrier frequency carries multiple handset streams between the handsets and the base station in both directions.

1.3.2   Cellular Telephone Networks

The wide area wireless access to the PSTN started with the analog cellular telephone. The technology for the analog cellular first generation (1G) systems was developed at the AT&T Bell

Laboratories in the early 1970s. However, the first deployment of these 1G systems took place in the Nordic countries using Nordic Mobile Telephone (NMT) technology about a year earlier than the deployment of the Advanced Mobile Phone Services (AMPS) in the United States. Since the United States was a large country, frequency administration and other regulation in the process was slower, so it took a longer time for the deployment. All 1G systems used traditional analog FM transmission and frequency division multiple access (FDMA) to share the medium among different users. The digital cellular networks in the Nordic countries started with the formation of the Groupe Speciale Mobile or GSM standardization group. The GSM standards group was originally formed to address international roaming, a serious problem for cellular operation in the European Union countries. The standardization group shortly decided to go for a new digital TDMA technology because it could allow integration of other services to expand the horizon of wireless applications [Hau94]. In the United States, however, the reason for migration to digital cellular was that the capacity of the analog systems in major metropolitan areas such as New York City and Los Angeles had reached their peak value and there was a need for increasing the capacity in the existing allocated bands. Although the Nordic countries, led by Finland, maintained the highest rate of cellular penetration in the early days of this industry, the United States was by far the largest market. By 1994, there were 41 million subscribers worldwide, 25 million of them in the United States. The need for higher capacity motivated the study of CDMA technology in the United States that was originally perceived to provide a capacity up to two orders of magnitude higher than other alternatives such as analog band splitting or digital TDMA. The CDMA technology used in the cellular network utilizes direct sequence spread spectrum (DSSS) technique for transmission and different users are differentiated using different spread-spectrum codes.

In the early 1990s, while the debate between TDMA and CDMA capacity for 2G cellular networks was in progress in the United States, deployment of the GSM technology started in the European Union that was looking for a technology to solve the roaming problem between a number of countries ready to form the European Union. Around the same period, developing countries, which had not yet deployed any cellular networks, started their planning for cellular telephone networks and most of them adopted the 2G GSM TDMA digital cellular technology over the legacy analog cellular systems. Soon after, GSM had penetrated into more than 180 different countries. An interesting phenomenon in the evolution of the cellular telephone industry was the unexpected rapid expansion of this industry in developing countries. In these countries, the growth of the infrastructure for wired telephone was slower than the growth of the demand for the new subscriptions and always there were long waiting times to acquire a wired telephone line. As a result, in most of these countries, telephone subscriptions were sold on the black market at a price several times its actual value. Penetration of cellular telephone in these countries was much easier because people were already prepared for a higher price for a telephone subscription. In addition, original deployment, maintenance, and expansion of cellular networks could be done in a manner that was much faster than that with wired telephones, resulting in a rapid penetration of cellular telephones in the world market.

In the beginning of the race between the TDMA and CDMA, CDMA technology was deployed only in a few countries. Besides, the experimentation had shown that the capacity improvement factor of CDMA was smaller than expected. In the mid-1990s when the first deployments of CDMA technology started in the United States, most companies were subsidizing the cost to stay in race with TDMA and analog alternatives. However, from day one, the quality of voice using CDMA was superior to that of TDMA systems installed in the United States. As a result, CDMA service providers under the banners like "you cannot believe your ears" started marketing this technology in the United States that soon become very popular with the users. Meanwhile, with the huge success of digital cellular telephony, all manufacturers worldwide started working on 3G cellular

International Mobile Telephone IMT-2000 wireless networks. Most of these manufacturers adopted wideband CDMA (W-CDMA) as the technology of choice for IMT-2000, assuming that W-CDMA eases integration of services, provides better quality of voice, and supports higher data rates of up to 2 Mbps for wireless Internet access.

In the early 2000s, Internet access to the home opened a new horizon for wireless data applications as WLAN became the choice for local access in home, office, and hot spots. The sporadic nature of WLAN coverage demanded the attention of cellular providers to provide wide-area high-speed wireless data access. The technology adopted in these efforts was no longer CDMA and these technologies evolved out of OFDM and MIMO technologies that had matured in the WLAN industry. This movement first started with WiMax (Worldwide Interoperability for Microwave Access) that was an extension to the WLAN technologies originally specified by IEEE 802.11. WiMax did not gain the expected market success and, at the time of writing, long-term evolution LTE has been more successful. Further, LTE provides an enhancement more suited to fragmented frequency bands used by cellular networks compared to WiMax. These technologies are referred to as fourth generation (4G) wireless technologies. LTE-advanced is the latest standard emerging in this area aiming at maximum data rates on the order of a gigabit per second. The basic physical layer of these technologies is similar to the physical layer of WLANs that better suits emerging data applications. The implementation of medium access control is adjusted to the specifics of cellular networks that are designed for comprehensive wide-area coverage and a fragmented bandwidth allocations. Figure 1.13 provides a chart depicting the evolution of different technologies for

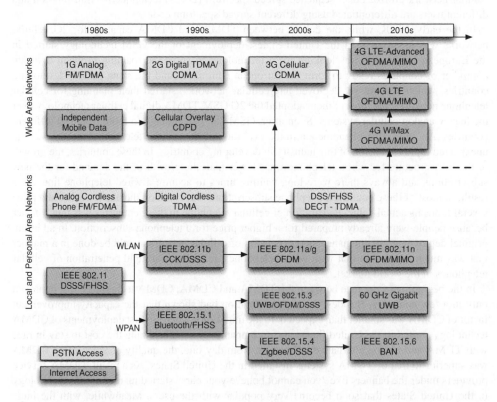

Figure 1.13 Evolution of networking technologies for wireless access to the PSTN and the Internet.

the original first four applications for local and wide-area voice and data applications shown in Figure 1.9.

1.4 Evolution of Wireless Access to the Internet

The main difference between wireless access to the Internet and wireless access to the PSTN is that the PSTN is a connection-based network primarily designed for telephone voice applications and the Internet is a connectionless network primarily designed for data applications. The telephony application generates massive amounts of data for two-way communication for a few minutes of real-time streaming at a low speed of around 10 Kbps per user that demands a steady quality of service during the conversation. These features require a simpler transmission technique but a more predictable medium access control method. Therefore, the focal points for the design of wireless access to these networks were the medium access control techniques and how they can support more numbers of users with a better quality of voice over a given bandwidth. Data applications are best described by bursts of information and these bursts of information need to percolate over the network in the fastest way, to incur minimal delay, and for that reason the highest achievable data rate dominates the design of the data-oriented networks.Therefore, the evolution of wireless data networks has focused on achieving the highest data rate in a given bandwidth that is achieved by innovations at the physical layer used for transmission of the bit streams in the wireless medium.

Table 1.3 provides the chronology of data-oriented wireless networks. As shown in Figure 1.9 (b) data-oriented wireless networks are divided into wide-area wireless mobile data networks and wireless local and personal area networks. Wireless local networking started with WLANs and later extended into WPAN technologies. WLANs originally emerged to support high data rates of above 1 Mbps (at that time) compared to mobile data networks that would support data rates of less than 10 kbps. In three decades of evolution, the WLAN industry changed significantly in developing innovative technology and marketing to the extent that, today, the WLAN standardization committee, IEEE 802.11, is evaluating gigabit wireless for the future of this technology and the sales of the WLAN chip sets exceeds several billions units per year. WPAN technology evolved under the IEEE 802.15 standardization process as a complement to WLAN technology to address application areas that are not fully covered by WLAN technology. They have already introduced popular standards such as Bluetooth for short-range communications between devices and ZigBee for low-power sensor networking applications. This group has worked on 3.4–10.6 GHz bands and more recently the 60 GHz spectrum for the implementation of higher data rates (gigabit wireless technologies) for short-distance connection such as wireless Universal Serial Bus (USB), three dimensional (3D) gaming, or cable replacement for multichannel high-definition television applications.

1.4.1 Local Wireless Data Networks

The lower part of Figure 1.13 illustrates the evolutionary path of WLAN and WPAN technologies. The concept of using the spread spectrum and infrared technologies to implement WLANs was first introduced around 1980 [Pah85], the first IEEE 802.11 standard using these technologies released in 1997 and the IEEE 802.15 for WPAN was formed in the following year. However, the sizeable market of several billion chipsets per year, which is comparable to the cellular chip set market, emerged only in recent years. A key feature of the WLAN and WPAN industry is operation in the unlicensed bands. The first unlicensed bands were the Industrial, Scientific and Medicine (ISM) bands at 900 MHz, 2.4, and 5.2 GHz, released in the United States in 1985 [Mar85]. Later in 1994 the unlicensed PCS bands and in 1997 unlicensed National Information Infrastructure

Table 1.3   Chronology of wireless data networks

Diffused Infrared WLAN: 1979 (IBM Rueschlikon Laboratories, Switzerland)
Spread Spectrum WLAN: 1980 (HP Laboratories, California)
ARDIS: 1983 (Motorola/IBM)
ISM bands for commercial spread spectrum applications: 1985
Mobitex: 1986 (Swedish Telecom and Ericsson)
IEEE 802.11 for WLAN standards: 1987
HIPERLAN-1: (High PERformance LAN) in Europe: 1992
Release of 2.4, 5.2, and 17.1–17.3 GHz bands in the European Community: 1993
PCS licensed and unlicensed bands for PCS: 1994
CDPD: 1993 (IBM and nine operating companies)
U-NII bands released, IEEE 802.11 completed, GPRS started: 1997
IEEE 802.11b with 11 Mbps: 1998
IEEE 802.15 for WPAN with Bluetooth as 15.a: 1998
IEEE 802.11a/HIPERLAN-2 started: 1999
IEEE 802.16 Metropolitan Area Networks: 1999
GPRS (General Packet Radio Services): Late 1990s
EDGE (Enhanced Data rates for GSM Evolution): Early 2000s
EV-DO (Evolution-Data Optimized or Evolution-Data Only): Mid-2000s
IEEE 802.15.3a for UWB Technology: 2003
IEEE 802.15.4 ZigBee: 2003
Mobile Wi-Max (Worldwide Interoperability for Microwave Access): 2005
HSPA (High Speed Packet Data): 2008
LTE (Long Term Evolution): 2009
Femtocell technology: Early 2010s
IEEE 802.11ad and IEEE 802.15.3c for Gigabit wireless: Early 2010s

---

(U-NII) bands were also released in the United States. The original legacy IEEE 802.11 used direct sequence spread spectrum (DSSS) and frequency hopping spread spectrum (FHSS) operating at 2.4 GHz, as well as a diffused infrared technology with 1 and 2 Mbps options. The DSSS and FHSS became the popular options and the enhanced versioned of the DSSS option using Complementary Code Keying (CCK) became the 802.11b standard, completed in 1998, that could support up to 11 Mbps with backward compatibility to the legacy DSSS option. The IEEE 802.11a standard, completed in 1999, operated in the ISM and U-NII bands at 5 GHz and used orthogonal frequency division multiplexing (OFDM) for the first time in a standard for wireless networking to support up to 54 Mbps.

The IEEE 802.15 WPAN standardization committee evolved from the IEEE 802.11 WLAN community and indeed it started in that community and then turned to an independent committee in 1998 [Hei98]. The first successful standard of this community was the IEEE 802.15.1 standard, popularly known as Bluetooth, which was released in the first year of the formation of IEEE 802.15. Bluetooth was designed for low-power ad hoc sensor networking using lower transmission power and FHSS technology. The FHSS transmission technology of Bluetooth is very similar to the FHSS technology used in legacy 802.11 but its medium access control is centralized, that better suits voice applications. As a result, Bluetooth has captured the market for wireless telephone connections over very short distances, such as the connection of cellular handsets or headphones to a base computer or a telephone base inside a car. The ZigBee technology was introduced by IEEE 802.15.4 in 2003 as a low-power WPAN technology for simple short data packet transmission applications in sensor networking. ZigBee, similar to legacy 802.11 DSSS option, uses DSSS for transmission with a contention-based medium access control that is a simplified version of IEEE 802.11 and better suits bursty and low-power connection applications. One can classify Bluetooth and ZigBee WPAN technologies as the low-power, smaller coverage area extensions of legacy IEEE 802.11 that have evolved for voice and data applications, respectively.

In addition to the low-power sensor complement of IEEE 802.11, IEEE 802.15 has also addressed higher data rate technologies at higher frequencies of operation as well. The first of these efforts was initiated by the IEEE 802.15.3a subcommittee, defining standard specifications for WPANs operating in the 3.1–10.6 GHz unlicensed ultra-wideband (UWB) bands released by the FCC in 2002. The entire band for the IEEE 802.11 community at 2.4 GHz is 84 MHz and at around that time the popular IEEE 802.11b WLANs specifications occupied a bandwidth of 26 MHz per carrier to support a maximum data rate of 11 Mbps. That had an enormous amount of overhead, leaving the highest achievable data rates at approximately 5.5 Mbps. The UWB system specifications could be envisioned to occupy a bandwidth on the orders of a few gigahertz and support data rates on the orders of gigabits per second. In practice, IEEE 802.15.3 began by completing a preliminary standard for 11 and 55 Mbps operation in the UWB spectrum in 2003. Fairly soon after this, the IEEE 802.15.3a group was formed that was aiming at using the UWB band and the technology to increase the effective data rate to several hundred Mbps to Gbps (that was at that time an order of magnitude higher than the corresponding data rates with IEEE 802.11b operating at 11 Mbps). In IEEE 802.15.3a, several options were evaluated for UWB communications, among which were the historical impulse radio technology, as well as Direct Sequence UWB (DS-UWB) and Multi Band OFDM (MB-OFDM) systems. This standard lost its momentum later and was dissolved in January 2006. However, the technical work produced by the committee was later transferred to activities in the 60 GHz band.

In 2001, the FCC released unlicensed bands at 57–66 GHz and this band opened another wave of standardization activities for gigabit wireless applications. The research and development community in this field argues that the millimeter wave frequencies have several advantages over the UWB bands. First, the UWB regulation in different countries is not the same. Second, the UWB spectrum overlaps with other technologies, causing potential interference with popular applications such as 802.11a WLANs and other consumer products such as cordless telephones operating at 5.2 GHz. The IEEE 802.15c group was formed to specify another PHY alternative to the original 802.15.3 WPAN standard released in 2003, which uses millimeter waves at 60 GHz. This group was formed in March 2005 and defined new MAC and PHY layer standards for the 60 GHz channel in September 2009 [IEEE09]. This technology is expected to accommodate coexistence with other WPANs and support high definition video streaming and other streaming applications that need data rates larger than 2 Gbps. Yet another alternative approach for Gbps wireless is being pursued by the IEEE 802.11ad group, which is building on the legacy of IEEE 802.11 and its commercial success in using MIMO/OFDM technology to achieve similar data rates. The IEEE 802.11ad group was expecting to complete its standard by December 2011 [Per10].

The lower parts of Figure 1.13 illustrate the evolution of local and personal area networking standards and technologies from the original concept of the WLAN for wireless Internet access. The evolution of wireless local Internet access in the past three decades is an excellent example of evolution for a successful and complex technology. This evolution experienced a huge swing from the early days of desperations for WLAN market in the early 1990s to the unexpected worldwide prosperity and growing popularity of Wi-Fi technology as the dominant technology for the wireless Internet access in the emerging smart phones, tablets and other devices at the time of this writing. Throughout this evolution, spread spectrum, OFDM, and MIMO technologies found their first popular applications in the wireless networking industry. Spread spectrum has emerged as the technology of choice for low-power applications and MIMO/OFDM for supporting the highest possible data rates. Another useful application of WLANs is the use of this technology for opportunistic wireless localization in indoor areas [Pah10]. WiFi localization is expected to emerge as one of the most popular wireless localization approaches for tens of thousands of location-dependent applications in smart phones and wireless tablets.

1.4.2   Wide Area Wireless Data Networks

In the wide area, mobile packet switched wireless data services were first introduced with the Advanced Radio Data Information Service (ARDIS) project between Motorola and IBM in 1983 that covered a number of metropolitan areas in the United States at a data rate of 4800 bps [Pah94]. The purpose of this network was to allow the IBM field crew to operate their portable computers wherever they wanted to deliver their services. In 1986, Ericsson introduced the Mobitex technology that was an open architecture implementation of ARDIS at a data rate of 8000 bps. In 1993, IBM and the nine Bell operating companies in the United States started the Cellular Digital Packet Data (CDPD) project operating at 19,200 bps, expecting a huge market by the year 2000. ARDIS and Mobitex were independent networks that owned their infrastructure and frequency bands and were using their own transmission technique and contention based medium access control. The CDPD infrastructure, however, was overlaid over the existing AMPS antennas and frequency bands with a separate physical layer and contention-based medium access control. This arrangement would allow a comprehensive coverage to the CDPD and would save in the cost of infrastructure that is heavily influenced by the cost of the land, tower and the frequency band. In the early 1990s CDPD was perceived to be the future of the mobile data industry, which at that time supported low-speed applications such as remote access to computers, file transfer, and email.

In the mid-1990s, the Internet started to penetrate the home market and the speed of operation and memory size of personal computers connected to the Internet opened a new horizon for evolution of a number of bandwidth-hungry applications that needed much higher data rates than those supported by the emerging wireless mobile data services such as CDPD at that time. This movement directed the cellular telephone community to consider integration of high-speed packet switched data networks into the 3G cellular networks and the IMT-2000 specifications for the 3G cellular networks set a data rate of 2 Mbps on a wireless packet switched network as its goal.

While 3G standards were under progress, 2G standardization activities responded to the market demand for higher-speed packet switched data with extensions to existing 2G digital cellular networks. The popular 2G networks included circuit switched packet data applications for low speeds of less than a couple of tens of kilobits per second and were interfacing to the PSTN. These new services would modify the infrastructure with new hardware elements that would direct the packet switched data to the Internet and would focus on increasing the data rate transmission over the air interface. In the late 1990s, General Packet Radio Services (GPRS) that was integrated in the successful GSM cellular systems was introduced with a theoretical data rate of up to around hundred Kbps. This is an order of magnitude higher data rate than previous technologies used for mobile data services. The GPRS system used the same physical air interface as the GSM but assigned more TDMA slots to a single user. In the early 2000s, Enhanced Data rates for GSM Evolution (EDGE) was released as a standard that used more bandwidth efficient transmission techniques to increase the data rate by several times to reach several hundred Kbps. EDGE still used the GSM infrastructure and TDMA framing. The real limitation of the data rate for GPRS and EDGE was due to the bandwidth of the carrier, that is 200 KHz for GSM.

The bandwidth per carrier of the 2G cdmaOne (IS-95) and the 3G CDMA2000 (IS-2000) is 1.25 MHz, allowing theoretical data rates of around several Mbps that was the goal of 3G cellular networks. In the mid-2000s, the Evolution-Data Optimized or Evolution-Data Only (EV-DO) standard was introduced that could achieve these data rates over the original QUALCOMM 2G CDMA technology by dedicating a 1.25 MHz carrier to packet data services. To achieve higher data rates, in a manner similar to GPRS and EDGE, multiple voice user channels (now separated in code) are devoted to a data user to transmit its burst of data at high speed. The data occupies the same bandwidth as voice but, similar to EDGE, can use a more bandwidth-efficient transmission to achieve even higher data rates. The same idea, when applied to Universal Mobile Telecommunications System (UMTS) using wideband CDMA technology with a bandwidth of 5 MHz, can allow another growth in the data rate for data burst transmissions. In the late 2000s, the High Speed Packet Data (HSPA) was introduced that could achieve several tens of Mbps. When these technologies are combined with MIMO technology, literally allowing multiple streams of the signal over the same bandwidth, another increase in data rate allows the achievement of several hundred Mbps for wireless mobile data access that is the highest one can expect from CDMA-based technologies.

Another path of evolution for cellular data networks came as an offspring of the IEEE 802.11 WLAN technology. In the early 1990s, while the European Union industry, lead by Nokia and Ericsson, was dominating the cellular telephone industry by capitalizing on the success of the GSM standard and its worldwide adoption, a smaller group initiated the idea of the High PERformance LAN (HIPERLAN) [Wil95a,b] to achieve higher data rates than the other LANs at that time. HIPERLAN-1 was the first standard to employ 5 GHz unlicensed bands with non-spread spectrum technologies for the next generation of WLANs. Spread spectrum technology that was mandated by FCC for the original unlicensed ISM bands sacrifices bandwidth to achieve resistance to interference and with lower power consumption. The focal point of the LAN industry, however, has always been the achievement of higher transmission rates. HIPERLAN-1 was considering an adaptive equalization of the channel that could achieve data rates an order of magnitude higher than those achievable with spread spectrum technologies adopted by the legacy IEEE 802.11 at that time [Sex89]. This standard did not gain popularity in commercialization and it was continued by HIPERLAN-2 that merged its physical layer activities with IEEE 802.11a. As mentioned earlier, they adopted OFDM technology that can achieve the highest data rates in multipath-rich indoor environments. An early comparison of all these transmission technologies for WLAN applications is available at [Fal96]. The differentiation between HIPERLAN-2 and IEEE 802.11a was medium access control. HIPERLAN-2 was devoted to centralized medium access, rather than the contention-based medium access used in IEEE 802.11, which is better in supporting quality of service in heavy traffic and was favored by cellular manufacturers making all their income at that time from cellular telephone applications.

HIPERLAN-2 did not meet commercial success either. However, close to the end of the 1990s, the idea of centralized medium access for better quality of service and using transmission techniques similar to WLANs extended to the IEEE 802.16 for metropolitan area networking for multipoint fixed access as a backbone for wireless networking. These activities later evolved into the mobile Wi-Max standard using OFDM with MIMO technology with a centrally assigned medium access control that gained significant attention in the late 2000s and received reasonable worldwide deployments. The next step in this evolutionary path was the Long Term Evolution (LTE) that at the time of writing is being followed by LTE-Advanced. The objective of these wireless technologies is to achieve gigabit per second data rates for wide area coverage.

The stimulus behind these technologies in the past few years has been the enormous success of the iPhone, followed by other smart phones and tablet computers demanding tremendous amount of bandwidth for data applications. The popularity of these devices is shifting the habits of people from using the telephone as the main medium for communications to email, short messaging, and social networking. New multimedia applications such as YouTube access are gaining rapid increases in usage. These data applications with high data rate demands have forced the service providers to re-design their networks for the emerging world of data communications and have been the force behind the emergence of WiMAX and LTE standards. Adapting these technologies for the emerging smart device is the key for the future success of these networks. However, smart devices prefer using WiFi because it is mostly free of charge, provides higher data rates per user, and consumes less battery. To take advantage of these features for cellular applications, this industry is considering Femtocell technology that uses traditional cellular technologies for integrated voice and data in an all-IP environment, such as WiMAX and LTE, to design base stations with a smaller coverage similar to WLANs. The future of the local wireless access industry could be a struggle between Femtocell technology and the existing WLAN technologies for local wireless access. Femtocells may be deployed by the service provider or by a user and they have their own feature of better quality of service for voice applications. WLANs are mostly deployed by individuals randomly but at a low-cost. If it happens, it would be one of those rare cases that a successful trend for the growth of a data communication network technology is reversed in favor of a voice-oriented technology. For wide area coverage, 3G and 4G technologies are competing to win their adoption by smart phone, tablet, and laptop manufacturers.

Figure 1.14 sketches a comparison among different wide, local, and personal area wireless networks. The vertical axis shows the degree of mobility and the horizontal axis the data rate per user. Since the coverage of a cell tower may span several tens of kilometers, while a WLAN access point covers less than 100 m and WPANs cover around10 m, the number of users sharing a carrier is much higher for the larger areas that restrict the data rate delivered to each user. Another hidden factor in this comparison is the issue of battery life for the mobile terminal. Larger areas demand higher power consumption and consequently lower battery life for the mobile terminal.

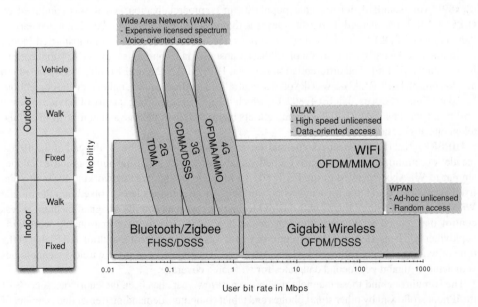

Figure 1.14   Overview of wireless technologies from a user perspective.

These issues are critical in the adoption of technologies for the future of this industry. For example, a smart phone manufacturer may consider 3G with lower data rates but lesser power consumption over 4G with higher data rates, but more power consumption.

1.5   Evolution of Wireless Localization Technologies

The RF localization industry started by addressing the problem of locating mobile radios used for military operations during World War II to locate soldiers in emergency situations. About 20 years later, during the Vietnam War, the United States Department of Defense launched a series of Global Positioning System (GPS) satellites to support localization during military operations in combat areas. In 1990, the signals from GPS satellites were made accessible to the private sector for commercial applications such as fleet management, navigation, and emergency assistance. Today, GPS technology is widely available in the civilian market for personal navigation applications. GPS receivers are designed to determine the locations of boats, planes, or mobile vehicles in open areas such as waterways, skyways, and highways. However, the accuracy of GPS positioning is significantly impaired in urban and indoor areas, where received signals can suffer from extensive multipath effects and additional path loss. In the past decade or so, to remedy this situation wireless localization technologies using signals other than GPS have emerged. We may refer to these localization technologies using signals of opportunity. A full description of the GPS system is beyond the scope of this book, but the interested reader can find much information in the open literature [Kap96]. In this book we address wireless localization techniques using time-of-arrival (TOA) and received signal strength (RSS) of the signals of opportunity.

1.5.1   TOA-based Wireless Localization

The GPS receivers measure the TOA of the received signal to measure the distance to satellite landmarks and from that locating itself on the Earth within a few meters. However, it does not work properly in urban and indoor areas where a number of computer-related applications can benefit from the location information. In the mid-1990s the Defense Advanced Research Projects Agency (DARPA) launched its small unit operation situation awareness system (SUO/SAS) program aiming at one-meter accuracy for indoor geolocation in military and public safety operations. About the same time, venture capitalists started funding startup companies such as PinPoint (in Woburn, Massachusetts) and WhereNet (based in Santa Clara, California), both seeking to develop and implement indoor geolocation technologies with accuracies comparable to those required for SUO/SAS.

The success of TOA-based techniques used in GPS positioning led military and commercial researchers to think in that direction. The idea sounded very straightforward. Using the operating frequency, bandwidth, and signal strength found in GPS systems, accuracies around several meters can be achieved within the course of a few minutes. If we want to extend this technology to practical indoor geolocation, we must overcome four challenges:

1. We need a positioning accuracy of better than a few meters to identify objects in different rooms of a building.
2. We need to cope with around 20–30 decibels of additional path loss to penetrate into the building.
3. We need algorithms to cope with multipath conditions.
4. We need to reduce the time to first fix to just a few seconds.

The pioneering military and commercial TOA-based systems designed in the late 1990s did not meet these challenges. DARPA had to compromise on its accuracy requirements and the commercial startups simply failed.

Radio propagation studies conducted for the SUO/SAS project revealed that the primary source of the problem for indoor geolocation was severe multipath conditions in obstructed line of sight (OLOS) environments that frequently caused large ranging errors [Pah98]. To remedy the situation, the developers of military and public safety applications resorted to such methods as UWB (ultrawideband), super-resolution, multipath diversity, and cooperative localization [Pah06]. More recently, inertial navigation systems and other sensors have been added to some systems in an effort to overcome the deficiencies of RF indoor geolocation; and hybrid localization using a variety of sensors complementing TOA-based RF localization is under development [Moa11]. The specifics of technologies for military and public safety applications are that these technologies are designed for first responder situations where minimal information about the environment is available because such environments are either unknown, such as a military target, or under rapid change, such as a disaster sight. In commercial applications, maps of the buildings are known and the site can be surveyed for the RF signature of the existing infrastructure in the building. The results of these RF surveys of the areas can be used later to localize the devices using less complex RSS-based localization.

1.5.2   RSS-based Localization

For commercial applications, other major problems include the cost of new proprietary hardware and the deployment of infrastructure. These cost factors led the industry to develop indoor geolocation techniques leveraging existing Wi-Fi and cellular networking infrastructures, which were growing rapidly in a variety of indoor environments. The accuracy of RSS-based localization is on the order of the coverage of the access points or base stations. Since the coverage of Wi-Fi is on the order of a few tens of meters while base stations cover several kilometers, Wi-Fi localization techniques using RSS have become a very important opportunistic localization technique to complement the GPS shortcoming in indoor areas. At the time of writing, RSS-based Wi-Fi localization is the most popular localization engine in smart devices such as iPhone, iPad, Kindle, and other similar devices. The database of leading Wi-Fi localization companies for smart devices, Skyhook (Boston, Massachusetts), receives several hundreds of millions of location request from smart devices per day.

The concept of using Wi-Fi infrastructure for TOA- and RSS-based localization in indoor areas was first introduced in 2000 [Li00; Bah00]. RSS-based localization needs a sight survey to create a data base as a reference for the RSS pattern of Wi-Fi devices at different locations in a building [Roo02a,b]. In the early days of this industry the data base was collected in indoor areas using a sight map. The locations where each data point was collected was marked on the map. This approach had three drawbacks: (1) the maps of indoor areas were not publicly available, (2) the data collection was very time-consuming, and (3) the coverage of the system was limited to a few uncoordinated buildings. The market for these application was certain buildings such as hospitals, museums, and warehouses in which localizing needs for equipment, merchandise, or people were in high demand.

In the mid-2000s Wi-Fi localization in urban areas emerged in the market. In these systems the data collection consisted of driving the streets of an urban area, with the receiver tagging the location where the RSS of access points were measured by the GPS reading in those locations [Pah10]. These systems are referred to as wireless positioning systems (WPS) and they cover the metropolitan areas. The WPS systems are those that were first adopted in the iPhone, the leading popular smart device, which revolutionized communication applications and networks to support those applications. The advantage of these systems is their comprehensive coverage using Google satellite maps. More recently Google started a collection of indoor maps for popular public places such as airports and shopping malls; and this has stimulated a number of new startup companies working on indoor geolocation applications in those areas.

1.6   Structure of this Book

Wireless access and localization is a very complex multidisciplinary systems engineering discipline. To describe these networks we need to divide their details into several categories to create a logical organization for the presentation of important material. From the material presented in the previous section on the evolution of wireless networks, we observe that the essential material needed for students to understand these systems in a comparative manner are the details of applied transmission techniques and the medium access control methods. Comparison of these techniques needs an understanding of radio propagation in urban and indoor areas that suffers extensively from the existence of multipath conditions. To make the book suitable for teaching in an engineering or science curriculum we need to present useful quantitative and analytical examples that are relevant for a comparative evaluation of these systems. In textbooks and associated courses in specific fields such as digital communications or signal processing, it is common to present the details of the derivation of techniques for transmission or design of filters and transforms that are useful for processing the signal. Multidisciplinary fields such as wireless networking, robotics engineering, or bio-engineering however *use* the results of several other disciplines and merge them for the creation of a new field. The analytical examples used for students in such fields are more diversified and they have to be selected carefully to avoid excessive complexity and yet remain useful and non-trivial to carry educational values. Therefore, a clear organization of the presented material and depth of discussion on the variety of issues is needed because the multidisciplinary nature of the material plays a significant role in the properness of the book.

This book is intended to describe technologies presented in the previous section and summarized in Fig. 1.13 during the evolution of wireless access techniques and to provide an understanding of the principles of emerging wireless localization techniques. It is intended to provide an overview of standards and technologies, the basic fundamental science behind these technologies, and examples of popular standards using these fundamental technologies to explain the "why" and "where" of their applications. The book is organized into an introduction chapter and five parts. The introductory material presented in Chapter one defines the meanings and sketches the evolutionary path of wireless networks. This chapter also provides an overview of the important wireless systems and outlines the details of the rest of the book. The material presented in Chapter one identifies different sectors of the wireless market, familiarizes the reader with the forces behind the growth of these sectors, and provides an overview of the standards developed to address them. The material presented in Chapter one motivates the reader to study the details provided in the remainder of the book.

The five parts of this book each consists of several chapters directed toward a description of certain aspects of wireless networks. The first and second parts of the book are devoted to the principles of air–interface design and the principles of wireless network operations. These parts provide the technical background needed for an understanding of wireless networks. The technical aspects are either related to the design of the air–interface or issues related to the deployment and operation of the infrastructure. The first part consists of three chapters describing technical aspects of the air–interface. The second part comprises three chapters devoted to the technical aspects of the wireless network infrastructure. The third and fourth parts of the book are devoted to a description of the details of typical wireless networks in a comparative manner. The third part comprises three chapters on local broadband and ad hoc wireless networks, describing popular WLANs and WPAN technologies. The fourth part consists of three chapters describing technologies used for the implementation of wireless wide-area networks. The fifth part of the book is devoted to RF localization and includes three chapters that describe system aspects, analytical bounds, and practical implementation, respectively.

1.6.1   Part I: Principles of Air–Interference Design

Wired terminals are powered and connect to transmission lines; and wired access to information networks is reliable, fixed, and relatively simple. Wireless mobile terminals are battery operated and wireless access is through the air that is unreliable and band-limited. The design of the physical layer connection and access method and an understanding of the behavior of the medium for wireless operation is far more complicated than for wired operation scenarios. Design of the air–interface for wireless connections needs a far deeper understanding of the behavior of the channel and more complex physical and medium access control mechanisms. The behavior of the wireless medium is more complex than that of the wired medium because in a wireless channel the received signal strength suffers from extensive power fluctuations caused by temporal and spatial channel dynamics. Transmission techniques used for wireless access are more complex because they have to be power- and bandwidth-efficient and they need to employ techniques to mitigate the received power fluctuations caused by the medium. Part I of this book is devoted to the analysis of the behavior of the channel in Chapter 2, an overview of the applied wireless transmission techniques in Chapter 3, and a description of the medium access control techniques used for wireless access to the PSTN and the Internet in Chapter 4.

Chapter 2 describes path-loss modeling, fluctuation of the channel, and the multipath arrivals of the signal. Path-loss models describe the relation between the average received power in a mobile station and its distance from a base station. These models are used in the deployment of networks to determine the coverage of a base station. In communication over the wireless medium the received power is not constant and it changes in time as the mobile moves or the environment changes. Models for variations of the channel are used to design the adaptive elements of the receiver, such as synchronization circuits or equalizers, to cope with variations of the channel. Models for multipath characteristics allow the design of a receiver that can handle the interference from signals arriving along different paths at the receiver.

The second chapter related to the air–interface is Chapter 3, which describes the digital transmission techniques used for the implementation of a variety of wireless networks. The diversity and complexity of transmission techniques in a wireless system is greater than that observed in wired networks. This chapter provides an overview of the principles of transmission techniques that are used for wireless access to the PSTN and the Internet. We first provide a brief description of the effects of multipath on the performance of wireless networks. Then we describe traditional transmission technologies, multipath resistant transmission techniques, the coding techniques used in wireless networks, and a brief description of cognitive radio and dynamic spectrum management.

The third chapter related to the air–interface is Chapter 4, devoted to the applied multiple access alternatives for packet transmission over the wireless medium. This chapter starts with a description and comparison of the assigned access schemes, such as TDMA, and CDMA, evolved for wireless access to the PSTN. The second part of this section is devoted to CSMA and ALOHA based random access techniques, such as ALOHA and Carrier Sense Multiple Access (CSMA), used for wireless access to the Internet. The last part of this section analyzes the applied access methods for the integration of voice and data that has evolved for wireless access to the PSTN and the Internet.

1.6.2   Part II: Principle of Network Infrastructure Design

In Part II of the book we address technical aspects of the design of the fixed infrastructure of wireless networks. This part consists of Chapters 5, 6 and 7, respectively addressing the deployment, operation, and security of wireless networks. Wireless networks share the medium and when they operate at the same or an overlapping frequency band, they interfere with each other. As a result, an understanding of the nature of the interference is essential for the deployment of a wireless network infrastructure. The frequency bands used by different wireless networks are either unlicensed, commonly used by WLANs and WPANs, or they are licensed to a service provider in a region, which is popular in cellular telephone networks. Unlicensed bands are open to the public, and as a result deployment of the network and interference between devices are not practically controllable. In licensed bands the service provider owns the bands as well as the network infrastructure; therefore, it can control the interference and use it to its advantage to provide an efficient and comprehensive coverage. In Chapter 5, we discuss interference between WLAN and WPAN products operating in uncontrolled unlicensed bands as well as frequency management in licensed bands to control the interference. Service providers using licensed bands often start with a minimal infrastructure and few antenna sites, to keep the initial investment low. As the number of subscribers grows, the service providers expand their wireless infrastructure to increase capacity and improve quality. The technology related to the deployment and expansion of the cellular infrastructure is also discussed in Chapter 5. This chapter is the first chapter related to the technical aspects of network infrastructure. It discusses interference in general, different topologies for wireless networks, frequency management in cellular infrastructure deployment, and issues related to expansion and migration to new technologies.

Chapters 6 and 7 are devoted to the functionalities of the fixed network infrastructure to support a mobile operation. These functionalities include mobility, radio resource, power and security managements. These issues are addressed in two separate parts of Chapter 6 and in Chapter 7. The mobility management part of Chapter 6 describes how a mobile terminal registers with the network at different locations and how the network tracks the location of the mobile as it changes its access to the network from one antenna site to another. The radio resource and power management part of Chapter 6 is devoted to the technologies used for controlling the quality and transmitted power of the terminals. Voice-oriented networks control the transmitted power of the mobile station to minimize interference with other terminals using the same frequency and to maximize battery life. Data-oriented networks use the sleep mode to avoid unnecessary consumption of power. Explanations of the methodologies and examples of how to implement power control and sleeping modes are provided in this part of Chapter 6. Chapter 7 is devoted to security in wireless networks. Wireless connections are inherently vulnerable to fraudulent connections and eavesdropping and need security features to avoid them. The security of wireless networks is provided by authentication and ciphering. When a wireless terminal connects to a network, an authentication process between the network and the terminal checks the authenticity of the terminal. When the connection is established, the transmitted bits are scrambled with a ciphering mechanism to prevent eavesdropping. Algorithms used for these purposes are discussed in the last part of Chapter 7.

1.6.3   Part III: Wireless Local Access

After completion of the overview of the standards in Chapter one and a study of the overview of technical aspects in Parts I and II, we provide detailed descriptions of practical wireless access techniques in Parts III and IV. These detailed descriptions are divided into two parts, addressing wireless local access (Part III) and wide area wireless access (Part VI).

Part III of the book describes WLAN and WPAN technologies used for wireless local access in unlicensed bands, and it consists of three chapters. The first chapter of these three is Chapter 8, which provides an overview of the WLAN industry and the IEEE 802.11 standardization committee. This chapter provides the details of IEEE 802.11 to demonstrate all aspects of the packet switching wireless standards operating in unlicensed bands. The medium access technology for IEEE 802.11 is CSMA/CA (CSMA with Collision Avoidance) that sets this standard as a connectionless packet switched standard. This feature eases Internet access, either by direct connection or by connection through an existing wired LAN. The contents of this chapter describe the objective of the standard, explain the specifications of packet framing, physical and medium access control layer alternatives supported by this standard, and provides the details of mobility support mechanisms, such as registration, handoff, power management, and security.

Chapter 9 is devoted to low-power WPAN technologies operating at data rates that are lower than those of WLANs, with particular emphasis on the details of Bluetooth and ZigBee technologies. This chapter starts by describing the IEEE 802.15 standards committee and low-power versus gigabit wireless technologies for WPANs. Then details of Bluetooth and ZigBee technology are described in reasonable depth. These types of WPANs are ad hoc networks designed for operating over short distances to connect personal equipment or sensors to one another. These personal devices or sensors ultimately provide wireless communications for voice- or data-oriented applications. The physical layer and medium access control methods for these systems are designed to suit one of these applications. Bluetooth access is centralized, which is more suited for the quality control needed for real-time streaming for voice applications and consequently it has gained more popularity in applications such as wireless headphones or wireless microphones. The ZigBee access method is a light version of the contention-based CSMA/CA, originally used in IEEE 802.11, which is better suited for data burst applications leading this technology towards sensor networking applications for low-power low-rate packet transmissions. After a description of Bluetooth and ZigBee as established WPAN standards, we conclude Chapter 9 with a short description of another low-power emerging standard, IEEE 802.15.6 for Body Area Networking (BAN), which is an active area of research at the time of writing.

Chapter 10 is devoted to technologies to support data rates beyond the existing WLANs and refers to these technologies as gigabit wireless networks for local and personal area networking. At the time of writing, a super high-speed WPAN industry is not as established as low-power WPANs and it is experiencing its evolutionary path waiting for a sizable market to emerge in the near future. However, those engaged in research and development in this industry believe that it will serve a number of applications, including high-definition video streaming, wireless gigabit Ethernet, wireless docking stations, desktop point to multipoint connections, wireless back haul, and wireless ad hoc networks [Guo07; Per10; Dan10; Kum11]. We start the Chapter 10 with the UWB technology for 3.4–10.6 GHz bands and we continue this discussion into gigabit wireless technologies evolving for 57–64 GHz bands for short-distance gigabit wireless. The objective of this chapter is to provide an overview of the evolution of this industry and to provide some detailed examples of the fundamentals for the design and operation of these networks.

1.6.4   Part IV: Wide Area Wireless Access

The fourth part of the book is devoted to a description of the important wide area networking technologies used for wireless access to the PSTN and the Internet. As explained in Section 1.2, cellular networks were originally designed for cellular telephone applications and connection to the circuit switched PSTN. Therefore, the center of attention for the design of 1G, 2G, and 3G cellular networks was connection to the PSTN for cellular telephone applications. The cellular telephone needs a low data rate of around 10 kbps for each of numerous simultaneous users engaged is two-way conversations. The quality of service for mobile users as they change their connection from one to different antennas in the network is very important for users. These requirements demand simple transmission techniques and more robust and centrally controlled medium access techniques. The cellular network technologies evolved for cellular telephone application using FDMA, TDMA, and CDMA techniques for 1G, 2G, and 3G cellular networks, respectively, to address this demand. In the late 2000s, and with the beginning of the popularity of smart phones, the cellular network providers started paying attention to all-IP 4G networks for Internet connection, and consequently OFDM and MIMO technologies evolved for wireless Internet access. In this part of the book we have three chapters to explain the principles of operation of TDMA and CDMA cellular networks primarily designed for wireless access to the PSTN and the differences between the 4G cellular networks and the WLANs that are designed for wireless Internet access.

In Chapter 11 we describe the overall infrastructure of a wide-area wireless access network to connect to the PSTN and details of the TDMA air–interface used in the popular 2G TDMA digital cellular network, the GSM. In Chapter 12 we describe the details of wireless access to the PSTN using CDMA technology used in original 2G cdmaOne and its extension to 3G cellular networks. The obvious reason for these selections is the worldwide popularity of GSM and the emergence of CDMA technology as the choice for 3G cellular systems. The last chapter in this part of the book is Chapter 13, devoted to the implementation of OFDM and MIMO technologies for 4G cellular networks.

Details of the architecture, the mechanisms to support mobility, and the layered protocols in a cellular telephone network are described in Chapter 11. This description is completed by including the specification of the air–interface of GSM as an example of a TDMA air–interface. Other TDMA digital cellular standards such as DECT, currently popular in cordless telephone applications, are very similar in nature to the GSM. In Chapter 11 we first describe all the elements of a cellular network architecture. Then, we address mobility support mechanisms with details of registration, call establishment, handoff, and security. The last part of this chapter provides details of how packets are formed and transmitted over the TDMA air–interface. The study of this chapter introduces the reader to the complexity and diversity of the issues involved in the development of a wireless cellular network.

Chapter 12 is devoted to CDMA and WCDMA technology that is adapted for 3G cellular systems. Since the wired backbone of the TDMA and CDMA systems are very similar, most attention is paid to the air–interface that is completely different from TDMA systems such as GSM. In this chapter, we describe the reasons for employing CDMA, the way users are separated within and across cells using spread spectrum codes, and how spread spectrum helps in exploiting multipath diversity. Details of how spread spectrum influences the design of the air–interface in 2G and 3G systems, as well as special challenges with CDMA such as soft handoffs and power control are also explained in this chapter. The evolution to HSPA through changes in the network architecture and air–interface concludes this chapter.

The last topic in this part is a description of 4G cellular networks that is provided in Chapter 13. This chapter is devoted to details of Wi-Max and LTE technologies for cellular networks. In this chapter, we explain the reasons for using OFDM as the primary transmission scheme with MIMO technology for higher data rates. We also describe the flattening of the network architecture with WiMax and LTE compared to the traditional hierarchical architecture used in GSM and CDMA networks. We finally discuss the evolution of this technology to LTE-Advanced.

1.6.5   Part V: Wireless Localization

The fifth and last part of the book is devoted to wireless localization techniques. A fundamental element of localization is a map and a means to find the distances from landmarks that can be identified on the maps. If we use an RF signal to find the distance from a landmark, we may call the system an RF localization system. The first popular RF localization system was GPS, designed originally for military applications and later available for commercial applications. GPS does not work properly indoors and in highly dense urban areas, it has a relatively long warm-up time, and it is a piece of hardware that has its own toll on the battery life of mobile devices. To extend the coverage of GPS to indoor and urban areas for both military and commercial applications wireless localization techniques using signals other than the GPS signals started in more recent years. The technical connection between wireless localization techniques and wireless networks is an understanding of channel behavior and its impact on waveform transmission on different media for wireless transmission and the fact that wireless localization techniques often use wireless infrastructures and signals for the purpose of localization. In Part V we have three chapters addressing different aspects of wireless localization. Chapter 14 provides an introduction to wireless localization systems, Chapter 15 covers the fundamentals of wireless localization, and Chapter 16 addresses the practical aspects of wireless localization.

Chapter 14 is devoted to an introduction to indoor geolocation and cellular positioning as emerging technologies to complement local and wide-area wireless services. This chapter provides a generic architecture for wireless geolocation services, describes alternative technologies for the implementation of these systems, and gives examples of evolving location-based services that are becoming necessary for a variety of applications.

Chapter 15 starts with modeling of the RF sensors used for localization technologies. This behavior modeling is important to understand the complexity of the technologies that evolved for wireless localization in indoor and urban areas. The next part of this chapter is devoted to the calculation of the Cramer–Rao Lower Bound (CRLB) for performance evaluation of wireless localization techniques. These bounds enable us to compare the performance of different wireless sensors in a quantitative manner. The last section of Chapter 14 is devoted to a survey of fundamental algorithms used for localization using different RF sensing techniques.

Chapter 16 is devoted to practical aspects of wireless localization techniques. We begin this chapter by addressing issues related to RSS-based Wi-Fi localization, which is the most popular wireless localization technique currently used in smart devices for numerous popular location-based applications. Then we address more precise indoor positioning using TOA-based localization by introducing challenges in measurement of the TOA with particular emphasis on the large errors that are caused when the direct path between the transmitter and the receiver is blocked. This discussion is followed by introducing localization using multipath diversity and cooperative localization for situations when the direct path between the transmitter and the receiver is blocked. The last section of this chapter is devoted to localization inside the human body as the latest localization technique that is emerging at the time of writing. The human body is a non-homogeneous liquid-like medium for RF propagation with extensive path loss and a blurred map because the objects inside it are constantly moving. As localization transforms from outdoors to indoors and then to inside the human body, we explain the challenges facing the research community to dicover this medium for RF localization needed for emerging wireless health applications.

Questions

1. Why is a wired network usually part of the wireless infrastructure?
2. How is a wireless network different from a wired network? Explain at least five differences.
3. What is the difference between a licensed and an unlicensed band? Give one example wireless technology standard that operates in licensed and one that operates in unlicensed bands.

4. What are the differences among WLAN and 4G (WiMax and LTE) data services in terms of regulation for frequency of operation, data rate, coverage, and cost charging mechanism?
5. What is Wi-Fi localization and how does Wi-Fi localization complement GPS technology?
6. Name the original four categories of markets that have evolved for the wireless networks.
7. Explain why standardization is important for wide, local, and personal area wireless networks.
8. What are the main differences between the characteristics of technologies needed for wireless access to the PSTN and the Internet?
9. How does an standard evolve for wireless networks and what type of organizations are involved in the process?
10. What are the differences among connection-based and connectionless networks? Give an example for each of the two networks.
11. What is WPAN? What is the difference between WPANs and WLANs? Name the two major standardization organizations writing the draft standards for these networks.
12. When the ISM bands were released, what was new about them, and what are the available ISM bandwidths at 0.9, 2.4, and 5.7 GHz?
13. What is ZigBee technology and how does it differ from Bluetooth technology? Which standardization organization is working on these technologies?
14. What does BAN stands for? Name two popular applications using BAN technology and one standards organization that is devoted to this technology?

Project 1.1

Search Chapter 1 and the Internet (IEEE Explore, Wikipedia, Google Scholar, ACM Digital Library) to identify one area of research and one area of business development which you think are the most important for the future of the wireless information networking industry. Give your reasoning why you think the area is important and cite at least one paper or a website to support your statement.

Project 1.2

Go over the technical program presentations and synapses of the WPI workshop on Body Area Networking Technology and Applications on 20 June, 2011: http://www.cwins.wpi.edu/workshop11/program.html. Do a Google scholar search for each category of fundamental challenges itemized in the synapses and find one paper in that area that you think is interesting. List the authors, title, publisher, and date of your selected papers for each category. Also, print and attach one paper that you enjoyed most and read through. Explain why you liked that attached paper.

Project 1.3

Read the following articles:

http://www.intomobile.com/2011/12/23/were-top-mobile-trends-2011-infographic/

http://www.businessweek.com/magazine/map-apps-the-race-to-fill-in-the-blanks-01122012.html

http://www.insidegnss.com/auto/may10-Pahlavan.pdf

and write a one-page essay on the current and future directions in Wi-Fi localization. You need to address the business aspects, technology challenges, and future directions of this area.

Part I

Principles of Air–Interference Design

2

Characteristics of the Wireless Medium

2.1 Introduction

An understanding of radio propagation is essential for coming up with appropriate design, deployment, and management strategies for any wireless network. In effect, it is the nature of the *radio channel* that makes wireless networks far more complicated than their wired counterparts. In wireless access and localization, which is the focus of this book, we use electrical signals or waveforms to transmit information for communications or to measure the distance between the transmitter and the receiver for localization applications. In open space waveform transmission for applications such as satellite communications the changes in the transmitted waveform are caused by the characteristics of the antennas and they are predictable. The designers of the antennas design them so that the vital characteristics that are essential at the receiver are preserved. The only unpredictable received signal is the noise that is collected through the medium or is produced during the implementation of the transmitter and the receiver devices. In these applications a system designer is mostly concerned with the ratio of the received signal strength and the noise at the receiver to determine the required transmission power to accommodate the desired coverage.

Radio propagation in open areas is very different from radio propagation indoors and in urban areas for which wireless networking applications are designed. Analysis of radio propagation in open areas across small distances or free space is relatively simple and deterministic because the signal arrives through one path between the transmitter and the receiver. Indoors and in urban areas the signal arrives from different paths and, as we show in this chapter, the multipath arrival causes signal fading over time, frequency selective fading, and distortions in the waveforms used for wireless access and localization. The signal fading over time is caused by movements of the transmitter or the receiver and by movements of objects close to the transmitter and the receiver. Frequency selective fading is caused in certain frequencies in which multipath components add to one another in a destructive manner. The resulting changes in the shape of the transmitted waveform cause interference among the transmitted symbols carrying the information bits or measurement of the time of flight between a transmitter and a receiver for localization purposes. Understanding the nature of multipath indoors and in urban areas is essential for understanding all these features of wireless transmission that affect the design and deployment of wireless networks.

Principles of Wireless Access and Localization, First Edition. Kaveh Pahlavan and Prashant Krishnamurthy.
© 2013 John Wiley & Sons, Ltd. Published 2013 by John Wiley & Sons, Ltd.

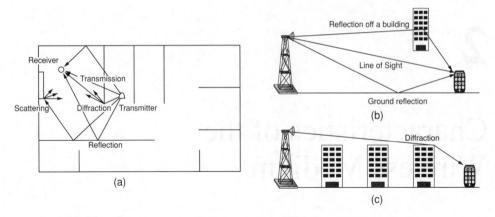

Figure 2.1   Examples of propagation paths and the mechanisms causing them.

2.1.1   Causes of Multipath Propagation

Electromagnetic waves at frequencies used in wireless networks can be treated as rays [Ber94], each representing a propagation path between a transmitter and a receiver. In free space there is only one direct line of sight propagation path connecting the transmitter and the receiver. Indoors and in urban areas multiple propagation paths carry the signal in different directions. Figure 2.1 shows examples of propagation paths in three different wireless communication scenarios. In Figure 2.1a four mechanisms (namely: transmission, reflection, diffraction, scattering) are demonstrated in the layout of an indoor area. Figure 2.1b demonstrates a scenario in an urban area that includes direct line-of-sight (LOS) transmission and reflection from buildings and ground. Figure 2.1c shows a dense urban area with diffraction mechanism. Therefore, we can divide the mechanisms causing multipath reception into direct LOS and four more mechanisms:

> *Reflection and transmission*: Upon reflection or transmission, a ray attenuates by factors that depend on the frequency, angle of incidence, material used in construction, and thickness of the walls. Reflection and transmission mechanisms often dominate radio propagation in indoor applications. In outdoor urban area applications, the transmission mechanism often loses its importance because it involves transmissions through multiple walls that reduce the strength of the signal to negligible values.

> *Diffraction*: Rays that are incident upon the edges of buildings, walls, and other large objects can be viewed as exciting the edges to act as a secondary line source. Diffracted fields are generated by this secondary wave source and propagate away from the diffracting edge as cylindrical waves. In effect, as shown in Figure 2.1(c), this phenomenon results in propagation into *shadowed* regions allowing the diffracted field to reach a receiver, which is not in the line of sight of the transmitter. Since a secondary source is created, it suffers a loss much greater than that experienced via reflection or transmission. Consequently, diffraction is an important phenomenon for outdoor applications in high-rise urban areas where signals transmission through buildings is virtually impossible. But it is less consequential indoors, where a diffracted signal is extremely weak compared to a reflected signal or a signal that is transmitted through a relatively thin wall.

Scattering: Irregular objects such as wall roughness and furniture in indoor areas and vehicles and foliage in outdoor areas scatter rays in all directions in the form of spherical waves. This occurs especially when objects are of dimensions that are on the order of a wavelength or less of the electromagnetic wave. Propagation in many directions results in reduced power levels, especially far from the scatterer. As a result, this phenomenon is not that significant unless the receiver or transmitter is located in a highly cluttered environment. For example, this mechanism dominates diffused infrared or any other light propagation when the wavelength of the signal is so small that the roughness of the wall results in extensive scattering. In satellite and mobile radio applications tree leaves often cause scattering. Scattering can however be beneficial in providing diversity and the ability to transmit and separate multiple data streams in the same bandwidth at the receiver using MIMO technology.

In the rest of this chapter, we analyze how multipath arrival causes problems in wireless networks and what methods have been adopted to model the behavior of multipath channels in different environments and frequencies of operation. A more detailed discussion can be found in [Pah05].

2.1.2   *Effects of Multipath Propagation*

In wireless networks, we transmit a signaling waveform or electronic *symbol* from a transmitter antenna, and at the receiver, we analyze the received waveform either to extract certain information about the channel and transmitted symbol or to estimate the distance between the transmitter and the receiver. The multipath nature of the channel results in several waveforms arriving at the receiver along different paths, each having a different amplitude, phase, and time of arrival. Therefore the received signal changes according to the multipath structure of the environment. Information in transmission schemes (discussed in more detail in Chapter 3) is usually embedded in the *amplitude* or *phase* of the transmitted signal. The estimation of the distance between the transmitter and receiver is obtained from the time of arrival or the strength of the received signal. Therefore, wireless communication and localization is highly affected by the multipath nature of the channel caused by the different mechanisms introduced in the last section. We need to understand these effects to design algorithms for reliable wireless communications and accurate localization. This analysis can be very complex for a static environment, and as a mobile user moves or people move in the vicinity of the transmission link, the multipath structure changes, which further adds to the complexity of the situation. Without getting into the details, first we explain what are the difficulties that are caused by multipath and what type of modeling of the channel is needed to help those working with wireless networks to achieve their goals.

Figure 2.2 shows an example of received signal strength in a multipath environment as a mobile terminal moves away from an access point (transmitter). When the transmitter is close to the receiver the power of the direct line of sight (LOS) path is dominant and the power in all other paths (caused by the various mechanisms) is negligible. As the terminal moves away from the access point, the power in the first path becomes comparable to those of the other paths and we see several multipath components with similar signal strength. When some objects such as walls, the human body, or furniture, block the direct LOS between the transmitter and receiver, the direct path receives significant attenuation and its power may even be less than that of some of the other multipath components (this is the effect of transmission through objects). Further, the power of the received signal fluctuates significantly; this is caused by fading that is caused by movement of the mobile terminal and the associated change in the multipath structure. This is in contrast to the

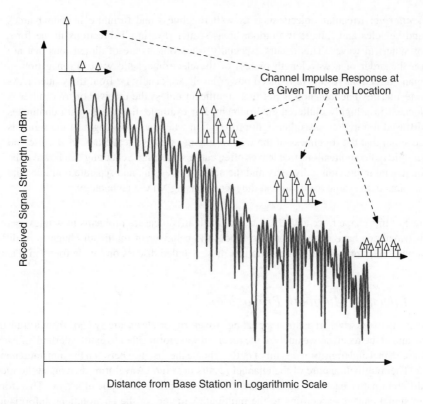

Figure 2.2   Relation between power in multipath components and received signal strength fluctuations with distance in wireless channels.

change in the frequency of channel that we will discuss next. We refer to this time and distance varying fading as *multipath fading* because it is caused by changes in the multipath behavior. An important factor we observe from Figure 2.2 is that the average received power is reduced as the distance between the access point and the terminal is increased. We need to model: (a) the average power reduction with distance, and (b) the fluctuations in the instantaneous power in time.

Figure 2.3 illustrates the measurement results of the channel *impulse response* between two locations inside a building and the *frequency response* of the channel, which is related to the Fourier Transform of the channel impulse response. We can measure the impulse response or frequency response characteristics either by: (a) sweeping the desired frequency band used for wireless communication or localization and taking the inverse Fourier Transform of the received waveform to form the impulse response, or (b) transmitting a narrow pulse and taking the Fourier Transform of the received waveform that consists of pulses arriving from different paths. In Figure 2.3 we have swept the channel in the frequency domain and we have obtained the channel impulse response from the measured frequency response. The impulse response illustrates the fact that a number of received pulses have arrived along different paths. The frequency response shows a huge variation in the signal strength at different frequencies and in particular certain specific frequencies in which signal strength is strongly attenuated (i.e., they cause *fading* in the received signal). In other words, the multipath nature of the channel causes *frequency selective fading* in the received signal. So, in

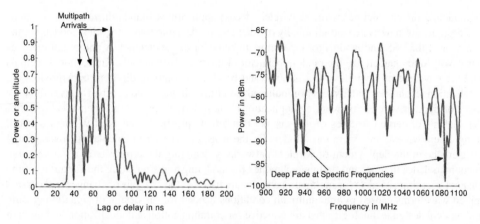

Figure 2.3   Typical (a) time and (b) frequency response of a radio channel.

addition to the models for received signal strength and time variations, to understand the nature of the channel we need to design models to explain the frequency selective behavior of the channel.

As shown in Figure 2.2, in wireless networking applications, when a receiver goes away from a transmitter, the multipath profile and the received signal strength vary randomly with the distance. The statistical behavior of these variations depends on the frequency of operation, bandwidth of the system, the architectural setting of the environment, and the scenario for the movement of the user and the application that it runs. The *frequency* of operation could go from hundreds of MHz for early cellular telephone applications up to several to tens of GHz for multichannel high-definition video transmission. The *bandwidth* used for the implementation of these systems can vary from tens of KHz for a cell phone to several GHz for multichannel high-definition video operation. The *architectural* setting can vary from a wide area cellular network operation along a highway that is made up of a simple, wide, and open area to streets of an urban area with a number of buildings compressed in a small area and to general indoor areas where Wi-Fi devices are installed inside a room or a car for a Bluetooth signal or even inside the human body for an endoscopy capsule operation. The speed of movement of the mobile terminal can be as different as the quasi-stationary applications running on a laptop residing in an office desk to a mobile cell-phone operating inside a high-speed train. Multipath characteristics and the rate of variations of the statistics of the channel behavioral characteristics are widely different in these scenarios. To accommodate such a wide variety of situations for modeling the statistics of the radio channel characteristics, standardization organizations and the designers of new systems for wireless access and localization spend a considerable amount of time in modeling the behavior of the channel for an emerging wireless technology and its evolving applications. We may not be very far off if we claim that the inception of wireless networks is based on application and propagation.

2.1.3   *Applied Channel Models for Wireless Communication Applications*

The three most important radio propagation characteristics used in the design, analysis, and installation of wireless information networks are the achievable signal coverage, the maximum data rate that can be supported by the channel, and the rate of fluctuations in the channel [Pah05]. Standardization committees for wireless networking usually provide models for the behavior of the power or received signal strength (RSS) and the typical channel impulse response caused by multipath.

In modeling the channel behavior for wireless access applications, standardization organizations and designers of new systems traditionally divide these models into two categories: models for the behavior of the RSS and models for the multipath behavior of the channel. A number of methods have evolved to characterize the random behavior of the wireless channels. These models help us find the coverage of a wireless access point and be able to compare different transmission and localization techniques under variety of multipath conditions indoors and in urban areas commonly used for popular wireless networking applications. These models provide mathematical tools for the analysis of certain features of the channel that affects practical aspects of wireless network design, deployment, and operation. Models for the RSS are used when calculating coverage and interference in the deployment of wireless networks, calculating the error rate as a wireless link encounters fades in time, and finding techniques for modeling and simulating the variations of the channel over time. Models for multipath behavior are used for the analysis of changes in the trans-mitted waveforms caused by the multipath conditions needed for the design of signal processing and modem design technologies that are effective for communication and localization in multipath channels.

The achievable signal coverage for a given transmission power determines the size of a cell in a cellular topology and the range of operation of a base station transmitter. This is usually obtained via empirical *path* loss models obtained by measuring the received signal strength as a function of distance. Most of the path loss models are characterized by a distance-power or path-loss *gradient* and a random component that characterizes the fluctuations around the average path loss due to shadow fading and other reasons. For efficient data communications, the maximum data rate that can be supported over a channel becomes an important parameter. Data rate limitations are influenced by the *multipath structure and spread* of the channel and the fading characteristics of the multipath components. This will also influence the signaling scheme and receiver design. Another factor that is intimately related to the design of coding and interleaving as well as the adaptive parts of the receiver, such as timing and carrier synchronization, phase recovery, and so on is the rate of fluctuations in the channel, usually caused by movement of the transmitter, receiver, or objects in between. This is characterized by the *Doppler spread* of the channel. We consider path-loss models and provide a summary of the effects of multipath and Doppler spread in subsequent sections of this chapter.

Depending on the data rates that need to be supported by an application and the nature of the environment, certain characteristics are much more important for wireless system design than others. This is because the channel characteristics impact the transmitted *symbols*, as we will see in Chapter 3. If the symbols last longer in time, the data rates are lower, but the impact of multipath is restricted to the rate of fluctuations of the channel. Here, the bandwidth of the transmitted signal is small and the signal fits within one of the *flat* regions in frequency seen in Figure 2.3 (i.e., there is no frequency selectivity – the entire signal faces fading in the same way). Thus the signal faces what we call narrowband or flat fading and the fluctuations in time matter most for narrowband cellular telephony. If the symbols last for very short periods of time, the data rates are larger, but the multipath now causes inter-symbol interference. Thus the multipath delay spread becomes important for high data rate wideband systems especially those that employ spread spectrum and OFDM technologies adopted in 3G and 4G cellular networks and in WLAN and WPAN applications. Another way of looking at this is that the bandwidth of the signal is large, and different parts of the signal in frequency face different frequency selective fading characteristics. Thus, both the fluctuations in time and the different fading characteristics in frequency become important here.

Other areas where the properties of the radio channel become important are in determining battery consumption, the design of tolerance of transmitter and receivers to the speed of mobile terminals, the design of medium access control protocols, design of adaptive and smart antennas, link-level monitoring for higher layer protocol performance, and the design of wireless system protocols for handoffs, power control, and interference cancellation.

2.2 Modeling of Large-scale RSS, Path Loss, and Shadow Fading

Figure 2.4 illustrates the variation in RSS as a function of the logarithmic distance between the transmitter and receiver as the receiver moves away from the transmitter and how we approach modeling them for different purposes. The RSS in a multipath environment always varies with time and with small local changes, on the order of the wavelength of the carrier frequency, in the location of the transmitter and receiver or the movement of the objects around them. However, the *average* received power over a small area is related to the distance from the transmitter to the center of the receiving area. To model this behavior for wireless networking applications we divide them into two classes of models: (1) models for average RSS change with respect to distance, and (2) models for RSS local fluctuations. In this section, we focus on the large-scale average RSS dependence on distance.

2.2.1 General Features of Large-Scale RSS

To model average RSS versus distance we first fit the empirical data as a function of distance on a logarithmic scale. As we will explain later, in wireless communications, the received power is exponentially related to the distance and that is the reason why we use a logarithmic scale for distance – to derive a simple linear relationship between the power and distance. In other words, we

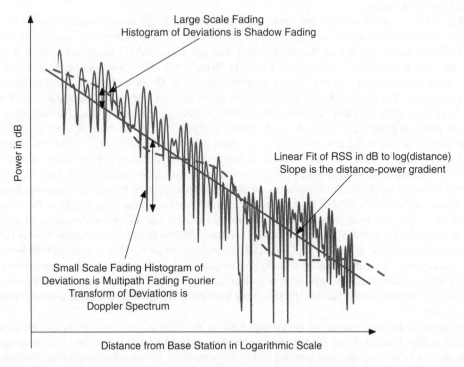

Figure 2.4 Modeling the average RSS for a mobile user with a linear fit path loss and a random shadow fading component; and modeling the statistics of fast fading and Doppler spectrum.

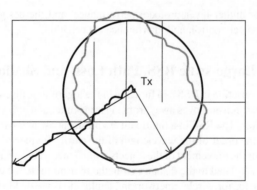

Figure 2.5 Variations in average RSS due to shadow fading when the distance is the same and when the distance is increasing.

could say that the large-scale average RSS in dB drops linearly with the logarithm of the distance from the transmitter. The drop in RSS is called the *path loss* for this reason (it is the loss in signal strength due to distance or path travelled). In simple words, the path loss is the difference between the transmitted power of the signal and the large-scale average RSS. If we consider the path loss instead of the RSS, the path loss *increases* linearly with the logarithm of the distance between the transmitter and receiver.

But, when we are in a large open area where the transmitter and the receiver are always in a LOS condition (they can see each other without obstructions) and away from obstacles like walls or furniture, the best fit line to the average RSS does not necessarily fit the actual average RSS around a given distance. For example, as shown in Figure 2.5, let the transmitter be in the middle of a room and let us draw a circle with a distance *d* around it. Obstacles between the transmitter and the receiver are different for different locations of the receiver, and so, the average RSS on the circle deviates from the calculated value of the average RSS using a best-fit line. This difference in the power is observed at locations in different directions, but at the same distance from the transmitter, because as the distances grow larger there is no longer a LOS between the transmitter and the receiver. Further, in each different direction, we have different objects obstructing the LOS. Deviations of the average RSS from the best-fit line is caused by the changes in the pattern of obstructions *shadowing* the direct transmission and for that reason it is referred to as *shadow fading*. As shown in Figure 2.5, shadow fading not only causes variations in the average RSS at the same distance, as we move away from the transmitter along a straight line, the power deviates randomly from the power predicted by the best fit line. Therefore, shadow fading causes fluctuations in the average RSS when the distance is kept fixed and deviations from the best linear fit to the RSS as we move away from the transmitter. Shadow fading is associated with the average short-time fading of the signal observed over a large scale of distances, so it is also referred to as the *long distance* or *long time or large-scale fading*. The term long time comes to mind for a mobile user traveling on a route.

In summary for the modeling of the average RSS, as shown in Figure 2.4, the slope of the best fit line to the observed data represents the exponential rate of variation of average RSS with the distance. As we will explain later, we refer to this exponent as the distance–power gradient. Shadow fading is the long-term average changes in the received signal strength caused by changes in the relative position of large objects between the transmitter and the receiver (e.g., buildings in urban areas, walls and furniture in indoor areas). The probability density function of the variations of the average amplitude of the RSS is the shadow fading characteristics of the channel.

Now that we have an intuitive understanding of the features of large-scale RSS, we begin introducing quantitative methods for calculation of the RSS in wireless communication applications. We start with path-loss modeling in free space using Friis' equation and its extension towards empirical path-loss modeling in multipath environments, then we provide example of models used for path-loss calculation, shadow fading, multipath fading, and Doppler spectrum.

2.2.2   Friis Equation and Path-Loss Modeling in Free Space

Harald T. Friis was a Bell Laboratory scientist born in Denmark (1893–1976). He developed a simple and practical model for the RSS in free space [Fri46], when there is only one path between the transmitter and the receiver. This simple model has been very helpful for understanding and empirical modeling of radio propagation. Free space implies that the transmitter and receiver are in a vacuum without even the Earth in between. Further, they both are only points in space and employ ideal isotropic radiating and receiving antennas, when an isotropic antenna radiates signal strength in all directions at the same power, the receiver antenna has an ideal effective area of $\lambda^2/4\pi$ where λ is the wavelength of the carrier of transmission. At a distance of d from the transmitter, the *density of signal strength* corresponds to the total radiated signal strength divided by the area of a sphere of radius d which is $4\pi d^2$. In other words, the signal strength density is $P_t/4\pi d^2$ on this sphere. An antenna with an effective area of $\lambda^2/4\pi$ receives a power $P_r = (\lambda^2/4\pi) \times (P_t/4\pi d^2)$ (see Figure 2.6 for a visualization). Thus, the transmitted power P_t and the received power P_r in free space are related by:

$$\frac{P_r}{P_t} = \frac{\lambda^2}{4\pi} \times \frac{1}{4\pi d^2} = \left(\frac{\lambda}{4\pi d}\right)^2 \qquad (2.1)$$

where d is the distance between the transmitter and receiver, $\lambda = c/f$ is the wavelength of the carrier, c is the speed of light in vacuum (3×10^8 m/s), and f is the frequency of the radio carrier. If the

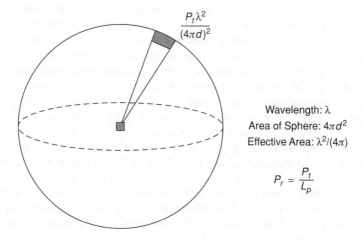

Figure 2.6   Visualization of the free space model.

transmitter antenna has a gain of G_t, and the receiver antenna has a gain of G_r in the direction from the transmitter to the receiver, then Equation 2.1 is modified to be:

$$\frac{P_r}{P_t} = G_t \times G_r \times \left(\frac{\lambda}{4\pi d}\right)^2 \tag{2.2}$$

We can simplify this equation to $P_r = P_0/d^2$, where $P_0 = P_t \times G_t \times G_r \times \left(\frac{\lambda}{4\pi}\right)^2$ is the received signal strength at the first meter ($d = 1$ m). Now, we can see that the received power falls exponentially, that is, as the square of the distance from the transmitter. The exponent associated with d is often called the path-loss gradient as it determines how rapidly the average RSS falls with distance.

We can rewrite Equation 2.2 in logarithmic or decibel form as follows:*

$$10\log\left(\frac{P_r}{P_t}\right) = 10\log G_t + 10\log G_r + 20\log\left(\frac{\lambda}{4\pi}\right) - 20\log d.$$

The first term is the path loss in decibels given by $L_p = 10\log_{10}(P_r/P_t)$, and if we define the path-loss at the first meter distance between the transmitter and the receiver as:

$$L_0 = 10\log G_t + 10\log G_r + 20\log\left(\frac{\lambda}{4\pi}\right) \tag{2.3}$$

we have:

$$L_p = L_0 + 20\log d \tag{2.4}$$

Equations 2.2 and 2.4 provide simple and practical descriptions of Friis' equation in linear and logarithmic forms, respectively, which are commonly used in variety of wireless networking applications discussed in this book. The received signal strength at the first meter, P_0, can be calculated or it can be measured in practice, when we set up the receiver antenna at a one-meter distance from the transmitter. The value of the received signal strength in free space at any distance is obtained by dividing the power at one meter by the square of the distance between the transmitter and the receiver. In other words, for a single-path radio propagation, the received signal strength decays with a distance power gradient of "two". We can use this simple equation for a variety of useful practical considerations to analyze the behavior of signals transmitted over wireless channels.

Equation 2.4 provides a simple relation between the path loss and the distance between transmitter and receiver. This equation shows that the path loss increases at a rate of 20 dB per decade of distance. In another words, if we plot pathloss or received signal strength (i.e., the inverse of path loss in dB) as a function of distance on a logarithmic scale, the plot is a line with a slope of 20. As we will see later, these plots are very popular in the empirical modeling of path loss for different wireless networks. As we mentioned earlier, from Equation 2.4 we can observe that the received signal strength along a path in free space loses 20 dB of strength per decade or 6 dB per octave of distance. We can use these equations to calculate the received signal strength at any distance. In the following example, we use Friis' equations to address a simple practical problem to calculate the received signal strength in free space when a receiver is located one mile away from the transmitter.

---

*Note that all logarithms are to base 10 unless otherwise mentioned.

Example 2.1: Coverage and Path Loss in Free Space

The transmitter power of a Wi-Fi device is 100 mW (20 dBm) and the transmitted power of Bluetooth devices is 1 mW (0 dBm). The receiver sensitivity for these devices is −90 dBm. Both devices operate in the 2.45 GHz ISM bands.

(a) Determine the coverage of the Wi-Fi device in free space when we use antennas with a gain of one (0 dBi).
(b) Repeat (a) for a Bluetooth device.
(c) Repeat (a) if each antenna has a gain of 3 dBi.

Solution:

(a) Assuming that the receiver needs an average RSS of at least −90 dBm to reliably decode the signal, the maximum allowable path loss is $L_p = 20$ dBm − (−90 dBm) = 110 dB.
   The device operates at 2.45 GHz, therefore, the path loss at the first meter can be calculated as follows:

$$\lambda = \frac{c}{f} = \frac{3 \times 10^8}{2.45 \times 10^9} = 0.122 \ (12.2 \text{ cm}); \ L_0 = -20 \log\left(\frac{\lambda}{4\pi}\right) = -20 \log\frac{0.122}{4\pi} = 40.2 \text{ dB}$$

Then the path-loss and coverage can be calculated as: $L_p = L_0 + 20 \log d \Rightarrow 110 = 40.2 + 20 \log d \Rightarrow d = 10^{\frac{69.8}{20}} = 3090$ m

(b) For a Bluetooth device, the maximum allowed path loss is $L_p = 0$ dBm − (−90 dBm) = 90 dB. Proceeding as before: $L_p = L_0 + 20 \log d \Rightarrow 90 = 40.2 + 20 \log d \Rightarrow d = 10^{\frac{49.8}{20}} = 309$ m. Note that the coverage is one decade of distance less than that of Wi-Fi which was transmitting with 20 dB more power compared to Bluetooth.

(c) If we add a gain of 3 dBi per antenna, $L_p = 3$ dB + 3 dB + 20 dBm − (−90 dBm) = 116 dB. Proceeding as before, $L_p = L_0 + 20 \log d \Rightarrow 116.8 = 40.2 + 20 \log d \Rightarrow d = 10^{\frac{75.8}{20}} = 6165.95$ m.
Note that this is one octave of distance more than that of Wi-Fi with a total of 0 dBi antenna gain rather than the 6 dBi gain.

The above example shows a simple but useful application of Friis' equation to relate the power to the distance in free space where the path-loss gradient is two. This relationship holds in wide open indoor areas and over short distances in outdoor areas where the first path is the strongest path and it dominates the received signal strength. As the distance between the transmitter and the receiver increases, other multipath components become stronger and the direct LOS ray may get obstructed and the path-loss gradient no longer stays at two. Statistical models describing these situations assume a reverse exponential relation similar to free space transmission, between the distance and power. However, the exponent of d in the denominator is not necessarily two and it can take different values in other communication scenarios. This exponent value is commonly referred to as the path-loss or distance-power gradient, α. That is, the linear fit to the large-scale average received signal power, P_r, after a distance d in meters from the transmitter, is represented by:

$$P_r = \frac{P_0}{d^\alpha} \tag{2.5}$$

The distance-power gradient in the single-path free space scenario for communications was $\alpha = 2$, but in other environments with multipath conditions it can take different values. In logarithmic form, this relation will change Equation 2.4 to a general equation of the form:

$$L_p = L_0 + 10\alpha \log d \tag{2.6}$$

This equation presents the total path-loss as the path-loss at the first meter plus the power loss relative to the power received at the one-meter distance. Variations of this equation are occasionally used in the wireless communications literature as path-loss models to represent the distance–power relationship in different environments and scenarios of operation.

To have a quantitative intuition to relate the change in the distance-power gradient on the coverage of a wireless base station or access point we provide a simple example next.

Example 2.2: Coverage and Distance–Power Gradient

A base station covers an area of radius 1 km in an area where the radio propagation can be modeled as a two-ray channel (discussed later) with a distance–power gradient of 4. What would be the coverage if we instead consider a satellite's transmission assuming free-space propagation.

Solution: In the base station's area, the path loss gradient is 40 dB per decade of distance. Therefore, in covering 1 km of distance the signal strength is reduced by:

$$\text{three decades of distance} \times 40 \text{ dB/decade} = 120 \text{ dB}$$

The three decades of distance is because 1 km $= 1000$ m $= 10 \times 10 \times 10$ m. In free-space communication for satellites, the loss will be 20 dB per decade of distance, which allows six decades of distance or 1000 km before the signal strength reduces by 120 dB. This huge difference in coverage is due to the change in distance–power gradient and this should be expected because we have doubled the exponent of the distance in Equation 2.5 for the received power.

Using calculations based on the simple path loss model given by Equation 2.6 and power specifications of the transmitter and the receiver, we can determine the coverage of different wireless devices in a variety of environments. A simple example can provide further clarification.

Example 2.3: Coverage and path loss at the first meter

The transmitted power of an 802.11 device is 20 dBm and the receiver sensitivity is −90 dBm. Determine the coverage for distance power gradient, α, of 2, 3, and 4. Assume the antenna gain is 1 and the frequency of operation is 2.45 GHz.

Solution: The maximum allowable path loss is $L_p = 20$ dBm $- (-90$ dBm$) = 110$ dB. The device operates at 2.45 GHz, and therefore, the path loss at the first meter is $L_0 = -20\log\left(\frac{\lambda}{4\pi}\right) = -20\log\frac{0.122}{4\pi} = 40.2$ dB as previously derived in Example 2.1. Then the coverages for different values of α are calculated as follows:

$$L_p = L_0 + 10\alpha \log d \Rightarrow 110 = 40.2 + 10\alpha \log d \Rightarrow d = 10^{\frac{69.8}{10\alpha}}$$

For $\alpha = 2$, $d = 3$ km, for $\alpha = 3$, $d = 212$ m, and for $\alpha = 4$, $d = 56$ m.

As the frequency increases, the path loss at the first meter also increases resulting in greater overall path-loss. However, higher frequencies may appear better for certain applications because of availability of more spectrum and need to smaller antennas.

Example 2.4: Path loss at the first meter at different frequencies

Consider a base station tower with two service providers. The first operates at a frequency of 875 MHz and the other at twice the frequency, that is, at 1750 MHz. What is the path loss observed at 1 km in each case assuming free-space propagation.

Solution: The wavelengths at the two frequencies are $(3 \times 10^8)/(875 \times 10^6) = 0.3429$ m and $(3 \times 10^8)/(1750 \times 10^6) = 0.1714$ m. Therefore, the path loss at the first meter in each case will be $L_0 = -20 \log \left(\frac{\lambda}{4\pi} \right) = -20 \log \frac{0.3429}{4\pi} = 31.3$ dB and $L_0 = -20 \log \left(\frac{\lambda}{4\pi} \right) = -20 \log \frac{0.1714}{4\pi} = 37.3$ dB. Then the path-loss at 1 km in each case will be: $L_p = L_0 + 20 \log d = 31.28 + 60 = 91.3$ dB at 875 MHz and 97.3 dB at 1750 MHz, which is 6 dB higher.

As we will see later, the path-loss models used in practical wireless networking applications are far more complex than the simple models discussed here, however, they all follow the same principles. Before we give some examples of applied path-loss models we discuss how we can empirically calculate the distance–power gradient in a wireless communications application.

2.2.3 Empirical Determination of Path Loss Gradient

In very simple situations, such as the two-path modeling for mobile radio applications in open areas (which we show later), one may be able to determine the distance–power gradient using mathematical approximations. In most other indoor and urban areas these calculations are not practical because multipath scenarios are more complex and they change drastically in different sites and application scenarios. As a result, researchers and standard organizations resort to empirical calculations of the distance–power gradient using measurements of the received signal strength. They employ empirical calculations to define path-loss models for different application scenarios. To measure the gradient of the distance–power relationship in a given area, the receiver is fixed at one location and the transmitter is placed at a number of locations with different distances between the transmitter and the receiver. The received power or the path loss (in dB) is plotted against the distance on a logarithmic scale. The slope of the best-fit line through the measurements is taken as the gradient of the distance–power relationship.

Figure 2.7 shows a set of measured data taken in an indoor area at distances of 1–20 m, together with the best-fit line through the measurements using linear regression. In this particular scenario we have a 24 dB change in signal strength over a decade of distance (from 1 to 10 m). Using Equation 2.6 we have $10 \times \alpha = 24$, resulting in a distance power gradient of 2.4. Using the best-fit line in Figure 2.7 we can also determine the estimated received power at the 1 m distance, which is approximately -5.7 dBm. Knowing the transmitted power and the estimated received power, we can also estimate the path loss in the first meter. Table 2.1 shows several examples of calculations of the distance–power gradients and path losses in the first meters of distance at a variety of center frequencies, environments, and scenarios normally used for WLAN applications [Pra92; Gue97; McD98].

2.2.4 Shadow Fading and Fading Margin

As shown in Figure 2.5, depending on the environment and the surroundings, and the location of objects, the received signal strength *for the same distance* from the transmitter will be different.

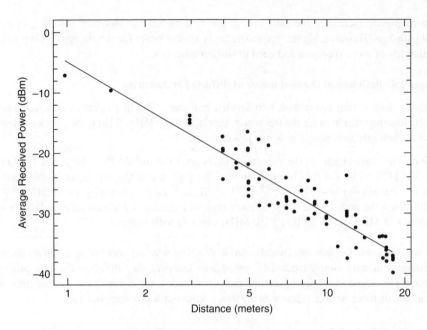

Figure 2.7 Measured received power and a linear regression fit to the data.

In effect, Equation 2.6 provides the mean value of the signal strength that can be expected if the distance between the transmitter and receiver is d. The actual received signal strength will vary around this mean value. As we described earlier, this variation of the signal strength due to location is often referred to as *shadow fading* or *slow fading*. The reason for calling this shadow fading is that very often, the fluctuations around the mean value are caused due to the signal being blocked from the receiver by buildings (in outdoor areas), walls (inside buildings), and other objects in the environment. It is called large-scale or slow fading because the variations are much slower with distance than multipath fading that we explained earlier and discuss in detail later. It is also found

Table 2.1 Distance–power gradients in sample environments at 2.4 GHz and 5.0 GHz

Center frequency f_c (GHz)	Environment	Scenario	Path loss at the first meter (dB)	Path-loss gradient α
2.4	Indoor office	LOS	41.5	1.9
		OLOS	37.7	3.3
5.1	Meeting room	LOS	46.6	2.22
		OLOS	61.6	2.22
5.2	Suburban residences	LOS and same floor	47.0	2 to 3
		OLOS and same floor		4 to 5
		OLOS and room in the higher floor directly above the Tx		4 to 6
		OLOS and room in the higher floor *not* directly above the Tx		6 to 7

that shadow fading has less dependence on the frequency of operation than multipath fading or fast fading, as we will discuss later. To include the effects of shadow fading the path-loss model of Equation 2.6 will have to be modified to include a random component representing the effects of shadow fading:

$$L_p = L_0 + 10\alpha \log d + X. \tag{2.7}$$

Here X is a random variable representing the effects of shadow fading on the path-loss calculations. If we use this equation to calculate the average RSS, then X represents the actual shadow fading. Figure 2.7 illustrates the average received power at different distances and the best-fit line for calculation of the distance–power gradient. In this figure the deviation of the actual measurements from the best-fit line approximation represents the sample values of this random variable. Several measurements and simulations indicate that this variation can be expressed as a log-normally distributed random variable. In decibel form, a log-normal random variable has a normal distribution. Therefore (in dB) the probability density function of this random variable representing the effects of shadow fading is given by:

$$f_{SF}(x) = \frac{1}{\sqrt{2\pi}\sigma} \exp\left(-\frac{x^2}{2\sigma^2}\right) \tag{2.8}$$

where σ is the standard deviation of the shadow fading. Statistical models for path loss recommended by a standards organization provide the model for the distance–power relation as well as the standard deviation of the shadow fading which can be used in conjunction with a model similar to Equation 2.7.

If we use Equation 2.7 without its shadow fading component and we find a value d for the coverage of a base station, similar to what we were doing in previous examples, this is only the mean coverage radius from a base station. In reality at that distance d we have a 50% probability of having adequate signal strength (RSS that is larger than the receiver sensitivity). This is because the normally distributed (in dB) shadow fading random variable can have a positive value with 50% probability and that positive value increases the path loss beyond what was used for calculation of the distance d. Figure 2.8 visually explains this situation. For a terminal located in distance d from

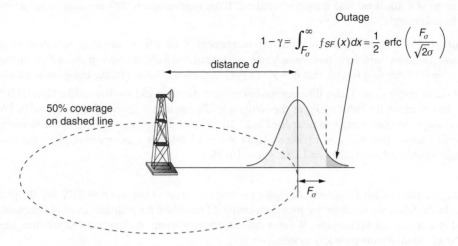

Figure 2.8 Calculation of coverage and its relation to fade margin.

the base station, we have a 50% probability to operate with the required minimum signal strength. To increase this probability one may add transmitting power to increase the probability of coverage at distance d. This additional power is referred to as *fading margin*, and it is represented by F_σ.

For $\gamma\%$ coverage, the base station should have an additional fade margin of F_σ such that:

$$1 - \gamma = \int_{F_\sigma}^{\infty} f_{SF}(x)dx = 0.5 \text{ erfc}\left(\frac{F_\sigma}{\sqrt{2}\sigma}\right) \tag{2.9}$$

The fade margin is the additional signal power that can provide a certain additional fraction of the locations at the edge of a cell (or near the fringe areas) with the required signal strength. Thus, for computing the coverage with certain assurance for coverage, we first determine the fade margin using the variance of the shadow fading. Then we employ the following equation to calculate the coverage:

$$L_p = L_0 + 10\alpha \log d + F_\sigma \tag{2.10}$$

where F_σ is the fade margin associated with the path loss to overcome the shadow fading component. This fading margin can be applied by increasing the transmit power and keeping the cell size the same, or reducing the cell size by setting a higher RSS threshold for making a handoff (see Chapter 6). For commonly employed probabilities of coverage such as 95 or 90%, we can use the following formulas to compute the fade margins. The value of F_σ for 90% coverage at the fringe is $F_\sigma = 1.282\sigma$ and for 95% coverage at the fringe, it is $F_\sigma = 1.654\sigma$. The following example provides clarity for this concept.

Example 2.5: Computing the fading margin

A mobile system is required to provide 95% successful communication at the fringe of coverage with a location variability (shadow fading) having a zero mean Gaussian distribution with standard deviation of 8 dB. What fade margin is required? If the requirement is 90% coverage, what would be the fade margin?

Solution: Note that the location variability component X (in dB) in this case is a zero mean Gaussian random variable. In this example, the variance of X is 8 dB. We have to chose F_σ such that $1 - \gamma = 0.5 \text{ erfc}(\frac{F_\sigma}{\sqrt{2}\sigma})$ is 0.05, that is, 95% of the locations will have a fading component smaller than the tolerable value. Using the complementary error function and a software like MATLAB®, we can determine the value of F_σ as the solution to the equation $0.05 = 0.5 \text{ erfc}(F_\sigma/\sigma\sqrt{2})$. For this example, the fade margin to be applied is 13 dB. We can also calculate this using the formulas specified above, that is, $F_\sigma = 1.654\sigma = 1.654 \times 8 = 13$ dB. For a coverage of 90%, the fade margin would be $F_\sigma = 1.282\sigma = 1.282 \times 8 = 10$ dB.

So far, we have discussed achievable signal coverage in terms of the received RSS and the path loss. In the following sections, we provide a couple of examples for path-loss models commonly used in wireless communications. We also discuss, where relevant, the important factors that lead to the adoption of these path-loss models.

2.2.5   Popular Models for Path Loss and Shadow Fading

Today, one of the essential parts of all wireless standardizations committees is to define channel models for the particular applications considered as the focal point of an emerging technology. We categorize these technologies in later chapters of this book and we address the details of some of the suggested channel models for these technologies in the relevant chapters. In this section we provide examples of path-loss models used in indoor and outdoor environments for cellular and WLAN applications. We begin with two simple models for indoor environments for which the models are designed for a single frequency band, the transmitter and the receiver are in the same floor and the height of antennas does not play an important role.

The first model is a mathematically simple wall-partitioned model, which assumes free space propagation in the indoor environment and adds to it, the path-loss associated with walls between the transmitter and the receiver, which depends on the material used for construction of the building. In this model we need to know the number of walls, the material used for construction of the wall and the path loss associated with each wall along the straight line that connects the transmitter and receiver. Then we discuss the distance-partitioned model recommended by the IEEE 802.11 WLAN standardization committee. This model assumes two different distance power gradients for calculation of the path loss. It identifies the appropriate breakpoint distance at which there is a change of gradient in different buildings. Both these models are applicable to indoor areas.

The last example used for path-loss modeling is the Okumura–Hata model that is used for outdoor cellular network planning. This empirical model provides the path loss as a function of frequency of operation and the heights of the antennas in different city environments. The worldwide frequency bands used for cellular networks are fragmented and distributed from hundreds of megahertz to a couple of gigahertz. In these applications the antennas can be mounted on top of towers and tall buildings as high as several hundred meters and the mobile antenna may operate on the ground in an urban canyons at a very small height. The Okumura–Hata model accommodates these factors (frequency of operation and height of antennas) in calculating the path loss.

A Simple Wall-Partitioned Model for Indoor Areas: The wall-partitioned model is very simple and it builds on Equation 2.6 for a free-space path-loss calculation by introducing losses for each partition that is encountered along a straight line connecting the transmitter and the receiver. This path loss model is given by:

$$L_p = L_0 + 20 \log d + \sum_{i=1}^{N} w_i. \tag{2.11}$$

Here w_i is the loss (in dB) attributed to each wall in between the transmitter and the receiver. The model requires some visual knowledge of the building architecture to determine the number of walls and the information on material used in the construction of these walls and their associated path-loss measurements. Table 2.2 shows some decibel loss values measured by Harris semiconductors at 2.4 GHz for different types of walls. Several other loss values, w_i, have been reported in [Rap02] that vary between 1 dB for dry plywood to 20 dB for concrete walls, depending on the carrier frequency. In calculating the coverage, appropriate fading margins have to be included to account for the variability in path loss for the same distance d. If we want to use this wall-partitioned model to calculate the coverage, we need knowledge of (or we need to assume the number of) the walls between the transmitter and the receiver.

Table 2.2   Partition-dependent losses

Signal attenuation of 2.4 GHz through:	dB
Window in brick wall	2
Metal frame, glass wall into building	6
Office wall	6
Metal door in office wall	6
Cinder wall	4
Metal door in brick wall	12.4
Brick wall next to metal door	3

Example 2.6: Coverage using the Wall-Partitioned Model

The transmitter power of an 802.11 device is 20 dBm and the receiver sensitivity is −90 dBm. Determine the coverage in an office building with an average of one wall every 17 m. Assume the antenna gains are one and the frequency of operation is 2.45 GHz. Use other numbers from Example 2.3.

Solution: From Example 2.3 we know that the maximum allowable path loss is $L_p = 110$ dB. The path loss at the first meter at 2.45 GHz is $L_0 = 40.2$ dB. From Table 2.2 we have a loss of 6 dB per office wall attenuation. Then the coverage is given by $L_p = L_0 + 10\alpha \log d + N \times 6$ where $N = \lfloor \frac{d}{17} \rfloor$. The sign $\lfloor \ \rfloor$ stands for the closest integer representing floor number for the value determined inside the brackets. To obtain the coverage, we need to solve the following two sets of nonlinear equations with two unknowns (here $\alpha = 2$):

$$\begin{cases} N = \left\lfloor \dfrac{d}{17} \right\rfloor \\ 69.8 = 10\alpha \log d + 6N \end{cases}$$

In other words, we need to find:

$$d = 10^{69.8 - 6N/20}$$

for which $N = \lfloor \frac{d}{17} \rfloor$. By inspecting different values of N we find that, for $N = 5$, the coverage is $d = 97.7$ m for which $N = \lfloor \frac{d}{17} \rfloor = \lfloor \frac{97.7}{17} \rfloor = \lfloor 5.74 \rfloor = 5$.

The results of Example 2.6 are between the 56 and 212 m coverage results of Example 2.3, where we used a single path-loss gradient model with distance–power gradients of 4 and 3, respectively. The above example also reveals the difficulties of this conceptually simple method in calculating the coverage that adds to the other difficulties we mentioned before.

Distance-Partitioned Model for Indoor Areas: In indoor areas, the area between the transmitter and the receiver is often not homogeneous with a single distance–power gradient. For small distances, the transmitter and the receiver are often in the same room where the dominant received signal power arrives from the direct LOS path between the transmitter and the receiver and the power arriving from other paths are negligible. In these situations the distance power gradient closely follows the free space LOS propagation. As the distances become larger and obstacles break the LOS between the transmitter and the receiver, reception of the signal over multiple paths contributes to the overall

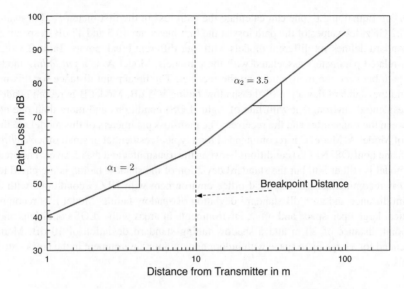

Figure 2.9   The IEEE 802.11 distance-partitioned path-loss model.

received signal and the distance power gradient increases to a higher value. Most of the recent path-loss models developed for WLAN and WPAN applications consider this phenomenon by partitioning the path-loss behavior in different segments with multiple distance–power gradients, each associated with a segment of the path between the transmitter and the receiver. Distance-partitioned models with two segments are the most popular in the industry.

In this section we describe the IEEE 802.11 recommended distance-partitioned path-loss model as our example for distance-partitioned path-loss modeling. Figure 2.9 shows the basics of this model (for model D in Table 2.3) in which the distance between the transmitter and the receiver is divided into two segments at a break-point distance d_{bp}. The distance-power gradient in the two segments are $\alpha_1 = 2$ and $\alpha_2 = 3.5$, respectively. The equations representing these models can be expressed as:

$$L_p = L_0 + \begin{cases} 10\alpha_1 \log d, & d < d_{bp} \\ 10\alpha_1 \log d_{bp} + 10\alpha_2 \log \left(\dfrac{d}{d_{bp}} \right), & d > d_{bp} \end{cases} \qquad (2.12)$$

Table 2.3   Parameters for different IEEE 802.11 recommended path-loss models for six environments

Environment	d_{bp} (m)	α_1	α_2	Shadow fading Std. Dev. (dB)
A	5	2	3.5	5
B	5	2	3.5	5
C	5	2	3.5	8
D	10	2	3.5	8
E	20	2	3.5	10
F	30	2	3.5	10

Where, using Equation 2.6, one can calculate the path loss in the first meter, L_0. For example, at 2.4 and 5.2 GHz, the values of the path loss at the first meter are 40.5 and 47 dB, respectively.

The standard defines six different models with four different break points. Table 2.3 shows the path loss related parameters associated with these models. Model A, is a *flat fading* model with a single path between the transmitter and the receiver. The breakpoint distance for this model is at 5 m and the standard deviation of the shadow fading is 5 dB. Model B is recommended for a typical residential environment with line of sight (LOS) conditions and more than one effective path between the transmitter and the receiver. The path-loss parameters of this model are the same as those of Model A. Model C is recommended for a typical residential or small office environment with LOS and nonLOS (NLOS) conditions between the transmitter and the receiver. The breakpoint for this model is still at 5 m but the standard deviation of the shadow fading is increased to 8 dB. Model D is recommended for a typical office environment with NLOS conditions with a 10 m breakpoint distance and an 8 dB standard deviation of shadow fading. Model E is recommended for a typical large open space and office environments in areas with NLOS conditions and it has a breakpoint distance of 20 m and a shadow fading standard deviation of 10 dB. Model F is recommended for a large open space with indoor and outdoor environment in the areas with NLOS conditions.

Example 2.7: Coverage using the IEEE 802.11 Distance Partitioned Model

The transmitter power of an 802.11 device is 20 dBm and the receiver sensitivity is −90 dBm. Determine the coverage in a small office building with LOS and NLOS using Model C. Assume antenna gains are one and the frequency of operation is 2.45 GHz.

Solution: We use Equation 2.12 and the parameters of Model C shown in Table 2.3 to calculate the IEEE 802.11 coverage. The maximum path loss allowed is 110 dB and the path loss at the first meter is 40.2 dB, just as they were in Examples 2.3 and 2.6. In LOS/NLOS small office areas and with a breakpoint distance of 5 m the coverage is calculated using the following expression:

$$110 = 40.2 + 20 \log 5 + 35 \log \left(\frac{d}{5} \right)$$

The coverage with 50% confidence is:

$$d = 5 \times 10^{\frac{69.8 - 14}{35}} = 195 \text{ m}$$

If we increase the confidence to 95%, (with an 8 dB standard deviation of shadow fading as was the case in Example 2.5), we need an additional 13.2 dB fade margin that reduces the coverage to:

$$d = 5 \times 10^{\frac{69.8 - 14 - 13.2}{35}} = 82.4 \text{ m}$$

Note that the distance determined by using the IEEE 802.11 distance-partitioned model in Example 2.7 is twice as much as that determined by the wall-partitioned model of Example 2.6. In Example 2.6, if we had one additional wall between the transmitter and the receiver, the calculation would result in a similar coverage. Also note that the coverage of a WLAN and in general all networks originally designed for wireless data applications are a function of data transmission rate and for higher data rates we usually have smaller coverage. We can use these path-loss models and the power requirement for different data rates to determine the coverage for all data rates. The relation between power and data rate is one of the topics that we will study in Chapter 3 when we discuss transmission techniques.

Okumura–Hata Model for Outdoor Areas: Outdoor models for the path-loss of wireless communication systems are mostly designed for cellular telephone applications and they are more detailed than models used for indoor areas. This is because of a few important factors. Cellular networks were originally deployed for the telephone application that demands a steady quality of service over a wide area. The cost of base stations for cellular networks is very high and such base stations operate in licensed bands that are very expensive and fragmented and spread over a large variety of frequencies. Therefore, the efficient deployment of the network is very important and considering that, in outdoor areas the terrain is not flat but includes hills and urban canyons with very sharp differences in elevation, as compared with the mostly flat interior of a building floor. As a result the path-loss models designed for outdoor applications are more detailed and they include the heights of antennas and the frequency of operation.

There have been extensive measurements in a number of cities and locations of RSS from base stations in cellular networks and these have been reported in the literature. The most popular of these measurements corresponds to those of Okumura who came up with a set of path loss curves as a function of distance in 1968 for a range of frequencies between 100 and 1920 MHz. Okumura also identified the height of the base station antenna h_b and the height of the mobile antenna h_m as important parameters. Masaharu Hata [Hat80] came up with empirical models that provide a good fit to the measurements taken by Okumura for transmitter–receiver separations d of more than 1 km. The expressions for path loss developed by Hata are called the Okumura–Hata models or simply the Hata models. Table 2.4 provides these models. The general formulation of Equation 2.13 is the same as the general path-loss model shown in Equation 2.6. However, the distance between the mobile and base station, d, is given in kilometers. Therefore, the first four terms in the equation provide for the path loss in the first kilometer rather than the first meter. Since this empirical

Table 2.4 Okumura–Hata models for macro-cellular path loss

General formulation:

$$L_p = 69.55 + 26.16 \log f_c - 13.82 \log h_b - a(h_m) + [44.9 - 6.55 \log h_b] \log d \qquad (2.13)$$

where f_c is in MHz, h_b and h_m are in m, and d is in km

Range of values

Center frequency f_c (MHz)		150–1500 MHz
h_b, h_m (m)		30–200 m, 1–10 m
$a(h_m)$ in dB Large city	$f_c \leq 200$ MHz	8.29 [log $(1.54\, h_m)]^2 - 1.1$
	$f_c \geq 400$ MHz	3.2 [log $(11.75\, h_m)]^2 - 4.97$
Medium–small city	$150 \geq f_c \geq 1500$ MHz	1.1 [log $f_c - 0.7$] $h_m - (1.56 \log f_c - 0.8)$

Suburban areas formulation
Use Equation 2.12 and subtract a correction factor given by:

$$K_r(\text{dB}) = 2[\log(f_c/28)]^2 + 5.4 \qquad (2.14)$$

Where f_c is in MHz

equation is calculated for different frequencies (in MHz) there is a frequency-dependent term for the path loss in the first kilometer. Another additional adjustment is the inclusion of the height of the antennas (in meters) in the path-loss for the first kilometer. This dependence is illustrated in Example 2.10 in Section 2.3.1, where we show that, in a two-path mobile radio operation, the received signal is dependent on the height of the antennas. The two terms in parenthesis in Equation 2.13 represent the $10 \times \alpha$ term which results in a distance-power gradient between 4.49 (for $h_b = 1$ m) and 3.18 (for $h_b = 100$ m). These numbers are around 4.0 which, as we will show in Section 2.3.1, is the distance power gradient for a cellular network application in an open area.

Example 2.8: Coverage using the Okumura–Hata model

The receiver sensitivity of a cell phone is -126 dBm and it operates at 900 MHz. For a base station height of 100 m and a mobile height of 2 m, determine the minimum transmit power for the base station to support a 30-km coverage.

Solution: We calculate the terms in the Okumura–Hata model as follows:

$$a(h_m) = 3.2[\log(11.75 h_m)]^2 - 4.97 = 1.05 \text{ dB}.$$

The path loss is:

$$L_p = 69.55 + 26.16 \log f_c - 13.82 \log h_b - a(h_m) + [44.9 - 6.66 \log h_b] \log d$$

$$= 69.55 + 26.16 \log 900 - 13.82 \log 100 - 1.05 + [44.9 - 6.55 \log 100] \log 30$$

$$= 165.11 \text{ dB}.$$

Therefore the base station transmitting power should be:

$$P_t(\text{dBm}) = L_p(\text{dB}) + P_r(\text{dBm}) = 165.11 + (-126) = 39.11 \text{ dBm}.$$

In watts, the transmit power is

$$10^{39.11/10} = 8147 \text{ mW} \approx 8 \text{ W}.$$

2.3   Modeling of RSS Fluctuations and Doppler Spectrum

In the last section, we discussed the behavior of the average RSS and its application to path-loss modeling and analysis of the effects of shadow fading. These models provide an insight into understanding of the relation between power and distance that is essential for deployment of wireless networks. Such a characterization of the RSS corresponds to a *large-scale* average value. In reality, as shown in Figure 2.4, the RSS is rapidly fluctuating in time and locally in space due to the mobility of the mobile terminal or the movement of other objects close to the transmitter and the receiving antennas causing changes in multiple signal components arriving via different paths.

Two effects contribute to the rapid fluctuations of the signal amplitude. The first, caused by the movement of the mobile terminal towards or away from the base station transmitter and is called *Doppler Spectrum*. The second is caused by the addition of signals arriving via different paths, which is referred to as *short time, small-scale* or *multipath fading*.

The short distance or short time variations of the RSS, referred to as *small-scale fading* is the rapid instantaneous changes in the received signal power caused by fast changes in the phase of the received signal from different paths due to small movements. As shown in Figure 2.4, for the analysis of the short-term variations of the channel we are interested in the statistics of short term multipath-fading and finding the shape of the Doppler spectrum of the signal. The statistics of the short-term variations in RSS allow us to calculate the *error rate* of different transmission techniques over the wireless medium. The statistics of the temporal multipath fading is characterized by the probability density function of the sampled values of the fast variations of the channel. As we will see later in this chapter, the most popular distribution for this variation is the Rayleigh distribution and for that reason sometimes this type of fading is referred to as Rayleigh fading.

The Doppler Spectrum is the Fourier transform of the samples of the variation of the signal. It is very important to model this spectrum because, if we want to simulate these variations, we will need a good model. If we know the spectrum of a random signal, we can regenerate it by designing a filter with that spectrum and stimulating that filter with a noise like random signal. The Doppler spectrum allows us to learn how to simulate variations of the channel in time to examine its impact on transmission of *packets* over the channel before we actually implement the expensive hardware for a particular transmission technique.

The cause of all problems in the wireless channel is multipath. In this section we use Friis' equation and a technique that we refer to as geometric ray tracing with simple models to show how multipath causes changes in the distance-power gradient and causes fluctuations of the RSS. Then we provide some examples of models for small-scale multipath-fading used in practical applications in wireless networking.

2.3.1   Friis' Equation and Geometric Ray Tracing

To model the path-loss we started with Equation 2.4 that represented Friis' equation in logarithmic form and, based on that equation, we showed how we develop empirical path-loss models for different wireless systems in multipath conditions and we described how we come up with the concept of large-scale shadow fading. To demonstrate how multipath causes short-term or small-scale fading and to discover the meaning of Doppler spectrum we start with Friis' equation in linear form given by Equation 2.1.

As we discussed in Section 2.1, electromagnetic waves at the high frequencies used in wireless networks can be treated as rays and if we have the geometry of the area, we can relate the length of these rays to the geometry using principles from geometric optics, which have been in use for a couple of millennia to describe the imaging inside mirrors. Using geometric optics, we can *trace* the paths that waves travel between transmitter and receiver. If for each path, we can find the amplitude and phase of the received signal, we will be in a position to analyze the received signal and explain how it causes counter intuitive observations such as change in the distance-power gradient or multipath fading.

We start this discussion by an example in which we use Friis' equation to calculate the magnitude and phase of a single tone cosine waveform, when it is transmitted over a single path wireless channel.

Example 2.9: Single-tone and single-path transmission

What are the amplitude and phase of a received signal if a single frequency cosine signal $x(t) = \sqrt{P_t} \cos 2\pi f t$ (where P_t is the transmitted power and f is the frequency of the signal) is transmitted along a single path free-space medium?

Solution: If we denote the received signal by $y(t)$, the amplitude of the received signal is obtained by taking the square root of the received signal power. The received signal power can be determined from simplified Friis' equation in linear form described by Equation 2.2, that is:

$$P_r = \frac{P_0}{d^2} \Rightarrow \sqrt{P_r} = \frac{\sqrt{P_0}}{d}$$

In this case, since the radio transmission environment forms a linear time-invariant system, if the transmitted signal is a cosine, the received signal will also be a cosine at the same frequency with a delay of:

$$\tau = \frac{d}{c}$$

That results in a phase value of:

$$\phi = 2\pi f \tau = \frac{2\pi f d}{c} = \frac{2\pi d}{\lambda}$$

where, $\lambda = \frac{c}{f}$ is the wavelength of the transmitted cosine. The received signal is simply:

$$y(t) = \sqrt{P_r} \cos 2\pi f (t - \tau) = \frac{\sqrt{P_0}}{d} \cos(2\pi f t - \Phi)$$

Figure 2.10 illustrates the basic concept behind single-tone transmission in a free-space single-path wireless transmission medium that was described in Example 2.9. Since the channel is linear, the

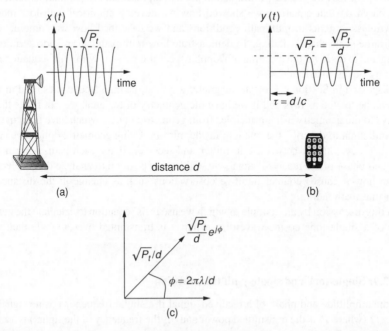

Figure 2.10   Visualization of the basic concept behind single-tone transmission in a free-space single-path wireless transmission link.

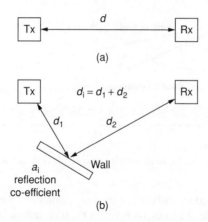

Figure 2.11 Comparison of direct and reflected paths: (a) direct LOS path, (b) a reflected path with reflection coefficient.

transmitted sinusoid is received as a sinusoid after propagation over a distance d. The *magnitude* corresponds to the amplitude $\frac{\sqrt{P_0}}{d}$, and the *phase* $\phi = \frac{2\pi d}{\lambda}$, of the received sinusoid are both functions of the distance. The amplitude changes slowly with the inverse of the distance and the phase rotates rapidly at the speed of one rotation which is 2π radians every $\frac{d}{\lambda}$ m. For example, with a Wi-Fi device operating at 2.45 GHz, the wavelength is 12.2 cm (see Example 2.1) and the phasor representing a path has one full rotation every 12.2 cm. For the same motion, the amplitude of the signal does not change significantly. As we will see later, this observation helps us to develop a clear intuition of the cause of multipath fading.

To analyze the effects of multipath on single tone transmission, we extend the results of single-tone transmission provided in Example 2.9, to a scenario where we have a single path arriving after reflection on a wall. For a reflected path from a wall, with minor modifications, we can use the same equations for calculation of the received signal amplitude and phase. As shown in Figure 2.11, the difference between a direct LOS and a reflected path is that after reflection, the incident propagated wave from the wall has a loss in its amplitude according to the reflection coefficient and it changes the polarity of its phase. Therefore, if a signal arriving along a path reflected from a wall with a length of d_i and a reflection coefficient of a_i, the amplitude and phase of the received signal are $\frac{a_i \sqrt{P_0}}{d_i}$ and $\phi_i = \frac{2\pi d_i}{\lambda} + \pi$. If we represent the reflection coefficient by a negative number, then the addition of π to the phase shift is not necessary. This simple observation allows us to explain the effects of multipath, when in addition to the direct LOS path, we also have reflected paths.

The usefulness of this simple analysis can be further clarified by using some examples. As our first example we use our simple techniques to calculate the amplitude and phase of the signal arriving from a direct and a reflected path in a two-path environment to demonstrate how a simple reflected path can modify the distance-power gradient drastically.

Example 2.10: Distance-power gradient for a two-path environment

Figure 2.12 represents the propagation environment for a cellular telephone application in an open environment, where the communication between the cell tower and the mobile involves two paths. Here, the base station and the mobile terminal are both assumed to be at elevations above the earth,

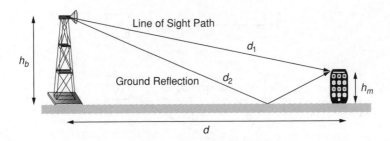

Figure 2.12   Two-path model for land mobile radio wireless communications in an open area.

which is modeled as a flat surface in between the base station and the mobile terminal. There is a line-of-sight (LOS) component that exists between the base station and the mobile terminal and carries the signal in a manner similar to free-space. There is also one path reflected off the flat surface of the earth. The two paths travel different distances based on the heights of the base station antenna, h_b, and the height of the mobile terminal antenna, h_m. We assume that the distance d between the transmitter and receiver is much larger than either antenna heights and the reflection coefficient is $a_i = -1$, which means that the ground acts as an ideally lossless reflector. Show that the distance-power gradient for this environment is four.

Solution: Considering Figure 2.12 and assuming that the distance between the two terminals is much larger than the heights of the antennas, the length of the two paths used for calculation of the amplitude of the received signal will be approximately the same as the *distance* between the transmitter and the receiver and we have $d_1 \approx d_2 \approx d$. However, the phase of the two paths will be $\Delta\phi$ apart, where $\Delta\phi$ is a very small value representing the difference between the phases of the two paths. If we represent the paths in complex form (phasors), for the signals arriving along the first and the second path we have amplitudes: $\frac{\sqrt{P_0}}{d_1}e^{j\phi_1} \approx \frac{\sqrt{P_0}}{d}e^{j\phi_1}$ and $a_i\frac{\sqrt{P_0}}{d_1}e^{j\phi_2} \approx \frac{\sqrt{P_0}}{d}e^{-j\phi_2}$. Then the amplitude of the received signal is the complex sum of these two phasors similar to those shown in Figure 2.10c which can be calculated as:

$$\frac{\sqrt{P_0}}{d}e^{j\phi_1} + \frac{\sqrt{P_0}}{d}e^{-j\phi_2} = \frac{\sqrt{P_0}}{d}e^{j\phi_1}\left(1 - e^{j\Delta\phi}\right)$$

where $\Delta\phi$ is the difference between the two phases. The received signal power is the square of this amplitude. That is:

$$P_r = \frac{P_0}{d^2} \times \left|1 - e^{j\Delta\phi}\right|^2 \approx \frac{P_0}{d^2} \times |\Delta\phi|^2$$

Because for small values of $\Delta\phi$, we have:

$$\left|1 - e^{j\Delta\phi}\right| \simeq |1 - (1 + j\Delta\phi)| \simeq |\Delta\phi|$$

The value of phase difference is $\Delta\phi = 2\pi f \Delta d/c = (2\pi/\lambda) \times \Delta d$ with Δd being the difference between the two path lengths. Since:

$$d_1 = \sqrt{(h_b + h_m)^2 + d^2} \simeq d + \frac{(h_b + h_m)^2}{2d}$$

and

$$d_2 = \sqrt{(h_b - h_m)^2 + d^2} \simeq d + \frac{(h_b - h_m)^2}{2d}$$

We have:

$$\Delta d = d_1 - d_2 \simeq \frac{(h_b + h_m)^2}{2d} - \frac{(h_b - h_m)^2}{2d} = \frac{2h_b h_m}{d}$$

and:

$$\Delta \phi = \frac{2\pi}{\lambda} \times \Delta d \simeq \frac{2\pi}{\lambda} \times \frac{2h_b h_m}{d}$$

Substituting this phase difference value into the equation for calculation of power we have:

$$P_r \approx \frac{P_0}{d^2} \times |\Delta\phi|^2 = P_0 \times \left(\frac{2\pi}{\lambda}\right)^2 \times \frac{4h_b^2 h_m^2}{d^4}$$

Which shows that the gradient of the distance–power relationship is four. Thus the power in this scenario for mobile radio operation will decrease 40 dB per decade of distance, in contrast with the 20 dB per decade found for the case of LOS transmission in free space.

As we will discuss in Chapter 5 (discussing the deployment of wireless networks), a distance-power gradient of four is often used to determine the coverage and interference in cellular networks. This example also reflects the effects of the height of an antenna in calculation of the average RSS in cellular networks and the reason why in WLANs, the path-loss models do not consider the height of an antenna, while the Okumura–Hata model for cellular networks has adjustment elements for the heights of the antennas.

In Example 2.10, we used the principles obtained from Friis' equation in free space and ray optics to show how multipath impacts the distance-power gradient for links operating in a wide area cellular network. Our next example uses the same methodology to demonstrate how multipath causes short-term fading in the RSS for a local indoor application. In this example we consider a scenario with two mobile devices in a large open area, shown in Figure 2.13a, in which we have three paths, the direct LOS path and the paths reflected from the ceiling and the floor of the building. In this situation we can make a reasonable assumption that the effects of other paths are negligible.

Example 2.11: Multipath fading in an open indoor area

Consider the three-path open indoor area scenario of operation shown in Figure 2.13a, assuming the height of the ceiling is 5.0 m, the antennas are located in 1.5 m above the floor and reflection coefficient is $a_i = -0.7$. For this scenario:

(a) Give equations for calculation of the RSS from all paths if the transmitted power and the antenna gains are normalized to 1.
(b) Use MATLAB to plot the RSS in dB for $1 < d < 100$ m to demonstrate the formation of multipath-fading. Assume the frequency of operation is 1 GHz.
(c) Repeat (b) for a frequency of 10 GHz.

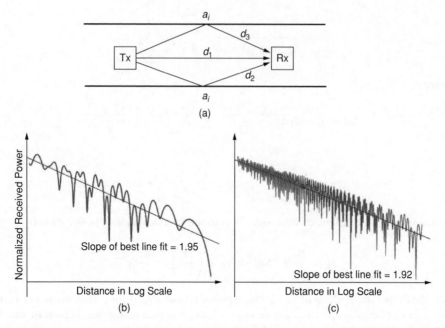

Figure 2.13 Ray tracing in an open area. (a) Scenario of operation. (b, c) Normalized received power versus distance at: (b) 1 GHz, (c) 10 GHz.

Solution:

(a) The amplitude and Phase of the i-th path is given by $a_i \frac{\sqrt{P_0}}{d_i} e^{j\phi_i}$ where for the LOS path $a_i = 1$ and for the other two paths $a_i = -0.7$. The amplitude and phase of the received signal is the complex sum of the three phasors associated with the three paths, given by

$$\sqrt{P_0} \sum_{i=1}^{3} \frac{a_i}{d_i} e^{j\phi_i}$$

the RSS is the magnitude square of this complex received signal:

$$P_r = \left| \sqrt{P_0} \sum_{i=1}^{3} \frac{a_i}{d_i} e^{j\phi_i} \right|^2 = P_0 \left| \sum_{i=1}^{3} \frac{a_i}{d_i} e^{j\phi_i} \right|^2$$

where

$$P_0 = P_t G_t G_r (\lambda/4\pi)^2 = (\lambda/4\pi)^2$$

In this example, the ceiling height is assumed to be 5.0 m and the antennas are 1.5 m above the floor. The reflection coefficients are assumed to be $a_1 = +1$ for the LOS path and $a_2 = a_3 = -0.7$ for the other two paths. The distance of the direct path is actual distance between the transmitter and the receiver, and the distances for the path from the ground and from the ceiling are given by:

$$d_2 = 2 \times \sqrt{\frac{d_1^2}{4} + (1.5)^2}$$

and

$$d_3 = 2 \times \sqrt{\frac{d_1^2}{4} + (3.5)^2}$$

(b) The following MATLAB code can be used to determine the normalized amplitude and phase of the received signal for variety of parameters. Figure 2.13b shows the normalized received power versus distance calculated for distances ranging from 1 to 100 m using MATLAB. The plot shows power in decibels, and distance on a logarithmic scale. This figure demonstrates the power fluctuations due to multipath fading.

(c) Figure 2.13c shows the results for the frequency of 10 GHz.

```
%Define parameters
c = 3e8;
Pt = 1; Gr = 1; Gt = 1;
a = [1, -0.7, -0.7];
fc = [100e6 1e9 10e9];
lambda = c./fc;
d = logspace(0,2,1000);

%Define NLOS distance vectors d1 and d2
d1 = 2*sqrt(0.25*(d.^2)+(1.5^2));
d2 = 2*sqrt(0.25*(d.^2)+(3.5^2));

%Part 1abcd:
for i=1:length(fc)

    %Calculate P0 for fc(i)
    P0 = Pt*Gr*Gt*((lambda(i)/(4*pi))^2);

    %Calculate phases for fc(i)
    phi1 = -(2*pi*fc(i)*d)/c;
    phi2 = -(2*pi*fc(i)*d1)/c;
    phi3 = -(2*pi*fc(i)*d2)/c;

    %Calculate received power for fc(i)
    Vr =
    (a(1)*(exp(j*phi1)./d)+a(2)*(exp(j*phi2)./d1)+a(3)*(exp(j*phi3)./d2));
    Pr_dB = 10*log10(P0*abs(Vr.^2));

    %Find the best-fit curve for the received power plot
    bf = polyfit(10*log10(d),Pr_dB,1);
    bf_val = polyval(bf,10*log10(d));

    %Plot power vs. distance
    figure(i)
    semilogx(d,bf_val,'r:',d,Pr_dB,'b-');

    %Labels
    xlabel('Distance [m]'); ylabel('Received Power [dB]');
    title('Received Power vs. Log Distance');
end
```

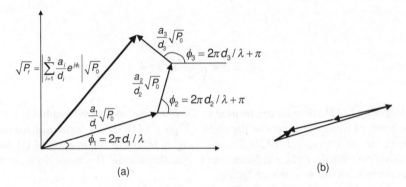

(a)   (b)

Figure 2.14   (a) Phasor diagram representing how multipath components adds up together. (b) Occurrence of a multipath fading.

Figure 2.13 illustrates the calculations in the above example. This figure indicates that while the average power decreases with distance, the power also fluctuates by as much as $20-30$ dB at a rate proportional to the frequency of operation. To explain what causes the fluctuations we resort to a phasor diagram in the complex plane. Figure 2.14 shows the three paths in the complex plane and the resulting vector sum of the three path-vectors each represented by a magnitude and a phase. As a mobile user moves away from the transmitter these three vectors constantly change their amplitude and phase. However, the amplitude of the paths changes slowly (proportional to the inverse of the distance), but the phase changes rapidly at a rate of $2\pi/\lambda$ radians per meter. This means that for a mobile with a carrier frequency of 1 GHz, we have a 360 degrees change in the phase every $\lambda = \frac{1}{3}$m. Therefore, in order to visualize the received signal strength in a multipath environment, one should consider Figure 2.14 when all the amplitudes and phases are changing, amplitudes very slowly and phases very rapidly. Each vector is shrinking in its amplitude slowly while it rotates like a road runner. The received signal amplitude is the vector sum of all path amplitudes and phases. When all paths are in line they add up and result in a strong amplitude. When they are aligned against one another, the result is very small amplitudes registering the fading in RSS. Therefore, as mobile moves, it observes extensive fluctuations in its amplitude caused by different combinations of the phases that add up or degrade the overall amplitudes. The rate of these variations and occurrence of fading is proportional to speed of rotation and consequently the wavelength of the signal. When we increase the frequency to 10 GHz as in Figure 2.13c, the rate of occurrence of the fades increases by ten times. The rate of variations of the phase of the multipath components is proportional to the frequency.

To connect this example to path-loss modeling, the reader should note that the slope of the best fit line to the RSS is the distance–power gradient that is shown in Figure 2.13b, c. This slope remains very close to 2 (1.95 and 1.92), that is consistent with the free space and IEEE 802.11 model for the path loss. The IEEE 802.11 model assumes a distance-power gradient of two for open indoor areas. The average power, however, does not show any shadow fading because the details of furniture or other objects contributing to the shadowing is not included in the scenario and there is no wall between the transmitter and the receiver.

In Example 2.11 we showed that as the distance between the transmitter and the receiver increases, the rapid changes in the phase of the multipath components causes rapid multipath fading. In a typical environment in indoor or urban areas, even when we move along a circle and keep the distance constant, similar to Figure 2.5, the multipath components change their phases and cause

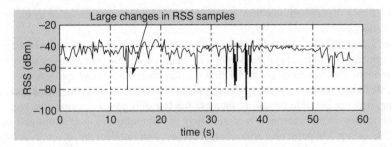

Figure 2.15 Measurement of the received signal strength from an IEEE 802.11b/g access point for the fixed location of a laptop.

rapid multipath fading. Even if we keep the distances constant and people or vehicles move close to the transmitter and the receiver, the phase of the multipath components changes and multipath fading is observed. Figure 2.15 shows the measured average RSS values in a laptop of a signal transmitted by an IEEE 802.11 access point. Notice the large variations in the RSS samples caused by multipath fading, even though the RSS samples are averaged values.

In this section we explained how multipath causes fluctuations in the RSS. As we explained in this section, these fluctuations in RSS of the received signal in wireless links is modeled by the statistics of multipath fading and the Doppler spectrum. Next, we provide some examples of popular models used for these purposes.

2.3.2 Modeling of Small-Scale Fading

Multipath fading results in fluctuations of the signal amplitude because of the addition of signals arriving with different *phases*. This phase difference is caused due to the fact that signals have traveled different distances by traveling along different paths. Since the phases of the arriving paths are changing rapidly, the received signal amplitude undergoes rapid fluctuation. This rapid fluctuation is often modeled as a random variable with a particular distribution.

To model these fluctuations one can generate a histogram of the received signal strength samples in time. The density function formed by this histogram represents the distribution of the fluctuating values of the received signal strength. The most commonly used distribution for multipath fading is the Rayleigh distribution whose probability density function is given by:

$$f_{ray}(r) = \frac{r}{\sigma^2} \exp\left(-\frac{r^2}{2\sigma^2}\right), r \geq 0 \tag{2.15}$$

Here, it is assumed that all signals suffer nearly the same attenuation, but arrive with uniformly different phases. Theoretical considerations indicate that the sum of such signals will result in the amplitude having the Rayleigh distribution of Equation 2.15. The random variable corresponding to the signal amplitude is r. This is also supported by measurements at various frequencies [Pah05]. When a strong LOS signal component also exists, the distribution is found to be Ricean, and the probability density function of such a distribution is given by:

$$f_{ric}(r) = \frac{r}{\sigma^2} \exp\left(-\frac{r^2 + K^2}{2\sigma^2}\right) I_0\left(\frac{Kr}{\sigma^2}\right), r \geq 0, K \geq 0 \tag{2.16}$$

Here, K is a factor that determines how strong the LOS component is relative to the rest of the multipath signals.

Equations 2.15 and 2.16 can be used to determine what fraction of time a signal is received such that the information it contains can be decoded or what fraction of area receives signals with the requisite strength. The remainder of the fraction is often referred to as *outage*.

Small-scale fading results in very high bit error rates. In order to overcome the effects of small scale fading, it is not possible to simply increase the transmit power because this will require a humungous increase in the transmit power. A variety of techniques are used to mitigate the effects of small-scale fading – in particular error control coding with interleaving, diversity schemes, and using directional antennas. These techniques will be discussed in Chapter 3.

2.3.3   Modeling of Doppler Spectrum

Equations 2.15 and 2.16 provide the distributions of the amplitude of a radio signal that is undergoing small-scale fading. In general, it is also important to know for what time a signal strength will be below a particular value (duration of fade) and how often it crosses a threshold value (frequency of transitions or fading rate). This is particularly important to design the coding schemes and interleaving sizes for efficient performance. We see that this is a second order statistic and it is obtained by what is known as the *Doppler spectrum* of the signal.

The Doppler spectrum is the spectrum of the fluctuations of the received signal strength. Figure 2.16 [How90] demonstrates the results of measurements of amplitude fluctuations in a signal and its spectrum under different conditions. In the top of Figure 2.16, the transmitter and receiver are kept fixed and there is nothing moving in the vicinity of the set up. The received signal has

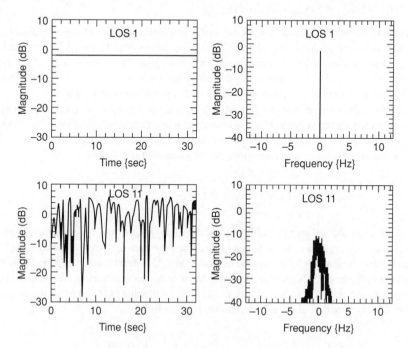

Figure 2.16   Measured values of the Doppler.

a constant envelope and its spectrum is an impulse for this reason. In the bottom of Figure 2.16, the transmitter is randomly moved, resulting in fluctuations of the received signal. The spectrum of this signal is now expanded (it is not an impulse) and spans around 6 Hz reflecting the *rate of variations* of the received signal strength. This spectrum is referred to as the Doppler spectrum.

In the mobile radio applications the Doppler spectrum for a Rayleigh fading channel is usually modeled by:

$$D(\lambda) = \frac{1}{2\pi f_m} \times \left[1 - \left(\frac{\lambda}{f_m} \right)^2 \right]^{-\frac{1}{2}} \quad \text{for } -f_m \leq \lambda \leq f_m \tag{2.17}$$

Here, f_m is the maximum Doppler frequency possible and is related to the velocity of the mobile terminal via the expression $f_m = v_m \times c/f$ where v_m is the mobile velocity and f is the center wavelength of the radio signal. Note here that we use λ as the Doppler frequency rather than the wavelength. This spectrum, commonly used in mobile radio modeling, is also called the *classical Doppler spectrum* and is shown in Figure 2.17. Another popular model for the Doppler spectrum is the bell-shaped distribution that is used for indoor applications [Pah05].

From the shape of the Doppler spread, it is possible to obtain the fade rate and the fade duration for a given mobile velocity [Pah05]. These values can then be used in the design of appropriate coding and interleaving techniques for mitigating the effects of fading. Diversity techniques are useful to overcome the effects of fast fading by providing multiple copies of the signal at the receiver. Since the probability that all of these copies are in fade is small, the receiver is able to correctly decode the received data. Frequency hopping is another technique that can be used to combat small-scale fading. Because all frequencies are not simultaneously under fade, transmitting data by hopping to different frequencies is an approach to combat fading. This is discussed in Chapter 3.

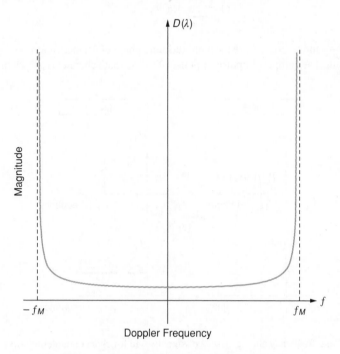

Figure 2.17   The classical Doppler spectrum.

2.4   Wideband Modeling of Multipath Characteristics

The phasor analysis based on Friis equation that we presented in Section 2.3 for a geometric optical explanation of the cause of multipath fading for the RSS was based on the assumption that we transmit a sinusoid signal at a given frequency. In circuits and systems, this analysis is sometimes referred to as frequency domain analysis because as shown in Figure 2.18a, in the frequency domain, the spectrum of a sinusoid has only a single impulse representing a frequency tone.

A sinusoid is an ideal signal in frequency domain. The equivalent of the ideal signal in time domain is an ideal impulse, shown in Figure 2.18b, that we use to find the time domain characteristics of a system. The impulse in time has a constant spectrum in frequency (i.e., it has all of the frequency components with the same power). Using an ideal impulse one can determine the impulse response of a channel that can be used to find out what happens to a waveform if it is transmitted over a multipath channel. Waveforms are used in communications to carry bits of information (see Chapter 3) and in localization (see Chapter 14) to determine the distance between the transmitter and the receiver. Knowing what happens to a transmitted waveform is essential for the design of efficient wireless access and localization systems.

2.4.1   Impulse Response, Multipath Intensity, and Bandwidth

If we consider the geometric optics based ray tracing technique that we have developed so far, we realize that in ideal conditions, if we are able to transmit an impulse, at the receiver we receive several impulses each arriving along a different path. In other words the impulse response of the a multipath channel is a discrete function of the form:

$$h(\tau) = \sum_{i=1}^{L} A_i \delta(\tau - \tau_i) e^{j\Phi_i} \qquad (2.18)$$

where $A_i = \frac{a_i \sqrt{P_0}}{d_i}$ and $\Phi_i = \frac{2\pi d_i}{\lambda}$ are the amplitude and phase of i-th path and $\tau_i = \frac{d_i}{c}$ is the time of arrival of the path. The complex impulse response of a three-path channel is shown in Figure 2.19a.

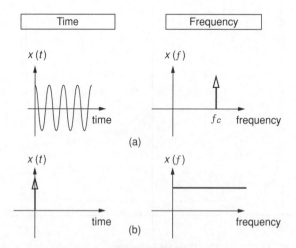

Figure 2.18   Time-frequency characteristics of signals used for channel modeling: (a) a sinusoid as an ideal narrow band signal, (b) an impulse, the ideal wideband signal.

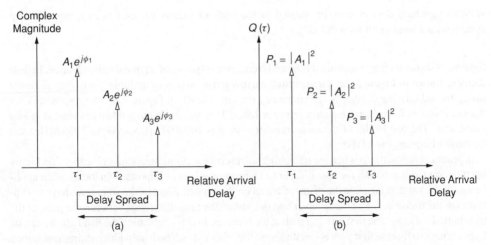

Figure 2.19   (a) Complex channel impulse response. (b) Delay power spectrum.

The magnitude square of the complex channel impulse response is called Delay Power Spectrum, shown in Figure 2.19b:

$$Q(\tau) = |h(\tau)|^2 = \sum_{i=1}^{L} P_i \delta(\tau - \tau_i) \qquad (2.19)$$

where $P_i = |A_i|^2$ is the power of the signal arriving along the i-th path. The physical meaning of the Delay Power Spectrum is that it represents the received power arriving along different paths as a function of the delay of arrival of these paths. Therefore, the horizontal axis is not the real time, it is the delay or lag between the various multipath components using arrival time of the first path as the reference.

To measure a typical channel impulse response we need to transmit a narrow pulse that resembles an impulse. The narrower the pulse, the wider is the required bandwidth for transmission of the pulse. To measure (or *resolve*) all multipath components the bandwidth should be wide enough that its inverse is proportional to the difference in the delay of arrival of different paths $\Delta\tau$. This delay reflects the intensity of the multipath arrivals. Since delay is a function of distance, the difference between arrivals or intensity of the paths is related to the difference in the path lengths Δd. The bandwidth requirement to isolate the paths with a measurement system is approximately $W = \frac{1}{\Delta\tau} = \frac{c}{\Delta d}$.

Example 2.12: Resolving multipath components in indoor and urban areas

In indoor areas used for WLAN applications, distances on the order of meters separate walls and other objects. Therefore, it is reasonable to assume that a measurement system used for measuring the indoor multipath should be able to resolve multipath components that are up to 1 m apart. For this resolution we need a bandwidth of $W = 3 \times 10^8 = 300$ MHz because difference in distances is 1 m. If we consider outdoor areas where buildings are around of tens of meters apart our measurement system may require bandwidth on the order of 30 MHz. If we need a measurement system for

WPANs in which devices may be located in fractions of meters we may need a measurement system with a bandwidth around 1 GHz.

Figure 2.3 shows a sample measured time and frequency response of a typical radio channel. In time domain, shown in Figure 2.3a, a transmitted narrow pulse arrives as multiple paths with different strengths and arriving delays. In the frequency domain, shown in Figure 2.3b, the response is not flat and it suffers from deep frequency selective fades. This measurement is taken in a typical indoor office area. The bandwidth of the measurement system is 200 MHz (from 900 to 1100 MHz) and the center frequency is 1 GHz.

In practical applications when we transmit a waveform for communication or localization knowing the required bandwidth for operation in an environment is very important. In localization applications if we want to measure the distance based on the time of flight of the signal, we have to synchronize the transmitter and the receiver and determine the time difference between the peak of the transmitted pulse and the first arriving path at the receiver. In this application the bandwidth must be large enough so that the first path is isolated from the other paths. The bandwidth requirement is then very similar to bandwidth requirements for the multipath channel measurement systems described earlier in this section. The bandwidth requirement for wireless communications is different because for reliable communications, the emphasis is on the *symbol transmission rate*. For a given multipath scenario we want to increase the rate of transmission of our symbols, but increasing the symbol transmission rate or bandwidth beyond certain values will cause inter-symbol interference (ISI).

2.4.2   Multipath Spread, ISI, and Bandwidth

In Figure 2.19 the delay between the arrival of the first and the last path is referred to as the excess multipath delay spread or simply the *delay spread* of the channel. In wireless communications the inverse of the width of the transmitted symbols approximately represents the required bandwidth for data transmission. If the multipath delay spread is comparable to or larger than the symbol duration, the inverse of the bandwidth, the received waveform representing one symbol spreads into the waveform representing adjacent symbols significantly and produces considerable ISI. If the ISI distorts the symbols so that the receiver cannot distinguish between the possible transmitted symbols regardless of how much power we use for transmission receiver can not differentiate the transmitted symbols and makes errors in detection process.

To explain this phenomenon, consider Figure 2.20, in which we want to transmit a rectangular waveform for communication purposes over a three-path channel. The information is coded in the amplitude of this waveform and every T_s seconds we transmit a symbol. The data rate or the required bandwidth for implementation of this transmission system is $W = 1/T_s$. Because of the multipath, at the receiver we have three waveforms that add together and the resulting waveform stretches beyond T_s, the duration allocated to the symbol, and interferes with the transmission of the next symbol causing ISI. As we increase the transmission data rate, the duration of the waveform T_s is reduced and the amount of ISI is increased. The increase in the ISI causes degradation in the performance of the transmission system. In data communications over single path channels, degradation of transmission performance is due to the background noise. When we encounter such a situation, we can increase the transmission power so that the received signal is stronger than the fixed background noise. In an ISI environment, increasing the transmitter power does not solve the problem because it increases the amount of ISI as well.

The amount of ISI depends on the delay spread and the strength of the individual paths with respect to the first path. The second central moment of the multipath intensity profile is called the

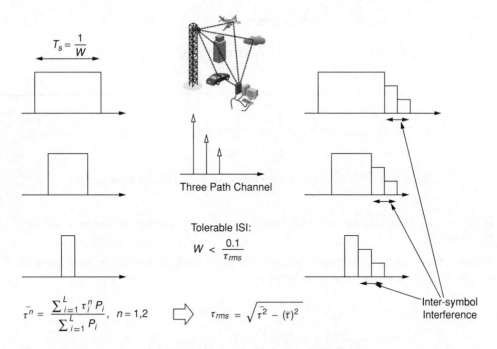

Figure 2.20 Relationship between among channel impulse response, ISI, and symbol transmission rate or bandwidth.

root mean square (rms) delay spread and it is used as a measure of the ISI. The basic definition of the second central moment is given by:

$$\tau_{rms} = \sqrt{\overline{\tau^2} - (\overline{\tau})^2}$$

where the *n*-th moment is defined by:

$$\overline{\tau^n} = \frac{\sum_{i=1}^{L} \tau_i^n P_i}{\sum_{i=1}^{L} P_i}, n = 1, 2$$

in which τ_i and P_i are the arrival delay and the power of the *i*-th path. If we combine the last two equations, we will have:

$$\tau_{rms} = \sqrt{\frac{\sum_{i=1}^{L} \tau_i^2 P_i}{\sum_{i=1}^{L} P_i} - \left(\frac{\sum_{i=1}^{L} \tau_i P_i}{\sum_{i=1}^{L} P_i} \right)^2} \qquad (2.20)$$

Example 2.13: Calculation of the rms delay spread

Consider a two path channel, with an impulse response comprising multipaths with arrival delays of $\tau_1 = 0$ ns and $\tau_2 = 50$ ns and path powers of $P_1 = 1$ (0 dBm) and $P_1 = 0.1$ (−10 dBm), shown

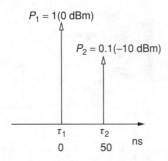

Figure 2.21   The delay power spectrum of the channel used in Example 2.9.

in Figure 2.21. Determine the rms delay spread from direct calculation of moments and using Equation 2.20.

Solution: For the direct calculation of the rms delay spread, using the expression for the moments, we can calculate the first and second moments as:

$$\bar{\tau} = \frac{0 \times 1 + 50 \times 0.1}{1 + 0.1} = 4.55(n \sec)$$

$$\overline{\tau^2} = \frac{0 \times 1 + 2500 \times 0.1}{1 + 0.1} = 227.27(n \sec)$$

Then we can determine the rms delay spread:

$$\tau_{rms} = \sqrt{\overline{\tau^2} - (\bar{\tau})^2} = \sqrt{227.27 - (4.55)^2} = 14.37(n \sec)$$

We could also determine the same value using Equation 2.20 in one step:

$$\tau_{rms} = \sqrt{\frac{0 \times 1 + 2500 \times 0.1}{1 + 0.1} - \left(\frac{0 \times 1 + 50 \times 0.1}{1 + 0.1}\right)^2} = 14.37(n \sec)$$

Considering Figure 2.20, if we want to have an intuition on the signal to noise ratio in presence of multipath, the strength of the signal relative to ISI noise is related to the duration of the transmitted pulse. The longer the length of the pulse, T_s, the smaller will the effects of the ISI be. Therefore, $\frac{T_s}{\tau_{rms}} = \frac{1}{W \times \tau_{rms}}$ is a measure of the signal to ISI noise. As we will see in Chapter 3, most basic digital transmission systems operate at a reasonable error rate when their signal to noise ratio is above 10 dB. Considering this value, one can conclude that reliable operations of a simple digital communication link over a multipath channel is possible if the symbol transmission rate or the bandwidth of the system is less than 10% of the inverse of the rms delay spread or:

$$W < \frac{0.1}{\tau_{rms}} \tag{2.21}$$

If the bandwidth of the transmission system over a multipath channel follows the above rule, the shape of the transmitted symbol is preserved without significant distortions in the shape of the waveform. As inverse of the rms delay spread plays an important role in calculation of the data transmission rate over multipath channels, in the literature this inverse is sometimes referred to as the *coherence bandwidth* B_c of the channel.

The rms delay spread varies depending on the type of environment. In indoor areas, it could be as small as 30 ns in residential areas or as large as 300 ns in factories [Pah05]. In urban macrocellular areas, the rms delay spread is on the order of a few microseconds. This means that the maximum data rates that can be supported by a simple binary modem in indoor areas can be as high as 6.7 Mbps (at 30 ns) and in wide area cellular networks it can be as low as 50 kbps (at 4 μs). This observation indicates that as we increase the range of coverage of a wireless network, the distances of objects causing multipath become larger and the rms delay spread increases resulting in a lower supportable data rate with simple schemes using a single frequency carrier.

In order to support higher data rates, different receiver techniques are necessary. Equalization is a method that tries to cancel the effects of multipath delay spread in the receiver. Direct sequence spread spectrum enables resolving the multipath components and exploiting them to improve performance. Orthogonal frequency division multiplexing (OFDM) uses multiple carriers, spaced closely in frequency, each carrying low data rates to avoid ISI. Beamforming with multiple input multiple output (MIMO) antenna systems reduce the number of multipath components, thereby reducing the total delay spread itself. We discuss these topics in Chapter 3. Here we have discussed a simple method to relate the bandwidth to the ISI.

2.4.3   Wideband Channel Models in Standardization Organizations

In reality, delay power spectrum, $Q(\tau)$, is a two-dimensional function, $Q(\tau, t)$, of *delay* and *time* of arrival. However, it is a slow-time varying function or channel and for practical purposes, we can represent it as a function of only delay [Pah05]. The physical meaning of slow-time varying channel is that when we are transmitting a waveform for communication or localization purposes, during the transmission of the symbol, the channel is stable or invariant. We have already modeled the rate of variations of the channel by the Doppler Spectrum function $D(\lambda)$ that reflects the effects of movement in the area of operation. In a typical wireless application, the rate of variations of the channel are at most several hundred hertz while typical transmission rates are on the order of mega symbols per second. In this situation, a slow-time varying model is very reasonable. Under this assumption we divide the channel behavior into a static and a dynamic component. The static behavior is represented by the delay power spectrum, $Q(\tau)$, and the dynamic behavior by the Doppler spectrum $D(\lambda)$. In classical radio channel modeling [Pah05], a combination of these two functions is called the *scattering function* and it is defined by:

$$S(\tau, \lambda) = Q(\tau) \times D(\lambda) \tag{2.22}$$

If we complement specifications for Equation 2.22 with a path-loss model, we have a complete set of analytical models to simulate a channel for both coverage and communication performance evaluation purposes.

Standardization organizations usually provide channel models that identify the scattering function and path-loss models for different scenarios. One of the major challenges for wireless standardization organizations is to compare and select the best modem design for the physical layer implementation among multiple proposed systems. To have a fair comparison among these proposed alternatives, a commonly accepted channel model is needed. After the completion of the standard, these models are used by manufacturers for the design and performance evaluation of

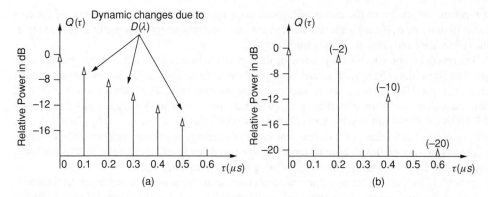

Figure 2.22   Two options for delay power spectrums recommended by the GSM committee: (a) six paths, (b) four paths.

their products. Since the bandwidth and environments in which these channel models are used are different, most standardization groups come up with their own standard model(s). Since we have different channel models for the wide area and local area networks we treat their details in separate sub-sections in the relevant chapters.

We proceed to explain the wideband model recommended by GSM as a simple example to understand the channel models recommended by standardization organizations.

Example 2.14: Wideband model for TDMA cellular networks

The GSM standardization group defines a set of channel profiles with discrete delay power spectra of different types for rural areas, urban areas, and hilly terrains [GSM91]. The basic difference between these models is the value of the rms delay spread and the number of multipath components that are used to represent the channel profile. These models is made for a TDMA cellular network, where the bandwidth of the system is 200 KHz. It is useful for any wireless access method operating in urban areas with similar bandwidths.

In this example, we only describe the model used for rural areas. This model defines a delay power spectrum with two options for implementation with six or four multipaths. Figure 2.22 shows the two delay power spectra recommended for rural areas. In the six-path model, the multipaths are 0.1 μs apart and they cover a delay spread of 0.5 μs. The power of the received signal at each path starts at 0 dB and decays by 4.0 dB with each successive path. In the four-path model, the delay spacing is 0.2 μs and it covers a delay spread of 0.6 μs. The relative powers of the paths are 0, −2, −10, and −20 dB, respectively. Both the six-path and four-path models roughly provide the same rms delay spread. If we want to evaluate the effects of multipath on the design of different modems, both models should provide similar results. The six-path option provides for a more refined model at the expense of additional hardware for implementation of two more paths. The bandwidth of the GSM channels is 200 KHz resulting in pulses with approximately 5 μs width. The delay spread of the channel is around 10% of this value that follows our bandwidth constraint for manageable ISI defined by Equation 2.21.

The first path in both profiles is assumed to have a Rician distribution given by Equation 2.16 because they are assumed to be along a direct LOS path. The rest of the paths are assumed to have classical Rayleigh distributions given by Equation 2.15. The Doppler spectrum choices for each path or tap of the model are either Rician or the classical Rayleigh. In a manner similar to the simulation of narrowband signals, the Doppler power spectrum for the classical Rayleigh model is the one given by Equation 2.17 and illustrated in Figure 2.17. The Rician spectrum is the sum of the classical Doppler spectrum in Equation 2.17, and one direct path, weighted so that the total multipath power is normalized and it is given by:

$$D(\lambda) = \frac{0.41}{2\pi f_m} \left[1 - (\lambda/f_m)^2\right]^{-1/2} + 0.91\delta(\lambda - 0.7 f_m), \quad -f_m < \lambda < f_m. \tag{2.23}$$

If we complement this multipath model with the Okumura–Hata path-loss model we have a complete model for behavior of the channel in the specified area. Using these models, we can predict the coverage of the system and simulate the effects of fluctuations of the channel as well as the effects of multipath on the waveforms transmission.

2.4.4 Simulation of Channel Behavior

In terms of hardware or software simulations, the separation of the static and dynamic behavior allows us to implement the behavior in the delay variable with a tapped delay line. Spacing of the taps are at the value given by delay power spectrum, $Q(\tau)$, and short term fading fluctuations of each path are implemented by using a filtered complex Gaussian noise with the shape of Doppler spectrum, $D(\lambda)$, as illustrated in Figure 2.23. In Figure 2.23b we have paths with different arrival delays, identified by $Q(\tau)$, implemented in parallel branches. Figure 2.23a shows the implementation of the amplitude and phase of each path using a filtered Gaussian noise with an spectrum shape prescribed by $D(\lambda)$. The simulated complex channel fluctuations in Figure 2.23a are scaled by the strength of the path so that the overall channel response in Figure 2.23b provides for the delay power spectrum defined in Equation 2.19. In general, the delay τ_i is a random variable as well, but for simplicity of implementation traditional standardization organizations assume fixed values for the delay and try to fit the rms delay spread of the multipath profile with the typical measurements in the environment that model is designed for.

The main objective in the development of a model for wideband characteristics of the channel is to develop a foundation for design and comparative performance evaluation of wireless modems. The analysis of the performance was traditionally performed using analytical equations and the calculation of the analytical equations using digital computers. As the speed of computers and digital hardware in general increased, the models were also used for real-time hardware and computer software simulations of the channel behavior.

2.5 Emerging Channel Models

In this section we discuss some new radio channel models that are gaining importance for different applications. Position location is becoming important for emergency and location-aware applications (see Chapter 14–16) and models developed for communication systems are no longer sufficient to address the performance of geolocation schemes. The use of smart antennas and adaptive antenna arrays require knowledge of the *angle of arrival* of the multipath components in order to steer antenna beams in the right directions. We provide a brief discussion of these models below.

2.5.1 Wideband Channel Models for Geolocation

With the advent of widespread wireless communications during the 1990s, location of people, mobile terminals, pets, equipment and the like by employing radio signals gained importance as

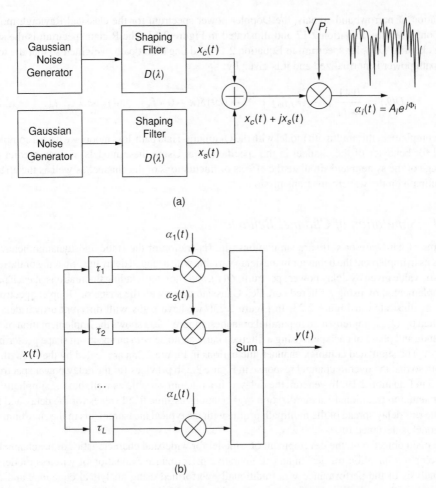

Figure 2.23 Elements of a complete channel simulator: (a) simulation of the tap gains with dynamic behavior of $D(\lambda)$, (b) simulation of static multipath using $Q(\tau)$.

well. Several new position location applications [Pah98] emerged in the market. Civilian applications included intelligent transportation systems, public safety (enhanced 911 or E-911 services), automated billing, fraud detection, cargo tracking, accident reporting, and so on. It was possible to employ position location for additional benefits such as cellular system design and futuristic *intelligent office* environments [War97]. Most tactical military units, however, were also heavily reliant on wireless communications. Ad hoc connectivity among individual warfighters, for instance in the Small Unit Operations, in restrictive radio frequency propagation environments requires *situation awareness systems* to enable the individual warfighters to determine their location and associated information. This situation occur in such environments such as inside buildings, tunnels, other urban structures, caves, mountainsides, and double canopy coverage in jungles and forests. In either case, the position location service had to operate within an environment where traditional geolocation techniques such as GPS fail due to lack of sufficient signal power and the harsh multipath conditions.

While RF propagation studies in the past have focused on telecommunications applications, position location applications require a different characterization of the indoor radio channel [Pah98]. For position location applications, accurately detecting the *direct* LOS path between the transmitter and receiver is extremely important. The direct LOS path corresponds to the straight line connecting the transmitter and receiver even if there are obstructions like walls in between. Detecting the direct LOS path is important because, the time of arrival or the angle of arrival of the direct LOS path correspond to the distance between the transmitter and receiver (or to the direction between them). This information is used in conjunction with multiple such measurements to locate either the transmitter or the receiver as the case may be. This is in contrast to telecommunications applications where the emphasis is on how data bits can be sent over a link efficiently and without errors. Another issue in positioning systems is the relation between the bandwidth of the transmitted signal and the required accuracy in ranging. An error of 100 ns in estimating the delay of an arriving multipath component could result in an error of 30 m in calculation of the distance between a transmitter and a receiver. Therefore, positioning system using time of arrival often require wide bandwidths to resolve multipath components and detect the arrival of the first path.

In wideband indoor radio propagation studies for telecommunication applications often channel profiles measured in different locations of a building are divided into line-of-sight and obstructed line-of-sight because the behavior of the channel in these two classes have substantially different impacts on the performance of a telecommunications system.

A logical way to classify channel profiles for geolocation applications is to divide them into three categories as shown in Figure 2.24. The first category is the dominant direct path case in which the direct LOS path is detected by the measurement system and it is the strongest path in the channel profile. In this case, traditional GPS receivers [Get93; Eng94; Kap96] designed for outdoor applications where multipath components are significantly weaker than the direct LOS path, lock on to the direct LOS path and detect its time of arrival accurately. The second category is the non-dominant direct path case where the direct LOS path is detected by the measurement system but it is not the dominant path in the channel profile. For these profiles traditional GPS receivers, expected to lock to the strongest path, will make an erroneous decision on the time of arrival that leads to an error in position estimation. The amount of error made by a traditional receiver is the distance associated with the difference between the time of arrival of the strongest path and the time of arrival of the direct LOS path. For the second category, locations with non-dominant direct path profiles, a more complex receiver [Pah98] can resolve the multipath and make an intelligent decision on the time of arrival of the direct LOS path. The third category of channel profiles is the undetected direct path profiles. In these profiles the measurement system cannot detect the direct LOS path and therefore traditional GPS or other receivers both cannot detect the direct LOS path. In reality direct LOS always exists, however, when its strength goes under the threshed of deductibility of the receiver and we have other paths strong enough to be detected by the receiver an undetected direct path occurs. In these circumstances no receiver can detect the direct path and an erroneous location error is guaranteed. To imagine how such a situation may occur assume that a large metallic object, such as an elevator, obstruct the direct LOS between the transmitter and a receiver in an indoor environment. The direct path becomes undetectable but other paths reflecting from surrounding walls will be detected by the receiver. More details on this topic are discussed in Chapter 15.

Figure 2.25 shows the results of ray tracing simulations of regions in the first floor of Atwater Kent Laboratories at Worcester Polytechnic Institute with different types of multipath profiles for a centrally located channel sounder.

In the same way that the bit error rate is the ultimate measure for comparing performance of different digital communication receivers, the error in the measurement of the time of arrival or

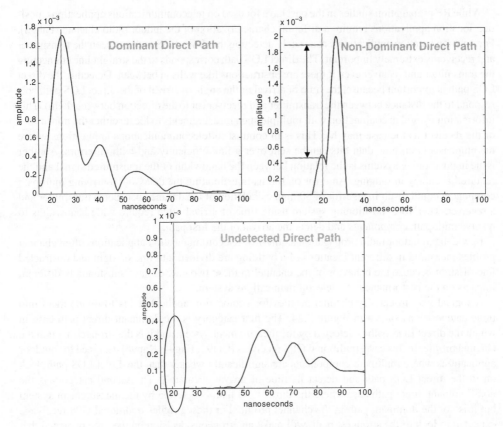

Figure 2.24   Multipath profiles for indoor geolocation.

angle of arrival of direct LOS path is a measure of the performance of the geolocation receivers. Traditional radio frequency studies consider the path loss and τ_{rms} as mentioned above and these are not sufficient for the geolocation problem. The relative power and delay of the signal arriving via other paths, the channel noise, the signal bandwidth and interference all influence the detection of the direct LOS path and thus the error in estimating the range (distance) between the transmitter and receiver.

Efforts to model the indoor radio channel for geolocation based on the error in the detection of the direct LOS path are reported in [Pah98; Kri98; Kri99a, b, c; Ala06a, b]. In these efforts, parameters of importance and development of a model are based on measurements of the radio channel used as input to software simulations.

2.5.2   SIMO and MIMO Channel Models

Recently, there has been a lot of attention placed on *spatial wideband channel models*, that not only provide the delay-power spectrum discussed in Equation 2.19, but also the angle of arrival of the multipath components. The advent of antenna array systems that are used for interference cancellation and position location applications has made it necessary to understand the spatial properties of the wireless communications channel.

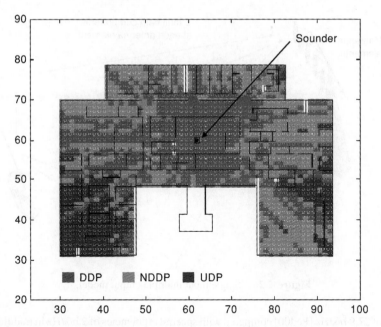

Figure 2.25 Simulated regions in the first floor of Atwater Kent Laboratories showing regions with different types of multipath profiles for a centrally located channel sounder.

We start with *single input multiple output* (SIMO) radio channel models [Ert98]. In these models, a typical cellular environment is considered where it is assumed that the mobile transmitters are relatively simple and the base station can have a complex receiver with adaptive smart antennas with M antenna elements. As shown in Figure 2.26, the multipath environment is such that up to L signals arrive at the base station from different mobile terminals (l) with different amplitudes (α) and phases (φ) at different delays (τ) from different directions (θ). These are in general time-invariant and as a result, the channel impulse response is usually represented by:

$$\vec{h}(t) = \sum_{l=1}^{L(t)} \alpha_l(t) e^{j\phi_l(t)} \delta(t - \tau_l(t)) \vec{a}(\theta_l(t))$$

Note that the channel impulse response is now a *vector* rather than a scalar function of time. The quantity $\vec{a}[\theta(t)]$ is called the array response vector and will have M components if there are M antenna array elements. Thus, there are M channel impulse responses each with L multipath components. A variety of models are available in [Ert98]. The amplitudes are usually assumed to be Rayleigh distributed although they are now dependent on the array response vector $\vec{a}(\theta(t))$ as well.

An extension of this model to the scenario where there are N mobile antenna elements and M base station antenna elements [Ped00] and this is called a *multiple-input multiple output* (MIMO) channel. In this case, the channel impulse response is an $M \times N$ matrix that associates a *transmission coefficient* between each pair of antennas for each multipath component. Experimental results and models are considered in [Ker00; Ped00].

There appears to be tremendous potential for improving capacity using MIMO antenna systems. Capacity increases of 300–500% are possible in cellular environments. Spectrum efficiency is also increased. For example, a 4×4 antenna array system over the MIMO channel can provide a spectral

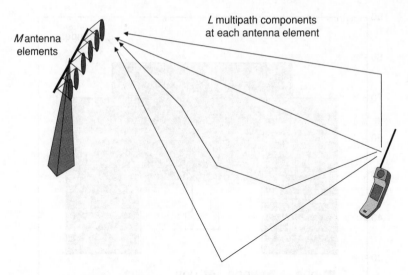

Figure 2.26   Single input multiple output model.

efficiency of 27.9 b/s/Hz [Ped00] compared with spectral efficiencies of 2 b/s/Hz in traditional SISO radio systems.

Appendix A2: What Is the Decibel?

The decibel (dB) is usually the unit employed to compute the logarithmic measure of power and power ratios. The reason for using dB is that all computation reduces to addition and subtraction rather than multiplication and division. Every link, node, repeater, or channel can be treated as a *black box* (see left side of Figure A2.1) with a particular dB gain. The dB gain of such a black box is given by:

$$\text{dB gain} = 10 \log \left(\frac{\text{power of output signal}}{\text{power of input signal}} \right) = 10 \log \left(\frac{P_{out}}{P_{in}} \right) \qquad (A2.1)$$

This corresponds to the *relative* output power with respect to the input power. The logarithm is always to the base 10. If the ratio in Equation A2.1 is negative, it is a decibel loss.

The decibel gain relative to an absolute power of 1 mW is denoted by dBm and that relative to 1 W is denoted by dBW. For example, if the input power is 50 mW, relative to 1 mW, the input power is 10 log (50 mW/1 mW) = 16.98 dBm. If this is followed by a link having a loss of 10 dB,

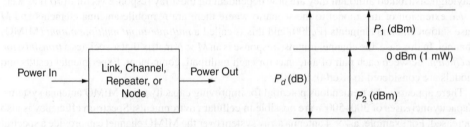

Figure A2.1   The decibel: (a) overall concept (b) relation between dB and dBm.

the absolute power at the output of the link will be $16.98 - 10 = 6.98$ dBm. Relative to 1 W, these values will be $10 \log (50 \times 10^{-3}/1) = -13$ dBW and -23 dBW respectively. Note that the difference between two power levels in dBm, P_1 and P_2, is in dB: P_1 (dBm) $- P_2$ (dBm) $= P_d$ (dB) – see right side of Figure A2.1.

Antenna gains are represented similarly with respect to an *isotropic* antenna (that radiates with a gain of unity in all directions) or a *dipole* antenna. The former gain is in dBi units and the latter in dBd units. The dBi units are 2.15 dB larger than the dBd units.

Questions

1. What are the three important radio propagation mechanisms at high frequencies? Which of them are predominant for wireless indoor applications?
2. Explain what path-loss gradient means. Give some typical values of the path-loss gradient in different environments.
3. Explain the meaning of the expression "a loss of 37 dB per decade of distance" in terms of the path loss gradient α.
4. Why does multipath in wireless channels limit the maximum symbol transmission rate? How can we overcome this limitation?
5. What is the Doppler spectrum and how can one measure it?
6. What are the differences between multipath, shadow and frequency-selective fading? Give an example distribution function that is used to model multipath fading and an example that is used to model shadow fading.
7. What is fading margin? Explain its meaning in terms of the fraction of coverage at the edge of a cell.
8. What are the situations where the Okumura–Hata model is applicable? Can it be directly applied for transmissions in the 2.4 GHz ISM bands?
9. What distribution is used to model the amplitude of a signal under multipath fading in LOS situation? In OLOS situations?
10. How is the symbol duration important when we consider radio propagation? Explain in terms of the inter-symbol interference caused by multipath.
11. What is the relationship between the coherence bandwidth and the rms multipath delay spread?
12. What techniques can be used to combat the effects of frequency selective fading (multipath delay spread)?
13. For position location applications using time of arrival, how are wideband radio channels classified? How is this classification useful?
14. What is the difference between a SIMO and a MIMO radio channel?
15. Explain ray tracing and its relation to geometric optics.

Problem 2.1

Assuming wireless devices use an antenna length of one-fourth of the wavelength of the transmitted frequency, what are the typical antenna lengths for cellular phones operating in the 900 and 1800 MHz bands and WLANs operating at 2.4 and 5.2 GHz?

Problem 2.2

What is the received power (in dBm) in free space of a signal whose transmit power is 1 W and carrier frequency is 2.4 GHz if the receiver is at a distance of 1 mile (1.6 km) from the transmitter? What is the path loss in dB? Assume antenna gains are one.

Problem 2.3

The path loss in a building was discovered to have two factors adding to the free space loss: a factor directly proportional to the distance and a floor-attenuation-factor (FAF). In other words, the path loss = free space loss + βd + FAF. If the FAF is 24 dB, and the distance between transmitter and receiver is 30 m, determine what should be the value of β so that the path loss suffered is less than 110 dB.

Problem 2.4

The modulation technique used in the an analog cellular radio systems is analog FM. The transmission bandwidth is 30 kHz per channel and the maximum transmitted power from a mobile user is 3 W. The acceptable quality of the input SNR is 18 dB and the background noise in the bandwidth of the system is -120 dBm (120 dB below the 1 mW reference power). In this cellular operation, we may assume that the strength of the signal drops 30 dB at the first meter of distance from the transmitter antenna and have an attenuation of 40 dB per decade of distance for distances beyond 1 m.

a. What is the maximum distance between the mobile station and the base station at which we have an acceptable quality of signal? Use Okumura-Hata model for your calculations.
b. Repeat (a) for a digital cellular system for which the acceptable SNR is 14 dB.

Problem 2.5

The transmitted power of an IEEE 802.11g device operating at 2.45 GHz is 100 mW. When the terminal is close to the access point (AP), the maximum data rate that is possible is 54 Mbps that requires an average received signal strength (RSS) of -72 dBm. The minimum supported data rate is 6 Mbps that requires a minimum RSS of -90 dBm.

a. Determine the coverage of the AP for data rates of 54 and 6 Mbps in a small office using the IEEE 802.11 channel model C.
b. Repeat (a) assuming a single distance-power gradient of 2.5 and compare your results with the results of part (a).

Problem 2.6

The transmitted power of an IEEE 802.11a device is 100mW and it operates at 5.2 GHz. When the terminal is close to the AP the maximum data rate is 54 Mbps that requires an RSS of -72 dBm. As the distance from the AP increases, the data rate reduces to values that require smaller RSS to operate. The minimum supported data rate is 6 Mbps that requires a minimum RSS of -94 dBm.

a. Determine the path-loss in the first meter in dB.
b. Determine the coverage of the AP for 54 and 6 Mbps in a residential or small office environment with LOS/NLOS using the IEEE 802.11 channel model D in Table 2.3.
c. Determine the fade margin for the system for 90% success at the fringe of coverage, again for the same model as in (b).
d. Repeat (b) and (c) for a large open indoor and outdoor space with NLOS conditions.

Problem 2.7

IEEE 802.11 WLANs operate at a maximum transmission power of 100 mW (20 dBm) using multiple channels with different carrier frequencies. The IEEE 802.11g devices use the 2.402–2.480 GHz bands and the IEEE 802.11a devices use the 5.150–5.825 GHz bands. Both standards use OFDM modulation with a bandwidth of 20 MHz.

a. Calculate the received signal strength in dBm at 1 m distance of an IEEE 802.11g access point for the smallest and the largest possible carrier frequencies in the band. Assume that the transmitter and receiver antenna gains are one (0 dBi) and at the first meter, signal propagation follows the free-space propagation rules.
b. Repeat (a) for the IEEE 802.11a WLANs.
c. Compare the received signal strengths at the one-meter distance for the IEEE 802.11g and IEEE 802.11a devices. Use the middle of the allocated band for each standard as the carrier frequency in your calculations.
d. Compare the rate of the received signal fluctuations (maximum Doppler shift f_m), due to the change in the frequency of operation, for both the IEEE 802.11g and the IEEE 802.11a devices. Use the middle of the allocated band for each standard as the carrier frequency in your calculations. Assume that the velocity of movement in the environment is 1 m/s.

Problem 2.8

Table P2.1 provides the minimum required RSS for an IEEE 802.11b device to operate at different rates.

a. Calculate the coverage associated with each data rate in the table using IEEE 802.11 channel model B.
b. Plot the staircase function of the data rate versus RSS.
c. Plot the staircase function of the data rate versus distance. This should be a plot that shows the data rate as a fixed number in the respective distance range.
d. If a mobile terminal moves away from an 802.11 AP, it goes out of the coverage area. Calculate the *average* data rate that terminal observes during the movement in the coverage area of the AP. Assume that the mobile terminal is moving at a constant speed and it is always receiving (or transmitting) packets. Also assume there is no interference and ignore any MAC layer effects.

Table P2.1   Data rate and minimum power requirement for IEEE802.11b

Data rate (Mbps)	RSS (dBm)	Coverage (m) using IEEE 802.11 path-loss model D
11	−82	
5.5	−87	
2	−91	
1	−94	

Problem 2.9

Use the Okumura–Hata model to determine the maximum radii of cells at 900 MHz and 1900 MHz respectively having a maximum acceptable path loss of 130 dB. Use

$a(h_m) = 3.2[\log(11.75h_m)^2 - 4.97]$ for both cases. Assume $h_b = 200$ m and $h_m = 10$ m. Here the frequency f_c is in MHz (use 1900), the antenna heights are in meters, and the distance d is in km.

Problem 2.10

In a mobile communications network, the minimum required signal to noise ratio is 12 dB (i.e., the RSS has to be 12 dB larger than the noise power level). The background noise at the frequency of operation is -115 dBm. If the transmit power is 10 W, the transmitter antenna gain is 3 dBi, the receiver antenna gain is 2 dBi, the frequency of operation is 800 MHz and the base station and mobile antenna heights are 100 m and 1.4 m respectively, determine the maximum in building penetration loss that is acceptable for a base station with a coverage of 5 km if the following path-loss models are used:

a. Free space path loss model
b. Two-ray path loss model
c. Okumura–Hata model for a small city.

Problem 2.11

A mobile system is to provide 95% successful communication at the fringe of coverage with a location variability having a zero mean Gaussian distribution with standard deviation of 8 dB. What fade margin is required?

Problem 2.12

Signal strength measurements for urban microcells in the San Francisco Bay area in a mixture of low-rise and high-rise buildings indicate that the path loss L_p in dB as a function of distance d is given by the following linear fits:

$$L_p = 81.14 + 39.40 \log f_c - 0.09 \log h_b + [15.80 - 5.73 \log h_b] \log d, \text{ for } d < d_{bk}$$

$$L_p = [48.38 - 32.1 \log d_{bk}] + 45.7 \log f_c + (25.34 - 13.9 \log d_{bk}) \log h_b$$
$$+ [32.10 + 13.90 \log h_b] \log d + 20 \log(1.6/h_m), \text{ for } d > d_{bk}$$

Here, d is in kilometers, the carrier frequency f_c is in GHz (that can range between 0.9 and 2.0 GHz), h_b is the height of the base station antenna in meters, and h_m is the height of the mobile terminal antenna from the ground in meters. The *breakpoint* distance d_{bk} is the distance at which two piecewise linear fits to the path loss model have been developed and it is given by $d_{bk} = 4h_b h_m/100\ \lambda$, where λ is the wavelength in meters. The shadow-fading component in dB is given by a zero mean Gaussian random variable with a standard deviation of 5 dB.

a. If 90% of the locations at the cell edge need coverage, what should be the fading margin applied? What percent of locations would be covered if a fading margin of 5 dB is used?
b. What would be the radius of a cell covered by a base station (height 15 m) operating at 1.9 GHz and transmitting a power of 10 mW that employs a directional antenna of gain 5 dBi? The fading margin is 7.5 dB and the sensitivity of the mobile receiver is -110 dBm. Assume that $h_m = 1.2$ m. How would you increase the size of the cell?

Note that the path losses predicted by the two equations are very close, but not exactly the same at $d = d_{bk}$. You can use either value in your calculation.

Problem 2.13

The path-loss at a frequency of 1.9 GHz in 95 existing suburban macro-cells was measured by AT&T Wireless in New Jersey, Seattle, Chicago, Atlanta, and Dallas (see [Erc99]). The numbers used in this problem are roughly for a flat terrain with low tree density. Based on their measurements, they proposed a path-loss model given by:

$$L_p = A + 10\alpha \log \frac{d}{d_0} + \sigma X$$

where $A = 78$ dB, α is a normally distributed random variable with mean value given by:

$$m = a - bh_b + c/h_b$$

where h_b is the height of the base station antenna. The standard deviation of α is empirically computed to be σ_α. The reference distance d_0 is 100 m. X is a normally distributed random variable with mean zero and standard deviation 1 and represents the shadow-fading component. Furthermore the standard deviation of shadow fading σ (when considered across different macro-cells) is itself normally distributed with a mean p and standard deviation q. So the overall path-loss model can be written as:

$$L_p = \left[A + 10 \left(a - bh_b + \frac{c}{h_b} \right) \log \frac{d}{d_0} \right] + Z$$

a. Show that Z is a random variable given by:

$$Z = 10Y\sigma_\alpha \log \frac{d}{d_0} + Up + UVq$$

Here Y, U, V are normally distributed with mean zero and standard deviation 1.
b. If $a = 3.6$, $b = 0.005$, $c = 20$, and $h_b = 25$ m, compute the median path-loss at a distance of 1 km from the base station.
c. Assume that the path-loss computed in (b) is the acceptable path-loss. Let $\sigma_\alpha = 0$ and $q = 0$. What should the fade-margin be to ensure that 95% of locations at the edge of the cell have acceptable path-loss given $p = 8.2$ dB?
d. Let $\sigma_\alpha = 0.6$, $p = 0$, $q = 0$. At a distance of 1 km from the base station, what fraction of locations have acceptable path-loss? Assume that the path-loss computed in (b) is the acceptable path-loss.

Problem 2.14

Assume that the path loss L_p is of the form $K + 10\alpha \log d$. At any given distance d from the transmitter, the actual path loss also has a shadow fading component that is normally distributed with mean zero and variance σ^2. If the maximum path loss that can be accepted in the system is $K + 10\alpha \log R$, compute the fraction of users in the cell (not at the cell edge) that have an acceptable path loss. You can express your answer as an integral.

Problem 2.15

A multipath channel has three paths at 0, 50, and 100 nsec with relative strengths of 0, -10, and -15 dBm, respectively.

a. What is the multipath spread of the channel?
b. Calculate the rms multipath delay spread of the channel.
c. What would be the difference between multipath spreads and rms multipath spreads of this three-path channel and a two-path channel formed by the first and the third path of this profile?
d. What would be the coherence bandwidths of the two channels?

Problem 2.16

Sketch the power-delay profile of the following wideband channel (Table P2.2) and calculate its excess delay spread, mean delay, and rms delay spread. A channel is considered "wideband" if its coherence bandwidth is smaller than the data rate of the system. Would the channel be considered a wideband channel for a binary data system at 25 kbps? Why?

Table P2.2   Data for wideband channel

Relative delay (in ms)	Average relative power (in dB)
0.0	-1.0
0.5	0.0
0.7	-3.0
1.5	-6.0
2.1	-7.0
4.7	-11.0

Problem 2.17

In the 1900 MHz bands, measurements [Bla92] show that the rms delay spread increases with distance. An upper bound on the rms delay spread is given by the equation $\tau = e^{0.065L(d)}$ in ns where $L(d)$ is the mean path loss in dB as a function of distance d between the transmitter and receiver. The path loss itself is given by the equation $L(d) = L_0 + 10\alpha \log_{10}(d/d_0)$ where $L_0 = 38$ dB and $\alpha = 2.2$ for $d < 884$ m and $\alpha = 9.36$ for $d > 884$ m. The standard deviation of shadow fading is 8.6 dB. Assume that you are using a transmission scheme that has a symbol rate of 135 ksps without equalization. If the maximum allowable path-loss is 135 dB, what limits the size of the cell – the rms delay spread, or the outage at 90% coverage at the cell-edge? Explain clearly all of your steps.

Problem 2.18

In Equation 2.22 we presented the *scattering function*. The scattering function of an indoor radio channel is defined as the product of a time function $Q(\tau)$ that represents the delay-power spectrum and a frequency function $D(\lambda)$ that represents the Doppler Spectrum (see Equation 2.22), that is:

$$S(\tau; \lambda) = Q(\tau)D(\lambda)$$

The scattering function provides a better description of the radio channel. Suppose for τ (in ns) we have:

$$Q(\tau) = 0.4\,\delta\,(\tau - 50) + 0.4\,\delta\,(\tau - 100) + 0.2\,\delta\,(\tau - 200)$$

and for λ (in Hz):

$$D(\lambda) = 0.1\,U\,(\lambda + 5) - 0.1\,U\,(\lambda - 5)$$

with $\delta(.)$ and $U(.)$ functions representing the impulse and unit step functions.

a. Determine the rms delay spread of the channel assuming $Q(\tau)$ is similar to Equation 2.19 and the mean square values are specified here.
b. What is the maximum Doppler spread of the channel?
c. Compute the coherence bandwidth of the channel.

Projects

Project 2.1: Simulation of Multipath Fading

Figure 2.13a shows two mobile devices communicating in a large open indoor area with a ceiling with a height of 5 m and two antennas that are 1.5 m above the ground. Communication between the terminals is taking place through three paths: the direct path, the path reflected from the ground and the path reflected from the ceiling. The reflection coefficients from the ground and the ceiling are 0.7 and each reflection causes an additional 180 degrees phase shift.

a. If the transmitted power is 1 mW, derive an expression for the calculation of P_o, the free space received signal strength at 1 m distance from the transmitter, as a function of frequency of operation, f_c.
b. Derive an expression for calculation of the amplitude, delay, and phase of each of the arriving paths as a function of distance, d, and the frequency of operation, f_c.
c. Derive an expression for calculation of the received signal strength (RSS) as a function of d and f_c.
d. Use MATLAB to plot the RSS versus distance for 1 m $< d <$ 100 m for center frequencies of 900 MHz, 2.4 GHz, and 5.2 GHz.
e. Discuss the relation between the received signal strength, rate of fluctuations, and frequency of operation.
f. Use the results for 2.4 GHz to design a path-loss model for the RSS by determining a suitable break-point and two distance power gradients in the two regions. Compare your model with the IEEE 802.11 models and discuss your observations.

Project 2.2: Ray Tracing Simulation of Multipath Fading

Figure P2.1 shows an access point and a laptop communicating with one another in a 20 × 50 m room. Communication between the terminals takes place through five paths: the direct path and four paths (reflections from each wall). The reflection coefficient from the walls is 0.7 and each reflection causes an additional 180° phase shift. The access point is located at a 2 m distance from the left wall and an 8 m distance from the lower wall of the room. The laptop moves from a distance

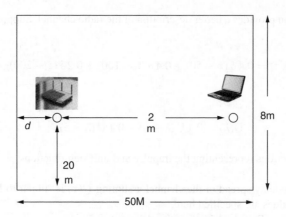

Figure P2.1   Scenario for Project 2.

of 1 m from the access point towards the right-hand-side wall until it is at a 2 m distance from the right wall.

1. If the transmitted power is 100 mW, derive an expression for P_o, the free space received signal strength at a 1 m distance from the transmitter, as a function of frequency of operation, f_c.
2. Explain a geometric method for calculation of the length of each of the five paths as a function of distance between the transmitter and the receiver, d, and dimensions of the room shown in the above figure.
3. Derive an expression for calculation of the amplitude, delay, and phase of each of the arriving paths as a function of distance, d, and the frequency of operation, f_c.
4. Sketch the channel impulse response for $d = 5$ m and $f_c = 2.4$ GHz and calculate the rms delay spread of the channel. What is the coherence bandwidth of the channel and how does it relate to the symbol transmission rate?
5. Repeat (4) for $d = 10$ m and $f_c = 5.2$ GHz. Explain the differences between the results of (4) and (5).
6. Derive an expression for calculation of the received signal strength (RSS) as a function of d and f_c.
7. Use MATLAB to plot the RSS versus distance for 1 m $< d <$ 46 m for center frequencies of 2.4 and 5.2 GHz.
8. Discuss the relation between the RSS rate of fluctuations and the frequency of operation.
9. Use the results for 2.4 GHz to design a path-loss model for the RSS and compare your model with the IEEE 802.11 models. Based on your observations, what is a suitable break-point for the two piece IEEE 802.11 model if we use it for our test scenario?
10. How can we use this basic concept to develop a ray tracing software which finds the channel impulse response using electronic map of a building?

Project 2.3: Multipath, Wireless Access, and Localization

In this project we illustrate the highlights of applications of radio channel propagation studies for wireless access and localization techniques.

Figure P2.2 shows the floor plan of a 20 × 50 m room and two mobile devices communicating in that room. The ceiling height is 5 m and two antennas are mounted on the devices at heights of

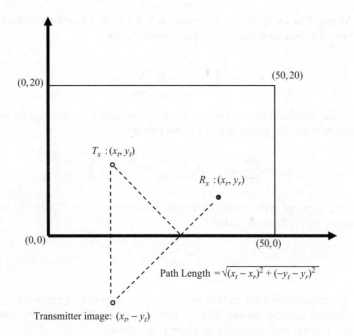

Figure P2.2   Geometry for calculation of path lengths for a Ray Tracing program in a square room.

1.5 m above the ground. The communication between the terminals is taking place through seven paths: the direct path, the path reflected from the ground, the path reflected from the ceiling and four paths bounced through the four walls. Assume paths that have more than one reflection have negligible strength and the reflection coefficient from the walls, ground and the ceiling is 0.7 and each reflection causes an additional 180° phase shift.

a. If the transmitted power is 1 mW, derive an expression for calculation of P_o, the free space received signal strength at a distance of 1 m from the transmitter, as a function of frequency of operation, f_c.

b. Calculate the path length associated with each of the seven paths from the geometrical relations shown in Figure P2.2.

c. Calculate the path amplitude, phase and delay for each path using Figure P2.3.

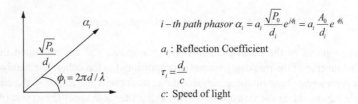

Figure P2.3   Phasor diagram to calculate magnitude, delay, and phase of a path of length d.

d. Determine the total narrowband received power calculated from Equation P2.1 below and plot it when transmitter is at $T_x : (25,15)$ and the receiver moves from $R_x : (1,10)$ to $R_x : (49,10)$ in

1 m steps. Assume that the frequency of operation is 2.4 GHz. Describe a method that can be used to estimate the distance using the received signal strength.

$$P_{NB} = \left| \sum_{i=1}^{7} \alpha_i \right|^2 = P_0 \left| \sum_{i=1}^{7} \frac{a_i}{d_i} e^{j\phi_i} \right|^2 \tag{P2.1}$$

e. Repeat (d) for the wideband received power given by Equation P2.2 and explain the reason for the differences between the results of part (d) and part (e).

$$P_{WB} = \sum_{i=1}^{7} |\alpha_i|^2 = P_0 \sum_{i=1}^{7} \left| \frac{a_i}{d_i} e^{j\phi_i} \right|^2 = P_0 \sum_{i=1}^{7} \left| \frac{a_i}{d_i} \right|^2 \tag{P2.2}$$

f. Plot the impulse response of the channel when the receiver is at $R_x : (10,10)$. The radio channel impulse response $h(t)$ is represented by:

$$h(t) = \sum_{i=1}^{N} \alpha_i e^{j\phi_i} \delta(t - \tau_i) \tag{P2.3}$$

where $N = 7$ represents the total number of paths in the response. A graphical representation of the radio channel impulse response $|h(t)|$ is shown in Figure P2.4. Note that in this case, the information as it relates to the phase ϕ_k of a path is not shown.

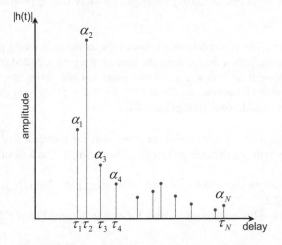

Figure P2.4   A typical channel impulse response.

g. The rms delay spread of the radio channel impulse response can be computed using Equation P2.4. The inverse of the rms delay spread is proportional to the coherence bandwidth of the channel that represents a limit on the achievable data rate for wireless access in the channel. Plot the coherence bandwidth as the transmitter moves along the route specified in part (d).

$$\tau_{rms} = \sqrt{\frac{\sum_{k=1}^{N} \tau_k^2 \alpha_k^2}{\sum_{k=1}^{N} \alpha_k^2} - \left(\frac{\sum_{k=1}^{N} \tau_k \alpha_k^2}{\sum_{k=1}^{N} \alpha_k^2} \right)^2} \tag{P2.4}$$

h. Use a raised cosine pulse given by Equation P2.5 in which W is the bandwidth of the system to plot the channel impulse response of the system for the transmitter and receiver of the locations defined in part (f). Assume the bandwidth is $W = 1, 5, 10, 50, 100, 500$, and 1000 MHz and for each bandwidth use the peak of the first transmitted pulse to calculate the distance estimate. Plot the distance measurement error versus bandwidth of the system.

$$f(t) = \frac{\sin(\pi W t)}{\pi W t} \frac{\cos(\beta \pi W t)}{1 - (4\beta W t)^2} \qquad \text{(P2.5)}$$

i. Repeat (h) assuming direct path between the transmitter and the receiver is blocked by a human subject.

Project 2.4: Simulation of fast envelope fading

The simulation of the fluctuations of the radio channel can be used in larger programs to evaluate the optimum system design parameters such as code lengths for error recovery and training times for adaptive equalizers to combat the harsh nature of the radio channel.

To simulate the fluctuations of the radio channel we need to generate a random process with specific envelope fading density function and a specific Doppler spectrum. The random variable is to be complex where the magnitude follows a Rayleigh fading distribution while the phase follows a uniform distribution. The power spectral density or spectrum of the random variable should follow the classic Doppler spectrum, given by:

$$D(f) = \frac{1}{2\pi f_m} \times \frac{1}{\sqrt{1 - \left(\frac{f}{f_m}\right)^2}} \qquad |f| \leq f_m$$

Where f_m is the maximum Doppler frequency. An example of this Doppler spectrum shape is shown in Figure 2.17.

One way to implement these fluctuations is shown in Figure P2.5. In this model, two independent Gaussian (Normal) random variables are filtered using a digital IIR filter that approximates the Classic Doppler Spectrum and added using an I and Q configuration (In-Phase and Quadrature). If

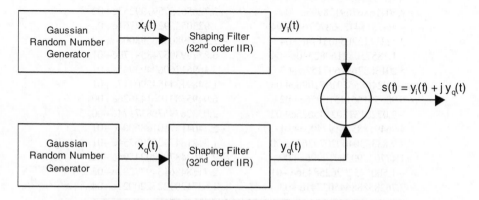

Figure P2.5 Simulation of channel fluctuations using a filtered Gaussian noise.

y_i and y_q are two independent Gaussian distributed variables, then $s = \sqrt{y_i^2 + y_q^2}$ will be a Rayleigh distributed random variable. Therefore, the magnitude of the output signal of the system shown below will follow a Rayleigh distribution and the power spectrum follows approximately the shape of the classic Doppler.

Using MATLAB, generate two sequences of independent Gaussian random numbers x_i and x_q of the length 5120 with function randn(1, 5120). Compute filtered Gaussian vectors y_i and y_q, shown in Figure P2.5, as the output of the filters using filtfilt function in MATLAB with the coefficients of the IIR filter provided in Table P2.3.

Table P2.3   Classic spectrum IIR filter coefficients

Denominator coefficients	Numerator coefficients
1.0000000000000000e+00	6.5248059900135200e−02
−1.2584602815172037e+01	−5.6908289014580038e−01
8.3781249094641240e+01	2.7480451166883220e+00
−3.8798703729842964e+02	−9.4773135180288293e+00
1.3927662726637102e+03	2.5786482996126544e+01
−4.1039030305379210e+03	−5.8241097311312117e+01
1.0278517997545167e+04	1.1247173657687033e+02
−2.2393748634049065e+04	−1.8904842233132774e+02
4.3133809439790406e+04	2.7936237305345003e+02
−7.4319282567554124e+04	−3.6418631194112885e+02
1.1554604041649372e+05	4.1715604202981109e+02
−1.6315680006218722e+05	−4.1320604132753033e+02
2.1026268214607492e+05	3.3901659663025242e+02
−2.4818342600838441e+05	−2.0059287960205506e+02
2.6898038693500403e+05	2.3734545818966293e+01
−2.6809721585952450e+05	1.5363912802007360e+02
2.4593366073473063e+05	−2.9424154728837402e+02
−2.0763108908648306e+05	3.7359596060374486e+02
1.6120527209223103e+05	−3.8642988435890055e+02
−1.1492103434104947e+05	3.4521505714177903e+02
7.5041686769138993e+04	−2.7265055759799253e+02
−4.4731841330872761e+04	1.9230535924562764e+02
2.4231115205405174e+04	−1.2153980630698008e+02
−1.1857508216082340e+04	6.8773930574859179e+01
5.2013837692697152e+03	−3.4696126060493945e+01
−2.0246855591971096e+03	1.5489134454590417e+01
6.9005516614518956e+02	−6.0495383196143626e+00
−2.0220131802145625e+02	2.0332679679817174e+00
4.9649188538197400e+01	−5.7404157101686004e−01
−9.8333304002079363e+00	1.3121847123296254e−01
1.4770279039919996e+00	−2.2867487042024594e−02
−1.5005452926258436e−01	2.7118486134987282e−03
7.7628588864503741e−03	−1.6371291227220021e−04

For all questions, assume $s(t)$ is sampled at four times the maximum Doppler frequency f_m.

a. Provide a plot of magnitude (in dB) and phase (in degrees) of $s(t)$ as a function of time.
b. Provide the histogram of the magnitude (in linear form, not in dB) and phase (in degrees) of $s(t)$. Examine whether they fit the expected Rayleigh and uniform distributions.
c. Provide the plot of the power spectral density of $|S(f)|^2$, as a function of normalized frequency f/f_m, where $S(f)$ is the Fourier transform of the $s(t)$. Compare the results of your simulation with the expected spectrum shown in Figure 2.17.

3

Physical Layer Alternatives for Wireless Networks

3.1 Introduction

In this chapter we describe and overview of transmission technologies, which have been adopted in many of the developing standards and products for wireless information networks. In principle, these techniques are applicable to all wired and wireless modems because the basic design issues are common to both systems. This is because the general objective in all communication systems is to be able to transmit data with the highest achievable data rate with the minimum expenditure of signal power, channel bandwidth, and transmitter and receiver complexity. In other words, we would like to maximize both *bandwidth efficiency* and *power efficiency* and minimize the transmission system *complexity*. However, the emphasis on these three objectives varies according to the application requirement, availability of bandwidth and medium for transmission. There are also certain constraints and details that are specific to particular applications. For example, the cellular telephone application requires a stable quality of service for a relatively narrow data rate of around 10 kbps over a wide area of coverage for a user that may be inside a moving vehicle at fairly large speeds. On the other hand, data applications using WLANs are mostly designed for transmission of a burst of data at the fastest data rate available over smaller coverage areas. Yet another important factor that impacts design, discussed in Chapter 2, is the characteristics of the medium of transmission (the *channel*). These design objectives are often conflicting and the engineering tradeoffs decide what factors are considered more important than others. As a result of this heterogeneity in objectives and channel behaviors a number of transmission techniques have emerged and are employed in various wireless standards and products. Covering the details of these techniques rigorously is beyond the scope of this book and the reader can refer to books addressing the details of digital transmission techniques (e.g., see [Pro08]). In this chapter we focus on a systematic comparative understanding of the transmission techniques that are applied to popular cellular, local, and personal area wireless networks.

Transmission techniques used in wireless networks build on what has evolved for the more *reliable wired communication channels*. In most wired data applications, such as local area networks, transmission schemes were developed for use over twisted pair telephone wires, coaxial cable, or optical fiber, and they are simpler. The received data from the higher layers are coded into the

---

Principles of Wireless Access and Localization, First Edition. Kaveh Pahlavan and Prashant Krishnamurthy.
© 2013 John Wiley & Sons, Ltd. Published 2013 by John Wiley & Sons, Ltd.

electrical voltages or optical light signals that are applied *directly* to the medium. These transmission techniques are often referred to as baseband transmission schemes. On longer telephone lines or coaxial cable modems for wide area applications, the transmitted signal is modulated over a sinusoidal *carrier* of a certain frequency. The amplitude, frequency, phase of the carrier, or a combination of these is used to carry data. The purpose of modulation in some of these applications is to eliminate the direct current component from the transmission spectrum, and allow the usage of more bandwidth efficient modulation to support higher data rates. For other data communication applications, modulation is used to allow *multiple channels* in a single medium. Specific impairments seen on telephone channels are amplitude and delay distortion, phase jitter, frequency offset, and effects of non-linearity. Many of the practical design techniques of wired modems have been developed to efficiently deal with these categories of impairments. In the next section, we discuss the very basics of physical layer techniques – those of data rate, bandwidth, and power.

3.2   Physical Layer Basics: Data rate, Bandwidth, and Power

The basic responsibility of the physical layer in a wireless network is to transfer a bit stream of information from a transmitter to a receiver through a wireless channel. The transmitter processes the bit stream of data arriving from the higher layers using variety of *coding* and *modulation* techniques. Through the coding process, the transmitter manipulates the bit stream and adds parity bits for reliability of transmission. The coded data stream is then mapped on to electrical waveforms, used as *symbols* for transmission of information, each carrying a set of coded data bits. Figure 3.1 shows this basic concept when we use a simple pulse shape with different amplitudes for transmission. In this case, each symbol carries two bits of information. The transmitted symbols or waveforms are fed to the transmitter antenna and sent over the channel. At the receiver, the received symbol is detected and used to decide on the transmitted bits. The ultimate objective of designing a digital communication technique is to find waveforms that *efficiently* and *reliably* transfer the information bits from the transmitter to the receiver. Efficient means that for a given bandwidth we can support higher data rates. Reliable means that the detected received data stream has a low symbol and bit error rate. To design such waveforms we need to establish a quantitative performance evaluation technique that allows us to compare performance of different modulation

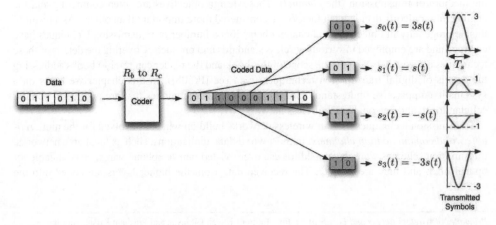

Figure 3.1   Basic concept of digital communication and its relation to coding and modulation.

and coding techniques. In an ideal situation, we want to increase the data rate, R_b, and reduce the average transmitted energy per symbol, E_s, while maintaining a low bandwidth, W. In this section, we provide a general picture of the basics related to the physical layer. In later sections of this chapter and some later chapters, details of applied modulation and coding schemes are considered.

3.2.1 Data Rate and Bandwidth

In Figure 3.1, if we use m coded bits for a symbol, the relation between m and the number of symbols, M, is $M = 2^m$. If we assume that the symbol duration is T_s, the symbol transmission rate, $R_s = 1/T_s$, is also a representative of the bandwidth of the channel, W, that is, $W \approx R_s$. In information theory and related applications, we often define the bandwidth to be the symbol transmission rate. In practice, the government agency responsible for regulating telecommunications or the telecommunications standardization activities define the bandwidth based on actual implementation of the filters in the frequency domain. The bandwidth may thus differ from the value of the symbol transmission rate. We will use the inverse of the symbol rate as a measure of the bandwidth in this book.

The bit rate of the coded data stream, R_c, is related to the symbol transmission rate by $R_c = m \times R_s$. Therefore, the relation between the application data (bit) rate, R_b, and the bandwidth is:

$$R_b = r \times R_c = r \times m \times R_s \approx r \times m \times W \tag{3.1}$$

where r is the *coding rate*, that is the ratio of the number of coded bits of data to the number of information bits (see Section 3.3.6 for more details).

Example 3.1: Relation between data rate and symbol transmission rate in 802.11g

The IEEE 802.11g standard for WLANs defines a number of data rates for a fixed symbol transmission rate of 12 MSps by providing for different number of bits per symbols and different coding rates. With four bits per symbol, $m = 4$, and a coding rate of $r = 1/2$, from Equation 3.1 we have the data rate as:

$$R_b = 12\,(\text{MSps}) \times 4\,(\text{bps}) \times \frac{1}{2} = 24\,\text{Mbps}$$

3.2.2 Power and Error Rate

To bring power and error rate performance into consideration and relate them to the bandwidth and the data rate we need to examine the *shape* of the transmitted symbol and the energy that it carries for transmission of each bit of information. If we represent the transmitted waveform for the i-th symbol by $s_i(t)$, the transmitted energy per symbol is:

$$E_i = \int_{-\infty}^{\infty} |s_i(t)|^2 dt \tag{3.2a}$$

and the amplitude of the received signal is proportional to $\sqrt{E_i}$. If we assume equally probable symbol transmissions, then the average transmitted symbol energy is given by:

$$E_s = \frac{1}{M} \sum_{i=0}^{M-1} E_i \tag{3.2b}$$

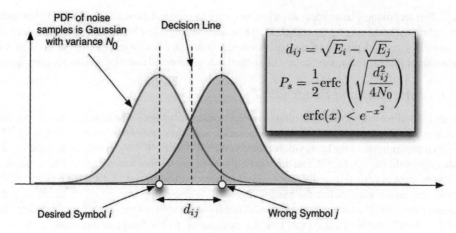

Figure 3.2   Relation between symbol error rate, signal energy and noise level.

which is referred to as the average energy per symbol. This average energy per symbol is a representative of the transmission power (when divided by the symbol duration) and it is directly related to the performance of the system. The measure of performance is the *symbol* or *bit error rate* and determination of the quantitative value of the error rate is possible if we assume that the cause of errors is the background noise that interferes with the detection process at the receiver. If we make the reasonable assumption that samples of the background noise has a Gaussian distribution with a zero mean and a variance of N_0, the relationship between the error rate and power becomes relatively simple. This model of the transmission channel where errors are caused only by background noise is often called the additive white Gaussian noise (AWGN) channel.

Figure 3.2 explains the relation among the energy of the symbols, variance of the noise and the symbol error rate. In this figure we have a desired symbol with desired received amplitude of $\sqrt{E_i}$ that is arriving at the receiver with a Gaussian distributed noise with a variance of N_0. We also have a second symbol with amplitude of $\sqrt{E_j}$. We can define the difference or distance between the amplitude of the two symbols as:

$$d_{ij} = \sqrt{E_i} - \sqrt{E_j}$$

An error in detecting the desired symbol occurs when the Gaussian noise level is more than half of this distance and the probability of this event happening is the same as half of the overlapping area shown in Figure 3.2. This area is the tail area of the distribution of a Gaussian random variable that is determined through the complementary error function (erfc), previously defined in Equation 2.8. Since the complementary error function is asymptotically bounded by an exponential function, we can determine the symbol error rate and its bound from:

$$P_e = \frac{1}{2}\mathrm{erfc}\left(\frac{d_{ij}}{2\sqrt{N_0}}\right) \le \frac{1}{2}e^{-\frac{d_{ij}^2}{4N_0}}$$

In a multisymbol transmission scheme, we have to calculate the error rate over all possible combinations of the transmitted symbols and take the average of all of them. However, since the relation between the error rate and the distance between the symbols are exponential, the average error rate

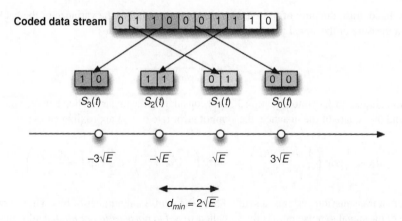

Figure 3.3   Signal constellation representation of the system in Figure 3.1.

will be dominated by the error rate associated with the *minimum distance* between the symbols and we have:

$$P_e \approx \frac{1}{2}\mathrm{erfc}\left(\frac{d_{min}}{2\sqrt{N_0}}\right) \leq \frac{1}{2}e^{-\frac{d_{min}^2}{4N_0}} \tag{3.3}$$

where d_{min} is the minimum distance between any pair of symbols in the transmission set. To visualize the distances between the symbols, often we map the signal amplitudes of all of the signals used in the transmission scheme in a diagram along multiple dimensions and refer to it as the *signal constellation*.

Example 3.2: Signal constellation and error rate

Figure 3.3 shows the signal constellation for the system described in Figure 3.1.

Let us suppose that the energy of the first signal (representing one symbol) is:

$$E = E_1 = \int_{-\infty}^{\infty} |s_1(t)|^2 dt$$

The minimum distance between signals in the constellation is given by $d_{min} = 2\sqrt{E}$. Note that the average energy in the constellation is:

$$E_s = \frac{1}{4}\sum_{i=0}^{3} E_i = \frac{9E_1 + E_1 + E_1 + 9E_1}{4} = 5E_1 = \frac{5}{4}d_{min}^2$$

For all practical purposes, the square of the minimum distance is linearly related to the average transmitted energy per symbol, defined by Equation 3.2, and we have $d^2 = a \times E_s$ where a is some constant. In contrast, the ratio of the average energy per symbol to the variance of the background noise is a measure of the signal to noise ratio of the system defined as:

$$\gamma_s = \frac{E_s}{N_0} \tag{3.4a}$$

Substituting Equation 3.4a into Equation 3.3 and considering the linear relation between the average energy and the square of the distance, the symbol error rate can be approximated by:

$$P_e \approx \frac{1}{2}\mathrm{erfc}\left(\frac{1}{2}\sqrt{a \times \frac{E_s}{N_0}}\right) = \frac{1}{2}\mathrm{erfc}\left(\frac{1}{2}\sqrt{a \times \gamma_s}\right) < \frac{1}{2}\exp\left(-\frac{a \times \gamma_s}{4}\right) \tag{3.4b}$$

For different transmission systems, a symbol may carry different number of bits. Thus, a more fair measure of the signal to noise ratio is the so-called *signal to noise ratio per bit* that allocates energy to each transmitted bit of information:

$$\gamma_b = \frac{E_b}{N_0} = \frac{E_s}{m} \times \frac{1}{N_0} = \frac{\gamma_s}{m} \tag{3.5a}$$

The symbol error rate in terms of normalized signal to noise ratio per bit is then given by:

$$\frac{1}{2}\mathrm{erfc}\left(\frac{1}{2}\sqrt{a \times m \times \gamma_b}\right) < \frac{1}{2}\exp\left(-\frac{a \times m \times \gamma_b}{4}\right) \tag{3.5b}$$

Before we proceed to give an example of a typical plot of this probability as a function of the signal to noise ratio per bit, we bring up a couple of interesting points. In Figure 3.3, bits are assigned to symbols so that if we make a symbol error between adjacent symbols, only one of the two bits is in error. This type of coding of the signal constellation is referred to as Gray coding, and with that the bit error rate (BER) and the symbol error rate are the same when symbol errors are mostly between adjacent symbols. In Equations 3.3–3.5, if we have several symbols with the same distance, then the factor of 1/2 preceding the exponential will change to another constant. However, the reader should be also aware of the fact that the signal to noise ratio affects the error in an exponential form but this factor is only linear and consequently, it can be mostly neglected when it is compared to the effects of the signal to noise ratio. More accurate error rate calculations in the literature also take into account the changes in the linear factor.

Example 3.3: A sample plot of probability of error versus signal to noise ratio

Figure 3.4 shows the signal constellation of a transmission system with two symbols each having the same energy level but transmitted with different polarities. This is a binary system (there are only two symbols with one bit per symbol, $m = 1$) and the relation between the minimum distance and average energy is $d = 2\sqrt{E} \Rightarrow a = 4$. Then the probability of bit error and its asymptotic bound versus the signal to noise ratio per bit (from Equation 3.5b) for this transmission system are:

$$P_s \approx \frac{1}{2}\mathrm{erfc}\left(\sqrt{\gamma_b}\right) < \frac{1}{2}e^{-\gamma_b}$$

Figure 3.4 shows the plots of the symbol error rate and its bound as a function of signal to noise ratio in dB. The basic plot is generated by the MATLAB® code in Appendix A3.

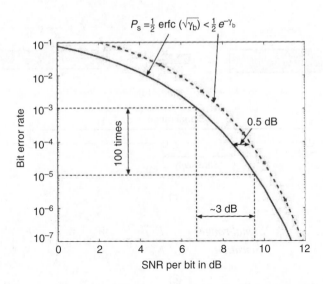

Figure 3.4   Typical probability of error and its bound as a function of signal to noise ratio per bit in dB.

A modem designer uses curves such as those in Figure 3.4 to translate an application or customer's error rate requirement to the signal to noise ratio. From Figure 3.4, if the customer demands a BER of 10^{-5}, the designer should implement the system so that the ratio of the received signal power to the background noise power at the receiver is such that the signal to noise ratio per bit is higher than 10 dB.

Figure 3.4 illustrates a couple of interesting observations. If we compare the performance in terms of the bit error rate, a 100 times reduction in the error rate requirement approximately reduces the signal to noise ratio per bit requirement by only 3 dB. In other words if we reduce the power by two times (3 dB) we will have hundred times more errors. This general observation is due to the fact that the relation between the error rate and signal to noise ratio per bit is exponential. The other interesting observation is that the simple exponential bound provides results within approximately 0.5 dB accuracy. These useful and simple facts are clearly demonstrated in the figure.

3.2.3   Shannon–Hartley Bound on Achievable Data Rate

In our discussions thus far, we have shown how the design of the signal constellation relates to the achievable data rate and the bandwidth and energy efficiency. At this time, it is useful to consider the ultimate limits on data rates and efficiency that are theoretically achievable. This is best done by examining Shannon–Hartley's well-known formula:

$$R_b \leq W \log_2 \left(1 + \frac{E_s}{N_0} \right) \text{ bits/s} \tag{3.6}$$

where R_b is the transmission data rate. The right side of this equation provides a bound on the maximum achievable information transfer rate in bps in a bandwidth of W hertz, where the signal-to-noise ratio is E_s/N_0 [Sha48].

To have an intuitive idea about this equation, consider Figure 3.5a, where a number of symbols with average energy E_s arrive at the receiver, disturbed by an additive white Gaussian noise with

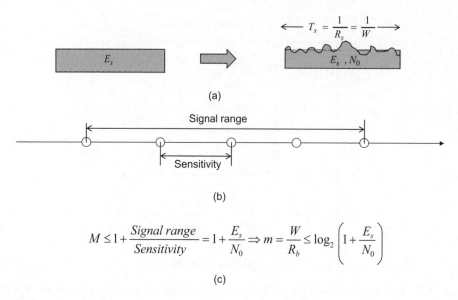

(a)

(b)

(c)

Figure 3.5 Shannon–Hartley bound relating power to bandwidth and data rate: (a) transmission parameters, (b) Hartley's observation, (c) the bound.

a variance of N_0. The duration of the symbol is T_s and the corresponding bandwidth is $W = R_s = 1/T_s$. The signal constellation for this system is shown in Figure 3.5b. Hartley had first observed that if the sensitivity of the receiver to differentiate two symbols from each other is known, then the number of symbols that can be sent through this communication channel is always less than the signal range divided by sensitivity plus one, that is:

$$M \leq 1 + \frac{Signal\ Range}{Sensitivity} = 1 + \frac{E_s}{N_0}.$$

Later on Shannon showed that the ratio of the signal range to sensitivity is indeed the signal to noise ratio per symbol of the channel E_s/N_0. Since the number of bits per symbol and the total number of symbols are rated by $M = 2^m$, and the number of bits per symbol is $m = R_b/R_s = R_b/W$, we can then derive the equation:

$$m = \frac{R_b}{W} \leq \log_2\left(1 + \frac{E_s}{N_0}\right) = \log_2\left(1 + m\frac{E_b}{N_0}\right) \tag{3.7}$$

that provides the basis for Equation 3.6. While this is not a rigorous derivation of the bound, it provides an intuitive sense of the relationships we explore next.

Example 3.4: A sample plot of the probability of error

The signal to noise ratio of a conditioned telephone line can be as high as 30 dB (1000 times). Therefore, we can say:

$$m \leq \log_2(1 + SNR) = \log_2(1001) = 9.96.$$

Since the bandwidth of the voice band telephone channel is 4 KHz, the data rate cannot exceed 40 Kbps or so. To support higher data rates, there is a need to increase the SNR. In 56 Kbps modems, the SNR is increased in the down stream channel by eliminating the analog circuitry in the connection between the user and the telephone network.

The Shannon–Hartley bound in Equation 3.6 reveals a fundamental relation between the power, bandwidth, and the data rate. By inspecting this equation we can conclude that if the bandwidth is much more than the data rate, we can design systems that operate at a *low signal to noise ratio* allowing wider coverage. This is the feature of a class of transmission techniques referred to as spread spectrum systems. The original legacy IEEE 802.11 WLAN is an example of such spread spectrum systems. The data rate of this WLAN technology was 2 Mbps, but it operated over a 26 MHz bandwidth. The spatial coverage of this system was better than the higher speed IEEE 802.11 systems, which evolved later. If we have a small bandwidth and we need high data rates, Equation 3.6 suggests that we need to use multiple symbols and a high signal to noise ratio per bit. Example 3.4 illustrates an example of such a communication system.

As we observe from Equation 3.4, the signal to noise ratio is also related to the required error rate. If we reduce the error rate requirement, the required signal to noise ratio reduces allowing us to achieve high data rates. For a give transmission technique we can get closer to the Shannon–Hartley bound if we reduce our error rate requirement as illustrated by Example 3.5.

Example 3.5: Error rate and bounds for QPSK

The QPSK modulation technique with $m = 2$ bpS needs a 10 dB (10 times in linear scale) signal to noise ratio per bit to provide an error rate of 10^{-5}. Using Equation 3.7, the bound for the number of bits for this transmission technique is:

$$m \leq \log_2(1 + 2 \times 10) = 4.39 \text{ bpS}.$$

This indicates that we have 2.39 bits per symbol more than QPSK. If we reduce the error rate requirement to 10^{-3}, we need smaller signal to noise ratio (3 dB less) because we have reduced our error rate requirement by 100 times. In this case the signal to noise ratio is 7 dB (five times in a linear scale) and:

$$m \leq \log_2(1 + 2 \times 5) = 3.47 \text{ bpS}.$$

By reducing the error rate requirement, the performance of QPSK becomes closer to the maximum achievable bound.

Similarly, if we can increase the number of bits per symbol, we can get closer to the bound. Therefore, constellations with larger number of points can be more efficient in terms of being closer to the Shannon–Hartley bound.

3.3   Performance in Multipath Wireless Channels

The two main characteristics of the wireless medium affecting the performance of a modem are large fluctuations of the received power level (called fading) and arrival of the received signal via delayed multiple paths referred to as multipath propagation. In the rest of this section, we give an overview of the performance of basic amplitude, phase, and frequency modulation techniques and we analyze their performance as Rayleigh fading and simple multipath conditions affect it. This will provide the reader with an intuitive understanding of the measures that are used to evaluate the performance of a modulation technique and how this measure is affected when we increase the power level or when we get exposed to fading or multipath conditions.

3.3.1 Effects of Flat Fading

One of the main characteristics of the wireless medium affecting the performance of a transmission technique is the large fluctuation of the received power level, referred to as *flat fading*. As opposed to wired channels, the received signal from wireless channels suffers from strong amplitude fluctuations (on the order of 30–40 dB) in specific time or frequencies that are referred to as fading in the received signal. During periods of signal fading, the error rate of the transmission system increases substantially and when the system is out of fade, the error rate becomes negligible.

Figure 2.15 shows a typical measurement of the received signal strength in multipath fading caused by movements around the transmitter or receiver. The signal to noise ratio follows the same pattern of fluctuations because it is the ratio of the received signal strength that fluctuates over the noise power that is a fixed value. Therefore, as shown in Figure 3.6, due to the fading effect, the signal strength and consequently the signal to noise ratio per bit, γ_b, randomly fluctuates in time. If we define the average signal to noise ratio per bit by $\bar{\gamma}_b$, the error rate when the signal is close to this average is very small (close to zero) and when the sinal level is in a deep fade, it is very high (close to 0.5). Therefore, most errors occur during deep fades when the signal level is very low and the frequency of occurrence of the fade governs the overall average error rate.

To evaluate the performance over a fading radio channel we resort to the average bit error rate $\bar{P}_e$, versus the average received SNR per bit, $\bar{\gamma}_b$. To calculate this average error rate we need a

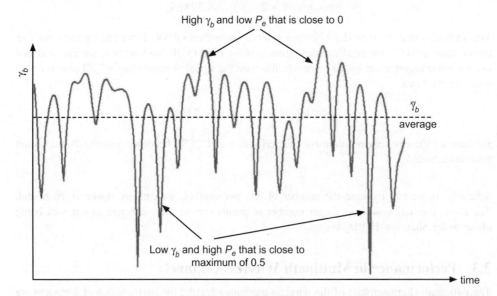

Figure 3.6 Fluctuations of the signal to noise ratio per bit and its relation to the error rate in and out of fading.

model for the statistics of the variations of γ_b. As we discussed in Section 2.4.3, the most common model used for the variations of signal amplitude in fading channels is the Rayleigh distribution given by Equation 2.14. The probability density function of the *square* of the Rayleigh amplitudes has an exponential distribution. Since the signal power is related to the square of the amplitude and signal to noise ratio follows the same statistics as the power, the signal to noise ratio in a Rayleigh fading channel follows an exponential distribution. As a result, to compute the average signal to noise ratio $\bar{\gamma}_b$, we use the distribution function of the of the signal to noise ratio that follows an exponential distribution given by:

$$f_\Gamma (\gamma_b) = \frac{1}{\gamma_b} e^{-\frac{\gamma_b}{\bar{\gamma}_b}}$$ (3.8)

in which γ_b is the instantaneous SNR per bit and $\bar{\gamma}_b$ is the average SNR per bit in the vicinity of a location. Equation 3.5 gives the relation between the SNR per bit and the probability of error for a given specific value of SNR per bit. As the channel fluctuates, the error rate also fluctuates, and the average error rate is the integral of the error rate function over the distribution of the SNR per bit, and it is given by:

$$\bar{P}_e = \frac{1}{2\bar{\gamma}_b} \int_0^\infty \text{erfc} \left(\frac{1}{2} \sqrt{a \times m \times \gamma_b} \right) e^{-\frac{\gamma_b}{\bar{\gamma}_b}} d\gamma_b \approx \frac{1}{2} \left[1 - \sqrt{\frac{a \times m \times \bar{\gamma}_b}{1 + a \times m \times \bar{\gamma}_b}} \right] \approx \frac{1}{am\bar{\gamma}_b}$$ (3.9a)

If we instead consider the exponential asymptotic bound for errors described in Equation 3.5 the average BER, $\bar{P}_e$, over all possible values of the SNR per bit is given by:

$$\bar{P}_e = \frac{1}{2\bar{\gamma}_b} \int_0^\infty \exp \left(-\frac{a \times m \times \gamma_b}{4} \right) e^{-\frac{\gamma_b}{\bar{\gamma}_b}} d\gamma_b = \frac{1}{2} \left[\frac{1}{4(1 + a \times m \times \bar{\gamma}_b)} \right] \approx \frac{1}{2am\bar{\gamma}_b}$$ (3.9b)

In a multipath fading environment, the average BER, $\bar{P}_e$, is a function of inverse of the average SNR per bit $\bar{\gamma}_b$. To compare this relation with the relation between the error rate, P_e, and SNR per bit, γ_b, for non-fading wired channels, as we show in Figure 3.4, in non-fading channels a 3 dB change in transmission power decreased the bit error rate by two orders of magnitude (hundred times). Equation 3.9 reveals that for fading channels we need 20 dB (100 times) more average power to increase the average bit error rate by two orders of magnitude. The upper part of Figure 3.7 shows the average probability of error in logarithmic form versus average signal to noise ratio per bit in dB for $a \times m = 4$. This relation is approximately a line with slope of one. Comparing this curve with the waterfall like plots of bit error rate versus signal to noise per bit, shown in lower part of Figure 3.7, we need much more power to overcome the effects of fading to achieve the same error rates. For example to achieve an error rate of 10^{-5} we need a γ_b of less than 10 dB on a non-fading channel compared with an average signal to noise per bit, $\bar{\gamma}_b$, of close to 45 dB needed to achieve the same average error rate on a multipath fading wireless channel. This difference means that achievement of the same error rate for the radio channel requires more than 3000 times more power (35 dB). Designers of the radio modems have worked hard in the past half a century to close this gap between the performance over non-fading and fading wireless channels. They have come up with a number of innovative solutions such as error control coding, multiple receive antennas, space–time coding (STC), and multiple-input multiple-output (MIMO) systems, which are part of the latest advancement in design of the physical layer of wireless networks. All of these solutions take advantage of the so-called diversity techniques that we will discuss later.

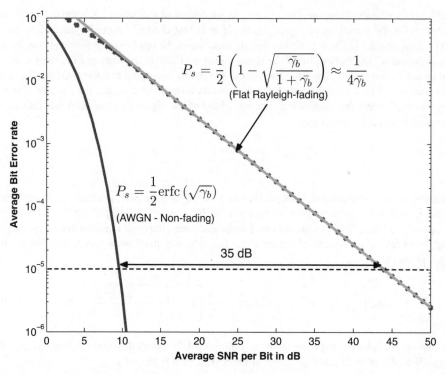

Figure 3.7 Average BER in logarithmic form versus average SNR/bit in dB for $a \times m = 4$ over a Rayleigh fading wireless channel and over a non-fading wired channel.

3.3.2 ISI Effects Due to Multipath

One of the main differences between wireless and wired channels is that the wireless channel suffers from multipath propagation. As we described in Section 2.5.2, the shape of the received pulse and its time duration of the signaling waveform are both changed due to multipath arrivals. The difference between the first and the last arriving pulses is the *delay spread* of the channel. If the symbol duration is much larger than the multipath spread, this means that the data rate is much smaller than the coherence bandwidth of the channel (see Equation 2.20 in Section 2.5.2), and all pulses received via different paths arrive roughly on top of one another causing only amplitude fluctuations and the fading phenomenon that was discussed in the previous section (called flat fading). If the ratio of the delay spread to the pulse duration becomes considerable, the received pulse shape is severely distorted, and it also interferes with neighboring symbols causing inter-symbol-interference (ISI). In addition to SNR fluctuations due to fading effects, this interference power also degrades the performance. However, the ISI effect of multipath degrades the performance in a different manner than flat fading. The effects of flat fading can be compensated via an increase in the transmit power by a fading margin (even though it may be as large as 20 dB). Increase of the transmit power cannot compensate for the adverse effects of ISI. This is because an increase in the transmit power increases the signal as well the ISI interference power keeping the signal to interference ratio at the same level. A simple example will clarify the situation.

Example 3.6: ISI effects of multipath

Assume that we have a multipath channel that causes ISI energy per bit that is 10% of the energy per bit of the transmitted symbols. Further assume that we can take into account the ISI interference as a Gaussian noise. The signal to noise ratio including the ISI is then given by:

$$\gamma_{b-ISI} = \frac{E_b}{N_0 + 0.1 E_b} = \frac{1}{\frac{1}{\gamma_b} + 0.1}$$

Using Equation 3.5 for a transmission techniques with $a \times m = 4$ we have:

$$P_e \approx \frac{1}{2}\mathrm{erfc}\left(\sqrt{\frac{1}{\frac{1}{\gamma_b} + 0.1}}\right)$$

Figure 3.8 compares this performance with and without ISI. When the signal to noise ratio is low, the associated ISI is less than the background noise and the performance is close in both cases.

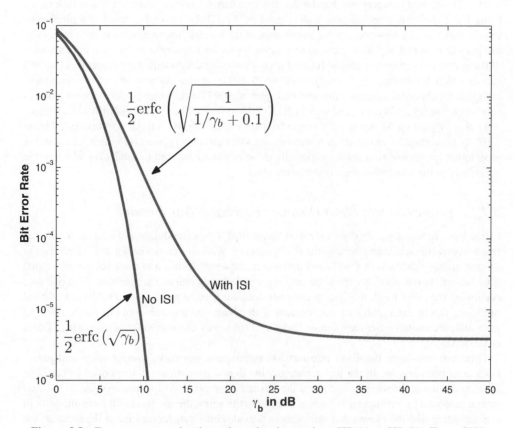

Figure 3.8   Error rate versus signal to noise ratio when we have ISI: (a) no ISI, (b) effects of ISI.

As the signal strength is increased the ISI increases as well and gradually it dominates over the background noise. From that point onwards, as we increase the signal strength the overall signal to noise ratio will remain almost flat resulting in a flat error rate curve. This flat value of the error rate is sometimes referred to as the irreducible error rate caused by ISI.

3.4   Wireless Transmission Techniques

The first generation wireless cellular and cordless telephone systems used analog frequency modulation that we do not consider in this book. With the emergence of second-generation wireless networks, digital techniques replaced analog modulation to increase the capacity. Analog voice was source coded into digital format at the mobile terminal for digital transmission. Speech coding at the terminal also facilitated the integration of voice and data services in a single terminal. After the emergence of second-generation systems, digital transmission has become the dominant choice for wireless communication networks. Therefore, in the rest of this chapter we describe the fundamentals of digital transmission techniques applied to modern wireless networks.

Popular digital wireless transmission techniques can be divided into three categories according to their applications. The first category consists of power efficient baseband pulse transmission techniques used for short-range wireless networking for optical wireless and ultra wideband networks. The second category are bandwidth efficient carrier modulated transmission techniques using basic modulation techniques, widely used in 2G TDMA cellular as well as a number of mobile wireless data networks. As the importance of the wireless Internet became obvious, a third category of multipath resistant transmission technologies have emerged to increase the data rates of the systems to support broadband Internet access and other data oriented applications. This third category includes multiple-input multiple-output (MIMO) antenna systems, spread spectrum technology and orthogonal frequency division multiplexing (OFDM) techniques. Spread spectrum was first adopted in legacy WLANs and later in 3G CDMA systems. The OFDM and MIMO techniques were also adopted by WLAN standard organizations and later by 4G cellular networks (see Figure 1.10). In the remainder of this chapter we consider some elementary aspects of these transmission techniques. Implementation details, especially those of more advanced transmission schemes, are considered in the later technology-oriented chapters.

3.4.1   Power Efficient Short Distance Baseband Transmission

In baseband transmission, the digital signal is transmitted without modulation of a carrier at a higher frequency and the spectrum is around the zero frequency. Without a carrier signal, the system cannot support multiple channels in a frequency division multiplexing format and consequently one signal (that belongs to one user) occupies the entire available bandwidth of the medium. With one user occupying the entire band, the system designer does not need to pay attention to the out of band radiation. To support a multiuser environment with a baseband transmission, one has to resort to innovative techniques other than simple traditional frequency division multiplexing (e.g., TDMA or CDMA).

There are two basic baseband transmission techniques, *line-coded transmission* and *pulse-modulated transmission*. In the first technique, the digital data stream is line-coded to facilitate synchronization at the receiver and avoid the direct current offset during transmission. If the data stream produced by a computer is applied directly to the wires, the receivers will have difficulty in synchronizing with the transmitted symbols. To provide better synchronization at the receiver, the format of the incoming data stream is modified before transmission. This modification process is often referred to as line coding. Baseband line coded signaling is commonly used in short distance wireless applications. For example, WLANs using infrared wireless optical signals often employ line-coded baseband transmission. In pulse modulation, the transmitted information is coded into amplitude, location, or duration of a pulse shape. Pulse-modulated baseband transmission is commonly used for low speed infrared data communications such as remote controls or connections between personal computers and the printers or keyboards. Ultra wideband pulse modulation has also been considered for very low power short-range radio communications.

In wired applications, baseband signaling is used in Ethernet, the dominant wired LAN technology. In wireless applications baseband transmission with line coding is popular in high-speed diffused and directed beam infrared WLANs.

Example 3.7: Manchester encoding in an IR transmitter

Figure 3.9 shows a Manchester code implementation of an infrared transmitter used in infrared WLANs. The digital data stream is first line coded in a format known as Manchester coding. In this format we always have a transition in the middle of the pulse. If the data stream is "1" we start from a high voltage value and if it is "0" we start from a low level voltage. The line-coded signal is then *intensity* modulated by the emitted infrared light by simply turning the transmitted light to on and off positions. The receiver consists of a simple photo-sensitive diode for detecting the presence and lack of presence of the light. This will either produce or not produce an electric signal that is then amplified and used as the received signal. The Manchester coded data doubles the transmission rate (each bit consists of two pulses) but provides one transition per bit. These transitions are important for the receiver because it uses them to synchronize its clock with that of the transmitter clock. In the original data stream without transitions, if we have a long stream of "0"s or "1"s, the receiver loses the reference timing of the transmitter.

In the baseband pulse transmission technique that we described in Example 3.7, the coded stream uses the intensity or amplitude of the signal to communicate between the transmitter and the receiver. Baseband pulse modulation techniques can use the *duration* or *position* of a pulse for communications. Since wireless channels suffer from extensive amplitude fluctuations caused by fading and near-far problems, communications using the pulse width or position are very popular in wireless baseband communications. The following example provides a practical implementation to illustrate this concept.

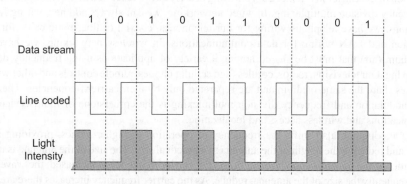

Figure 3.9 Data stream and line coding for a light intensity infrared transmission system.

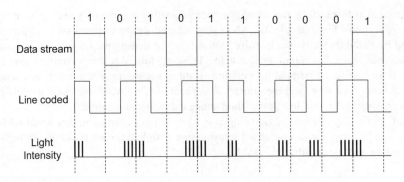

Figure 3.10   Data stream and line coding for a pulse position transmission system.

Example 3.8: Using position of pulses for communications

A practical implementation of an infrared communication system using the position of a pulse connecting a keypad to a computer is shown in Figure 3.10. The received data stream is first Manchester encoded and then the falling edge of the coded signal is used to generate a pulse position communication signal. Rather than one single pulse, multiple narrow pulses are transmitted that code the transmission of digitized information. When detected by the photosensitive diode at the receiver, the multiple narrow pulses will produce a single continuous pulse that is close to what would have been received if a single pulse were transmitted. However, with multiple narrow pulses, the required transmission power is smaller because the light is on for a shorter period of time. Therefore, multiple pulse transmissions saves the life of the battery at the transmitter. A disadvantage of using several pulses per symbol is the large bandwidth occupied by each pulse. However, this is not too important in infrared applications, where one user occupies the entire band.

3.4.2   Bandwidth Efficient Carrier Modulated Transmission

In carrier modulated signaling, the message signal is mixed with a *carrier signal at a higher frequency* before transmission. Carrier modulation shifts the spectrum of the transmitted signal to the location of the carrier in the spectrum allowing orderly coexistence of a number of transmissions via frequency division multiplexing. In wired networking, we can always add new wiring for new applications, we have telephone wiring for our telephones, coaxial or fiber-optic cable for video distribution and LAN wiring for data communications. In wireless networking, we have only one medium (air) that must be shared among a variety of applications using frequency division multiplexing. Our television, radios, cordless and cellular phones, remote controls and other wireless appliances share the same medium and are separated only by their carrier frequencies. Therefore, carrier modulation and frequency division multiplexing is the cornerstone of multiservice radio communications and wireless access and localization.

Carrier modulation also shifts the frequency of operation to higher values, providing higher capacity and reducing the length of the antenna to a practical size. The size of the antenna is usually on the order of the transmission wavelength. As the carrier frequency is increased, the wavelength and consequently the size of the antenna, reduce. As the carrier frequency increases there are wider frequency bands available to support higher data rates. However, with increasing carrier frequency, path-loss in the first meter increases, the design of circuits becomes more challenging and also in-building penetration of the signal becomes smaller. Designers of wireless modems need to make a compromise among availability of frequency of operation, required bandwidth, coverage of indoor and outdoor areas, and the cost of implementation. Another advantage of carrier modulation is that we can implement two-dimensional (2D) transmission in the same frequency band. To understand what a 2D transmission means, we start with example of a simple 1D digital carrier transmission.

Example 3.9: An implementation for Example 3.3

Figure 3.11 shows a simple example of a carrier-modulated implementation of the binary communication system that was described in Example 3.3. In this transmission system, the baseband information signal is encoded first and then multiplied with a cosine carrier at a frequency f_c at the transmitter before transmission with an antenna.

The received signal is multiplied with an identical locally generated cosine carrier and passed through an integrator to recover the transmitted signal. The frequency of the carrier is selected so that during the transmission of a symbol, we have an integral number of cycles of the carrier frequency (i.e., there is always a complete cycle of cosines in T_s seconds). In this situation we have:

$$\frac{2}{T_s} \int_0^{T_s} [\cos 2\pi f_c t]^2 dt = \frac{1}{T_s} \int_0^{T_s} [1 + \cos 4\pi f_c t] dt = \frac{1}{T_s} \int_0^{T_s} dt + \frac{1}{T_s} \int \cos 4\pi f_c t dt = 1 + 0 = 1$$

Thus, the detected received signal in the absence of noise is the same as the transmitted signal. As shown in Figure 3.11 the difference between "1" and "0" signal in this transmission technique is a 180° shift in the carrier frequency. For this reason this modulation technique is referred to as binary phase shift keying or BPSK modulation. The signal constellation and error rate plots of this system are identical to that of Example 3.3.

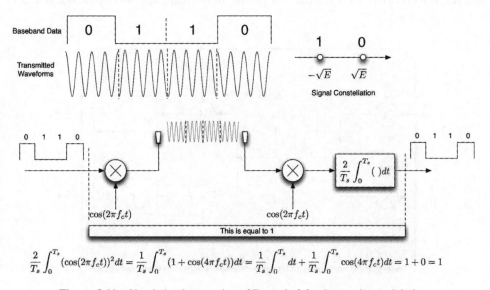

Figure 3.11   Simple implementation of Example 3.3 using carrier modulation.

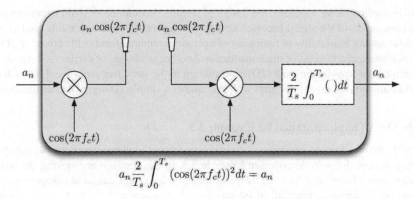

$$a_n \frac{2}{T_s} \int_0^{T_s} (\cos(2\pi f_c t))^2 dt = a_n$$

Figure 3.12 Simple concept behind carrier transmission.

We can extend this concept to implement any signal constellation with multiple symbols, such as the constellation with four symbols shown in Figure 3.3. Figure 3.12 shows a simple extension of the implementation of a binary transmission to multisymbol transmission, where a_n represents the amplitude of different symbols. This type of modulation is usually referred to as pulse amplitude modulation (PAM) and as shown in Figure 3.3 the corresponding constellation has a 1D representation.

In wireless communications we can also create a 2D transmission system over a single carrier. Such transmission techniques are usually referred to as quadrature amplitude modulation (QAM) and their constellations have a 2D representation. To understand the concept of 2D transmission, consider Figure 3.13. Here we have two streams of data symbols, a_n and b_n, which are mixed with a cosine and a sine carrier separately. The streams with the carrier are then added together and sent over the wireless channel. The receiver now consists of two multipliers and integrators, similar to

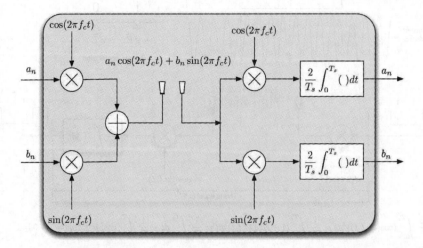

Figure 3.13 Basic concept behind 2D carrier transmission.

the single one in Figure 3.12, one synchronized to the transmitted cosine and the other one to the transmitted sine. Since a cosine and a sine at the same frequency are orthogonal signals we have:

$$\frac{2}{T_s} \int_0^{T_s} [\cos 2\pi f_c t]^2 dt = 1, \quad \frac{2}{T_s} \int_0^{T_s} [\sin 2\pi f_c t]^2 dt = 1,$$

$$\frac{2}{T_s} \int_0^{T_s} \cos 2\pi f_c t \sin 2\pi f_c t dt = \frac{1}{T_s} \int_0^{T_s} \sin 4\pi f_c t dt = 0$$

Thus, the cosine component of the signal is detected in the upper branch and eliminated in the lower branch that is designed for sine signal and visa versa. As a result the two output streams are separated at the receiver without interference. With a wireless transmission system that is capable of carrying two independent stream of symbols with different amplitudes, a_n and b_n, we can establish a 2D communication system with a two-dimensional signal constellation.

Example 3.10: QPSK transmission

Figure 3.14 shows the 2D signal constellation of a four-symbol (4-QAM) transmission system also known as quadrature phase shift keying (QPSK). It is a realization of the system shown in Figure 3.13, where each orthogonal branch carries a binary information stream mapping two bits to each point in the constellation. The 1D counterpart of this constellation was described in Example 3.3 and Figure 3.3. For the same minimum distance of $d_{min} = 2\sqrt{E_1}$, the average energy per symbol is:

$$E_s = \frac{1}{4} \sum_{i=0}^{3} E_i = \frac{2E + 2E + 2E + 2E}{4} = 2E = \frac{1}{2} d_{min}^2$$

Considering that the minimum distance is the governing factor in the error rate, we see that with the same error rate or minimum distance, the 2D QPSK modem needs 2.5 times or approximately 4 dB less average power to achieve the same error rate as that of 4-PAM, described in Example 3.2, which also employs four symbols, but in one dimension.

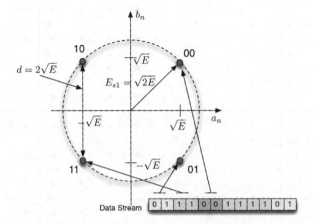

Figure 3.14 The 2D 4-QAM or QPSK constellation with two bits per symbol.

If we compare the QPSK constellation with the BPSK constellation of Figure 3.11 the average energy per symbol for both transmission schemes is the same. However, in the case of QPSK, we are sending two bits per symbol and in that context, the QPSK energy per bit is two times or 3 dB smaller than BPSK. In general it can be shown that if the number of symbols in the constellation is large, in a 2D constellation, we will need 3 dB (four times) more power to transmit one more bit per symbol with the same error rate. This 3 dB for one additional bit per symbol rule is very useful one for system engineering designs, where we want to have an intuitive quantification of comparative performance of complex systems with large signal constellations.

Wireless data networks use a variety of signal constellations to adjust the data rate as the distance between a mobile device and an access point or base station changes. When the user is close to an access point, the signal to noise ratio is high and a larger signal constellation is used. As the distance between the user and access point increases, the number of points in the constellation is reduced. As we described in Chapter 2, power and distance are related in a logarithmic form and power in dB is proportional to $10\alpha \log d$, where α is the path-loss gradient and d is the transmitter-receiver distance. If we want to double the coverage, we need $10\alpha \log 2 \approx 3\alpha$ more power in dB. If we want to increase the coverage by 10 times, we need 10α more power in dB. A numerical example can provide some intuitive engineering insight.

Example 3.11: Data rate coverage in QAM systems

Consider the QPSK constellation with two bits per symbol shown in Figure 3.14 and the 16-QAM (four bits per symbol) and 64-QAM (six bits per symbol) signal constellations, shown in Figure 3.15. Also assume that the error control coding rate before transmission for all of these constellations is $r = 3/4$ and the symbol transmission rate for these constellations are all 12 MSps. In a manner similar to Example 3.1, we can determine the bit rate of QPSK to be $R_b = 12(\text{MSps}) \times 2(\text{bpS}) \times \frac{3}{4} = 18$ Mbps. For 16-QAM and 64-QAM we have the data rates as 36 and 54 Mbps respectively. In other words this system supports (18 Mbps for QPSK, 36 Mbps for 16QAM and 54Mbps for 64-QAM). In terms of coverage, according to the 3 dB per bit rule described above, 16- and 64-QAM need 6 and 12 dB more power than the QPSK, respectively. That means, if the 64-QAM covers D meters, because at this distance its power is not adequate to

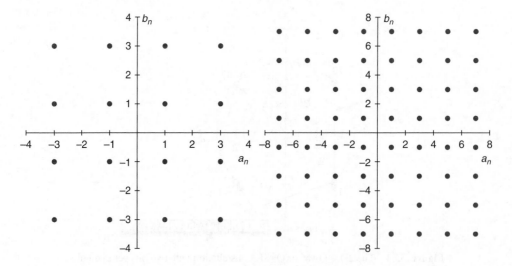

Figure 3.15 16- and 64-QAM constellations for four and six bits per symbol.

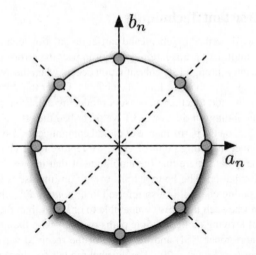

Figure 3.16   8-PSK signal constellation used in some wide area wireless data networks.

provide an acceptable error rate the error rate of the 16-QAM is still acceptable because it operate up to 6 dB less level of power. To have a quantitative number consider an environment with a distance power gradient of $\alpha = 2$. In this environment the coverage of the 16-QAM, d, with respect to the coverage of the 64-QAM, D, will be:

$$10\alpha \log \frac{d}{D} = 6 \Rightarrow d = 10^{6/(20)} D = 2D$$

Similarly, in the same environment, a QPSK system that has a 12 dB less power advantage, will assures a coverage of 4D.

The above example provides a simple and useful method to relate the data rate, coverage and the environment in a wireless network. Using the 2D implementation of Figure 3.13 we can implement most signal constellations. We have specific names for some of these constellations. As we have discussed previously, the rectangular constellations shown in Figure 3.15 are referred to as QAM and the four-point constellation of Figure 3.14 by QPSK. Figure 3.16 shows another popular constellation called 8-PSK that is commonly used for wide area high-speed wireless data communications. The advantage of an M-PSK over M-QAM is that the amplitudes of all the symbols are the same. Therefore, the receiver does not need to have a very accurate value of the amplitude to detect the received symbol because the transmitted information is embedded only in the phase of the signal.

Based on the relation between the minimum distance and the energy we can determine the error rate associated with these constellations. The error rate for M-QAM is given by:

$$P_e \approx \frac{1}{2}\mathrm{erfc}\left(\sqrt{\frac{3}{2M-1}m\gamma_b}\right) \tag{3.10}$$

and for M-PSK by:

$$P_e \approx \frac{1}{2}\mathrm{erfc}\left(\sqrt{\sin^2\left(\frac{\pi}{M}\right)m\gamma_b}\right) \tag{3.11}$$

3.5   Multipath Resistant Techniques

In the previous section, we described baseband and carrier modulation techniques used for wireless networks assuming that multipath propagation does not impact the error rate or change the shape of the waveform significantly through ISI. This situation occurs when the symbol transmission rate is much lower than the coherence bandwidth of the channel. Under this circumstance, the arriving signals along multiple paths add together as phasors producing extensive power fluctuations that we refer to as small-scale fading. In Section 3.3.1 we described the effects of Rayleigh multipath fading, shown in Figure 3.6, on the performance of digital communication systems (see Figure 3.7). From the multipath channel modeling point of view, we represent the channel impulse response with only one impulse and we declare that the amplitude of that impulse fluctuates according to a Rayleigh or Rician distribution. The Fourier transform of an impulse is a flat line in frequency domain, so this type of fading channel is also referred to as a *flat fading* channel. As we increase the symbol transmission rate such that it is comparable to or larger than the coherence bandwidth of the channel, a symbol arriving along multiple paths overlaps with the next transmitted symbol causing inter symbol interference (ISI) and the shape of the received waveform is not the same as the transmitted symbol (see Figure 2.20). The channel models for these situations have several impulses, each representing a cluster of arriving paths and their associated path fluctuations. In the frequency domain the spectrum is not flat any more and it suffers from *frequency selective fading*. Figure 3.17 shows a typical multipath profile for arrival of the paths and typical frequency domain characteristics that is not flat anymore and is affected by frequency selective fading. The cause of frequency selective fading, in a manner similar to multipath fading, is the changes in the phase of the signal arriving along different paths. An intuitive understanding of how the multipath structure is related to the symbol transmission rate is essential for understanding how different modem design technologies have evolved.

With low data rate transmissions, we are exposed to flat fading. Then, at the receiver we only need to cope with power fluctuations since the transmitted waveform preserves its shape. In a flat fading channel, the huge number of errors during the fade dominates the error rate performance of any transmission technique. Any mitigating technique that reduces the number of fade hits would improve the performance of the system. If we increase the data rate, we get exposed to ISI caused by multipath arrival and frequency selective fading. Any technique that handles the ISI or mitigates

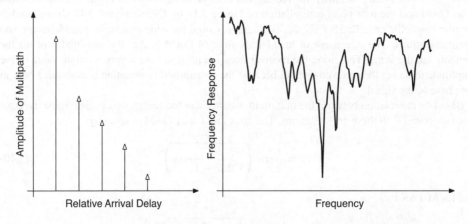

Figure 3.17   Multipath arrival and the associated frequency selective fading.

frequency selective fading is desirable for high speed or wideband wireless networking in multipath channels.

In the past few decades wireless networking evolved as the most popular access technique for traditional telephone networks originally designed for voice applications and became also popular for Internet access using mobile devices for mostly data oriented applications. A number of multipath fading mitigating technologies have evolved during this period. In principle, we can divide these techniques into three categories: (1) antenna diversity techniques addressing flat fading, (2) direct sequence spread spectrum (DSSS) techniques addressing multipath arrival and ISI, (3) frequency hopping spread spectrum (FHSS) domain addressing frequency selective fading.

3.5.1   Flat Fading, Antenna Diversity, and MIMO

As we observed in the previous sections (see Figure 3.6), flat fading is manifested as signal amplitude fluctuations over a wide dynamic range. In particular, during short periods of time, the channel goes into deep fades causing significant numbers of errors that virtually dominate the overall average error rate of the system. In order to compensate for the effects of fading when operating with a fixed-power transmitter, the power must typically be increased by several orders of magnitude relative to non-fading operation (see Figure 3.7). This increase of power protects the system during the short intervals of time when the channel is deeply faded.Instead of increasing the transmit power significantly, a very effective method of counteracting the effects of fading is to use diversity techniques in transmission and reception of the signal. The concept here is to provide multiple copies of the received signals whose fading patterns are different (and hopefully independent). With the use of diversity, the probability that all the received signals are in a fade at the same time reduces significantly, which in turn can yield a large reduction in the average error rate of the system. Diversity can be provided spatially by using multiple antennas, in frequency by providing signal replicas at different carrier frequencies separated by the coherence bandwidth, or in time by providing signal replicas with different arrival times separated by the channel coherence time. It is conventional to refer to the diversity components as diversity branches. We assume that the same symbol is received along *different branches*, with each branch exposed to a separate random fluctuation. This has the effect of reducing the probability that the received signal will be faded simultaneously on all the branches; this in turn reduces the overall outage probability, as well as the average BER.

Figure 3.18 shows fluctuations in two branches of a diversity channel in time and how they help in the reduction of overall error rates. When one of the branches is in deep fade, causing a large number of errors, the correct data can be retrieved from the other branch (see the dashed box on the left). In a diversity channel, large numbers of errors can occur when all branches are in deep fade at the same time. Since the probability of a deep fade occurring in all branches is much lower than in only one branch, the error rate on a diversity channel is much less than on a single-branch fading channel. The occurrence of deep fading on all branches is a function of correlation among different branches and of the number of diversity channels. As the correlation among the diversity branches decreases and they become independent and the number of branches increases, the error rate decreases.

A variety of techniques are available for reception of the diversity signals. In the most popular and the optimum method of combining, called *maximal-ratio combining*, the diversity branches are weighted prior to summing them, each weight being proportional to the received branch signal amplitude.

Let us assume that the amplitudes of the signals received on different branches are all uncorrelated Rayleigh-distributed random variables and all branches of diversity have the same average received

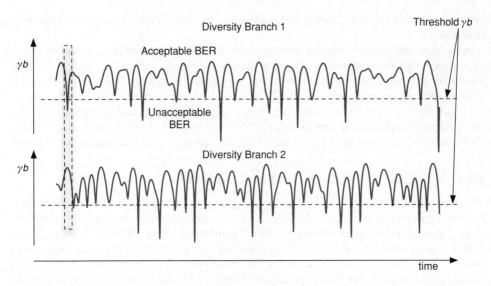

Figure 3.18   Fading in two branches of a diversity channel.

signal power and the average SNR on each branch is denoted by $\bar{\gamma}_b$. The probability distribution function of the maximum ratio combiner signal to noise ratio per bit is then given by the gamma function:

$$f_\Gamma(\gamma_b) = \frac{1}{(D-1)!} \frac{\gamma_b^{D-1}}{\bar{\gamma}_b^D} e^{-\gamma_b/\bar{\gamma}_b} \tag{3.12}$$

where D is the order of diversity. It can be shown that using Equation 3.12 instead of exponential function in Equation 3.9 the average probability of error for the maximal-ratio combiner output is given by [Pah05; Pro08]:

$$\bar{P}_e = \int_0^\infty f_\Gamma(\gamma_b) P_b(\gamma_b) d\gamma_b \approx \left(\frac{1}{8\frac{m}{a}\bar{\gamma}_b}\right)^D \binom{2D-1}{D} \tag{3.13}$$

where we have used the standard notation for a binomial coefficient:

$$\binom{N}{k} = \frac{N!}{(N-k)!k!}$$

The expression in Equation 3.13 shows that average BER performance at the maximal-ratio combiner output improves exponentially with increasing D, the order or number of branches of diversity. Figure 3.19 shows the average probability of error $\bar{P}_e$ versus average signal to noise ratio per bit $\bar{\gamma}_b$, for different orders of diversity. Included in the figure is the error rate curve for steady-signal reception without fading. As we saw earlier, with a single antenna, we lose 30–35 dB in performance relative to steady-signal reception at reasonable levels of error rate. With two independent diversity branches, the performance loss is reduced to about 25 dB, and with four orders of diversity the signal to noise penalty is reduced to around 10 dB. With additional orders of diversity the penalty relative to non-fading can be further reduced. There will of course be a practical limit to the order of

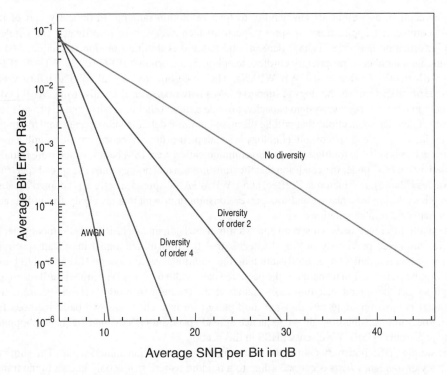

Figure 3.19   Average BER versus average SNR per bit for different orders of diversity.

diversity implemented because, for example, one cannot put an arbitrarily large number of antennas into a small smart phone used as a wireless communications terminal.

In wireless networks MIMO is referred to the systems using multiple antennas at both the transmitter and the receiver to improve communication performance. It is one of several forms of *smart antenna* technologies build to allow beamforming with combining the transmitted or received signals from different antennas with different weight combinations. In the CDMA digital cellular telephone networks, smart antenna systems adjust the spatial antenna pattern of the base station and the mobile device of a specific user to reduce interference from other users. Reduction of interference among the mobile users allows more simultaneous users in the area that increases the capacity of the network. In the wireless data applications, if the antenna pattern is focused to the direct path between the transmitter the receive signal from other paths reduces, resulting in significant reduction of the rms multipath spread of the channel. Since the maximum data rate is inversely proportional to the rms delay spread of the effective channel between the transmitter and a receiver, MIMO system increase the maximum data transmission rate or throughput of the wireless data networks. Because of these benefits, MIMO technology has been considered for 3G CDMA cellular telephone standards and has been adopted by latest OFDM-based wireless data networking standards such as the IEEE 802.11n for WLANs and 4G long-term evolution (LTE) systems.

3.5.2   Frequency Hopping Spread Spectrum Transmissions

Spread spectrum technology was first invented during the second-world war and it has dominated military communication applications, where it is attractive because of its resistance to interference and interception, as well as its amenability to high resolution ranging. In the early part of the 1980s, commercial applications of spread spectrum technology were investigated for wireless office information networks [Pah85, Pah88a] and today it is the transmission technique used in 3G cellular, a number of proprietary cordless telephones, the original IEEE 802.11 WLAN, IEEE 802.15 Bluetooth, ZigBee, and UWB WPANs. The voice-oriented digital cellular industry has selected spread spectrum technology to support CDMA networks as an alternative to TDMA/FDMA networks in order to increase system capacity, provide a more reliable service, and to provide soft handoff of cellular connections that will be discussed in more detail in subsequent chapters. In the WLAN industry, spread spectrum technology was adopted primarily because the first unlicensed frequency bands suitable for high speed radio communication were ISM bands, which were initially released by the FCC under the condition that the transmission technology must use spread spectrum techniques [Mar85]. In Bluetooth, ZigBee, and UWB WPANs, spread spectrum is adopted because of simplicity of implementation and low power consumption, which is necessary for ad hoc and sensor networks implementations.

The main difference between spread spectrum transmission and traditional radio modem technologies discussed previously in this chapter, is that the transmitted signal in spread spectrum systems occupies a much larger bandwidth than the traditional radio modems. Compared to baseband impulse transmission techniques, the occupied bandwidth by spread spectrum is still restricted enough so that the spread spectrum radio can share the medium with other spread spectrum and traditional radios in a frequency division multiplexed format. There are two basic methods for spread spectrum transmission: direct sequence spread spectrum (DSSS) and frequency hopping spread spectrum (FHSS). We discuss FHSS in this section.

The simple FHSS transmitter shifts the center frequency of the transmitted signal. The shifts in frequency or *frequency hops* occur according to a random pattern that is only known to the transmitter and the receiver. If we move the center frequency randomly among one hundred different frequencies, then the required transmission bandwidth is hundred times more than the original transmission bandwidth. We call this new technique a *spread spectrum technique* because the spectrum is spread over a band that is hundred times larger than original traditional radio. FHSS can be applied to both analog and digital communications but it has been applied primarily for digital transmissions.

Example 3.12: FHSS and LFSR codes

Figure 3.20 shows the hopping pattern and associated frequencies for a frequency hopping system transferring data packets over the air. A three state recursive machine is used for generation of the code that determines the frequency to which the transmitter will hop next. This machine, shown in Figure 3.20a, generates a sequence of seven states, representing all seven non-zero values which can be formed by three binary digits, periodically. Figure 3.20b shows all seven states of the machine and their associated decimal number, which is used as the frequency index for FHSS. Each packet is transmitted using one of these frequencies. The sequence of frequencies is $f_4, f_2, f_5, f_6, f_7, f_3, f_1, f_4$ before returning to the first frequency f_4. Figure 3.20c shows transmission of the information packets in time and frequency in this FHSS system.The random sequence used to pick the frequencies in these situations are referred to as pseudo noise (PN) sequences. The particular code used here is generated using a Linear Feedback Shift Register (LFSR) code (more details of which are available later).

In FHSS, the hopping of the carrier frequency does not affect the performance in the presence of additive noise because the noise level in each hop remains the same as the noise level of

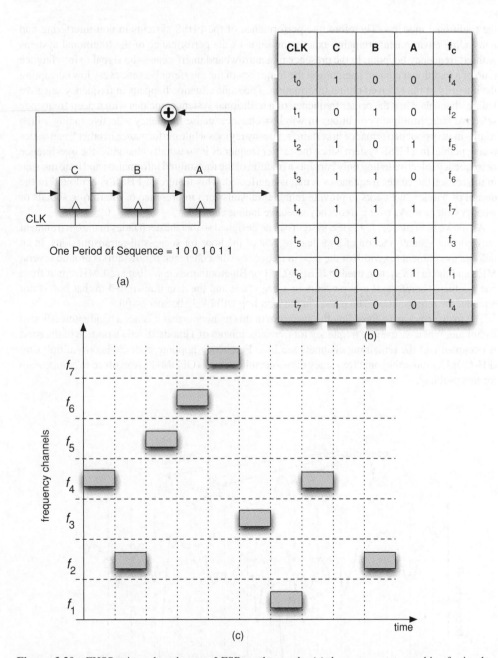

CLK	C	B	A	f_c
t_0	1	0	0	f_4
t_1	0	1	0	f_2
t_2	1	0	1	f_5
t_3	1	1	0	f_6
t_4	1	1	1	f_7
t_5	0	1	1	f_3
t_6	0	0	1	f_1
t_7	1	0	0	f_4

One Period of Sequence = 1 0 0 1 0 1 1

(a)

(b)

(c)

Figure 3.20 FHSS using a length seven LFSR random code: (a) three stage state machine for implementation of the LFSR code, (b) states of the memory in time and frequency, (c) transmission of information bits in time and frequency in FHSS.

the traditional modems. Therefore, the performance of the FHSS systems in non-interfering and non-fading environments remains exactly the same as the performance of the traditional systems without frequency hopping. In the presence of a narrowband interference, the signal to interference ratio of a traditional modem operating at the frequency of the interferer becomes very low corrupting the integrity of the received digital information. The same situation happens in frequency selective fading channels when the center frequency of a traditional system coincides with a deep frequency selective fade. Multipath conditions in wireless channels cause frequency selective fading which results in very poor performance in certain frequency regions while performance in other frequencies is acceptable. In a FHSS system, since the carrier frequency is constantly changing, the interference or frequency selective fading only corrupts a fraction of the transmitted information and transmission in the rest of the center frequencies remains unaffected. This feature of FHSS is exploited in the design of wireless networks to provide reliable transmissions in presence of interfering signals or when a system works over a frequency selective fading channel.

As shown in Figure 3.21, a FHSS system can be designed so that the deep fades in the environment only corrupt a small fraction of hops leaving rest of the hops for successful retransmissions. In an indoor environment the width of the fade in frequency (the coherence bandwidth) is around several MHz and the FHSS system used in IEEE 802.11 or Bluetooth uses hops that are 1 MHz apart from one another. Therefore, if a hop occurs in a deep fade and the data transmitted in that hop is not reliable, the retransmitted data packet in the next hop will likely be successful.

The main criticism to spreading the spectrum in this manner is that it is not a bandwidth efficient technique. When we use one frequency for transmission out of a hundred, only a portion of the band is occupied and the remaining channels are idle. Frequency hopping code division multiplexing (FH-CDMA) and orthogonal frequency division multiplexing (OFDM) have evolved to compensate for this problem.

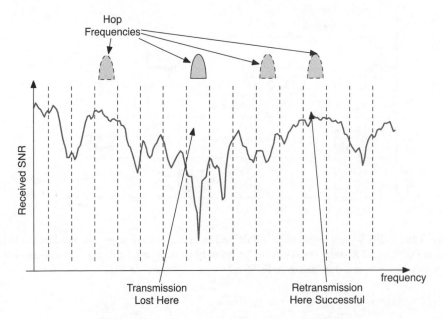

Figure 3.21 Frequency selective fading and FHSS.

3.5.3 FH-CDMA and OFDM

Frequency hopping spread spectrum allows the coexistence of several transmissions in the same frequency band using different hopping codes. Different users could be members of the same network following a coordinated hop pattern or two different networks each using their own pattern. For example, IEEE 802.11 FHSS and Bluetooth both use the same bandwidth of 1 MHz and the same 78 channels in the 2.4 GHz ISM bands, shown in Figure 3.22. Members of each network coordinate their transmissions and the hopping patterns in their own network while there is no coordination among the users in the two different network and they simply co-exist. The multiuser interference occurs when two different users transmit on the same hop frequency. If the hopping codes are random and independent from one another, the "hits" will occur with some calculable probability. If the codes are synchronized and the hopping patterns are selected so that two users never hop to the same frequency at the same time, multiuser interference is eliminated. When we have multiple users with lower data rates, this approach allows them to share the medium.

Basic OFDM can be considered as a coordinated FH-CDMA system in which all channels are assigned to a single user. It follows the same features of FHSS but it can support high data rates for data rate sensitive applications such as wireless Internet access. The basic OFDM can also be considered a multicarrier modulation system in which all carriers are assigned to a single user so that the user has high-speed access for bursty data applications. OFDM is then an implementation of multicarrier modulation using orthogonality of the adjacent carriers. OFDM is an implementation of such multicarrier operation that takes advantage of the orthogonality of the channels and employs a computationally efficient implementation based on the Fast Fourier Transform algorithm.

Multicarrier Modulation was first evaluated for high-speed voice band modems in the early 1960s and it was augmented with an FFT implementation for the same application in the early 1980s [Pah88b]. It found its way into WLANs (IEEE 802.11), DSL modems, cable modems (IEEE 802.14), WMANs (IEEE 802.16), WPANs (IEEE 802.15), and many other wired and wireless applications in the 1990s. The concept here is very simple. Instead of transmitting a single stream

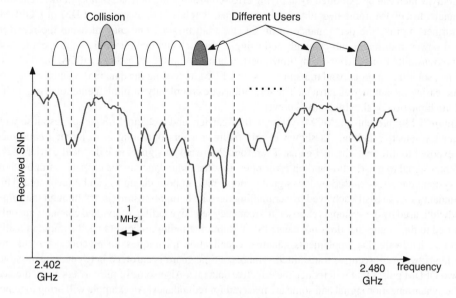

Figure 3.22 FH-CDMA in legacy IEEE 802.11 and Bluetooth systems.

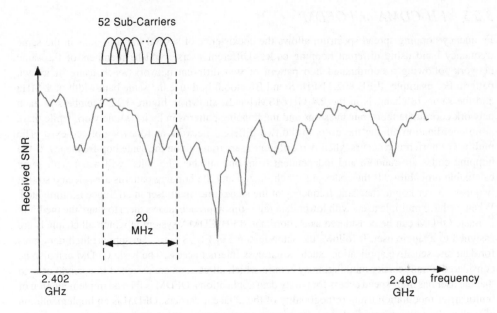

Figure 3.23   Frequency selective fading and multicarrier modulation.

at a rate of R_s symbols/s, we use N streams over N carriers spaced by about R_s/N Hz each carrying a stream at the rate R_s/N symbols/s. The primary advantage of MCM is its ability to cope with severe channel conditions such as high attenuations at higher frequencies in long copper lines or the effects of frequency-selective fading in wireless channels. Sub-channels provide a form of frequency diversity, which can be exploited by applying error-control coding *across symbols* in different sub-channels. In OFDM implementations, this technique is referred to as coded OFDM or COFDM. To further improve the performance of an OFDM transmission, one may measure the received signal power in different sub-channels and using a feedback channel may adjust the specification of the transmitted sub-carriers to optimize performance. With these features, OFDM has become an ideal solution for broadband transmissions. In OFDM increasing the data rate is simply a matter of increasing the number of carriers. The limitations are complexity of implementation using FFTs and the limitation on the transmitted power.

Figure 3.23 represents an MCM system with N carriers and a channel frequency response with frequency selective fading. As shown in this figure, carriers operating at different frequencies are exposed to different channel gains. Therefore, the received signals in individual carriers have different signal to noise ratios and bit error rates. If some redundancy in the carriers is imposed on the system, the errors caused by low signal to noise ratio in poor channels can be recovered. This redundancy can be easily achieved by scrambling the data before transmission. Although the entire bandwidth used by the system is exposed to frequency selective fading, individual channels are only exposed to flat fading that does not cause ISI. Without flat fading, receivers need computationally intensive hardware that implements adaptive equalization techniques. In practice to avoid any overlap between consecutive transmitted symbols, a time guard is enforced between transmissions of two OFDM pulses that will reduce the effective data rate. Also some of the carriers are dedicated to the synchronization signal and some are reserved for redundancy. An example will bring together all the details of implementation in OFDM systems.

Example 3.13: Implementation of OFDM in IEEE 802.11g

The IEEE 802.11g transmission has 48 effective sub-carriers, with each sub-carrier carrying data at a symbol rate of 250 KSps. The user symbol transmission rate is 48×250 KSps $= 12$ MSps. The bit transmission rate depends on the number of bits per symbol (i.e., the constellation used for transmission) and the rate of the convolutional code used for transmission. With BPSK, an $r = 1/2$ coding rate, and 1 bpS, the data rate is 12 (MSps) $\times$ 1/2 $\times$ 1 (bpS) $= 6$ (Mbps) and with 64-QAM and 6 bpS and convolutional coding of rate $r = 3/4$ we have 12 (MSps) $\times$ 3/4 (bpS) $\times$ 6 (bpS) $=$ 54 (Mbps). As the distance between the transmitter and the receiver is increased, the data rate is reduced by adjusting the coding rate and the symbol transmission rate (size of the constellation) to maintain reliable communications.

3.5.4   Direct Sequence Spread Spectrum Transmission

In a DSSS, each transmitted information symbol is coded into N narrower pulses referred to as chips using a PN sequence. At the receiver, the transmitted chips are first demodulated and then passed through a correlator to calculate their autocorrelation function (ACF). For now, we define the ACF of a PN sequence $\{b_i\}$ of length N as:

$$R(k) = \sum_{i=0}^{N-1} b_i b_{i-k} = \begin{cases} N, k = mN \\ -1, \; otherwise \end{cases} \qquad (3.14)$$

The PN sequence is a periodic sequence of length N and therefore the correlation is calculated over one period of the sequence. The autocorrelation function of a PN sequence has a very high peak of height N at $k = 0$ at integral multiples of N, which is usually referred to as the processing gain of the receiver. The value of this autocorrelation function for $k \neq 0$ is far below the peak value (it is -1 for maximal length sequences). Therefore, DSSS systems use the peak of the autocorrelation function to detect the transmitted bit.

Example 3.14: DSSS using 7-chip LFSR code

The output of the LFSR circuit shown in Figure 3.20c is the sequence $\{-1 -1\ 1\ 1 -1\ 1\ 1\ 1\}$. Figure 3.24 shows a sample data bit "1" in a binary communication DSSS system, the transmitted code for the data bit, and the autocorrelation function at the receiver with its high peak and low sidelobes. LFSR codes are generated with simple state machines with feedback and usually produce maximum length PN sequences following $2^m - 1$ values. Such maximum-length PN sequences are widely used in CDMA cellular networks and the time of arrival based geolocation techniques. More details are available in Chapter 12.

As shown in Figure 3.24, for any transmitted bit the output of the correlator function at the receiver produces a narrow pulses with a height of N and a base that is twice the chip duration. Therefore, a DSSS system can be thought of as a *pseudo pulse transmission technique*, where we receive a narrow pulse designating each transmitted bit. In a multipath radio channel environment, as shown in Figure 3.25, the signal arriving along different paths will carry different pulses at the receiver at different times. An intelligent receiver can use these pulses as a source of *diversity* to improve its performance. These receivers are referred to as RAKE receivers, which are commonly implemented in DSSS wireless transmission systems to achieve reliable communications. Figure 3.25 also

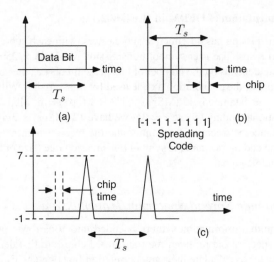

Figure 3.24 Seven-chip LFSR-7 code used in DSSS: (a) a transmitted bit, (b) the seven-chip code, (c) circular autocorrelation function of the code.

compares the DSSS transmission with traditional symbol transmission. In the traditional systems, multipath causes ISI that is very damaging. The DSSS system with a RAKE receiver over the same channel enjoys multipath diversity.

Example 3.15: Multipath reception in CDMA

The original Qualcomm CDMA system uses a chip rate of 1.25 Mcps and the symbol transmission rate of 9600 Sps. Therefore, it can resolve multipath components that are on the order of $1/(1.25$ Mcps$) = 800$ ns apart. A multipath spread of up to $1/(9800$ bps$) = 1.04$ ms does not cause ISI in the system. The multipath spread in the outdoor micro-cellular environment is on the order of

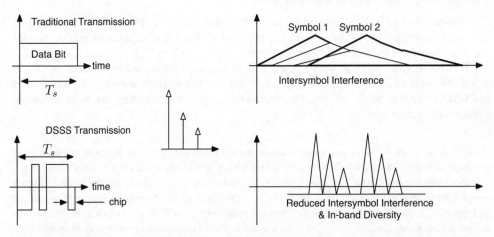

Figure 3.25 Comparison of traditional systems and DSSS system used for communication over a three-path channel.

several tens of microseconds and in indoor pico-cell areas it is on the order of several hundreds of nanoseconds. Therefore, this system does not suffer from ISI in any of the indoor or outdoor environments. However, in an indoor pico-cellular environment, it is unlikely that the system resolves the multipath components since the spread is smaller than 800 ns. Some 3G Wideband-CDMA (W-CDMA) systems offer similar bit rates with chip durations that are up to an order of magnitude shorter (ten times larger bandwidth). With a pulse resolution of around 80 ns for W-CDMA systems, one can expect the receiver to be able to resolve several multipath components even in the indoor pico-cellular environment.

The bandwidth of any digital system is inversely proportional to the duration of the transmitted pulse or symbol. Since the transmitted chips are N times narrower than data bits, the bandwidth of the transmitted DSSS signal is N times larger than a traditional system without spreading. As a result, in a manner similar to FHSS, the DSSS transmission is not a bandwidth efficient system by itself because it occupies a much wider bandwidth than its symbol transmission rate. To increase the bandwidth efficiency of the DSSS in cellular telephone applications, where we have a number of users each demanding a relatively low data rate, we use CDMA by assigning different codes to each user. For the data applications, where we need very high data rates for transmission of a burst by a single user, we resort to M-ary orthogonal coding, where a single user uses multiple orthogonal codes for transmission.

3.5.5 DS-CDMA and M-ary Orthogonal Coding

In a multiuser DS-CDMA environment for cellular telephone application, different codes are assigned to different users. In other words, each user has its own unique "key" code that is used to spread and despread only his messages. The codes assigned to other users are selected so that during the dispreading process at the receiver, they produce very small signal levels (like noise) that are on the order of the sidelobes of the autocorrelation function. Consequently, they do not interfere with the detection of the peak of the autocorrelation function of the target receiver. In this manner each user is only a source of noise for the detection of other users signal even though all users occupy the same bandwidth at the same time.

Figure 3.26 shows an example of CDMA operation for two users with different key codes. The actual symbol transmission rate or bandwidth of each user is R_s Hz, but when their spectrum is spread N times, the transmission bandwidth is spread to $W = N R_s$ Hz. In this figure the top code belongs to the desired user and the receiver multiplies that with its key code. The key code returns the transmitted signal to its original despread form and smaller bandwidth, but the undesired signal is not despread to its original form. It remains with the same transmission bandwidth of $W = N R_s$. As a result, the peak of the desired signal at the receiver is N times higher than the interfering signals from other users that are now only a source of noise.

As the number of users increases the multiuser interference increases for all of the users. This phenomenon continues up to a point when the mutual interference among all terminals stops the proper operation for all of them.

PN sequences have good auto-correlation properties but the cross-correlations are non-zero and cause interference to other users operating over the same channel. In order to increase capacity, CDMA systems sometimes employ orthogonal sequences. The cross-correlation between two orthogonal sequences is zero by definition and consequently, if two users employ orthogonal sequences for spreading, at the receiver, it is possible to completely separate the two signals by correlating them with the signal replicas, without interference. The problem with orthogonal sequences is that the users must be *synchronized* since orthogonal sequences do not have good

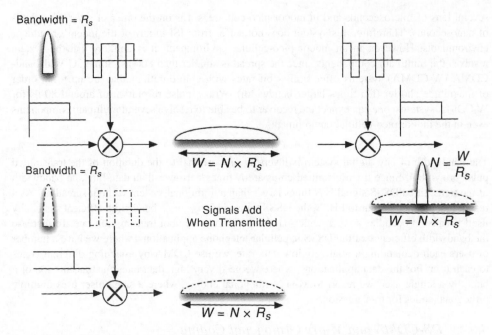

Figure 3.26   Basic concept of DS-CDMA for two users explained in frequency domain.

auto-correlation properties. Consequently, orthogonal sequences are employed on the downlink where the base station is able to synchronize transmissions directed to all users. PN sequences are preferred for spreading on the uplink.

Example 3.16: Walsh orthogonal sequences for CDMA

Walsh sequences are generated by the rows of a Hadamard matrix. A Hadamard matrix is a square matrix of the order $n \times n$ where all pairs of rows are orthogonal. The following matrix is an example of a Hadamard matrix.

$$\begin{bmatrix} 0 & 0 & 0 & 0 \\ 0 & 1 & 0 & 1 \\ 0 & 0 & 1 & 1 \\ 0 & 1 & 1 & 0 \end{bmatrix}$$

The row [0 0 0 0] corresponds to the all-zero Walsh code. Each row of the Hadamard matrix corresponds to a unique Walsh code. It is easy to construct Hadamard matrices of order $2^m \times 2^m$. The first CDMA system designed by Qualcomm used Walsh codes of order 64 generated by a Hadamard matrix of the order 64×64 to separate the data stream send from the base station to different mobile users. Details of implementation of that system will be discussed in Chapter 12 when we discuss CDMA cellular technology.

Using a set of orthogonal codes to separate different users on the downlink is commonly used in CDMA cellular telephone networks. In wireless data networks one may desire to assign a group of codes to a single user so that the data burst can be transmitted at high data rate. If all of the orthogonal codes are assigned to a single user, the transmitted system is referred to as M-ary orthogonal coding. From a different point of view, M-ary orthogonal coding is a coding technique that provides redundancy in the transmitted bit streams and as such increases the reliable of transmission. In wireless indoor data networks it can be used to increase the data rate [Pah90]. In the CDMA cellular telephone application, since it is difficult to synchronize signals on the uplink, the orthogonal codes are sometimes used differently. Instead of using them to separate the signals of users through their orthogonality, they are instead used to encode the signals thereby spreading them through the redundancy therein. This improves the reliability of the signal even under interference from other signals. An example clarifies this transmission scheme that is referred to as M-ary orthogonal coding.

Example 3.17: Walsh code for M-ary orthogonal coding

Considering the four orthogonal codes forming rows of the matrix described in Example 3.16, we can encode two-bits of data into one of the four-bit rows of the code. This way we create a redundancy in decoding process that adds to the integrity of the signal. In Example 3.16, we have four code words that are each four-bits in length. The total number of 4-bit words is 16 and therefore the selected orthogonal set only uses four out of 16 possibilities. This redundancy allows the receiver to correct some of the erroneous bits to add to the integrity of the transmission. In this rate 1/2 coding technique, if we receive one of the 12 possibilities for received sequence that are not used in our coding scheme we can find the closest possible code for that transmission and correct the error bit. This system is a 4-ary orthogonal coding system. Specific example of these transmission schemes are described in chapters 8 and 12 when we describe CDMA systems and the physical-layer of the IEEE 802.11b WLAN, respectively.

3.5.6   Comparison of DSSS, FHSS and OFDM

In this section, following [Fal96], we compare the bandwidth and power requirements of spread spectrum techniques with that of OFDM operating in a test indoor area, shown in Figure 3.27. An indoor area is considered because broadband communications for applications such as WLANs are very important. However, the overall conclusions can be extended to any other wireless fading multipath channel. The test area consists of seven rooms in the second floor of the Atwater Kent Laboratories at the Worcester Polytechnic Institute. Results of seven hundred wideband channel measurements in these rooms [How90] are used to calibrate a ray-tracing algorithm [Yan94] that is then used to generate several hundreds of thousands of channel multipath profiles in this indoor area. These profiles are then used for performance evaluation of different modem design technologies operating in this area. The basic modulation for all techniques was QPSK and acceptable performance for a modem was considered to be an error rate that is better than 10^{-5} in 99% of the locations in the area. The purpose of this exercise was to examine the relationship between bandwidth and power requirements to the maximum data rate of each technique in a realistic indoor area with a considerable amount of multipath.

First we assume a fixed bandwidth of 10 MHz and we compare the minimum required radiation power for each transmission technique to cover the test area. This part will address the important issue of power consumption for battery operated mobile terminals. It was found that with a maximum transmission power of 100 mW, DSSS, FHSS, and OFDM transmission techniques considered are able to cover the test area. If we assume the transmission power is maintained at 100 mW and there is no restriction on the bandwidth, we can determine the maximum data rate that can be achieved with any of the transmission techniques in the test area. This exercise addresses the transmission technologies in the context of demand for higher data rate that has been an extremely important factor in the evolution of the WLAN industry.

Example 3.18: Comparison between DSSS, FHSS, and OFDM

Figure 3.28 shows the minimum radiation power required to achieve 90% coverage of the test area shown in Figure 3.27 for a 10 MHz channel using DSSS with a four-tap RAKE receiver and processing gains of 15 (bars in the middle), as well as FHSS (bars on the left) and OFDM with 15 carriers (bars on the right). There are two bars for each technology representing the power requirement and the data rate supported with that technology. Therefore, using this figure we can find the power requirements and supported data rate for the three different techniques in a fixed bandwidth (10 MHz) environment. From this figure we can draw a number of practical conclusions.

In 10 MHz of bandwidth, the DSSS system with a processing gain of 15 and a 4-tap RAKE receiver, using time diversity, needs −9 dBm to cover the area with a data rate of 1.33 Mbps. The FHSS system with 15 hopping frequencies, taking advantage of frequency diversity, provides a 1.33 Mbps system with −9 power consumption. Since we have a processing gain of 15, the symbol transmission rate is $10/15 = 0.665$ MSps. With two bits per symbol for QPSK modulation, the data transmission rate for both systems is 1.33 Mbps. The DSSS scheme consumes less power because it isolates the multipath components and with a 4-tap RAKE receiver, takes advantage of the in-band multipath diversity. In comparison, with OFDM operating in the same 10 MHz bandwidth, the data rate is 13.3 Mbps because here we have separated the carriers 1.5 times more than with FHSS. This is necessary for reduction of adjacent channel interference in the OFDM system in which all

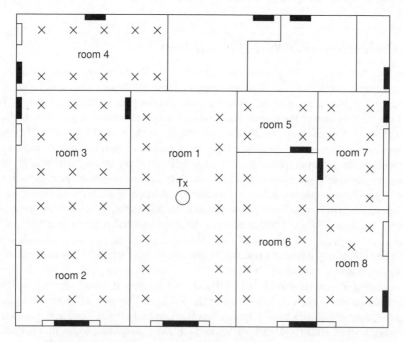

Figure 3.27   Indoor test area in WPI's Atwater Kent Laboratories.

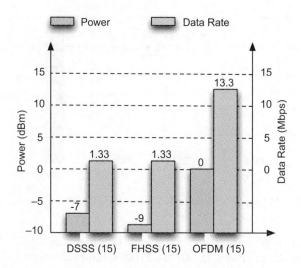

Figure 3.28   Power requirements of different transmission schemes.

carriers are sending data at the same time. The power consumption of the OFDM system is 9 dB more than DSSS and 7 dB more than the FHSS system.

The above example leads us to the conclusion that for fixed bandwidth channels OFDM provides higher data rates at the expense of higher power consumption. The spread spectrum systems provide a better coverage at the expense of lowering the operating data rate.

The quantitative study of radiated power consumption is important for two reasons:

1. Given the restriction on the maximum radiated power by the frequency administration agencies we need to know how the coverage of various transmission techniques compare with one another.
2. Considering the increasing expansion of the market for battery-operated WLANs, we need to examine the total power consumption of a modem.

The results of Figure 3.28 are extended to a variety of modem design technologies in [Fal96] and they can be used directly to analyze the coverage of various transmission techniques in a typical indoor area. This allows a designer or a standards regulator to justify the selection of a particular transmission technique for a specified system description. Since the power consumption varies substantially with the design and fabrication technology used in the implementation of a WLAN, the designer can use these results with their estimation of the electronic power consumption for their own implementation of a system.

Figure 3.29 presents the maximum attainable data rate for different transmission techniques in the test area shown in Figure 3.27. The transmission power is maintained at 100 mW and there is no constraint on the bandwidth; a situation similar to most unlicensed bands where several hundreds of MHz of bandwidth is available for implementation of a broadband service. In this figure, the maximum achievable data rate and the required bandwidth for each technology are indicated on two bars. In a manner similar to Figure 3.28, a number of interesting practical conclusions can be drawn from this figure as well.

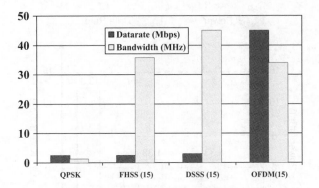

Figure 3.29   Maximum achievable data rates with different transmission schemes.

An OFDM system, using frequency diversity, with 15 carriers can achieve data rates close to 40 Mbps in a bandwidth of around 35 MHz. A FHSS scheme achieves the same data rate as a QPSK modem (2.4 Mbps) with a bandwidth consumption that is 15 times larger (36 MHz). The DSSS scheme using a RAKE receiver can increase the data rate slightly to 3 Mbps using 45 MHz of bandwidth.

From the above example one can observe that by appropriate selection of the bandwidth of the individual carrier one may compensate for the effects of multipath fading for the system. As a result, if there is no power or bandwidth constraint one can achieve any data rate with an OFDM modem. This property is not shared by other techniques where the increase of data rate finally reaches to a point that the effects of frequency selectivity of the channel are dominant and an increase in the transmit power is no longer effective. The restriction in OFDM is implementation complexity that increases with an increase in the number of carriers. In the practical band- and power-limited applications, the above performance is expected to improve even further if a smart frequency diversity receiver uses multirate transmission or exploits measurement of the channel characteristics to adjust the power in different carriers. Because of these features OFDM transmission is dominating the wireless data communication for Wi-Fi and 4G cellular data services where maximum achievable data rate is more important than power consumption.

Example 3.19: Data rates with OFDM

Obviously, spread spectrum provides lower data rates than OFDM. However, bandwidth efficiency is only one side of the equation; the processing gain that spread spectrum provides can serve to reduce the transmitter's power requirement. This is an important consideration if the wireless network is being used to connect portable, battery-operated sensors. Because of these features, spread spectrum technology is commonly adopted in low power ad hoc wireless sensor networking technologies such as Bluetooth or ZigBee.

3.6   Coding Techniques for Wireless Communications

Coding of bits is a common technique that is employed in wireless networks for a variety of reasons. Indeed, we have already mentioned Manchester coding for baseband signaling as well as PN-sequence coding and Walsh codes for CDMA and M-ary orthogonal coding applications. The traditional reason for coding is error control. Codes are also employed to convert voice into bits. The perceived quality of voice often depends on the speech coding employed, and there is a tradeoff between this quality and the bandwidth requirement for digitized voice.

In Section 3.5, we discussed a variety of diversity techniques that essentially employ *redundancy* in time, frequency or space to improve the reliability of reception of transmitted data. In some sense, error control coding is also a diversity scheme since it introduces redundancy in the transmitted bits to correct errors that may be introduced by a channel and if correction is not possible, to provide the capability to detect the occurrence of errors.

Error control coding, as the name suggests, is a technique to *code* the transmitted bits to control the error rate. This becomes increasingly important in radio communications because of the harsh channel conditions. Errors in wireless channels usually occur in bursts. That is, a string of data bits are subject to fading or other harsh impairments like interference, resulting in several consecutive bits (up to 50% of the bits in the burst) arriving at the receiver in error. This is in contrast to wired communications where errors usually occur at random, a bit at a time. Consequently, error control coding schemes are different for wired and wireless channels.

Error control coding is also dependent upon the application under consideration. Voice packets can usually tolerate error rates as high as 1 in 100 bits or 10^{-2}. Such error rates are generally unacceptable for data packets and messaging systems that require error rates as low as 10^{-5}. In some cases, it is impossible to achieve such low error rates and in such cases, it is reasonable to *retransmit* the data packets that are lost. Such schemes are referred to *as automatic repeat request* schemes. In order to determine whether or not a packet (or a block of data bits) has been received in error, *block-coding* schemes are employed and the process is called error detection. Block codes can also be used to correct errors and this is called *forward error correction*. Another coding scheme that can be used for forward error correction is convolutional coding that employs some memory of previously transmitted bits to determine a most *likely* sequence of transmitted bits. Powerful turbo-coding schemes have been developed in the last decade that further improves upon the performance of block and convolutional codes. Space–time coding was another significant technique for use in multiple antenna systems. Coding techniques are also used for block interleaving, scrambling and speech coding used in wireless netwroks. In the remainder of this section, we provide a brief overview of these coding techniques.

3.6.1   Block Codes

Block coding, as the name suggests, involves encoding a *block* of bits into another block of bits, with some redundancy to combat errors. Block coding in its simplest form consists of a *parity check* bit. An extra bit is added to each block of k bits and the extra bit is selected so that each new block of $k + 1$ bits has either an even number of ones or an odd number of ones. The extra bit is called the parity check bit. The result is that if the channel introduces a single bit error in a block of $k + 1$ bits, the number of ones in the block will no longer be even (or odd), and the receiver can *detect* the error. The simplicity of this code is clear because if there is an even number of errors in the block, the errors cannot be detected since the parity of number of ones is maintained.

Using a variety of algebraic techniques, efficient encoding rules have been obtained that calculate a set of $n–k$ parity check bits that apply parity checks to a group of bits in the block of k bits. Together with these parity check bits, the size of the encoded block is $k+(n–k) = n$ bits. The block code is called an (n, k) block code and the code rate is $r = k/n$. This means that if the raw data rate is R bps, only kR/n bps corresponds to actual data. The rest of the bits do not contain useful information and are included only for error control purposes.

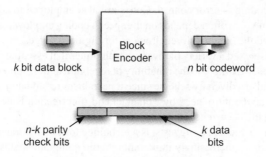

Figure 3.30   Operation of a block code.

Example 3.20: Code rate for the GSM control channel

In GSM, the most popular TDMA 2G cellular system, a block of 184 bits is encoded into 224 bits of codeword on the control channel before it is sent to a convolutional encoder. The number of parity check bits is 40. The code rate of this block encoder is $184/224 = 0.82$.

Block codes use finite-field arithmetic based properties to encode and decode blocks of bits or symbols. Most operations are based on linear feedback shift registers that are easy and inexpensive to implement. Most block codes are created in *systematic* form where the k data bits are retained as is and the $n - k$ parity check bits are either prepended or appended to them (see Figure 3.30). The parity check bits are generated via a generator matrix or generator polynomial. The encoded block of n bits is called the *codeword* and this is transmitted over the channel. Codes generated by a polynomial are called cyclic codes and block codes of this nature are called *cyclic redundancy check* codes and are employed in a variety of data transmission schemes for error correction and detection.

The received word may be identical to the codeword in the case of error-free transmission or may have been modified due to channel errors. The modifications may result in another valid codeword, in which case, it is not possible either to detect or correct the errors. The probability of such a false detection is upper bounded [Wol82] by:

$$P_{FD} \leq 2^{-(n-k)} \tag{3.15}$$

The idea behind the design of block codes is to have a large *distance* between any pair of codewords. This distance is measured in terms of the number of positions (bits or symbols) in which the codewords differ, and is called the Hamming distance. The *minimum* Hamming distance between the set of all codewords of a block code determines its error detection or correction capability. A block code with a minimum distance of d_{min} can detect blocks of errors that have a "weight" of less than d_{min} and can correct blocks of errors that have a weight up to t_{max} where:

$$t_{max} = \left\lfloor \frac{d_{min} - 1}{2} \right\rfloor \tag{3.16}$$

Here, $\lfloor x \rfloor$ refers to the largest integer less than or equal to x. An error block is also represented in a manner similar to a block of data bits. Those bits that are not changed are represented by zeros and those that are, are represented by ones. The weight of the error block corresponds to the number of ones in the block, or the number of bits that are changed and thus in error. Intuitively, we can see why a block code with a minimum distance of d_{min} can correct up to t_{max} errors. Given two codewords in the set, the distance between them is greater than or equal to d_{min}. An error block modifies a codeword into the received word. If its distance from the correct codeword is less than half the distance between the correct codeword and any other codeword, we can associate the received word as being *closest* to the original codeword, and correct it accordingly. If however, the error block modifies the received word to make it closer to some other codeword, the error correction procedure will not work.

For sensitive data transfer, it is still possible to *detect* errors with weights larger than t_{max}. The detection of errors is performed by determining whether or not the received word is a valid codeword. This can be done by computing the parity check bits again from the k data bits and comparing it with the parity check bits received over the channel. It could be possible that the data bits were received correctly but the parity check bits were in error, but the two cases are indistinguishable.

3.6.2   Convolutional Codes

Convolutional codes unlike block codes do not map individual blocks of bits into blocks of codewords. Instead, they accept a continuous stream of bits and map them into an output stream introducing redundancies in the process. Usually, a code rate can be defined for convolutional codes as well. If there are k bits per second input to the convolutional encoder and the output is n bits per second, the code rate is once again close to k/n. The redundancy is however dependent on not only the incoming k bits, but also several of the preceding k bits. The number of the preceding k bits used in the encoding process is called the constraint length m that is similar to the *memory* in the system.

Example 3.21: Convolutional coding in wireless systems

In the Qualcomm CDMA network, a convolutional encoder is employed in both the forward and reverse links. In the forward link, a rate 1/2 convolutional encoder is used that has a constraint length of $m = 9$. On the reverse link, a rate 1/3 convolutional encoder is employed with the same constraint length.

GSM, the 2G time multiplex cellular system also employs a convolutional encoder. Digitized voice is broken up into 182 class I bits and 78 class II bits. The most significant 50 bits of the class-I bits is enhanced with a block code that adds three parity bits. The sum total of 182 class I bits plus the three parity bits, puls four tail bits (189 bits in all) are passed through a rate 1/2 convolutional coder to produce 378 bits. The 78 class II bits are added to these 378 bits to produce 456 bits of encoded data.

In general, convolutional codes are more powerful than block codes in terms of forward error correction, but are not useful for error detection or automatic repeat request schemes. At the receiver, forward error correction is performed using a maximum-likelihood decoding algorithm that determines what sequence was most likely transmitted given the received sequence of bits. The Viterbi algorithm is the most common algorithm of all and several implementations of this algorithm are available [Pro08].

Most receivers simply decide whether a received bit is a zero or a one and send this information to the channel decoder. The decoder employs its knowledge of the coding scheme to either detect or correct errors at this stage. Such a procedure is called *hard decision*. In soft decision decoding schemes, the receiver will convert the received signal into several *levels* of output. Usually there are Q quantized levels, where Q is larger than the number of alphabets. For example, in a binary system, there may be 8 levels of quantized demodulator outputs instead of the usual two levels with hard decision. The decoder will use the additional information now available in order to make a decision on the received block. The Viterbi algorithm can be used for soft decisions in both convolutional encoding and block coding techniques.

3.6.3   Turbocodes and Other Advanced Codes

Over the last decade, error control codes have been designed that can reduce the error rates significantly even at low SNRs. Some of these codes like the low-density parity check codes have been available for several decades, but they have implementation complexity issues that has precluded them from being used in communication systems till very recently. Others such as turbo-codes have made their way into 3G and 4G cellular telephone standards.

Turbo-codes make use of concatenation of truncated convolutional codes. The incoming data stream is encoded using a truncated convolutional code. It is scrambled and then encoded using another convolutional code. The idea is that bits that are lost with low probability will get additional redundancy because of the scrambling prior to encoding. Turbo-codes employ iterative decoding and soft decisions to improve performance. One of the drawbacks of turbo-codes is that they display an error floor once a certain SNR per bit is reached. This is typically on the order of 10^{-5} or so, which is still useful for practical applications.

3.6.4   Space–Time Coding

Space–time coding (STC) techniques are used for wireless communication systems with multiple transmit antennas and a single or multiple receive antennas. STC techniques are realized by introducing temporal and spatial correlation into the signals transmitted from different antennas. In traditional multiple antenna systems with or without MIMO features the same transmitted stream of data in time is sent and received over all the antennas. The transmitter and the receiver antenna gains are adjusted to optimize the performance of the detected received data stream. Therefore, benefits of those system are achievd through manipulation of the signal propagation in the space. In the STC, the transmitted stream is coded into different streams and each antenna carries one of these different streams of data. The decoder at the receiver combines different received streams according to decoding scheme to reconstruct the original stream of data. This way we add time diversity to the the existing spatial diversity of the systems with multiple antennas and that is why they are referred to as space-time coding techniques.

Using STC does not require increasing the total transmitted power or transmission bandwidth. The overall diversity gain of the STC technique results from combining the time diversity obtained from coding with the space diversity obtained from using multiple antennas. In wireless networks, a number of antennas can be deployed in the access point or base station while the mobile terminal's receiver usually has one main antenna with some possible support from other antennas depending on the size of the terminal. In traditional multiple-antenna access point or base station systems all transmit antennas carry the same signal and the signal received at a receiver antenna is the summation of the received signals from different transmit antennas. The mobile station combines the diversified signal from different antennas to optimize the performance. The basic principle of STC is to encode the transmitted symbols from different antennas at the base station and modify the receiver to take advantage of the space and time diversity of the arriving signal from multiple transmitter antennas. Using STC at the base station, we can improve the performance of the downlink (base to mobile) channel significantly to support asymmetric applications such as Internet access, where the downlink data stream operates at a much higher rate than does the uplink data stream.

Using STC, significant increases in throughput over single antenna system are possible with only two antennas at the base station and one or two antennas at the mobile terminal. It can be implemented like block [Ala98] or convolutional codes [Nag98; Tar98] with simple receiver structures. The basic concept of STC with the simple two transmit and one received antenna block coding system as well as a simple MIMO scheme for two transmit and two received antennas is described in [Ala98]. The simulation results show that the performance is identical to a system that uses maximal ratio combining with one transmit and four receiver antennas. In other words, Alamouti has shown that with his simple block coded STC with two transmit and one and two receiver antennas, one can obtain the same diversity performance as achieved with optimum MRC with two and four receiver antennas. More elegant approaches that combine transmit diversity with channel coding, similar to Trellis Coded Modulation (TCM), are also available in [Nag98; Tar98]. A good overall overview of STC and its applications is provided in [Dha02].

3.6.5   Automatic Repeat Request Schemes

In voice networks, if a packet is received in error, it is either dropped or replaced with an attenuated version of the previous packet to provide a semblance of continuity and retransmissions of damaged packets are not done because of the sensitivity of voice to delay. Automatic repeat request or ARQ schemes essentially are used in data networks where reliability of received information is of paramount importance and delay is less of a problem compared to real-time multimedia applications. If a block of data is received in error, the receiver requests retransmission of the block of data. This request may be explicit or built into several protocols already operational in the system for flow control or other purposes. Usually, an acknowledgement packet is employed to indicate correct reception of one or more transmitted packets. If an acknowledgement is not received in a certain time frame, or a negative acknowledgement is received, the transmitter will retransmit the packet. Such mechanisms are commonly employed with random-access protocols discussed in Chapter 4.

There are three basic ARQ schemes. The stop and wait ARQ scheme waits for an acknowledgement for each individual packet before sending the next one. This is especially inefficient if the round-trip times are large because the transmitter spends a lot of time waiting for the acknowledgement. In order to improve upon this scheme, the Go-back-N ARQ scheme transmits up to N packets at a time and waits for acknowledgements. Multiple packets can be acknowledged with one response. Depending on the receipt of acknowledgements, the transmitter will back up to the last correctly received packet and retransmit the following ones (N or less packets). It is possible that some of the subsequent packets are received correctly, but they will be discarded. In order to eliminate this inefficiency, a selective repeat ARQ scheme can be employed. Here, only those packets that are received in error are retransmitted.

Example 3.22: Acknowledgements and retransmissions in IEEE 802.11

In IEEE 802.11, every transmitted packet is acknowledged because the channel is unreliable. It is similar to a stop and wait protocol in that sense. Because round-trip times are small, and both

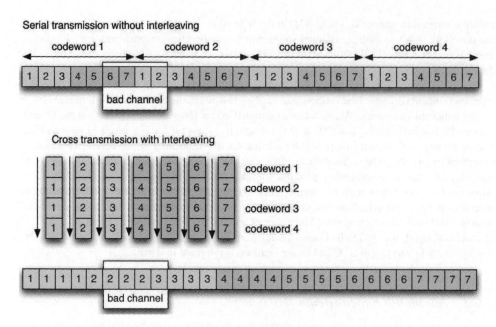

Figure 3.31 Block interleaving.

the access point and the mobile stations share the channel, the inefficiency is limited. It is often possible to piggyback the acknowledgements over data packets transmitted in the other direction.

3.6.6 *Block Interleaving*

Block interleaving is a technique used in wireless systems to spread the errors out over a large number of codewords. For instance, consider a Hamming code that can correct single bit errors over codewords of size seven bits. This means that if there is a single error over a block of seven bits, the coding scheme can correct it. However, a burst of five errors cannot be corrected by this code. If we can however spread these errors over five codewords so that each codewords "sees" only one error, it is possible to correct each of the errors. The way this works is shown in Figure 3.31. Codewords are arranged one below the other and bits are transmitted vertically. At the receiver, the codewords are reconstructed and the bits are decoded horizontally. Since the burst of errors affects the serially transmitted vertical bits that are spread over several codewords, the errors can be corrected. Block interleaving introduces delay because several codewords have to be first received before the voice packet can be reconstructed. There is only so much delay that is acceptable for normal voice conversations and the interleaving process should not create an unacceptable value of delay in the process.

Example 3.23: Block interleaving in 2G Cellular

In GSM, a 2G cellular technology, the output of the convolutional encoder consists of 456 bits for each input of 228 bits. The 456 bits are split into eight blocks of 57 bits each. The 57 bits are spread over eight frames so that even if one frame out of five is lost, the voice quality is not affected.

The delay in reconstructing the codewords corresponds to the reception of eight frames that takes 37 ms. A delay of less than 100 ms is usually tolerable for voice conversations.

3.6.7 Scrambling

Scramblers are used in older wireless networks to provide some resilience to eavesdropping when data is broadcast over the air. This way, for example, a cellular network provider can protect the customer's data on the air and a broadcast satellite service provider can make sure that only its subscribers can use their channels. Another example application for scramblers is to avoid transmission of large sequences of similar bits, either "0"s or "1"s. A scrambler in this context is different from encrypting the data to make it unintelligible by unwanted users. In this context, a scrambler replaces the data stream by another stream removing the possibility of occurrence of strings of similar bits not suitable for synchronization process at the receiver. However, they are very useful in randomizing the data in particular when nothing is transmitted and the transmission medium is idle. To ensure reception of data with adequate transitions in the stream, as we explained in previously, line coding techniques are used.

A scrambler is a pseudo randomizer device that manipulates a data stream before transmission. The manipulations are reversed by a descrambler at the receiver to recover the original data stream. Simple scramblers are implemented using linear feedback shift registers (similar to those shown in Figure 3.20) and the implementation of the scrambler and the descramblers are very similar. An example will further clarify the basic implementation of a scrambler and a descrambler.

Example 3.24: Scrambler and Descrambler Implementation

Figure 3.32 represents a typical scrambler and descramblers implemented with shift register delay lines and binary adders. Figure 3.33 represents the general block diagram of the IEEE 802.11g transmitter with the above scrambler placed before the convolutional encoder. Block interleaving is performed after forward error correction and before data is sent to the OFDM modem.

3.6.8 Speech Coding

Encoding analog voice into digital format has received much attention since it influences not only the quality of voice, but also the performance and capacity of the system. The important parameters of a speech code are the transmitted bit rate, the speech quality, the robustness in the presence of transmission errors, and complexity of implementation. Speech quality is usually subjective and is determined by the "mean-opinion-score" or MOS [Jay84]. A low rate speech coder will essentially require less bandwidth for transmission (which is beneficial in wireless environments) but usually compromises on the quality of speech.

There are two bit rates associated with a speech coder: the uncoded or raw bit rate, and the encoded bit rate to account for error correction. Where the bandwidth efficiency is not the most important criterion and the voice quality is, a higher rate encoder scheme is employed for speech coding. Resistance to channel errors is also an important issue. Voice codes may give a poor performance when the error rates are as large as 10^{-2}. This is the reason why some error control coding is applied with low rate voice coders. The voice compression scheme removes redundancies in digitized voice, and the error-coding scheme introduces some structured redundancy to provide better performance. Block interleaving techniques are sometimes used to improve performance.

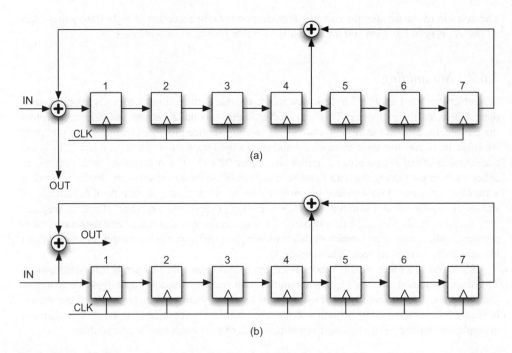

Figure 3.32 Implementation of a typical (a) scrambler and (b) descrambler, using shift register codes.

The important speech coding techniques include waveform encoding techniques like pulse code modulation, model-based speech coders, regular-pulse excitation and code-excite linear predictive techniques, and hybrid schemes. The complexity of implementation is low in the case of waveform encoding schemes and the quality of voice is extremely high. The bit rates are also correspondingly larger making them unattractive for wireless applications. Linear prediction coding techniques can provide good voice quality at bit rates as low as 2400 bps compared to the 64 kbps rate of pulse code modulation at the expense of burdensome computation. GSM uses a version of regular-pulse excitation that has an acceptable implementation complexity and delay. It operates at 13 kbps and utilizes a speech frame that lasts for 20 ms. For still further low bit rates, the quality of GSM coder voice is not adequate and code excited linear prediction techniques are preferred. A version of this coding technique is used by QUALCOMM in their original CDMA network to achieve voice-coding rates of 8 kbps and less. A detailed table of sample speech coding techniques and their relative data rates is available in Chapter 4 of [Pah09].

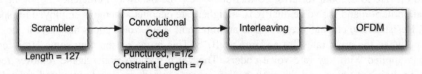

Figure 3.33 Overall block diagram of the IEEE 802.11g using a scrambler to randomize data, convolutional codes for FEC, and block interleaving to spread blocks of errors.

3.7 Cognitive Radio and Dynamic Spectrum Access

Receiver techniques discussed so far have considered traditional ways of implementation. With the emergence of a large number of standards, software implementation of the mobile terminal that can dynamically adapt itself with time to the radio environment in which it is located is an attractive solution. This concept is generally referred to as *software radio* [Bur00]. Software radio provides the impetus for fast rollout of new services; mix-and-match services offered by a variety of standards, provide choice to the customer, and increase the hardware lifetime of mobile terminals. In the literature, software radio has several definitions that include: (a) a software controllable and flexible transmitter/receiver architecture, (b) replacement of radio functionality by signal processing as much as possible, (c) the ability to download an air–interface architecture and dynamically reconfigure the user terminal, (d) multi-mode or multistandard support, and (e) a transceiver that can define the frequency bands, the modulation and coding schemes, radio resource and mobility management, and the user applications to be used in the software. DSP technology and reconfigurable hardware technology is driving efforts towards actual implementation of software radio.

From a mobile terminal's point of view, software radio needs to have limited circuit complexity, low cost, low power consuming, and have a small form factor. Ideally, the analog components of a software radio are limited and most of the radio functionality should be implemented digitally to enable software reconfigurability. However, Analog to Digital conversion at RF is extremely difficult and instead a programmable downconverter appears promising. However the limited bandwidth of the converters, the jitter introduced by the digital operations, and intermodulation products are problems with the sampled signal that are yet to be completely addressed. The processing power becomes an issue of significance in mobile terminals operating via battery power. Special-purpose DSPs, that are required for real-time computation, are expensive and complex.

For adaptive reconfigurability several solutions have been considered. Some of them are as follows. Each manufacturer could have proprietary software for a variety of hardware platforms. This provides the ability to differentiate products in the market, but creates a problem for network operators especially when mobile terminals are required to "download" the air–interface. A standard hardware platform would eliminate the numerous proprietary solutions, but would restrict the differentiability of products. A third solution proposed includes a real-time compiler that would compile a common source code into solutions for different hardware platforms. In [Bur00], Java is suggested as a language for implementing the third option. This is because Java already possesses the ability to have a uniform "bytecode" for all hardware platforms requiring only an interpreter for the bytecode for each platform.

Connecting the terminal to a PC, by using smart cards, or over the air, could do the download of the air-interface. Using a PC is not a feasible solution especially while the user is on the move. Potentially, smart cards could provide a fast solution for changing the air–interface, but there are technology limitations as of today. Over the air downloads is preferable and requires no effort by the user and intelligent updates are possible. In this case, there is a suggestion of a *universal control channel* for accessing the radio personalities over the air. The problems with such a solution are the security of such a download, the possibility of the radio channel introducing errors during the download, delay and slowness of the download procedure, and the need for protocols, resources, and bandwidth to assist the procedure.

Appendix A3

MATLAB code for Example 3.3

```
gb_db = linspace(0, 15, 100);
```

```
gb = 10.^(gb_db/10);
pe_0 = 1e-5;

% m/? = 1
a = 0.5;
b = 1;
pe = a * erfc (sqrt(b * gb));
%pe_bpsk_fading = 1 ./(2.*gb);
pe_bound = a * exp (-b * gb);
gb_0 = (1/b) * (erfinv(1-(pe_0/a)))^2;
gb_bpsk_db = 10 * log10(gb_0);
figure(1)
semilogy(gb_db, pe)
hold on
grid off
semilogy(gb_db, pe_bound)
%semilogy([0 15], [1e-5 1e-5],'r')
%semilogy([0 15], [1e-3 1e-3],'r')
axis([0 12 1e-7 1e-1])
set(gca, 'XTick', [0:1:15])
xlabel('SNR/bit (dB)')
ylabel('BER')
```

Questions

1. How does number of symbols, symbol transmission rate, bandwidth, and data rate relate to one another?
2. How do we implement a system to carry the symbols of a 2D signal constellation over a wireless media?
3. Why bandwidth and power efficiency is so important for wireless networking?
4. How do data rate and power in a physical layer implementation relate to one another?
5. Why is the Shannon-Hartley equation important for understanding transmission techniques?
6. What is the additional power for transmission of one additional bit per symbol in a QAM system?
7. In a multipath channel, explain how inter-symbol interference (ISI) is formed.
8. Why is out of band radiation an important issue in designing modulation schemes for cellular telephone networks?
9. Use Figure 3.5 to explain the error rate pattern in a wireless fading channel.
10. Explain why: In Figure 3.7, for the same bit error rate, we need a much higher signal to noise ration in the case of the fading channel.
11. Use Figure 3.19 to explain why diversity techniques are so effective in improving the performance over fading channels.
12. Differentiate between frequency hopping and direct sequence spread spectrum.
13. How can we increase the bandwidth efficiency of a spread spectrum system when it is used for a cellular telephone application? Explain why.
14. How can we increase the bandwidth efficiency of a spread spectrum system when it is used for a wireless data application? Explain why.
15. Name four diversity techniques.
16. Explain how spread spectrum receivers can exploit multipath diversity using RAKE receivers.
17. What is the difference between a MIMO system and a traditional system using multiple antennas to obtain space diversity?

18. For a fixed given bandwidth, which transmission technique among OFDM, DSSS, and FHSS provides the highest data rate and which one consumes the minimum power?
19. For a fixed transmission power and unlimited available bandwidth which transmission techniques among OFDM, DSSS, and FHSS provides the highest data rate?
20. Differentiate between block codes and convolutional codes.
21. What is block interleaving? How is it useful in combating the effects of fast fading?
22. What is the difference between a STC antenna system and a traditional system using multiple antennas to obtain space diversity?
23. What is a turbo code?
24. Why dynamic spectrum access has attracted more attention in cellular telephone applications?

Problems

Problem 3.1

The Wi-Fi devices use multiple modulation techniques in the same transmission bandwidth to provide different data rates. When the mobile terminal is close to the access point a 64-QAM modulation is used and as modem goes to the edges of the coverage of the access point a BPSK modulation is used that requires substantially lower received signal strength to operate.

a. If the data rate for the BPSK users is 6 Mbps and for this data rate the modem uses a rate $r = 1/2$ convolutional code, what is the symbol transmission rate of the system?
b. If the 64-QAM modem uses the same symbol transmission rate and the user data rate is 54 Mbps, what is the coding rate for the modem?

Problem 3.2

a. Determine the average energy per symbol of the BPSK and 64-QAM and from that give the difference between the received signal strength requirement of the two modulation techniques in the context of Wi-Fi application explained in Problem 1.
b. If the coverage with *64-QAM* is D meters, what is the coverage with a BPSK modem when we operate in a large indoor open area with a distance–power gradient of $\alpha = 2.$?
c. Repeat (b) for an indoor office area with a distance–power gradient of $\alpha = 3.5$.

Problem 3.3

Use Shannon–Hartley bounds to determine the maximum data rate achievable over a channel with a bandwidth of 150 MHz and minimum signal to noise ratio requirement of 10 dB.

Problem 3.4

a. Plot the signal constellations for 16-PSK and 16-QAM modulations.
b. Determine the average energy in each constellation, E_s, as a function of the minimum distance between the points in the constellation, d_{min}.
c. Starting with Equation (3.3):

$$P_s \approx \frac{1}{2} erfc \left(\frac{d_{min}}{\sqrt{N_0}} \right) \geq \frac{1}{2} e^{-\frac{d_{min}}{\sqrt{N_0}}},$$

give an approximate equation for calculation of error rate of each of these modems which relate the probability of the symbol error P_s to the average energy in the constellation, E_s, and the variance of the background noise, N_0.

d. Plot the probability of symbol error versus signal to noise ratio of the two modulation techniques. Use Figure 3.4 as a guideline for the plotting.

e. Comment on the results. Which modulation scheme is more power efficient? What is the advantage of using the other one?

Problem 3.5

a. Determine and plot the error rate curves as a function of the SNR per bit for 4-, 16-, and 64-QAM.

b. Determine the SNR per bit required to obtain an error rate of 10^{-5} for each of the above modulation schemes. You can use the plots or the MATLAB erfinv function for your calculation.

Problem 3.6

a. Give the symbol error rate of the QPSK as a function of the signal to noise ratio and plot the error rate versus signal to noise ratio in dB.

b. Repeat (a) if we have intersymbol interference (ISI) of 5, 10, and 20%.

c. What is the maximum achievable error rate for the QPSK without ISI and with each level of ISI?

Problem 3.7

a. Give the symbol error rate of the 8-PSK as a function of the signal to noise ratio over a channel disturbed only by white Guassian noise (WGN).

b. Plot the error rate versus signal to noise ratio in dB. Give the signal to noise ratio requirement if the error rate is 10^{-5}.

c. Give the average symbol error rate of the 8-PSK as a function of the average signal to noise ratio over a flat fading channel.

d. Plot the average error rate versus the average signal to noise ratio in dB. Give the average signal to noise ratio requirement if the average error rate is 10^{-5}.

e. How much additional average signal power in dB is needed so that the error rate in WGN becomes the same as the average error rate over the fading channel?

Problem 3.8

a. Give the symbol error rate of the BPSK as a function of the signal to noise ratio over a channel disturbed only by white Guassian noise (WGN).

b. Plot the error rate versus signal to noise ratio in dB. Give the signal to noise ratio requirement if the error rate is 10^{-5}.

c. Give the average symbol error rate of the BPSK as a function of the average signal to noise ratio over a flat fading channel, when the system has integrated two orders of independent diversity.

d. Plot the average error rate versus the average signal to noise ratio in dB. Give the average signal to noise ratio requirement if the average error rate is 10^{-5}.

e. How much additional average signal power in dB is needed so that the error rate in WGN becomes the same as the average error rate over the fading channel?

Figure P3.1 A sample differential Manchester coded waveform.

Problem 3.9

In the differential Manchester coded signal, shown in Figure P3.1, information is coded in transitions between the bits. A "low-to-high" means next bit is a "0" and a "high-to-low" means "1".

a. Show the beginning and the end of each bit.
b. Identify all the bits in the data sequence.

Problem 3.10

This problem illustrates the concept of using orthogonal waveforms in CDMA. Figure P3.2 shows the waveforms used by three users, Al, Bill, and George, to send messages from a wireless text transmitter device to a base station. The chip duration is 100 ns and the bit duration is 400 ns. Each of them is transmitting eight bits each, corresponding to the ASCII codes for A, B, and G respectively. They transmit their waveform if the ASCII bit is a zero, and the negative of the waveform if it is a one.

a. Draw the signals corresponding to their ASCII code transmitted by each of them.
b. Draw the composite transmitted signal that will be the sum of the individual signals if they are transmitting at the same time and are synchronized perfectly.

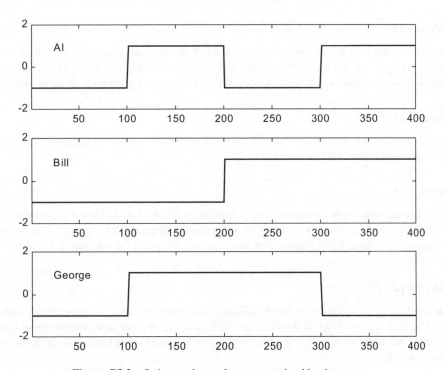

Figure P3.2 Orthogonal waveforms transmitted by three users.

c. Let us refer to the transmitted signal in (b). We shall call the part of the signal during the durations 0–400 ns, 400–800 ns, and so on, as "symbols". Compute the cross-correlation at zero lag of each of the symbols in the transmitted waveform with the waveforms of Al, Bill, and George. Comment on the results.

Problem 3.11

Assume instead of the waveforms shown above, Al, Bill, and George use the following waveforms:

$$\text{Al: } \cos(\pi t/T), 0 \le t \le T; \text{ Bill: } \cos(2\pi t/T), 0 \le t \le T; \text{ George: } \cos(3\pi t/T), 0 \le t \le T$$

Draw these three waveforms for $T = 1$ μs. Show that they are orthogonal. Draw also the spectrum of the three waveforms. Note that these waveforms are at baseband. If a single user is transmitting all the three waveforms on a single carrier, we have OFDM with three sub-carriers.

Problem 3.12

In MATLAB, the function conv performs the convolution of two vectors. Samples of a signal can be represented as vectors (as in the case of spread spectrum pulses). Suppose that the M-sequence [1 −1 −1 1 1 1 −1 1 1 1 −1 −1 −1 −1 1] is used as the basic waveform for transmitting a zero and its negative is used to transmit a one. The matched filter will have the flipped version of the M-sequence as its impulse response, that is, [1 −1 −1 −1 −1 1 1 1 −1 1 1 1 −1 −1 1]. You convolve the input to the matched filter with this vector to get the output. Let us suppose we are transmitting four bits 0, 1, 1, 0. Assume also that the channel is a three-path channel with inter-path delays of $5T_c$ and $8T_c$ respectively. Plot the output of the matched filter.

Problem 3.13

Show that the sequences shown in Example 3.16 are orthogonal to one another. *Hint: Represent the zeros by −1s in the sequences. Orthogonality is demonstrated as follows. Multiply the sequences element-wise and sum the resulting elements. If the sum is zero, the sequences are orthogonal.*

Problem 3.14

Show the steps to generate the periodic M-sequence of period 7 from the linear feedback shift register shown in Figure 3.20.

Problem 3.15

Suppose the maximum fade duration over a radio channel is 0.001 ms. Assume that all the bits are in error when a signal encounters a fade. What is the maximum number of consecutive bits in error for a transmission through this channel if the data rate is 10 kbps? If the data rate is 11 Mbps?

Problem 3.16

Block interleaving is a solution to enable simple error correcting codes to correct long bursts of errors. For both the situations in Problem 15, determine the number of codewords over which interleaving has to be performed if the length of the codeword is seven bits and a single bit error can be corrected.

Problem 3.17

If codewords have to be received in sequence for message delivery, determine the delay encountered by the block interleaving scheme of Problem 16. How does this impact voice transmission?

Problem 3.18

Use the results of data rate versus power consumption for the central part of the Atwater Kent building (see Figures 3.28 and 3.29) to answer the following questions:

a. For 10 MHz of bandwidth, what is the power requirement and maximum data rate supported by DSSS-15 and OFDM-15?
b. What is the maximum data rate and required bandwidth for DSSS-15, OFDM-15 modulations to cover this area with 100 mW power?
c. Name three standards using DSSS, and and three using OFDM for implementation of a wireless network.

Projects

Project 3.1: Error rate and phase jitter in QPSK modulation

a. Sketch a typical QPSK signal constellation and assign the two-bit binary codes to each point in the constellation. Define the decision line for the received signal constellation so that the receiver can detect received noisy symbols from each other.
b. In Section 3.3 we discussed that the probability of symbol error for a multi-amplitude, multi-phase modem with coherent detection can be approximated by $P_s = 0.5\,erfc\,(d/2\sqrt{N_0})$, where d is the minimum distance between the points in the constellation and N_0 is the variance of the additive Gaussian noise. Use this equation to calculate the probability of error of the QPSK modems. Observe that if we consider the signal constellation and the decision lines of part (a), this equation can be modified to $P_s = 0.5\,erfc\,(\delta/8\sqrt{N_0})$, where δ is the minimum distance of a point in the constellation from a decision line.
c. Use MATLAB or an alternative computation tool to plot the probability of symbol error versus signal to noise ratio in dB. What are the signal to noise ratios (in dB) for the probability of symbol error of 10^{-2} and 10^{-3}? Let us refer to these two SNRs as SNR-2 and SNR-3.
d. Simulate transmission of the QPSK signal corrupted by additive Gaussian noise for 10,000 transmitted bits. Generate random binary bits and use each two bits to select a symbol in the constellation of part (a), add complex additive white Gaussian noise to the symbol so that the signal to noise in dB is SNR-2, and use the decision lines to detect the symbols. Find the number of erroneous symbol decisions and divide it by the total number of symbols to calculate the symbol error rate. Compare the error rate with the expected error rate of 10^{-2}.
e. Repeat (d) for SNR-3 and error rate of 10^{-3}.
f. Assume that a channel produces a fixed phase error θ. Give an equation for calculating the probability of error for a QPSK modem operating over this channel. Use the minimum distance from a decision line, δ, and the equation $0.5\,erfc\,(\delta/8\sqrt{N_0})$ for this calculation.
g. Assume that the received signal-to-noise is 10 dB, and sketch the probability of error versus the phase error $0 < \theta < \pi/4$.
h. Repeat (c), (d) and (e) for a channel with a phase shift of $\theta = \pi/8$.

Project 3.2: Error rate in fading channels for QPSK modulation

This project combines the results of Project 2.6 in Chapter 2 and Project 3.1 to analyze the performance of a QPSK modem over a fading channel.

a. Give the equation for calculation of the average error rate of the QPSK as a function of the average signal to noise ration. Plot the average bit error rate versus average signal to noise ratio for this modulation technique.
b. Use MATLAB or an alternative computation tool to plot the average probability of symbol error versus signal to noise ratio in dB. What are the signal to noise ratios (in dB) for the probability of symbol error of 10^{-2}? Let us refer to these two SNRs as SNR-2 and SNR-3.
c. Simulate transmission of the QPSK signal over a Rayleigh fading channel corrupted by additive Gaussian noise for 10,000 transmitted bits. For simulation of the channel use the results of Project 6 in Chapter 2 and for the implementation of the transmitted symbols and the detection process use the results of Project 1. Run the simulation for the average signal to noise ratios of found in part (b). Compare the error rates observed in the simulation with the expected error rates of 10^{-2}. Explain the differences between the two values.

4

Medium Access Methods

4.1 Introduction

This chapter presents an overview of the medium access methods commonly used in networks. Access methods form a part of Layer 2 of the TCP/IP protocol stack and Layer 3 of the IEEE 802 standard for local area networks that are responsible for interacting with the medium to coordinate successful operation of multiple terminals over a shared channel. Most multiple access methods were originally developed for wired networks and later adapted to the wireless medium. However, requirements on the wired and wireless media are different, thereby demanding modifications in the original protocols to make them suitable for the wireless medium. Today, the main differences between wireless and wired channels are availability of bandwidth and reliability of transmissions. The wired medium includes optical media with enormous bandwidth and very reliable transmission (with error rates very close to zero all the time). Bandwidth in wireless systems is always limited because the medium (air) cannot be duplicated and also the medium is shared between all wireless systems that include multi-channel broadcast television, and a number of other bandwidth demanding applications and services. In the case of wired operation, we can always lay additional cables to increase the capacity, as needed even if it is an expensive proposition. In a wireless environment, we can reduce the size of *cells* to increase capacity as will be discussed later in Chapter 5. With the reduction of the size of the cells, the number of cells increases, and with this, the need for improvements in the wired infrastructure to connect these cells increases. Also the complexity of the network for handling additional handoffs and mobility management increases posing a practical limitation upon the maximum capacity of the network. As far as transmission reliability is concerned, as we saw in Chapter 2, the wireless medium suffers from multipath and fading that causes serious threat to reliable data transmission over the communication link. Since the wireless channel is so unreliable, as discussed in Chapter 3, people have developed a number of signal processing and coding techniques to improve transmission reliability over the wireless channel. In spite of these techniques, the reliability of the wireless medium is lower than that of the wired medium used as the backbone of the wireless networks.

Although in practice we prefer to have the same access method and the same frame structure for wired backbone and the wireless access, wireless networks often use different packet sizes and a modified access method to optimize the performance to the specifics of the unreliable wireless medium, as the following two examples illustrate.

Principles of Wireless Access and Localization, First Edition. Kaveh Pahlavan and Prashant Krishnamurthy.
© 2013 John Wiley & Sons, Ltd. Published 2013 by John Wiley & Sons, Ltd.

Example 4.1: IEEE 802.3 and IEEE 802.11

The IEEE 802.3 standard (based on Ethernet) is the successful and dominant standard for local wired communications. Consequently, the IEEE 802.11 wireless local area network standard (see Chapter 8), under ideal circumstances, desired to use the same access method when its development started in the early 1990s. Carrier sense multiple-access with collision detection (CSMA/CD) is the protocol used in the Ethernet. However, collision detection in wireless channels is difficult for several reasons and the IEEE 802.11 standard had to resort to carrier sense multiple-access with collision avoidance (CSMA/CA) that can be viewed as a wireless adaptation of IEEE 802.3.

Example 4.2: ATM and wireless ATM

In the mid- to late-1990s ATM was perceived to be the transmission scheme for all future networking. In the mid-1990s when wireless solutions were considered, a wireless ATM working group was formed to extend the ATM short packet solution with quality of service (QoS) naturally for wireless access. The group had to make significant changes to the wireless version because ATM was designed for error free and reliable transmission over optical channels.

To avoid substantial overlap with existing literature, we use as examples the access methods used in wireless networks with justification of why and how they are employed in different wireless networks. We allude to wired networks where appropriate.

The access methods adopted by voice oriented and data oriented networks were traditionally quite different. Voice oriented networks were designed for wireless access to the Public Switched Telephone Network (PSTN) with relatively long telephone conversations as the major application exchanging several megabytes of information in a *full duplex*[1] mode. A signaling channel that exchanges short messages between two calling components sets up the call by obtaining resources (such as the link, switches, etc.) in the telephone network at the beginning of the conversation and terminates these arrangements by releasing the resources at the end of the call. The wireless access methods evolved for interaction with these networks assign a slot of time, a portion of frequency or a specific code to a user preferably for the entire length of the conversation. We refer to these techniques as centralized assigned-access methods. Data networks were originally designed for bursts of data mainly for the wireless access to the Internet, for which the supporting network does not have a separate signaling channel. In packet communications each packet carries some "signaling information" related to the address of the destination and the source. We refer to the access methods used in these networks as random-access methods accommodating randomly arriving packets of data. Certain local area data networks also *take turns* in accessing the medium as in the case of token passing and polling schemes. In some other cases, the random access mechanisms are used to temporarily *reserve* the medium for transmitting the packet. In recent years, the use of Voice-over-IP (VoIP) for telephone conversations has blurred the distinction between voice-oriented and data-oriented networks. However, differentiation between the types of access schemes, namely assigned-access and random-access, still exists.

---

[1]Duplexing refers to how communication is possible between two parties in both directions. In simplex communications, the information flow is unidirectional, in half-duplex communications, it is bi-directional, but only in one direction at a time (time division duplexing – TDD), and in full duplex, the communication is bi-directional simultaneously. For full duplex operation, two separate physical channels are usually required [like two frequency bands (frequency division duplexing – FDD) or two separate cables].

In the next two sections of this chapter we provide a short description of assigned-access techniques used for wireless access to the PSTN and random access-methods used for wireless access to the Internet respectively.

4.2   Centralized Assigned-Access Schemes

Wireless access standards and technologies for connection to the connection-based PSTN, such as cellular telephony or PCS services, use assigned-access schemes or channel partitioning techniques for the medium access control. In assigned-access schemes, a fixed allocation of channel resources, frequency, time, or a spread spectrum code is made available on a predetermined basis to a single user for the duration of the communication session. The three basic assigned multiple-access methods are frequency division multiple access (FDMA), time division multiple access (TDMA) and code division multiple access (CDMA). Circuit switched wired telephone networks started with FDMA and evolved into using TDMA in the mid-twentieth century. The choice of the multiple-access method has a great impact on the capacity and quality of the service provided by a cellular telephone network. The impact of multiple access schemes is so important in this case that we commonly refer to various voice oriented wireless systems by their channel access method that is only a part of the layer two specification of the air interface of the network.

Example 4.3: Common terminology for digital cellular systems

The global system for mobile communications (GSM) and the North American IS-136 digital cellular standards are commonly referred to as digital TDMA cellular systems and the IS-UMTS are called digital CDMA cellular systems.

In reality, different systems use different modulation techniques as well. However, as we will see in the rest of this book, the impact of the choice of access method on the capacity and overall performance of the network is much more profound in cellular wireless systems. Consequently, the system is really distinguished by its access method. As we will see in our examples of cellular networks, a network that is identified with an access-technique often uses other random or fixed assignment techniques as a part of its overall operation. However, it is identified by the access techniques employed for transferring the main information source for which the network is designed to carry.

Example 4.4: Random access techniques in cellular networks

GSM uses slotted-ALOHA (a random access method) to establish a link between the mobile terminal and the base station. It also has an optional frequency-hopping pattern that improves the system performance when there is fading of the radio signal. However, the GSM network is built for voice communications and each session uses TDMA as the access method.

Another important design parameter related to the access method is the differentiation between the carrier frequencies of the forward (downlink – communication between the base station and mobile terminals) and reverse (uplink – communication between the mobile terminal and the base station) channels. If both forward and reverse channels use the same frequency band for communications, but the forward and reverse channels employ alternating time slots, the system is referred to as employing Time Division Duplexing (TDD). If the forward and reverse channels use different carrier frequencies that are sufficiently separated, the duplexing scheme is referred to as Frequency

Division Duplexing (FDD) – see Figure 4.1 for examples. With TDD, since only one frequency carrier is needed for duplex operation, we can share more of the RF circuitry between the forward and the reverse channels. The reciprocity of the channel in TDD allows for exact open-loop power control and simultaneous synchronization of the forward and reverse channels. TDD techniques are used in systems intended for low-power local area communications where interference must be carefully controlled and where low complexity and low power consumption are very important. Thus TDD systems are often used in local area pico- or micro-cellular systems deployed by PCS networks. FDD is mostly used in macro-cellular systems designed for coverage of several tens of kilometers where implementation of TDD is more challenging.

4.2.1   Frequency Division Multiple Access

In an FDMA environment all users can transmit signals simultaneously and they are separated from one another by their frequency of operation. The *Frequency-Division Multiple-Access* (FDMA) technique is built upon *Frequency Division Multiplexing* (FDM). FDM is the oldest and still a commonly used multiplexing technique in the trunks connecting switches in the public switched telephone network (PSTN). It is also the choice of radio and TV broadcast as well as cable TV distribution. FDM is more suitable for analog technology since it is easier to implement. When FDM is used for channel access it is referred to as FDMA.

Example 4.5: FDMA in AMPS with FDD

Figure 4.2a shows the FDMA/FDD system commonly used in first generation analog cellular systems and a number of early cordless telephones. In FDMA/FDD systems, forward and reverse channels use different carrier frequencies and a fixed sub-channel pair is assigned to a user terminal during the communication session. At the receiving end, the mobile terminal filters the designated channel out of the composite signal. The early analog cellular telephone system called Advanced Mobile Phone System (AMPS) allocated 30 kHz of bandwidth for each of the forward and reverse channels. The result is a total of 421 channels in 25 MHz of spectrum assigned to each direction – 395 of these channels were used for voice traffic and the rest for signaling.

Example 4.6: FDMA in CT-2 with TDD

Figure 4.2b shows an FDMA/TDD system used in the CT-2 digital cordless telephony standard. Each user employs a single carrier frequency for all communications. The forward and reverse transmissions take turns via alternating time slots. This system was designed for distances of up to 100 m and a voice conversation is based on 32 kbps Adaptive Differential Pulse Code Modulation (ADPCM) voice coding. The total allocated bandwidth for CT-2 is 4 MHz supporting 40 carriers each using 100 kHz of bandwidth.

The designer of an FDMA system must pay special attention to adjacent channel interference, in particular in the reverse channel. In both forward and reverse channels, the signal transmitted must be kept confined within its assigned band, at least to the extent that the out of band energy causes negligible interference to the users employing adjacent channels. Operation of the forward channel in wireless FDMA networks is very similar to wired FDM networks. In forward wireless channels, in a manner similar to that of wired FDM systems, the signal received by all mobile terminals has the same received power and interference is controlled by adjusting the sharpness of the transmitter and

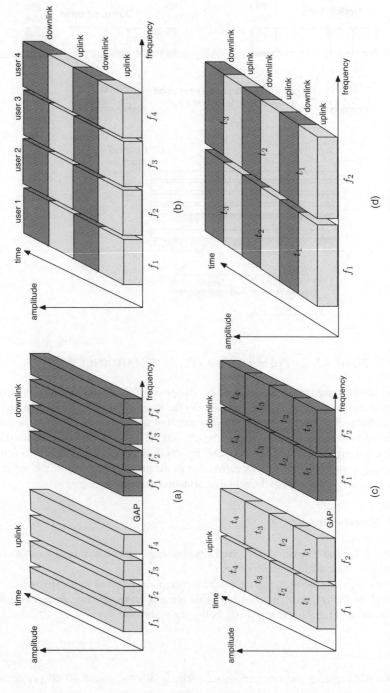

Figure 4.1   (a) FDMA/FDD. (b) FDMA/TDD. (c) TDMA/FDD. (d) TDMA/TDD with multiple carriers.

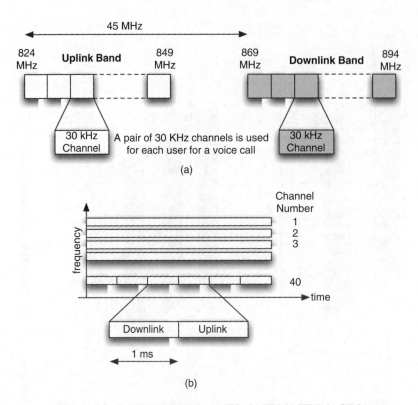

Figure 4.2   (a) FDMA/FDD in AMPS. (b) FDMA/TDD in CT-2.

receiver filters for the separate carrier frequencies. The problem of adjacent channel interference is much more challenging on the reverse channel. On the reverse channel, mobile terminals will be operating at different distances from the BS. The received signal strength (RSS) at the BS, of a signal transmitted by a mobile terminal close to the BS, and the RSS at the BS of a transmission by a mobile terminal at edges of the cell are often substantially different causing problems in detecting the weaker signal. This problem is usually referred to as the near–far problem. If the out of band emissions are large, they may swamp the actual information-carrying signal.

Example 4.7: Near–far problem

(a) What is the difference between the received signal strength of two terminals located at 10 m and 1 km from a base station in an open area?
(b) Explain the effects of shadow fading on the difference in the received signal strengths.
(c) What would be the impact if the two terminals were operating in two adjacent channels? Assume out of band radiation that is 40 dB below the main lobe.

Solution:

(a) As we saw in Chapter 2, the received signal strength falls by around 40 dB per decade of distance in open areas. Therefore, the received powers from a mobile terminal at 10 m from a BS and another, at a distance of 1 km, are 80 dB apart.

(b) In addition to the fall of the RSS with distance, we also discussed the issues of multipath and shadow fading in radio channels that cause power fluctuations on order of several tens of dBs. Therefore, the difference in the received powers due to the near-far problem may exceed even 100 dB.

(c) If the out of band emission is only 40 dB below that of the transmitted power, it may exceed the strength of the information-bearing signal by almost 60 dB.

To handle the near–far problem, FDMA cellular systems adopt two different measures. First, as we discuss in Chapter 5, when frequencies are assigned to a cell, they are grouped such that the frequencies in each cell are as far apart as possible. The second measure employed is power control that will be discussed in Chapter 6. In addition, whenever FDMA is employed, *guard bands*[2] are also used in between frequency channels to further reduce adjacent channel interference. This however has the effect of reducing the overall spectrum efficiency.

4.2.2   Time Division Multiple Access

In TDMA systems, a number of users share the same frequency band by taking assigned turns in using the channel. The *Time-Division Multiple Access* (TDMA) technique is built upon the *Time Division Multiplexing* (TDM) scheme commonly used in the trunks for telephone systems. The major advantage of TDMA over FDMA is its format flexibility. Because the format is completely digital and provides flexibility of buffering and multiplexing functions, time-slot assignments among multiple users are readily adjustable to provide different access rates for different users. This feature is particularly adopted in the PSTN and the TDM scheme forms the backbone of all digital connections in the heart of the PSTN. The hierarchy of digital transmission trunks used in North America is the so-called T-Carrier system that has an equivalent European system (the E-carriers) approved by the ITU. In the hierarchy of digital transmission rates standardized throughout North America, the basic building block is the 1.544 Mbps link, known as T-1 carrier. A T-1 transmission frame is formed by time-division multiplexing 24 pulse code modulation (PCM)-encoded voice channels each carrying 64 kbps of user data. Service providers often lease T-carriers to interconnect their own switches and routers and for forming their own networks.

Example 4.8: The use of T-carriers in cellular networks

Cellular service providers often lease T-carriers from the long-haul telephone companies to interconnect their own switches referred to as Mobile Switching Centers (MSCs). The difference between the MSC and a regular switch in the PSTN is that the MSC can support mobility of the terminal. The details of these differences are discussed in later chapters when we provide examples of cellular networks. The end user subscribes to the cellular service provider.

Example 4.9: The use of T-1 lines in the Internet

The routers in the Internet are sometimes connected through leased T-carrier telephone lines to form part of the Internet. The difference between a router and a PSTN switch is that the router can handle packet switching while the PSTN switch uses circuit switching. The end-user subscribes to an Internet Service Provider (ISP) in this case.

---

[2]When we say each AMPS channel is 30 kHz wide in Example 4.5 this also includes guard bands.

With TDMA, a transmit controller assigns time slots to users, and an assigned time slot is held by a user until the user releases it. At the receiving end, a receiver station synchronizes to the TDMA signal frame and extracts the time slot designated for that user. The heart of this operation is synchronization that was not needed for FDMA systems. The TDMA concept was developed in the 1960s for use in digital satellite communication systems and first became operational commercially in telephone networks in the mid-1970s [Pah94].

In cellular and cordless systems, the migration to TDMA from FDMA took place in the second-generation systems. The first cellular standard adopting TDMA was GSM. The GSM standard was initiated to support international roaming among Scandinavian countries in particular and the rest of Europe in general. The digital voice adoption in TDMA format facilitated the network implementation, resulted in improvements in the quality of the voice and provided a flexible format to integrate data services in the cellular network. The FDMA systems in the United States very quickly observed a capacity crunch in major cities and among the options for increasing capacity, standards for North American TDMA (IS-54/IS136) evolved and deployed but were ultimately replaced either by the IS-95 CDMA or the GSM. TDMA was also adopted in second-generation cordless telephones such as CT-2 and Digital Enhanced Cordless Telephony (DECT) to provide format flexibility and to allow more compact and low power terminals. At the time of this writing, DECT is very popular in cordless telephones.

Example 4.10: TDMA in GSM

Figure 4.3 shows an FDMA/TDMA/FDD channel used in second-generation digital cellular of Europe (GSM). The particular example shows the eight-slot TDMA scheme used in the GSM system. Forward and reverse channels use separate carrier frequencies (FDD). Each carrier can support up to eight simultaneous users via TDMA, each using a 13 kbps encoded digital speech, within a 200 MHz carrier bandwidth. A total of 124 frequency carriers (FDMA) are available in the 25 MHz allocated band in each direction. 100 kHz of band is allocated as a guard band at each edge of the overall allocated band.

Example 4.11: TDMA in digital enhanced cordless telephone (DECT)

Figure 4.4 shows an FDMA/TDMA/TDD system used in the Pan-European digital PCS standard DECT. Since distances are short, a TDD format allows using the same frequency for forward and reverse operations. The bandwidth per carrier is 1.728 MHz that can support up 12 ADPCM coded speech channels via TDMA. The total allocated band in Europe is 10 MHz that can support five carriers (FDMA). Figure 4.4 shows the details of the TDMA/TDD time slots use in the DECT system. The frame duration is 10 ms, with 5 ms for portable-to-fixed stations and 5 ms for fixed-to-portable. The transmitter transfers information in signal bursts which it transmits in slots of duration $10/24 = 0.417$ ms. With 480 bits per slot (including a 64-bit guard time), the total bit rate is 1.152 Mbps. Each slot contains 64 bits for system control (C, P, Q, and M channels) and 320 bits for user information (I channel).

Example 4.12: North American TDMA Cellular

Figure 4.5 shows the frame format for the TDMA/FDD with six slots considered for the North American TDMA standard (IS-136) both for the forward (base to mobile) and reverse (mobile to base) channels. In this standard each 30 kHz digital channel has a channel transmission rate of

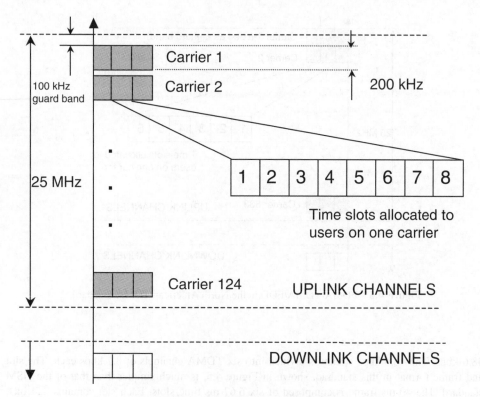

Figure 4.3 FDMA/TDMA/FDD in GSM.

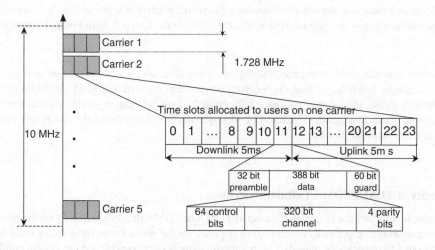

Figure 4.4 FDMA/TDMA/TDD in DECT.

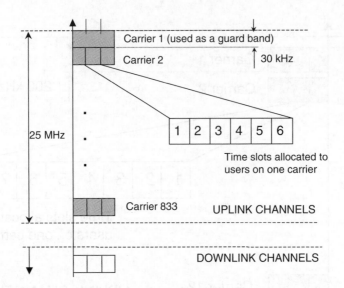

Figure 4.5   FDMA/TDMA/FDD in the North American TDMA Standard.

48.6 kbps. The 48.6 kbps stream is divided into six TDMA channels of 8.1 kbps each. The slot and frame format in this standrad, shown in Figure 4.5, is much simpler than that of the GSM standard. The 40-ms frame is composed of six 6.67-ms time slots. Each slot contains 324 bits, including 260 bits of user data, and 12 bits of system control information in a slow associated control channel. There is also a 28-bit synchronization sequence, and a 12-bit digital verification color code used to identify the frequency channel to which the mobile terminal is tuned. In the mobile-to-base direction, the slot also contains a guard time interval of 6-bit duration when no signal is transmitted, and a 6-bit ramp interval to allow the transmitter to reach its full output power level.

Due to the near–far problem the received signal on the reverse channel from a user occupying a time slot can be much larger than the received power from the terminal using the adjacent time slot. In such a case, the receiver will have difficulties in distinguishing the weaker signal from the background noise. In a manner similar to FDMA systems, TDMA systems also use power control to handle this near–far problem.

Capacity of TDMA/FDMA Cellular Systems

The capacity of a cellular system in the case of TDMA or FDMA depends primarily on the number of channels available per cell, which in turn depends on the reuse factor (how many cells apart can the same frequencies be reused). Let us suppose there is only one cell and the system employs TDMA. Then, the number of simultaneous users that can be supported will be simply the number of users per carrier m multiplied by the total number of carriers (in the case of FDMA, one user is supported per carrier). The total number of carriers is given by the total bandwidth W, divided by the bandwidth of one carrier B. So, the number of simultaneous users supported will be $M = m \times W/B$. If the frequency reuse factor is N_f, then the number of simultaneous users per cell will be:

$$M = \frac{m}{N_f} \left(\frac{W}{B} \right)$$

We will revisit this in our discussion of CDMA next.

4.2.3   Code Division Multiple Access (CDMA)

With the growing interest in the integration of voice, data and video traffic in telecommunication networks, CDMA appears increasingly attractive as the wireless access method of choice. Fundamentally, integration of various types of traffic is readily accomplished in a CDMA environment since coexistence in such an environment does not require any specific coordination among user terminals. In principle, CDMA can accommodate various wireless users with different bandwidth requirements, switching methods and technical characteristics without any need for coordination. Of course, since each user signal contributes to the interference seen by other users, power control techniques are essential in the efficient operation of a CDMA system.

To illustrate CDMA and how it is related to FDMA and TDMA, it is useful to think of the available band and time as resources we use to share among multiple users. In FDMA, the frequency band is divided into slots and each user occupies that frequency through out the communication session. In TDMA, a larger frequency band is shared among the terminals and each user uses a slot of time during the communication session. As shown in Figure 4.6, in a CDMA environment, multiple users use the same band at the same time and the user are differentiated by a code that acts as the key to identify that users. These codes are selected so that when they are used at the same time in the same band a receiver knowing the code of a particular user can detect that user among all the received signals. In CDMA/FDD (Figure 4.7a) the forward and reverse channels use different carrier frequencies. If both transmitter and receiver use the same carrier frequency (Figure 4.7b), the system is CDMA/TDD.

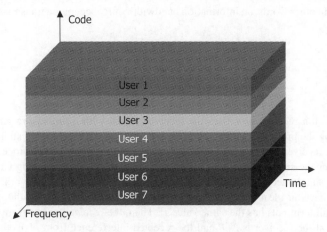

Figure 4.6   Simple illustration of CDMA.

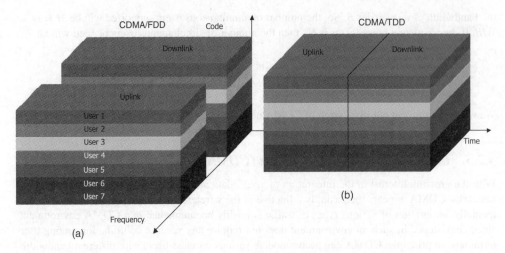

Figure 4.7   (a) Frequency. (b) Time division duplexing with CDMA.

In CDMA, each user is a source of noise to the receiver of other users and *if* we increase the number of users beyond a certain value, the entire system collapses because the signal received in each specific receiver will be buried under the noise caused by many other users. An important question is: how many users can simultaneously use a CDMA system before the system collapses? We investigate this answer below.

Capacity of CDMA

CDMA systems are implemented based on direct sequence spread spectrum transmission at the physical layer that was presented in Chapter 3. In its most simplified form, a spread spectrum transmitter spreads the signal power over a spectrum N_p times wider than the spectrum of the message signal. In other words, an information bandwidth of R_b occupies a transmission bandwidth of W, where:

$$W = N_p R_b \tag{4.1}$$

The spread spectrum receiver processes the received signal with a *processing gain* of N_p. This means that during the processing at the receiver, the power of the received signal having the code of that particular receiver will be increased N_p times beyond the value before processing.

Let us consider the situation of a *single cell* in a cellular system employing CDMA. Assume that we have M simultaneous users on the reverse channel of a CDMA network. Further let us assume that we have an ideal power control enforced on the channel so that the received power of signals from all terminals has the same value P. Then, the received power from the target user after processing at the receiver is $N_p P$ and the received interference from M-1 other terminals is $(M-1)P$. If we also assume that a cellular system is interference limited and the background noise is dominated by the interference noise from other users, the received signal to noise ratio for the target receiver will be:

$$S_r = \frac{N_p P}{(M-1)P} = \frac{N_p}{M-1} \tag{4.2}$$

All users always have a requirement for the acceptable error rate of the received data stream. For a given modulation and coding specification of the system, that error rate requirement will be supported by a minimum S_r requirement that can be used in Equation 4.2 to solve for the number of simultaneous users. Then, solving Equations 4.1 and 4.2 for M, we will have:

$$M = \frac{W}{R_b}\frac{1}{S_r} + 1 \cong \frac{W}{R_b}\frac{1}{S_r} \tag{4.3}$$

Example 4.13: Capacity of one carrier in a single-cell CDMA system

Using QPSK modulation and convolutional coding the IS-95 digital cellular systems require 3 dB $< S_r < 9$ dB. The bandwidth of the channel is 1.25 MHz and transmission rate is R = 9600 bps. Find the capacity of a single IS-95 Cell.

Solution: Using Equation 4.3 we can support from:

$$M = \frac{1.25 \text{ MHz}}{9600 \text{ bps}}\frac{1}{8} \approx 16 \quad \text{to} \quad M = \frac{1.25 \text{ MHz}}{9600 \text{ bps}}\frac{1}{2} \approx 65 \text{ users}$$

Practical Considerations

In the practical design of digital cellular systems, three other parameters affect the number of users that can be supported by the system as well as the bandwidth efficiency of the system. These are the number of sectors in each base station antenna, the voice activity factor, and the interference increase factor. These parameters are quantified as factors used in the calculation of the number of simultaneous users that the CDMA system can support. The use of sectored antennas is an important factor in maximizing bandwidth efficiency. Cell sectorization using directional antennas reduces the overall interference, increasing the allowable number of simultaneous users by a *sectorization gain factor,* which we denote by G_A. With ideal sectorization the users in one sector of a base station antenna do not interfere with the users operating in other sectors, and $G_A = N_{sec}$ where N_{sec} is the number of sectors in the cell. In practice antenna patterns cannot be designed to have ideal characteristics, and due to multipath reflections, users in general communicate with more than one sector. Three-sector base station antennas are commonly used in cellular systems, and a typical value of the sectorization gain factor is $G_A = 2.5$ (4 dB). The voice activity interference reduction factor G_v is the ratio of the total connection time to the active talkspurt time. On average, in a two-way conversation, each user talks roughly *50%* of the time. The short pauses in the flow of natural speech reduce the activity factor further to about *40%* of the connection time in each direction. As a result, the typical number used for G_v is $1/0.4 = 2.5$ (4 dB). The interference increase factor H_0 accounts for users in other cells in the CDMA system. Since all neighboring cells in a CDMA cellular network operate at the same frequency they will cause additional interference. This interference is relatively small due to the processing gain of the system and the distances involved; a value of $H_0 = 1.6$ (2 dB) is commonly used in the industry.

Incorporating these three factors as a correction to Equation 4.3, the number of simultaneous users that can be supported in a CDMA cell can be approximated by:

$$M = \frac{W}{R_b}\frac{1}{S_r} + 1 \cong \frac{W}{R_b}\frac{1}{S_r}\frac{G_A G_v}{H_0} \tag{4.4}$$

If we define the *performance improvement factor* in a digital cellular system as:

$$K_p = \frac{G_A G_v}{H_0} \tag{4.5}$$

assuming the typical parameter values given earlier, the performance improvement factor is $K_p = 4$ (6 dB).

Example 4.14: Capacity of one carrier in a multi-cell CDMA system with correction factors

Determine the multi-cell IS-95 CDMA capacity with correction for sectorization and voice activity. Use the numbers from Example 4.13.

Solution: If we continue the previous example with the new correction factor included in, the range for the number of simultaneous users becomes $64 < M < 260$.

4.2.4   Comparison of CDMA, TDMA and FDMA

CDMA was by far the most successful multiple access scheme in second-generation cellular wireless systems in the United States. With wideband CDMA adopted as the multiple access scheme of choice in third-generation cellular networks, one wonders why CDMA has become the favorite choice for wireless access in voice-oriented networks. Spread spectrum technology became the favorite technology for military applications because of its capability to provide a low probability of interception and strong resistance to interference from jamming. In the cellular industry, CDMA was introduced as an alternative to TDMA to improve the capacity of second-generation cellular systems in the USA. As a result, much of the early debates in this area were focused on calculation of the capacity of CDMA as it is compared with TDMA. However, capacity is not the only reason for the success of the CDMA technology. As a matter of fact, calculation of the capacity of CDMA using the simple approach provided above is *not* very conclusive and is subject to a number of assumptions such as perfect power control that cannot be practically met. The first CDMA service providers in the United States were using slogans such as "you can not believe your ears!" to address the superior quality of voice for the CDMA. However, the superiority of voice is partially dependent on the speech coder and it is not a CDMA versus TDMA issue. In order to provide a good explanation for the success of a complex and multi-disciplinary technology, such as a cellular network, addressing consumer market issues has always been very important. Those of us involved in this debate for the past decade have seen the discussion of the ups and downs of CDMA in variety of forums. One of the most interesting events that the principal author remembers was in 1997 in a major wireless conference in Taipei where one the most famous figures in this debate in his keynote speech at the opening of the conference declared that "we have seen in the past that the VHS which was not a better technology defeated BETA". In his perception, at that time, CDMA was similar to BETA. In less than a year or so after that, CDMA was selected by a number of different communities around the world as the technology of choice for 3G and IMT-2000.

In the rest of this section we bring out a number of issues that may enlighten the reader towards a deeper understanding of the technical aspects of CDMA systems as they are compared to TDMA and FDMA networks. We hope that this may lead the reader to her/his own conclusion about the success of CDMA.

Format Flexibility: As we discussed before, telephone voice was the dominant source of income for the telecommunication industry up to the end of the past century. In the new millennium strong emergence of Internet and cable TV industries has created a case for other popular multi-media applications. The cellular phones that were designed for telephony applications are now being used for other applications and need support for multimedia applications. To support a variety of data rates with different requirements, a network needs format flexibility. As we discussed earlier, one of the reasons for migrating from analog FDMA to digital TDMA was that TDMA provides a more flexible environment for integration of voice and data. The time slots of a TDMA network designed for voice transmission can be used individually or in a group format to transmit data from users and to support different data rates. However, all these users should be time synchronized and the quality of the transmission channel is the same for all of them. The chief advantage of CDMA relative to TDMA is its flexibility in timing and the quality of transmission. In CDMA, users are separated by their codes, unaffected by the transmission time relative to other users. The power of the user can also be adjusted with respect to others to support a certain quality of transmission. In CDMA each user is far more liberated from the other users allowing a fertile setting to accommodate different service requirements to support a variety of transmission rates with different quality of transmission to support multi-media or any other emerging application.

Performance in multipath fading: As we saw in Chapter 2, multipath in wireless channels causes frequency selective fading. In frequency selective fading, when the transmission band of a narrowband system coincides with the location of the fade, no useful signal is received. As we increase the transmission bandwidth, fading will occupy only a portion of the transmission band providing an opportunity for a wideband receiver to take advantage of the portion of the transmission band not under fade and provide a more reliable communication link. In Chapter 3 we introduced DFE, OFDM, sectored antennas, and spread spectrum as technologies that can be employed in wideband systems to handle frequency selective fading. The wider the bandwidth, the better is the opportunity for averaging out the faded frequency.

These technologies are not used in the first-generation analog cellular FDMA systems because they were analog systems and these techniques are digital. The Pan European GSM digital cellular system uses 200 kHz of band and the standard recommends using DFE. The North American digital cellular system, IS-136, uses digital transmission over the same analog band of 30 kHz of the North American AMPS system and does not recommend equalization because the bandwidth is not very large. An equalizer needs additional circuitry and some power budget at the receiver that was one of the drawbacks considered in IS-136. The bandwidth of IS-95 CDMA system is 1.25 MHz and W-CDMA systems for 3G networks use bandwidths that are as high 5 MHz. RAKE receivers are used to increase the benefits of wideband transmission by taking advantage of the so-called in-band or time diversity of the wideband signal (see Chapter 12 for more details). This is one the reasons for having a better quality of voice in CDMA systems. As we mentioned earlier, quality of voice is also affected by the robustness of the speech-coding algorithm, coverage of service, methods to handle interference, handoffs, and power control.

System Capacity: Comparison of the capacity depends on a number of issues including the frequency reuse factor, speech coding rate, and the type of antenna. Therefore a fair comparison would be difficult unless we go to practical systems. The following simple example compares the capacity of FDMA, TDMA, and CDMA used in debates to evaluate alternatives for the first-/second-generation North American cellular systems to replace the first-generation analog.

Example 4.15: Comparison of the capacity of different 2G systems

Compare the capacity of 2G CDMA with 1G FDMA and 2G TDMA systems. For the CDMA system, assume an acceptable signal to interference ratio of 6 dB, data rate of 9600 bps, voice duty cycle of 50%, effective antenna separation factor of 2.75 (close to the ideal three-sector antenna), and neighboring cell interference factor of 1.67.

Solution: For the 2G CDMA system, using Equation 4.4 for each carrier with $W = 1.25$ MHz, $R_b = 9600$ bps, $S_r = 4$ (6 dB), $G_v = 2$ (50% voice activity), $G_A = 2.75$, and $H_0 = 1.67$, we have:

$$M = \frac{W}{R_b} \frac{1}{S_r} \frac{G_A G_v}{H_0} = 108 \text{ users per cell}$$

For the 2G TDMA system with a carrier bandwidth of $B = 30$ kHz, number of users per carrier of $m = 3$, and frequency reuse factor of $N_f = 4$ (commonly used in these systems), each $W = 1.25$ MHz of bandwidth provides for:

$$M = \frac{W}{B} \frac{m}{N_f} = 31.25 \text{ users per cell}$$

For the 1G analog system with carrier bandwidth of $B = 30$ kHz, number of users per carrier of $m = 1$ and frequency reuse factor of $N_f = 7$ (commonly used in these systems), each $W = 1.25$ MHz of bandwidth provides for:

$$M = \frac{W}{B} \frac{1}{N_f} = 6 \text{ users per channel}$$

Another example of this form is instructive to compare these systems with the 2G TDMA system – Global System of Mobile Communications (GSM) – that originated in Europe and since then has also become a presence in the United States.

Example 4.16: Comparison of NA systems with GSM

Determine the capacity of GSM for $N_f = 3$.

Solution: For the GSM system with a carrier bandwidth of $B = 200$ kHz, number of users per carrier of $m = 8$, and frequency reuse factor of $N_f = 3$ (commonly used in these systems), each $W = 1.25$ MHz of bandwidth provides for:

$$M = \frac{W}{B} \frac{m}{N_f} = 16.7 \text{ users per cell.}$$

Handoff: As we discuss in Chapter 6, handoff occurs when a received signal in a MS becomes weak and another BS can provide a stronger signal to the MS. The first generation FDMA cellular systems often used the so-called hard decision handoff in which the base station controller monitors the received signal from the BS and at the appropriate time switches the connection from one BS to another. TDMA systems, use the so-called mobile-assisted hand-off (MAHO) in which the mobile station monitors the received signal from available BSs and reports it to the base station controller which then makes a decision on the handoff. Since adjacent cells in both FDMA and TDMA use different frequencies, the MS has to disconnect from and re-connect to the network that will appear as a click to the user. Handoffs occur at the edge of the cells when the received signals from both BSs are weak. The signals also fluctuate anyway because they are arriving over radio channels. As a result, decision making for the handoff time is often complex and the user experiences a period of poor signal quality and possibly several clicks during the completion of the handoff process. Since adjacent cells in a CDMA network use the same frequency, a mobile moving from one cell to another can be make "seamless" handoff by the use of signal combining. When the mobile station approaches the boundary between cells, it communicates with both cells. A controller combines the signals from both links to form a better communication link. When a reliable link has been established with the new base station, the mobile stops communicating with the previous base station and communication is fully established with the new base station. This technique is referred to as soft handoff. Soft handoff provides a dual diversity for the received signal from two links that improves the quality of reception and eliminates clicking as well as the ping-pong problem.

Power Control: As we discussed earlier in this chapter, power control is necessary for FDMA and TDMA systems to control adjacent channel interference and mitigate the unexpected interference caused by the near–far problem. In FDMA and TDMA systems, some sort of power control is needed to improve the quality of the voice delivered to the user. In CDMA however, the capacity of the system depends *directly* on the power control and an accurate power control mechanism is needed for proper operation of the network. With CDMA, power control is the key ingredient in maximizing the number of users that can operate simultaneously in the system. As a result, CDMA systems adjust the transmitted power more often and with smaller adjustment steps to support a more refined control of power. Better power control also saves on the transmission power of the MS that increases the life of the battery. The more refined power control in CDMA systems also helps in power management of the MS that is an extremely important practical issue for users of the mobile terminals.

Implementation Complexity: Spread spectrum is a two-layer modulation technique requiring greater circuit complexity than conventional modulation schemes. This in turn will lead to higher electronic power consumption and larger weight and cost for mobile terminals. Gradual improvements in battery and integrated circuit technologies, however, have made this the issue transparent to the user.

4.2.5   *Performance of Assigned-Access Methods*

Fixed assignment access methods are used with circuit switched cellular and PCS telephone networks. In these networks, in a manner similar to the wired multi-channel environments, the performance of the network is measured by the blockage rate of an initiated call. A call does not go through for two reasons: (1) when the calling number is not available, (2) when the telephone company is out of resources to provide a line for the communication session. In a plain old telephone service (POTS), for both cases the user hears a busy tone signal and cannot distinguish between the two types of blockage. In most cellular systems, however, type (1) blockage results in a response that is a busy tone and type (2) with a message like "all the circuits are busy at this time; please try your call later". In the rest of this book we refer to blockage rate only as type (2) blockage rate. The statistical properties of the traffic offered to the network are also a function of time. The telephone service providers often design their networks so that the blockage rate at the peak traffic is always below a certain percentage. Cellular operators often try to keep this average blockage rate below 2%.

The blockage rate is a function of the number of subscribers, number of initiated calls, and the length of the conversations. In telephone networks, the Erlang equations are used to relate the probability of blockage to the average rate of the arriving calls and the average length of a call. In wired networks the number of lines or subscribers that can connect to a multi-channel switch is a fixed number. The telephone company monitors the statistics of the calls over a long period of time and upgrades the switches with the growth of subscribers so that the blockage rate during peak traffic times remains below the objective value. In cellular telephony and PCS networks, the number of subscribers operating in a cell is also a function of time. In the downtown areas, everyone uses their cellular telephones during the day and in the evenings they use them in their residential area that is covered by a different cell. Therefore, traffic fluctuations in cellular telephone networks are much more than the traffic fluctuations in POTS. In addition, telephone companies can easily increase the capacity of their networks by increasing their investment on the number of transmission lines and quality of switches supporting network connections. In wireless networks, the overall number of available channels for communications is ultimately limited by the availability of the frequency bands assigned for network operation. To respond to the fluctuations of the traffic and cope with the bandwidth limitations, cellular operators use complex frequency assignment strategies to share the available resources in an optimal manner. Some of these issues are discussed in Chapter 5.

Traffic Engineering using the Erlang Equations

The Erlang equations are the core of the traffic engineering for telephony applications. The two basic equations used for traffic engineering are the Erlang B and Erlang C equations. The Erlang B equation relates the probability of blockage $B(N,\rho)$ to the number of channels N_u and the normalized call density in units of channels ρ. The Erlang B formula is:

$$B(N_u, \rho) = \frac{\rho^{N_u}/N_u!}{\sum_{i=0}^{N_u}(\rho^i/i!)} \qquad (4.6)$$

where $\rho = \lambda/\mu$, λ is the call arrival rate and μ is the service rate of the calls[3].

Example 4.17: Call blocking using the Erlang-B formula

We want to provide a wireless public phone service with five lines to a ferry crossing between Helsinki and Stockholm carrying 100 passengers where on the average each passenger makes a 3-min telephone call every 2 h. What is the probability of a passenger approaching the telephones and none of the four lines are available?

---

[3]The equation assumes that the arrivals are Poisson and the service rate is exponential. For details, see [Ber87].

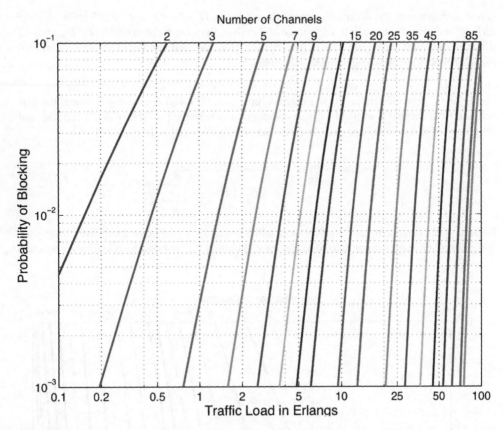

Figure 4.8 Erlang B chart showing the blocking probability as a function of offered traffic and number of channels.

Solution: In practice, often the probability of call blockage is given and we need to calculate the number of subscribers. Here we need an inverse function for the Erlang equation that is not available. As a result, a number of tables and graphs are available for this inverse mapping. Figure 4.8 shows a graph relating the probability of blockage $B(N_u, \rho)$ to the number of channels N_u and the normalized traffic per available channels ρ. From this graph, we can estimate the blocking probability. The traffic load is 100 users $\times$ 1 call/user $\times$ 3 minutes/call per 120 min $= 2.5$ Erlangs. Since there are five lines available and the traffic is 2.5 Erlangs, the blocking probability is roughly 0.07.

Example 4.18: Capacity using the Erlang B formula

A North American TDMA cellular phone provider owns 50 cell sites and 19 traffic carriers per cell each with a bandwidth of 30 kHz. Assuming each user makes three calls/h and the average holding time per call of 5 min, determine the total number of subscribers that the service provider can support with a blocking rate of less than 2%.

Solution: The total number of channels is $N_u = 19 \times 3 = 57$ per cell. For $B(N_u, \rho) = 0.02$ and $N_u = 57$, Figure 4.8 shows that $\rho = 45$ Erlangs. With an average of 5 calls/min, the service rate is $\mu = 1/5$ min, and the acceptable arrival rate of the calls is $\lambda = \rho \times \mu = 1/5$ (min^{-1}) $\times$ 45 (Erlang) = 9 (Erlang/min). With an average of 3 calls/h, the system can accept 9 (Erlang/min) / 3 (Erlang) / 60 (min) = 180 subscribers per cell. Therefore the total number of subscribers are 180 (subscribers/cell) $\times$ 50 (cells) = 8000 subscribers.

The Erlang C formula relates the waiting time in a queue if a call does not go through but it is buffered till a channel is available. These equations start with the probability that a call does not get processed immediately and gets delayed. The probability a call is delayed is given by:

$$P(\text{delay} > 0) = \frac{\rho^{N_u}}{\rho^{N_u} + N_u! \left(1 - \dfrac{\rho}{N_u}\right) \displaystyle\sum_{k=0}^{N_u-1} \dfrac{\rho^k}{k!}} \tag{4.7}$$

Because of the complexity of the calculation, tables or graphs are again used to provide values for this probability based on normalized values of ρ. Figure 4.9 illustrates the relationship between

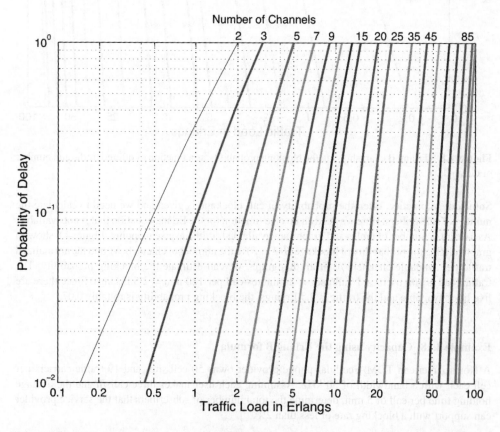

Figure 4.9   Erlang C chart relating the offered traffic to the number of channels and the probability of queuing.

probability of delay, number of channels N_u and the normalized traffic per available channel ρ. The probability of having a delay that is more than a time t is given by:

$$P[delay > t] = P[delay > 0]e^{-(N_u - \rho)\mu t} \tag{4.8}$$

This indicates the exponential distribution of the delay time. The average delay is then given by the average of the exponential distribution:

$$D = P[delay > 0]\frac{1}{\mu(N_u - \rho)} \tag{4.9}$$

Example 4.19: Call delay using Erlang C formula

For the ferry described in Example 4.17 answer the following questions:

(a) What is the average delay for a passenger to get access to the telephone?
(b) What is the probability of having a passenger waiting more than a minute for the access to the telephone?

Solution:

(a) Using Equation 4.7 for $N_u = 5$ and $\rho = 2.5$, we have $P[\text{delay}>0] = 0.13$. Using Equation 4.9 the average delay is $0.13/(5.0–2.5)/3 = 0.17$ min.
(b) Using Equation 4.8 $P[\text{delay}>1 \text{ min}] = 0.13 \exp[–(5.0–2.5)1/3] = 0.13 \exp(–0.83) = 0.0565$.

4.3   Distributed Random Access for Data Oriented Networks

Random access methods have evolved around bursty data applications for computer communications. In our discussion of fixed-assignment access methods, we noted that such methods make relatively efficient use of communications resources when each user has a steady flow of information to be transmitted. This would be the case for example with digitized voiced traffic, data file transfer, or facsimile transmission. However, if the information to be transmitted is intermittent or bursty in nature, fixed-assignment access methods can result in communication resources being wasted for much of the duration of the connection. Furthermore, in wireless networks, where subscribers pay for service as a function of channel connection time, fixed-assignment access can be an expensive means of transmitting short messages and will also involve large call set up times. *Random-access* methods provide a more flexible and efficient way of managing channel access for communicating short-burst messages. In contrast to fixed-assignment access schemes, random access schemes provide each user station varying degrees of freedom in gaining access to the network whenever information is to be sent. A natural consequence of randomness of user access is that there is contention among the users of the network for access to a channel, and this is manifested in collisions of contending transmissions. Therefore these access schemes are sometimes called contention-based schemes or simply *contention-schemes*.

   Random access techniques are widely used in wired local area networks and the literature in computer networking provides adequate description of these techniques. When applied to wireless applications these techniques often are modified from their original wired version. The objective of the rest of this section is to describe the evolution of random access techniques that are used in wireless networks. We first discuss the random access methods used in wireless data networks and then we provide some details of the access methods used in wireless LAN applications.

4.3.1   Random Access Methods for Data Services

The random access methods used data networks can be divided into two groups. The first group consists of ALOHA-based access methods for which the terminals transmit their packets without any coordination between them (they contend for the medium). The second class is the carrier-sense based random access techniques for which the terminal senses the availability of the channel before it transmits its packets.

ALOHA-based Random Access Techniques

The original *ALOHA protocol* is sometimes called *pure-ALOHA* to distinguish it from subsequent enhancements of the original protocol. This protocol derives its name from the ALOHA system, a communications network developed by Norman Abramson and his colleagues at the University of Hawaii and first put into operation in 1971 [Abr70]. The initial system used ground-based UHF radios to connect computers on several of the island campuses with the university's main computer center on Oahu, by use of a random-access protocol which has since then been known as the ALOHA protocol. The word ALOHA means hello in Hawaiian.

The basic concept of ALOHA protocol is very simple. A terminal transmits an information packet upon when the packet arrives from the upper layers of the protocol stack. Simply put, terminals say "hello" to the medium interface as the packet arrives. Each packet is encoded with an error-detection code. The BS/AP checks the parity of the received packet. If the parity checks properly, the BS sends a short acknowledgement packet to the MS. Of course, since the MS packets are transmitted at arbitrary times, there will be collisions between packets whenever packet transmissions overlap by any amount of time, as indicated in Figure 4.10a. Thus after sending a packet, the user waits a length of time more than the round-trip delay for an acknowledgment (ACK) from the receiver. If no acknowledgment is received, the packet is assumed lost in a collision and it is transmitted again with a randomly selected delay to avoid repeated collisions.

The advantage of ALOHA protocol is that it is very simple and it does not impose any synchronization between mobile terminals. The terminals transmit their packets as they become ready for transmission and if there is a collision they simply retransmit. The disadvantage of the ALOHA protocol is its low throughput under heavy load conditions. If we assume that packets arrive randomly, they have the same length, and are generated from a large population of terminals, the maximum through put of the pure-ALOHA is 18%.

Example 4.20: Throughput of pure ALOHA

(a) What is the maximum throughput of a pure ALOHA network with a large number of users and a transmission rate of 1 Mbps?
(b) What is the throughput of a TDMA network with the same transmission rate?
(c) What is the throughput of the ALOHA network if only one user was effective?

Solution:

(a) For a large number of mobile terminals each use a transmission rate of 1 Mbps to access a BS/AP using ALOHA protocol, the maximum data rate that successfully passes through to the BS is 180 kbps.

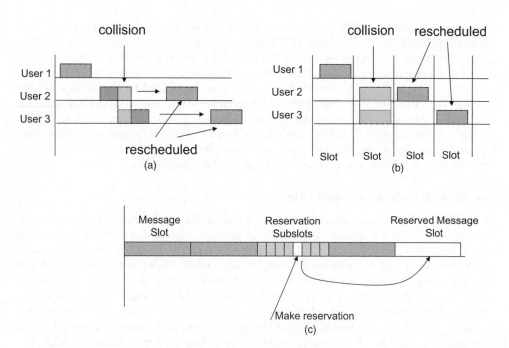

Figure 4.10   (a) Pure ALOHA protocol (b) Slotted ALOHA protocol (c) Reservation ALOHA.

(b) If we have a TDMA system with negligible overhead (long packets) the throughput defined this way is 100% and throughput is 1Mbps.

(c) The 1 Mbps can be attained in an ALOHA system only if we have one user (no-collision) who transmits all the time.

In wireless channels where bandwidth limitations often impose serious concerns for data communications applications, this technique is often changed to its synchronized version, referred to as slotted-ALOHA. The maximum throughput of a slotted ALOHA system under the conditions mentioned above is 36% that is double the throughput of pure-ALOHA.

In slotted-ALOHA protocol, shown in Figure 4.10b, the transmission time is divided into time slots. The BS/AP transmits a beacon signal for timing and all MSs synchronize their time slots to this beacon signal. When a user terminal generates a packet of data, the packet is buffered and transmitted at the start of the next time slot. With this scheme we eliminate partial packet collision. Assuming equal length packets, either we have a complete collision or we have no collisions. This doubles the throughput of the network. The report on collision and retransmission mechanisms remains the same as in pure ALOHA. Because of its simplicity, the slotted-ALOHA protocol is commonly used in the early stages of registration of a MS to initiate a communication link with the BS.

Example 4.21: Slotted ALOHA in GSM

In the GSM system, the initial contact between the mobile station and the base station tower to establish a traffic channel for TDMA voice communications is performed through a Random Access Channel using slotted-ALOHA protocol. Other voice oriented cellular systems adopt similar approaches as the first step in the registration process of a mobile station.

Throughput of slotted-ALOHA protocol is still very low for wireless data applications. This technique is sometimes combined with TDMA systems to form the so-called reservation-ALOHA (R-ALOHA) protocol, shown in Figure 4.10c. In R-ALOHA, time slots are divided into contention periods and contention free periods. During the contention interval MSs use very short packets to contend for the upcoming contention free intervals that will be used for transmission of long information packets. The R-ALOHA protocol was used in the ALTAIR wireless LANs that was developed in the early 1990s to operate in licensed frequency bands around 18–19 GHz. The detailed implementation of R-ALOHA can take a variety of forms and for that reason sometimes it is used under different names. The following example provides some details of the so-called Dynamic Slotted ALOHA protocol that is used in the Mobitex mobile data networks.

Example 4.22: Dynamic slotted-ALOHA

Mobitex has a full-duplex communication capability (simultaneous transmissions on the uplink and downlink) and employs a *dynamic* slotted ALOHA protocol. Suppose that there are three mobile stations MS_1, MS_2, and MS_3 in a cell. The situation is such that the BS has two messages to send to MS_3, MS_1 has a short status update that requires one slot, MS_2 has a long message to send, and MS_3 has nothing to transmit. A MS can transmit only during certain "free" cycles consisting of several slots of equal length that are periodically initiated by the BS using a FREE frame on the downlink. In this example, shown in Figure 4.11, the BS indicates that there are six free slots for contention, each of a certain length. This can change depending on the traffic and hence the term "dynamic". Also note that MSs cannot transmit whenever they want as in slotted ALOHA. The MSs with traffic to send that is, MS_1 and MS_2 select one of the six slots at random. In this case, MS_1 selects slot 1 and MS_2 selects slot 4. Hence there is no collision. MS_1 is able to transmit its short status update in slot 1 after which it ceases transmission. MS_2 transmits in slot 4 requesting access to the channel using a message called ABD. Simultaneously, the BS would have transmitted its message to MS_3. Upon receipt of the message, MS_3 acknowledges it. The free slots are designed to be of the duration of the downlink message to MS_3 so that the ACK from MS_3 can be received without contention. The BS also acknowledges the status report from MS_1 and sends an access grant (ATD) to MS_2. As MS_2 transmits its long message on the uplink, the BS can simultaneously send the second message to MS_3. After proper ACKs are transmitted and received, a new FREE cycle is started.

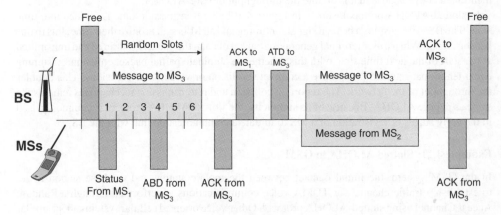

Figure 4.11 Dynamic slotted-ALOHA used in Mobitex.

Example 4.23: Packet reservation multiple access (PRMA)

An example of a system that uses reservation for integrating voice and data services is the work done by David Goodman and his colleagues in developing the concept of Packet Reservation Multiple Access (PRMA) [Goo89]. PRMA is a method for transmitting, in a wireless environment, a variable mixture of voice packets and data packets. The PRMA system is closely related to Reservation-ALOHA, in that it merges characteristics of slotted ALOHA and TDMA protocols. PRMA has been developed for use in centralized networks operating over short-range radio channels. Short propagation times are an important ingredient in providing acceptable delay characteristics for voice service. Our description here closely follows [Goo89] and [Goo91].

The transmission format in PRMA is organized into frames, each containing a fixed number of time slots. The frame rate is identical to the arrival rate of speech packets. The terminals identify each slot as either "reserved" or "available", in accordance with a feedback message received from the base station at the end of the slot. In the next frame, only the user terminal that reserved the slot can use a reserved slot. Any terminal, not holding a reservation that has information to transmit, can use an available slot.

Terminals can send two types of information, referred to as periodic and random. Speech packets are always periodic. Data packets can be random, if they are isolated, or periodic if they are contained in a long unbroken stream of information. One bit in the packet header specifies the type of information in the packet. A terminal having periodic information to send starts transmitting in contention for the next available time slot. Upon successfully detecting the first packet in the information burst, the base station grants the sending terminal a reservation for exclusive use of the same time slot in the next frame. The terminal in effect "owns" that time slot in all succeeding frames as long as it has an unbroken stream of packets to send. After the end of the information burst, the terminal sends nothing in its reserved slot. This in turn causes the base station to transmit a negative acknowledgment (NACK) feedback message indicating that the slot is once again available.

To transmit a packet, a terminal must verify two conditions. The current time slot must be available, and the terminal must have permission to transmit. Permission is granted according to the state of a pseudorandom number generator, permissions at different terminals being statistically independent. The terminal attempts to transmit the initial packet of a burst until the base station acknowledges successful reception of the packet or until the terminal discards the packet, because it has been held too long. The maximum holding time, D_{max} (seconds), is determined by delay constraints on speech communication, and is a design parameter of the PRMA system. If the terminal drops the first packet of a burst, it continues to contend for a reservation to send subsequent packets. It drops additional packets as their holding times exceed the limit D_{max}. Terminals with periodic data (as opposed to voice) packets to send store packets indefinitely while they contend for slot reservations (equivalent to setting D_{max} to infinity). Thus as a PRMA system becomes congested, both the speech packet dropping rate and the data packet delay increase.

In [Goo91], Goodman and Wei analyzed PRMA efficiency, which they quantified as the maximum number of conversations per channel that the system can support within a chosen constraint on packet-dropping probability. In their work they adopted a constraint of $P_{drop} < 0.01$. They used a speech source rate of 32 kbps and a header length of 64 bits in each packet. Using computer simulation methods, they investigated the effects of six system variables on PRMA efficiency: (1) channel rate, (2) frame duration, (3) speech activity detector, (4) maximum delay, (5) permission probability, and (6) number of conversations. Over the range of conditions examined, they found many PRMA configurations capable of supporting about 1.6 conversations per channel and found that this level of efficiency could be maintained over a wide range of conditions. Their overall conclusion was that PRMA shows encouraging potential as a statistical multiplexer of speech packets. However, they judged that there are still many questions to be answered in order to verify that PRMA will perform properly in short-range radio systems. Issues requiring further investigation include the effects of mixing random information packets (data) and periodic information packets (speech) in PRMA and the effects of packet transmission errors on PRMA efficiency.

Example 4.24: Reservation in general packet radio service (GPRS)

A single 200 kHz carrier in GSM has eight time slots, each capable of carrying data at 9.6 kbps (standard), 14.4 kbps (enhanced), or 21.4 kbps (if forward error correction is completely omitted). The raw data rate can thus be as high as $8 \times 21.4 = 171.2$ kbps. The same time slots can be reserved for data access using slotted ALOHA. Medium access is based on a slotted ALOHA reservation protocol. In the contention phase a slotted-ALOHA random access technique is used to transmit reservation requests, the BS then transmits a notification to the MS indicating the channel allocation for an uplink transmission, and finally the MS can transfer data on the allocated slots without contention. On the downlink, the BS transmits a notification to the MS indicating the channel allocation for downlink transmission of data to the MS. The MS will monitor the indicated channels, and the transfer occurs without contention.

Reservation in OFDMA

Reservation in OFDMA systems behaves in a manner similar to reservation in TDMA based systems described in Examples 4.23 and 4.24 above. Instead of reserving time slots for individual mobile stations, *resource blocks* are reserved on the downlink in Long Term Evolution based cellular systems. A resource block comprises of a set of frequency sub-carriers and time slots. In Long Term Evolution, a resource block is 1 ms long (two time slots) and 180 kHz of bandwidth comprising of 12 sub-carriers, each 15 kHz wide. Resource blocks on the downlink can be allocated in a contiguous manner or in chunks that are not together. On the uplink in Long Term Evolution, resource blocks have to be contiguous because of the single carrier nature of transmission (described in Chapter 13).

CSMA based Random Access Techniques

The main drawback of ALOHA-based contention protocols is the lack of efficiency caused by the collision and retransmission process. In ALOHA, users do not take into account what other users are doing when they attempt to transmit data packets and there are no mechanisms to avoid collisions. A simple method to avoid collisions is to sense the channel before transmission of a packet. If there is another user transmitting on the channel, it is obvious that a terminal should delay the transmission of the packet. Protocols employing this concept are referred to as Carrier-Sense Multiple-Access (CSMA) or Listen-Before-Talk (LBT) protocols. Figure 4.12 shows the basic concept of the CSMA protocol. Terminal "1" senses the channel first and then sends a packet. This is followed by a sensing and packet transmission by terminal "1" again. During the second transmission time of terminal "1", terminal "2" senses the channel and discovers that another terminal is also using the medium. It then delays its transmission for a later time using a back-off algorithm. The CSMA protocol reduces the packet collision probability significantly compared to ALOHA protocol. However it cannot eliminate the collisions entirely. Sometimes, as shown in Figure 4.12, two terminals sense the channel busy and reschedule their packets for a later time but their transmission time overlap with each other causing a collision. Such situations do not cause a significant operational problem because the collisions can be handled in the same way as they were handled in ALOHA. However,

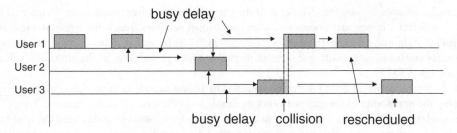

Figure 4.12   Basic operation of CSMA protocol.

if the propagation time between the terminals is very long such situations happen more frequently thereby reducing the effectiveness of carrier sensing in preventing collisions. As a result, several variations of CSMA have been employed in local area applications while ALOHA protocols are preferred in wide area applications.

Example 4.25: Examples of wireless networks that employ ALOHA and CSMA

As described earlier, Mobitex uses a variation of ALOHA protocol while the IEEE 802.11 standard for wireless LANs employs a version of CSMA protocol. ALOHA protocol is also used in the random access logical channels in cellular telephone and satellite communication applications (wide area networks).

A number of strategies are used for the sensing procedure and retransmission mechanisms that have resulted in a number of variations of the CSMA protocol for a variety of wired and wireless data networks. Figure 4.13 depicts the key elements of distinction among these protocols. If after sensing the channel, the terminal attempts another sensing only after a random waiting period the carrier-sensing mechanism is called "non-persistent". After sensing a busy channel, if the

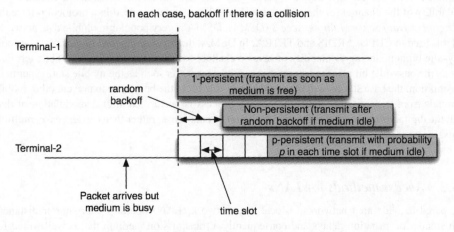

Figure 4.13   Retransmission alternatives for CSMA.

terminal continues sensing the channel until the channel becomes free, the protocol is referred to as a "persistent". In persistent operation, after the channel becomes free, if the terminal transmits its packet right-away it is referred to as "1-Persistent" CSMA and if it runs a random number generator and based on the outcome transmit its packet with a probability "p" the protocol is called p-persistent CSMA.

In a wireless network, due to multipath and shadow fading as well as the mobility of terminals, sensing the availability of the channel is not as simple as in the case of wired channels. Typically in a wireless network, two terminals can each be within range of some intended third terminal but out of range of each other, because they are separated by excessive distance or by some physical obstacle that makes direct communication between the two terminals impossible. This situation, where the two terminals cannot sense the transmission of each other, but a third terminal can sense both of them, is referred to as the *hidden terminal problem*. This is a more likely situation in cases of radio networks covering wider geographic areas in which hilly terrain blocks some groups of user terminals from sensing other groups. In this situation the CSMA protocol will successfully prevent collisions among the users of one group but will fail to prevent collisions between users in groups hidden from one another.

To resolve the hidden terminal problem we need to facilitate the sensing procedure. In multi-hop ad hoc networks, where there is no centralized station or infrastructure, a protocol called *Busy-Tone Multiple-Access (BTMA)* has been used in packet radio for military applications. A brief summary of BTMA is given in [Tob80], where a number of packet communication protocols are discussed and compared. In the BTMA scheme, the system bandwidth is divided into two channels, a *message channel* and a *busy-tone channel*. Whenever a station sends signal energy on the message channel, it transmits a simple busy-tone signal (e.g., a sinusoid) on its busy-tone channel. When any other terminal senses a busy-tone signal, it turns on its own busy tone. In other words, as a terminal detects that some user is on the message channel, it sounds the alarm on the busy tone channel in an attempt to inform every user including those hidden to the transmitting terminal. A user station with a packet ready to send first senses the busy-tone channel to determine if the network is occupied.

Most cellular mobile data networks use different frequencies for forward (downlink) and reverse (uplink channels). The messages in the forward channel are transmitted from the mobile data base station that are designed and deployed to provide a comprehensive and reliable coverage. In another words, the base stations are not hidden to the mobile terminals while the mobile terminals may be hidden from one another. In this situation one may use the forward channel to announce the availability of the channel for the mobile terminals. This concept is used in a protocol referred to as *Digital* or *Data Sense Multiple-Access (DSMA)*. DSMA is very popular in mobile data networks and it is used in CDPD, ARDIS and TETRA. In DSMA, the forward channel broadcasts a periodic busy-idle bit announcing availability of reverse channel for data transmission. A mobile terminal checks the busy-idle bit prior to transmission of its packet. As soon as the mobile station starts its transmission, the base station will change the busy-idle bit to the busy state to prevent other mobile terminals from transmission. Since the sensing process is performed after demodulation of data from the digital information it is referred to digital or data sense, rather than carrier sense, multiple access.

4.3.2   Access methods for LANs

Compared to wide area networks, a local area network (LAN) operates over shorter distances with smaller propagation delays and consequently a transmission medium that is well suited for variations of the CSMA protocol. Low speed wide area networks are developed for communicating shorter messages while local area networks are designed to facilitate large file transfers at high data rates. As a result, the length of the packets in LANs is much larger than the length of the packets in a low speed mobile data networks. When the length of the packets is long it would be very useful to pay further attention to packet collisions. Local area networks often employ variations of the CSMA protocol that either stop transmission as soon as a packet collision is detected or add additional features to avoid collisions.

Example 4.26: Packet sizes in wide and local area wireless networks

(a) Determine the transfer time of a 20 kB file with a mobile data network with a transmission rate of 10 kbps.
(b) Repeat for an 802.11 WLAN operating at 2 Mbps.
(c) What is the length of the file that the WLAN of part (b) can carry in the time that mobile data service of part (a) carries its 20 kB file.

Solution:

(a) The early mobile data networks, such as ARDIS and Mobitex, limited the length of a file to around 20 kB. For a data rate of around 10 kbps it would take $20 \times 8 / 10 = 16$ s to transfer such a file.
(b) An IEEE 802.11 network operating at 2 Mbps would transfer this file in 80 ms.
(c) In a 16 s time interval the same WLAN transfers a 4 MB file.

The most popular version of the CSMA for wired LANs is CSMA with Collision Detection (CSMA/CD) adopted in the IEEE 802.3 (Ethernet) standard, the dominant standard for wired LANs supporting data rates that can be up to several gigabits per second. The basic operation of CSMA/CD is the same as CSMA implementations discussed earlier. The defining feature of CSMA/CD is that it provides for detection of a collision shortly after its onset, and each transmitter involved in the collision stops transmission as soon as it senses a collision. In this way colliding packets can be aborted promptly, minimizing the wastage of channel occupancy time by transmissions destined to be unsuccessful. More details of Ethernet are provided in Chapter 8. Other LAN multiple access protocols employed the idea of taking turns (as in polling or token based schemes), but such schemes are not widely deployed. Ethernet pretty much dominates the wired LAN deployments of today.

Unlike plain CSMA, which requires an acknowledgment (or lack of an acknowledgment) to learn the status of a packet collision, CSMA/CD requires no such feedback information, since the collision-detection mechanism is built into the transmitter. When a collision is detected, the transmission is immediately aborted, a jamming signal is transmitted, and a retransmission back-off procedure is initiated, just as in CSMA [Pah94]. As is the case with any random access scheme, proper design of the back-off algorithm is an important element in assuring the stable operation of the network.

Example 4.27: Binary exponential back-off

The back-off algorithm recommended by IEEE 802.3 Ethernet is referred to as the binary exponential back-off algorithm that is combined with 1-persistence CSMA protocol with collision detection. When a terminal senses a transmission it continues sensing (persistent) until the transmission is completed. After the channel becomes free, the terminal sends its own packet. If another terminal was also waiting, a collision occurs because of the 1-persistence and the two terminals re-attempt transmission with a probability of 1/2 after a time slot that spans twice the maximum propagation delay allowed between the two terminals. A time slot that spans twice the maximum propagation delay is selected to ensure that in the worse case scenario the terminal will be able to detect the collision. If a second collision occurs the terminals re-attempt with the probability of 1/4 that is half of the previous retransmission probability. If collision persists the terminal continues reducing its retransmission probability by half up to 10 times and after that it continues with the same probability six more times. If no transmission is possible after 16 attempts, it reports to the higher layers that the network is congested and transmission shall be stopped. This procedure exponentially increases the back-off time and gives the back-off strategy its name. The disadvantage of this procedure is that the packets arriving later have a higher chance to survive the collision that results in an unfair first come last serve environment. It can be shown that the average waiting time for the exponential back-off algorithm is 5.4 T where T is the time slot used for waiting [Sta00; Tan10].

The CSMA/CD scheme is also used in many infrared LANs, where both transmission and reception are inherently directional. In such an environment a transmitting station can always compare the received signal from other terminals with its own transmitted signal to detect a collision. Radio propagation is not directional posing a serious problem in determining other transmissions during your own transmission. As a result, collision detection mechanisms are not well suited for radio LANs. However, compatibility is very important for WLANs and therefore designers of these networks have had to consider CSMA/CD for compatibility with the Ethernet backbone LANs that dominate the wired-LAN industry.

While collision detection is easily performed on a wired network, simply by sensing voltage levels against a threshold, such a simple scheme is not readily applicable to radio channels because of fading and other radio channel characteristics. The one approach that can be adopted for detecting collisions is to have the transmitting station demodulate the channel signal and compare the resulting information with its own transmitted information. Disagreements can be taken as an indicator of collisions, and the packet can be immediately aborted. However, on a wireless channel the transmitting terminal's own signal dominates all other signals received in its vicinity, and thus the receiver may fail to recognize the collision and simply retrieve its own signal. To avoid this situation the station's transmitting antenna pattern should be different from its receiving pattern. Arranging this situation is not convenient in radio terminals because it requires directional antennas and expensive front-end amplifiers for both transmitters and receivers.

The approach called *CSMA with Collision Avoidance* (CSMA/CA), shown in Figure 4.14, is actually adopted by the IEEE 802.11 wireless LAN standard. The elements of CSMA/CA used in the IEEE 802.11 are Inter Frame Spacing (IFS), Contention Window (CW), and a back-off counter.

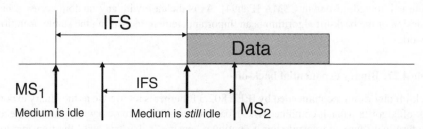

Figure 4.14   CSMA/CA adopted by the IEEE 802.11.

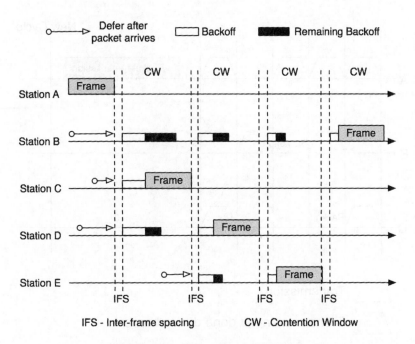

Figure 4.15   Illustration of CSMA/CA.

The CW intervals are used for contention and transmission of the packet frames. The IFS is used as an interval between two CW intervals. The back-off counter is used to organize the back-off procedure for transmission of packets. The method of operation is best described by an example.

Example 4.28: Operation of Collision Avoidance in IEEE 802.11

Figure 4.15 provides an example for the operation of the CSMA/CA mechanism used in the IEEE 802.11 standard. Stations A, B, C, D, and E are engaged in contention for transmission of their packet frames. Station A has a frame in the air when stations B, C, and D sense the channel and find it busy. Each of the three stations will run its random number generator to get a back-off time by random. Station C followed by D and B draws the smallest number. All three terminals persist on sensing the channel and defer their transmission until the transmission of the frame from terminal A is completed. After completion, all three terminals wait for the IFS period and start their counters immediately after completion of this period. As soon as the first terminal, station C in this example, finishes counting its waiting time it starts transmission of its frame. The other two terminals, B and D, freeze their counter to the value that they have reached at the start of transmission for terminal C. During transmission of the frame from station C, station E senses the channel, runs its own random number generator that that in this case ends up with a number larger than remainder of D and smaller than the remainder of B, and defers its transmission for after the completion of station C's frame. In the same manner as the previous instance, all terminal wait for IFS and start their counter. Station D runs out of its random waiting time earlier and transmits its own packet. Stations B and E freeze their counters and wait for the completion of the frame transmission from terminal D and the IFS period after that before they run start running down their counters. The counter for terminal E runs down to zero earlier and this terminal sends its frame while B freezes its counter.

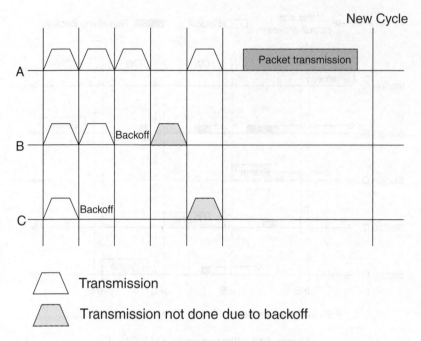

Transmission

Transmission not done due to backoff

Figure 4.16   Illustration of combing.

After the IFS period following completion of the frame from station E the counter in station B counts down to zero before it sends its own frame. The advantage of this back-off strategy over the exponential back off used in IEEE 802.3 is that the collision detection procedure is eliminated and the waiting time is fairly distributed in a way that on average a first come first serve policy is enforced.

Another related technique considered for collision avoidance in wireless LANs is the *combing* method [Wil95a, b]. As shown in Figure 4.16 the time is divided into comb and data transmission intervals. During the comb period each station alternates between transmission and listening periods according to a code assigned to the station. All stations will continue advancing in their code until they sense a carrier during their listening period. If they do not sense a carrier at the end of the code, they transmit their packet. If they sense a carrier they postpone their transmission until the next comb interval. A simple example will further clarify this method.

Example 4.29: Combing for collision avoidance

Figure 4.16 shows three stations with five digit codes 11101 (terminal A), 11010 (terminal B), and 10011 (terminal C). All three terminals transmit their carrier during the first slot, because all codes have 1 in that slot. In the second slot terminal C will listen and after sensing other two withdraws from contention and waits for the next combing. In the third period station B goes to listening state and after sensing the carrier of terminal A defers its transmission until the next cycle. Terminal A continues its sequence of alternating transmissions and listening until the end of the comb period when it transmits its data packet (as it heard no other terminal). After completion of the data transmission from station A, the other two terminals will wait for an inter-packet spacing

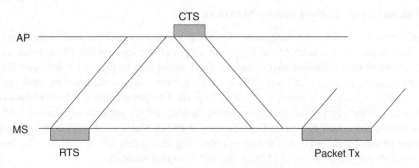

Figure 4.17 RTS/CTS in IEEE 802.11.

(IPS) for a new contention after which station B transmits its packet. Station C will transmit after the second transmission cycle.

In CSMA/CA, as we will see later, priority is assigned by dividing the IFS into several different sized intervals associated with different priority levels. In combing, priority can be arranged by assigning different classes of numbers to the codes. The lower priority packets will receive earlier zero codes and higher priority packets will have a zero in their codes in the later intervals.

Another access method used in wireless LANs is the Request To Send / Clear To Send (RTS/CTS) mechanism shown in Figure 4.17. A terminal ready for transmission sends a short RTS packet identifying the source address, destination address, and the length of the data to be transmitted. The destination station responds with a CTS packet. The source terminal will then send its packet with no contention. After acknowledgement from the destination terminal the channel will be available for other usage. IEEE 802.11 supports this feature as well as CSMA/CA (see Chapter 8 for more details). This method provides a unique access right to a terminal to transmit without any contention.

Quality of Service in WLANs

With the emergence of VoIP as an important application over random access networks, it became important to provide support for quality of service for voice and other real-time traffic over such networks. The IEEE 802.11e standard is an attempt to provide quality of service at the MAC layer in WLANs. The basic idea in IEEE 802.11e is to provide different priorities for different classes of traffic. Priority here refers to the ability of frames from certain classes of traffic to access the channel earlier than other classes of traffic. This helps in reducing the delay for such classes of traffic, as well as in providing higher throughput. One way of providing different priorities to different classes of traffic in IEEE 802.11 is to enable different waiting times (inter-frame spaces) or different back-off intervals for different classes of traffic. This way, voice traffic, for example, could wait for a much shorter period of time than web traffic reducing latency and improving throughput for voice packets.

This however implies that access to the channel is not fair (flows from some classes of traffic may be starved of bandwidth). Fair access to the medium while assuring throughput and latency is possible by using techniques that are distributed and yet provide fair access to the medium. Several techniques have been proposed to address this issue (see [Pat03] for more details).

MAC Protocols for Wireless Sensor Networks

Wireless sensor networks are a new class of networks that may include several thousand low-cost devices deployed in a sensor field for monitoring and sensing specific phenomena. In such networks, sensors may actually transmit and received useful data (typically sensed quantities that have been processed locally) only very infrequently. Sensors may be deployed in areas that may be inaccessible to humans. Consequently, it is important to ensure that the limited batteries in such devices last for years. Medium access schemes designed for high-speed wireless LANs are quite unsuitable for sensor networks. Coordinating sleep schedules of sensors in order to reduce the amount of idle listening is one method of enhancing the battery life in sensors. The reader is referred to [Dem06] for a survey of MAC protocols for sensor networks.

Enhancements for High-Speed Operation

One of the drawbacks of the CSMA/CA protocol employed in WLANs is that the waiting times employed severely limit the throughput of the network. Recall that no useful data is sent for the inter-frame spacing time, during the backoff slots, or during the transmission of the acknowledgment frames. These times form an overhead for the transmission of every single MAC layer frame in IEEE 802.11. Especially as the data rates increase, the waiting times become an increasingly large part of the transmission of a frame. Consider that the actual size of a frame reduces as the data rate increases. If the waiting times remain fixed, eventually they dominate the transmission of frames so that the throughput in a WLAN cannot exceed a certain threshold irrespective of how high the physical transmission rate is. To overcome this problem, in IEEE 802.11n (the high speed standard for WLANs), additional MAC layer features have been introduced to reduce the overhead. These include aggregating several frames for transmission and block acknowledgments for several frames [Xia05, Sko08].

4.3.3   Performance of Random Access Methods

In voice oriented circuit switched networks, performance is measured by the probability of blockage (blockage rate) of initiating a call. If the call is not blocked, a fixed rate full-duplex channel is allocated to the user for the entire communication session. In other words, interaction between the user and the network takes place in two steps. First, during the call establishment procedure, the user negotiates the availability of a line with the network and if successful (not blocked) the network guarantees a connection with a certain Quality of Service (QoS; data rate, delay, error rates) to the user. For real-time interactive applications such as telephone conversations or video conferencing, if the user does not talk the resource allocated to the user is wasted. If these facilities, originally designed for two-way voice application, are used for data application: (1) for bursty data file transfers during the idle times between transmission of two packet bursts allocated resources are wasted, (2) large file transfers suffer a long delay or waiting time for the transfer because resources allocated to each user is more restricted.

Users of packet switched networks are always connected and there is neither an initiation (negotiation) procedure to be blocked nor a fixed QoS to be allocated. In this situation, analysis of the performance for real time interactive applications such as telephone conversations is complicated and will be addressed later. Performance of these networks for data applications is often measured by the average throughput, S, and average delay, D, versus the total offered traffic, G. The *channel throughput, S*, is the average number of successful packet transmissions per time interval T_p. The offered traffic, G, is the number of packet transmission attempts per packet time slot T_p that includes

Table 4.1 Throughput of various random access protocols

Protocol	Throughput
Pure ALOHA	$S = Ge^{-2G}$
Slotted ALOHA	$S = Ge^{-G}$
Unslotted 1-persistent CSMA	$S = \dfrac{G[1 + G + aG(1 + G + aG/2)]e^{-G(1+2a)}}{G(1 + 2a) - (1 - e^{-aG}) + (1 + aG)e^{-G(1+a)}}$
Slotted 1-persistent CSMA	$S = \dfrac{G[1 + a - e^{-aG}]e^{-G(1+a)}}{(1 + a)(1 - e^{-aG}) + ae^{-aG(1+a)}}$
Unslotted non-persistent CSMA	$S = \dfrac{Ge^{-aG}}{G(1 + 2a) + e^{-aG}}$
Slotted non-persistent CSMA	$S = \dfrac{aGe^{-aG}}{1 - e^{-aG} + a}$

new arriving packets as well as retransmissions of old packets. The average delay D, is the average waiting time before successful transmission, normalized to the packet duration T_p. The standard unit of traffic flow is Erlang that can be thought of as the number of the packets per packet duration time T_p. The throughput is always between zero and one Erlang while the offered traffic, G, may exceed one Erlang.

The analyses of the relationships between S, G, and D for a variety of medium access protocols have been a subject of research for a few decades. This analysis depends on the assumptions on the statistical behavior of the traffic, number of terminals, relative duration of the packets, and the details of the implementation. Assuming a large number of terminals generating fixed length packets with a Poisson distribution[4], Table 4.1 summarizes the throughput expressions for ALOHA, and 1-persistent and non-persistent CSMA protocols, including the slotted and un-slotted versions of each. The expressions for p-persistent protocols are very involved and are not included here. The interested reader should refer to [Kle75; Tob75; Tak85], where the derivations of the other CSMA expressions can also be found. The expressions in the table are also derived in [Ham86] and [Kei89]. The parameter a in this table corresponds to the normalized propagation delay defined as $a = {}^\tau/_{T_p}$ where τ is the maximum propagation delay for the signal to go from one end of the network to the other end.

Example 4.30: Calculation of the normalized propagation delay

Determine the parameter a in IEEE 802.3 (Ethernet) 10 Mbps LANs and IEEE 802.11 2 Mbps LANs.

Solution: The IEEE 802.3 standard for star LANs allows a maximum length of 200 m between two terminals. The propagation speed in the cables is usually approximated by 200 000 km/s resulting in $\tau = 1 \mu s$. The IEEE 802.11 allows maximum distance of 100 m between the AP and the MS. The radio propagation is at the rate of 300 000 km/s resulting in $\tau = 0.33 \mu s$. For a star LAN operating at 10 Mbps with 1000 bit packets, the value is $a = 0.01$. For an IEEE 802.11 operating at 2 Mbps with the same packet size $a = 0.00066$.

---

[4]Poisson distribution assumes packets are generated independent from one another and the inter-arrival time between the packets forms an exponentially distributed random variable.

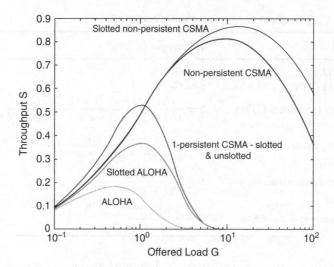

Figure 4.18   Throughput S versus offered traffic load G for various random access protocols.

Figure 4.18 shows plots of throughput S versus offered traffic load G for the six protocols listed in Table 4.1, with normalized propagation delay of $a = 0.01$. All curves follow the same pattern. Initially as the offered traffic G increases the throughput S also increases up to a point where it reaches a maximum S_{max}. After the throughput reaches its maximum value, an increase in the offered traffic actually reduces the throughput. The first region depicts the stable operation of the network in which an increase in aggregating traffic G that includes arriving traffic as well as retransmissions due to collisions, increases the total successful transmissions and thus S. The second region represents unstable operation where an increase in G actually reduces the throughput S because of congestion and eventually halting of the operation. In practice, as we saw in the last section, retransmission techniques adopted for the real implementation include back off mechanisms to prevent operation in unstable regions.

The throughput curves for the slotted and unslotted versions of 1-persistent CSMA are essentially indistinguishable. It can be seen from the figure that for low levels of offered traffic, the 1-persistent protocols provide the best throughput, but at higher load levels the non-persistent protocols are by far the best. It can also be seen that the slotted non-persistent CSMA protocol has a peak throughput almost twice that of persistent CSMA schemes.

The equations in Table 4.1 can also be used to calculate capacity, which is defined as the peak value S_{max} of throughput over the entire range of offered traffic load G [Ham86]. An example is helpful to show how to relate the curve to a particular system.

Example 4.31: Relating throughput and offered traffic to data rates

To relate throughput and offered traffic to data rates assume that we have a centralized network that supports a maximum data rate of 10 Mbps and serves a large set of user terminals with the pure ALOHA protocol.

(a) What is the maximum throughput of the network?
(b) What is the offered traffic in the medium and how is it composed?

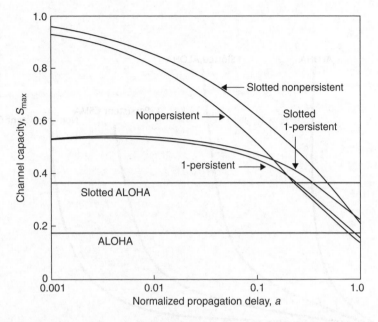

Figure 4.19 Capacity versus normalized propagation delay for various random access protocols.

Solution:

(a) Since the peak value of the throughput is S = 18.4%, the terminals contending for access to the central module can altogether succeed in getting at most 1.84 Mbps of information through the network.

(b) At that peak the total traffic from the terminals is 5 Mbps (because the peak occurs at G=0.5), which is composed of 1.84 Mbps of successfully delivered packets (some mixture of new and old packets) and 3.16 Mbps of packets doomed to collide with one another.

Plots of capacity versus normalized propagation delay are plotted in Figure 4.19 for the same set of ALOHA and CSMA schemes. The curves show that for each type of protocol the capacity has a distinctive behavior as a function of normalized propagation delay a. For the ALOHA protocols, capacity is independent of a, and is the largest of all the protocols (compared when a is large). As we discussed earlier this is the case where the area of coverage is large and propagation delays are comparable to the length of packets. The plots in Figure 4.19 also show that the capacity of 1-persistent CSMA is less sensitive to the normalized propagation delay for small a, than is non-persistent CSMA. However, for small a, non-persistent CSMA yields a larger capacity than does 1-persistent CSMA, though the situation reverses as a approaches the range 0.3–0.5 [Ham86].

Another important performance measure for packet data communications is the delay characteristics of the transmitted packets. For real-time applications such and voice conversations, if the delay is more than a certain values (several hundred milliseconds) packet is not useful and it is dropped. Therefore, we need to analyze the delay characteristics of the channel to determine the capacity of the access method. In the data transfer applications the delay characteristics is usually related to the throughput of the medium and it usually follows a hockey-stick shape. At low traffic

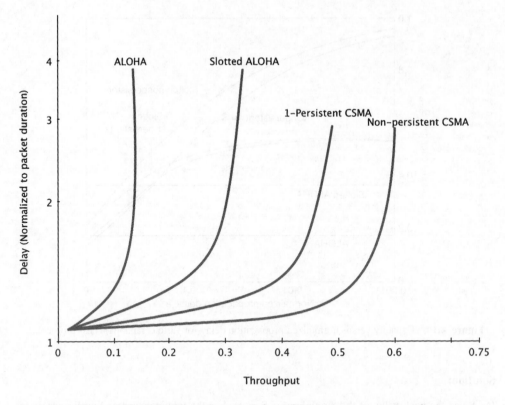

Figure 4.20   Delay versus throughput for various random access protocols.

when a small fraction of the maximum throughput is utilized, the delay often remains the same as the transmission delay. As the throughput increases the number of retransmitted packets increases, resulting in higher average delay for the packets. Around the maximum throughput the delay retransmissions grow rapidly pushing the network toward unstable condition where the channel is dominated with retransmissions and the packet delays grow extremely large. Figure 4.20 shows the delay–throughput behavior of the ALOHA, S-ALOHA, and CSMA protocols [Tan10].

Practical considerations

The analysis provided in the previous chapter is abstract and is used to provide an intuitive framework for the operation of different classes of access methods. In practice implementations deviate considerably from the abstract and the performance is evaluated by analysis or simulation of case-by-case situations. Examples of this type of analysis for the CSMA/CD with exponential back-off algorithm used in the IEEE 802.3 Ethernet and performance of token ring access method used in IEEE 802.5 are available in the last chapter of [Sta00].

Complications caused by wireless channel

Three factors that are effective in throughput analysis in wired environment are propagation delay, users' idle periods (not transmitting), and packet collisions. In a wireless environment, analysis of

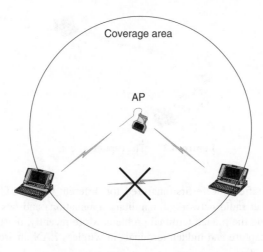

Figure 4.21   Hidden terminal problem.

the real throughput of a protocol is much more complicated because it involves hidden terminal and capture effects. To analyze these effects, let us assume we have a centralized AP with a number of terminals connected to it and communicating via a random access method.

Figure 4.21 demonstrates the basic concept behind the hidden terminal problem. The two terminals contending to communicate with the AP are both in the coverage area of the AP but they are out of the coverage area of each other. Limited antenna range and shadowing are two major causes for hidden terminal degradation. A hidden terminal problem does not affect the performance of the ALOHA-type protocols but it degrades the performance of CSMA protocols. In a CSMA environment affected by the hidden terminal problem, some terminals cannot sense the carrier of the transmitting terminal and their transmitted packets have a higher probability of colliding degrading the overall throughput.

In real installations, the coverage area of the AP is usually larger than that of the mobile terminals, because the AP is installed in a selected location to optimize the coverage (high on the walls or on the ceiling) that will increase the negative impacts of the hidden terminal problem. Assuming that the coverage area of the AP and the mobile terminals are the same, there is still no guarantee that all the terminals in the coverage area of the AP can hear one another. This is because two terminals at the maximum distance L from the AP could be as far as $2L$ apart. Therefore, the hidden terminal problem is unavoidable and natural to the operation of the centralized access systems using CSMA protocol that are common in wireless LAN operations.

Another phenomenon impacting the throughput of a radio network is capture. In radio channels, sometimes collision of two packets may not destroy both packets. Because of signal fading or the near–far effect, packets from different transmitting stations can arrive with different power levels, and the strongest packet may survive a collision. Figure 4.22 shows the basic concept of the capture phenomenon. The received power from the terminal closer to the access point is much larger than the received power from the terminal located at a distance. If two packets collide in time, the packet with the weaker signal will appear as a background noise and the AP captures (detects) the packet from the closer terminal successfully. The capture effect increases the throughput of the radio network because in calculating the throughput, we always assume that the colliding packets are destroyed (not detected).

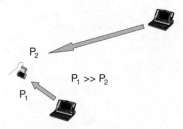

Figure 4.22   The capture effect.

The hidden terminal effects were first analyzed for different types of CSMA protocols used in rapidly moving packet radio networks for military applications and busy tone signaling was suggested for eliminating the hidden terminal problem. More recently, there have been efforts to analyze the effects of capture and hidden terminals in wireless LAN environment using various assumptions [Zha92; Zah97].

In reality, the capture of a packet is a random process, which is a function of the modulation technique used for transmission, received signal to noise ratio, and the length of the packet.

Example 4.32: Capture effect and throughput

Figure 4.23 [Zha92] shows the effects of capture on the throughput of the conventional slotted ALOHA and the CSMA systems for variety of packet lengths of 16, 64, and 640 bits. Also shown

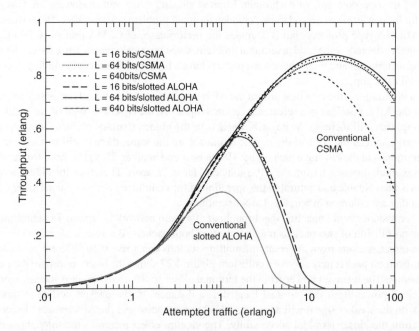

Figure 4.23   Effects of packet length on throughput for CSMA and slotted ALOHA with capture. The modulation is BPSK, and the SNR = 20 dB.

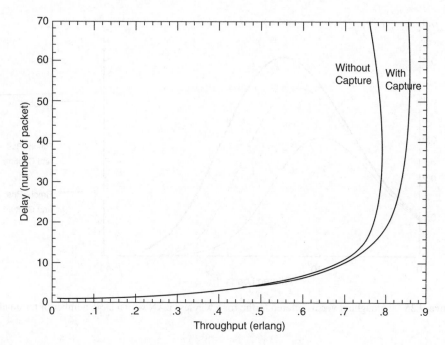

Figure 4.24 Delay versus throughput of CSMA for BPSK modulation and SNR = 20 dB, with and without capture.

for comparison are the curves for conventional non-persistent CSMA and slotted ALOHA without capture. With capture, the maximum throughput of CSMA with packet length 16 bits is 0.88 Erlang, which is 0.065 Erlang more than the case without capture. The maximum throughput for slotted ALOHA with the same packet length is 0.591 Erlang, which is 0.231 Erlang higher than the case without capture.

In slow-fading channels, if the terminal generating the test packet is in a "good" location, the interference from other packets is small and all the bits of the test packet survive the collision. In contrast, for a test packet originating from a terminal in a "bad" location, all the bits are subject to high probability of error and the packet does not survive the collision. As a result, the system shows minimal sensitivity to the choice of packet length, which is consistent with our assumption of slow fading. Figure 4.24 [Zha92] shows the delay-throughput for the CSMA protocol with and without capture for a 640-bit packet network. The packet delay is normalized to the length of the packet. For both cases as the throughput reaches around its maximum value system becomes rapidly unstable causing unacceptable delays for the delivery of the packets. When the capture effects are included maximum throughput is increased and the instability occurs at a slightly higher value of the throughput.

Example 4.33: Effect of capture and hidden terminals in a WLAN environment

Figure 4.25 [Zah97] shows the throughput versus offered traffic curves for a wireless LAN access point using CSMA protocol and surround by a large number of terminals uniformly distributed within the AP's coverage area. In this scenario, as shown in Figure 4.26, each terminal senses a

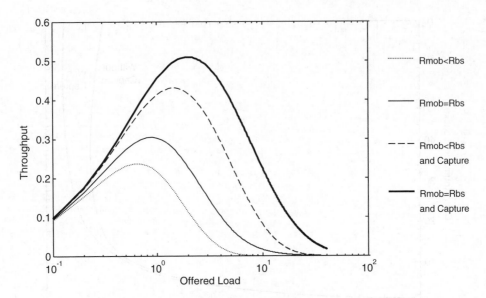

Figure 4.25   Throughput versus offered traffic for a wireless LAN with a large number of terminals.

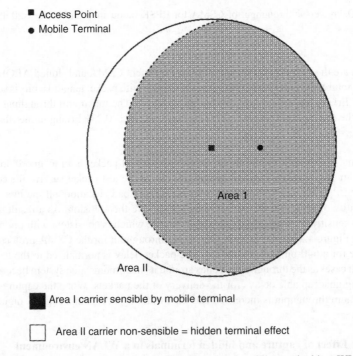

Figure 4.26   Coverage areas of an AP and a tagged mobile terminal in a WLAN.

group of terminals within its coverage area (area I in the figure) and cannot sense those that are out of its coverage area but are still within the coverage area of the AP (area II in the figure). The throughput of the target terminal with respect to terminals in the area I is the same as the throughput of a CSMA system. However, the throughput of the target terminal with respect to terminals in the area II is the same as the throughput of ALOHA networks because carrier sensing does not work and the terminals transmit their packets without knowledge of the transmission from terminals in the area II. Using these facts in [Zor97] the throughput at each point in the area of the coverage of the AP is calculated and then it is averaged over the entire coverage area of the AP for different coverage areas for the mobile terminals. Obviously, this average throughput will always remain between the throughput of CSMA and that of ALOHA. The lowest curve in Figure 4.22 shows the throughput when the hidden terminal problem is considered and the coverage of each terminal is 70% of the coverage of the AP. Because a number of terminals can not sense the transmission of others, the peak throughput has declined to less than 25% that is slightly higher than 18% maximum throughput of the ALOHA and far below the maximum of CSMA. The second curve from below, with a peak value of around 30%, represents the same results where the coverage of the AP and the mobile terminals are the same. The third curve depicts the performance when the both effects of hidden terminal and capture are considered and the coverage of the mobile terminals is smaller than that of the AP. The capture effect increases the throughput to more than 40%. The top curve is the same as the third curve where the coverage of the AP and the mobile terminals are the same. This situation has increased the throughput by another 10% to above 50% that is getting closer to the performance of conventional CSMA.

4.4   Integration of Voice and Data Traffic

As the wireless communications industry moves toward 3G and 4G networks, one of the important objectives is the use of a single wireless system for multi-media applications to support a variety of communications services, including voice, data and voice in various forms and combinations. A key technical problem to be dealt with in such integrated systems is that of multi-user access. As we saw earlier in this chapter, an access method that efficiently supports one category of service may be unsuitable for another category of service. In 1G and 2G networks, as we saw in Chapter 1, the wireless industry evolved around two separate paths for voice-oriented and data-oriented applications. If a data service can be efficiently integrated with a voice service, transmission resources that are otherwise wasted (because there is no voice transmission) can be used for data, which typically does not have stringent delay requirements. Firstly, the voice-oriented networks evolved into supporting data. More recently with the popularity of voice over IP over the Internet and PSTN, supporting voice in wireless LANs has become attractive as well.

4.4.1   Access Methods for Integrated Services

As we saw earlier, in a packet communication environment, voice and data have different requirements. Voice packets can tolerate errors and even packet losses (a loss of 1–2% of voice packets has insignificant effect on the perceived quality of reconstructed voice [Kum74]), while data packets are sensitive to loss and errors but can generally tolerate delays. Also, the rate at which information is transmitted is constant in the case of voice, thereby making circuit-switching a viable and efficient approach whereas information generated for data transmission is very bursty. As a result, voice and data oriented networks use different multiple access methods. In a wireless environment, the simplest approach is to assign different frequency bands to isochronous (voice) and asynchronous (data) packets. However, integration in one frequency band will result in a more efficient usage of the bandwidth, a simpler radio interface, and an environment that provides a better control for synchronizing voice and video (e.g., lip-sync).

4.4.2   Data Integration in Voice-Oriented Networks

Fixed access methods such as FDMA, TDMA, and CDMA were basically designed for access to the circuit switched voice-oriented networks. Later, as we saw in Chapter 1, several data services evolved around these systems. The economical incentive for using this medium for mobile data services is to take advantage, either partially or fully, of the existing infrastructure, the terminals, and the frequency bands designed for the voice oriented networks. This way the mobile data service provider saves in the major costs of deployment that includes the cost of real estates and the installation of the antenna and there no longer is a need to obtain new frequency bands for operation of the data service. If possible, using the same terminal for voice and data will reduce the cost and facilitates marketing of the service.

Example 4.34: Mobile data over FDMA analog cellular

The Cellular Digital Packet Data (CDPD) was introduced in the early 1990s and it used available frequency channels in the existing analog FDMA cellular telephone network (AMPS) to provide an overlaid packet data service supporting data rates similar to voice-band modems (up to 19.2 kbps). This system did not exploit the pauses between talk spurts but simply took advantage of the frequency bands temporarily unused by mobile telephone users in each cell area. CDPD used the unused AMPS channel to develop a communication link between a mobile data unit and a mobile data base station. Ideally, a CDPD terminal could use the RF and antenna of an AMPS terminal to communicate packet data bursts. However, the most important issue for the CDPD network was that it could use the same antenna site and the antenna towers and the frequency bands of an existing AMPS network. Since real estate, installation of the antenna post, and the frequency bands are perhaps the most expensive parts for implementation of a network, CDPD was perceived to provide a cost efficient solution for a mobile data service with a comprehensive coverage. The air interface protocol and modulation technique used in the CDPD were however different from that of the AMPS system.

Example 4.35: Mobile Data over TDMA systems

The GPRS packet data network, introduced in the late 1990s, uses the air interface and the infrastructure of the GSM network to provide a mobile packet data service that can support data rates of up to a couple of hundred kilobits per second. GPRS uses the same physical packet format and modulation technique as GSM. The logical channels used in GPRS do not use the dialing procedure used in the GSM. In a manner similar to CDPD, through the wired infrastructure of the network, the packets of data in GPRS are routed to the Internet rather than being switching through the PSTN. GPRS is designed to take advantage of the unused time slots of a TDMA voice-oriented GSM network.

Example 4.36: Mobile data over CDMA networks

In TDMA systems and FDMA systems the data users may use free time slots and free channels, respectively, as they become available. In a CDMA system the situation is somewhat different. The structure of CDMA is such that all active users use the entire bandwidth-time space simultaneously. The resource to be managed is signal power. With the application of efficient power control algorithms, the signal levels transmitted by mobile stations and the base station are continually adjusted in response to the changing locations of mobiles and the number of users on the system at any given time. In a CDMA network, the integration of data calls with voice calls is straightforward in principle, since various numbers of both categories of calls are readily mixed together, with each call accessing the channel with its unique user signal code. Therefore in a CDMA system no modification need be made to the channel access scheme to accommodate integration of voice and data "channels", and the information rate for voice or data traffic in any one channel can in principle be varied by a variable-rate scheme such as is used for voice service in the IS-95 standard. The integration of voice and data services in a single user channel is not necessarily straightforward.

From a technical point of view, there are two incentives for integration of data into voice-oriented fixed-assignment access methods:

1. The fixed-assignment access methods used in voice-oriented networks are designed to support a certain number of simultaneous users. When the number of active users falls below that number, some portion of the transmission resources is wasted.
2. A typical two-way conversation does not make full use of the call connection time, since only one of the parties talks at a time. Furthermore the flow of natural speech is actually composed of *talk spurts* with intervening short pauses. It is generally estimated that in a two-way voice connection, the average *voice activity factor* for each party is in the vicinity of 40% and thus about 60% of available transmission time remains unused.

Assume that we have N voice channels available for a given area (e.g., the coverage area of a sectored antenna in a cellular deployment) to accommodate newly originated calls as well as calls handed off from other areas. Further assume that the overall calls in the area are generated according to a Poisson process with normalized rate of $\rho = \lambda/\mu$ calls/unit channel and length of call holding is generally distributed with a unit mean. The number of idle channels, N_{idle}, according to renewal theory [Bud97] is given by:

$$N_{idle} = N_u - p[1 - B(N_u, \ p)] \tag{4.10}$$

where $B(N_u, \rho)$ represents the blocking probability for the M/G/1 queue calculated from the Erlang B equation given by:

$$B(N_u, \rho) = \frac{\rho^{N_u}/N_u!}{\sum_{i=0}^{N_u} (\rho^i/i!)} \tag{4.11}$$

Figure 4.27 shows the average number of the idle voice channels per area N_{idle} versus the number of available channels N for variety of call blocking rates. At call blocking rates of around 2% that is desirable for most cellular systems, a number of free channels are available for data communications. As the network accepts larger values of blocking rates, because most of the time all channels are in use, regardless of the number of channels, only a few channels will be available for data. If the integrated system uses the idle channels for data transmission, on the average, the maximum throughput available to the data user is N_{idle} times the encoding rate of the voice channel. If the

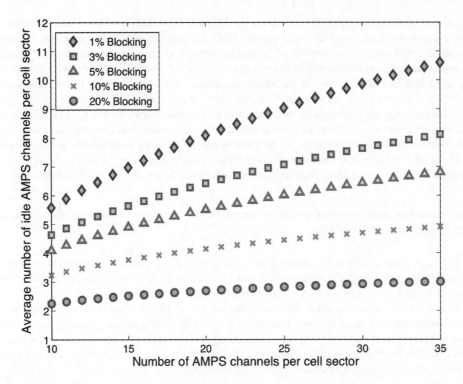

Figure 4.27 Average number of idle channels per area as a function of the number of available channels for a given call blocking rate.

system can take advantage of the silence periods in the two-way telephone conversations, as we indicated earlier, an additional throughput of up to 60% is available for data applications.

Another important parameter is the idle time for a voice channel. The idle time is the period of time that a voice channel is not occupied by a voice user. In an N channel voice-oriented network, assuming that each channel receives an equal fraction of the call load, one may calculate T_I, the average length of time a channel is idle, by [Bud97]:

$$T_1 = \frac{N_u - \rho \times [1 - B(N_u, \rho)]}{\rho \times [1 - B(N_u, \rho)]} \tag{4.12}$$

Figure 4.28 shows the normalized average idle period per channel T versus the number of available voice channels for different blocking rates. For a typical holding time of around a couple of minutes and a blocking rate of around 2%, the average idle periods are fairly long implying that a data network has a reasonable time to detect the availability and send its bursts of data.

There are periods of time for which all the voice channels are occupied for the voice users and there is no channel available to the data service. If these periods are short and infrequent some data applications may accept the situation. Otherwise, specific channels should be allocated for data only usage. In these situations the data users have their own channel as well as unused portions of the voice channels.

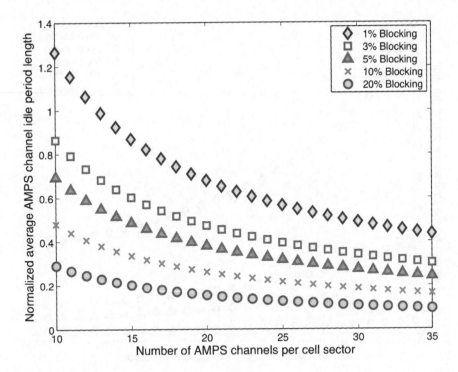

Figure 4.28 Normalized average idle period per channel as a function of the number of channels and call blocking probability.

Assuming that we accept the block out periods and assign no dedicated channel resources to the data application, we will have periodic operation between the available and blocked out periods. Assuming that the holding time for the telephone conversations is exponentially distributed, it can be shown [Bud97] that the average active period where a channel is available for data, T_a, is given by:

$$T_a = \frac{1 - B(N_u, \rho)}{N_u B(N_u, \rho)} \qquad (4.13)$$

The mean length of the blackout period, T_b, for this case is independent of the call load and hence blockage rate and it is given by:

$$T_b = \frac{1}{N_u} \qquad (4.14)$$

Figure 4.29 shows the normalized mean length of the active period T_a versus the number of channels. As the call blocking rate increases, the active period shortens leaving the system in more blackout periods. For a blocking rate of 5% or less and with fewer than 25 channels, the system has the equivalent of one dedicated channel for data applications. At higher blocking rates and data applications that cannot tolerate blackout periods, the data service must be deployed with at least one dedicated channel. Some practical examples are helpful at this stage.

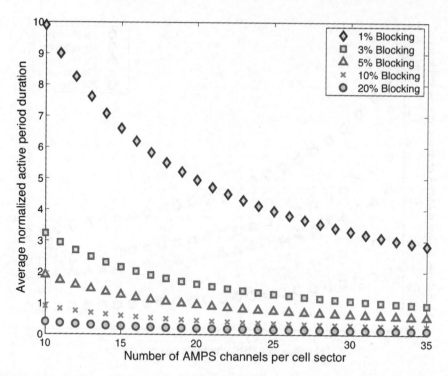

Figure 4.29   Mean length of available time for data as a function of the number of voice channels.

Example 4.37: Data overlay in FDMA systems – CDPD

The above analysis was actually developed for CDPD and all of the above analyses and discussions are directly applicable to it. CDPD operated over analog FDMA cellular networks at the rate of 19.2 kbps per channel that corresponds to digital transmission using GMSK modulation over 30 kHz AMPS channels. CDPD supported a channel hopping feature that allows a mobile data terminal to move to another channel during a communication session releasing the current channel for voice telephone conversation. This feature helps maintain the blockage probability at its nominal value and allows continual operation during the handoffs. The weakness of CDPD data overlay is that it does not assign several voice channels for one data user to support higher data rates. In general, in an FDMA system, the assignment of multiple voice channels to a single data user involves simultaneous operation of several RF channels by one terminal that is not practically attractive. For the same practical reason, data overlay in FDMA systems encounters difficulties in taking advantage of the silence periods during telephone conversations.

Example 4.38: Data overlay in TDMA systems – GPRS

An example of TDMA data overlay is GPRS. All of the above analysis is applicable to GPRS applications as well. However, the format flexibility of TDMA allows multi-slot assignment to support higher data rates. GPRS also does not take advantage of silence periods in two-way telephone conversation.

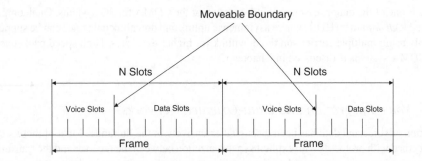

Figure 4.30   Frame structure in a moveable boundary frame-polling system.

An efficient method of integrating voice and data packets is a *movable boundary TDMA* scheme with *silence detection*. This method has been applied in the time-assignment speech interpolation (TASI) system used in T1-carrier telephone networks [Fis80] to maximize the number of voice users carried and to integrate data transmission into the channel. Using this basic idea, it is possible to design a *TDMA/framed-polling* protocol to integrate voice and data packets in a WLAN. This system consists of a number of voice and data terminals and a central station, which coordinates all the transmissions [Zha90]. The protocol for integration of voice and data packets is a movable boundary TDMA scheme shown in Figure 4.30. A frame is divided into two regions with a boundary between them. The first region is used for both voice and data traffic where the voice traffic has priority. If not, voice packets occupy all the slots in this region and the remaining slots are used for data traffic. The second region is reserved exclusively for data traffic. The boundary between the voice and data regions moves in accordance with the number of active voice packets in each frame. The maximum number of voice packets per frames is N_1 which is assigned an appropriate value to ensure some minimum data traffic capacity and to keep the blockage of voice packets below a selected value (2% in [Zha90]).

The result of extensive analysis and simulation in [Zha90] provides a simple experimental relation between the capacity that can be allocated for voice and data applications: $D = R_T - 0.032 \times N_v - 0.29$ where D is the data rate in Mbps available for data applications, N_v is the number of active 64 kbps PCM encoded telephone conversations and R_T is the transmission rate available on the medium. We can apply this equation to the TASI system that uses this protocol to accommodate 30 voice users with some additional data in a traditional 24-voice channels system (T1 carrier). The transmission rate is $R_T = 0.064 \times 24 = 1.536$ Mbps to support 24 voice users at 64 kbps. Using the equation for $N_v = 30$ active users, the data throughput with maximum delay of 10 ms (used in the simulation) is 286 kbps. The new protocol supports 6 more voice users plus 280 kbps data. With the same 24-voice users 487 kbps would be available for data applications that are around 30% of the overall transmission rate.

Example 4.39: Data overlay on CDMA

As we saw earlier in this chapter, integration of bursty data with voice in the CDMA system is very simple and CDMA systems already take advantage of the voice activity factor. Therefore with the same infrastructure and terminals, data services can be overlaid on CDMA. If higher data rates are needed one can either reduce the processing gain of the data channel or assign several parallel channels for one data link. Indeed the natural flexibility of CDMA to accommodate a variety of data services is one of the major reasons behind selection of the CDMA for 3G systems. Qualcomm has suggested *high data rate* (HDR) where asymmetric uplink and downlink data rates can be supported by simply using multiple carriers on the downlink for higher data rates. High speed packet access in 3G CDMA systems is discussed in Chapter 12.

4.4.3   Voice Integration into Data-Oriented Networks

Integration of voice and data has been discussed extensively in the literature. Most of these studies are concerned with protocols with explicit synchronization between the receiver and the transmitter. These approaches use assignment-based protocols for integration of voice and data, which allocate a fixed reference time such as a slot for transmission of packets. Synchronous systems provide more control on delay for voice traffic but less flexibility for bursty data traffic. Another approach that does not need explicit synchronization between the receiver and the transmitter is the asynchronous approach. The asynchronous packet approach mostly uses protocols extended from packet data networks, which are more suited for bursty data traffic. The voice traffic in this approach requires relatively complicated handling to limit the delay.

Contention-based packet communications protocols such as ALOHA and CSMA are used for data oriented wireless networks. They are especially well suited to networks comprising of many user stations each with low average data rate and potentially high peak rates. These protocols can operate with little or no centralized control, and can generally accommodate variable numbers of users in the network. However, contention-based schemes can become very inefficient in sharing the communications resources when the traffic load is heavy, as the system throughput degrades and the transmission delays increase. The unpredictability of throughput and time delays make these access methods unattractive for a voice dominated communication services, where a minimum throughput and delay is essential for user acceptance of the quality of service. Up to recent times (the Internet and wireless age), wired telephone services and the PSTN were producing the dominant source of income for the telecommunications industry. In the past century, telephone users have accepted the quality of the PSTN wired voice services as a normal standard for telephone conversations.

QoS in Voice Services

In the language of digital packet communications, the QoS of the PSTN voice user is specified by a guaranteed 64 kbps PCM (or 32Kbps ADPCM) coded data rate and a maximum delay of around 100 ms. We refer to this QoS as *wire-line-voice quality*. With the introduction of cellular telephone services in the late twentieth century, users accepted a lower QoS that suffered from the effects of fading due to the radio channel and dropped calls due to handoff, lack of coverage, or other reasons. As we saw in Chapter 1, cordless telephony and PCS services were aimed at bringing their QoS close to that of wire line quality and 3G cellular, that is a merger of cellular telephony and PCS services, follows the same pattern. If the quality of voice in a cordless telephone were far below that of wireline quality, users would have rejected the service. This is because they have the choice to receive a better QoS (no drops or fading effects) with their wired telephone that is also available at home or in the office. However, users had to accept the lower quality QoS with cellular telephony because there was no other alternative service provision for vehicles and other mobile applications.

Another recent event deviating from the wireline QoS in the emergence of voice over Internet or as most people refer to, the voice over IP, phenomenon. Popularity of the Internet, its penetration into the home market, its capability to support multi-media, and the most important advantage, its uniform cost for local and long haul communications encouraged development of Internet telephony. Operating on a packet switched environment with contention-access, the QoS of these services are not guaranteed at all and in its present stage of technology, the quality of voice calls is well below that of wire-line quality. However, free international calls through an Internet connection have been an incentive for some users to try this option as well.

In wireless networks, voice over IP (VoIP) does not make sense for mobile data applications because these services provide low data rates and after all they are evolving as an auxiliary network over already existing voice-oriented networks. Nevertheless, in 3G networks, recently, the use of VoIP over their high-speed packet access protocols is being investigated. However, voice over IP can be considered for wireless LAN environments. Imagine a wireless LAN installation in a stock market hall supporting wireless terminals for the users working in the hall. It would be useful and beneficial if they had a voice over IP service at the same terminal to use it for their telephone conversations as well. This incentive has initiated preliminary work on voice over IP in a wireless LAN environment using contention based access methods. The work in [Fei99] determines the number of supported voice terminals in a wireless LAN environment under a variety of conditions.

Capacity of a Wireless LAN with Voice and Data

Wireless LANs are becoming very popular in indoor applications such as in stock exchange halls where mobile users demand a high-speed wireless data access to the network and voice capabilities for telephone conversations. To deploy such a network a mathematical framework is very helpful to compare the capacity performance of wireless LANs with voice and data services in different scenarios. Therefore, for an asynchronous wireless LAN using the TCP/IP protocol suite, we need to find an answer to two questions:

1. What is the number of network telephone calls that can be carried with a given amount of data traffic?
2. What is the maximum data traffic per user for a given number of voice users?

A mathematical framework to answer these two questions is provided in [Zah00] where the integration of voice and data with TCP/IP protocol that operates in an asynchronous CSMA access environment is analyzed.

To integrate voice packets in a TCP/IP environment, the first step is to select a speech coder. A variety of speech coding algorithms exist with different rates as discussed in Chapter 3. We first present some theoretical results and then discuss some experimental results reported in the literature. To reduce the load on the network generated by voice traffic, [Zha90] adopted IMBE that is a popular low data rate vocoder (4.8 kbps) with an acceptable quality of service. This vocoder has been used in INMARSAT-M and AUSSAT mobile satellite communication systems and is proposed for APCO-25 standard for narrow-band digital land mobile radio.

Using TCP/IP will provide two options for sending packets in the network: the transmission control protocol (TCP) and the user datagram protocol (UDP). As a streaming protocol, UDP has no support for error correction, acknowledgment, sequencing, and flow control. Under high traffic conditions, the lack of flow control in UDP may cause bandwidth saturation in the Internet that should be prevented by the application program. In contrast, TCP has an error-correcting mechanism, uses acknowledgment, and guarantees in-order packet delivery. These requirements demand additional overhead that increases the average delay and reduces the overall throughput of the network. In general data packets can tolerate delay but cannot tolerate packet loss while the voice packets can accept packet loss of the order of 1% but cannot afford delays of more than 200 ms between consecutive packets. In the system described in [Zha90] TCP is used for the data packets

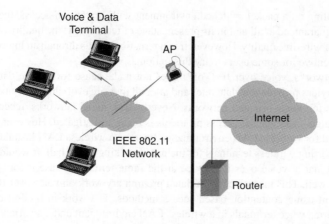

Figure 4.31   Schematic of a system employing voice and data terminals in a WLAN.

to guarantee accuracy of information and UDP for voice packets to handle the delay requirement. This approach is adopted in several products available in the market while other products use TCP for both voice and data. Although in our analysis TCP is selected for data transmission, there are some products, which use UDP for data transmission. In this case the upper layers are responsible for delivery accuracy.

Figure 4.31 represents a general overview of the system where several voice and data terminals communicate with an access point (AP) of a WLAN. Since the human ear is sensitive to time delays larger than 200 ms (T_{th}) in a voice conversation, wireless terminals should provide some facilities to minimize voice time delay. Allocating higher priority for transmitting voice over the data traffic is one way to decrease the voice packet time delay. Therefore, voice and data packets should be stored in a queue and wait for transmission as shown in Figure 4.32. The total delay for each packet transmission consists of a *queuing delay* at the terminal and *channel transfer delay*. The work in [Zha90] uses an *M/G/1* queue[5] with two priorities and a single server for node modeling. The arrival is modeled by a Poisson random variable. Figure 4.33 shows the system capacity and delay for T_{th} = 100 and 50 ms with 1 and 2 Mbps channel bandwidths. The maximum number of the voice users declines with reducing T_{th}.

Example 4.40: Capacity of a WLAN with voice and data users

Using Figure 4.30, find the number of users for T_{th} = 100 ms and T_{th} = 50 ms when the channel bandwidth is 1 Mbps.

Solution: From the figure a maximum of 18 voice users are supported with T_{th} = 100 ms and decreasing T_{th} to 50 ms will reduce the maximum voice users to 14. In this example the data traffic is less than 10 kbps.

In practice, there are a number of VoIP software packages and services, such as Skype, Line2, Asterisk, Speakfreely, Net2Phone, DialPad etc. that can be used to implement a real-time test bed

---

[5] An M/G/1 queue implies that packet arrivals are Poisson and the service rate of the queue has a general distribution with known mean and variance (see [Ber87]).

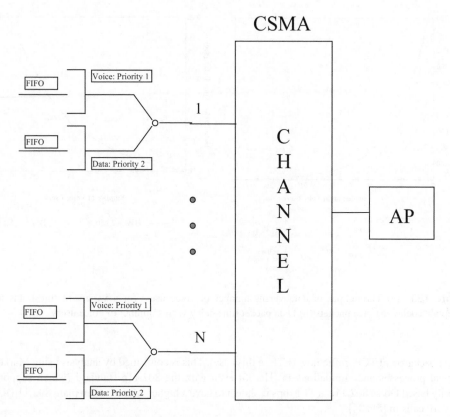

Figure 4.32 Queuing model for prioritizing voice traffic.

to analyze the behavior of voice in an IEEE 802.11 wireless LAN environment. The purpose of setting such experimental test beds or simulations is to determine the number of voice users that can be supported and relate it to the design parameters.

So how do real wireless LANs do in terms of carrying VoIP calls? A simple analysis that was performed in [Gar03] indicates that a single 802.11b cell can support only 3 to 12 VoIP calls with a G711 codec and 20 ms audio payloads depending on the average transmission rates of the mobile station. The analysis in [Gar03] was backed by limited experimentation where calls were gradually added to an 802.11b cell. When a seventh call was added, it caused all calls to result in unacceptable quality. Experiments revealed that the downlink (access point to mobile stations) was the affected part. The waiting times (inter frame spacing and backoff) and the packet overhead (headers and acks) in IEEE 802.11 drastically limit the capacity of a WLAN to carry VoIP calls. Similar conclusions under different scenarios were reached in [Ela04a, b] in an experimental testbed. The work in [Med04] confirms the poor capacity of 802.11 WLANs for carrying VoIP calls and extends this to 802.11g and 802.11a systems. Packet losses were ignored as also VoIP quality considerations in this work. If packets are lost, the capacity drops further.

The work in [Wan05] addresses the degradation in capacity due to the downlink by *aggregating VoIP payloads* at the access point into a single multicast packet intended for many MSs. With this scheme, up to 22 VoIP calls could be supported using a GSM 6.10 codec instead of 12 calls without this scheme. This work also demonstrates the degradation of VoIP calls in the presence of even a

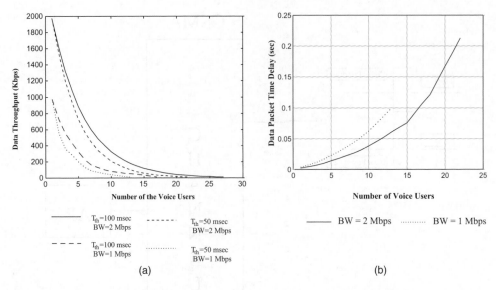

Figure 4.33 (a) Throughput of data versus number of voice users for a variety of thresholds for acceptable delay in voice packets (b) Data packet time delay versus number of voice users.

single background TCP traffic flow (FTP in this case). This is confirmed by analysis using Markov renewal processes and simulations in [Har06] even with the 802.11e standard (which supports priority based QoS). 802.11e at 11 Mbps is shown to have a higher capacity than plain 802.11 DCF for VoIP calls in [Sha04].

IP Telephony Using a Wireless LAN

The principles of operation of voice over contention based packet switched networks and assigned access circuit switched networks are very different. In a circuit switched network, a fixed connection between the two terminals is established during the call establishment procedure. This connection supports end-to-end communications with a fixed data rate and a controlled delay dominated by propagation delay and a negligible delay jitter. In packet voice communications with contention access and packet switched networks, delay and jitter are the dominant sources of the performance degradation. To regulate jitter, the receiver has a buffer to store the received packets at different delays but pump them to the user at constant intervals to reconstruct real-time voice. The performance of the system is then related to the size of the buffer at the receiver. Next, we provide a summary of the experimental work presented in [Fei99] to relate the throughput to the buffer size.

The first step is to describe the overall scenario in a practical situation for the implementation of the voice over IP in a wireless LAN environment. Figure 4.34 describes the arrival of packets. Because of random access and packet switched networking, the packets sent at fixed intervals arrive at a variety of delays.

The overall delay is the minimum network delay plus the individual jitter per packet that will be regulated with the jitter compensation buffer at the receiver. The packets arriving at different delays are stored in a buffer and the application at the receiver reads this buffer periodically. When the packet with the right sequence number is available the receiver reads the packet and plays it

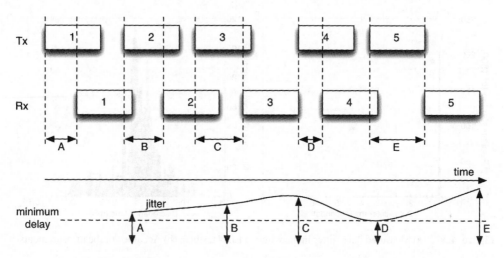

Figure 4.34 Illustration of the arrival of voice packets transmitted at a constant rate.

through the speaker. When the packet is not available the application software at the receiver skips that packet. A simple example further clarifies the operation.

Example 4.41: Jitter in VoIP on WLANs

Figure 4.35 illustrates the details of the operation of the receiver and the relationship between the jitter compensation buffer, arrival and playing time of the packets. When the first packet arrives it is delayed in the receiver's jitter compensation buffer and after the maximum allowed delay, it is delivered to the user application to be played in the first slot. The second packet is discarded because it has arrived after the deadline for playing. The third packet arrives normally before the

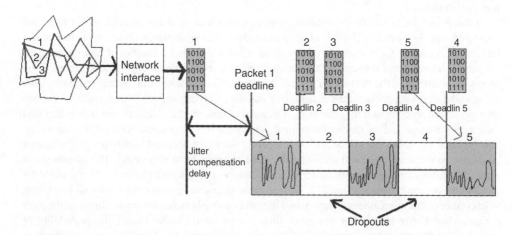

Figure 4.35 Reception of voice packets and buffering to maintain the appearance of no jitter.

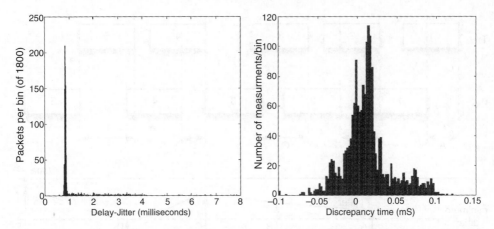

Figure 4.36 (a) Measured delay jitter in the wireless LAN testbed. (b) Accuracy of the measurements.

deadline and it is delivered at its appropriate time to the speaker. The fourth and the fifth packets arrive off-sequence. The fourth one is late (arriving after its deadline) and it is discarded. The fifth packet has arrived before the deadline of the fourth packet and it is shifted to its own time slot.

As we have discussed before, the user can accept a packet drop rate of around 1%. To reach the goal of 1% packet drop, the receiver has the choice of increasing the length of the jitter compensation buffer at the expense of additional overall delay at the receiver. Therefore, the length of the jitter compensation buffer is an important parameter in voice over IP applications. This parameter is adjusted by changing the length of the buffer at the receiver. The delay observed by the user is the minimum network delay plus the jitter compensation buffering delay. The above example also illustrates that in voice over IP applications, in addition to transmission packet losses occurring in the network, we have packet losses due to late arrival at the receiver (that is a function of the length of the jitter compensation buffer). Therefore, the length of the jitter compensation buffer and packet loss are interrelated.

To determine the relationship between the jitter compensation delay and the packet loss rate, a test bed was developed in [Fei99] to implement the scenario shown in Figure 4.31. In this test bed an infrastructure for wireless LAN operation using an AP and a number of laptops are used for measurement of the statistics of the delay jitter in a voice over IP application. Figure 5.36a shows the statistics of the delay jitter for 1800 packets. The measurement system transmits, time stamps, and stores the packet at the transmitting and the receiving laptops. The stored files are then post processed to eliminate the effects of differences between the clocks of the transmitter and the receiver laptops and we extract the refined delay jitter measurements. Figure 4.36b shows the accuracy of the system (that is measured by comparing the results obtained from two separate laptops connected to the same point and receiving the same message form a receiver). The measurement error (difference of measurements in two identical laptops) has a mean of around 0.01 ms, while the mean of the measurements is around 1.0 ms restricting the measurement error to around 1%. Using a delay jitter distribution, one can simply find the relationship between the packet loss and the jitter compensation buffer length. For any given jitter compensation buffer length, the probability of packet loss is the same as the probability of having a delay jitter larger than the jitter compensation buffer length. Figure 4.37 shows the experimental results for up to five stations operating in the

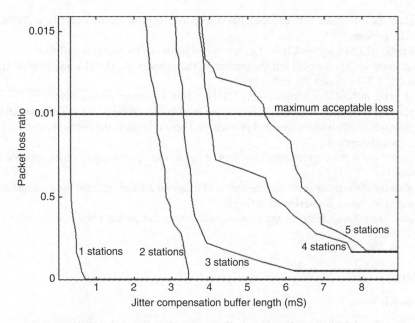

Figure 4.37   Packet loss versus jitter compensation buffer length.

wireless LAN test bed. If we fix the acceptable packet loss rate to 1%, the minimum buffer length increases from 0.5 to 7.0 ms when we increase the number of users from one to seven. The details of algorithms for the implementation of the test bed and the results of OPNET simulation for large number of voice users are available in [Fei99].

Questions

1. Name two duplexing methods and one example standard that uses each of these technologies.
2. What are the popular access schemes for data networks? Classify them.
3. Name a cellular telephony standard that employs FDMA.
4. What is binary exponential back-off algorithm, which standard uses that and what is the purpose of using it? What is its weakness?
5. What is the purpose of the IEEE 802 standard committee and what are the steps taken to make its recommendations an international standard?
6. What is the difference between the access techniques of the IEEE 802.3 and IEEE 802.11?
7. Why do most PCS standards use TDD and most cellular standards use FDD?
8. Why in the PSTN backbone hierarchy the FDM multiplexing lost its popularity to TDM multiplexing?
9. Why did 2G cellular systems shift from analog FDMA to digital TDMA and CDMA technologies?
10. Name three standards using TDMA/TDD as their access method.
11. What are the advantages of the CDMA access technique?
12. What is the difference between performance evaluation of voice-oriented fixed assignment and data-oriented random access methods?

13. Explain the difference between the effects of power–control on the capacity of TDMA and CDMA systems.
14. In a radio ALOHA network how does a terminal learn that its packet is collided?
15. What is the difference between the maximum throughputs of ALOHA and slotted-ALOHA networks? What causes this difference?
16. What is the difficulty of implementing CSMA/CD in a wireless environment?
17. Explain the difference between carrier sensing mechanisms in the wired and wireless channels.
18. Explain what is the hidden terminal problem and how it impacts the performance of a CSMA based access method.
19. Explain what is the capture effect and how it impacts the performance of the random access methods.
20. Explain the differences between integration of data into a voice-oriented network and integration of voice into a data-oriented network.
21. Explain the relation between the receiver buffer size and packet error rate in voice over IP applications.

Problems

Problem 4.1

To provide public telephone access to commercial ferries a telephone company installs a multi-channel wireless telephone system in a ferry. This wireless radio system connects to a base station in the shore through the air. The base station is connected to the PSTN using wires.

a. If the telephone company installs a four-channel system what is the probability of having a person come to the telephone and non of the lines are available? Assume that the average length of a telephone call is 3 min and 150 passengers of the ferry, on the average make 1 call/h.
b. What is the average delay for accessing the telephones?
c. How many channels were needed to keep the blockage probability below 2%?

Problem 4.2

a. Neglecting the frequency spectrum used for control channels, what is the maximum number of two-way voice channels that can fit inside the frequencies allocated to the AMPS system.
b. What is the number of channels in each cell. Note that $N_f = 7$ was originally used in the AMPS.
c. Repeat (b) for North American TDMA in which $N_f = 4$ and the number of slots per TDMA channel is three.
d. Repeat (b) for IS-95 CDMA assuming the minimum required Eb/No is 6 dB. Include the effects of antenna sectorization, voice activity, and extra CDMA interference.
e. Repeat (d) for broadband CDMA where a 5-MHz band is used in each direction.

Problem 4.3

a. Sketch the throughput versus offered traffic G for a mobile data network using slotted non-persistent CSMA protocol. The packets are 20 ms long and the radius of coverage of each BS is 10 km.

   Assume the radio propagation speed is 300 000 km/s and use the worse delay for calculation of the "a" parameter.

b. Repeat (a) for slotted ALOHA protocol.

c. Repeat (a) for 1-persistent CSMA protocol.

d. Repeat (a) for a wireless LAN with access point coverage of 100 m.

e. Repeat (a) for a satellite link at a distance of 20 000 km from the earth.

Problem 4.4

Use the equations given in [Bud97] to reproduce Figures 6 and 8 in that paper.

Problem 4.5

A cellular carrier has established 100 cell sites using AMPS with 395 channels and $K = 7$.

a. Use the provided graphs to calculate the total number of subscribers for a blocking probability of 0.02, average of 2 calls/h, and average telephone conversation of 5 min.

b. Use a mathematical software tool to calculate the same values using the Erlang B equation directly.

c. Determine (either from plot or calculation) the average delay for a call.

d. Repeat (a) for a blocking probability of 0.01.

e. Repeat (a) if North American TDMA cellular was used with $N_f = 7$.

f. Repeat (a) if North American TDMA cellular was used with $N_f = 4$.

Problem 4.6

a. Sketch the throughput versus offered traffic G for a mobile data network using slotted nonpersistent CSMA protocol. The packets are 20 ms long and the radius of coverage of each BS is 10 km.

   Assume the radio propagation speed is 300 000 km/h and use the worse delay for calculation of the "a" parameter.

b. Repeat (a) for the slotted-ALOHA protocol.

c. Repeat (a) for the 1-persistent CSMA protocol.

d. Repeat (a) for a wireless LAN with an access point coverage of 100 m.

e. Repeat (a) for a satellite link at a distance of 20 000 km from the earth.

Problem 4.7

We want to use a GSM system with sectored antennas ($N_f = 4$) to replace the exiting AMPS system ($N_f = 7$) with the same cell sites. In the existing AMPS system the service provider owns 395 duplex voice channels.

a. Determine the number of voice channels per cell for the AMPS system.

b. Determine the number of voice channels per cell for the GSM system.

c. Repeat (b) if we were using a W-CDMA system with the bandwidth of 12.5 MHz for each direction. Assume a signal to noise ratio requirement of 4 (6 dB) and include the effects of antenna sectorization (2.75), voice activity (2), and extra CDMA interference (1.67).

Problem 4.8

We provide a wireless public phone with four lines to a ferry crossing between Helsinki and Stockholm carrying 100 passengers where on the average each passengers makes a 3 min telephone call each 2 h.

a. What is the probability that none of the four lines is available when a passenger approaches the telephone.
b. What is the average delay for a passenger to get access to the telephone.
c. What is the probability of having a passenger waiting more than 3 min for access to the telephone.
d. What would be the average delay if the ferry had 200 passengers.

Problem 4.9

A WLAN hop accommodates 50 terminal running the same application. The transmission rate is 2 Mbps and the terminals are using slotted ALOHA protocol. The commutative traffic produced by the terminals are assumed to form a Poisson process.

a. Give the throughput versus offered traffic equation for the system and determine the maximum throughput in Erlangs.
b. What is the maximum throughput in bits per second.
c. What is the maximum throughput in bits per second for each terminal.

Problem 4.10

A local 3-h tour boat with 50 passengers has one AMPS radio phone to connect to the shore. On the average each user places one call per tour and the average holding time for the calls is 3 min.

a. What is the probability that a person attempts to use the phone and he/she finds it occupied?
b. Repeat (a) if the AMPS phone is replaced by three NA TDMA cellular phones using the three slots of the NA TDMA cellular over the same band.
c. Repeat (a) if this phone is replaced by six upgraded NA TDMA cellular phones using 6-slot upgraded IS-54 TDMA over the same band.

Problem 4.11

In a datagram packet switched network (Figure P4.1) with:

P: packet size in bits

N: number of hops between two given systems

B: data rate in bps on all links

H: overhead (header in bits per packet)

T: end to end delay

N_p: number of packets

L: message length in bits

D: propagation delay per hop

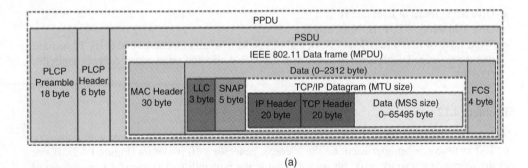

(a)

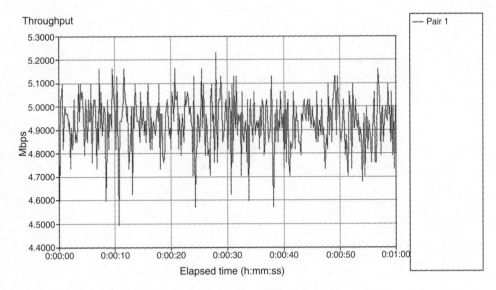

(b)

Figure P4.1 Packet transmission in the IEEE 802.11b. (a) Overheads for the formation of a packet. (b) Overheads for successful transmission of a TCP packet.

Figure P4.2 Throughput variations in one location for 802.11b.

a. Give Np in terms of L, P, and H
b. Give T in terms of L, P, H, N, B, and D.
c. What value of P, as a function of N, B, and H, results in minimum end-to-end delay T? Assume that the message length is much larger than the packet size and propagation delay is negligible $(D = 0)$.

Problem 4.12

An ad hoc 2 Mbps wireless LAN using ALOHA protocol connects two station with a distance of 100 m from one another each, on the average, generating 10 packets/s (Figure P4.2). If one of the terminals transmits a 100 bit packet, what is the probability of successful transmission of this packet. Assume that the propagation velocity is 300 000 km/s and the packets are produced according to a Poisson distribution.

Part II

Principles of Network Infrastructure Design

5

Deployment of Wireless Networks

5.1 Introduction

Wireless networks come in all shapes and sizes. While many aspects of these diverse networks have common foundations, there are several other aspects that differ widely and need separate treatment. It is thus important to classify and sample example wireless networks to obtain a better understanding of the issues discussed in this and the next few chapters. In this part of the book, we consider the general deployment and operational aspects of wireless networks. In this chapter, we will start with a discussion of where wireless networks fit in with respect to their wired counterparts. The rest of the chapter will focus on interference and frequency reuse for deployment of wireless networks. In Chapter 6, we will consider specific examples of different wireless networks – wide area, local area, and personal area networks. In the process, we will illustrate the challenges and issues in designing, analyzing, deploying and managing these wireless networks. Chapter 7 examines the security aspects of wireless networks.

In this chapter, we start by considering the place of wireless networks in a general networked architecture and then provide classifications of wireless networks that takes into account the topology and coverage of the network. The primary goal of networked systems is to enable connectivity between different devices. The underlying network enables connection of computers to computers, sensors to computers, cell phones to telephones, computers to telephones, and so on. Wired networks have usually been prevalent prior to the corresponding wireless networks (e.g., a wired local area network was implemented prior to a WLAN and wired telephony existed for several decades before cellular telephony became a reality). Backbone or core networks primarily make use of fiber, although microwave point to point links and satellite links have been used for long distance communication. So there is already this *fixed network* of devices and links that carries bits from some source to some destination, perhaps separated across continents. This network can be the public switched telephone network (PSTN) or the Internet although the boundary between the two has blurred significantly. We will just call this the "fixed network".

When we consider wireless networks today, we are not very interested in the point to point wireless links. What makes wireless networks interesting is the fact that they offer *portable* or *mobile* access to networks and applications. By portable, we mean the ability to carry the device *anywhere* and still be able to connect. By mobile, we mean the ability to stay connected at *anytime* even if the user is moving with the device at speeds of up to 100 km/h. Another type of wireless network of interest is one that is made up of *low-power* devices with small form-factor that may be fixed

Principles of Wireless Access and Localization, First Edition. Kaveh Pahlavan and Prashant Krishnamurthy.
© 2013 John Wiley & Sons, Ltd. Published 2013 by John Wiley & Sons, Ltd.

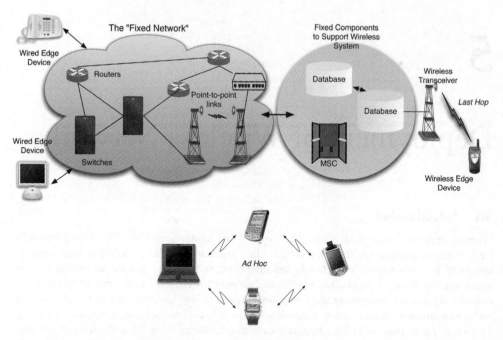

Figure 5.1 Positioning of the wireless network infrastructure in relation to the wired network infrastructure.

[such as sensor networks or radio frequency identification (RFID) tags]. Note that portability and mobility also mean that devices operate on battery power and have a reasonably small form-factor.

In order to modify the traditional fixed network infrastructures to support portable or mobile wireless connections a new infrastructure is needed as an interface between the backbone wired network infrastructure and the mobile communication terminals. The mobile communication terminals need to be equipped with wireless front-ends to communicate with the wired backbone through the new wireless infrastructure. Figure 5.1 shows the positioning of the wireless infrastructure in relation to the wired infrastructure.

In addition to switches, routers, and point to point links, the wireless network infrastructure also needs wireless transceivers to communicate with the wireless communication terminals and act as *points of entry* to the fixed part of the wireless network infrastructure. These transceivers are referred to as *base stations* (BS) or *access points* (AP). Any wireless base station has a limited coverage area. If the coverage area is less than the desirable coverage area for the wireless service, we need multiple base stations to cover the service area. In the case of multiple base stations operating in an area, the wireless network infrastructure needs to coordinate the continuity of a wireless connection as the mobile communication terminal moves through the coverage areas of different base stations.

5.2 Wireless Network Architectures

We have already qualified some types of wireless networks above by using adjectives such as "portable", "mobile", "last hop", "infrastructure", and "ad hoc". Now, we will try to formally come

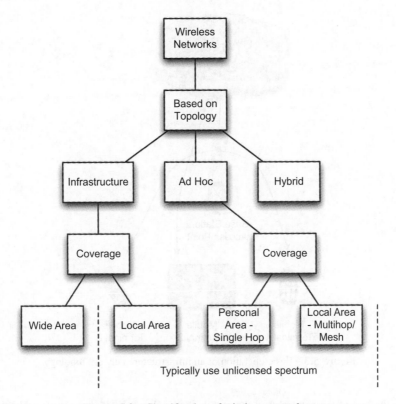

Figure 5.2 Classification of wireless networks.

up with a classification of wireless networks with specific example technologies and standards. We first consider the end wireless device. All devices like cell phones, laptops, sensors, RFID tags, and other devices that use wireless transmissions will be called *mobile stations* or MSs with exceptions where the context requires us to identify the device exactly (e.g., as a sensor). Note that some of these devices may not be mobile, but we will still call them MSs.

Figure 5.2 provides a classification of wireless networks. At the top of this classification, wireless networks have been classified into three broad categories based on their *topology*. We refer to *topology* as the configuration in which a mobile terminal communicates with another.

5.2.1 Classification of Wireless Networks Based on Topologies

An *infrastructure* topology is what we primarily encounter today. Wireless networks with infrastructure topology are the most well-developed with specific applications. Here, there is a fixed radio transceiver that connects to the fixed network. MSs in an infrastructure topology have to connect to each other or to other entities in the fixed network using this fixed radio transceiver. It is common to refer to this fixed radio transceiver as a *base station* (BS). There are other names used for the fixed radio transceiver. It is called an *access point* (AP) in WiFi because it is the point of access to the fixed network. A general term used in some standards is "co-ordination point" (CP). Figure 5.3 shows the basic operation of an infrastructure network with a single BS/AP. The BS/AP

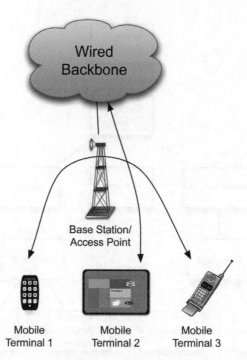

Figure 5.3   Basic operation of an infrastructure network topology.

serves as the hub of the network and the mobile terminals are located at the ends of the spokes. Any communication from one wireless user station to another, that is, between peers, has to be sent through the BS/AP. The hub station usually controls the mobile stations and monitors what each station is transmitting. Thus the hub station is involved in managing user access to the network.

In any case, it is possible for MSs to connect to the network through a given BS/AP over some area near the BS/AP. This area is called a *cell*. Thus, a number of "cells" cover the area and enable connectivity of MSs. The BSs or APs are connected over a wired or fixed network so that they can provide coverage over a given geographical area. Thus, in the infrastructure topology, there is a fixed (wired) infrastructure that supports communication between mobile terminals and between mobile and fixed terminals. The infrastructure networks are often designed to support large coverage areas and multiple base station or access point operations. Most of the discussion in this chapter is around this type of operation.

Example 5.1: Systems that employ infrastructure network topology

All standardized cellular mobile telephone and wireless data systems (1G, 2G, 3G, 4G) use an infrastructure network topology to serve mobile terminals operating within the coverage area of any BS. The IEEE 802.11 standard and most of the wireless LAN products support infrastructure operation. Typical architectures of cellular systems are described in Chapter 6.

The *ad hoc or distributed topology* is common with personal digital accessories like digital cameras, printers, headsets, cell phones, and the like that connect to one another as and when needed. For

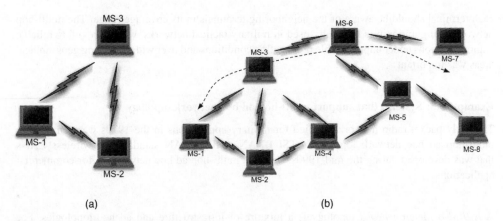

Figure 5.4 Ad hoc networking. (a) Single-hop peer to peer topology. (b) Multi-hop ad hoc networking topology.

example, a cell phone may connect to a headset when a call is made and disconnect from it after the call is completed. It may connect to a computer at a later time to synchronize calendars and address books. The computer itself may be simultaneously connected to a digital camera that is uploading its pictures. The word ad hoc implies that the wireless network has been formed or arranged for a particular purpose only. Once the purpose is completed, a MS may not be part of the network anymore. Ad hoc topologies are common in the case of sensor networks. A number of sensors may be randomly deployed to perform some activities over a given area. They may need to communicate with one another as needed. Such an architecture also applies to reconfigurable networks that can operate without the need for a fixed infrastructure. These networks are primarily used by the military and also in a few commercial applications for voice and data transmission. Such a topology is suitable for rapid deployment of a wireless network in a mobile or fixed environment. Figure 5.4 shows two variations of the ad hoc network topology. Figure 5.4a shows a single-hop ad hoc network where, as the name implies, every user terminal has the functional capability of communicating directly with any of the other user terminals.

Example 5.2: Systems that support single hop ad hoc network topology

The IEEE 802.11 wireless LAN standard supports single hop peer to peer topology for ad hoc networking. When a terminal is turned on, it first searches for a beacon signal from an AP or another terminal announcing the existence of an ad hoc network. If a beacon is not detected, the terminal takes the responsibility of announcing the existence of an ad hoc network. Several PCS services, such as PHS and NEXTEL satellite, support peer to peer walkie-talkie type communication among voice terminals. Bluetooth supports peer to peer connections between cellphones and computers or headsets.

In some ad hoc networking applications, where users may be distributed over a wide area, a given user terminal may be able to reach only a portion of the other users in the network, due to transmitter signal power limitations. In this situation, user terminals will have to cooperate in carrying messages across the network between widely separated stations. Networks designed to function this way are called multi-hop ad hoc networks, and are illustrated in Figure 5.4b. In an ad hoc multi-hop network, each terminal should be aware of the neighboring terminals in its coverage range. The multi-hop network configuration was originally used in military tactical networks, where providing reliable communications under unpredictable propagation conditions and over widely varying geographical areas was important.

Example 5.3: Systems that support multi-hop ad hoc network topology

The early packet radio networks studied for military applications in the 1970s were employing multi-hop ad hoc network topology. ETSI BRAN's HIPERLAN standard for wireless LANs that was developed during the mid-1990s, supports multi-hop ad hoc networking for commercial applications.

A *hybrid* or *heterogeneous* topology is a mixture of infrastructure and ad hoc topologies. For example, an ad hoc network of MSs may be created for emergency purposes in a disaster area. Some of these MSs may also be able to simultaneously connect to a cellular network forming a bridge between the ad hoc network and the fixed network. Sometimes, the words hybrid and heterogeneous are used to indicate networks with mixed technologies such as a cellular telephone network and WiFi network [Pah00]. In such networks, a MS should be able to seamlessly roam from a WiFi network to the cellular network or vice versa.

Comparison of ad hoc and infrastructure network topologies: A number of attributes can be used to compare infrastructure and ad hoc network topologies.

Scalability: In peer to peer single hop networks, expansion is always limited to the coverage of the radio transmitter and receiver and there is no simple way to scale up the network coverage or capacity (wireless traffic) that can be supported by the network. In multi-hop ad hoc networks as the number of terminals increases the potential coverage of the network is increased. However, the traffic handling capacity of the network remains the same. To connect an ad hoc network to the backbone wired network one needs to use a proxy server with a wireless connection as a member of the ad hoc network. In practice, all terminals supporting ad hoc networking operate in a dual mode that also supports infrastructure operation. Wireless infrastructure networks are inherently scalable. To scale up a wireless infrastructure network the number of BSs or APs is increased to expand the coverage area or to increase the capacity while using the same available spectrum. Therefore, for wide area coverage and for applications with variable traffic loads, infrastructure networks are always used. We look at how the available capacity can be expanded in an infrastructure topology later in this chapter.

Flexibility: Operation of infrastructure networks requires deployment of a network infrastructure which is very often time consuming and expensive. Ad hoc networks are inherently flexible and can be set up instantly. Therefore, ad hoc networks are always used for temporary applications where flexibility is of prime importance.

Controllability: To coordinate proper operation of a radio network, we need to centrally control certain features such as time synchronization, transmitted power of the mobile stations operating in a certain area, and so on. In an infrastructure network all these features are naturally implemented in the BS or AP. In an ad hoc network, implementation of these features requires more complicated structures demanding changes in all terminals.

Routing complexity: In multi-hop peer to peer networks each terminal should be able to route messages to other terminals. This capability requires each terminal to monitor the existence of other terminals and be able to connect to those available in the immediate neighborhood. For this, there is need for a routing algorithm that directs information to the next appropriate terminal. Implementation of these features adds to the complexity of the terminal and the network operation. In infrastructure and peer to peer single-hop ad hoc networks, this problem does not exist.

Coverage: In wireless LANs, coverage of the network is an issue of concern since it has an effect on the selection of the topology. In peer to peer single hop network topology, the maximum distance between two terminals is the range of coverage of the wireless interface used in the terminal. In an infrastructure network, two wireless terminals communicate through an AP or a BS. The maximum distance between two terminals is thus twice the range of coverage of a single wireless modem since the communicating terminals maybe located at the edge of the coverage area of the BS or AP. In practice, often APs or BSs are fixed in opportunistic locations using elevated mountings that increases the coverage of the wireless modem. This usually results in a maximum coverage distance between two terminals that is greater than twice the coverage distance of the same modem in an ad hoc configuration.

Reliability: Another issue of concern in small-scale wireless LAN operation and military applications in battlefields is resistance to failure. Infrastructure networks are "single failure point" networks. If the AP or BS fails, the entire communications network is destroyed. This problem does not exist in ad hoc peer to peer configurations.

Store and forward delay and media usage efficiency: In peer to peer single-hop networks information is transmitted only once and there is no store and forward procedure. In the infrastructure topology we have transmission of data twice, once from the source to the AP/BS and once from the AP/BS to the destination. The AP/BS also should store the message and forward it later. This adds to the delay encountered by the data packets. Multi-hop ad hoc networks may have several transmissions and several store and forward delays that depend on the instantaneous topology and number of hops required to send the data from the source to the destination.

5.2.2   Classification of Wireless Networks Based on Coverage

If we move a level down the classification shown in Figure 5.2, we see that both infrastructure and ad hoc networks can be further classified based on *coverage.* Coverage can be loosely defined as the contiguous geographical area where a MS can communicate with the wireless network of interest. If the geographical area is as large as a city, the wireless network is called a "Wide-Area" network. An example of a Wireless Wide Area Network (WWAN) is the cellular telephone network that provides coverage over cities and sometimes across States. For geographical areas smaller than a city, but large enough to include a building or several buildings in a campus, the wireless network is called a "Local Area" network. Wireless Local Area Networks (WLANs) based on WiFi are examples of local area networks. A "Personal Area" network is one that exists in a small space around a person and his devices. Usually, the communication is only across a single hop. That is, a device only communicates with another device if it is in range. For example, a cell phone connecting to a computer and a headset via Bluetooth creates a Wireless Personal Area Network (WPAN). A device typically does not relay packets or messages between two other devices by acting as an intermediary. A second type of ad hoc network allows relaying of packets by devices across multiple hops so that the effective communication area is larger (for example a local area). Note that ad hoc networks that cover local areas have been clubbed into the same category as "mesh" networks. The reason for this is that ad hoc networks are neither widespread nor standardized although prototype implementations and working sensor networks do exist. We will revisit this topic later. A last comment on Figure 5.2 is on the use of *licensed* and *unlicensed* spectrum by wireless networks. Most local and personal area networks use unlicensed spectrum with some regulations on the transmit power levels and fair usage. WWANs, in contrast, use licensed spectrum that is allocated to the service provider by the Federal Communications Commission (FCC) in the United States or another regulatory authority elsewhere.

There are other classifications of wireless networks that are possible – based on the mobility of devices and based on the power consumption of devices. We have already discussed the difference between portability and mobility. Sometimes, fixed and stationary wireless connections are also included in this mix. Power consumption is usually proportionally related to the coverage – a device needs a higher transmit power to communicate over larger one-hop distances. Transmissions in WWANs require higher power than those in sensor networks with WLANs being somewhere in between. However, transmission power and radio propagation also impacts the deployment of wireless networks because of the interference that may be caused to transmissions in adjoining geographical areas. We explore this next in the context of unlicensed spectrum and local wireless networks.

5.3   Interference in Wireless Networks

When two wireless network overlap in their coverage and operate at the same frequency at the same time without any access coordination, they will obviously interfere with one another. For example, several cordless phones, Bluetooth, ZigBee WPANs, and Wi-Fi WLANs operate at the 2.4 GHz unlicensed bands and they are subject to interference from each other. The literature on military communication systems offers many detailed analyses of the performance of communication systems in the presence of various intentional interferers or jammers [Sim85]. These jammers are designed to disrupt the operation of a system and they can employ relatively sophisticated techniques such as multi-tone jamming and pulsed jamming. In civilian applications, interference is neither intentional nor sophisticated. Most often, the interferer is simply another system designed to operate in an adjacent band or a portion of or the entire band of operation of a given system and the users are generally willing to cooperate so as to minimize the mutual interference. WWANs have to maintain specific quality of service due to regulatory restrictions and to satisfy paying subscribers. As we will see later, deployment of such networks is more complex and involves careful frequency reuse or interference management (we devote Sections 5.5–5.8 for the deployment of WWANs). WLANs and WPANs, due to the use of unlicensed bands and the relative simplicity of architecture are usually deployed with minimal planning unlike cellular telephone and WWANs. Depending on the level of coordination of the overlapping wireless network since the early days of the IEEE 802.11 [Hay91], the WLAN industry has specified three types of overlapping networks: interference, coexistence, and interoperation.

Multiple wireless networks are said to *interfere* with one another if collocation causes significant performance degradation of any of the devices. Multiple wireless networks are said to *coexist* if they can be collocated without significant impact on the performance of any of the devices. *Coexistence* provides for the ability of one system to perform a task in a shared frequency band with other systems that may or may not be using the same set of rules for operation. *Interoperability* provides for an environment for multiple overlapping wireless systems to perform a given task using a single set of rules. In an interoperable environment multiple wireless networks exchange and use the information among each other. Interoperability is an important issue for wired as well as wireless networks.

Coexistence and interference are issues that are significant primarily for wireless network designers and it becomes more important for the case of ad hoc networks. This terminology for unlicensed bands was first discussed in the IEEE 802.11 community [Hay91]. Later on when FCC released unlicensed PCS bands the issue of *etiquettes* or rules of coexistence in unlicensed PCS bands attracted attention [Pah97] that ultimately lost its momentum, as the unlicensed bands did not gain significant popularity. Around year 2000, the IEEE 802.15 WPAN group was engaged in interference analysis in one of its task groups. They performed preliminary interference analysis between Bluetooth and IEEE 802.11 devices operating in 2.4 GHz unlicensed bands working on practical coexistence and interoperability methods [IEE01; Enn98].

Bluetooth is a fast frequency hopping (1600 hops/s 1 Mbps) wireless system operating in the 84 MHz of bandwidth that is available in the 2.4 GHz unlicensed bands that are also used for IEEE 802.11 systems mostly using variety of technologies and IEEE 802.15.4 ZigBee using DSSS. Therefore, the interaction between a Bluetooth system and a co-located 802.11 WLAN or 802.11.4 ZigBee system needs an analysis of the interference between the different radio systems. In what follows, we use randomly deployed devices of different types at 2.4 GHz to illustrate the general concept of interference in wireless networks.

5.3.1   Interference Range

The first issue in interference is the *interference range*, that is the distance between two terminals that may cause them to interfere, in case they operate at the same frequency and at the same time. The range of interference is related to the radio propagation characteristics of the environment, the transmitted power from different devices and the sensitivity level of the transmission technique to the interference. Figure 5.5 illustrates a general interference scenario between an interference source, a desired transmitter, and a target receiver co-located in the area of coverage of the interference source. The interference takes place when the target receiver is receiving information from the desired source while the interference source is transmitting for its own objective of communicating with its paired receiver.

Consider the general Friis equation given by Equation 2.4. If we consider that, P_0, the power at first meter distance given in Equation 2.2b, is a linear function of the transmitted power, we have $P_0 = K P_t$ where K is a constant related to wavelength of the signal and antenna gains. Therefore, in general for the received signal strength at distance d in an environment with a distance–power gradient of α, we have:

$$P_r = \frac{KP_t}{d^\alpha} = KP_t d^{-\alpha} \tag{5.1}$$

Then, at the time that the target receiver is receiving a signal from the desired source and interference source is also transmitting, *the signal to interference level* at the target receiver is given by:

$$S_r = \frac{P_{r-d}}{P_{r-I}} = \frac{KP_{t-d}R^{-\alpha}}{KP_{t-I}D^{-\alpha}} = \frac{P_{t-d}}{P_{t-I}}\left(\frac{D}{R}\right)^\alpha \tag{5.2}$$

where R and D are the distances between the source and desired transmitter and the target receiver and interference source respectively. Also, P_{t-d} and P_{t-I} represent the transmitted power by the

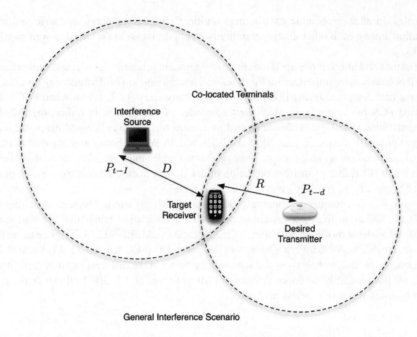

Figure 5.5   The basic interference scenario between two devices.

desired transmitter and the interference source respectively. Therefore, the *range of interference* between the interference source and the target receiver is given by:

$$D_{int} = R \times \sqrt[\alpha]{\frac{S_{min}P_{t-I}}{P_{t-d}}} \tag{5.3}$$

where D_{int} is the maximum distance at which the two terminals interfere and S_{min} is the *minimum acceptable received signal to interference ratio* needed for proper operation of the target receiver. In other words, the range of interference of the interfering device to the target receiver is directly related to the distance to the desired transmitter, required signal to noise ratio for proper operation of the target receiver, and transmit power of interfering source, and it is inversely related to the transmit power of the desired transmitter. In general, as we discussed in Chapter 2, the value of α may change from less than two in hallways and open areas up to around six in building with metal partitioning. Depending on the location of the devices, the path loss gradients between the interference source and target receiver may be different from α as well. In open areas with no walls, that include a number of scenarios involved with short-range devices, the environment is close to free space propagation and α is often close to 2.

Example 5.4: Range of a WPAN Interfering with a WLAN

Consider an IEEE 802.11 AP WLAN with transmitting power of 20 dBm and a minimum signal to noise requirement of 10 dB is located at a distance of 20 m of a Wi-Fi enabled device. If a ZigBee or Bluetooth device with transmitted power of 0 dBm is operating in that area, what is the minimum distance between the target Wi-Fi device and the interfering WPAN so that the signal from WPAN can corrupt the Wi-Fi packets?

(a) Assuming an open area with $\alpha = 2$, $S_{min} = 10$ (10 dB), $P_{t-d} = 100$ mW (20 dBm), $P_{t-I} = 1$ mW (0 dBm), and d $= 20$ m, from Equation 5.3, we have $D_{int} = 6.4$ m. That means if the frequencies of the WPAN and Wi-Fi are the same and WPAN transmits at the same time that the Wi-Fi device receives, the WPAN device that is closer than 6.4 m to the Wi-Fi device will interfere and destroy the received packets.

(b) In a partitioned environment with $\alpha = 4$ this interfering distance will increase $D_{int} = 10.2$ m.

(c) If the WPAN device is at its maximum transmit power of 100 mW (20 dBm) in the same partitioned area, then $D_{int} = 17.7$ m which is an order of magnitude larger than the value in the 0 dBm mode.

Example 5.5: Range of a WLAN Interfering with a WPAN

Consider two low-power IEEE 802.15 WPANs such as Bluetooth or ZigBee with transmitting power of 0 dBm and a minimum signal to noise requirement of 10 dB are communicating with each other in a typical distance of 2m. If a Wi-Fi device with transmitted power of 20 dBm is operating in that area, what is the minimum distance between the target WPAN device and the interfering WLAN so that the signal from WLAN can corrupt the Wi-Fi packets?

(a) Using Equation 5.3 with $R = 2$ m for the distance between the two WPAN devices, $P_{t-d} = 1$ mW (0 dBm), $S_{min} = 10$ (10 dB), and $P_{t-I} = 100$ mW (20 dBm) we will have $D_{int} = 63.2$ m. This is because the 802.11 device is radiating 100 times more power.

(b) If the WPAN device operates at 20 dBm, with the same power as the 802.11 then $D_{int} = 6.32$ m.

If the transmission technology is DSSS, used in ZigBee and one of the options of legacy IEEE 802.11, as we discussed in Chapter 3, the minimum required received signal to interference ratio at the target receiver is reduced by a factor equivalent to the value of the processing gain of the DSSS, N. Then, the interference range of narrowband FHSS will reduce significantly.

Example 5.6: Interference between Bluetooth and DSSS based IEEE 802.11

Consider the scenario described in Example 5.5 where the IEEE 802.11 AP WLAN uses DSSS with the processing gain of 11 that was originally recommended by the standard.

(a) Assuming an open area with $\alpha = 2$, $S_{min} = 10$ (10 dB), $P_{t-d} = 100$ mW (20 dBm), $P_{t-I} = 1$ mW (0 dBm), and $R = 20$ m. For a processing gain of $N = 11$, the effective minimum signal to noise ratio becomes $S_{min} = 10/11$ and the interference range will reduce to around $D_{int} = 1.9$ m (from 6.4 m).

(b) For $P_{t-I} = 100$ mW (20 dBm) we have an interference range of $D_{int} = 19$ m.

The conclusion from these simple examples is that the DSSS reduces the interference of the narrowband systems. This results in a $\sqrt[\alpha]{1/N}$ reduction in the range of interference compared to a FHSS system. However, the spectrum of DSSS is much wider and the probability of frequency coincidence of the DSSS and FHSS is much higher than the probability of hit of a two FHSS systems. In the next section we quantify this statement further.

5.3.2 Probability of Interference

In the last section we analyzed the range of interference between different devices and gave examples of systems using FHSS and DSSS. Using these examples, we showed that the range of interference of FHSS to DSSS or any other system that uses a fixed bandwidth is smaller than the range of interference of two FHSS systems. However, FHSS is a narrowband signal that changes its frequency of operation randomly while other systems always use the same spectrum. A narrowband FHSS such as a Bluetooth transmitter will interfere with the reception of another system with a greater probability than it will with the reception of another FHSS system. Therefore, the probability of interference between two devices differs and it plays an important role in the inference analysis. In particular probability of interference becomes substantially different when we consider two FHSS system.

To further analyze the interference we first pay attention to interference between two FHSS system and we use Bluetooth and FHSS 802.11 devices as our example interfering systems. Both Bluetooth and FHSS 802.11 are frequency-hopping systems using the 79 carrier frequencies in the 2.4 GHz ISM bands shown in the vertical axis of Figure 5.6. Bluetooth packets are normally shorter than 802.11 packets and hop at a much slower rate of 2.5 hops/s. When a terminal is in the

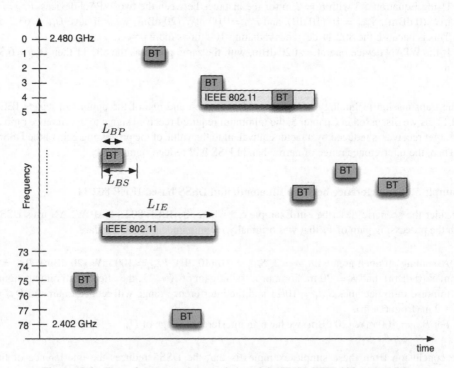

Figure 5.6 Time–frequency characteristics of the FHSS IEEE 802.11 and BT.

interference range of the other terminal and the hopping frequencies are the same, packets collide and get destroyed. To analyze this situation we need to find the probability of collision in time and in frequency.

Since Bluetooth packets are shorter than 802.11 packets, during transmission of one 802.11 packet, the co-located Bluetooth device hops and sends one packet per hop several times. Assuming L_{IE} is the length of the IEEE 802.11 packet and L_{BS} the length of a Bluetooth slot, the minimum number of Bluetooth hops occurring during transmission of one 802.11 packet is: $n = \lceil L_{IE} / L_{BS} \rceil$ where $\lceil x \rceil$ represents the smallest integer greater than or equal to x. The maximum number of Bluetooth hops occurring in duration of an 802.11 packet is $\lceil L_{IE} / L_{BS} \rceil + 1$. It can be easily shown [Enn98] that the probability of an 802.11 packet overlap with $n = \lceil L_{IE} / L_{BS} \rceil$ Bluetooth dwell periods of duration L_{BS} is:

$$P_n = L_{IE}/L_{BS} - \lceil L_{IE}/L_{BS} \rceil$$

The probability that it overlaps with $n + 1 = \lceil L_{IE} / L_{BS} \rceil + 1$ dwell periods is:

$$P_{n+1} = 1 - L_{IE}/L_{BS} + \lceil L_{IE}/L_{BS} \rceil .$$

These simple equations can be used to analyze a number of different practical scenarios.

Example 5.7: Probability of interference between Bluetooth and FHSS IEEE 802.11

If $L_{IE} / L_{BS} = 4.3$ the probability of overlap of 802.11 packet with $n = 4$ Bluetooth dwell periods is 30% and the probability of overlap with $n + 1 = 5$ dwell periods is 70%.

Considering these expressions, the probability of an 802.11 packet surviving BT interference, $P_{survive}$, is approximated by:

$$P_{survive} = (1 - P_{hit})^n P_n + (1 - P_{hit})^{n+1} P_{n+1}$$

where P_{hit} is the probability of having the same frequency for both 802.11 and Bluetooth. The probability of collision is given by $P_{collision} = 1 - P_{survive}$.

Example 5.8: Collision probability between IEEE 802.11 FHSS and BT

The probability of a Bluetooth hop to occur at the operating frequency of the FHSS system is $P_{hit} = 1/79 = 0.013$. For a 1000 byte 802.11 packet at 2 Mbps:

$$L_{IE} = \frac{1000(\text{bytes}) \times 8(\text{bits/byte})}{2(\text{Mbits/s})} = 4 \text{ ms}$$

If Bluetooth is sending 1-slot packets $L_{BS} = 625$ μsec. Therefore:

$$n = \left\lceil \frac{4 \text{ ms}}{625 \text{ μs}} \right\rceil = 6$$

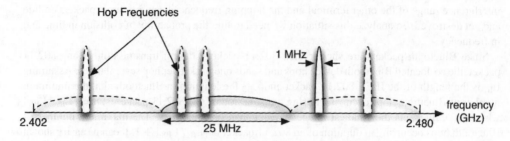

Figure 5.7 Overlapping DSSS IEEE 802.11 and FHSS Bluetooth spectrum.

and $P_n = 0.4$ that result in $P_{n+1} = 0.6$. Therefore:

$$P_{survive} = (1 - 0.013)^6 \times 0.4 + (1 - 0.013)^7 \times 0.6 = 0.92$$

Therefore the collision probability is 0.08 or 8%.

Example 5.9: Collision probability between IEEE 802.11 DS-SS and BT

Figure 5.7 shows the mechanism with which the frequency-hopping pattern of BT and the spectrum of the DSSS 802.11 or CCK 802.11b hit one another. The probability of a Bluetooth hop to occur at the operating frequency of the DSSS system is $P_{hit} = 26/78 = 0.33$. For a 1000 byte 802.11 packet at 2 Mbps all the other parameters remain the same as the last example and we will have:

$$P_{survive} = (1 - 0.33)^6 \times 0.4 + (1 - 0.33)^7 \times 0.6 = 0.072$$

The probability of collision is 0.928 or 92.8% as compared with 8% for the FHSS 802.11 example.

Example 5.10: BT interference with IEEE 802.11b

The IEEE 802.11b uses the same band as 802.11 DSSS to transmit at 11 Mbps. Therefore, again we have $P_{hit} = 26/79 = 0.33$. However, for a 1000 byte 802.11 packet at 11 Mbps we have

$$L_{IE} = \frac{1000(\text{bytes}) \times 8(\text{bits/byte})}{11(\text{Mbits/s})} = 727 \ \mu s$$

With Bluetooth 1-slot packets we have:

$$n = \left\lceil \frac{727 \ \mu s}{625 \ \mu s} \right\rceil = 1$$

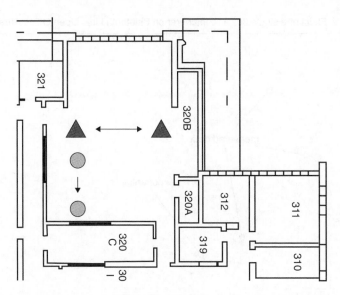

Figure 5.8   Scenario for experimental analysis of interference between Bluetooth and IEEE 802.11b.

and $P_n = 0.16$ that results in $P_{n+1} = 0.84$. Therefore:

$$P_{survive} = (1 - 0.33)^1 \times 0.16 + (1 - 0.33)^2 \times 0.84 = 0.49$$

The collision probability is 0.51 or 51% that is substantially better than 802.11 DSSS and much worse than the 802.11 FHSS.

5.3.3   Empirical Results

The analysis in the last section is at the PHY layer, but a more thorough analysis including the effects of all layers should be done experimentally. A group of undergraduate students at Worcester Polytechnic Institute developed a testbed for the experimental analysis of the interference between the IEEE802.11b and Bluetooth voice and data channels for their senior undergraduate project [Cha01]. In this project they considered a number of scenarios and measured the overall packet loss, throughput, and delay characteristics of the interfering Bluetooth and 802.11 devices as well as cordless telephones. In this section we provide some of their results and conclusions that are related to the scenarios described in Figure 5.8 that relates the performance of interfering 802.11b and BT terminals to the distance between the devices.

Example 5.11: PLR in BT with interfering IEEE 802.11b devices

Figure 5.8 shows the floor plan and one of the measurement scenarios in which two 20 dBm Bluetooth equipped laptops (triangles) are separated by 10 m and an 802.11b laptop (circle) is moved from a distance of 1 to 10 m from the Bluetooth laptop. The 802.11b station is communicating with another laptop that is far away and does not interfere significantly with the Bluetooth device.

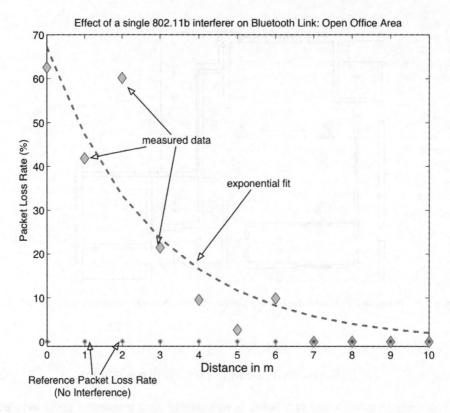

Figure 5.9 Packet loss rate (%) over a Bluetooth link with and without interference from an 802.11b terminal.

Figure 5.9 shows the packet loss rate (PLR) of the Bluetooth device. As the distance of the interfering 802.11 device increases the packet loss reduces. When the BT and 802.11b devices are next to one another the PLR is 70%, but as the distance increases to five meters, there is no effect from the interference. Note that 802.11 packets are much longer than Bluetooth transmission slots, but they are also transmitted at higher raw data rates. Figure 5.10 shows the delay characteristics evaluated using ping messages (that measures the round trip delay).

Example 5.12: PLR in IEEE 802.11b with an interfering BT device

Figure 5.11 shows the PLR of 802.11b in a scenario that is the opposite of the last example shown in Figure 5.10. In this example two 802.11b devices are located at a distance of 10 m and an interfering BT terminal is moved from 1 to 10 m from them. At close distances, the PLR is close to 45%, and as the distance of the interfering BT device increases beyond 3 m the effects of interference is negligible.

The general conclusion of these studies is that the interference between FHSS 802.11 and BT devices is negligible, however, DS-SS 802.11 devices will interfere significantly with the BT devices.

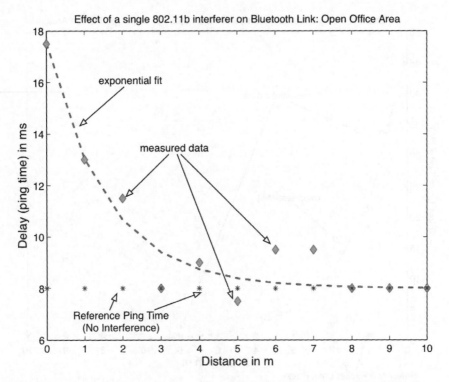

Figure 5.10   Delay characteristics over a Bluetooth link with and without 802.11b interference.

5.4   Deployment of Wireless LANs

We first describe the deployment of WLANs, which has a simpler architecture an less complicated operational procedures compared to cellular networks (see Chapter 6). Let us consider the differences among issues related to deployment of WLANs and cellular telephone networks (details of which are discussed in the succeeding sections). WLANs operate in unlicensed bands while cellular telephone networks use licensed bands. As we discussed in Chapter 4, the network capacity in FDMA and TDMA based cellular networks depend on the frequency reuse factor during deployment, which is determined by calculations of the interference from the neighboring cells. In the case of CDMA networks, again we had a parameter for the calculation of capacity that was related to interference from neighboring cells, which again relates capacity to interference. All these calculations are based on the assumption that the band is licensed to one operator, which technically means that the network planner has control on the interference. WLANs operate in unlicensed bands in which a network planner does not have control on the interference. Network managers in a university campus or a corporation may restrict students or employees in the deployment of WLANs other than those owned by the university or the corporation to control interference, which does not comply with the government regulations on using these bands and can be considered illegal.

Unlike WLANs, for optimal deployment, cellular networks use relatively accurate statistical coverage prediction models, such as Okumura–Hata model discussed in Chapter 2. Statistical channel models for coverage in indoor areas are much less accurate and this poses a challenge in the

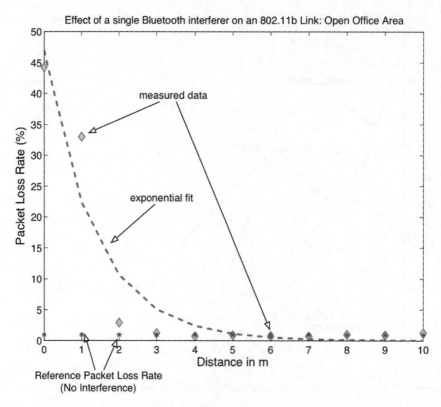

Figure 5.11   Packet loss rate (%) over an 802.11b link with and without an interfering Bluetooth device.

analysis of coverage of WLANs unless we resort to labor intensive empirical measurements or use computationally intensive ray tracing algorithms. However, APs for WLANs are very inexpensive, around a few hundred dollars or lesser at the time of this writing, and they do not need expensive antenna towers and site landscape for installation. BSs are orders of magnitude more expensive than access points and for large area coverage (macrocells), they need expensive towers and appropriate land for cell sites. Radio resource management in WLANs is much simpler because WLANs have limited numbers of non-overlapping frequency bands (for example three non-overlapping channels in the case of the popular 802.11g standard), cellular networks have an order of magnitude more channels and users/cell/carrier to handle. Mobility management for cellular networks is much more complex than WLANs because most popular WLAN applications are quasi-stationary while cellular networks were designed to operate inside a vehicle. In addition, traditional WLAN data traffic is in bursts, it is non-symmetric and location- and time-selective. Therefore the medium access control is not performed centrally through radio resource management techniques. As a result of all of these differences, an 802.11 AP is simple and connects to the Internet backbone directly through a wired LAN or cable/DSL modem while a cellular BS is connected to the PSTN using a hierarchy that includes a base station controller (BSC) and a mobile switching center (MSC). The BSC and MSC are needed because in cellular networks we have more complex radio resource management, mobility management, and connection management techniques. Cells are larger, voice connections need more quality control, and cellular telephones have higher mobility.

When WLAN standardization activities started in the late 1980s, the main issue related to installation was the selection of topology, not large-scale cellular deployment. At the time, WLANs were perceived as an extension to wired LANs that avoids wiring challenges. With the popularity of Internet access starting in the mid-1990s, the corresponding growth of the WLAN industry, and thoughts on integrating WLANs into 3G cellular systems, more attention was paid to large scale WLAN deployment for wide area coverage. Since commercially successful cellular telephone networks were deployed very carefully based on relatively accurate channel models for path-loss, research efforts in systematic deployment of WLANs in large areas using automated coverage prediction software attracted attention at that time.

However, in practice, WLANs were installed in small areas, such as a residential home or a small shop, in a random manner by users in the most convenient location where the WLAN access point (or router) could connect to the backbone Internet connection points such as a cable or a DSL modem. In large area applications, such as wireless mobile access inside of a large office building or a warehouse area, or outdoor deployments for wireless Internet access, WLANs were deployed in grid formation [Unb03]. Access points are installed every 20–30 m inside a building in convenient locations such as corridors and large open areas where larger traffic was expected. In outdoor applications, APs are installed in much wider grids on top of the utility posts, high on the outside walls of multi-floor buildings, or on the roof of buildings to optimize coverage. The grid installation, deployed by building owners or independent service providers (for instance in the hallways of shopping malls), provide excellent coverage with large overlap between adjacent cells and a very low outage probability. The basic difference between user deployment and grid installation is that in the grid installation some primitive network planning by visual inspection, measurements of the RSS in selected locations or a study of the construction drawings of the building is performed to provide for a better coordination among the locations of the APs. In user deployments the AP positions is mainly decided based on the convenience of installation. Figure 5.12 illustrates these differences. This figure also shows optimal deployment, which uses software tools to determine the best places for deploying access points (perhaps minimizing interference or optimizing for user loads).

For outdoor grid installations in a street in a typical downtown or industrial area, we need specialized technicians and coordination with the city and utility companies for the installation of the APs on top of the lampposts. Besides, extensive wiring would be needed to connect the APs to a backbone network. The physical installation using coverage optimization technique will look very similar to the grid installation.

It is true that at the time of this writing the dominant bands for using WLANs are the 2.4 GHz ISM bands. However, as the popularity of WLANs increase many researchers envision that at some points this industry will resort to higher frequencies where wider portions of bandwidth are

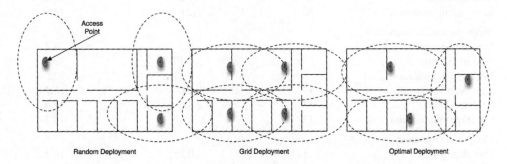

Figure 5.12   Random, grid and optimal deployment of large-scale WLANs.

available. The path of migration is first to 5 GHz and then to tens of GHz (around 60 GHz) and perhaps to the IR domain. Cellular planning for higher frequencies would involve other challenges. In some situations, the desired signal is attenuated by obstacles whereas an interfering signal arrives unobstructed, and there is a consequent degradation in the carrier to interference ratio (denoted by C/I). This situation may prevent the design of a logical layout for the cells in an indoor area. However, if the signal is contained in one room and does not penetrate the walls of the room, the walls can be used to define the boundaries of the cells. This situation exists for infrared and microwave (above 20 GHz) WLANs, wherein each room constitutes one cell of the network. A more quantitative example will further clarify this situation.

Example 5.13: WLAN coverage in different environments

A systematic comparative performance evaluation of random, grid, and coverage optimization techniques for deployment of WLANs in office buildings, shopping centers, and campus areas at 5, 17 and 60 GHz frequencies is analyzed in [Unb03]. In this work ray tracing software was used to generate channel profiles in different environments. Figure 5.13 shows the sample building layouts for an office, a shopping mall, and a campus area used by the ray tracing software. Considering different APs densities, the received SNR in different location is calculated to form the cumulative distribution function of the SNR for various conditions. Figure 5.14 shows two sets of cumulative distribution functions at 5 GHz in a manufacturing floor. Each set consists of user deployment, grid installation, and optimal network planning installation for 0.1 and 0.15 APs per 1000 square meter area. Table 5.1 provides the 10th percentile point for the density function of the SNR in different

Table 5.1   Required 10th percentile SNR in different environments and different frequencies

Deployment method	Frequency (GHz)	Minimum AP density/1000 m^2	10th percentile SNR (dB)
Office environment			
User deployment	5	1.85	14.6
Grid installation	5	1.85	17.5
Coverage optimization	5	1.85	21.2
User deployment	17	14.8	16.1
Grid installation	17	7.4	16.9
Coverage optimization	17	7.4	19.9
User deployment	60	74.0	18.0
Grid installation	60	52.0	15.2
Coverage optimization	60	52.0	14.9
Shopping mall environment			
User deployment	5	0.5	21.3
Grid installation	5	0.17	19.0
Coverage optimization	5	0.17	22.2
User deployment	17	6.0	16.2
Grid installation	17	3.0	16.0
Coverage optimization	17	3.0	18.5
Campus environment			
User deployment	5	0.15	14.2
Grid installation	5	0.15	21.0
Coverage optimization	5	0.15	22.0

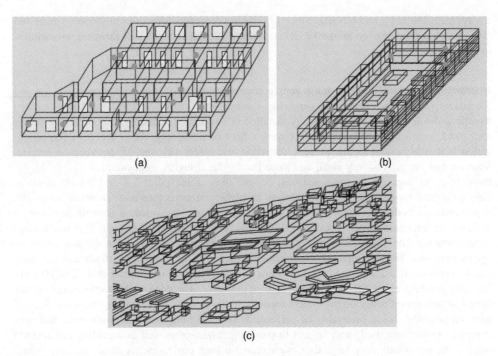

Figure 5.13 Typical building layouts used for the performance analysis of WLAN using ray tracing software in (a) an office (b) a shopping mall, and (c) a campus area. © 2003 IEEE. Reprinted, with permission, from [Unb03].

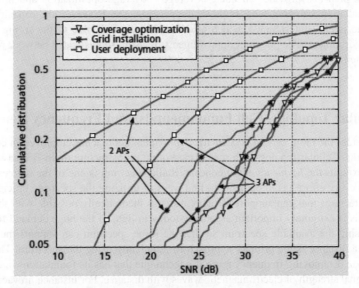

Figure 5.14 Cumulative distribution function of the received SNR at 5 GHz in a shopping area. © 2003 IEEE. Reprinted, with permission, from [Unb03].

environment and 5, 17, and 60 GHz. Since 17 GHz has coverage difficulties in shopping areas and 60 GHz is not suitable even unlike 17 GHz operation in open areas, these values are not included in the table.

In general, office buildings provide a simple environment for WLAN deployment. For 2.4 and 5.0 GHz systems, since the signal penetrates through walls, the coverage is relatively unproblematic and a few APs opportunistically installed in convenient areas, such as corridors, sitting areas, and large lecture halls, typically covers a floor of a building. At 17 and 60 GHz, propagation is mostly limited to line of sight operation and essentially one AP per room is needed. Since size of the cells is very small and more bandwidth at higher frequencies are available, the throughput per user can be increased well beyond those achieved at 2.4 or 5.0 GHz systems. This additional throughput is at the expense of higher infrastructure cost. In office areas, the deployment methodology is not very critical and performing sophisticated network planning or coverage optimization does not result in substantial performance gain. At such high frequencies deployment must be dense and in fact, user deployment can outperform grid installation or coverage optimization. In the campus environment, coverage with an acceptable infrastructure density can be provided if a 2.4 or a 5.0 GHz system is used. The small cell sizes at 17 and 60 GHz basically preclude an outdoor installation of such networks. The deployment method does in fact have a significant influence on the system capacity in the campus environment, and some form of network planning should preferably be performed. The shopping mall is a problematic and very complex environment, both with respect to providing coverage as well as regarding interference issues. The potentially very large user populations in such environments cause not only strong shadowing, but also result in a relatively low average throughput per user. The strong fragmentation of the environment and the fact that the network layout needs to be done in three dimensions (the shopping mall comprises of multiple floors) makes it very difficult to devise a reasonably good network plan that provides adequate coverage. Coverage optimization and in some cases also grid installation, can somewhat improve the performance. However, the capacity is mainly limited by the intense interference, which also fluctuates heavily due to the strong fragmentation of the environment. The potential improvements by using sophisticated network planning are however limited.

5.5   Cellular Topology, Cell Fundamentals, and Frequency Reuse

Next we consider the more careful, systematic deployment approaches employed in cellular telephone networks. Cellular topology is a special case of an infrastructure multi-BS network configuration that exploits the *frequency reuse* concept. Radio spectrum is one of the scarcest resources available and every effort has to be made to find ways of utilizing the spectrum efficiently and to employ architectures that can support as many users as theoretically possible with the available spectrum. This is extremely important especially today in light of the huge demand for capacity. Spatially reusing the available spectrum so that the same spectrum can support multiple users separated by a distance is the primary approach for efficiently using the spectrum. This is called *frequency reuse*. Employing frequency reuse is a technique that has its foundations in the attenuation of the signal strength of electromagnetic waves with distance. For instance, in vacuum or free space, the signal strength falls as the square of the distance. This means that the same frequency spectrum may be employed without any interference for communications or other purposes, provided the distance separating the transmitters is sufficiently large, and their transmit powers are reasonably small (depending on the reuse distance). This technique has been used for example in commercial radio and television broadcast where the transmitting stations have a constraint on the maximum power they can transmit so that the same frequencies can be used elsewhere. The cellular concept is an intelligent means of employing frequency reuse. Cellular topology is the dominant topology used in all large-scale terrestrial and satellite wireless networks. The concept of cellular communications was first developed at the Bell Laboratories in the 1970s to accommodate a large number of users with a limited bandwidth [Mac79].

5.5.1   The Cellular Concept

By cellular radio, we mean deploying a large number of low power base stations for transmission, each having a limited coverage area. In this fashion, the available capacity is multiplied each time a new base station or transmitter is set up since the same spectrum is being *reused* several times in a given area. The fundamental principle of the cellular concept is to divide the coverage area into a number of contiguous smaller areas each served by its own radio base station. Radio channels are allocated to these smaller areas in an intelligent way so as to minimize the interference, provide an adequate performance, and cater to the traffic loads in these areas. Each of these smaller areas is called a *cell*. Cells are grouped into *clusters*. Each cluster utilizes the entire available radio spectrum. The reason for clustering is that adjacent cells cannot use the same frequency spectrum because of interference. So, the frequency bands have to be split into chunks and distributed among the cells of a cluster. The spatial distribution of chunks of radio spectrum (which are called sub-bands) within a cluster has to be done in a manner such that the desired performance can be obtained. This forms an important part of network planning in cellular radio.

Two types of interference are important in such a cellular architecture. The interference due to using the same frequencies in cells of different clusters is referred to as *co-channel interference*. The cells that use the same set of frequencies or channels are called *co*-channel cells. The interference from frequency channels used within a cluster whose side-lobes overlap is called *adjacent channel interference*. The allocation of channels within the cluster and between clusters must be done so as to minimize both of these.

The cellular concept can increase the number of customers that can be supported in the available frequency spectrum as illustrated by the following examples by deploying several low power radio transmitters.

Example 5.14: Cellular concept

Consider a single high power transmitter (see Figure 5.15) that can support 35 voice channels over an area of 100 square km with the available spectrum. If seven lower power transmitters are used so that they support 30% of the channels over an area of 14.3 km^2 each, a total of $\approx$80 voice channels are now available in this area instead of 35. In reality, channels will have to be allocated to base stations in such a way as to prevent interference between one base station and another. In Figure 5.13, base stations 1 and 4 could use the same channels, as their coverage areas are sufficiently far apart and so also base stations 3 and 6. Suppose the cells labeled 1, 2, 5, 6, and 7 use disjoint frequency bands and the channels used in 1 and 6 are reused in 3 and 4. The set of cells $\{1,2,5,6,7\}$ forms a cluster. Cells 3 and 4 form part of another cluster. In the limiting case, the density of base stations can be made so large that the capacity is infinite. However, in practice this is impossible for several reasons that include drastic increases in the network and signaling load, number and frequency of handoffs, and cost of infrastructure and planning.

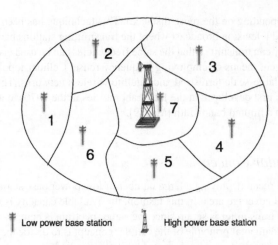

Low power base station   High power base station

Figure 5.15   The cellular concept.

Example 5.15: Importance of cellular topology

We want to provide a radio communication service to a city. The total bandwidth available is 25 MHz and each user requires 30 KHz of bandwidth for voice communication. If we use one antenna to cover the entire town we can only support 25 MHz/30 KHz = 833 simultaneous users. Now let us employ a cellular topology where twenty lower power antennas are opportunistically located to minimize both kinds of interference. We divide our frequency band into four sets and assign one set to each cell. Each cell has a spectrum of 25/4 = 6.25 MHz allocated to it. We have a *cluster* of four cells in this example. The number of simultaneous users supported per cell is 6.25 MHz/30 KHz = 208.

The number of users per cluster is $4 \times 208 = 832$. The total number of simultaneous users is now $832 \times 5 = 4160$ since we have five clusters of four cells each. The new capacity is roughly five times the capacity with a single antenna.

Examples 5.14 and 5.15 illustrate the main benefits and elements of a cellular network planning by relating the bandwidth, number of cells, frequency reuse factor, and capacity of the network. If W is the total available spectrum, B is the bandwidth of a carrier, N is the cluster size or frequency reuse factor, and m is the number of users per carrier, the number of simultaneous users is given by:

$$M = \frac{m}{N}\left(\frac{W}{B}\right) \tag{5.4}$$

In particular we observe that the capacity of the network can be increased by: (a) increasing m, that is, the number of users that can be supported on a carrier (by changing the modulation or transmission scheme), (b) decreasing the frequency reuse factor (number of clusters N). A major remaining question at this point is how to assign the groups of sub-bands to individual cells so that interference between different users using the same sub-bands is acceptable. We address this issue in subsequent sections. Let us now consider some other important issues related to a cellular topology.

A cellular topology reduces the coverage requirements of both the mobile terminal and the BS. The reduction of the size of coverage lowers the required transmitted power by the mobile terminal since mobile terminals are located closer to the base stations and they require less power to communicate with the network. This increases the battery lifetime and reduces the size of a terminal. These issues are extremely important to the user of a hand-held terminal. Therefore, the larger the number of cells, the larger the capacity and the smaller the size of the hand-held terminal. However, we need a fixed network infrastructure to interconnect the cells and ensure that the entire system works in a coordinated manner. As we increase the number of cells, the cost and the time for deploying the network increases. In addition, the smaller the size of the cell, the larger is the frequency of a mobile changing its connection from one BS to another. Therefore, a reduction in the size of the cells increases the complexity of the design and deployment of the network as well as the signaling load in the fixed part of the infrastructure. The art of designing a cellular topology involves striking a balance between all these elements and this is the subject of details that follow in this chapter.

Another important factor in deployment of wireless cellular networks is provision for expansion. The main investment of a wireless service provider is towards the cost of the fixed infrastructure, which includes the base station and connections between them. When a service provider starts an operation, he needs to minimize the cost of infrastructure while continuously increasing the number of subscribers. As the number of subscribers increases, new income is generated and the service provider can afford to expand the network by increasing the complexity of its infrastructure to support a further larger population of subscribers. Therefore, there is a need for a plan to take into account the growth of the subscriber base and thus the entire wireless network.

In summary, we need to address to the following technical issues for planning a cellular network:

- Selection of a frequency reuse pattern for different radio transmission techniques.
- Physical deployment and radio coverage modeling.
- Plans to account for the growth of the network.
- Analysis of the relationship between the capacity, cell size, and the cost of the infrastructure.

5.5.2   Cellular Hierarchy

There are three reasons to use a hierarchical cellular infrastructure supporting cells of different sizes. One is to extend the coverage to the areas that are difficult to cover by a large cell. For example, cells designed to cover suburban areas have antennas with tall towers and cover a large area. Signals from these antennas, however, cannot propagate sufficiently into urban canyons or indoor environments. For urban canyons we need to install antennas at lower heights and in indoor areas we may mount the antennas on walls to provide a comprehensive coverage. Antennas mounted in these locations are of low power and cover a smaller area resulting in the creation of a smaller sized cell. The second reason to have a cellular hierarchy is to increase the capacity of the network for those areas that have a higher density of users. Imagine the number of cellular phone users in the world trade center and compare it with the number of mobile users on an interstate highway. To support the larger subscriber demand and higher traffic in smaller areas we need to increase the number of cells by reducing their sizes. The third reason is that sometime an application needs certain coverage. Consider the increasing number of wireless devices that we are carrying in our bag these days and the increasing need for communication between these devices. This necessitates extremely small sized cells that provide a wireless network for connecting laptops or notepads to cellular phones.

In a modern deployment of a cellular network a number of cell sizes are used to provide a comprehensive coverage supporting traffic fluctuations in different geographical areas and supporting a variety of applications. One way of dividing the cells into a hierarchy is to define the following cell sizes:

Personal-cells: These are the smallest unit of the cellular hierarchy used for connection of personal equipment such as laptops, notepads, and cellular telephones. These cells need to cover only a few meters where all these devices are in physical range of the user.

Pico-cells: These are small cells inside a building that support local indoor networks such as wireless LANs. The size of these networks is in the range of a few tens of meters.

Micro-cells: These cells cover the inside of streets with antennas mounted at heights lower than the rooftop of the buildings along the streets. They cover a range of hundreds of meters and are used in urban areas to support personal communication services (PCS).

Macro-cells: These cover metropolitan areas and are the traditional cells installed during the early phases of the cellular telephony. These cells cover areas on the order of several kilometers and their antennas are mounted above the rooftop of typical buildings in the coverage area.

Mega-cells: These cells cover nationwide areas for with ranges of hundreds of kilometers and are mainly used with satellites.

Figure 5.16 illustrates the relationship between different cells with example applications. An ideal network has a hierarchy of these cells to cover airplane travelers with mega-cells, car drivers in

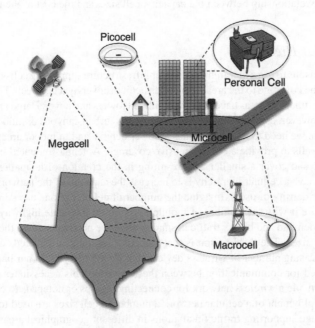

Figure 5.16   Cellular hierarchy.

sub-urban areas with macro-cells, pedestrians in the streets via micro-cells, indoor users with pico-cells, and connect personal equipment with femto-cells.

5.5.3 Cell Fundamentals and Frequency Reuse

Having looked at the cellular topology and the concept of employing a cellular architecture to increase the communications capacity and to cater to a large subscriber demand in hotspots, we now consider quantitative means to characterize the interference in a cellular topology. This in turn leads to quantitative means for determining the best cluster size and simple techniques for allocating the sub-bands of spectrum and within a cluster.

Even though in practice cells are of arbitrary shape (close to a circle) because of the randomness inherent in radio propagation, it is easier to obtain insight and understanding for system design by visualizing all cells as having the same shape. Also, it is easier to mathematically analyze a cellular topology by assuming a uniform cell size for all cells. Once some insight is obtained as to what the effects of interference are, measurements, simulation and a combination of these can be employed in actually determining the planning of a network.

For cells of the same shape to form a tessellation so that there are no ambiguous areas that belong to multiple cells or to no cell, the cell shape can be of only three types of regular polygons: equilateral triangle, square, or regular hexagon, as shown in Figure 5.17.

A hexagonal cell is the closest approximation to a circle of these three and has been used traditionally for system design (see Figure 5.18). The argument for a hexagonal shape comes from the fact that among the three shapes mentioned, for a given radius (largest possible distance between the polygon center and its edge), the hexagon has the largest area.

In most of the literature and in back of the envelope design, the hexagonal cell shape is chosen as the default cell shape. In particular cases that consider continuous distributions of traffic load and interference between different transmission schemes, a circular cell shape is employed for tractable calculation.

In order to investigate the effects of interference, which changes with distance, there is a need to come up with an elegant way of determining distances and identifying cells. Fortunately, it is possible to do this easily in the case of hexagonal cells [Mac79]. In order to maximize the capacity, co-channel cells must be placed as far apart as possible for a given cluster size. It can be shown that there are *only* six co-channel cells for a given reference cell at this distance. The distance between

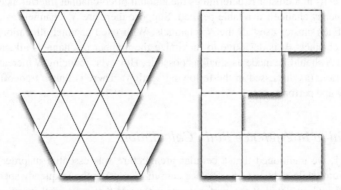

Figure 5.17 Triangular and rectangular cells.

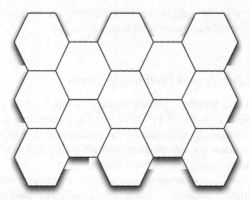

Figure 5.18 Arranging regular hexagons that can cover a given area without creating ambiguous regions.

the co-channel cells can be shown to be $D_L = \sqrt{3}R_L$. Here R_L is the radius of a cell. The relationship between the distance between co-channel cells, the cluster size, and the cell radius is given by:

$$\frac{D_L}{R} = \sqrt{3N} \tag{5.5}$$

This quantity is also referred to as the *co-channel reuse ratio*. Values for N can only take on values of the form $i^2 + ij + j^2$ where i and j are integers.

Example 5.16: Cluster size of $N = 7$

As described above, i and j can only take integer values. If we take $i = 2$ and $j = 1$, we see that $N = 4 + 2 + 1 = 7$. Selecting a cell A, we can determine its co-channel cell by moving two units along one face of the hexagon and one unit in a direction 60° or 120° to this direction. Proceeding in this fashion, clusters of size $N = 7$ can be created as shown in Figure 5.19. A value of $N = 7$ is employed in the United States in the advanced mobile phone service (AMPS).

The number of cells in a cluster N determines the amount of co-channel interference and also the number of frequency channels available per cell. Suppose there are N_c channels available for the entire system. Each cluster uses all the N_c channels. With fixed channel allocation, each cell is allocated N_c/N channels. It is desirable to maximize the number of channels allocated to a cell. This means that N should be made as small as possible. However, reducing N increases the signal to interference ratio (as discussed in the following section). There is thus a tradeoff between the system capacity and performance.

5.5.4 Signal to Interference Ratio Calculation

In Section 5.5.1, we mentioned that a cellular architecture was essential in order to reuse the available spectrum while reducing interference caused by reusing the frequency spectrum. In this section, we will look in detail at the performance measures that are useful in system design, in

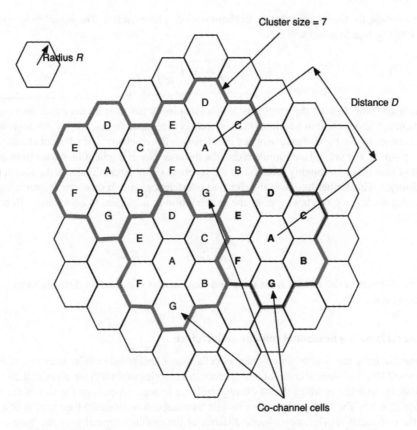

Figure 5.19   Hexagonal cellular architecture with a cluster size of $N = 7$.

particular the signal to interference ratio and its relationship with the path loss, and the grade of service. Recall the signal to interference ratio calculations in the unlicensed frequency bands in Section 5.3 here.

In general, the signal to interference ratio is calculated in a manner similar to that in Section 5.3 and can be written as follows:

$$S_r = \frac{P_{desired}}{\sum_i P_{interference,i}} \tag{5.6}$$

The signal strength falls as some power of the distance α called the power–distance gradient or path-loss gradient. That is, if the transmitted power is P_t, after a distance d in meters, the signal strength of a radio signal will be proportional to $P_t d^{-\alpha}$. In its most simple case, the signal strength falls as the square of the distance in free space ($\alpha = 2$). Suppose there are two base station transmitters BS_1 and BS_2 located in an area with the same transmit power P_t and a mobile terminal is at a distance of R from the first and D from the second. If the mobile terminal is trying to communicate with the first base station, the signal from the second base station is interference. The *signal to interference* ratio for this mobile terminal will be:

$$S_r = \frac{KP_t R^{-\alpha}}{KP_t D^{-\alpha}} = \left(\frac{D}{R}\right)^{\alpha} \tag{5.7}$$

The larger the ratio D/R is, the greater is S_r (the smaller is the interference) and the better the performance. The objective in a cellular radio system is to allocate frequencies or channels to cells within a cluster so that the distance between interfering cells (co-channel or adjacent channel) is as large as possible. For urban land mobile radio, the distance power gradient increases from two (in the case of free space) to roughly four so that the received signal strength falls as the fourth power of the distance. This further improves the signal to interference ratio. If there are J_s interfering base stations surrounding a given base station, the general form of the signal to noise ratio will be:

$$S_r = \frac{d_0^{-\alpha}}{\sum_{n=1}^{J_s} d_n^{-\alpha}} \tag{5.8}$$

where the distance of the mobile from the given base station is d_0 and its distance from the n-th base station is d_n.

Example 5.17: S_r in a hexagonal cellular architecture

Recalling that there are exactly six co-channel cells with a hexagonal cellular structure, it is clear that they will all cause similar levels of interference to a mobile terminal in the given cell. So $J_s = 6$ here. Also, the distance at which the co-channel cells are located depends on the size of the cluster from Equation 5.5. The farthest distance a mobile terminal can be from the base station of a given cell is the cell radius R. The approximate distance of the mobile terminal from the base stations of each of the co-channel cells is D_L. For land mobile radio, if only the six co-channel cells that make up the first tier of interferers are considered, $J_s = 6$, and the signal to interference ratio can be approximated as:

$$S_r \approx \frac{R^{-4}}{J_s D_L^{-4}} = \frac{R^{-4}}{6 D_L^{-4}} = \frac{1}{6}\left(\frac{D_L}{R}\right)^4 = \frac{3}{2} N^2 \tag{5.9}$$

In terms of dB, we can write the signal to interference ratio as:

$$S_r = -7.78 + 40\log\left(\frac{D_L}{R}\right) = 1.76 + 20\log N \tag{5.10}$$

Figure 5.20 shows how the signal to interference ratio given by Equation 5.9 varies with the cluster size N. Equation 5.9 is commonly used to determine the cluster size for an adequate performance. Note that the signal to interference ratio is influenced by the co-channel reuse ratio D_L/R in that a given D_L/R has to be maintained for a particular S_r. However, it is an approximation since different base stations may employ different transmit powers and the path loss model may not be as simple as the d^{-4} model used here. The S_r calculation will be different for the uplink (mobile terminal to base station communication) compared to the downlink (base station to mobile terminal communication).

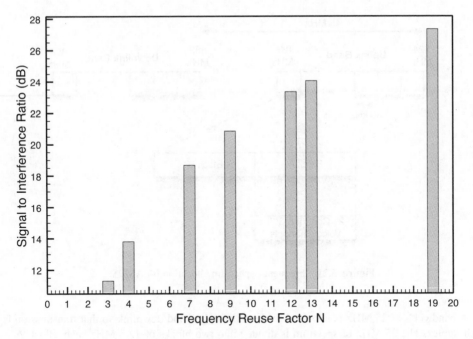

Figure 5.20 S_r as a function of N.

We have so far assumed that the received signal strength falls as the fourth power of the distance for land mobile radio. In dB, this translates to a path loss model of the form:

$$P_r(d)(\text{dB}) = P_t(\text{dB}) - 40\log d + 10\log K \tag{5.11}$$

The factor $10\log K$ usually corresponds to the path loss at the first meter or kilometer as the case may be and d is in the same units. This path loss model may not be appropriate, especially since measurements of the received signal strength indicate that the path loss is dependent not only on the distance between the base station and the mobile, but also the radio frequency of operation and the antenna heights. The path loss is also dependent upon the scenario, whether the cellular architecture corresponds to land mobile radio or to a micro-cellular personal communication services (PCS) application. However this simple model is appropriate for first-cut approximations in system design.

Let us consider an example of a real cellular system that tries to bring together many of the concepts that have been considered so far in this chapter.

Example 5.18: Cellular architecture of AMPS

As an example of cellular architecture, we consider the very first cellular radio telephone system in the United States, called *Advanced Mobile Phone Service* (AMPS) based on analog FM modulation scheme. Each voice channel in AMPS occupies 30 kHz of bandwidth and uses frequency modulation (FM). Figure 5.21 shows the spectrum allocations for AMPS.

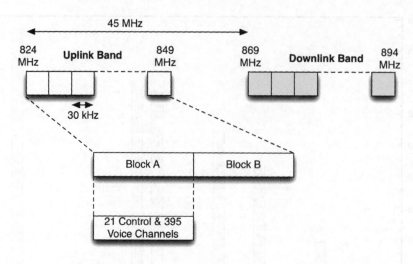

Figure 5.21   Frequency spectrum allocation for AMPS.

A bandwidth of 25 MHz is allocated for both the uplink and downlink so that transmission is full duplex. The 25 MHz of spectrum is divided into two blocks of 12.5 MHz each. Block A is allocated to carriers who are not traditional telephone service providers. Block B is allocated to traditional telephone service providers. Each 12.5 MHz of spectrum can support 416 channels each of which is 30 kHz wide. Of these, 395 are dedicated channels for voice and 21 are dedicated for call control.

Based on subjective voice quality tests, it was determined that a signal to interference ratio of 18 dB can be tolerated while providing a good voice quality to the user. From Equation 5.9, this means that the cluster size has to be $N = 7$. Figure 5.19 shows the cellular architecture with this cluster size. Cells with the same label use the same frequency spectrum. They are separated by a distance $D_L = 4.58\,R$ in this case which ensures that the signal to interference ratio is around 18 dB.

Let the 395 voice channels available for a service provider be numbered from 1 to 395. For example, on the downlink, 869.0–869.030 corresponds to channel 1, 869.030–869.060 to channel 2 and so on. Channels 1, 8, 15, ... are allocated to cells labeled A. Channels labeled 2, 9, 16, ... are allocated to cells labeled B and so on. This ensures that there is a sufficient separation between channels used within a cell so that adjacent channel interference is minimized. In practice, the numbering scheme is different since the entire 25 MHz of bandwidth was not available for AMPS initially. However, a separation of seven adjacent channels is maintained between channels used within a cell. It was also found in some cases that, since cells actually do not subscribe to a hexagonal shape and because of the assumptions made in coming up with the value for N, in reality, a cluster size of $N = 12$ has to employed for good voice quality.

5.6   Capacity Expansion Techniques

In the past decade, the dominant source of income for the wireless telecommunication industry has been the cellular telephone service. This industry has grown exponentially during the last decade of the past millennium. Numerous companies are in fierce competition to gain a portion of the income of this profitable and prosperous industry. The main investment in deploying a cellular network is the cost of the infrastructure that includes the cost of base station and switching equipment, property (land for setting up the cell sites), installation, and links connecting the BSs. This cost is proportional to the number of BS sites. The income of the service is directly proportional to the number of subscribers. The number of subscribers should grow with time and a cellular service provider has to develop a reasonable deployment plan that has a sound financial structure to account for many of these aspects. All service providers start their operation with the minimum number of cell sites to cover a service area that requires the least initial investment. As the number of subscribers increases, it generates a source of income for the service provider. At such a point of time, they can increase the investment on the infrastructure to improve service and increase the capacity of the network to support additional subscribers. Therefore, a number of methodologies have evolved to facilitate the expansion of cellular telephone networks.

There are basically four methods to expand the capacity of a cellular network. The simplest method is to obtain additional spectrum for new subscribers. This is a very simple but expensive approach. The so-called PCS bands were sold in the United States for around $20 billion. If we assume that each new subscriber generates a profit of approximately $1000 per year, we will still need 20 twenty million additional subscribers to recover this amount in a year. With the fierce competition to provide the lowest cost to the customer, this has proved to be suicidal. A case in example is that the top three companies that purchased the PCS bands have already filed for bankruptcy. The reader should not however conclude that this is not an acceptable method. With our pessimistic scenario we are accentuating the vital importance of the need for other alternatives to expand capacity in addition to this simple approach of getting additional spectrum.

The second method to expand the capacity of a cellular network is to change the cellular architecture. Architectural approaches include cell splitting, cell sectoring using directional antennas, Lee's micro-cell zone technique [Lee91], and using multiple reuse factors (called reuse partitioning [Hal83]). These techniques, described in detail in the rest of this section, change the size and shape of the coverage of the cells by adding cell sites or modifying the nature of antennas to increase the capacity. These techniques do not need additional spectrum or any major changes in the wireless modem or access technique of the system that will require the user to purchase a new terminal. These features of architectural approaches distinguish them as one of the more practical and less expensive solutions to expand the network capacity.

The third method for capacity expansion is to change the frequency allocation methodology. Rather than distributing existing channels equally among all cells it is possible to use a non-uniform distribution of the frequency bands among different cells according to their traffic need. The traffic load of each cell is dynamically changed by the geography of the service area and with time depending on the traffic load. In most downtown areas we have the largest traffic loads during rush hours and a relatively light traffic load in the evening hours and weekends. This situation is reversed in residential areas. If allocate channels dynamically to different cells we can increase the overall capacity of the network. These techniques do not need any change in the terminal or physical architecture of the system and they are implemented somewhere inside the computational devices used for network control and management.

The fourth and the most effective method to expand the network capacity is to change the modem and access technology. The cellular industry started with analog technology using FM modulation and has now evolved towards TDMA and then a CDMA air interface using digital modems. Digital technology increases the network capacity and also provides a fertile environment for integration of voice and data services. However, this migration requires the user to purchase new terminals and the service provider to install new components in the infrastructure.

5.6.1   Architectural Methods for Capacity Expansion

Cell Splitting: As the number of subscribers increase within a given area, the number of channels allocated to a cell is no longer sufficient for supporting the subscriber demand. It then becomes necessary to allocate more channels to the area that is being covered by this cell. This can be done be *splitting* cells into smaller cells and allowing additional channels in the smaller cells.

Consider Figure 5.22. In this figure, we have a cellular architecture where a cluster size of seven is employed. When the traffic load increases, a smaller cell is introduced such that it has half the area of the larger cells. This will ultimately increase the capacity fourfold (since area is proportional to the square of the radius). However, in practice, only a single small cell will be introduced such that it is midway between two co-channel cells. In this case, these are the larger cells labeled **A**. It is logical to thus reuse the channels allocated to these cells in the smaller cell to minimize the interference.

This approach gives rise to some problems. Let us suppose that the radius of the smaller split cell (labeled **a**) is $R/2$. Let the transmit power of the base station of the small cell be the same as the transmit powers of the larger cells. As far as the smaller cell is concerned, the signal to interference ratio is maintained because the maximum distance the mobile can be from the base station in this cell is $R/2$. So though the distance between this cell and the co-channel cells **A** is reduced by half, the value of S_r remains the same. On the other hand, this is not the case for the cells labeled **A** since the co-channel reuse ratio for these cells is now $D_L/2R$ with respect to the smaller split cell. In order to maintain the same level of interference, the transmit power of the base station in the smaller cell should be reduced. But this will increase the interference observed by the mobiles in the smaller cell. The other alternative is to divide the channels allocated to cells labeled **A** into two parts: those used by **a** and those not used by **a**. The channels used by **a** will be used in the larger cells only within a radius of $R/2$ from the center of the cell so that the co-channel reuse ratio will be maintained as far as these channels are concerned. This is called the *overlaid cell concept* where a larger *macro-cell* co-exists with a smaller *micro-cell*.

The downside of this approach is that the capacity of the larger cells is reduced which will ultimately lead to introducing split cells in their area, till such time a chain reaction will result in the entire area being served by cells of a smaller radius. Also, the base stations in cells labeled **A** will become more complex and there will be a need for handoffs between the overlays.

Using Directional Antennas for Cell Sectoring: The simplest and the most popular scheme for expanding the capacity of cellular systems is cell sectoring using directional antennas. This

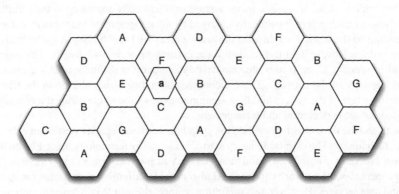

Figure 5.22   Cell splitting.

technique attempts to reduce the signal to interference ratio and thus reduce the cluster size, thereby increasing the capacity. The idea behind using directional antennas is the reduction in co-channel interference that results by focusing the radio propagation in only the direction where it is required. In order to achieve this, the coverage of a base station antenna is restricted to part of a cell called a *sector* by making the antenna directional. In implementing this technique cell site locations remain unchanged and only the antennas used in the site will be changed. The main objective here is to increase the signal to interference ratio to a level that enables us to use a lower frequency reuse factor. A lower frequency reuse factor allows larger number of channels per cell increasing the overall capacity of the cellular network.

As we discussed earlier (see Equation 5.10) the signal to interference ratio is given by:

$$S_r = \frac{1}{J_s} \left(\frac{D_L}{R} \right)^4 = \frac{9}{J_s} N^2 \tag{5.12}$$

where J_s is the number of interfering cell sites. Using a sector antenna reduces the factor J_s resulting in the interference and an increase in S_r. The most popular directional antennas employed in cellular systems are $120°$ directional antennas. In some cases $60°$ directional antennas are also employed. In the following two examples we evaluate the impact of these antennas that enables the reuse factor to be reduced from $N = 7$ to $N = 4$ and $N = 3$ respectively.

Example 5.19: Three-sector cells and a reuse factor of $N = 7$

Consider a seven-cell cluster scheme with $120°$ directional antennas shown in Figure 5.23. Channels allocated to a cell are further divided into three parts, each used in one sector of a cell. As shown in the figure, the number of co-channel interfering cells is reduced from six to two. Thus, there is an improvement in the signal to interference ratio. For omni-directional antennas (see Examples 5.17 and 5.18), the value of S_r for a cluster size of $N = 7$ is 18.66 dB. In this case, in a manner similar to (7), the signal to interference ratio is given by:

$$S_r \approx \frac{R^{-4}}{J_s D_L^{-4}} = \frac{R^{-4}}{6 D_L^{-4}} = \frac{1}{2} \left(\frac{D_L}{R} \right)^4 = \frac{9}{2} N^2 \tag{5.13}$$

For $N = 7$, this will give us $S_r = 23.43$ dB. To see the importance of this gain note that the required signal to noise ratio for AMPS systems is 18 dB which suggests $N = 7$. However a larger S_r is required because of non-ideal situations.

Example 5.20: Three-sector cells and a reuse factor of $N = 4$

Equation 5.13 remains unchanged in this case as there are only two interfering cells again (see Figure 5.24). With omni-directional antennas, $J_s = 6$, and for $N = 4$, we end up with $S_r = 13.8$ dB which is woefully inadequate for AMPS.

It can be seen that the signal to interference ratio with three-sector cells is substantially better compared to omni-directional antennas and no cell sectoring. With $N = 4$, the signal to interference ratio is 19.9 dB. This value is larger than the requirement of 18 dB based on subjective mean-opinion-score (MOS) tests of voice quality.

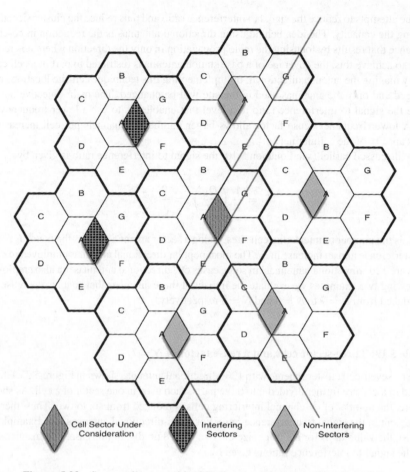

Figure 5.23   Seven-cell reuse with 120° directional antennas (3-sector cells).

Example 5.21: Six-sector cells and a reuse factor of $N = 4$ and $N = 3$

With 60° directional antennas, we have six sectors within a cell. The number of interfering co-channel cells reduces to one and the signal to interference ratio can be written as:

$$S_r \approx \left(\frac{D_L}{R}\right)^4 = 9N^2 \qquad (5.14)$$

It is possible to employ a cluster size of four or three with six-sector cells since the signal to interference ratio will be 21.58 or 19.1 dB, respectively, which has a sufficient margin for AMPS. The cellular layout and relation to sectors in this case is left as an exercise for the readers.

In practice we cannot ideally sector a cell because ideal antenna patterns cannot be implemented. Therefore, the numbers obtained in above examples for ideal cell sectors are optimistic. However, our conclusion from the above examples is that the use of sectoring increases the signal to interference ratio at the terminal. We should emphasize that in the particular examples we could reduce the frequency reuse factor from $N = 7$ to $N = 4$ or even $N = 3$ by using three- and six-sector

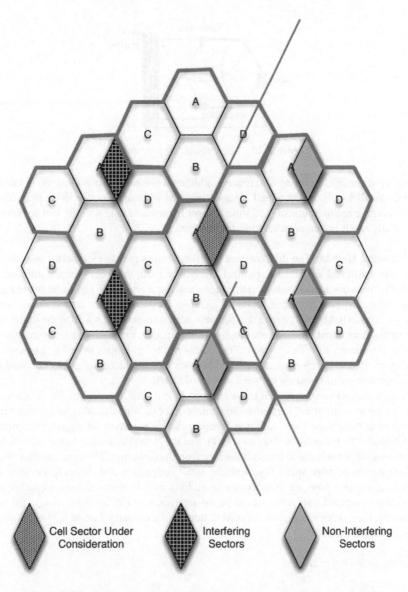

Figure 5.24 Four-cell reuse with 120° directional antennas (3-sector cells).

cells respectively. This reduction in frequency reuse from seven to four or even three would result in a capacity increase of 1.67 and 2.3, respectively, allowing an equal increase in the number of subscribers and consequently income of the service provider. The service provider needs to add these antennas to the base stations in the desired area. Compared to the cell-splitting method, using directional antennas is less effective in increasing capacity but it can be significantly less expensive. The cost of additional cell sites, needed in cell splitting, includes costs of the property and installing the antenna mounting tower that are usually far expensive compared to deploying

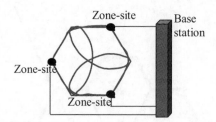

Figure 5.25   Lee's micro cell zone concept.

directional antennas. Cell splitting also requires additional planning efforts to maintain interference levels in the smaller cells. If directional antennas are used without reduction in the frequency reuse factor, the average required transmitted signal power from the mobile stations will be reduced that can potentially result in longer battery life for the user.

Lee's Micro-cell Method: The disadvantage of using sectors is that each sector is nothing but a new cell with a different shape, since channels have to be partitioned between the different sectors of a cell. The network load is substantially increased since a handoff has to be made each time a mobile terminal moves from one sector of a cell to another. Also, in all the discussion in the previous sections, it has been assumed that the base station antenna is located at the center of a cell, whether or not directional antennas are employed. In practice, employing directional base station antennas at the corners of cells can reduce the number of base stations [Mac79]. Lee's micro-cell zone technique [Lee91] exploits corner excited base stations to reduce the number of handoffs and eliminate partitioning of channels between sectors of a cell.

Figure 5.25 shows Lee's micro-cell zones concept. In this case, there is one base station per cell, but there are three "zone sites" located at the corners of a cell. Directional antennas that span 135° are employed at these zone sites. All three zone sites act as receivers for signals transmitted by a mobile terminal. The base station determines which of the zone sites has the best reception from the mobile and uses that zone site to transmit the signal on the downlink. The zone sites are connected to the base station by high-speed fiber links to avoid congestion and delay. Since only a single zone site is active at a time, the interference faced by a mobile terminal from a co-channel zone site is smaller compared with what would be the interference with an omni-directional antenna. Consequently, the cluster size can be reduced to three and a capacity gain of 2.33 is obtained over a seven-cell cluster scheme.

Consider the following example:

Example 5.22: Lee's Micro-Cell Zone Technique

Show that the co-channel reuse ratio D_Z/R_Z for the zones in Lee's micro-cell zone concept is larger than 4.6 if a cluster size of three is employed. Use Figure 5.26 for your calculations.

In this example, a cluster size of $N = 3$ is employed. Each "cell" is divided into three "zones". On the downlink, only one of the zones is active. Since the zone sites are directional, they cause interference only in corresponding zone sites in another cluster. The co-channel reuse ratio D_z/R_z in Figure 5.14 is clearly $6 \times \sqrt{3}/2 = 5.196$ which is larger than 4.6. Even if all six co-channel zones cause interference, the capacity is still larger by a factor of 2.33 since the cluster size is now $N = 3$ as compared to the usual value of $N = 7$.

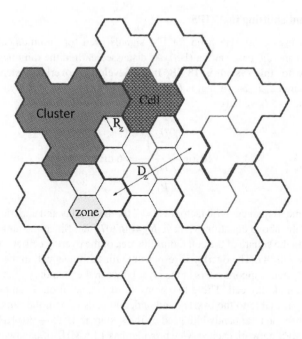

Figure 5.26   Example of Lee's micro-cell zone concept.

Using Overlaid Cells: The *overlaid cell concept* introduced in the section on cell splitting can be used to increase the capacity of a cellular network. Here, channels are divided among a larger macro-cell that co-exists with a smaller micro-cell contained entirely within the macro-cell. The same base station serves both the macro- and micro-cells. Figure 5.27 illustrates the basic concept for overlaid cell concept. There are four parameters R_1 and D_1 representing the radius of coverage and distance among co-channel cells for the macro-cells and R_2 and D_2 denoting the radius of coverage and the distance among co-channel cells for the micro-cells. The design is made such that D_2/R_2 is larger than D_1/R_1 and from Equation 5.7 the signal to interference ratio for the micro-cells will be substantially greater than that of the macro-cells. There are two methods to exploit this situation to increase the capacity of the network: using split-band analog systems and reuse partitioning. Often the micro-cells are said to belong to an *overlay* network that is overlaid on top of an underlying macro-cellular network referred to as the *underlay* network.

Split-Band Analog Systems: The split-band analog systems use a more bandwidth efficient modulation within the overlay cells. This technique is applied in analog cellular systems using frequency modulation (FM). In FM, the signal to noise requirement is inversely proportional to the square of the bandwidth. If we reduce the bandwidth to half the original value, the signal to noise ratio requirement will be increased four times (by 6 dB). If we arrange R_2 and D_2 to have a co-channel reuse ratio that is four times larger than usual, we end up with a signal to interference ratio (from Equation 5.6) that remains unchanged. The overlay system can then use FM with half the bandwidth of the underlay system, doubling the capacity within the overlay part of the network. An example will further clarify this situation.

Example 5.23: Band-splitting in AMPS

The AMPS system uses a 30 kHz band for FM signals used for communication between the mobile station (MS) and the base station (BS). As discussed earlier, the minimum required signal to interference ratio for this system is 18 dB. If we develop an overlay system with a 15 KHz bandwidth, the required S_r is 24 dB that is 6 dB more than the system employing a bandwidth of 30 kHz. From Equation 5.6 we have:

$$10\log \frac{\left(\dfrac{D_2}{R_2}\right)^4}{\left(\dfrac{D_1}{R_1}\right)^4} = 6\,\text{dB}$$

If we employ the same frequency reuse factor of $N = 7$ for the overlay and underlay networks, $D_1 = D_2$ and solving for the above equation we have $R_2 = 0.7079\,R_1$. Since the area covered by each cell is proportional to the square of the cell radius the area of the overlay cell, A_2, will be half of the area of the underlay cell, A_1. The overlay is responsible for terminals within the smaller hexagon, while the underlay system supports users in the layer between the boundary of the overlay cell and the boundary of the underlay cell. These two areas are the same in our example. Therefore, the number of channels available to the overlay and underlay cells remain the same. If we represent this number by M then the total bandwidth used by the system is $M(15 + 30)$ KHz.

In the original AMPS network each service provider has 12.5 MHz of bandwidth that is divided into 416 channels from which 395 channels are used for voice and 21 channels for control signaling. Therefore, 395×30 KHz of bandwidth was used for actual traffic. If we replace that system with a split-band underlay-overlay network we have:

$$M(15 + 30) = 395 \times 30 \geq M = 263$$

The total number of channels $M = 263$ for each of the overlay and underlay cells and $263 \times 2 = 526$ will be the total number of channels available. This number is 1.34 times larger than the original system, improving the capacity of the system by 34%.

Compared to cell splitting or using sectored cells, this technique provides a smaller improvement in capacity. However, it does not need any change in the hardware infrastructure. However, the MS and BS need minor changes to cope up with multiple bandwidths. To further improve the capacity of this technique, it is possible to use another layer of overlay system using even smaller cells. As we saw in Chapter 1, the Japanese analog systems use 25 kHz per user for underlay networks and 12.5 kHz (and even 6.25 kHz) for the overlay networks. The downside of underlay-overlay networks is the increased complexity at the base station for keeping track of which channel belongs to which overlay and increased number of handoffs when a mobile moves from one overlay to the next (or from a micro- to a macro-cell). This requires additional complexity at the base station and handoffs when a mobile terminal moves from a micro- to a macro-cell.

Reuse partitioning and Fractional Frequency Reuse: The overlaid cell concept described above can be used to increase the capacity of a cellular network through what is called the *reuse partitioning* concept [Hal83]. Here, channels are divided among a larger macro-cell and a smaller micro-cell contained entirely within the macro-cell. The bandwidth in both cells remains the same. Since the radius of the micro-cell is smaller, the signal to interference ratio (S_r) for the overlay is larger and it

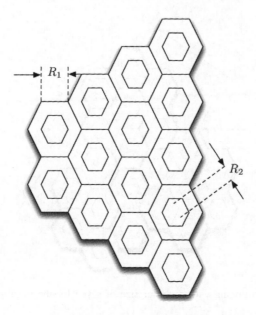

Figure 5.27 Reuse partitioning.

is able to employ a smaller co-channel reuse distance compared to the underlay or macro-cell. The channels allocated to the micro-cell for instance, may be reused in every third or fourth micro-cell whereas the channels allocated to the macro-cells can be reused in only every seventh or 12th cell, as the case may be. This requires additional complexity at the base station and handoffs when a mobile terminal moves from a micro- to a macro-cell. To explain this situation, consider the following example.

Example 5.24: Reuse partitioning of 7 and 3

Assume that in Figure 5.27 the radius of the underlay macro-cells is R_1 and the radius of the micro-cells of the overlay is R_2. If we have an AMPS network operating on this infrastructure, the required S_r for both networks is 18 dB. From Equation 5.7, both underlay and overlay networks should have $D_1/R_1 = D_2/R_2 = 4.6$. Since R_2 is smaller than R_1, D_2 can be made smaller than D_1 by a factor equal to the ratio of R_1 to R_2. The improvement in co-channel reuse ratio comes from the fact that the micro-cells in the overlay are *not* contiguous to one another.

Suppose the co-channel reuse ratio without reuse partitioning was $D_1/R = Q$. The cluster size N in this case is $Q^2/3$ (from Equation 5.7) and the number of channels available per cell is $N_c/N = 3N_c/Q^2$. With reuse partitioning, let the ratio of the macro-cell radius to the micro-cell radius be $\kappa = R_1/R_2$. From Example 5.14, the cluster size for the micro-cells can be reduced by a factor of κ^2 since the micro-cells are non-contiguous.

Example 5.25: Channel allocation to underlay and overlay cells

Consider Figure 5.28. Here we are using a cluster size of $N_1 = 7$ for the underlying macro-cells to ensure that $D_1/R_1 = 4.6$ to provide a suitable S_r for AMPS. We now overlay micro-cells with

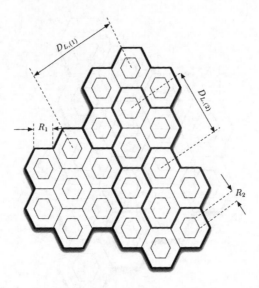

Figure 5.28 Reuse partitioning with a cluster size of seven for the macro-cells and three for the micro-cells.

a radius R_2 such that the cluster size of the micro-cells is $N_2 = 3$. If $N_2 = 3$, we can see from Figure 5.28, that $D_2 = 3R_1$. Clearly, $3R_1/R_2 = 4.6$ or $R_2 = 0.652R_1$. One way of allocating channels to the micro- and macro-cells is to distribute them by area occupied. This may not be the best case. However, for this example, we employ this technique. The area of a cell is proportional to the square of its radius. We see that the area of a micro-cell is either 0.652^2 times the area of a macro-cell or 0.425 times the area of a macro-cell. Let the total number of channels available be N_c, If channels are distributed according to area and there are L channels available per cell, let us assume that $0.425L$ channels are allocated to the micro-cell and $0.575L$ channels are allocated to the macro-cell.

Since the cluster sizes are 7 and 3 respectively, we have:

$$N_c = 7 \times 0.575L + 3 \times 0.425L => L = N_c/5.3$$

The total number of available channels for an AMPS operator is 395. Therefore, $L = 75$. The inner overlay uses approximately 32 channels and the underlay uses 43 channels. Originally we had 395 channels with $N = 7$, providing approximately 56 channels per cell. The increase in the capacity is $75/56 = 1.34$, a 34% increase in capacity. In reality, a larger capacity can be expected since the channels allocated to the macro-cells may also be used within the micro-cells.

Multiple overlays can provide an additional increase in capacity. As compared to the other expansion techniques, the advantages and disadvantages of the frequency reuse partitioning are very similar to frequency splitting. However, reuse partitioning does not need modification in the BS or MS radio equipment and it can be easily applied to other modulation techniques. The derivation of S_r for frequency splitting was highly correlated with how FM works and it cannot be extended to digital systems in a straightforward way.

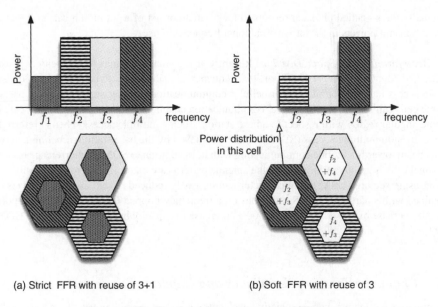

(a) Strict FFR with reuse of 3+1   (b) Soft FFR with reuse of 3

Figure 5.29   Fractional frequency reuse in LTE systems.

Reuse partitioning is employed in a different way in 4G LTE systems where OFDMA is the access method. In OFDMA, frequency sub-carriers and time slots are allocated to individual users for transmissions. Through what is called *fractional frequency reuse* or FFR, certain frequency sub-carriers are used at lower transmission power, so that they only service mobile devices that are at a closer distance to a base station, while other frequency sub-carriers are employed throughout a cell. This way, those low-power sub-carriers that are used in the center of a cell do not interfere with transmissions in neighboring cells at the same frequencies. The way these sub-carriers are selected depends on the cell sizes, loads in cells, tolerable interference, and other such factors. The primary difference between the way this is deployed in 4G LTE systems and in the past is that the power assignment and reuse can be done on a fine-grained time and frequency scale in 4G systems, especially because there are more powerful ways of predicting and coordinating interference in these systems. This is often called *inter-cell interference coordination* or ICIC in LTE systems [Ger10]. In contrast, the reuse partitioning approaches in 2G/1G cellular systems were mostly done on much longer time-scales and the frequency allocations were fairly static.

Figure 5.29 shows two approaches for fractional frequency reuse in LTE systems [Gho11]. In the case of strict fractional frequency reuse, the interior of a cell uses its own frequencies which are not reused in the cell exterior. As shown in Figure 5.29a, the interior of every cell use frequencies f_1. The cell exteriors use frequencies f_2, f_3, f_4, with a reuse factor of three for a total of four different frequencies. Thus, the transmit power with frequency f_1 is smaller than the transmit power associated with the other frequencies. In the case of soft fractional frequency reuse, the interior of a cell can use the frequencies that are not used in its own cell's exterior, but are used in the exterior of neighboring cells. For example, in the cell that uses f_4 in its exterior with higher power, the frequencies f_2 and f_3 are used with a lower power so that they can be used only in the cell interior and not cause interference to the users at the edges of neighboring cells. We note here that we have referred to frequencies, but as mentioned previously, and described in more detail in Chapter 13, it is actually the so-called physical resource blocks (that consist of a set of sub-carriers) that use different transmit powers in the case of fractional frequency reuse.

Using Beamforming and Space Division: Recently, using smart antennas for capacity expansion has attracted attention [Leh99]. Traditionally, frequency division, time division and code division multiple access has been employed for cellular communications. Using smart antennas, users in the same cells can use the *same* physical communication channel as long as they are not located in the same angular region with respect to a base station. Such a multiple access scheme, referred to as space division multiple access (SDMA), can be achieved by the base station *directing* a narrow antenna beam towards a mobile communicating with it. In addition to SDMA, interference between co-channel cells is greatly reduced since the antenna patterns are extremely narrow. Previously, we saw that using sectored cells, the reuse factor can be vastly reduced. Even larger advantages can be obtained with smart antennas. Simulations on a frequency-hopped GSM system have reported a capacity increase of 300%. A fivefold (500%) increase in capacity has been reported for CDMA [Leh99].

5.6.2   Channel Allocation Techniques and Capacity Expansion

In the previous section we were associating each cell with a group of channels that is assigned by the service provider to that cell. In the analysis of capacity we were assuming that all channels in a cell will be used and we were finding methods to increase the number of available channels per cell in a given geographical area. We examined a variety of architectural methods to increase the number of available channels per cell. In all of these schemes the number of channels in cells of equal size were assumed to be the same. This assumption would be valid if the distribution of the users in the area was stationary and uniform over all cells. In practice, during the day, a cell in downtown areas of a city carries a high traffic that peaks during rush hour. But this very same cell does not carry a lot of traffic during late evenings or weekends. A cell in a residential area may have traffic characteristics that are the opposite of that of the cell in the downtown area. Clearly, the number of terminals in a cell changes in time depending on the location of the cell, and this means there is need for a more complex methodology to allocate channels to the cells dynamically based on the traffic load at a given time. A number of channel allocation strategies have been developed to address this issue [Kat96].

To address channel assignment or allocation techniques in cellular networks we look at this problem from the angle of view of a user. For the user it is not important how many channels are available or how they are allocated. A circuit switched (voice) user will dial a number and if a channel is available he/she is happy. If the call is blocked because a free channel is not available, the user will be dissatisfied with the service provider. A measure of whether channels will be available for a user when he attempts a call is the probability of call blockage (we discuss this in detail in a later section). It depends on the number of available channels and the traffic load. This is thus the quantity that represents user satisfaction. Service providers believe that a probability of call blockage of around 2% will keep customers happy and they aim at this number. However, the probability of call blockage changes its value as mobile terminals move in and out of the boundaries of a cell. A service provider should maintain the number of subscribers under a particular value so that it is possible to accommodate fluctuations in the probability of call blockage over time across all the cells.

The main objective of channel allocation techniques is to stabilize the fluctuations in the probability of call blockage over the entire coverage area of a network over time. Reduction of the fluctuations in probability of call blockage allows service providers to accept a higher number of subscribers over the coverage area. This can be considered equivalent to expansion of the network through additional channels. In other words, as the number of subscribers increase, one way of accommodating such an expansion is to use a more efficient channel allocation technique to cope up with the situation. Service providers use a variety of proprietary algorithms for channel allocations. These techniques can be divided into three main categories: Fixed Channel Allocation (FCA), Dynamic Channel Allocation (DCA), and Hybrid Channel Allocation (HCA) techniques. We study these techniques in the following sections.

Fixed Channel Allocation: In cellular telephony, the number of chunks of frequency spectrum available for the service provider and the bandwidth required per user governs the number of available channels. Fixed channel allocation techniques, in their simplest form, divide the available spectrum by the cluster size to determine the number of radio channels per cell. That is, if the available spectrum is W Hz and each channel needs B Hz, the total number of channels is $N_c = W/B$. If the cluster size is N, the number of channels per cell is $C_c = N_c/N$. These C_c available radio channels are then distributed in the cells in a manner so as to minimize adjacent channel interference. One obvious distribution pattern for channels among the cells is to assign adjacent radio frequency bands to different cells. In analog cellular systems each radio channel corresponds to one user (one voice channel) while in digital TDMA or CDMA networks, each radio frequency channel carries several time slots or codes associated with voice channels.

Example 5.26: Fixed channel allocation in 2G Cellular TDMA (GSM)

In the GSM cellular system, we have a pair of 25 MHz bands allocated to the downlink or forward channel and uplink or reverse channel. Each radio carrier uses 200 KHz of bandwidth and each carrier contains 8 time slots capable of supporting 8 voice users. Potentially we have 125 carriers, but in practice only 124 of them are used. Let us number the channels as 1, 2, 3, ... , 124. If the cluster size is $N = 4$ the simple FCA technique results in four sets of frequencies given by:

$\{1, 5, \ldots , 120\}$, for the first set of cells

$\{2, 6, \ldots , 121\}$, for the second set of cells

$\{3, 7, \ldots , 122\}$, for the third set of cells

$\{4, 8, \ldots , 123)$ for the fourth set of cells.

Example 5.27: Fixed channel allocation in 1G Cellular FDMA (AMPS) and 2G NA-Cellular TDMA (IS-136)

In United States cellular systems, discussed in Example 8 earlier, each service provider has 12.5 MHz of spectrum available to him. In the downlink or forward channel, service providers use half of the 869–894 MHz band and in the uplink or reverse channel, half of the 824–849 MHz band. These bands are divided into 416 pairs of radio frequency bands each having 30 KHz chunks for forward and reverse channels. In AMPS each frequency band carries one voice user while in the IS-136 digital TDMA system, three users per 30 KHz of radio channel are supported via time slots. Of the 416 radio frequency channels, 21 are used for control channels and 395 for voice traffic. With FCA and a frequency reuse factor of seven we can create the following seven sets of frequencies to minimize the interference:

{1, 8, ... , 390} For the first set of cells

{2, 9, ... , 391} For the second set of cells

. . .

{7, 14, ... , 396} For the seventh set of cells.

The FCA strategy described above is simple to implement and if the traffic in the network is uniform, so that the number of active users in each cell is the same; and it remains constant with time, which is also an optimum channel allocation strategy. However, in practice, traffic in each cell changes with time due to the movement of mobile terminals from one cell to another. This results in higher probabilities of call blocking in some cells and lower values in others, which results in poor utilization of the available bandwidth. To equalize the utilization of channels in all cells, the obvious solution is that the cells with higher traffic load should somehow use the free channels available in low traffic cells. This is possible by a non-uniform allocation of channels to cells in the first place. When we assume that the traffic density in all the cells is the same, as illustrated by Examples 5.16 and 5.17, the channel assignment algorithm is very straightforward. We simply divide the total number of available channels by the cluster size of the system and allocate this number of channels to each cell. Using the traffic density and the number of channels in one cell, we can determine the call-blockage probability in that cell. The probability of call blockage in all other cells and consequently the average probability of call blockage in the entire network will be the same as that of one cell. Here the channel assignment algorithm and the calculation of probability of call blockage are performed in two independent steps. With a non-uniform channel allocation technique, we need to include the call blockage probability as a criterion for the channel allocation algorithm. Since the relation between the number of channels and the call-blockage probability is a complex function, this algorithm becomes significantly more complex. The following example helps in our understanding of the complexity involved.

Example 5.28: Probability of call-blockage with non-uniform traffic distribution

Assume that we have only four cells and one cluster. We simply divide our available channels N_c by four. That is, we assign $N_c/4$ channels per cell. Also assume that for a uniform distribution of the traffic, the calculated probability of blockage for each cell is the desired number (2%). If the traffic becomes non-uniform, the blockage rate of all four cells will be changed and the average over all of them does not remain at 2% anymore. The general idea is to increase the number of channels in cells with the higher traffic load and decrease it in cells with a lower traffic load so that the overall blockage rate of the network is minimized. There are four variables, namely the number of channels per cell and the cost function to be minimized is the probability of call blocking. This has a very complex expression involving these variables. The minimization process is far more complex than the uniformly distributed traffic case for which we know that the same number of channels per cell would provide the minimum blockage rate. This is only one dimension of the complexity of the problem. The other dimension of complexity emerges when we consider co-channel cells having different traffic densities. The optimization problem is now a function of N_c variables and in addition, the regular frequency reuse pattern studied earlier does not work any more. The relatively simple frequency reuse pattern strategies discussed earlier were based on the assumption that the channels are fixed and thus calculating the co-channel interference on the basis of having the same number of channels per cell. These patterns become much more complex as we start thinking of unequal number of channels per cell.

Algorithms that distribute channels among the cells according to their traffic load have been investigated. For example, a *non-uniform compact channel allocation algorithm* is discussed in [Zha91]. This algorithm first defines a set of patterns for non-uniform distribution of channels then it selects the pattern that minimizes the average call blocking probability in the system. Results of example simulations in [Zha91] show that non-uniform distribution of channels adopted by this algorithm provide better call blocking probabilities in the system. The reduction in call blocking probabilities allows an average of 10% and a maximum of 22% traffic to be added to the system while maintaining the same call blocking probability as that of uniform channel allocation. Note that channels are still permanently allocated to cells here and this still corresponds to fixed channel allocation.

Channel Borrowing Techniques: Non-uniform channel allocation is quite complicated. A simpler scheme enables high traffic cells to *borrow* channel frequencies from low traffic cells and maintain them until significant changes in traffic pattern is measured or predicted. In other words the high traffic cells *borrow* channels from low traffic cells. These techniques are usually referred to as *Channel Borrowing Techniques* [Kat96]. The technical issues are how can we relate the traffic distribution to the channel allocation? and which cell to borrow the channels from? There are two methods to borrow channels: *temporary channel borrowing* and *static channel borrowing*. In *temporary channel borrowing*, high traffic cells return the borrowed channels after the call is completed. In *static channel borrowing* channels are non-uniformly distributed among cells according to the available statistics of the traffic and changed in a predictive manner.

Temporary channel borrowing deals with short-term allocation of borrowed channels to cells. Once a call associated with the borrowed channel is completed, the channel is returned to the cell from which it was borrowed.

Example 5.29: Temporary Channel Borrowing

Let us suppose that we had forty-nine cells in a region. Let the cluster size be seven. In the case of uniform traffic density, we divide the total number of available channels, N_c, into seven groups and using prescribed reuse patterns studied in earlier sections, we assign channel groups to different cells (see Figure 5.30). The calculation of the probability of call blockage will remain the same as before and let this value be 2%. If we make a pool of the N_c channels and allocate them purely on the basis of demand, we have N_c different channels each with a different characteristic in terms of the traffic on them.

Some channels may not be in use and others may be continually in use. Depending on how often a channel is used and where it is being used, it may cause a high or low interference to its co-channel elsewhere. Suppose the cell A in the central cluster in Figure 5.30 borrows a channel from the solid shaded cell F within its cluster to accommodate extra traffic load. This means that the corresponding channel in three cells labeled F to the left in neighboring clusters are locked till this channel is released by the cell A. This is because the re-use distance for the borrowed channel has decreased since it has been moved from the solidly shaded cell F, to the cell A.

A number of methods for selecting free channels from a lightly loaded cell for borrowing by another cell are summarized in [Kat96]. These methods either make all channels available for borrowing (this is called simple borrowing schemes) or they partition the channels into borrowable and non-borrow-able (such schemes are referred to as hybrid borrowing schemes). Simple borrowing schemes are found to be better under light or moderate traffic loads. A quantitative comparison of some of these techniques is provided in [Kue92]. In specific examples addressed in this paper, some of the suggested borrowing techniques are found to be capable of supporting up

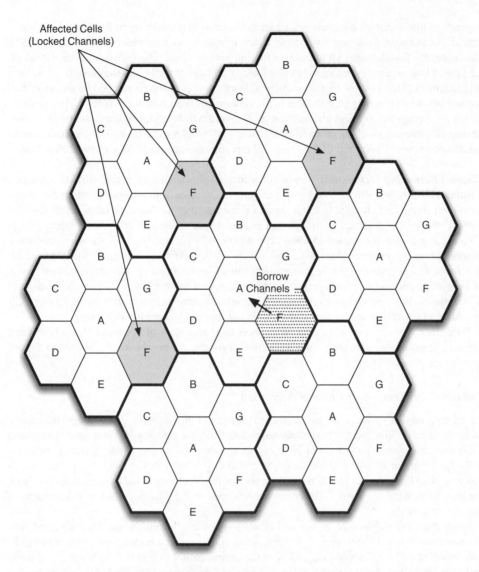

Figure 5.30 Temporary channel borrowing.

to 35% more traffic than uniformly distributed FCA. Channel borrowing schemes however require additional computational complexity, and frequent switching of channels. They may also affect handoff strategies.

Dynamic Channel Allocation: Many researchers have studied the shortcoming of FCA techniques to accommodate temporal and spatial traffic variations and have suggested various dynamic channel allocation (DCA) techniques in the past two decades. In DCA all channels are placed in a pool and they are assigned to a new call according to the overall signal to interference pattern in all cells. Each channel can be used in any cell as long as it satisfies the signal to interference ratio requirements for the system. The channel is returned to the pool after the cellular call is terminated. This technique adapts well to the temporal and spatial changes in traffic load. In fact, according to [Cox99], the capacity is maximized when the signal to interference ratio of every set of co-channel users (users in co-channel cells using the same frequency bands) is balanced around some level that is no larger than strictly necessary. The downside is that DCA is extremely complex and inefficient under high traffic load conditions. Although many claims have been made about the relative performance of each DCA schemes, the tradeoffs and the range of achievable capacity gains are still unclear and a number of questions remain unanswered [Kat96]. Micro-cellular systems of high density personal communication networks have shown to benefit most out of the DCA and results of simulations shows a near ideal performance with DCA algorithms [Kat96].

The basic idea of the DCA is straightforward. However, a number of DCA schemes have been studied. The question then arises as to how these schemes differ from one another. In DCA often several choices are available for assigning channels to a requesting cell. To devise a selection policy, we have to define a cost function to determine the appropriateness of the channel to be selected. This cost function quantitatively ranks the available channels based on the overall interference, average probability of call-blockage, or parameters that somehow relate to these quantities. The difference between DCA techniques lies in the selection and optimization of this cost function.

In reference [Kat96] a number of DCA schemes are introduced and they are divided into two categories: centralized and distributed schemes. In centralized schemes, a central pool of all channels exists. The various schemes differ in the way the cost function handles the priorities in selecting a candidate channel and returning it to the central pool upon termination of the call. Distributed DCA techniques are considered for micro-cellular systems where channel propagation is less predictable and traffic is denser. In these schemes, the BS decides on the frequency assignment locally. Distributed DCA techniques are further divided into two classes of techniques: cell-based and interference-based. Cell-based techniques require each BS to maintain a table of available channels in its vicinity and based on this table, the BS decides and assigns the channel to the users in its cell. This technique is very efficient, but the expense incurred is additional inter-BS communication traffic, that increases with the traffic in a cell. To avoid this situation, another sub-class of distributed DCA schemes has evolved. Here, each BS makes the channel assignment based on the received signal strengths (RSS) of the mobiles in its vicinity. In such schemes all channels are available to the BS and the BS make its decision based on the local information without any need to communicate with other BSs. These schemes are self-organizing, simple, efficient, and fast but they suffer from additional unwanted co-channel interference that may result in channel interruption and network instability.

Example 5.30: Centralized DCA

In [Zha89], a locally optimized dynamic assignment (LODA) strategy is discussed. Here the particular cell allocating channels considers candidate channels based on whether they are being used in the first, second, third tiers of co-channel cells and so on till the n-th tier. It then assigns the channel with minimum cost to the requested call from the mobile. Simulations of the LODA scheme with 49 cells indicate that the call-blocking probability can be reduced by 40% for light loads compared to fixed channel assignment.

Example 5.31: Distributed DCA (DDCA)

Examples of a signal strength-based distributed dynamic channel assignment strategies are simulated in [Goo93]. The DDCA schemes are implemented for micro-cellular environments or

Table 5.2   Comparison of three classes of algorithms for DCA [Kat96]

Algorithm	Centralized DCA	Distributed DCA Cell-based	Measurement-based
Advantages	Near optimum channel allocation	Near optimum channel allocation	Sub-optimal channel allocation Simple assignment algorithm Use of local information Minimum communication between base stations System capacity increases, efficiency, radio coverage Fast real time processing Adaptive to traffic changes
Dis-advantages	High centralized overhead	Extensive communication between base stations	Increased co-channel interference Increased interruption Deadlock probability Instability

one-dimensional cellular systems. Similar schemes are implemented in DECT [Kat96]. When a mobile requests a channel from the base station, the base station measures the interfering signal power on all channels not already assigned to mobiles in its cell. The channel with the maximum signal to interference ratio is assigned to the mobile. For the same mean traffic, the DDCA strategies provide a much lower probability (around 30-50% lower) of call blocking compared to fixed channel assignment schemes. The exact increase in capacity depends on the number of mobile stations and the offered traffic load in the given cell.

Table 5.2 compares all three classes of algorithms for DCA. It also summarizes in brief, the important characteristics of DCA strategies. Centralized DCA strategies do provide optimal or near optimal channel allocation. However, they require a lot of computational and signaling effort since the centralized location has to be aware of the available channel and the necessary parameters required to make an optimum decision on which channel to allocate to an incoming call. These parameters could be how channels are allocated in co-channel cells, what are the signal to interference ratio values for the channel under consideration, what is the expected traffic load in and around the region, and so on. Also, since the decision mechanism is centralized, it is not robust, and a failure here could lead to an entire system-wide shutdown. In that sense, distributed DCA strategies are better. Cell based DDCA strategies also can allocate channels optimally since base stations can communicate with each other to obtain knowledge of the entire system. While the base stations make the decisions to allocate channels, the need for frequent communication and update between base stations is a disadvantage. Signal-strength or measurement based DCA strategies do not provide optimal allocation of channels. However, the algorithm for channel allocation is truly localized and the speed with which call set-ups can be made is adequate. Such schemes do provide capacity increases and are implemented in digital cordless systems like DECT.

Comparison of FCA, DCA, and Hybrid Allocation: Overall, DCA techniques have shown 30-40% performance improvements over the simple FCA techniques [Goo93]. Table 5.3 compares fixed and dynamic channel allocation strategies.

Table 5.3   Comparison between FCA and DCA [Kat96]

Attribute	Fixed Channel Allocation	Dynamic Channel Allocation
Traffic Load	Better under heavy traffic load	Better under light/moderate traffic load
Flexibility in channel allocation	Low	High
Reusability of channels	Maximum possible	Limited
Temporal and spatial changes	Very sensitive	Insensitive
Grade of service	Fluctuating	Stable
Forced call termination	Large probability	Low/moderate probability
Suitability of cell size	Macro-cellular	Micro-cellular
Radio equipment	Covers only the channels allocated to the cell	Has to cover all possible channels that could be assigned to the cell
Computational effort	Low	High
Call set-up delay	Low	Moderate/high
Implementation complexity	Low	Moderate/high
Frequency planning	Laborious and complex	None
Signaling load	Low	Moderate/high
Control	Centralized	Centralized, decentralized, or distributed

Under low to moderate traffic loads, DCA strategies perform far better than FCA techniques. Since DCA is based on random arrivals of mobiles and random allocation of channels to them, unless maximizing the "packing" of channels is an optimization criterion, it is likely that distances larger than what is required may separate co-channels. This will prevent channels from being reused as often as they can be resulting in less capacity at larger loads. DCA however reduces the fluctuations in the call blocking probabilities as well as forced call terminations. FCA strategies require a lot of "offline" effort in frequency planning. DCA strategies need plenty of effort in real time for channel allocation. A unified framework for comparing all kinds of DCA strategies versus FCA is not available [Kat96] and it is hard to say which of these schemes are actually beneficial. In addition, DCA schemes that jointly optimize power control and handoff strategies have also been proposed.

Since DCA is better at lower traffic loads and FCA is better at higher traffic loads, the natural question of whether the two channel allocation techniques can be combined to provide both advantages arises. Indeed, *hybrid* channel allocation strategies have also been investigated. The total number of channels is partitioned into *fixed* and *dynamic* sets. The ratio of fixed to dynamic channels becomes important in the performance of the system. HCA schemes have been shown to perform better than FCA schemes for load increases up to 50% [Kat96].

5.6.3   Migration to Digital Systems

Analog cellular systems had inherent capacity crunch and also had the corresponding disadvantages of analog communication systems. Migration to digital cellular systems during the early 1990s resulted in advantages for both the service providers and end users by providing additional capacity, feature flexibility and control. In North American TDMA or IS-136, the same 30 kHz channels employed with AMPS were deployed in TDMA format with six time slots per 30 kHz channel increasing the capacity by a factor of up to six. With full-rate voice coding, each user is allowed access to two time slots so that three users can be accommodated on each AMPS carrier. The frequency planning discussed above is equally applicable to such TDMA systems since they use exactly the same carriers as the AMPS systems. However, it has been found that the interference that can be tolerated by digital TDMA systems is much larger than AMPS so that a much tighter reuse ratio could be employed. For instance, for the IS-136 systems, an S_r of 12 dB is sufficient as against 18 dB for AMPS. This further increases the capacity since a reuse factor of $N = 4$ is possible. In GSM, an S_r of 9 dB is sufficient with slow frequency hopping and this enables employing a reuse factor of $N = 3$.

Digital CDMA systems can provide an even larger capacity increase because of the various interference combating capabilities of CDMA. With CDMA, the *same frequencies* can be employed in adjacent cells thereby increasing the reuse factor to one. We discuss cellular network planning issues in CDMA below.

5.7 Network Planning for CDMA Systems

Code division multiple access presents some unique features that are not present in traditional TDMA and FDMA systems. In TDMA and FDMA systems, the users operating in one channel are completely isolated from the users operating in other channels. The only interference comes from the fact that the same frequency bands are employed in spatially separated cells and this interference is co-channel interference. Of course leakage of signal from adjacent bands causing adjacent channel interference is also a factor but intelligent design can reduce this effect greatly. However, in the case of CDMA, all users are operating on the same frequency channel at the same time resulting in everyone causing co-channel interference. This problem is reduced on the downlink by employing time-synchronized orthogonal codes. On the uplink, a combination of convolutional coding, spreading, and orthogonal modulation are employed to combat the effects of this interference. Network planning in the case of CDMA is far more complicated than in the case of TDMA/FDMA in that sense, but at the same time, using CDMA completely eliminates the concept of conventional frequency reuse since the same frequencies can be deployed in all cells.

Instead of defining an acceptable signal to interference ratio, in CDMA, it is necessary to define the *quality of the signal* [Hal96]. Usually this is expressed in terms of the acceptable energy per bit to total interference ratio E_b/I_t that results in roughly a 1% data frame error rate. The E_b/I_t is used in CDMA instead of the signal to interference ratio S_r. In this section only, we use E_b/I_t and S_r interchangeably. The reason for selecting this as a measure is that this frame error rate results in acceptable speech quality at the vocoder output. The value of E_b/I_t is usually between 6 and 11 dB depending on the speed of the mobile terminal, propagation conditions, the number of multipath signals that can be used for diversity etc. The value of I_t depends on the number of interfering signals and the transmit powers of the interfering users. Consequently, power control and thresholds play a very important role in the coverage of a CDMA cell and the soft handoff process associated with it. Details of power control and soft handoff in CDMA are discussed in later chapters.

The number of active traffic channels (or calls) that are possible at a given point of time is given in the case of CDMA by:

$$M = \frac{W}{R_b} \frac{1}{S_r} \frac{G_A G_v \alpha}{H_0} \tag{5.15}$$

where (W/R_b) is the spreading gain, H_0 is the additional interference from adjacent cells (usually 0.6, or 60% of the interference from within the cell), α is the power control accuracy (between 0.5

and 0.9), G_v is the voice activity factor (about 0.45) and G_A is the gain due to sectored cells (2.55 for a three sectored cell) [Gar00].

5.7.1 Issues in CDMA Network Planning

Many of the principles that apply to TDMA/FDMA systems also apply to CDMA systems but there are important differences. For example, the path loss is very similar to TDMA systems in that the signal strength drops roughly as the fourth power of the distance in macro-cells and is quite site specific and terrain dependent. However, the design issues that differ are described below.

Managing the noise floor: In CDMA, managing the noise floor is very important. If the number of users in a particular area increases beyond that dictated by Equation 5.15, the system is interference limited and increasing the transmit power will not benefit any user or set of users as the total interference also increases. It is quite possible that interference from many cells can raise the noise floor to such a level that holes may be created in the region where the coding/spreading gain is not sufficient to overcome the interference levels. This is illustrated in Figure 5.31 [Hal96]. If there is an isolated three sector cell, most of the cell has an E_b/I_t larger than 7 dB and in the regions where there is soft handoff (where the mobile terminal can connect to more than one base station), the E_b/I_t value from each base station is around 3 dB providing sufficient diversity gain to allow communication. If too many cells are deployed as shown in the figure, there may be some regions where the noise level is so high that it is impossible to communicate. It is often possible to cover the same area with fewer cells to reduce the total interference levels and it is usually not a very good idea to cover an area by more than three cells or cell sectors. The problem becomes more severe when terrain plays a role and in addition to site selection, it will be important to use the downtilt of antennas and the use of minimum radiated power levels to manage the noise floor.

Cell breathing: In CDMA, the boundary of a cell is not fixed and depends on where the E_b/I_t value is reached. For example, consider the uplink E_b/I_t value that is observed at a base station. As the number of traffic channels on the uplink is increased, this value also increases and it is clear from

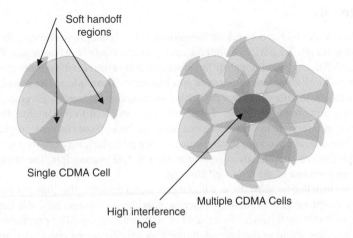

Figure 5.31 Noise floor management in CDMA.

Equation 5.15 that the handoff boundary (where the mobile terminal has to move from one base station to another) shifts closer to the base station. This effect is called cell breathing. In order to ensure that a correct handoff is performed, the transmit power of the pilot channel of the base station (see Chapter 12 for details) must also be reduced so that the forward link handoff boundary is also maintained at the same level as the reverse link boundary. In some cases, cell breathing can have a deleterious impact on the system performance and this should be taken into account while planning the system, either by deploying more cells or offloading capacity to other carriers.

5.7.2   Migration from Legacy Systems

Often, service providers have to migrate from a legacy technology to an emerging technology due to the benefits that the new technology provides. This causes additional deployment problems that are often specific to the technologies involved. Here we use the migration from 1G analog to 2G CDMA systems as an example.

2G CDMA systems typically operate on the same frequency bands allocated to 1G Analog systems and some 2G digital TDMA systems. For example, a single 2G CDMA channel requires removing 41 contiguous analog 1G FDMA channels. Since the CDMA carrier requires a different set of signal to interference constraints, when service providers were migrating from analog to CDMA systems, they had to take care to minimize the interference between these systems. There are three possible approaches that a service provider may employ to migrate from analog 1G to digital 2G CDMA systems [Gar00]. They are as follows.

1. Two independent systems, one based on 1G analog and the other on 2G CDMA.
2. An integrated 1G analog/2G CDMA system.
3. A partially integrated system with analog 1G providing coverage at the fringes and 2G CDMA in the core along with the analog system.

The characteristics, advantages and disadvantages of these approaches are summarized in Table 5.4.

5.8   Femtocells

Over the last few years, cellular network service providers have engaged in allowing the deployment of very small base station like devices in the residences of subscribers or in businesses. These devices are generally called *femtocells* although different service providers refer to their devices by different names such as microcells, miniature towers, or wireless network extenders. These devices are being marketed as solutions to either improve the quality of reception or provide coverage in otherwise dead spots inside and around buildings. Femtocells are connected to the Internet through wired services that may be available in a residence or a business through coaxial cable, digital subscriber lines, or fiber. Femtocells can communicate with a service provider's network through the Internet. Femtocells typically can provide coverage over an area of 5000 square feet. The reader is referred to [Cla08] for an overview of the femtocell concept.

Femtocells are installed by subscribers without the careful network planning aspects that accompany the deployment of base stations by service providers. Consequently, the deployment of femtocells requires that such devices be capable of self-configuration. The process usually works as follows. Upon connecting to the Internet through an existing wired connection such as cable,

Table 5.4 Approaches to migrating from analog 1G to 2G CDMA

Approach	Characteristics	Advantages	Disadvantages
Independent systems	CDMA uses a separate subset of the spectrum. CDMA may cover a larger area (because of greater capacity).	• Allows independent operation and vendor independence • All digital service everywhere • Smaller number of CDMA base stations can be deployed reducing capital costs	• Capacity loss due to spectrum segmentation • If there is dumping of analog terminals, the blocking rate for analog subscribers may increase • Operational complexity
Integrated systems	The same service provider provides both CDMA and AMPS over the entire service area.	• All digital service everywhere • High spectral efficiency • Operational simplicity • No need for dual mode phones	• Requires deployment of digital base stations everywhere and may underutilize the available capacity
Partially integrated systems	Part of the system is converted to support both analog 1G and 2GCDMA.	• CDMA capacity advantage is placed only where it is required • Simpler than the independent approach in terms of operation	• Needs a buffer zone where adjacent analog 1G channels are removed and cannot be operated • Digital service will not be available everywhere • Needs dual-mode phones • Voice quality changes are perceived when a handoff is made from 2G CDMA to analog 1G

a femtocell device will communicate with a server on the service provider's network. This server can authenticate the femtocell and upgrade its firmware and capabilities without intervention from the subscriber. Most femtocells are equipped with GPS to obtain an estimate of their location (to ensure that it is allowed to be used there), also based on which the service provider can assign some parameters to the femtocell, such as the frequencies to use and in the case of CDMA networks, the PN codes to be used by the femtocell. The femtocell monitors the environment to detect what network deployed base station transmissions can be heard by it. Based on such measurements, the femtocell can create a neighbor list for handoffs and adjust its transmit power levels.

There are a few challenges in the deployment of femtocells. There are many types of interference factors – femtocell downlink interfering with a regular base station's downlink and vice versa, a femtocell's downlink interfering with another femtocell's downlink, a mobile device connecting to a femtocell interfering on the uplink with a transmission to a regular base station and vice versa, and a mobile device connecting to a femtocell interfering with another mobile's transmission on the uplink to another femtocell. Further, the dynamic range of femtocell devices has to be large to account for the minimum transmit power limits on mobile devices and the short distances between mobile devices and femtocells.

Questions

1. Name any three advantages of an infrastructure topology over that of an ad hoc topology.
2. Compare peer to peer and multi-hop ad hoc topologies.
3. Name the five different cell types in the cellular hierarchy and compare them in terms of coverage area and antenna site.
4. Why hexagonal cell shape is preferred, over square or triangular cell shapes, to represent the cellular architecture?
5. What are the most popular frequency reuse factors for AMPS, GSM, and IS-95?
6. Of the following, what values are possible for a cluster size in a cellular topology? Why? Assume a hexagonal geometry: 8, 21, 23, 30, 47, 61, 75.
7. Name five architectural methods that are used to increase the capacity of an analog cellular system without increasing the number of antenna sites.
8. Explain why band splitting is not used in the second-generation cellular networks.
9. Explain why reuse-partitioning can be used for both first- and second-generation cellular networks.
10. What is the difference between band-splitting and underlay-overlay techniques for increasing the capacity of cellular networks? What is the effectiveness of each in improving the capacity and how they differ from one another?
11. Explain how smart antennas can improve the capacity of a cellular network.
12. Explain why in fixed channel allocation techniques neighboring frequency channels are assigned to different cells.
13. How does the high interference holes in CDMA deployment are created?
14. Compare FCA and DCA frequency assignment techniques.
15. What are femtocells? Why are they deployed?

Problem 5.1

Consider the general interference scenario of Figure 5.5 in which a Bluetooth device is collocated with an IEEE 802.11 device and interferes with its operation.

a. Shannon–Hartley bound given by Equation 3.7 gives the relation between minimum signal to noise, S_{min}, and the number of bits per transmitted symbol, m. Using that S_{min} and Equation 5.3 calculate the minimum distance for interference, D_{int}, as a function of distance-power gradient, α, power of the Bluetooth device, P_{BT}, the power of the desired access point, P_{AP}, and the distance between the AP and the 802.11 device, R.
b. Plot the minimum distance for interference, D_{int}, as a function of the transmission rate, $R_b = m R_s$, $\alpha = 2$, $P_{BT} = 0$ dBm, and $P_{AP} = 20$ dBm.
c. Repeat for $\alpha = 3$ and $\alpha = 4$.

Problem 5.2

A FHSS IEEE 802.11 and a Bluetooth device are operating in close vicinity of each other. Generate a computer plot illustrating the probability of collision of their packets versus the size of the FHSS packet. Using the results of computer plots explain the impact of packet length on the probability of collision between FHSS IEEE 802.11 and Bluetooth. Note that the maximum length of the 802.11 packets is specified by the standard.

Problem 5.3

Assume that you have six-sector cells in a hexagonal geometry. Draw the hexagonal grid corresponding to this case. Compute S_r for reuse factors of 7, 4, and 3. Comment on your results.

Problem 5.4

Assume that we wanted to deploy an analog FM AMPS system with half band of 15 KHz rather than the existing 30 KHz. Also assume that in analog FM the carrier to interference ratio (C/I) requirement is inversely proportional to the square of the bandwidth (four times the increase in C/I for dividing the band into two).

a. What is the required C/I in dB for the 15 MHz per channel if the required C/I for the 30 KHz systems was 18 dB?
b. Determine the frequency reuse factor K needed for the implementation of this 15 KHz per user analog cellular system.
c. If a service provider had 12.5 MHz band in each direction (up-link and down-link) and it would install 30 antenna sites to provide its service, what would be the maximum number of simultaneous users (capacity) that the system could support in all cells? Neglect the channels that are used for control signaling.
d. If we use the same antenna sites but a 30 KHz per channel system with K = 7 (instead of the 15 KHz system), what would be the capacity of the new system?

Problem 5.5

We have an installed cellular system with 100 sites, frequency reuse factor of $K = 7$ and a 500 overall two-way channels.

a. Give the number of channels per cell, total number of channel available to the service provider, and the minimum carrier to interference ratio (C/I) of the system in dB.
b. To expand the network we decide to create an underlay–overlay system where the new system uses a frequency reuse factor of $K = 3$. Give the number of cells assigned to inner and outer cells to keep a uniform traffic density over the entire coverage area.

Problem 5.6

Repeat Problem 5.3 with $N = 12$

Problem 5.7

a. What is the number of RF channels per cell in the GSM network described in Chapter 7? The frequency reuse factor of the GSM is $K = 4$.
b. What is the maximum number of simultaneous users per cell in this system?
c. Assume that we want to replace this GSM system with an IS-95 spread spectrum system in the same frequency bands. What is the maximum number of users per cell? Assume an ideal power control and use the practical considerations for the IS-95 system.

Problem 5.8

a. Determine the carrier to interference ratio, in dB, of a cellular system with frequency reuse factor of $K = 7$.
b. Repeat (a) for $K = 4$.
c. If we consider multi-symbol QAM modulation for the digital transmission of the information, how many more bits per symbol can be transmitted with $K = 4$ as it is compared with $K = 7$ architecture?

6

Wireless Network Operations

6.1 Introduction

In the previous chapters we provided the principles of radio propagation, wireless modem design, wireless access methods, and deployment of cellular systems. These issues were all related to the air interface design and physical characteristics of the wireless medium that is needed to connect a mobile terminal to a base station or access point that then connects to the backbone wired networks. To support mobile operation, the backbone (sometimes called the core network) has to add several new functionalities that do not exist for the operation of wired terminals, because they are not usually necessary. These functionalities include mobility and location management and radio resource and power management.

The very nature of mobile communications implies that the mobile terminal is constantly changing locations warranting a need for tracking the mobile and restructuring existing connections or packet flows as the mobile device moves. *Mobility and location management* handles the operations required for these purposes.

As we have discussed earlier, bandwidth is a scarce resource as also is the battery power of the mobile terminal. Also, we have seen that a consequence of employing a cellular topology to "multiply" the bandwidth results in wireless networks that are limited by interference. *Radio resource and power management* schemes are used to address operational aspects related to reducing interference, improving battery life and handling the scarce radio resources.

A number of algorithms and methodologies have been implemented in different wireless networks to implement the features necessary for managing mobility and radio resources. In this chapter we provide an overview of the techniques that are used for implementation of these features in wireless networks.

In this section we will consider examples of some types of wireless networks. Our goal in this section is to describe the architectural elements and explain, at a very high level, their functionality and operation. We also introduce some of the terminology used in wireless networks. Details and example approaches are presented in later sections and chapters (e.g., security is considered in Chapter 7, and 4G architecture in Chapter 13). We will also consider at a high level, the issues and challenges specific to each type of wireless network since they can appear to be substantially different across the different network types and yet be similar in nature.

Principles of Wireless Access and Localization, First Edition. Kaveh Pahlavan and Prashant Krishnamurthy.
© 2013 John Wiley & Sons, Ltd. Published 2013 by John Wiley & Sons, Ltd.

6.1.1   Operations in Cellular Telephone Networks

A *cellular telephone network* is an example of a WWAN with infrastructure topology. It also by far has the most complex architecture of all wireless networks. Let us consider a general architecture, not specific to any particular cellular telephone network standard, shown in Figure 6.1. Here, we describe a voice-oriented architecture. We can break up the network into three parts – *the radio subsystem, the network subsystem* and *the management subsystem*. While the names by themselves suggest the components of the network, we will look at some details of the subsystems below.

The radio subsystem comprises of the entities in the network that have a radio interface or air interface. From Figure 6.1, the two components in the radio subsystem are the Mobile Station (MS) and the Base Station (BS). The BS provides coverage in a given area called a cell as mentioned earlier. Cells are shown as ovals in Figure 6.1 although they are really irregular in shape as we saw in Chapter 5. When a MS powers up, it needs to determine the availability of services in the cell in which it is located. This process is called service discovery or cell search. For this purpose, all BSs periodically (or continuously) transmit some kind of *beacon* signals which have information related to the network on known frequency channels using known signaling schemes. These beacon signals have different names in different standards. MSs will first decode this information. If there are multiple beacon signals that are detected, algorithms within the MS will allow it to pick one of them based on one or more criteria such as communications quality, type of network, network capabilities, network ownership, and so on.

In any given cell, as this is an infrastructure topology, MSs always communicate with a BS whether the other party is another MS or a telephone connected to the PSTN. The communication between a MS and a BS is two-way (duplex). The MS can receive and send at the same time, usually

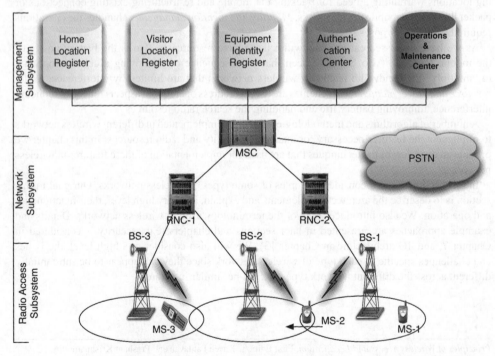

Figure 6.1   A general cellular network architecture.

by using two different frequencies. The transmission from the MS to the BS is called the *uplink* or *reverse channel*. The transmission from the BS to the MS is called the *downlink* or *forward channel*. Note that multiple MSs connect to the same BS. So a BS must be capable of handling many simultaneous uplink and downlink transmissions. Different MSs may be transmitting and receiving at the same time. These transmissions must not interfere with one another. Separation of these transmissions is achieved by using different frequencies, time slots or spreading codes for each MS as described in Chapter 4. Also, there will be simultaneous transmissions in adjacent and neighboring cells (e.g., cells served by BS-1 and BS-2). To ensure that these transmissions do not interfere, similar steps must be taken. It is common to use different frequencies or spreading codes in different cells, but not different time slots as it is harder to synchronize transmissions. Moreover, the propagation delays are unpredictable since they depend on the locations and physical separation between two MSs.

A technical challenge in the radio subsystem is the dynamic assignment of frequencies, time slots and/or spreading codes to MSs (see Chapters 11–13 for 2G systems based on TDMA, 2G/3G systems based on CDMA and 4G systems based on OFDMA). MSs can begin a call at any point of time and it does not make sense to statically allocate a frequency, time slot or spreading code to a MS. Sometimes, it is possible that certain frequencies are facing more interference resulting in poorer communications quality. In such a case, a MS may be asked to switch to a different frequency or increase its power. The management of the time slots, frequencies, spreading codes and transmit power of a MS (and the corresponding changes at the BS) is typically the task of the Radio Network Controller (RNC) which is part of the network subsystem. The time slots, frequencies, spreading codes, and transmit power are the so-called *radio resources* that need to be efficiently managed. Hence this task is called Radio Resources Management (RRM). An RNC controls many BSs and performs allocation of the radio resources through the BSs. In Figure 6.1, notice that MS-2 is moving away from BS-1 towards BS-2. As it moves in this direction, the communications quality of the link between MS-2 and BS-1 deteriorates and at some point of time, MS-2 must switch to BS-2 as the point of access to the network. This is called *handoff*. The decision of performing the handoff requires using metrics associated with the radio resources and is part of RRM. Note that measuring the metrics associated with radio resources (communications quality, received signal strength, and so on) is not a trivial task either. Often, the beacon signals are used as reference signals for comparison and determination of radio resource metrics.

For communicating with the external world, a *Mobile Switching Center* (MSC) is required. The external world includes the PSTN and other cellular telephone networks. One MSC usually controls a group of RNCs creating a tree-like hierarchy. Usually, the BS by itself is only a transceiver. To set up a communication between a MS and another party, the RNC and the MSC need to intervene. For example, if the MS is communicating with a telephone connected to the landline PSTN, the MSC needs to communicate with switches in the PSTN and set up a circuit between the last switch in the PSTN and the MS. This circuit will go through the RNC controlling the BS over the air to the MS. One can immediately see two problems. First, let us suppose the MS is not connected (it is idle or on stand-by) and someone somewhere wants to place a call to this MS. How does the network know where the MS is located even if it is powered up? To solve this problem, *location management* techniques are used in the network. A MS can use the beacon signals to determine whether it is in the same location area (essentially a group of cells) as it was sometime back or whether it has moved. In any case, it sends a location update message periodically to the network. These location update messages are carried to a database called the Visitor Location Register (VLR) that is connected to the MSC. Note that the VLR and the other databases discussed below are part of the management subsystem. When the MS powers up and sends a location update message, the VLR contacts another database called the Home Location Register (HLR) and exchanges this information with it. The HLR keeps a pointer to the VLR that is serving the MS currently. When a call arrives, it is first sent to a MSC connected to the HLR. The HLR uses the pointer to redirect the connection to the MSC connected to the VLR serving the MS. This MSC uses the RNC to *page* the MS in a group of cells where it was last reported to be and obtain a response.

If the MS is already connected to another party and moves from one BS to another, how is the circuit maintained? To solve this problem *handoff management* techniques are used in the network. The VLR and HLR exchange information about a MS as it moves keeping track of it. If the MS crosses to a new BS that is controlled by the same RNC as the previous BS, no changes need be made except for the RNC to set up a circuit between itself and the new BS. If the new BS is controlled by a different RNC, the MSC will have to get involved. If the handoff takes place to a BS that is controlled by a different MSC, a new VLR will have to get involved in migrating the circuit. MSs not in active use will often go to sleep or stand-by to save battery life. They wake up occasionally to see if there are any pages from incoming calls for them. The network and MSs have to maintain some synchronization and clocking to ensure that the MS indeed sees any messages intended for it when it wakes up. Protocols have to make use of this timing information to ensure reliable delivery of information as well. When the MS first powers up and tries to connect to the network based on some set of decoded beacon signals, an *authentication* and *key establishment* process needs to take place. This will ensure that the MS is authentic and will create keys in the MS and the RNC that will allow communications to be encrypted. The cryptographic keys used for this purpose are known only to the MS and a database called the Authentication Center (AuC). The AuC communicates temporary secret information to the MSC/RNC/BS to enable the authentication of the MS and encrypted communication between the entities. The Equipment Identity Register (EIR) is a database used to verify whether the MS is legitimate (not stolen, cloned or one that has not paid the subscriber fees). The Operation and Maintenance Center (OMC) handles billing, accounting (e.g., roaming, peak minutes and so on) and other operational tasks. The entities and messages change slightly for cellular data services, but the general concepts are similar.

Clearly, we can see that the cellular telephone network is a fairly complex network of many different entities handling many different tasks. In fact, this is by far the most complex wireless network. As we will see next, the IEEE WLAN is not as complex in its architecture although it has to perform almost the same number of tasks as a cellular network. Further, the network complexity is also reflected in the way these systems are deployed as described in this chapter.

6.1.2   Operations in Wireless Local Area Networks

Next, we will see how WLANs operate in general (details are available in Chapter 8). Once again, we will not try to consider standard-specific operation, but try to provide a general overview. First, we will consider an infrastructure topology shown in Figure 6.2. Notice that this architecture looks much simpler than the corresponding cellular network architecture in Figure 6.1.

Like the cellular network, we can think of a radio subsystem in WLANs comprising of MSs and BSs (that are called access points – APs). The APs connect to the fixed Local Area Network (LAN) which connects to the Internet through a router. APs cover areas called cells (which are also irregular in shape although they are shown as ovals in Figure 6.1) and transmit beacons periodically to enable service discovery. Such periodically broadcast packets are in fact called *beacons* in the WiFi standard and contain information about the *basic service set* (BSS) and the *extended service set* (ESS). The BSS is one AP, the cell it covers and all MSs connected to it. The ESS is the set of all APs on the same network and all the MSs connected to them. Note that APs in the same network are connected through a wired LAN segment as shown in Figure 6.2. The wired network that connects the APs is called a *distribution system*.

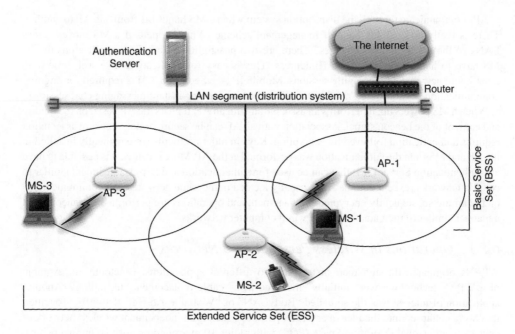

Figure 6.2 A general WLAN architecture.

In a LAN, packets are *broadcast* on the medium and all devices connecting to a LAN share the medium using a *Medium Access Control* (MAC) protocol. Devices on the LAN pick up packets if they are addressed to them and discard them otherwise. Note that entities such as the RNC and MSC do not exist in the WLAN architecture. Moreover, the AP and MSs have to be inexpensive. Thus, the MAC protocol has to be simple and distributed so as to not require central control or expensive components. Thus, there is no concept of having different frequencies or time slots to separate transmissions. MSs and APs transmit on the same frequency channel, but these transmissions have to be somehow kept separate. Carrier sensing is a common methodology used in LANs for medium access. Loosely speaking, each MS senses the medium to see if there is a transmission. If the medium is free, the MS can transmit its packet. The actual process is more complex because the MAC protocol has to handle *collisions* that occur when two MSs transmit at the same time.

RRM is minimal in a WLAN because there is really no need for allocation of time slots, spreading codes or frequency carriers. There is no transmit power control - retransmissions take care of collisions or packets lost due to poor communications quality. Determination of the need to handoff does exist. The MS makes a decision to handoff if it sees poor signal quality which is measured on beacon signals. Location management is simple. Every device on a LAN is supposed to receive packets and it is up to the device to take a packet or discard it. So all a sending party needs to know is the IP address of the destination device on a LAN. An IP packet destined to the device reaches the LAN through a router attached to the LAN. The packet is simply put on the LAN segment and the device will pick it up. In the case of a MS, things are a bit different. A MS registers with an AP on the LAN. The AP has to pick up the packet intended for the MS and transmit it on the air. All MSs in the BSS receive the packet. Only the intended MS will pick up the packet and the rest will discard the packet.

APs communicate through the distribution system when a MS hands off from one AP to another. There is really no complex handoff management scheme. What happens if a MS moves across LANs so that its IP address changes? There are two possibilities. The client applications on the MS have to handle the change in IP address. This is easy for applications like web browsing, e-mail, and new instant messaging sessions. Mobile IP (see Section 6.3.3) is required for ongoing communication sessions (e.g., a voice over IP call or on ongoing instant messaging chat session).

When a MS powers up, it performs an association with an AP. It picks the AP based on the quality and content of the beacon signal. Association with an AP can be secure or not. A secure association requires authentication followed by encryption. Keys in older protocols were manually installed in the AP and MSs and the authentication was performed at the AP. More recently, it is possible to have an Authentication Server (AS) that can be used for authentication and key establishment similar to cellular networks (except that the AS is usually located in the same network). All communications on the air are subsequently encrypted. Management and operations are performed extraneously in a manner similar to the management of wired computer networks.

6.1.3   Operations in Wireless Personal Area Networks

WPANs originated through their history in two different applications. Bluetooth, an example of a WPAN technology were initially envisaged as a "cable replacement" technology. Another application of interest was the so-called "BodyLAN" or "Wearable Ad Hoc Network". Together, they were collapsed into the idea of a WPAN [Bra00] which is an independent wireless network of devices in a Personal Operating Space (POS) with radius 10 m around one or two human beings. The MSs in this network are all within the communication range of one another (see Figure 6.3) A network is created spontaneously and unobtrusively as and when the MSs need to communicate with one another.

As the network is supposed to be plug and play, self-configuration and service discovery become important issues. Recall that there is no central fixed transceiver (such as a BS or AP) in this network. So MSs have to discover one another and also discover the capabilities of one another. This is done using specific self-configuration mechanisms and service discovery protocols that we will discuss in later chapters. In the case of Bluetooth, a Master device initiates communications and polls all responding Slave devices for data as and when required (see Figure 6.3). Most often, MSs

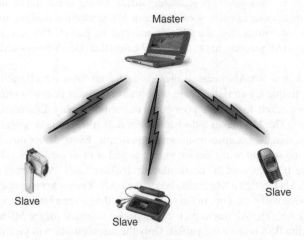

Figure 6.3   A general WPAN architecture.

operate using a single available frequency band for communications. Interference and coexistence with other networks (WPANs or WLANs) is a significant issue here. To avoid interference between co-located WPANs, frequency hopping within this band is adopted by Bluetooth. In the 802.15.4 WPAN standard, carrier sensing is employed as in the case of WLANs (see Chapter 9). A very important requirement in WPANs is to have the ability to conserve power to the maximum possible extent. MSs in WPANs should be able to operate for days. Mechanisms are also necessary to authenticate MSs to one another and establish keys for encrypted communications between MSs. In Bluetooth, this is achieved by installing PIN numbers in trusted MSs. The PIN numbers along with the MAC addresses and other random numbers are employed for key establishment.

Location management, handoffs, and RRM are not as important in WPANs as they are for cellular networks, especially because of the medium access control protocol, smaller coverage area, the use of unlicensed bands, and low transmission power.

6.2   Cell Search and Registration

When a mobile device powers up, it has to first monitor its environment to see what services are available. In an infrastructure wireless network, it is common for a mobile station to be able to "see" multiple base stations or access points. The way the mobile device selects an appropriate base station, and also tune itself to the parameters of the base station is often called cell search. In other words, cell search refers to the earliest procedures that are undertaken by a mobile device before it registers itself with the wireless network. The procedures enable a mobile station to determine what base station is serving an area, and what frequencies, time slots or codes it can use with a selected base station.

The general idea behind cell search, as briefly mentioned earlier, is for the mobile station to search for one or more "beacon signals". The beacon signals are typically the strongest signals (in terms of received signal strength) in the geographical area. Among the challenges that face a mobile station is the fact that a wireless network employs a variety of frequency carriers. With frequency reuse, the close base stations in the area will be only using a subset of the frequency carriers. The mobile station often has to sift through many frequency carriers to find the ones that are active in its geographical area. In wide area wireless networks, cell search depends significantly on the underlying technology and transmission scheme. 1G systems employed frequency modulation with FDMA as the multiple access scheme. 2G systems used either TDMA/FDMA or CDMA/FDMA, while 3G systems use CDMA/FDMA and 4G systems employ OFDMA. In each case, the challenges for cell search are different. We briefly describe some of the approaches taken in various wide area wireless networks to illustrate the general cell search process.

In 1G systems, voice traffic was carried on frequency carriers that were exclusive to the conversation. A simplified view of cell search and registration in an example 1G system is as follows. In Advanced Mobile Phone Service, 30 kHz channels were employed. In a band that spanned 12.5 MHz (one slice of 12.5 MHz existed for the forward link and another slice of 12.5 MHz for the reverse link), a service provider could have 416 frequency channels, of which 395 were used for duplex voice conversations and 21 channels were used for control data. The control data was transmitted using frequency shift keying. A mobile device had to search for the strongest of the 21 forward control channels that it could hear. Once it found the strongest forward control channel, it could decode the information on that channel which included the system ID (e.g., to discover which service provider was in the area). If the mobile device is unable to decode the control information because of interference or other reasons, it would select the next strongest control channel. On the reverse control channel, the mobile station would transmit its own ID and register with the network.

In 2G TDMA based cellular systems, once again, the mobile station scanned the air for the strongest frequency carrier. On a carrier, transmissions were multiplexed in time. For example, in GSM, on each 200 kHZ carrier, there exists a hierarchy of slots, frames, superframes, etc. that allows a mobile station to discover the relevant timing information. Also, certain time slots simply carry an unmodulated carrier (sometimes called the frequency correction channel) to allow the mobile station to synchronize itself to the frequency of the carrier. The beacon signal is then decoded by the mobile station to discover the cell ID, timing synchronization parameters, and other information related to the network operations. A registration procedure follows that enables the mobile to let the network know that it is active.

In CDMA systems, every cell reuses every frequency carrier. So a mechanism that can distinguish between the transmissions of various base stations is necessary. In 2G CDMA systems, a primary and secondary carrier (each with a bandwidth of 1.25 MHz) are pre-assigned in the system. The mobile always tries to find the primary carrier or the secondary carrier first. Each base station in cdmaOne, the common 2G CDMA system, is synchronized to a common time and transmits an unmodulated pilot signal (which is still spread using a scrambling code) which forms the beacon signal. The scrambling codes used by each base station is offset from a baseline time (that is a multiple of the even second in system time). When the mobile station performs a cyclic or periodic autocorrelation of the primary carrier with a locally generated unmodulated, but scrambled, carrier, it can detect peaks in the autocorrelation from the various base stations transmitting in the area. The mobile station picks the strongest peak in the autocorrelation. Timing and system parameters are decoded subsequently from a synch channel and paging channels. Some details are described in Chapter 12. In 3G CDMA systems based on UMTS, the carriers are much wider (5 MHz wide) and fewer in number. However, each cell has its own scrambling code which makes it cumbersome for a mobile station to search for the beacon signal. A hierarchical assignment of scrambling codes, discussed in brief in Chapter 12 allows the mobile station to rapidly sift through the possibilities and find the serving base stations in the area.

In 4G systems where systems may use varying bandwidths, the problem becomes more challenging. Because 4G systems employ OFDM, the central sub-carriers within any given band are used for transmitting the beacon signal.

In WLANs, the mobile station has to still sift through the various frequency channels to discover what networks are active. Each WLAN typically transmits a beacon packet that includes information about the network and also has other parameters that a mobile station may need for registering with the network. The mobile usually picks the access point that is transmitting the beacon packet with the largest signal strength in a given network. If multiple networks exist, users are often prompted to select the network based on the service set ID.

Connection States: In cellular networks, the mobile devices are placed in states depending on what the devices are doing. The common state when the mobile device is not making a voice call or transmitting packets is usually called the IDLE state. Upon cell search and registration, a mobile phone usually enters the IDLE state. In this state, a mobile device is not wasting its battery power by trying to continually maintain radio connectivity with a base station or network (see section 6.4.3). When the mobile is making a voice call or is engaged in packet data transfer, it is called a CONNECTED state. There are various possibilities for the CONNECTED state depending on the technology and service. This becomes important for packet data oriented services like LTE and HSPA. Here the delays in transitioning to a state where actual data transfer can take place becomes an issue for the quality observed by the user and in terms of the overhead. In WLANs, if there is no data transfer, mobile stations usually enter a sleep mode.

6.3   Mobility Management

The primary advantage of wireless communications is the ability to support tetherless access to a variety of services, whether voice-oriented as in the case of cellular radio and PCS or data services and access to the Internet as in the case of mobile data networks and wireless LANs. Tetherless access implies that the user has the ability to move around while connected to the network and continuously possess the ability to access the services provided by the system to which he is attached. This leads to a variety of issues because of the way in which most communications networks operate. First, in order for any message to reach a particular destination, there must be some knowledge of where the destination is (location) and how to reach the destination (route). In static networks, where the end terminals are fixed, the physical connection (wire or cable) is sufficient to indicate the destination. In wireless networks, where the terminal may be anywhere, there must be a mechanism to locate the terminal in order to deliver the communication to it.

Location management refers to the activities a wireless network should perform in order to keep track of where the mobile terminal is. As discussed in Chapter 5, the most common wireless topology uses multiple cells to provide coverage over a larger area. The location of the mobile terminal must be determined such that there is a knowledge of which point of access (base station or access point) is serving the cell in which the mobile terminal is located. Second, once the destination is determined, it is not enough to assume that the destination will remain at the same location with time. When a mobile terminal moves away from a base station, the signal level from the current base station degrades and there is a need to switch communications to another base station.

Handoff is the mechanism by which an ongoing connection between a mobile terminal and a correspondent terminal is transferred from one point of access to the fixed network to another. *Handoff management* handles the messages required to make the changes in the fixed network to handle this change in the location during an ongoing communication. Location and handoff management together are commonly referred to as *mobility management* [Aky98].

6.3.1   Location Management

Location management involves tracking of the location of the mobile terminal, as it moves, for delivery of voice or data communication to it. In the case of voice networks, when a call is made to a mobile number, a dedicated channel has to be set up from the calling party to the called party for the conversation to proceed. For this, a circuit has to be set up over the fixed part of the network and a pair of radio channels have to be allocated to the mobile terminal for the voice conversation. The mobile terminal has to be located to set up this dedicated channel. Note that this is before the actual conversation takes place. If the mobile terminal moves during the course of a conversation, the steps taken to handle the continuity of conversation is called handoff and handoff management. In the case of data networks, packets are addressed to a destination terminal. Routers, within the data network, will use the destination address to deliver the packet. The address information is usually hierarchical and fixed, which means that the address points towards a physical location. If the terminal is fixed, the packet is routed appropriately to the physical location of the terminal. In the case of a mobile terminal, some steps are required to determine where the mobile terminal is before the packet is routed to it. Another important functionality of location management is to determine the status of the mobile terminal. If the mobile terminal is switched off, the network should be aware that it is unreachable so that appropriate action may be taken depending on the service requested. For example short messages may be stored on a server for later delivery.

Location management in general has three parts to it: location updates, paging, and location information dissemination. *Location updates* are messages sent by the mobile terminal regarding its changing points of access to the fixed network. These updates may have varying granularity and frequency. Each time the mobile terminal makes an update to its location, a database in the fixed part of the network has to be updated to reflect the new location of the mobile terminal. Whether or not there is a change in the location, the update message will be transmitted over the air and over the part of the fixed network. Since the updates are periodic there will be some uncertainty in the location of the mobile terminal to something around a group of cells. In order to deliver an incoming message to the mobile terminal, the network will have to *page* the mobile terminal in such a group of cells. The paged terminal will respond through the point of access that is providing coverage in its cell. The response will enable the network to locate the terminal to within the accuracy of the cell in which it is located. Procedures can be then initiated to either deliver the packet or set up a dedicated communications channel for voice conversation. In order to initiate paging however, the calling party or the incoming message should trigger a location request from some fixed network entity. The fixed network entity will then access some kind of database that will contain the most current location information related to the particular mobile terminal and use this information to generate the paging request, as well as deliver the message or set up a channel for the voice call. *Location information dissemination* refers to the procedures that are required to store and distribute the location information related to the mobile terminals serviced by the network.

The basic issue in location management is the tradeoff between the cost of the nature, number, and frequency of location updates and the cost of paging [Won00]. If the location updates are too frequent and the incoming messages few, the load on the network becomes an unnecessary cost, both in terms of the usage of the scarce spectrum as well as network resources for updating and processing of the location updates. If the location updates are few and infrequent, a larger area and thus a larger number of cells will have to be paged in order to locate the mobile terminal. Paging in all the cells where the mobile is not located is a waste of resources. Also depending on the way paging is performed, there may be a delay in the response of the mobile terminal because the paging in the cell in which it is located might be performed much later than the cell in which it last performed its location update. For applications such as voice calls, this will result in unnecessary call dropping since the mobile terminal did not respond in a reasonable time. In the case of data networks, depending on the type of mobility management scheme implemented, packets might simply be dropped if the mobile terminal is not located correctly.

As we discussed earlier in this section, location management consists of three activities – location updates, paging, and location information dissemination. There are different types of location management schemes that employ a variety of location update mechanisms, paging schemes and dissemination architectures. We discuss these in the following sections.

Location Update Algorithms: Location update algorithms are usually of two types – static and dynamic [Won00]. In static location updates, the topology of the cellular network decides when the location update needs to be initiated. In dynamic location updates, the mobility of the user as well as the call patterns are used in initiating location updates.

In the most common form of static location updates, which is the case in most cellular networks, a group of cells is assigned a *location area* (LA) identifier, as shown in Figure 6.4. Each base station in the location area broadcasts this identification number periodically over some control channel. A mobile terminal is required to continually listen to the control channel for the location area identifier. When the identifier changes, the mobile terminal will make an update to the location by transmitting a message with the new identifier to the databases containing the location information. If there is an incoming message, paging is performed in the group of cells corresponding to the

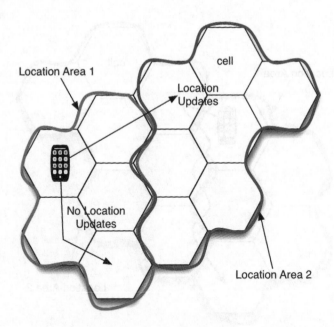

Figure 6.4   Location area (LA) based location updates.

location identifier stored in the database. The mobile terminal usually responds, unless the location area identifier has changed in the meanwhile) and the communication can be delivered successfully.

Example 6.1: Location update mechanism in cellular infrastructure

In cellular infrastructure, for example GSM, a location area identity, also called a paging area, is used for location updates. A location area usually consists of a group of cells controlled by a base station controller (BSC). A mobile terminal will perform a location update under three circumstances: (a) upon powering up, it compares the location area identity it previously had recorded with the one currently being broadcast. If the two location area ideates are different, a location update is performed (b) when the mobile terminal crosses the boundary of a location area, it performs a location update (c) after a period of time predetermined by the network, a location update is performed to ensure that the mobile terminal is available. In case (b), the MS detects a change in location area because the BS broadcasts the location area identity, which the MS is required to monitor and compare with the stored value. In case (c), the update mechanism might be costly if the MS does not leave a location area for long periods of time.

The primary problem with the static location area identifier approach is that if a mobile terminal is frequently crossing the boundary of two location area as shown in Figure 6.5, there will be a *ping-pong* effect of continually switching between two location areas. A solution to this problem is to employ a dwell timer that persists without a location update for sometime to ensure that the location update is worthwhile. Similar problems and associated algorithms to resolve the problem are encountered during handoff that we will see later in this chapter.

A variety of other static location update schemes are possible. These include distance-based – where the location update is performed after crossing a certain number of cells, timer-based – where

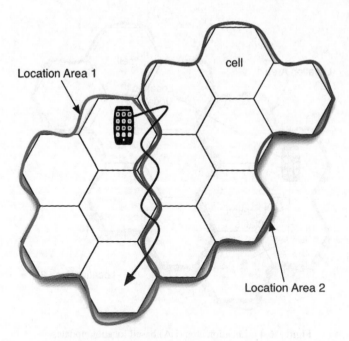

Figure 6.5   Location area ping-pong effect.

a location update is performed after a certain time elapses, and variations on these two schemes that take into account the signaling load on control channels and the location and velocity of the mobile terminal [Won00].

Examples of dynamic location update schemes are the *state-based* and *user-profile-based* location updates. In the state-based location update scheme, the mobile terminal makes a decision on when to perform an update based on its current state information. The state information can include several metrics that includes the time elapsed, the distance traveled, the number of location areas crossed, and the number of calls received. that could be changed based on the user's mobility and call patterns. The user-profile-based location update schemes maintain a sequential list of location areas that the mobile terminal may be located in at different points of time based on the history of the mobile terminal. A detailed comparative evaluation of several of these schemes has been discussed in [Won00].

Paging Schemes: Paging is broadcasting a message in a cell or a group of cells to elicit a response from the mobile terminal for which a call or message is incoming. Transmitting the page in only the cell in which the mobile terminal is located that makes the most accurate location estimate can reduce the cost of paging. The problem with paging in the most accurate cell location is that it is quite impossible to determine location accurately especially if the location update cost has to be kept low.

Example 6.2: Blanket paging in cellular infrastructure

Blanket paging refers to paging the mobile terminal in all the cells within a location area simultaneously. This means that, if the location area update is correct, in the very first paging cycle, the mobile terminal will receive a paging request and respond to it. The advantage of this system, employed in GSM, is that the delay of the response to paging is kept minimum. The disadvantage is that paging has to be done in several cells all of which have the same location area identity.

Another strategy for paging is to use "closest cells first" approach. Here, the cell where the mobile terminal was last seen is paged first followed by subsequent *rings* of cells that are equidistant from this cell in each paging cycle. If there are delay constraints, as in the case of voice calls, several rings may be polled simultaneously in a paging cycle. In general, if the first location estimate is not correct, the next page should be performed so that the probability of locating the mobile is the next largest and so on. The paging is performed in the area corresponding to the last location update and then subsequent pages are performed in most likely locations based on parameters that may include past history and distance. A timer is used to declare the mobile terminal as unreachable in a particular paging cycle. This is sometimes called *sequential paging*. Results indicate that blanket polling provides the lowest delay at small load, while sequential paging can sustain a higher paging request rate, especially when there are several incoming calls to a certain area. The behavior pattern of mobile terminals as well as the user profile may also be employed in paging algorithms in a manner similar to location update mechanisms.

Location Information Dissemination: When there is an incoming packet to the mobile terminal, there is the need for at least one fixed network entity, whose location and address is known, that can be reached to obtain information about the terminal. In general, this is often referred to as an *anchor*. The anchor has some information regarding the location and routing information of the mobile terminal. If a single anchor is used for all mobile terminals, not only is the load on this entity increased making it a bottleneck for communications, but also makes it a failure point that can result in the collapse of the network. Usually, multiple anchor points are employed as described below. What we describe below is in general terms how network entities and databases are employed for location and it can be also extended to other applications such as handoff management. Specific implementations are different.

Every mobile terminal is associated with a *home network* and a *home database*. The home database keeps track of the profile of the mobile terminal – such as the mobile identification, authentication keys, subscriber profile, accounting and location. The location of the mobile is maintained in terms of a *visiting network* where the mobile is located and a *visiting database* that keeps track of the mobile terminals in its service area. The home and visiting databases communicate with each other to authenticate and update each other about the mobile terminal. We will see more of this in the section on handoff management.

Example 6.3: Location information dissemination in cellular infrastructure

In most cellular network infrastructures, for example GSM, the home and visiting databases are called *home location register* (HLR) and *visiting location register* (VLR) respectively. When the mobile terminal observes a change in the location area identity, it transmits a location update message through the base station to a mobile switching center (MSC). The MSC contacts its VLR with the location update. The VLR does nothing if it serves both the old and new location area. If the VLR has no information about the mobile terminal, it contacts the HLR of the mobile terminal via a location registration message. The HLR authenticates and acknowledges the location registration, updates its own database and sends a message to the old VLR canceling the registration there.

Example 6.4: Call delivery in a cellular telephone system

When a call is made to a mobile telephone number, the *anchor* entity contacted is the MSC associated with the HLR of the mobile terminal. The HLR contacts the VLR that is associated with the mobile terminal and enables call set up. Detailed examples of the call delivery is presented in Chapter 11 for TDMA cellular systems.

Other Issues in Location Management: Access to the database and management of queries is very important in order to reduce delay and maintain quality. To reduce the load on a centralized database (like an HLR) local caches of the mobile terminal information can be maintained. Similar strategies are being considered in Mobile IP, as we will see later. Alternative location update strategies and paging algorithms are being investigated. An important factor that influences the performance of all of these techniques is traffic modeling that can accurately represent the nature of incoming calls, paging requests, and movement of the mobile terminal.

6.3.2   Handoff Management

Handoff management involves the entire gamut of issues and actions that are required to handle an ongoing connection when a mobile terminal moves from the coverage of one point of access to another. Handoff [Pol96] is extremely important in any mobile network because of the default cellular architecture employed to maximize spectrum utilization. Handoff (HO), in the case of cellular telephony involves the transfer of a voice call from one BS to another. In the case of WLANs, it involves transferring the connection from one AP to another. In hybrid networks, it will involve the transfer of a connection from one BS to another, from one AP to another, between a BS and an AP, or vice versa.

For a voice user, HO results in an audible click interrupting the conversation for each HO [Pol96] and because of handoff, data users may lose packets and unnecessary congestion control measures may come into play [Cac95]. Degradation of the signal level is however a random process and simple decision mechanisms such as those based on signal strength measurements result in the *ping-pong effect*. The ping-pong effect refers to several handoffs that occur back and forth between two base stations. This has a severe toll on both the user's quality perception and the network load. A way of eliminating the ping-pong effect is to persist with a base station for as long as possible. However if HO is delayed, weak signal reception persists unnecessarily, resulting in a lower voice quality, increasing the probability of call drops and/or degradation of quality of service (QoS). Consequently, more complex algorithms are needed to decide on the optimal time for handoff. HO also involves a sequence of events in the backbone network that include re-routing the connection and re-registering to the new point of access, which are additional loads on the network traffic. HO has an impact on traffic matching and traffic density for individual BSs (since the load on the air-interface is transferred from one point of access to another). In the case of random access techniques employed to access the air interface, or in the case of CDMA, moving from one cell to another impacts QoS in both cells since throughput and interference depend on the number of terminals competing for the available bandwidth.

While significant work has been done on handoff mechanisms in circuit switched mobile networks [Pol96; Tri98], there is not much of literature available for packet switched mobile networks. Performance measures such as call blocking and call dropping probabilities are applicable only to real-time traffic and may not be suitable for bursty traffic that exists in client-server type of applications. When a voice call is in progress, allowed latency is very limited, resource allocation has to be guaranteed and while occasionally some packets may be dropped and moderate error rates are permissible, retransmissions are not possible and connectivity has to be maintained continuously. On the other hand, bursty data traffic by definition needs only intermittent connectivity and can tolerate greater latencies and employ retransmission of lost packets. In such networks handoff is

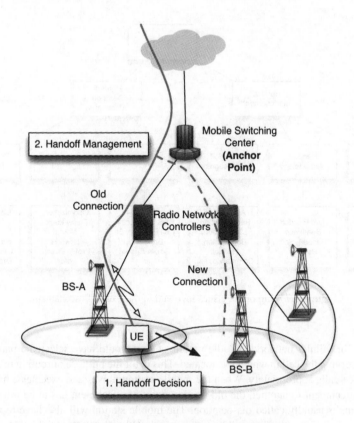

Figure 6.6 Two basic actions during handoff.

warranted only when the terminal moves out of coverage of the current point of attachment or the traffic load is so high that a handoff may result in greater throughput and utilization.

There are a variety of issues related to HO. In particular we can consider handoff as consisting of two different steps as shown in Figure 6.6. In the first step, the handoff management process determines that a handoff is required (handoff decision and initiation). In the second step, the rest of the network is made aware of the handoff and the connection is restructured to reflect the new location of the mobile terminal. Note that there is an *anchor* in the fixed part of the network that must be involved in the handoff management process in a manner similar to location management. Several issues arise during the handoff management process.

As shown in Figure 6.7, these issues are divided into two categories: architectural issues and HO decision time algorithms. Architectural issues are those related to the methodology, control, and software/hardware elements involved in re-routing the connection. Issues related to the handoff decision time algorithms are the types of algorithms, metrics used by the algorithms and performance evaluation methodologies.

Architectural Issues in Handoff: Handoff procedures involve a set of protocols to notify all the related entities of a particular connection that a handoff has been executed and that the connection has to be redefined. In data networks, the mobile device is usually registered with a particular point

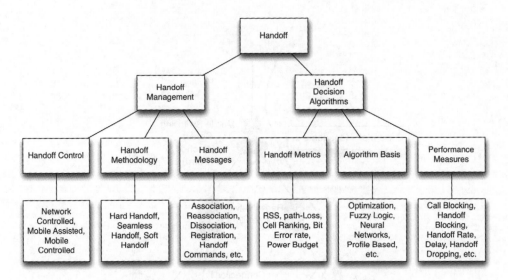

Figure 6.7   Important issues involved in the handoff mechanism.

of attachment. In cellular networks, an idle mobile station would have selected a particular base station that is serving the cell in which it is located. This is for the purpose of routing incoming data packets or voice calls appropriately. When the mobile station moves and executes a handoff from one point of attachment to another, the old serving point of attachment has to be informed about the change. This is usually called dissociation. The mobile station will also have to re-associate itself with the new point of access to the fixed network. Other network entities that are involved in routing data packets to the mobile station or in switching voice calls have to be aware of the handoff in order to seamlessly continue the ongoing connection or call.

Example 6.5: Connection in 4G LTE

The mobile device in the case of LTE can be in one of two states – an IDLE state and a CONNECTED state. They are defined as radio resource control states. In the idle state, the mobile station can only listen to broadcast parameters or paging messages from the network. But the mobile station autonomously decides which base station to select based on these parameters. Further, the mobile station is only tracked up to an area similar to a location area. In the CONNECTED state, the mobile station can be tracked up to a granularity of a cell. It can transmit measurements and data to the base station. But the network decides when the mobile station must make a handoff.

Depending on whether a new connection is created before breaking the old one or not, HOs are classified into hard and seamless handoff. Hard handoff occurs when the MS completely breaks connection with the old BS before connecting to the new BS and synchronizing itself to it. Seamless handoff refers to the case where the MS sets up a traffic channel with the new BS before breaking off from the old BS. However, communication is possible only through one BS at a time. In CDMA, the existence of two simultaneous connections during HO results in soft handoff [Tek91]. Soft handoff is discussed in more detail in Chapter 12.

The decision mechanism or *handoff control* could be located in a network entity (as in cellular voice) or in the MS (as in WLANs) itself. These cases are called network controlled handoff (NCHO) and mobile controlled handoff (MCHO) respectively. In GPRS, information sent by the MS can be employed by the network entity in making the handoff decision. This is called mobile assisted handoff (MAHO). As described in Example 6.5, the selection of the base station may also depend upon the connection state of the mobile station.

Example 6.6: Handoff control in different systems

In AMPS, the analog 1G standard, handoff decision is network controlled (NCHO). The mobile telephone switching office uses the RSS measurements from a MS at different base stations to initiate handoff. In the case of IEEE 802.11 LANs, the mobile station controls handoff decision (MCHO). It monitors the *beacon* of several access points (APs) to decide which AP to connect to. The network has no role in deciding when to make a handoff.

In any case, the entity that decides on the handoff uses some metrics, algorithms and performance measures in making the decision. The measurement and handoff decision are usually part of the radio resource management procedures (described in Section 6.3). However, we examine handoff decision mechanisms in this section under mobility management to keep the procedures for handoff together. These are discussed below.

Handoff Decision Time Algorithms: Several algorithms are being employed or investigated to make the correct decision to handoff [Pol96; Tri98]. Traditional algorithms employ thresholds to compare the values of *metrics* from different points of attachment and then decide on when to make the handoff. A variety of metrics have been employed in mobile voice and data networks to decide on a handoff.

Primarily, the received signal strength (RSS) measurements from the serving point of attachment and neighbouring points of attachment are used in most of these networks. Alternatively or in conjunction, the path loss, carrier-to-interference ratio (CIR), signal-to-interference ratio (SIR), bit error rate (BER), block error rate (BLER), symbol error rate (SER), power budgets, and cell ranking have been employed as metrics in certain mobile voice and data networks. In order to avoid the ping-pong effect, additional parameters are employed by the algorithms such as hysteresis margin, dwell timers and averaging windows. Additional parameters (when available) may be employed to make more intelligent decisions. Some of these parameters also include the distance between the MH and the point of attachment, the velocity of the MH, traffic characteristics in the serving cell, and so on.

Traditional handoff algorithms are all based on the received signal strength (RSS) or received power P. Some of the traditional algorithms [Pol96] are as follows:

1. *Received Signal Strength*: The base station whose signal is being received with the largest strength is selected (choose BS B_{new} if $P_{new} > P_{old}$).
2. *Received Signal Strength plus Threshold*: A handoff is made if the received signal strength of a new BS exceeds that of the current one and the signal strength of the current BS is below a threshold T (choose B_{new} if $P_{new} > P_{old}$ and $P_{old} < T$).
3. *Received Signal Strength plus Hysteresis*: A handoff is made if the received signal strength of a new BS is greater than that of the old BS by a hysteresis margin H (choose B_{new} if $P_{new} > P_{old} + H$).

4. **Received Signal Strength, Hysteresis and Threshold:** A handoff is made if the received signal strength of a new BS exceeds that of the current one by a hysteresis margin H and the signal strength of the current BS is below a threshold T (choose B_{new} if $P_{new} > P_{old} + H$ and $P_{old} < T$).

5. **Algorithm plus Dwell Timer:** Sometimes a dwell timer is used with the above algorithms. A timer is started at the instant when the condition in the algorithm is true. If the condition continues to be true till the timer expires, a handoff is performed.

Figure 6.8 illustrates these algorithms in the case of a mobile terminal traveling between two base stations along a straight line. Note that the RSS is not smooth as shown in this figure but more random as illustrated in Figure 6.9.

Other techniques such as hypothesis testing [Lio94], dynamic programming [Rez95], and pattern recognition techniques based on neural networks or fuzzy logic systems [Tri97] have been suggested for handoff decisions (for an excellent survey of various algorithms, see [Pol96; Tri97]). These complicated algorithms are necessitated by the complexity of the handoff problem especially in hybrid data or voice networks. The mobile terminal has to monitor the air for wireless data services

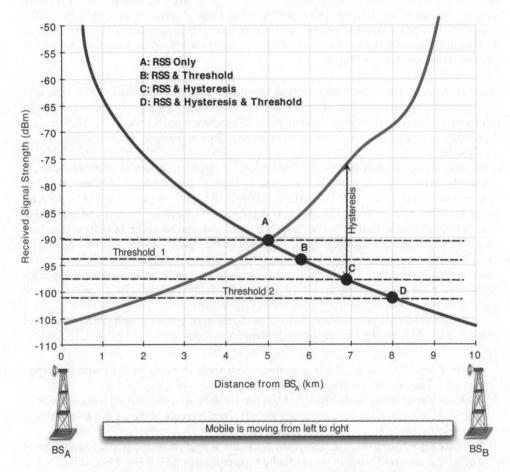

Figure 6.8 Traditional handoff algorithms using RSS thresholds and hysteresis.

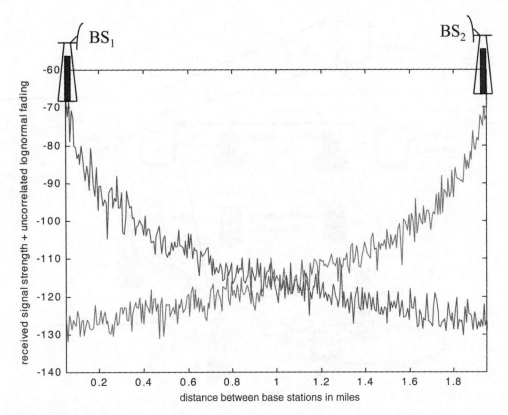

Figure 6.9 Sample RSS from two base stations as seen by a MS traveling in a straight line between them.

that may be available for attachment. As an example, consider a MS that could connect to either an 802.11 WLAN access point connected to a local area network or a GPRS base station subsystem (BSS) connected to a backbone GPRS network. There must be a mechanism or algorithm within the mobile terminal that will enable it to choose the best available service and switch to this service as soon as it is available. For example, the mobile terminal must be able to switch from the GPRS service to the WLAN access point as soon as it detects the availability of a connection to an access point. For data traffic (e.g., web browsing), the delay may be acceptable.

The performance of handoff algorithms is determined by their effect on certain performance measures. Most of the performance measures that have been considered such as call blocking probability, handoff blocking probability, delay between handoff request and execution, call dropping probability, and so on, are related to voice connections. Handoff rate (number of handoffs per unit time) is related to the ping-pong effect and algorithms are usually designed to minimize the number of unnecessary handoffs. While minimizing the handoff rate is important in mobile data networks, other issues include throughput maximization and maintaining QoS guarantees during and after handoff. However, these issues have not received sufficient attention in the literature.

Generic Handoff Management Process: In this section, we will show the different messages and processes that are required for handoff management in a generic wireless network. As in the case

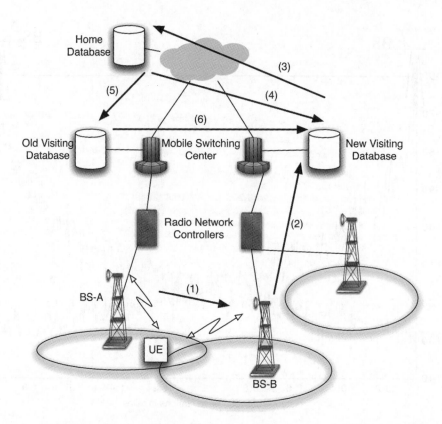

Figure 6.10   Generic handoff management process.

of location management, the specific implementations will be different. We will consider some specifics in later chapters.

In Figure 6.10, a generic architecture is shown for the handoff management process. There are two types of databases in the network – the home database that also acts as the anchor and the visiting database. Every mobile terminal is associated registered with a home database that keeps track of the profile of the mobile terminal. The visiting database keeps track of the mobile terminals in its service area. The home and visiting databases communicate with each other during the handoff management process as described below.

1. In the first step, a decision is made to handoff and handoff is initiated. This decision, as discussed above, maybe made in the network by some entity with or without the help of the mobile terminal, or at the mobile terminal. For this purpose the decision time algorithms are employed.
2. The mobile terminal registers with the "new" visiting database via a handoff announcement message. This is the first information to a network entity in the case of a mobile controlled handoff. In the case of a network controlled or mobile assisted handoff, the new visiting database may already be aware or expecting this message.
3. The new visiting database communicates with the home database to obtain subscriber profile and for authentication. This is the first information exchange between network entities about the changed location of the mobile in MCHO. In the case of MAHO or NCHO, these entities may already be in communication.

4. The home database responds to the new visiting database with the authentication of mobile. If the mobile is authenticated, in the case of circuit-switched connections, a pair of traffic channels that might be kept ready is allocated to the mobile terminal for continuing the conversation. In the case of packet data traffic, no such dedicated channels are required since the traffic is bursty. The two databases are updated for delivering new messages that may arrive to the mobile terminal. The new visiting database includes the mobile terminal in its list of terminals that are being serviced by it.
5. The home database sends a message to the old visiting database to flush packets intended for and registration information related to the mobile terminal. This is because packets that may have been routed to the old visited network while the mobile was making a handoff need to be dropped or redirected and the old visiting database needs to clear resources it had maintained for the mobile terminal since they are no longer required.
6. The old visiting database flushes or redirects packets to the new visiting database and removes the mobile terminal from its list.

Each of the above steps is important in order to correctly, securely, and efficiently implement handoff and release resources that are otherwise not used in the system. There are other architectural issues such as handoff between channels in a cell (intra-cell handoff), handoff between two base stations associated with the same database (inter-cell intra-domain handoff) and different databases (intra-cell inter-domain handoff), and so on. One such issue is examined in Example 6.7. These issues are also discussed later when we describe specific technologieslater.

Example 6.7: Two types of Handoff in 4G LTE

In LTE, there is expected to be a flat architecture (see Chapter 13 and the references therein) where the base station also performs the functionality of the radio network controller. It is expected that the base stations are connected to each other. When the base stations involved in a handoff are connected to each other, the current base station contacts the target base station to which a handoff is anticipated with a handoff request. The target base station allocates resources and acknowledges the current base sttaion which then asks the user equipment (mobile device) to make the handoff. Once the handoff is completed by the mobile, the target base station contacts the databases to update them about the new point of access and to switch the paths appropriately. It also asks the old base station to flush its resources as shown in Figure 6.11.

When the base stations are not connected, different mobility management entities are involved in the handoff process. In this case, the current base station has to contact the current mobility management entity, which in turn contacts the target mobility management entity which in turn contacts the base station as shown in Figure 6.12. Notice that the entities at a higher hierarchical layer (the mobility management entities) are now involved in the handoff.

Mobility management procedures have details that are specific to the respective systems. They need some description of the network entities since the nomenclature for the databases, and the controlling entities are different with different functionality. Descriptions of mobility management procedures are provided in subsequent chapters where individual technologies are described. Also procedures are adopted when a MS returns to a cell from which it had been handed off to simplify the connections in the backbone.

Handoff in Hybrid Networks: With data traffic far exceeding voice traffic in recent times, cellular service providers who were initially unfavorable to WLANs started to see WLANs as a

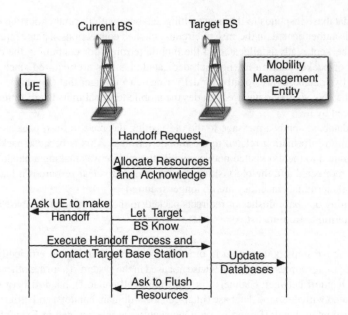

Figure 6.11   Simplified handoff in LTE when base stations are connected.

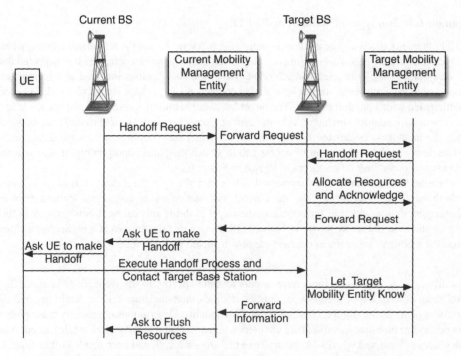

Figure 6.12   Simplified handoff in LTE when base stations are not connected.

complementary solution that would reduce the traffic load on their networks. In other words, users of smartphones and tablets would be better served if they would use WLANs when the WLAN service was available and use the cellular data services otherwise. This implies that mobile devices should, to the extent possible, seamlessly handoff from cellular networks to WLANs and vice versa. Today, most cellular service providers actively encourage users tos witch to WLANs which may provide higher data rates at lower cost, but perhaps not universal coverage. Consumers benefit from not having to pay higher subscriber fees for larger amounts of monthly data traffic, and service providers do not have to invest in expanding capacity rapidly.

In the process of switching between WLANs and cellular data services, it may become necessary for the use of Mobile IP (discussed in the next section) to maintain seamless network layer communications for real time traffic, but thisis often not even necessary for browsing the web or for streaming applications where the user can either retstart the application or tolerate the delay. When delay becomes an important issue, handoff decision time algorithms have to be more sophisticated to handle the switching between networks of different types. Some of the research work done in this area in the late 1990s and early 2000s is available in [Pah00] which considers architectural approaches and handoff decsion time algorithms for handing off between cellular networks and WLANs.

6.3.3   Mobile IP and IMS

The Internet Protocol (IP) that is the most popular network layer protocol for data networks was not designed with wireless or mobile networks in view. Mobile IP tries to address this issue by creating an "anchor" for a mobile host that takes care of packet forwarding and location management. As 3G and 4G cellular networks move towards an all IP backbone (fixed) infrastructure, Mobile IP becomes more important. As we discuss in Chapter 11, the earliest data services overlaid on GSM mimicked the behavior of Mobile IP. Further, when a mobile device switches its connection across LAN segments that belong to different IP sub-networks, mobile IP becomes important. In this section, we will consider Mobile IP as a specific handoff management scheme that has two drawbacks – it does not specify either the first step (handoff decision and initiation) or the last step (flushing and redirecting data). These are technology specific as far as Mobile IP is considered.

Mobile IP [Per97] is simplified because IP packets do not need mechanisms to set up dedicated bandwidth or channels as in the case of circuit switched connections. However, it solves a different kind of problem that IP created when the terminals were mobile. The IP address is used for dual purposes – for routing packets through the Internet and also as an end-point identifier for applications in end-hosts. The connections in an IP network use *sockets* to communicate between clients and servers. A socket consists of the following tuple:

<source IP address, source port, destination IP address, destination port>.

A transmission control protocol (TCP) connection cannot survive any address change because it relies on the socket to determine a connection. However, when a terminal moves from one network to another, its address changes. This is because the Internet uses domain names that are converted to an IP addresses. A packet addressed to one IP address gets routed to the *same place* always since the IP address also points to the location of a physical network.

An IP mobility working group of the Internet Engineering Task Force (IETF) is in charge of activities related to Mobile IP. Several standards and requests for comments (RFCs) related to mobile IP are available [RFC96]. The basic design criteria for a mobile IP were: (a) compatibility with existing network protocols, (b) transparency to higher layers (TCP through application) and to the user, (c) scalability and efficiency in terms of not requiring a great deal of additional traffic or network elements, and (d) security due to changing locations of the mobile node. The way Mobile IP handles location and handoff management is discussed below, starting with some terminology.

A *mobile node* (MN) is a terminal that can change its location and thus its point of attachment. The partner for communication is called the *correspondent node* (CN) that can be either a fixed or a mobile node. The IP network where the MN resides is called the *home network* and the IP network where the MN is visiting is called the *foreign network*. The *Home Address* of a MN is a long term IP address assigned to the MN that is part of the home IP network. It remains unchanged regardless of where the MN is and it is used for (domain name system) DNS determination of the MN's IP address. The *Care-of Address* (COA) is an IP address in the foreign network that is the reference pointer to the MN when it is visiting the foreign network. The *home agent* (HA) is the anchor in the home network for the MN. All packets addressed to the MN reach the HA first unless the MN is already in its home network. A *foreign agent* (FA) (only in the case of IPv4) acts as the reference point in the foreign network for the MN. The COA is usually the IP address of the Foreign Agent. The MN can act as its own FA in which case, it is called a **co-located** COA.

Location Management in Mobile IP: Location management in Mobile IP is achieved via a registration process and the so-called agent advertisement. Foreign agents and home agents periodically "advertize" their presence using ***agent advertisement*** messages. The same agent may act as both a HA and a FA Mobility extensions to ICMP messages are used for agent advertisements. The messages contain information about the COA associated with the FA, whether the agent is busy or not, whether minimal encapsulation is permitted, whether registration is mandatory, and so on. The agent advertisement packet is a broadcast message on the link. If the MN gets an advertisement from its HA, it **must** deregister its COAs and enable a gratuitous ARP. If a MN does not "hear" any advertisement, it must solicit an agent advertisement using ICMP. The entire connection search flow is shown in Figure 6.13.

Once an agent is discovered, the MN performs either a registration or deregistration with the HA, depending on whether the discovered agent is a HA or a FA. The MN sends a *registration request* using UDP to the HA through the FA (or directly if it is a co-located COA). The HA creates a ***mobility binding*** between the MN's home address and the current COA that has a fixed lifetime. The MN should re-register before the expiration of the binding. A *registration reply* indicates whether the registration is successful or not. A rejection is possible by either the HA or FA for reasons such as insufficient resources, the HA is unreachable, there are too many simultaneous bindings, for failed authentication etc. If a MN does not know the HA address, it will send a broadcast registration request to its home network called a *directed broadcast*. The response to this request is a reject by every valid HA. The MN uses one of the HA addresses in the reject message to make a valid registration request. The HA and FA maintain lists of MNs in what we can relate to as the home and visiting databases. Upon a valid registration, the HA creates an entry for a mobile node that has the mobile node's care of address, an identification field, and the remaining lifetime of the registration. Each foreign agent maintains a visitor list containing the following information: link layer address of the mobile node, mobile node's home IP address, UDP registration request source port, HA IP address, an identification field, the registration lifetime, and the remaining lifetime of pending or current registration.

Handoff Management in Mobile IP: Mobile IP enables datagrams addressed to the MN at the home address to be delivered wherever the MN is. As shown in Figure 6.14, the CN transmits a datagram to the MN that is routed to the MH home network as usual in step (1). The HA intercepts

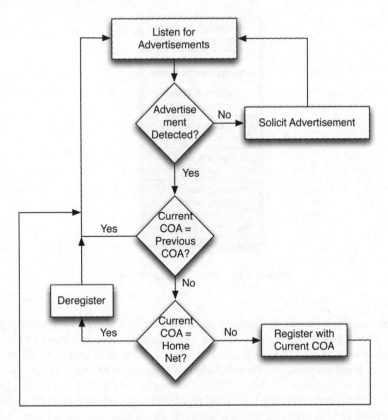

Figure 6.13   Agent discovery procedure.

the packet, encapsulates and tunnels it to FA in step (2). The FA decapsulates and forwards the packet to the MN in step (3). Packets from the MN to the CN are sent as usual (4). This procedure is called triangle routing.

In order to intercept packets addressed to the MN, the HA performs a proxy address resolution protocol (ARP) on behalf of the MN when it is away. The way ARP works is as follows. An

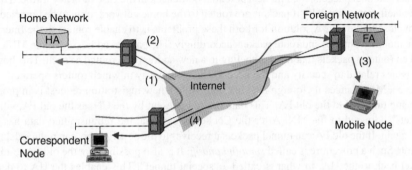

Figure 6.14   Triangle routing in Mobile IP.

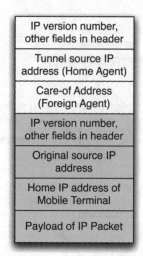

Figure 6.15   IP-in-IP encapsulation.

ARP request is a broadcast message seeking the MAC (physical) address of a terminal given its IP address. When a packet arrives to the MN, an ARP request is made to obtain its MAC address on the home network. If the MN is away, the HA will respond with its own MAC address. When it returns to the home network, the MN will perform a gratuitous ARP that is an unsolicited ARP reply broadcast to each node on the home network clearing the ARP caches. Forwarding packets is achieved by encapsulation (tunneling). A virtual pipe between tunnel entry point (HA) and tunnel termination point (FA) is created through a datagram that includes the packet from the CN as its payload. The mandatory implementation for Mobile IP is IP-in-IP encapsulation as shown in Figure 6.15 though more efficient implementations (called minimal encapsulation) are optional. As far as the IP packet from the correspondent node is concerned, it looks like a single hop within the Internet.

Figure 6.16a shows the sequence of events when a MN moves from the home network to a foreign network and Figure 6.16b shows the sequence of events when it returns to the home network.

Other Issues in Mobile IP: There are several issues in Mobile IP that are under consideration. Since the HA has to tunnel packets, it could be a potential bottleneck in the case of heavy traffic. Triangle routing is inefficient especially if packets are routed to the home network, only to be tunneled back to a point close to the CN. A solution for both these problems is to enable routers in the Internet to cache the mobility binding and route packets accordingly. The packets addressed to the MN can be detected en route as packets that need tunneling to a new address and routed as such. This however leads to issues related to security and the need to change the way in which routers operate.

Suppose a MN changes its foreign network and while a new registration request is in progress, data is being tunneled to the old FA. This data has to be resent by the CN, as the old FA will drop the packets addressed to the MN. After the CN sends the data, the retransmitted data has to be tunneled again. If the old FA can tunnel packets it receives to the new FA, this can reduce delay and congestion. Such a procedure is called *smooth handoff*. It is also possible that the old FA re-tunnels the packet back to the HA, in what is called a "special tunnel". This enables the HA to detect a "loop" if a new registration request has not been enabled.

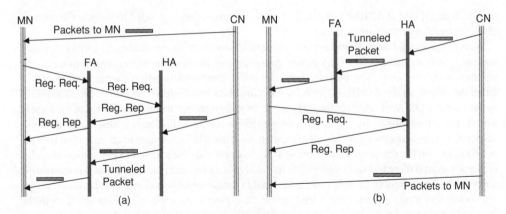

Figure 6.16 Sequence of events: (a) when the MN moves to a foreign network, (b) when the MN returns to the home network.

Sometimes packets will have to be tunneled through the HA. Two common reasons for this are that firewalls drop outgoing packets that have an IP address that corresponds to another network. So the packet cannot be directly sent from the MN to the CN. Also packets addressed by the MN to hosts on the home network usually have a small time-to-live (TTL) since they are supposed to be on the same network. A small TTL implies that the packet needs to sense the Internet as one hop. In both cases, the MN can tunnel the packet to the home agent for retransmission and this procedure is called *reverse tunneling*.

IP Multimedia Subsystem: In 4G networks and beyond, it is expected that all of the transfer of packets at the network layer will be based on IP. Since all traffic, including that over the air is based on IP, the provision of traditional voice conversations is also expected to be very different. One popular architecture that is being actively considered is the IP Multimedia Subsystem or IMS, that creates a domain behind the fixed part of the wireless network. This IMS domain provides a variety of application services such as Voice over IP, video calls, file transfers, and instant messaging. The IMS domain includes a *home subscriber server* or HSS (in some respects replacing the HLR) which incorporates user profiles and security features. A detailed discussion of IMS is beyond the scope of this book. The interested reader is referred to an early paper [Mag06] describing IMS.

6.4   Radio Resources and Power Management

The term "radio resource" is itself not obviously indicative of what exactly it is. In wireless networks, as we described in Part I of this book, the quality of a link is dependent upon several factors. For example, given a transmit power, a lower level modulation scheme like BPSK or a low rate coding scheme can keep the link reliable (i.e., the bit error rate at acceptable levels) at much longer distances than it may be possible with no coding or a higher level modulation scheme like 64-QAM. This would come at the cost of a lower data rate on the link. Alternatively, it may be possible to increase the transmit power to maintain the quality of a link, but this may cause interference to other parts of the network. Thus the term "radio resource" implicitly includes the transmission scheme and the transmit power. The link itself could be a frequency carrier, a time slot on a frequency carrier, or associated with a spread-spectrum code that separates users on a carrier. With OFDM and MIMO, the idea of *radio resources* gets expanded to include the antenna elements and frequency sub-carriers.

We also need to distinguish between power control, power saving mechanisms, energy efficiency, and radio resource management. By *power control*, we shall mean the algorithms, protocols, and techniques that are employed in a wireless network to dynamically adjust the transmit power of either the mobile terminal or base station for reducing co-channel interference, near-far interference in the case of CDMA, or other reasons. *Power-saving mechanisms* are employed to save the battery life of a mobile terminal by explicitly making the mobile terminal enter a suspended or semi-suspended mode of operation with limited capabilities. This is however done in co-operation with the network, so as to *not* disrupt normal communications or provide the user with a perception of such a disruption even if there was one. *Energy efficient design* is a new area of research that is investigating approaches to save battery life of a mobile terminal in fundamental ways such as in protocol design, via coding and modulation schemes, and in software. *Radio resources management* refers to the control signaling and associated protocols employed to keep track of relationships between signal strength, available radio channels, and other parameers (e.g., modulation and coding) in a system so as to enable a mobile terminal or the network to optimally select the best radio resources for communications.

Radio resource and power management is an important part of any wireless network because of several reasons. The scarcity of radio spectrum and the need to improve capacity in a serving area has resulted in frequency reuse with cells as discussed in Chapter 5. The installation of multiple base stations to provide service results in certain phenomena that need to be correctly addressed for proper operation of the wireless network.

First, signals from mobile terminals operating in the coverage area of one base station cause interference to the signals of mobile terminals operating in the coverage area of another base station on the uplink. There is a need to reduce such an interference by properly controlling the transmit powers of mobile terminals. Similarly, signals transmitted by one base station will interfere with the signals transmitted by another base station on the downlink. The transmit powers of base stations on interfering channels need to be controlled to minimize this interference.

Second, correctly controlling the transmit powers of mobile terminals can enhance their battery life and make mobile terminals lighter and handier to use. Since wireless terminals are mobile, they run on battery power which needs to be conserved as long as possible to avoid the inconvenience of requiring a fixed power outlet for re-charging. Most of this power is consumed during transmission of signals. Consequently, the transmit power of the mobile terminal must be made as small as possible. In turn, this requires reducing the coverage area of a point of access – whether a base station (BS) or an access point (AP) so that the received signals are of adequate quality. Also, as mobile terminals move, the ability to communicate with the current base station degrades and they will need to switch their connection to a neighboring base station. At some point during the movement, a decision has to be made to handoff from one base station to another. This decision will have to be made based on the expected future signal characteristics from several base stations that may be potential candidates for handoff, the capacities and available radio resources of such base stations, and interference considerations. For example, if a mobile terminal continues to communicate with a base station when it is deep into the coverage area of another base station, it will cause significant interference in some other cell that employs the same channel.

Third, there is a need for the wireless network to keep track of the radio resources, signal strengths and other associated information related to communication between a mobile and the current and neighboring base stations. All of these tasks are not undertaken by a single entity.

We already discussed the last issue of handoff decision in Section 6.2.2. As we will discuss next in this chapter, several schemes and technologies are employed for the first two issues of radio resources and power management in wireless networks. The radio resource and power management functionality is usually handled by a management entity which interfaces across the lower three layers of the OSI protocol stack. Alternatively, it may be viewed as an application layer on top of the lower three layers (as in the case of GSM). In any case, handling of this functionality requires knowledge of how exactly signals are behaving at a particular point of time. This automatically requires feedback from the lower layers of the protocol stack. As we discuss in Chapter 11, in the case of GSM, the Base Station Controller and the Mobile Switching Center need to communicate with the mobile to obtain information about the state of the radio channel and provide the instructions for power control and selection of channels. A *bi*-directional logical channel is required for the communication between the MS and the network. In data networks where the mobile station has autonomous operation and decides for itself what should be the appropriate action, power control is much harder to implement. Moreover, the selection of radio channels and entering sleep mode of operation is done at the mobile station. In this case a *unidirectional* channel may be sufficient. The protocol architecture and location of the RRM layers of some specific technologies are described in subsequent chapters.

6.4.1   Adjusting Link Quality

The quality of a wireless link depends on a variety of factors, as discussed in Chapters 2 and 3. The transmit power, the distance between the transmitter and receiver, the radio propagation conditions, the signal bandwidth and symbol durations, the modulation and coding schemes employed, inter-leaving of coded streams, the transmission scheme itself (e.g., spread spectrum), the complexity of the transceiver, and interference all influence the quality of the link.

As a rule of thumb, higher level modulation schemes and higher code rates result in larger bit error rates for the same received signal-to-interference ratios. Alternatively, if the received signal-to-interference ratio is larger, it may be possible to forego some coding overhead or switch to a modulation scheme with a larger spectrum efficiency. These are controlled by the radio resource management modules in wireless networks. Typically, both ends of the link monitor the bit error rates, received signal-to-interference ratios, packet or frame error rates and autonomously, by using feedback, or through central control switch between various modulation and coding schemes. Interleaving is usually fixed in a certain format to satisfy the delay constraints on the application because of the buffering of codewords needed with interleaving.

We do not delve into the details of the various protocols and schemes used for adjusting link quality in this chapter. In voice applications (real time), the modulation and coding schemes are not modified continually depending on the channel. Instead, the networks rely on power control to maintain acceptable voice quality. However, we note here that WLANs support a variety of data rates on a link based on varying the modulation schemes and the code rates (see Chapter 3 for examples). Link adaptation mechanisms exist in most cellular data networks, some of which are discussed in later chapters. For example, incremental redundancy in EDGE, a 2.5G data service overlaid on GSM sends frames with no coding initially. If the transmissions are successful, high bit rates are possible.However, if the error checks fail, the frames are retransmitted with lower code rates till the transmission is successful. Using the channel quality information and hybrid ARQ schemes is common in 3G and 4G data networks.

6.4.2   Power Control

In this section we will discuss basic power control mechanisms and why they are important through example implementations in cellular networks.

Basic Idea in Cellular Networks: Power control has been an issue of importance since the very first deployment of analog cellular systems. As discussed in Chapter 5, co-channel interference limits the capacity of a cellular network. Co-channel interference also causes the quality of a voice signal to deteriorate and an attempt has to be always made to ensure that co-channel interference is minimized. This translates into forcing a mobile terminal or a base station to operate at the transmit power that results in the *lowest acceptable received signal to interference ratio* (SIR) (such that the voice or communications quality is acceptable). This appears to be a paradox because one would expect that it is important to maintain a high signal-to-interference ratio for good communications quality. While this is true in ordinary communications systems, in wireless communications with a cellular topology, operating at a high signal to interference ratio implies that the transmit power of a mobile terminal or a base station is large. A large transmit power in one frequency channel in one cell results in a large co-channel interference in all the closest co-channel cells that employ the same frequency channel, albeit at a sufficient distance away from the given base station. This will reduce the communications quality all around and is not desirable.

Example 6.8: Minimum S_r operation in AMPS on the reverse link

Consider an AMPS network. As discussed in Chapter 5, usually a cluster of seven cells uses the entire spectrum allocated to an operator. The spectrum is then reused in neighboring clusters. The approximate distance between the centers of co-channel cells D_L is $4.58R$ where R is the radius of a cell. Consider a mobile terminal in one of the cells of the cluster. Without power control, let us suppose that it would transmit at some maximum power P_t, which is independent of distance. If it is at the cell edge, the received power at the base station will be proportional to $P_t R^{-4}$ if the path-loss in the cell has a distance-power gradient of 4. Consider a second mobile terminal that is located at a distance of $R/2$ from its own base station. The received power is proportional to P_t $(R/2)^{-4} = 16 P_t R^{-4}$. The transmit power must thus be reduced to $P_t/16$ since the base station would otherwise receive a signal that is 16 times stronger than a mobile that is at the edge of the cell. This will in turn reduce the interference it causes to co-channel cells as well as adjacent channel cells. By reducing the transmit power, the mobile terminal will also improve its battery life.

Example 6.9: Effect of large transmit power on the forward link in AMPS

Consider an AMPS network with a reuse factor of $N = 7$. Assume that channels 1, 8, 15, and so on, are allocated to cells labeled A in Figure 6.17 (see also Example 5.18 in Chapter 5). Channel 1 corresponds to the frequency band 869.0–869.030 MHz. Let the base station in the shaded cell transmit signal on Channel 1 at a transmit power six times as large as the other base stations of its co-channel cells. This is an increase in the transmit power by less than 8 dB. The effect on the signal to interference ratio observed by the mobile terminal in Figure 6.17 is as follows:

$$S_r \approx \frac{P_t R^{-4}}{5 P_t D_L^{-4} + 6 P_t D_L^{-4}} = \frac{1}{11}\left(\frac{D_L}{R}\right)^4$$

From Chapter 5, we know that the ratio D_L/R is 4.58 for the case where $N = 7$. Here D_L is approximately the distance of the mobile terminal from its co-channel cells. The signal to interference ratio calculated here is around 16 dB that is 2 dB lower than the required value for good communications quality. If all the base stations in the area are erratic, the signals received by the mobile terminals will all be of poor quality.

From the above examples, we can see that controlling the transmit power of both mobile terminals and the base stations is important. When power control is applied properly it can improve the quality

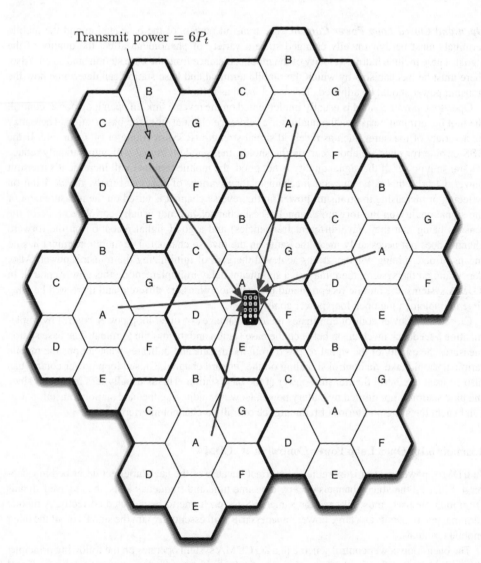

Figure 6.17 Effect of large transmit power.

of communications by increasing the signal-to-interference ratio. An alternative way to view this is in terms of increase in system capacity. If the signal-to-interference ratio can be increased, this will imply that a lower frequency re-use can be employed. As discussed in Chapter 5, this will increase the capacity accordingly. Further, in CDMA systems, where the near-far effect is critical, strict power control becomes mandatory as discussed in Chapter 12.

When examining power control, some of the aspects that are often considered are: (a) whether it is being applied to the forward or reverse links, (b) what are the power control step sizes and are they fixed or adaptive, (c) is the control decentralized or centralized, and (d) what measurements are used for power control – the received signal strength, the signal to interference ratio and/or the packet, frame, or bit error rates. We discuss some of these aspects below with examples.

Open and Closed Loop Power Control: The transmit powers of the base station and the mobile terminals must be dynamically changed since a variety of phenomena affect the quality of the signal. These include fading, velocity of the mobile, distance from the base station, and so on. Also, there must be mechanisms by which the mobile terminal and base station can determine how the transmit power should be adjusted. This is achieved as follows.

Open loop power control is usually implemented on the reverse link. In open loop power control, the mobile terminal measures the quality of a reference channel from the base station. There may be a variety of measures such as received signal strength (RSS) or frame or bit error rate. If the RSS or bit error rates are above certain thresholds, the mobile terminal will automatically reduce it's transmit power. If the signal quality is not good, the mobile terminal will increase it's transmit power. Clearly, this is not a good mechanism for a variety of reasons. Firstly, the decision on reducing or increasing the transmit power on the reverse channel is based on the measurement of the signal quality on the forward channel. These channels are not usually correlated (one of the reasons being that they have different frequencies) and a good signal reception on the forward channel does not necessarily mean the same on the reverse channel. The mobile terminal has no means of determining whether it has achieved the goal of minimizing the transmit power. Also, depending on the system, there may be a significant delay in implementing this power control. In TDMA systems, the mobile reception and transmission times are different and there will be a lag time in implementing open loop power control.

Closed loop power control eliminates the disadvantages of open-loop power control by implementing a feedback mechanism between the base station and the mobile terminal. The base station measures the quality of the signal received from the mobile and indicates what actions the mobile terminal should take via control signaling on the forward channel. Closed loop power control can also be used to control the transmit power at the base station. This is usually less important since the base station is not limited by battery power. However, adjusting the base station transmit power can benefit the system in terms of reducing the overall co-channel interference.

Example 6.10: Open Loop Power Control in 2G CDMA

In CDMA, power control is extremely important because of the near-far effect described in Chapter 4. Since all the voice channels occupy the same time and frequency slots, the received signals from multiple users must all have the same RSS for each one to be detected correctly. A mobile terminal that transmits at a large power unnecessarily will essentially jam the signals of all the other mobile terminals.

The open-loop power control scheme in a 2G CDMA system operates on the following principle. A mobile terminal that is closer to a base station should transmit less power because its signal suffers a smaller path-loss. Mobile terminals that are in deep fade or far away from a base station should transmit at a larger transmit power to overcome the loss in signal strength. Upon powering up, the mobile terminal adjusts its transmit power based on the *total* received power from all base stations on the forward channels [Gar00]. The base stations are not involved in the power control mechanism. The reference channel on the forward path used to determine the transmit power of the mobile terminal is the *pilot channel*. If the pilot channel is received very strongly, the mobile transmits a weak signal to the base station. Else it transmits a strong signal to the base station.

Example 6.11: Closed Loop Power Control in 2G TDMA

The closed loop power control on the reverse link of 2G TDMA systems such as GSM operates as follows [Gar99]. The mobile terminal measures the RSS and the signal quality of up to six neighboring base stations and reports its measurements to the base transceiver subsystem. The base transceiver subsystem also measures the RSS, signal quality, and distance to the each of the mobile terminals in its serving area. From these measured values, it determines the minimum required transmit power and informs the mobile terminal of this value via a five-bit field in the slow associated control channel. The power control is performed in steps of 2 dB. Note that the power control is performed at a far slower rate compared to that in CDMA systems.

Centralized and Decentralized Power Control: Open loop and closed loop power control mechanisms discussed above try to dynamically adjust the *transmit power* of the mobile terminals or base stations based on the thresholds for signal to interference ratio or bit error rate set in the network. The goal of any power control scheme should be to uniformly render the signal to interference ratio of all users to a value, which is usually the *maximum possible SIR* in the system. In terms of how such an optimization can be done are two approaches – centralized and distributed.

In a centralized power control (CPC) scheme, a central controller, in the base station controller or mobile switching center has knowledge of *all* the radio links in the system. That is, the transmit powers, received powers, SIRs, and bit error rates for all mobile terminal – base station combinations are known to this centralized controller. Assuming that the system is interference limited, an optimization algorithm can be implemented to maximize the minimum SIR in the system and minimize the maximum SIR in the system, thereby equalizing the SIR of all radio links [Gra93]. While this provides an optimum solution, this scheme is extremely hard to implement because the centralized controller has to dynamically keep track of all the links in the system and compute the transmit powers for each mobile terminal.

In distributed power control (DPC) [Gra94], the mobile terminals adjust their transmit powers in discrete steps. This is similar to what is actually done in practice. For theoretical simplicity, it is often assumed that the mobiles adjust their transmit powers synchronously. The power adjustments made by the mobiles result in the transmit powers iteratively converging to the optimum power control solution. Ideally this should result in all the mobile terminals having the same SIR (which should be the maximum possible in the system) as in the CPC scheme after a number of adjustments. In practice, the adjustment to the power levels is also discrete (in steps of a few dB).

Example 6.12: Power adjustment levels in example wireless networks

In GSM, each mobile terminal is required to either increase or decrease its power level by 2 dB depending on the message sent by the BTS. In 2G CDMA, mobile terminals can change their power levels in steps of 1 dB. The North American 2G TDMA standard required that a mobile terminal be able to change its transmit power by 4 dB in response to a command from the base station in 20 ms. In some older wireless data networks, the transmit power was set based on the signal power received and *not* the SIR. Consequently, power control is not based on signal quality. It is based only on the absolute signal strength.

6.4.3   Power Saving Mechanisms in Wireless Networks

In addition to dynamically changing the transmit power, there are several other mechanisms built into the operation of most wireless networks for saving the battery power of the mobile terminals. A variety of measurements have indicated that the most battery power is consumed during transmission of signals. A significant amount of power is consumed during active reception of signals (although this is lesser than the power consumed during transmission). A third mode of operation, called *standby* mode consumes nearly an order of magnitude less power than either during transmission or reception of signals [Agr98].

Example 6.13: Power consumption in Lucent WaveLAN

Lucent's WaveLAN (later Orinoco) was a wireless local area network product based on the early IEEE 802.11 WLAN standard. It operated in the 2.0 GHz ISM band. The power consumed by a 15 dBm WaveLAN radio was 1.825 W in the transmit mode, 1.8 W in the receive mode and 0.18 W in the standby mode [Agr98]. Clearly, the standby mode operates with very little power and operating in this mode can save the battery life. Many of these products have reduced their power consumption over time with improvements in technology.

The operation of wireless networks is often designed to ensure that the mobile terminal spends as much time as possible in either a standby or sleep mode in order to conserve power. Several techniques are employed in wireless networks to reduce the amount of time spent in transmitting or receiving signals. In the case of laptops and other data terminals, a *sleep* mode of operation is preferred where the radio transceiver is shut off to conserve power. In voice networks such as GSM and IS-95, the *voice activity* factor is used to reduce either the transmit power or completely stop transmission when there is no speech activity. We discuss these techniques in the following sections.

Discontinuous Transmission and Lower Transmit Power: A particularly attractive option for saving the battery power of a mobile terminal is not to send information unnecessarily. With real-time communications, it is often assumed that there is a constant stream of data to be transmitted and also such communications are sensitive to delay and jitter. Usually, this is not the case especially in a two-way conversation where one of the users is listening for some duration of time. Data communications are less affected by delay and jitter, it is possible to buffer data and transmit it at a later time. Discontinuous transmission is mostly employed in cellular telephone networks where additional hardware and algorithms are used to detect the presence or absence of voice. In ancient mobile telephones, some amount of information would be transmitted all the time, whether the person was actually speaking or not. With the use of *voice activity detection* (VAD) it is now possible for a mobile terminal to behave differently when no voice activity is detected. One of the possibilities (assuming ideal voice activity detection) is to not transmit any signal when the user is not speaking. A second alternative is to repeat data, but at a far lower signal power than usual. This will ensure that data is transmitted all the time, but the total power consumed corresponds to only that data which was actually generated by speech. VAD has its associated problems. In high-noise situations, the mobile terminal must be able to distinguish between the presence of useful signals in high noise and simply noise. Also it should be able to detect low-level voice activity. If VAD is not implemented correctly, there may be clipping of speech or an additional hangover after a talk spurt. Also, if there is absolutely no transmission, subjective tests indicate that silent gaps are extremely annoying. Consequently, systems usually insert a very low power *comfort noise* signal during silent gaps.

Example 6.14: Discontinuous transmission in 2G TDMA

In 2G TDMA cellular systems such as GSM, the device transmits a comfort noise signal when there is no speech activity. When a VAD determines that there is no speech activity, the mobile terminal enters a hangover state to prevent clipping of speech due to very short silence periods. If there is no speech activity after the hangover period has elapsed, a silence identifier frame is transmitted at larger intervals than voice frames. The receiver will insert comfort noise when it detects the presence of the silence identifier frame.

Example 6.15: Discontinuous transmission in 2G CDMA

In cdmaOne, the 2G CDMA cellular technology, speech coders operate at different rates depending on the voice activity. The number of bits generated per frame will be different depending on the rate of the speech coder. The data stream corresponds to 9600, 4800, 2400, or 1200 bps. If traffic is generated at 9600 bps, bits are transmitted at 100% of the transmit power. On the forward link, at lower data rates, the bits are repeated and then transmitted at half, one-fourth or one-eighth of the transmit power on the forward channel. On the reverse link, discontinuous transmission is employed. The bits are gated after repetition so that only one copy is transmitted.

Sleep Modes: A common approach for saving battery power is to allow the mobile terminal to enter into a *sleep mode* during periods of inactivity. The idea here is based on earlier discussion where we mentioned that the most power is consumed during signal transmission and significant power is consumed during signal reception as well. By entirely shutting off the RF hardware, it is possible to further reduce the battery consumption. However, there are problems associated with shutting off the RF hardware completely. What happens if a call or a packet arrives for a mobile terminal when it is shut off? The network should be able to make provisions for handling calls or packets that arrive for a mobile terminal that is in sleep mode.

Example 6.16: Sleep Mode in North American 2G TDMA

In the North American version of 2G TDMA, the *standby time* is defined as the time for which a mobile terminal can be powered on and is available for service on a control channel, before it needs recharging. The operation has been designed to allow the mobile terminal to enter a *sleep mode* for long periods of time when it is on standby. A mobile terminal is required to monitor the forward link only for a few time slots in order to determine whether or not there is a call addressed to it. The network may however require the mobile to monitor channels more frequently. The mobile also has to monitor neighboring channels for handoff and monitor broadcast information. These will affect the time for which it can enter the sleep mode. For the rest of the time slots, the mobile terminal can enter a sleep mode.

Example 6.17: Sleep mode in IEEE 802.11 WLANs

In IEEE 802.11 WLANs, a mobile terminal can enter the sleep mode and inform an access point of its decision. Since this is a local area network and handoffs are less frequent than in cellular systems, handoff is less of a problem. Because of the bursty nature of data traffic, arrival of packets addressed to the mobile terminal is a bigger concern. The access point buffers packets addressed to the sleeping mobile in 802.11. A beacon signal is transmitted periodically that contains information about buffered packets intended for sleeping mobiles. The mobile terminal wakes up at times when it expects the beacon and determines whether it should re-enter the sleep mode or awaken completely to receive packets.

6.4.4   Energy Efficient Designs

The most common technique for conserving energy in a mobile terminal is to use advanced hardware design. Low power digital CMOS, mobile CPU microprocessors, and other hardware design approaches that consume very little power are usually employed in laptops and handheld computers. Beyond the actual hardware design, there are approaches at other layers that can enable savings on power consumption thereby improving the lifetime of a battery. There are three approaches for improving battery life. The first approach is to tune the protocols employed in a wireless network to reduce power consumption. The second approach is to investigate power efficient modulation and coding techniques (see Chapter 3 for more details on modulation and coding). The final approach is in software design for mobile terminals. We discuss the first and third approaches below.

Energy Efficient Protocols: Most protocols in data and voice networks have some relationship with the OSI protocol model even though they do not exactly match the seven layers. In wireless networks, the more important layers are the physical layer that handles the actual transmission of symbols and the link-layer that handles transmission of data packets or voice packets in link-layer frames and also controls access to the wireless medium. With the emergence of TCP and IP as the most popular transport and network layer protocols, these two are usually seen in wireless networks, although they are more restricted to data networks currently. As voice over IP becomes popular, IP will make its presence in voice networks as well. As discussed earlier, the important power conservation principle in mobile terminals is to *minimize* the amount of time spent in transmitting signals. This principle can be applied to the design of different protocols in different layers.

Link Layer and MAC Design: At the link layer and in the design of medium access control techniques, power conservation factors should be taken into consideration. Two areas of design have been addressed in the literature. The first area is MAC design where the design goal is to eliminate unnecessary collisions and to employ better protocols for sleep mode and broadcast operation, as well as eliminating unnecessary processing at the MAC layer. The second area of design involves the link layer where automatic repeat request (ARQ) error control schemes are employed for retransmission of lost or damaged packets.

Design at the MAC layer has focused on the following issues. Techniques to avoid retransmission due to collision as far as possible have been incorporated. Collision is not an issue in cellular telephony, where a channel is dedicated to the voice call for the duration of the call. However, in most local wireless data networks, such as IEEE 802.11 or IEEE 802.15.4, collision is an issue. Even in reservation-based schemes like GPRS, the access to the network is achieved by a contention-based protocol.

Example 6.18: Collision avoidance mechanisms in wireless data networks

Collision avoidance mechanisms have been incorporated in WLAN standards such as IEEE 802.11 and HIPERLAN that are based on carrier sensing. In IEEE 802.11, there are two forms of carrier sensing – at the physical layer and at the MAC layer. At the MAC layer, the length of transmission of a signal is detected via a field in the MAC frame and a *net allocation vector* (NAV) is set so that no signal transmission or physical carrier sensing is attempted in this period, thereby reducing the chance of a collision and also reducing the energy spent in monitoring the channel. In HIPERLAN, multiple contention phases eliminate the chance of collision to a great extent. In some older wide area wireless data networks, the downlink carried a "status" flag that indicates whether the uplink is busy or idle. This once again reduces the possibility of collision.

Outside of collision avoidance, it is possible to use some intelligent techniques to further reduce unnecessary battery consumption. In local area networks, a mobile terminal will receive *all* packets irrespective of whether it is addressed to it or not. If the packet is not addressed to the mobile terminal, it is discarded. This results in unnecessary wastage of battery resources. One possibility for improving this situation is to simply look at only the header information and continue receiving the signal only if the packet is addressed to the mobile terminal.

Example 6.19: Intelligent processing in Carrier-Sensing based WLANs

In HIPERLAN, an early WLAN standard that did not commercially succeed, some header information was to be transmitted via a low bit rate transmission scheme to reduce battery consumption [Woe98]. The reason for this is as follows. As the data rate increases, the effects of multipath delay spread require the use of equalization techniques as discussed in Chapter 3. Equalization schemes consume a lot of battery power. At 23 Mbps, which is the data rate supported by HIPERLAN, equalization becomes very important. In order to reduce battery power consumption, the header information is transmitted at a lower data rate (1.4706 Mbps). The entire header is not transmitted at a low data rate. Only a 34 bit has value of the destination address is sent at this low data rate. The mobile terminal determines whether or not the received has value matches its own hash value. If the hash value does not match, the rest of the packet is not received. If the hash value matches, the equalization circuitry is switched on and the rest of the packet is decoded. Of course, there is a possibility that the packet still may not be intended for the mobile terminal, since the hash values are not unique. However, the chances of this happening are small. By not receiving signals or using the equalization circuitry unnecessarily, HIPERLAN terminals can save battery power significantly.

Other possibilities for intelligent packet reception are possible. Since the downlink is controlled in infrastructure networks by a base station or access point, this can schedule the broadcast of packets intended for different mobile terminals. The mobile terminal will then have to decode only those packets that arrive in the vicinity of its schedules reception time [Agr98].

Automatic repeat request (ARQ) schemes are employed at the link layer to retransmit data packets that are lost (see also Chapter 3). ARQ schemes are typically not useful for real-time traffic like voice. Packet losses can occur due to several reasons – collisions, interference, fading, and multipath delay spread. Collision avoidance mechanisms discussed earlier try to eliminate collisions to the extent possible. However, collisions cannot be entirely avoided as discussed in Chapter 4. Also, interference, fading, and other radio channel effects can result in errors in the received packet. Retransmission techniques are incorporated at the link layer based on error detection schemes. It is possible to reduce retransmissions if error recovery can be performed at the receiver via forward error correction. In fact several wireless systems include *block interleaving* as discussed in Chapter 3 to reduce the effects of burst errors and enable forward error correction. However, if channels conditions are very bad, none of these techniques can recover from errors and retransmission of packets will be necessary.

Retransmitting packets will not be useful if the channel conditions continue to be harsh. In fact this will result in unnecessary transmissions of packets that are bound to be lost or damaged. In [Zor97], the energy efficiency of an error control protocol is defined as follows:

$$\lambda = \frac{\text{total amount of data delivered}}{\text{total energy consumed}} \tag{6.1}$$

Zorzi and Rao argue that this metric, which corresponds to the average number of packets delivered correctly during the lifetime of a battery, will influence the choice of ARQ protocols, and that sub-optimal protocols may in fact be better as far as energy consumption is concerned.

Example 6.20: An adaptive energy efficient Go-Back-N ARQ protocol

A classic Go-back-N ARQ scheme will transmit up to M packets and wait for acknowledgements from the receiver. The receiver will only accept packets in order and will send a negative acknowledgement (NAK) if a packet is not received. The receiver may also acknowledge several packets received correctly and in sequence with a single acknowledgement for the last correctly received packet. If the sender times out while waiting for an acknowledgement for packet N in the set of M packets or receives a NAK for the packet N, it will retransmit all the packets starting from N till M once again. All of these packets may be lost if the channel conditions continue to be bad. Instead, the adaptive protocol suggested in [Zor97] operates as follows. A probe packet is transmitted when there is a NAK or a timeout. The probe packet is a small packet that has minimum payload or simply a header so that the mobile terminal does not waste resources transmitting large packets. Only if a positive acknowledgement is received for the probe packet will the sender resume normal transmission of packets. Under slow fading conditions and small energy consumption for probing packets, this scheme can increase the energy efficiency by three times. Such a scheme can be worse than the regular scheme if the channel conditions are varying rapidly because the channel may have degraded immediately after an acknowledgement is received for the probe.

Transport layer Design: The most common protocol employed at the transport layer in data networks is the transmission control protocol (TCP). As discussed in the previous section, as long as the channel conditions are bad, it is wasteful to transmit packets since they will not be delivered correctly. TCP has in-built mechanisms to back off when it detects packet losses. This backing-off of transmissions is initiated not because channel conditions are suspected to be bad, but because TCP assumes that there is congestion in the network, and transmissions should be reduced to ease congestion. However, it is possible that indirectly, this may also aid in reducing unnecessary energy consumption in wireless networks especially when there are correlated errors on the wireless channel. In [Zor99], the energy efficiency [defined in (6.1)] of TCP is investigated with this point of view. Analysis there indicates that depending on the nature of the channel, and the type of TCP implementation, TCP parameters can be tuned to increase energy efficiency significantly. In certain cases, it is possible to increase the energy efficiency by almost three times.

If parameters of TCP are not set correctly, the congestion avoidance mechanisms can degrade throughput and increase energy inefficiency. Split approaches for TCP [Agr98] introduce intermediate hosts in the network that keep track of missing packets and acknowledgements and handle TCP congestion avoidance mechanisms appropriately. These approaches can also increase the energy efficiency of the system.

6.4.5   Energy Efficient Software Approaches

Significant reduction of battery consumption can be obtained if the mobile terminal can be made to operate intelligently to reduce power consumption. Battery is consumed in mobile devices due to accessing of hard disks, operation of the central processing unit, and power consumption in the display in addition to the wireless communication unit and the communication protocols that we have considered so far. Energy management strategies are appropriate for each of these components, and this is usually achieved by the operating system (OS). These components usually have low power modes of operation and in many cases multiple modes of operation. A thorough discussion of these issues is provided in [Lor98]. The operating system will have to decide which mode of operation is appropriate at what time and when there should be a switch from one mode to another (*transition*). It should decide how a component's functionality can be modified to move it into low power modes

Table 6.1   Energy management in software in mobile terminals

Component	Secondary storage	Processor	Display
Power-saving features	Five power modes: active, idle, standby, sleep, off	Clock slowdown, shut off functional units, shut off processor	Color to monochrome, reduce update frequency, turn display off
Issues	Sleep and standby modes consume far less power, but moving from standby to idle mode consumes power	Clock speed reduction can increase task time thereby increasing power consumption; Shutting off processor works best	Affects readability; Can be annoying if there are flashes for updates
Transition strategies	Enter sleep mode when inactive for some fixed threshold of time (several seconds) Use dynamic thresholds based on previous samples of disk access Predictive spin-ups of the disk (has not worked very well)	Process scheduler knows whether processes are ready to run or running; When processes are blocked, the CPU can be turned off (UNIX and Windows) Predict the number of busy CPU cycles and set the CPU clock speed	Turn off or turn down the display if there is no user input after a period of time
Load change strategies	Increase number of disk accesses – increase cache size Pre-fetch data based on prediction of usage before spinning disk down Reduce paging and improve memory access locality	Use lower power instructions (energy efficient compilers) Reduce the time taken by low level tasks and pipeline tasks to units that can be turned off Reduce unnecessary tasks (block instead of busy-wait)	No formal load change strategies
Adaptation strategies	Switch to flash memory for cache and storage Offload access to the network Reduce disk speeds in low power modes, improve energy consumption of disks	Design motherboards that automatically power down all components when the processor does not present them with a load	Use sensors to determine if a user is looking at a display or not; Use lighter colors for display; Provide ability to display only active windows and dim the rest of the display

as often as possible (*load change*) and also how software can be employed to permit novel power-saving use of such components (*adaptation*). Important factors in deciding strategies for any of these decisions involve what effect a strategy may have on the *overall* power savings since saving power in one component may affect the performance or power consumption of other components in addition to introducing unnecessary overhead. In certain cases, the lifetime of a battery will not be the sole issue, but how much of productivity can be obtained from a mobile terminal.

Table 6.1 provides a summary of results from [Lor98] that considers energy management issues related to secondary storage (hard disk, etc.), the processing unit and the mobile terminal display units. A variety of power-saving strategies are possible as shown in Table 6.1. The improvement in power saving varies depending on the type of mobile device, secondary storage device, processor and display.

Significant power is consumed solely by access to hard disc and it is suggested that it may be better to offload some of the storage onto a network file server. This assumes that the power consumed with the wireless transceiver will be smaller than the power consumed by disc access. Since transmission and reception does not involve mechanically moving parts, this may be a viable option, especially if energy efficient protocols are employed. Flash memory is non-volatile and consumes less power. Against 0.9 W consumed only in idle mode for a hard disk, flash memory consumes 0.15–0.47 W for reading and writing. However, it is about twenty times more expensive than hard disks. Simulations indicate that flash memory used as secondary storage can reduce power consumption by 60–90%. Turning the processor off and reducing the voltage and clock speed of a processor can result in power savings that can be as large as 70% if the strategies are implemented correctly. The use of low-power states can further reduce power consumed by display units. Other components that can benefit from the OS employing low-power modes include sound cards, modems and main memory [Lor98].

Questions

1. What three operational issues are important in wireless networks compared with wired networks? Why?
2. What is cell search? How is it accomplished in cdmaOne?
3. Why are connection states important?
4. Name the two important issues in mobility management.
5. What is location management? What are the three components of location management? What are the tradeoffs between them?
6. Name three location update mechanisms.
7. Name three paging mechanisms. Explain blanket paging.
8. What are the two steps in handoff?
9. Explain three traditional handoff techniques.
10. Differentiate between mobile controlled and mobile assisted handoff.
11. Explain a general handoff procedure. Explain the entities involved with GSM as an example.
12. How is handoff in LTE different when the base stations are connected and when they are not connected?
13. What is agent advertisement? Why is it important in Mobile IP?
14. What are smooth handoffs? What application(s) may benefit from them?
15. How can the BER quality of a link be maintained as a mobile station moves from a point closer to a base station to a point that is farther from the base station?
16. Why is power control important in wireless networks?
17. What are the differences in power control for voice oriented and data oriented networks?
18. Differentiate between open-loop and closed-loop power control.
19. Differentiate between centralized and distributed power control.
20. Name two types of power-saving mechanisms.
21. Differentiate between sleep modes in TDMA cellular systems and IEEE 802.11 WLANs.
22. What intelligent protocol features are available in IEEE 802.11 and HIPERLAN to save battery power?
23. Describe some energy efficient software approaches.

Table P6.1   RSS from four base stations

Time(s)	0	2.5	5.0	7.5	10.0	12.5	15.0	17.5	20.0
BS_1	−47	−57	−52	−55	−60	−62	−60	−65	−64
BS_2	−59	−56	−55	−54	−52	−51	−49	−60.5	−52
BS_3	−70	−72	−75	−70	−58	−50	−60.5	−62	−75
BS_4	−72	−71	−65	−60	−55	−53	−50	−49	−56

Problem 6.1

A mobile terminal samples signals from four base stations as a function of time. The times and signal strengths (in dBm) from the samples are given in Table P6.1. Assume the mobile terminal is initially attached to base station 1 (BS_1). The mobile makes handoff decisions by considering the signals from the base stations after each sampling time.

For example, if just RSS is used, just after $t = 12.5$ s, the mobile terminal would be connected to BS_3. On a plot, show the handoff transitions between base stations for each of the following algorithms as a function of time. If a condition is met for more than one base station, assume the best one (strongest RSS) is selected.

a. Received signal strength (RSS)
b. RSS + threshold of −60 dBm
c. RSS + hysteresis of 10 dB
d. RSS + hysteresis of 5 dB + threshold of −55 dBm.

Problem 6.2

In Problem 6.1 above, which technique is the best in terms of reducing the number of unnecessary handoffs? What parameters will you change to reduce the number of unnecessary handoffs? If the minimum required RSS for good signal quality is -55 dBm, would your answers change?

Problem 6.3

A mobile node has a home address 136.142.117.21 and a care-of address 130.216.16.5. It listens to agent advertisements periodically.

a. The agent advertisement indicates that the care-of address is 130.216.45.3. What happens? Why?
b. The agent advertisement indicates that the care-of address is 136.142.117.1. What happens? Why?

Problem 6.4

Mobile terminals in seven co-channel cells (labeled A) of a cellular system are transmitting on the same frequency channel as shown in Figure P6.1. Due to a glitch in the handoff mechanism, the transmit power of the mobile in the center cell increases without control as it moves away from a base station such that it continues to be connected to it even after it moves out of the cell A into cell D as shown in the figure so that its transmit power is now three times as the rest. Assume that the reuse factor is $N_f = 4$. Determine the interference suffered by the other six mobiles. You

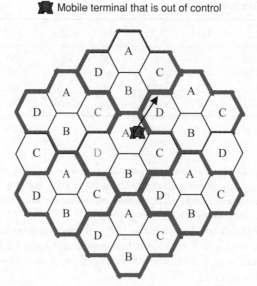

Figure P6.1 Impact of interference in a four-cell cluster.

can approximate the distance between mobiles with the distance between the cells they are in. Comment on your results. What are the implications of this in terms of power control?

Problem 6.5

The energy per bit in a transmission system is 10 dBm. Using Equation 6.1 Determine the energy efficiency of error correction codes that have a code rate of 1/5, 1/3, and 1/2 respectively if there are NO errors on the channel. Next consider a harsh radio channel where a stop and wait protocol is employed, i.e., packets are retransmitted if they are not received correctly one by one. About 30% of packets are damaged on this channel on average. The rate 1/5, 1/3 and 1/2 codes can respectively repair 80, 40, and 10% of the damaged packets. Determine the energy efficiency of the error correcting codes in this case. Assume that all events are independent.

Project: Simulation of shadow fading and handoff

In this project we simulate the fluctuations of average received signal strength due to shadow fading in a micro-cellular network and we use that to analyze a simple handoff algorithm.

The scenario of operation is shown in Figure P6.2, four base stations, BS-i ($I = 1, 2, 3, 4$) are located in four street-crossings in a micro-cellular network. The mobile station, moving from BS-1 toward BS-2 and in the figure is communicating through BS-1. As the mobile station moves away, the received signal strength (RSS) from BS1 decreases and the RSS values from BS-2, BS-3, and BS-4 increase. At certain points, the received power from BS-1 becomes weak and the mobile station starts to search for another BS that can provide a stronger signal and selects that base station as its point of connection this change of base stations is referred to as handoff. In an ideal system we would expect the handoff decision only once in the middle of the path between BS-1 and BS-2. In practice depending on the handoff algorithm, we may have several handoffs in different locations.

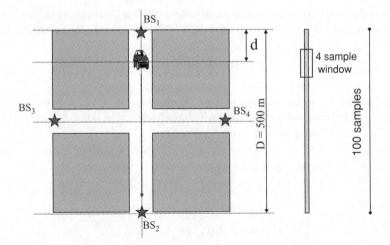

Figure P6.2   Four base stations scenario for a micro-cellular operation.

In this project we consider the simplest and the most obvious algorithm that simply connects the MH to the BS with strongest average RSS. To analyze the situation, we use a channel model to simulate the average RSS from various base stations and we observe the number and location of the handoffs.

We use a distance-partitioned model with two slopes to simulate the channel (see Chapter 2). In this model, the path loss increases with a slope of 2 to a break point at the distance 150 m, then the slope increases to 3. With this model, the RSS at a distance d in the LOS paths associated with BS1 and BS2 are given by:

$$RSS(d) = P_t - P_0 - \begin{cases} 20\log_{10}(d); & d \leq 150 \\ 20\log_{10}(150) + 30\log_{10}(d/150); & d > 150 \end{cases} + l(d)$$

where d is the distance between the mobile host and BS-1 in meters, $P_t = 20$ dBm is the transmitted power of the base stations, $P_0 = 38$ dB is the path loss at the first meter calculated for 1.9 GHz PCS bands, and $l(d)$ is the lognormal shadow fading with a standard deviation of 8 dB.

For the OLOS propagation associated with the RSS from BS-3 and BS-4, a LOS propagation is assumed up to the street corner and after the corner the propagation path loss is calculated by placing an imaginary transmitter at the corner with the transmit power equal to power received at the corner from the LOS base station (this is a simple way of simulating diffraction at an edge). As a result, the RSS at a distance $(d + R)$ from the OLOS base station is given by:

$$RSS(d) = P_t' - \begin{cases} 20\log_{10}(d); & d \leq 150 \\ 20\log_{10}(150) + 30\log_{10}(d/150); & d > 150 \end{cases} + l(d)$$

where P_t' is the RSS in LOS at the middle cross-section:

$$P_t' = P_t - P_0 - 20\log_{10}(150) + 30\log_{10}(250/150)$$

This model assumes all of the power arriving in the cross-section is diffracted in the direction of the mobile.

For the simulation of lognormal fading we assume a random Gaussian noise $N(0,1)$ with zero mean and variance of 1 is passed through a low-pass filter characterized by the transfer function:

$$H(z) = \frac{\sigma_2}{1 - \alpha z^{-1}}$$

where α designating the location of the pole of the filter is a number very close to the one to keep the bandwidth low. Then the samples of the shadow fading effects can be simulated from the following equations. The shadow fading sample is $s(i)$.

$$\alpha = e^{(-\frac{1}{85})}$$
$$\sigma_1^2 = 8$$
$$\sigma_2^2 = \sigma_1^2 \cdot (1 - \alpha^2)$$
$$s(1) = \sigma_1 \cdot N(0, 1)$$
$$s(i) = \alpha \cdot s(i - 1) + \sigma_2 \cdot N(0, 1)$$

In this simulation the first point of the simulation must be at $d = \sqrt{g} = \sqrt{150}$ and the last point should be at $d = 2R - \sqrt{g} = 500 - \sqrt{150}$

The following sample code in MATLAB® facilitates the simulations. This code generates one simulation of the RSS from the four base stations when the mobile moves from BS-1 to BS-2 and both path-loss gradients are assumed to be two.

```
% Declare the various variables used for distances
R = 250;
L = 2 * R;
speed = 1;
sample_time = 0.1;
step_distance = speed * sample_time;
g = 150;
min_distance = sqrt(g);
max_distance = L - sqrt(g);
d1 = [min_distance:step_distance:max_distance];
d2 = L - d1;
d3 = abs(R - d1);
d4 = abs(R - d1);
Ns = length(d1);

% Declare variables and compute RSS
% Part 1: Computations independent of the random variable
%          for shadow fading
Pt = 20;
Po = 38;
grad1 = 2;
grad2 = 2;
alpha = exp(-1/85);
sigma1 = sqrt(8);
sigma2 = sqrt(sigma1^2 * (1 - alpha^2));

RSS01 = Pt - Po - (10 * grad1 * log10(d1) + 10 * grad2 * log10(d1/g));
RSS02 = Pt - Po - (10 * grad1 * log10(d2) + 10 * grad2 * log10(d2/g));
```

```
RSS_corner = Pt - Po - (10 * grad1 * log10(R) + 10 * grad2 * log10(R/g));
RSS03 = RSS_corner - (10 * grad1 * log10(d3) + 10 * grad2 * log10(d3/g));
RSS04 = RSS_corner - (10 * grad1 * log10(d4) + 10 * grad2 * log10(d4/g));

for i=1:Ns
    if d3(i) < min_distance
        RSS03(i) = RSS_corner;
    end;
    if d4(i) < min_distance
        RSS04(i) = RSS_corner;
    end;
end;

% Part 2: Adding the random variable for shadow fading
s1(1) = sigma1 * randn(1);
s2(1) = sigma1 * randn(1);
s3(1) = sigma1 * randn(1);
s4(1) = sigma1 * randn(1);

for i=2:Ns
    s1(i) = alpha * s1(i-1) + sigma2 * randn(1);
    s2(i) = alpha * s2(i-1) + sigma2 * randn(1);
    s3(i) = alpha * s3(i-1) + sigma2 * randn(1);
    s4(i) = alpha * s4(i-1) + sigma2 * randn(1);
end;

RSS1 = RSS01 + s1;
RSS2 = RSS02 + s2;
RSS3 = RSS03 + s3;
RSS4 = RSS04 + s4;

% Plot the RSS values obtained
figure(1)
plot(d1, RSS1,'r')
hold on
plot(d1, RSS2,'b')
hold on
plot(d1, RSS3,'g')
hold on
plot(d1, RSS4,'c')
title('RSS versus distance along route')
xlabel('distance from BS1 in meters');
ylabel('dBm');
```

Figure P6.3 provides a sample result expected from this simulation if both gradients are fixed at two.

Using the above discussion do the following:

a. Write your simulation for the RSS from the four base stations with two different gradients described earlier and plot a sample result similar to Figure P6.3. Assume that the mobile station moves from the vicinity of BS-1 toward BS-2 with a constant walking speed of 1 m/s and the distance of a block shown in Figure P6.2, is $R = 250$ m. The sampling frequency is 10 Hz, that is, the mobile station measures the RSS every 0.1s.

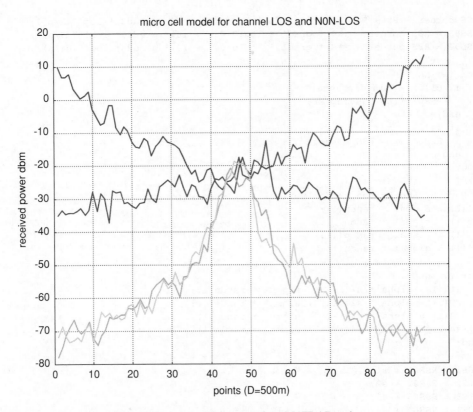

Figure P6.3   Sample output of the MATLAB code.

b. Assuming that the mobile station always connects to the BS with the strongest RSS. Expand you simulation to record the location and number of handoffs in each experiment when moving from BS-1 to BS-2. Run the program ten times to get ten random trials. Determine the number and location of handoffs for all ten trials. Suggest some rational modifications to the simple algorithm to reduce the number of handoffs.

7

Wireless Network Security

7.1 Introduction

Wireless access to the Internet is becoming pervasive with diverse mobile devices with varying capabilities. Before the past decade, the deployment of wireless devices was on a small scale and, typically, wireless devices accessed closed networks with specific applications (e.g., voice conversations on cell phone networks). The devices themselves were not very sophisticated and could not be remotely contacted easily. The associated security risks at that time were fairly low. As of today, with smart phones rivalling the capabilities of personal computers and connecting to the Internet using extremely high speed data networks, the security risks have significantly increased. Moreover, wireless networks and devices are becoming very critical for all kinds of communications. Thus the security and availability of wireless devices and networks is an issue of paramount importance.

The security threats and corresponding security requirements for wireless communications are similar to the wired counterparts but sometimes need to be treated differently because of the applications involved and the potential for fraud. Different parts of the wireless network need different security approaches. Over the air security has usually focussed on the privacy of voice conversations. This is changing with the increasing use of wireless data services where message authentication and integrity have also become important.

The widely varying features and capabilities of wireless communication devices and networks introduce several security concerns. The broadcast nature of wireless communications renders it extremely susceptible to malicious interception, malicious access and wanton or unintentional interference. Analog telephones in 1G cellular systems were extremely easy to tap and conversations could be eavesdropped using an RF scanner. 2G digital systems based on TDMA and CDMA are much harder to tap and RF scanners cannot do the trick anymore, but since the circuitry and chips are freely available, it is not hard for someone to break into a system that deploys no security. As long as the deployment was sparse and potential for harm to the consumer small, primitive security measures and even no security measuress were not a problem. As more people now use wireless access to the Internet and use wireless networks for e-commerce or credit-card transactions and store all kinds of private information on their smart phones, the potential for harm has increased significantly. At least some minimal security features are essential to prevent casual hacking into wireless networks. In this chapter, we address security issues in general. We will provide an overview of network security services and mechanisms and describe some specific examples related

---

Principles of Wireless Access and Localization, First Edition. Kaveh Pahlavan and Prashant Krishnamurthy.
© 2013 John Wiley & Sons, Ltd. Published 2013 by John Wiley & Sons, Ltd.

to wireless networks. In this chapter, we try to address some security threats and requirements that have been identified for wide, local, and personal area wireless networks.

The chapter is organized in terms of the various types of wireless networks (classified based on coverage). Section 7.2 provides an overview of the security issues and protocols used in wireless local networks. Section 7.3 presents the security issues and solutions for wireless personal networks. Section 7.4 considers wide area cellular networks; and Section 7.5 discusses miscellaneous issues.

7.1.1 General Security Threats

Security problems occur for a variety of reasons, but the open nature of communications over the Internet, as previously described, assists malicious entities in launching attacks because of the inherent vulnerabilities that exist. The emergence of very large cyber-crime operations has moved network security attacks from the realm of hobbyists to criminal organizations making them more dangerous with potential for great economic harm. Figure 7.1 summarizes some of the technical reasons why security attacks are possible. We will refer to a general malicious entity – a human, a criminal organization or software – as Oscar in this chapter.

Information Leaks: Just as a robber would stake out his physical target (e.g., a bank) first before attempting the actual robbery to determine how many security guards are there, what exits he could use, and so on, it is common for cyberattackers to stake out their victims networks. Many protocols aid attackers by "leaking" information about networks and services running in these networks. Information leaks can be used to launch social engineering attacks (discussed later), for cracking passwords, mapping network topologies, and determining open services provided by servers. Information leaks are part of the normal operation of some protocols and it may be difficult to prevent such leaks. Other leaks can and should be plugged, especially those that occur when Oscar actively probes the victim networks for information. The installation of malicious apps on smartphones that may capture information from other apps or from the operating system sunning on smartphones is an additional threat to information leaks.

Software Bugs: One common reason for successful security attacks is that servers listening at known ports have bugs in their implementation (e.g., buffer overflows). In such cases, it is possible for a malicious entity like Oscar to craft packets that can be sent to buggy services. When the service is compromised, it can enable Oscar to take control over the host. This means, Oscar can perhaps install malicious software on the host, use the host to launch malicious packets targeted towards other vulnerable hosts, steal files containing valuable information that may be stored on

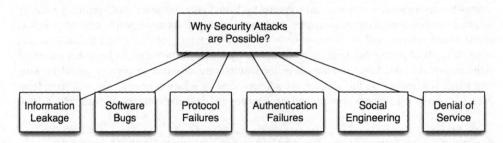

Figure 7.1   Reasons why security attacks are possible.

the compromised host (or on other hosts on the network that trust the compromised host), and so on. With smartphones running many kinds of software applications, software bugs, once not a big issue for wireless devices, is now becoming a security problem.

Authentication Failures: A question that arises is why services do not authenticate the origin (and content) of packets such that maliciously crafted packets do not make it and thus they do not exploit software bugs. In fact, many services do try to implement mechanisms to verify the legitimacy of the source of a command or request. Unfortunately, not all such authentication mechanisms always work. You cannot trust the "From" field in an e-mail address simply because it is too easy to forge. Similarly, many services implicitly trust the source IP address. Again the source IP address is easy to spoof. We will discuss cryptographically secure authentication and integrity mechanisms used in wireless networks later in the chapter.

Social Engineering: Over the centuries, malicious entities have exploited the lack of experience, wisdom, or judgment in people for their own benefits. One example is Oscar walking into a building wearing an official-looking uniform so that no one questions you and getting whatever it is he can lay his hands on. Social engineering is this approach that Oscar uses to exploit the naivete or carelessness of users. Crafted e-mails that make users click on links or open attachments, fabricated websites that look like their original counterparts, malicious wireless access points that are named like trusted one, are some examples of how Oscar may lure victims, social engineering is not restricted to the cyberdomain. It is quite possible that Oscar makes use of some information he has obtained to call up a system administrator and ask him to reset a password. Social engineering attacks are easier in many cases than breaking encryption or overcoming other technological hurdles.

Denial of Service: Denial of Service, sometimes called DoS, is primarily an attack against availability of resources. Resources could mean the bandwidth in the network, information flow or access to stored information, computing resources, or software at the client/server side. Thus a network denial of service typically involves flooding a target with packets at the link or network layers, application level service denial could include providing false information, interrupting information or crashing a server. DoS on the client side could mean crashing the client software. It is impossible to prevent denial of service, but it may be possible to mitigate its effects.

7.1.2 Cryptographic Protocols for Security

The appendix in this chapter provides a quick overview of cryptography and its related terminology and technical aspects, such as the importance of key sizes, the types of encryption algorithms, and how security services such as confidentiality, integrity, and authentication may be provided using cryptography.

As we will see in later sections, most wireless networks of today provide confidentiality (through encryption) and some authentication of messages. A good security protocol has two steps. In the first step, it typically starts with the authentication between the two parties that want to communicate (see Figure 7.2). It is common to call these two parties as Alice and Bob and the adversary as Oscar. This entity authentication is different from message authentication, and is more like identification schemes described in Appendix A7. It is preferable to have mutual authentication of the entities. That is, Alice is assured that she is talking with Bob and Bob is assured that he is talking with Alice. It is also common to exchange, distribute, or agree upon (establish) session keys during this entity

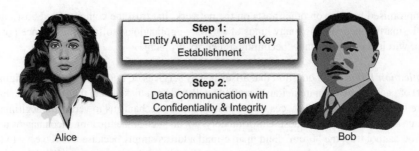

Figure 7.2   Typical sequence in a secure communication protocol.

authentication phase. Note that this process assumes that Alice and Bob *share* some long-term secrets already. These long-term secrets may be master keys, certificates, or other such constructs. The key establishment process uses some quantities called *nonces* that are used only once with the long-term secrets to create session keys at both Alice and Bob. Without knowledge of the long-term secrets, it is impossible for Oscar to re-create the session keys even with knowledge of the nonces. Typically, different session keys are used for confidentiality and integrity to ensure that using the same keys for different purposes does not impact security.

In the second step, when Alice and Bob have authenticated each other and established session keys, packets or frames are exchanged between them in a secure manner. "Secure manner" implies that the frames or packets are encrypted for confidentiality and also have an attached message authentication code or message integrity check. This ensures that Oscar cannot eavesdrop on the conversation, nor can he fabricate or modify packets. Further, he will not be able to masquerade as Alice to Bob or Bob to Alice in the middle of the session.

The way the entity authentication and subsequent data confidentiality and message integrity are implemented varies among networks and technologies. Further, the management and installation of keys in mobile devices and corresponding network entities can take on many different forms. While Appendix A7 has more details, it is beyond the scope of this book to consider all of the various aspects related to these cryptographic protocols.

7.2   Security in Wireless Local Networks

7.2.1   Security Threats

When IEEE 802.11 devices were early on in the market in the late 1990ss, they were quite expensive (on the order of hundreds of dollars). Thus their deployment was sparse and restricted to university campuses and some specific business organizations. Over time, improvements in technology brought down the cost of access points to tens of dollars. Instead of using an external PC card to connect to WiFi, laptops came built-in with the IEEE 802.11 WLAN interfaces (thereby hiding the cost to the consumer). During the early 2000s, this caused the rapid proliferation of IEEE 802.11 devices and networks. Most of these networks were deployed independently of others, used a fixed set of frequencies in unlicensed bands, and were deployed with the default settings since consumers did not really care or know about potential security issues.

Eavesdropping and Unauthorized Access: Security gradually became important for such private wireless data networks. For example, an organization that had installed a wireless LAN within its premises soon discovered that the coverage area extends into a neighboring street. With the reducing costs of wireless LAN PC-cards, anyone could buy a PC card and access the organization's WLAN from the street. Once a malicious entity obtained access to an organization's network, the vulnerabilities in hosts and servers in the network could be exploited easily. An attacker could gradually gather information about the topology of the network of the organization, the nature of operating systems running on hosts and servers, open services that may have weaknesses, and so on. This not only applied to organizations, but to general consumers who deployed wireless networks in their homes.

Initially, most of the 802.11 products came with the suggested security features, but by default, many consumers did not activate the features. Further, even when activated, the *wired equivalent privacy* or WEP protocol employed in the early versions of 802.11 had protocol flaws, as described in Section 7.2.2 below. Thus, it was easy for malicious entities to eavesdrop on WiFi links, and even access the Internet or inject packets into a private WLAN. Even though the security protocols used with WiFi have significantly improved, it is often recommended that the wireless LAN of an organization be placed outside the firewall so that traffic that may enter the WLAN has to further penetrate a firewall before it reaches the important assets (such as servers and regular hosts) in the wired network.

Social Engineering: The previous discussion makes it appear that threats to WLANs emerge from malicious mobile devices equipped with a WiFi interface. In fact, the creation of a malicious WLAN is not too difficult because of the inexpensive devices that are available. A special form of phishing attacks, called "evil twins," appeared a few years back [Rot08]. The fact that most smartphones and laptops have WiFi has made WiFi hotspots extremely popular. Typically, WiFi access points are placed in areas (e.g., hot spots like coffee shops or hotels). In the case of evil twins, malicious WiFi access points are placed close to where legitimate service is being provided by some service provider. When a legitimate user Alice tries to connect to such access points placed by Oscar, a web-page, similar to the one displayed by legitimate service providers is displayed. It is common for subscribers to enter credit-card and other sensitive information on these web-pages. When Alice enters this information on a web-page delived by the evil-twin, this enables Oscar to steal such information. The use of mutual authentication can help protect against this security problem.

7.2.2   Security Protocols

Till the early 2000s, WEP [Edn04] was the only security available for WiFi devices. It is still in use by legacy devices that cannot support the new options for WiFi security. As mentioned earlier, the expense and cost of deploying WLANs had kept the number of WiFi devices small and the rationale for using something like WEP, in an honest world seemed reasonable. The term "wired equivalent" meant that by using WEP, the designers envisaged that a WLAN would be as difficult to break into as wired LANs although it happened to be not the case. WEP used RC-4, a well-known fast and efficient stream cipher for encryption. The initial key sizes were set at 40 bits to allow for export regulations at that time. The system was self-synchronizing, in that each MAC layer frame was independently encrypted of other MAC layer frames. So a lost MAC frame would not impact subsequent MAC frames that were delivered correctly.

WEP and Problems with WEP: The early version of security in WLANs skipped the entity authentication and key establishment phase shown in Figure 7.2. Key establishment and management was very simple with WEP. There were two types of keys – default keys and key-mapping keys. Most devices used default keys where *every mobile device* and the access points in a WLAN shared the

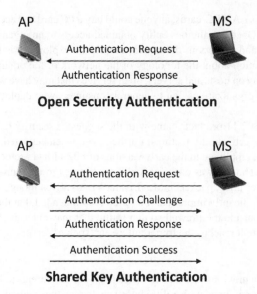

Figure 7.3 Two entity-authentication approaches in early IEEE 802.11 that were not secure.

same set of keys. Up to four different keys could be used, but it was common for all of the devices to employ a single key. Key-mapping keys allowed the use of different shared keys for different mobile devices. An access point would maintain a table or list of mobile devices and keys. But this option was not commonly used. Further, data confidentiality and integrity were accomplished using very simple methods that were not secure. We describe these below.

Example 7.1: Entity authentication in early IEEE 802.11

Some organizations filtered packets based on the 48-bit 802.11 MAC address. A list of authorized MAC addresses was maintained and if the MAC address did not match, the packet would be discarded. Once again, it is not too hard to spoof the physical address of the device once a legitimate address is discovered and this process for authenticating the entity was not secure.

If the devices had WEP enabled, there were two options for entity authentication (see Figure 7.3). In the first option, the access point accepted connections from all mobile stations, and mobile stations connected to any available access point. In the second option, a challenge-response protocol was used. As described in the appendix, this usually involves one party sending a challenge to a second party which responds with the encrypted challenge. This was futile in the case of WEP for the following reason. In the case of stream ciphers, a key is used to generate a key stream. The key stream is XOR-ed with the data during encryption. Thus, if *both* the plaintext and ciphertext are known, the key stream is also known. In the case of WEP, the challenge and response are transmitted over the air. An attacker can thus determine the key stream and use it to authenticate with the access point. This is also possible because all the mobile devices usually share the same single key with the access point.

As described in Example 7.1, it is easy to recover the key stream if the plaintext and ciphertext are known when encryption is performed with a stream cipher. For the same reason, the same key stream should not be used repeatedly when employing stream ciphers. To avoid this, in the case of WEP, a 24-bit *initial vector* or IV was concatenated with the key to generate the key stream.

Once again, this problem was created since all mobile devices shared the same key and there was no mechanism to establish a key during the entity authentication phase. Since the initial vector used by Alice will not be known to Bob, Alice has to transmit the initial vector in plaintext form. This increased the original key size in WEP from 40 bits to $40 + 24 = 64$ bits. In later versions, it increased the key size for generating the key stream from 104 bits to $104 + 24 = .128$ bits. But since the IV was transmitted in plaintext form, the key sizes were still smaller. The use of the IV was also rendered ineffective in terms of security as explained in Example 7.2.

Example 7.2: (In)effectiveness of the Initial Vector

Clearly, an initial vector should never be repeated. If is is repeated, the key stteam is repeated and this is never good when stream ciphers are used for confidentiality. A 24-bit initial vector implies that there are $2^{24} = 16,777,216$ different possibilities for the initial vector. By some estimates, a busy access point may receive up to 700 MAC layer frames every second. Since all mobile devices share the same key, it does not matter how many mobile devices are transmitting packetsdds. At 700 frames/s, it takes $2^{24}/700 = 23968s = 399$ minutes $= 6.65$ hours to see an IV repeated in the worst case. In reality, IVs get repeated more frequently. Many laptops start with the same IVs when they are restarted. Also, the generation of IVs was based on a pseudorandom generator making the sequence of IVs predictable.

The way confidentiality and integrity was implemented in WEP carries over the other flaws previously observed with WEP (i.e., the use of IVs, single shared keys, and stream ciphers). Data is broken into fragments suitable for use in a MAC layer frame. An *integrity check value* that is basically a CRC code is added to the fragment. The device selects an IV, appends it to the shared secret key, generates the key stream using RC-4, and encrypts the data fragment and concatenated CRC by XOR-ing it with the key stream. The MAC header and the initial vector are transmitted without encryption. A field in the MAC header informs the recipient of the packet that the packet is encrypted. The CRC is used as the integrity check (protection against fabrication or modification) but it is not secure.

Example 7.3: Message authentication in early IEEE 802.11

In early IEEE 802.11, the packet is encrypted using the RC-4 stream cipher. If the key stream used for encryption is not correct, upon decryption, the CRC (error detecting code) in the packet will fail and the access point can discard the packet. However, this implementation can be easily broken since the way the CRC works is known. So, an attacker can guess the bits that may be flipped when the CRC is checked by the access point and flip those bits only. This way, an encrypted packet can be modified and it can yet be accepted by an access point as legitimate.

WEP has no protection against replay. Further, there are some RC-4 keys that are weak and can be discovered because the IV is transmitted in plaintext form. Fluhrer, Mantin, and Shamir described an attack in 2001 (now called the FMS attack based on their names) that can discover the first 8 bits of a key with just 60 MAC frames. Better attacks were discovered over the years that made WEP really insecure [Tew09]. Several free and open-source tools incorported these attacks allowing even laymen to discover keys used with WEP, thereby allowing them access into a supposedly secure network.

WPA and 802.11i: The security flaws in WEP lead to a completely new approach to WLAN security resulting in the IEEE 802.11i standard [Edn04]. The IEEE 802.11i standard and its brand

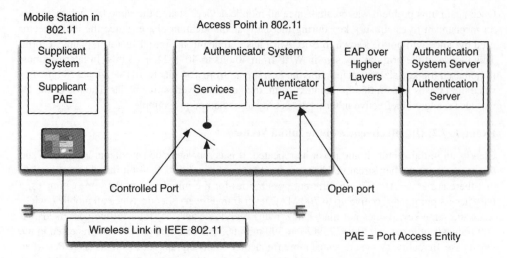

Figure 7.4 The 802.1X supplicant-authenticator-authentication server schematic.

equivalents called WiFi Protected Access (WPA) have some differences. We will not elaborate on these differences here, but treat them as mostly the same.

WPA/802.11i implements security using the general approach shown in Figure 7.2. The idea here is that entity authentication and key establishment are necessary whenever a mobile station joins a WLAN network. Once key establishment is performed and new session keys created, the link can provide confidentiality and integrity services to MAC layer frames. The entity authentication and key establishment process sometimes needs higher layer protocols or connections to servers on a network. Thus, *some initial access* to the network may be necessary to accomplish this part of the security protocol. However, access to other parts of the network must be prohibited till entity authentication and key establishment is completed.

Towards this, 802.11i adopts the 802.1X architecture comprising of a supplicant, authenticator, and authentication server shown in Figure 7.4. The mobile station acts as the supplicant which has a port access entity (PAE). It requests entity authentication by connecting to the authenticator. The authenticator presents two *ports* to all devices. The controlled port can be accessed only after entity authentication is successful. The open port can be accessed by all devices, but it only leads to an authentication server. Entity authentication can be achieved in many ways. There could be a simple password or pre-shared key between the authentication server and the mobile station. The authentication server could reside in the access point in this case. More complicated authentication protocols (such as those using transport layer security, Kerberos, Radius, or other approaches) are also allowed as options. They may need communication between the authenticator (the access point) and external servers or networks. In fact it may be possible to authenticate to an access point using credentials that reside in a cellular telephone network.

The process of entity authentication makes use of the *extensible authentication protocol* or EAP. The extensible authentication protocol is not really an authnetication protocol – it is more like a protocol specification that can be employed by various authentication mechanisms. It only specifies messages for excahnging the necessary information for authentication and either positively authenticating an entity or rejecting it. In general, it allows for authentication request, response, success and failure messages.

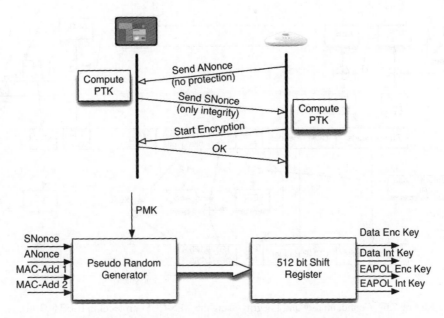

Figure 7.5   The four-way handshake in 802.11i and creation of keys.

At the completion of the entity authentication and key establishment process, the mobile station (supplicant) and the access point (authenticator) have a *pairwise-master key* (PMK) established. The way the PMK is established depends on the authentication protocol or the shared password between the mobile station and the access point. The PMK is not used directly for providing confidentiality or integrity. Instead, four different session keys are established through a *four-way handshake* as shown in Figure 7.5. The four keys, called data encryption key, data integrity key, EAPOL-key encryption key and EAPOL-key integrity key, are together referred to as the *pairwise-temporal key* or PTK. The PTK is generated using *nonces* that are exchanged at the beginning of the session and the MAC addresses of the communicating devices. The four-way handshake tells both the mobile station and the access point that they should install the PTK and use it for security. As the name suggests, the data encryption key is used for encrypting the MAC frame and the data integrity key is used for providing integrity checks. The PTK has to be re-created for new sessions.

With the above changes, it is possible to use RC-4 and provide a greater level of security while maintaining some backwards compatibility with WEP. This is one of the options of WPA. However, all of the new 802.11 devices eschew RC-4 and instead use the advanced encryption standard (AES) as the algorithm for providing confidentiality and integrity. This requires new hardware and thus cannot be used with legacy IEEE 802.11 devices.

Figure 7.6 shows how encryption and message integrity is provided using AES. An 802.11 MAC layer frame is broken up into blocks of 128 bits each. Each block is encrypted using AES in the counter mode (essentially a standardized stream cipher that is considered secure). In this case, a counter is incremented and encrypted using AES and the data encryption key to produce a key stream that is XOR-ed with data. Note that the MAC frame headerand packet number are not encrypted. To provide message integrity, the frame header and payload goes through the *cipher-block chaining* or CBC mode of operation of AES. Essentially, a 128-bit plaintext block is first XOR-ed with the previous ciphertext block and encrypted to create the current 128 bit ciphertext.

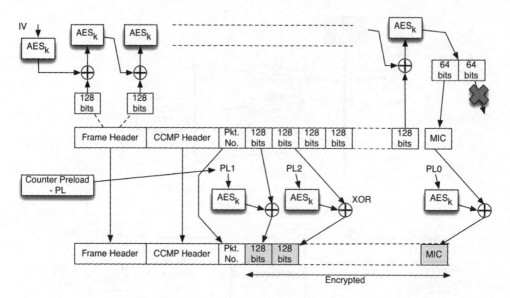

Figure 7.6   Confidentiality and Integrity using the AES-CCM protocol in IEEE 802.11i.

The most significant 64 bits of the last ciphertext block is used as the integrity check (to ensure that the packet has not been modified or fabricated). This protocol is called the *Counter mode with CBC MIC* or the CCM protocol. As we mention later, this approach has also been adopted by the IEEE 802.15.4 low-rate sensor networks as well.

7.3   Security in Wireless Personal Networks

7.3.1   Security Threats

Bluetooth: Bluetooth (see Chapter 9) is the most popular wireless personal network that allows devices of different types to connect to one another at short distances. As we discuss in Chapter 9, pairing of Bluetooth devices enables devices to communicate. If a malicious device gets paired with an honest device, the attacker may be able to obtain valuable information from the honest device (e.g., its addressbook containing e-mail addresses and other kinds of contact information) or use the Bluetooth connection to make expensive calls to services. This attack has been called "Bluesnarfing".

There have been several exploits of Bluetooth (see for example [Jak01]). The object exchange (OBEX) protocol in Bluetooth has been exploited to send spam messages to unsuspecting Bluetooth enabled devices. This attack has been called "Bluejacking". A tool called Bluetooth Sniper was developed to eavesdrop on unencrypted Bluetooth links, from far distances. Many mobile viruses have been known to spread using Bluetooth links. Warnibbling is an approach that has been discussed where a tool can be used to map all Bluetooth devices near a given location. This is an eavesdropping attack.

Low Rate Sensors: Security in sensor networks is an important topic although there are opinions that suggest that work in this area may be of academic interest only. In this section, we briefly discuss the security features provided in the IEEE 802.15.4 standard and briefly mention some of the research work describing security threats in sensor networks. Sensor networks are especially vulnerable to security attacks for the following reasons: (a) the transmission medium is air and so it is easy to obtain remote access to the medium using powerful antennas either for eavesdropping, jamming, or injecting malicious traffic (b) sensors may be deployed in huge numbers and are supposed to be of low cost with the implication that the possibility of some of the sensors being captured, tampered with, and compromised being very real. It is possible for adversaries to deploy their own sensors into a sensor field, but this requires physical protection of the sensed region. There are numerous security threats that have been considered in the research literature of wireless sensor networks making it difficult to consider them all together. Instead it is easier to group the threats into categories. While there are overlaps between them, we can classify these threats into the following categories.

Physical layer threats: At the physical layer of the communications protocol stack, common threats against sensor networks are disruptions to communications through jamming and node disabling. By jamming, an attacker may disrupt reliable communications by transmitting signals that interfere with the radio signals of sensor nodes. This may result in partitioning of the network, lower reliability of the sensed data because of the lack of availability of data from certain sensed regions, and ultimately result in the battery exhaustion of nodes repeatedly transmitting data till they are acknowledged or receiving bogus data. Jammers can be classified as those that may be outsiders employing a constant radio signal, deceptive jammers that inject regular packets into the network, random jammers that alternate between sleep and awake states and reactive jammers that cleverly disrupt communications upon sensing channel activity. Experimental studies indicate that packet delivery ratios are adversely impacted by all of these types of jammers. Jamming may adversely impact sensor nodes at the edges of a network or those that are towards vulnerable physical areas.

Eavesdropping threats: One of the most common threats in wireless sensor networks is information leakage where an adversary may obtain the sensed information by simply passively eavesdropping on the radio signals being transmitted by sensor nodes. Eavesdropping is especially problematic even with encryption because of the potential for sensor nodes possessing keys to be compromised or captured by adversaries. One model for computing the eavesdropping vulnerability is based on the adversary interested in predicting the behavior or aggregate output of the sensor network. In addition to information leakage, radio transmissions may reveal the location of sensors and the sink node and allow other kinds of analyses on the traffic patterns. An adversary may also be able to actively poll sensor nodes for information if there is no authentication of queries in the network.

Threats impacting routing: In networks, it is important for nodes to know where to send data packets so that they reach the destination in an efficient way. Such routing protocols in sensor networks are still evolving since attempts to directly use routing protocols designed for mobile ad hoc networks in sensor networks have faced challenges due to the scalability and energy requirements of sensor networks. Geographical and geometric routing that makes use of the knowledge of the Euclidean coordinates of sensors is proposed as an efficient means of routing data to the destination. However, it is likely that in general, routing in sensor networks faces the same threats as those in mobile ad hoc networks. Such threats include location disclosure, replay of old routing information, disruption by fabricating routing information, and route table poisoning. In addition, wormhole, blackhole, and Sybil attacks are possible. In blackhole attacks, malicious nodes advertise themselves as closer to the destination thereby making themselves part of most routes. They can then disrupt network operation by dropping packets or get information by eavesdropping. In Sybil attacks, a single malicious node claims to be more than one node. This way, it could claim a disproportionate amount of resources and also perform blackhole attacks. If there are collaborating nodes, they may create a wormhole (a tunnel) between them and create the impression of a false network topology.

Threats impacting position information: The position of a sensor node has importance in several applications. For example, temperature variations over a given area may have to be accurately characterized in which case, the position location of the sensor reporting the temperature reading needs to be known to certain accuracy. Such position information may be used for routing or even in security measures. Further, the location of a sensor monitoring a critical quantity may itself need to be kept secure (location privacy). Malicious nodes can interfere with the reporting of position location information in many ways. They can fabricate the position information or interfere with the support infrastructure using which sensors can determine their positions. In the latter case, there are many different approaches for determining the position of sensor nodes such as using beacons from nodes at known positions, determining the number of hops a node is away from a reference node and so on. Malicious nodes can interfere with such position determining activity.

Threats impacting data aggregation and in-network processing: Data aggregation and in-network processing is an important feature of data-intensive sensor networks. Because sensor nodes collect a huge amount of data and sometimes only aggregate information is necessary at the sink (e.g., average value or the sum of the sensed quantity), intermediate nodes can process the received data (in-network processing) or fuse data and forward those values. This reduces the communication costs and delays in the network. However, such functionality makes it extremely easy for malicious nodes to introduce false values that corrupt the processed or fused values. If a malicious node is responsible for fusing or aggregating data, the problem could be worse. If a Sybil attack is launched, a node can claim multiple identities and further skew the aggregated data by creating multiple false reports.

Threats against time synchronization: Sensor networks often require nodes in the network to be time synchronized for many reasons such as data fusion, scheduled transmissions for saving power, tracking duplicate sensed data and so on. Time synchronization can be achieved using reference broadcasts or sender-receiver synchronization. It is possible to disrupt the sensor network operation by misleading different nodes about the time at which they have to perform operations like sensing or transmissions of packets.

Miscellaneous threats: If sensor nodes are compromised, they can disrupt a sensor network in many ways. For instance, a compromised node may not follow the medium access protocol and hog the medium. If a node assumes several identities, as in the case of a Sybil attack, it could access the medium more often than it should normally have fair access. These may both deny access to radio resources by legitimate sensor nodes. In many types of sensor networks, a sink node is used to collect data after a query to many sensor nodes in the networks and for other types of network maintenance. In some cases, mobile sink nodes are employed to poll sensors or collect data from a set of static sinks. Compromise of sink nodes can lead to damages or disruption of a sensor network.

7.3.2   Security Protocols

Security Protocols in Bluetooth: Bluetooth considers devices that are paired with a given device as trusted. Usually, pairing requires manual intervention. For example, Figure 7.7 shows a window

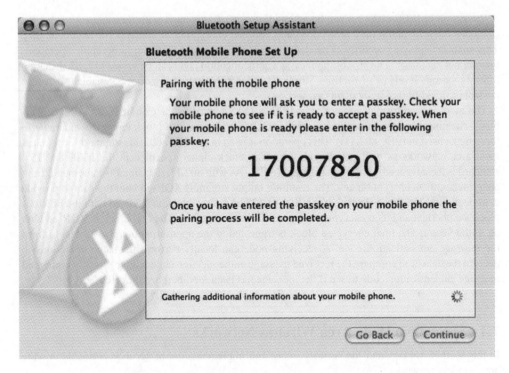

Figure 7.7   Pairing of a mobile phone with a computer.

that appears on a computer when a mobile phone is trying to pair itself with the computer. The human user has to manually enter the passkey shown on the computer into the mobile phone to enable pairing of the devices. This helps prevent unknown devices from connecting to a known device easily. The same process does not usually hold for interface-free devices like headsets or microphones. However manual intervention in terms of press-and-hold of a button or similar procedures are adopted to ensure security. However, some devices use no security whatsoever and can be vulnerable to malicious pairing.

Bluetooth specifies several functions and algorithms for encryption and authentication. Some use the SAFER+ algorithm while others are based on weaker linear feedback shift registers.

Security in IEEE 802.15.4: In IEEE 802.15.4, no attempt is made to address the many different vulnerabilities or threats that exist for sensor network applications described above. However, cryptographic protection of communications over links is part of the standard. The cryptographic operations that are part of the standard provide for confidentiality, integrity, and authentication of the communicated data using encryption and message authentication codes. The standard assumes that secrets that are to be shared between sensor nodes is an issue that is beyond its scope. So it is necessary to employ additional key establishment and key management schemes with IEEE 802.15.4 sensor devices. Keys may be pairwise or shared by a group of nodes. The rest of the section assumes that somehow pairs of communicating nodes share keys with each other (group or pairwise).

Protection in IEEE 802.15.4 can be adopted on a per frame basis with message authentication (includes integrity and replay protection) and optional encryption of contents for confidentiality. This enables deploying security as needed without expending energy for cryptographic operations that are not necessary. The size of the message authentication code can be varied (32, 64, 128 bits) offering various levels of protection. Similarly, encryption may or may not be enabled. It is also possible to send frames without any protection.

The most general form of protection in IEEE 802.15.4 involves the Counter mode with Cipher-block-chaining Message authentication code (CCM) operation of a block cipher (comparable to the scheme used in IEEE 802.11i). This operation is used in the IEEE 802.11i standard for wireless local area networks as well (see Figure 7.6). The block cipher specified in the IEEE 802.15.4 standard is the advanced encryption standard (AES). As with 802.11i, a counter is incremented and encrypted with an encryption key. The resulting output stream is XOR-ed with the data to provide confidentiality. The data is broken into bocks of 128 bits. Each block is XOR-ed with the previous block's ciphertext and then encrypted using an authentication key. The first block is XOR-ed with an initial vector. The final encrypted block is truncated to the appropriate number of bits to form the message authentication code. A receiving node can locally perform the same operations to decrypt the data or to compare the received message authentication code with the locally computed message authentication code to see if the message has been modified or fabricated.

7.4   Security in Wide Area Wireless Networks

In this section, we consider the security issues and implementation of security protocols in wide area wireless networks.

7.4.1   Security Threats

We can consider confidentiality, fraud, and integrity threats to data as the three major security issues in wide area wireless networks. The security problems with applications on smartphones is another category of security related problems in wireless networks.

Confidentiality: As we have discussed in detail in earlier chapters, voice conversations have been the primary driver of wide area cellular wireless networks. The confidentiality or privacy requirements for voice conversations and locations of mobile terminals are two-fold in wireless networks. Along with the air-interface, there is also a fixed infrastructure for handling the registration of mobiles, billing, mobility, power control and other issues. There are privacy requirements for the air-interface and others for the messages transmitted over the wired infrastructure. The air-interface is susceptibe to eavesdropping. The wired infrastructure has physical protection, but as the infrastructure moves towards an all IP network connecting to the Internet, vulnerabilities may make it accessible and subject to attacks from the Internet.

As we briefly discuss in Chapter 11, a variety of control information is transmitted over the air in addition to the actual voice or data. These include call set-up information, user location, user ID (or telephone number) of both parties etc. These should all be kept secure since there is potential for misusing such information. They may be used to track mobile devices. Calling patterns (traffic analysis) can yield valuable information under certain circumstances. A flurry of calls between the CEOs of two major companies may indicate certain trends if it was discovered, even if the actual information in the calls was secure.

In [Wil95a], various levels of privacy are defined for voice communications. At the bare minimum, it is desirable to have wireline equivalent privacy for all voice conversations. We commonly assume that all telephone conversations are secure. While this is not true, it is possible to detect a tap on a wireline telephone. It is impossible to detect taps over a wireless link. To provide privacy that is equivalent to that of a wired telephone, for routine conversations it may be sufficient to employ some sort of an encryption that will take more than simple scanning and decoding to decrypt. In order to alert wireline callers about the insecure nature of a wireless call that is *not at all* encrypted, a "lack of privacy" indicator may be employed. Wilkes [Wil95a] calls these two levels of security as levels zero and one respectively. Level 0 privacy is when there is no encryption employed over the air so that anyone can tap into the signal. Level 1 privacy provides privacy equivalent to that of a wireline telephone call, one possibility being encrypting the over-the-air signal. For commercial applications, a much stronger encryption scheme would be required that would keep the information safe for more than several years. Secret key algorithms with key sizes larger than 80 bits (preferably 128 bits) are appropriate for this purpose. This is referred to as Level 2 privacy. Encryption schemes that will keep the information secret for several hundreds of years are required for military communications and fall under Level 3 privacy.

For wireless data networks, a bare minimum level would be to keep the information secure for several years. The primary reason for this is that wireless electronic transactions are becoming common. Credit card information, dates of birth, social security numbers, e-mail addresses, and so on can be misused (fraud) or abused (e.g., junk messages). Consequently, such information should never be revealed easily.

Fraud: While privacy and confidentiality continue to be the important issue in wireless networks, other security requirements are becoming significant in recent times. There has been widespread fraud and impersonation of analog cellular telephones in the past. Since the advent of analog telephony, wireless service providers have suffered several billion dollars of losses due to fraud. Although this is more difficult with digital systems, it is not impossible. There is thus a need to correctly *identify* and *authenticate* a mobile terminal when it connects to a cellular network. It is important to ensure that stolen devices, cloned devices which were common in 1G systems, or subscribers who have not paid their bills do not get access to the network. We consider the security protocols used in example WWANs in the next section.

Integrity: As described in Chapter 6, there are several operational issues in wireless networks that require management messages to be sent to mobile devices to make handoffs, change tarnsmit powers, and so on. Such messages should have integrity checks. If spoofed, fabricated, or replayed messages are employed, they could disrupt the smooth operation of the cellular network.

Mobility: Unlike WLANs, mobile devices in cellular networks are highly mobile and users may travel across thousands of miles or across countries and yet use their cell phones as if they are connecting to their home network. This applies for incoming and outgoing calls. This adds another factor in terms of security. It is unlikely that a mobile device is known as a valid mobile device to every network. Even if a mobile device connects to a known service provider, the security credentials (such as shared keys) cannot be locally stored at every base station or base station controller. As a result, there is need for mechanisms to authenticate the network and mobile devices to one another remotely and to create session keys that can be locally employed for security.

Device Security: With smartphone operating systems capable of running applications and software just like computers, installation of malicious software is a real problem. As an example, in the past decades, Symbian was the major operating system on cell phones and there were many malicious software that exploited vulnerabilities. A worm called Cabir was capable of infecting Symbian phones made by many manufacturers. The Cabir worm could replicate using Bluetooth connections or through a shared application. It would ask users if they would be willing to receive a message over Bluetooth and then it would install itself if they accepted this request. At the minimum, this worm would constantly scan the air for other nearby Bluetooth devices thereby reducing the battey life of the device. There have been reported malware that access toll numbers causing unwanted charges to users.

Things have become more problematic today with malicious marketplaces for apps that can be installed on a smartphone. There have been a variety of reports on malware in Android phones. As smartphones can often carry a lot of valuable information (credit cards, bank accounts, passwords, pictures, location information), malware that could access these types of information and send it to cyber-criminals are problematic. To prevent security problems, the general recommendation is for users to avoid unknown marketplaces and choose to install only signed apps from known entities.

7.4.2   Security Protocols

There were few security features in 1G cellular systems. 2G cellular systems adopted challenge-response based entity authentication to identify whether the device connecting to the network was legitimate. This reduced fraud to a large extent. Example 7.4 briefly describes the entity authentication process in North American 2G cellular networks. 2G networks also adopted some encryption for voice traffic to prevent eavesdropping. The encryption algorithms were not public in many 2G systems and were subject to cryptanalytic attacks over the years.

Example 7.4: Challenge response schemes in 2G cellular networks

Figure 7.8 shows the architecture of North American 2G TDMA standard. This architecture is similar to the GSM architecture discussed in Chapter 11. The lower half of the figure shows part of the challenge response mechanism implemented in the standard for network operations. The network (Bob) generates a random number RANDU and sends it over the air to the MS (Alice). The MS computes a value AuthU using the encryption algorithm called CAVE (Cellular Authentication and Voice Encryption) algorithm. The value AUTHU is transmitted over the air. The network computes its version of AUTHU and compares the two values. If the values match, the MS is identified (authenticated in the terminology of the 2G standard). There was no authentication of the network itself.

In 3G systems such as UMTS (see Chapter 12 for some details on the architecture and physical layer of UMTS), mutual authentication was adopted. The protocols follow the two steps shown in Figure 7.2. In the first step, the entities (in this case, the mobile device and the connecting network) are authenticated to one another. At the same time, keys are established for communication during that session. Then control messages have integrity while data traffic is simply encrypted. We consider some details below. Further, there was no encryption of any kind in 1G analog cellular systems and as discussed earlier, it was easy to clone phones and evesdrop on conversations. In digital systems, the mobile device (equipment) identity is kept in the Equipment Identity Register (EIR). If a mobile device is reported as missing or stolen, the EIR can keep track of this to ensure

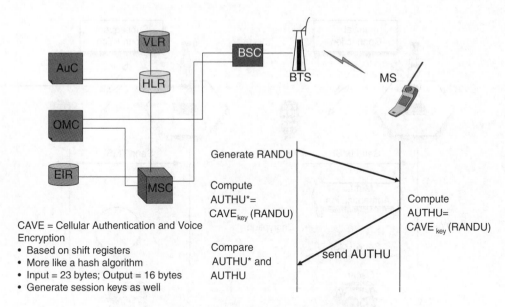

Figure 7.8 Challenge–response in 2G cellular.

that such devices do not connect to the network. Further tracking of stolen devices that attempt to connect to a network may be possible for law enforcement purposes.

As encryption is used on digital systems, the privacy problems are more related to location tracking. The prospect of mobile phones being tracked by malicious entities becomes an important security issue. As shown in Figure 7.9, this problem is addressed in 3G systems through the use of encryption. Initially, a mobile device transmits its so-called International Mobile Subscriber Identity or IMSI in plaintext form. This can be tracked by a sniffing malicious entity. But once the mobile device is authenticated and encryption starts, the network sends a *temporary mobile subscriber identity or TMSI* to the mobile device. Since the TMSI is encrypted when it is sent to the mobile device, the malicious entity cannot associate it with the IMSI. The next time the mobile device connects to the network, it uses the TMSI, in plaintext form. However, this TMSI, even when it is visible, cannot be linked to the IMSI and hence the privacy of the mobile device is maintained.

Security Protocols in 3G Systems: One of the critical aspects of WWANs is that the mobile device could be almost anywhere in a geographical area, and often may be served by different service providers. Clearly, the authentication credentials cannot be kept at every local point of access (either the base station or the radio network controller – see Chapter 6 for a description of the architecture of cellular systems). The shared master secret key is known only to the mobile station and the authentication center/home location register (AuC/HLR in Figure 7.10). When the mobile station connects to the visiting network, the serving GPRS support node (SGSN) or the visitor location register (VLR) contacts the HLR for authentication credentials. The first step in the security process is to obtain the authentication credentials from the HLR/AuC. In the second step, the VLR performs entity authentication. Unlike 2G systems, the mobile station can also ensure that it is connecting to a valid 3G network in this process. Also, session keys are established at the mobile station

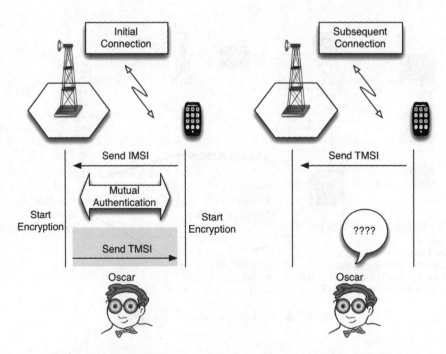

Figure 7.9   Maintaining privacy by using a temporary subscriber identity.

and at the radio network controller. Communications between the mobile station and the RNC are secured through encryption and integrity checks as described below. The communications in the core (wired) network employs a different approach that is discussed briefly in Section 7.5. More details of access security in 3G UMTS systems are available in [Koi04].

In the first step, the HLR/AuC sends a so-called *authentication vector* to the visiting network when it is contacted in response to a mobile station connecting to the visiting network. The VLR/SGSN of the visiting network contacts the HLR/AuC when it has no authentication vectors that correspond to a given mobile device. The authentication vector comprises of the following information:

- A 128-bit random number used as a nonce for challenge called RAND, which is generated from the internal state of the AuC
- An expected 32- to 128-bit response from the mobile device called XRES
- A 128-bit cipher key (CK) used for confidentiality and a 128-bit integrity key (IK) used for ensiring integrity of control messages
- An authentication token (AUTN) which has a 48-bit sequence number (SQN), a 16-bit authentication management field, and a 64-bit message authentication code (MAC-A).

Details of the second step are shown in Figure 7.11. The mobile station first verifies if the network is valid by locally computing an *expected* message authentication code X-MAC that is generated using the master key shared with the authentication center, and the quantities RAND, AMF, and SQN. If X-MAC matches MAC-A, the network is authenticated. Then, the mobile checks to make sure the sequence number is within an acceptable window and adjusts the window for the future.

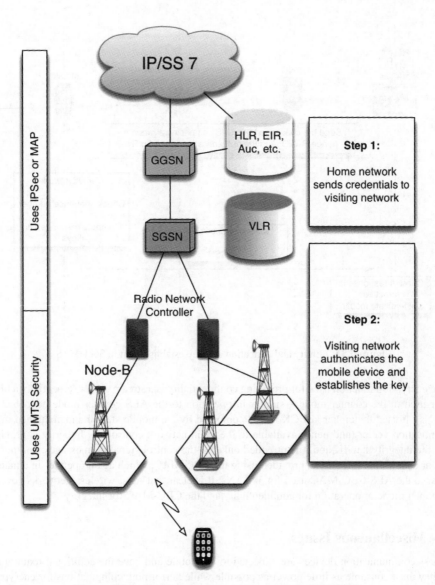

Figure 7.10   General view of security process in 3G UMTS.

The mobile station computes the expected response XRES, the cipher key CK and the integrity key IK which are derived from the master key and the nonce RAND. The mobile device then sends the expected response to the SGSN/VLR where it is compared with the XRES contained in the authentication vector. If the two match, the mobile station is authenticated. An anonymity key that hids the sequence number can also be derived. Various functions (made up of encryption algorithms such as AES or its variations or hash functions) are used for these purposes. Operators can pick encryption schemes of their choice although the standards offer some recommendations for encryption and creating the message authentication code.

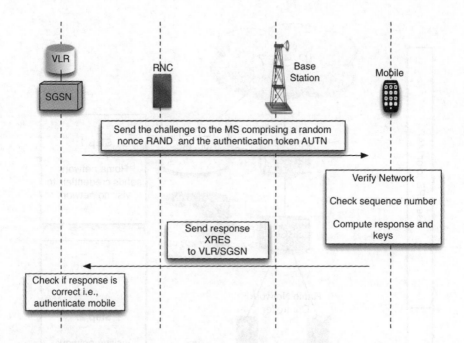

Figure 7.11   Entity authentication and key establishment in 3G UMTS.

The encryption function used for providing confidentiality is a stream cipher (essentially a block cipher used in the counter mode – similar in operation to the AES counter mode described for WLANs). It used the cipher key CK that is computed by the mobile station and contained in the authentication vector (and hence available to the radio network controller). Integrity is provided using the integrity key IK and it is provided only for the control signaling messages, and not for voice or data traffic. Integrity is provided using the CBC-MAC, which again operates in a manner similar to the AES CBC-MAC used for WLANs. It is clear that most wireless networks now use the counter mode of operation for confidentiality and the CBC-MAC for integrity.

7.5   Miscellaneous Issues

Wireless communication devices are expected to be mobile and have the additional requirement that they must consume as little power as possible while performing computations for encrypting or decrypting data to conserve battery power. This is a significant issue because cryptographic algorithms can be computationally intensive and may drain the battery of a mobile terminal quickly. Since the spectrum is scarce, cryptographic protocols should also not waste resources by requiring several handshakes between the mobile terminal and the fixed network. This requirement is usually contrary to security services as they are implemented in wired networks. The wireless channel is error-prone and may also result in messages being lost, duplicated, or damaged. It is also not clear what effects this may have on the overall performance of security protocols. Interference, fading, disconnections, handoff and other mobility-related procedures, and other peculiarities of the wireless network require robust security services that are at the same time resource efficient.

The security of information carried in the core (wired or fixed) part of the cellular network was typically based on lack of external access to the network in early systems. Signaling System-7

(SS-7) is used as the control signaling overlay network in 1G and 2G systems. The mobile application part or MAP is used as roughly the network layer protocol. Security was added on top of MAP to create MAPSec for use in the earlier systems. As 3G and 4G systems have migrated to IP as the network protocol over the fixed network, IPSec is the natural option for providing security for packets.

IPSec encrypts all IP traffic between two hosts or devices, or two networks, or combinations of hosts with possibly different terminating points for different security services. By performing encryption and authentication at the IP layer, no applications need be modified and security is transparent to applications and the user with IPSec. However, this approach makes it more difficult to authenticate users and provide them privileges appropriately. Keys may be manually established or a very complex protocol called *Internet Key Exchange* (IKE) [Kau02] can be used for authenticating entities to one another and establish keys. Keys are established as part of a unidirectional "security association" which specifies the destination IP address, keys, encryption algorithms and "protocol" to be used. Mechanisms are included to prevent denial of service attacks similar to TCP SYN floods where one side of the security association would otherwise waste resources for key establishment.

"Protocol" in the above paragraph corresponds to one of two specific security protocols provided by IPSec – *Authentication Header* (AH) and *Encapsulated Security Payload* (ESP). In AH, a message authentication code is created on the entire IP packet minus the fields in the IP header that change in transit. This enables the receiver to detect spoofed or modified IP packets. However, the payload is in plaintext and visible to anyone who may be capable of capturing the IP packet. ESP provides confidentiality and integrity to the payload of the IP packet, but not the header.

Use of the two protocols in a simple manner where a security association is set up between two hosts that directly apply either AH or ESP is called "transport mode" in IPSec. It is also possible to use a "tunnel mode" where the original IP packet is tunneled in another IP packet with a new IP header. This makes the original IP packet the payload, thereby protecting it completely with either a message authentication code in the case of AH or encryption and integrity in the case of ESP. Multiple security associations called "bundles" are also possible where both AH and ESP can be applied to the same IP packet. A security association database and a security policy database are used to decide what to do with an IP packet that is inbound or outbound from a host. IPSec is fairly complicated in all its details. We refer the interested reader to [Kau02] for more information.

Appendix A7: An Overview of Cryptography and Cryptographic Protocols

Security services such as confidentiality, entity and message authentication, integrity, and non-repudiation can be provided to communication protocols using cryptography. In this section, we provide a brief overview of the important topics in cryptography and cryptographic protocols. More details can be found in [Kau02; Ches03; Sta03; Sti02].

Cryptographic Primitives

Cryptographic protocols make use of cryptographic *primitives* that are used to provide the required security services. A classification of such primitives is shown in Figure A7.1. Cryptology is the broad discipline that includes the science of designing ciphers (cryptography) and that of breaking ciphers (cryptanalysis). Data that is encrypted is called "plaintext" and the result of encryption is called "ciphertext".

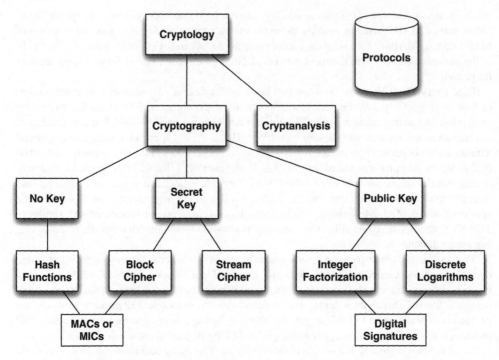

Figure A7.1 Classification of cryptographic primitives.

It is common to denote two communicating parties as Alice and Bob and the adversary or opponent as Oscar. Mathematically, the encryption of a plaintext x into a ciphertext y using a key k is written as:

$$y = e_k(x) \qquad\qquad (A7.1)$$

The corresponding decryption is written as:

$$x = d_k(y) \qquad\qquad (A7.2)$$

Ideally, we would like the encryption scheme to be such that it cannot be broken at all. Since there are no practical methods of achieving such an unconditional security, encryption schemes are designed to be computationally secure. The encryption scheme has to be powerful, in that given significant computational resources, an adversary must not be able to either find the key or decrypt the message in a reasonable time. Alternatively, if either the key or the plaintext can be determined in a short time, it should cost the adversary much more than what the value of the secret information would be to him. Usually, it is assumed that Oscar has knowledge of how the algorithm works, but not the key. Also, because of the standard formats of data packets and control messages in voice networks, Oscar usually has access to a limited number of plaintext-ciphertext pairs that he can use to perform a known plaintext attack to recover the key. Once the key is recovered, all subsequent ciphertext can be decrypted easily.

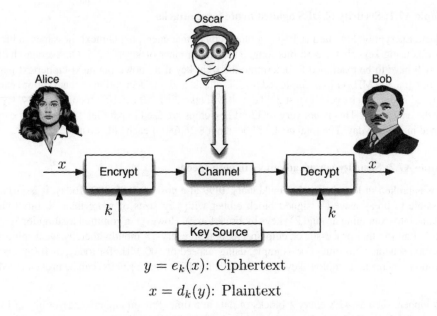

$$y = e_k(x): \text{ Ciphertext}$$

$$x = d_k(y): \text{ Plaintext}$$

Figure A7.2   Conventional encryption model.

So far, we have discussed security services and we have said that encryption can provide some of these services. What we have not discussed are the encryption algorithms that are employed or can be employed within these mechanisms. The details of these algorithms are beyond the scope of this book and the subject matter forms a wide area of interest in itself. In this section, we will briefly mention some of the algorithms that are in wide usage today. We will also discuss the key sizes that are required to make these algorithms secure.

Ciphers or encryption algorithms can be classified into *secret key* and *public key* categories. Encryption schemes have been available through the ages and have all been what are known as secret-key algorithms. Here, the communicating parties (Alice and Bob in Figure A7.2) share a secret key that they use to encrypt any communication between themselves. Usually, the encryption and decryption algorithms use the same key and hence, such algorithms are also called symmetric key algorithms. Block ciphers such as the advanced encryption standard (AES) also fall under this category. Figure A7.2 illustrates a schematic of a conventional encryption scheme. The opponent Oscar has access to the insecure channel and thus the ciphertext. However, he has no knowledge of the secret key k shared by Alice and Bob.

Secret-key algorithms such as AES are based on two principles: confusion and diffusion. The former introduces a layer of scrambling that creates confusion as to what exactly might be the transmitted message. The latter creates a randomness whereby the effect of changing a small part of the plaintext message will result in changing half of the encrypted ciphertext. This eliminates matching patterns or frequencies of occurrence of messages. Most secret-key algorithms are thus unbreakable except by brute force [Sil00]. If the length of the key of a secret-key algorithm is n bits, at least 2^{n-1} steps are required to break the encryption on average. Today, a key length of 80 bits is considered to be sufficiently safe from brute force attacks even though a key size of 128 bits is usually recommended.

Example A7.1: Security of DES against brute force attacks

The Data Encryption Standard (DES) is a block cipher that encrypts plaintext messages in blocks of 64 bits using keys that are 56 bits long. The total number of keys is 2^{56}. On average, half of them will have to be examined to determine the right key if a known plaintext-ciphertext pair is available. If a 500 MHz chip is employed for this attack, and one decryption (or encryption) can be performed in one clock cycle, to test 2^{55} keys, it will take $2^{55} / (500 \times 10^6)$ seconds = 834 days to break the encryption. This is not very secure if 834 chips are used in parallel, since the key can be obtained in a single day. The total cost will be about $ 16 680 if each chip costs $20!

Example A7.2: Security versus advances in chip speeds

DES was broken in less than a day in January 1999 at a cost of $500 000. Today, it is virtually impossible to break a well-designed block cipher with key sizes of more than 80 bits (which translates into examining around 2^{80} keys by brute force). However, a common assumption (called Moore's Law) is that processor or chip speeds double every 18 months thereby weakening any encryption scheme with time. For example, using a speed of 500 MHz for today, in 100 years, an encryption scheme that employs key sizes twice that of DES (i.e., 112 bits) can be broken in a day.

Block ciphers such as AES encrypt blocks of data at a time. Stream ciphers encrypt bits or bytes of data by XOR-ing the input data stream with a key strteam that has to be generated securely. The advantage of stream ciphers is that there is no need for buffering data up to the block size and for padding. Stream ciphers may also be more suitable for a jitter sensitive voice conversations. The disadvantage is that these have to be used carefully because encryption with stream ciphers uses simple XOR operation. Thus it becomes necessary to use different key streams for encryption each time (because a simple XOR can be inverted to obtain a previously used key stream).

DES was the secret-key encryption standard for over twenty years. The National Institute for Standards and Technology (NIST) examined proposals for the Advanced Encryption Standard in 1998. Of the five candidate algorithms, NIST selected Rijndael as the algorithm for AES in October 2000. A variety of factors were considered by NIST to determine the suitability of the algorithm for a standard:

1. Security – resistance to cryptanalysis, mathematical soundness, randomness of the algorithm output.
2. Cost – licensing requirements, computational efficiency on various platforms, memory requirements.
3. Algorithm implementation characteristics – the ability to handle variable key sizes and block lengths, implementation as stream ciphers and hash functions, hardware and software implementations, and algorithm simplicity are three categories used for evaluation of these algorithms.

In addition to these standards, several freeware and other secret-key algorithms are available such as IDEA, RC-4, and Blowfish [Sta98]. RC-4 in particular has been widely employed in web browsers as well as in wireless networks like IEEE 802.11.

The primary advantage of secret-key algorithms is that they are fast and at the huge data rates that are being supported by today's networks, it is virtually impossible to employ *public-key* algorithms (discussed below). However, since every pair of users has to have a key, for a communication system with N users, at least $N(N-1)/2$ keys need to be created and distributed. This is not a trivial exercise and has its own weaknesses.

Example A7.3: Number of keys with symmetric key encryption algorithms

Assume a small corporate network with 500 computers. A total of 124 750 keys are required (one between each pair of computers). Each computer needs to store 499 keys associated with the remaining computers. Suppose an employee gets a new handheld personal computing device. Not only will this handheld device need to load 500 new keys, but the remaining 500 old computers also need to be each updated with a key for the handheld computer.

There are techniques of key distribution for symmetric key algorithms such as the Needham-Schroeder key distribution scheme and Kerberos. All of these schemes need several handshaking steps and also an initial configuration of computers with *master keys*. Such master keys can be distributed physically in a secure manner. However, key distribution is still a potential weakness in the system. Another way of generating fresh keys for each communicating session is to use the master key and a one-time random number (called nonce) as inputs to a one-way hash function to generate a key. Alternatively, the nonce can be encrypted using the master key.

Public-key encryption is a radical shift in the way data is encrypted. Diffie and Hellman introduced the concept in 1977. With secret key algorithms, we have a situation that is similar to having a locked mailbox for *each pair of users*. Both users associated with a mailbox share a key that can unlock or lock the mailbox. Consider Figure A7.3. Here, if Alice desires to communicate with Dan, she unlocks the mailbox shared between her and Dan, deposits the message, and locks the mailbox again. The message is now accessible only to Alice and Dan who also has an identical key.

Clearly, the number of mailboxes required for N users like Alice and Dan is $N(N-1)/2$. For example, we have six mailboxes for four users as shown in Figure A7.3. The situation described

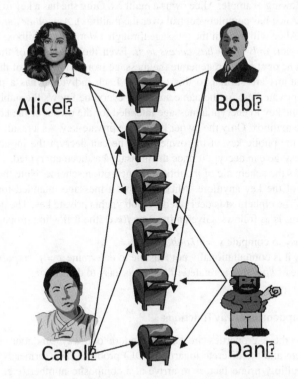

Alice␡   Bob␡

Carol␡   Dan␡

Figure A7.3   Multiple mailboxes with secret-key encryption.

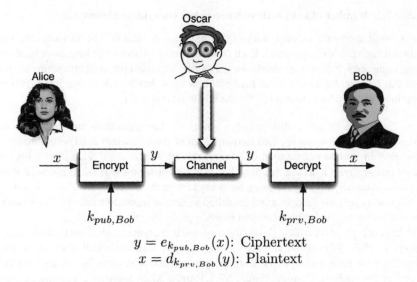

$$y = e_{k_{pub,Bob}}(x)\text{: Ciphertext}$$
$$x = d_{k_{prv,Bob}}(y)\text{: Plaintext}$$

Figure A7.4   Public-key encryption scheme.

above is not the natural way in which we employ mailboxes. Mailboxes are *associated with individuals* and not pairs of communicating parties. The natural way to employ a mailbox is as described in the following example. Alice owns a mailbox. Only she has a key to lock or unlock the mailbox (i.e., only Alice has complete control over the mailbox). *Any other person* who wishes to communicate with Alice will deposit the message through a *slot* in the mailbox. Once the message is deposited in the slot, *only Alice has access to it*. Even the originator of the message cannot retrieve it, although he or she may regenerate the message from knowledge of the contents.

Public-key algorithms are similar to this example. Each individual has a pair of keys – the public-key and the private-key. As the name suggests, everyone knows the public-key. So anyone can employ the public key to encrypt a message intended for the owner of the key. The public-key is like the slot in the mailbox. Only the owner knows the private-key. As a result, once the message is encrypted using the public key of the owner, only he can decrypt the message. Not even the originator of the message can decrypt it once the message has been encrypted.

Figure A7.4 shows the schematic of a public-key encryption scheme. Note that there is no need for secure transfer of the key anymore. Alice encrypts a message intended for Bob with Bob's public key $K_{pub,bob}$. The ciphertext is decrypted by Bob via his private key. The design criterion for public-key algorithms is as follows. Given a function $f(k,x)$, the following properties always hold.

- It is extremely easy to compute $y = f(kpub,x)$
- Given *kpub* and y it is computationally not feasible to determine $x = f^{-1}(kpub,y)$
- With a knowledge of *kprv* that is related to *kpub* it is easy to determine:
- $X = f^{-1}(Kpuby)$

Example A7.4: Trapdoor one-way functions

Functions that have these properties are called trapdoor one-way functions. Examples are the factorization problem and the discrete logarithm (DL) problem. The former is based on the fact that it is easy to multiply prime factors to arrive at a composite number (e.g., it is easy to find $7 \times 17 \times 109 \times 151 = 195\,821$, but it is quite a hard task to split $30\,616\,693$ into its prime number factors). The latter is based on the fact that it is easy to determine what 2^{23} mod 109 is (the answer is 77). It is quite hard to find out what u is given 2^u mod $109 = 68$. Note that with real number arithmetic, it would have been trivial to determine u as $u = \log_2 68$. The modulo function which reduces the operations to be on set of numbers that are non-negative integers less than 109 makes this problem very hard to solve. Integer factorization is employed in RSA and the DL problem is used in the Diffie–Hellman (DH) Key Exchange protocol and Digital Signatures.

Example A7.5: The Diffie–Hellman Key Exchange Protocol

The DH key exchange protocol is based on the discrete-logarithm (DL) problem discussed in Example A7.4. Let us suppose that Alice wishes to exchange a session key with Bob without sharing any secret with him. Alice chooses a base α and a large prime number p that are publicly known. She only chooses a random private number a. She computes $k_{pubA} = \alpha^a$ mod p which she sends to Bob. Note that given k_{pubA}, α, and p, it is computationally impossible to determine a. Similarly, Bob chooses a private random number b and computes $k_{pubB} = \alpha^b$ mod p which he transmits to Alice. Once again, it is extremely difficult to determine b. After obtaining the public keys of each other, Alice and Bob raise these public keys to the exponent corresponding to their private numbers respectively. That is, Alice will compute:

$$ks = k_{pubB}{}^a \bmod p = \alpha^{ab} \bmod p$$

Bob computes:

$$ks = k_{pubA}{}^b \bmod p = \alpha^{ab} \bmod p$$

This way both Alice and Bob have generated a common session key. An adversary Oscar cannot determine this key without solving the DL problem. At least, there is no known solution for obtaining the session key other than by solving the DL problem.

RSA has been the most popular public-key algorithm. It employs integer factorization. The Diffie–Hellman key exchange protocol based on discrete logarithms is also very popular in wireless networks. This protocol is described in Example A7.5 and is commonly employed for key exchange for web transactions, e-commerce, and IP security. The Digital Signature Standard (DSS) is also based on discrete logarithms. Signature schemes based on RSA are also widely employed.

However, in the case of public-key algorithms, Oscar, the opponent, is aware of Bob's public key and this adds an additional parameter to the problem. Since public-key algorithms are based on mathematical structures, for small key sizes, there are well-known results or tables that can be employed to break the encryption. As such, the key sizes are extremely large compared to secret-key algorithms. Today, for good security, public-key algorithms need keys that are 3–15 times larger than their secret-key counterparts. Because the mathematical bases on which public-key algorithms work are well known, they are susceptible to analytical attacks and require much larger key sizes compared to secret key algorithms. The mathematics of elliptic curves is also being employed in encryption schemes since they need smaller key lengths compared to RSA.

Table A7.1 presents the key lengths and the time required to break some of the well-known public and secret-key algorithms [Sil00]. The values in this table are based on the assumption that $10 million is available for computer hardware. The key sizes in each row are equivalent.

The mathematical operations for public-key algorithms are quite computationally intensive. Consequently, the encryption rates are quite small and public-key algorithms are rarely used

Table A7.1   Cost equivalent key lengths (in bits) of various encryption schemes

Secret-key algorithm	Elliptic curve	RSA	Time to break	Memory
56	112	430	<5 min	Trivial
80	160	760	600 months	4 Gb
96	192	1020	3×10^6 years	170 Gb
128	256	1620	10^{16} years	120 Tb

for bulk data transfer. Instead, they are employed to exchange a *session key* between a pair of communicating entities who will then use the session key with secret-key algorithms for the duration of that communication (bulk data encryption). This ensures that a new session key is employed each time a communication is initiated thereby reducing the possibility of an adversary breaking the encryption scheme.

Although the public key of an honest party like Alice can be made public, its authenticity needs to be verified since Oscar can claim to be Alice and publish his key as hers. It is common to use *digital certificates* signed (see digital signatures below) by one of a few trusted certification authorities to verify the authenticity of the public key. This approach is used in modern web browsers for e-commerce applications.

We include hash functions in the classification in Figure A7.1. Hash functions are not strictly encryption schemes. They map any sized data to a fixed size digest. Given the digest, it is considered infeasible to obtain any data that maps to the digest if the size of the digest is at least 160 bits. Popular hash functions in use today are MD-5 and SHA.

Providing Security Services Using Encryption

Encryption schemes and hash functions are widely employed in password protection schemes and access control lists that are used for access control – the ability to allow or deny people access to certain resources based on their identification. Identification or entity authentication by itself is an important security service that needs to be provided for a variety of applications. Access to an automatic teller machine (ATM), logging on to a computer, identifying the user of a cellular telephone to the network, and so on. involve identification schemes. Note that there is a difference between identification and message authentication. When we talk about message authentication (discussed next), there is usually some information-containing message that is exchanged between the parties and one or both parties need to be authenticated. Identification schemes (sometimes referred to as *entity authentication*) involve real time verification of a party's identity and *need not* involve exchanging information-bearing messages.

Weak identification schemes are based on passwords or pin numbers that are time invariant. Usually the password or pin value is compared with a securely stored hash value. Such schemes are easily susceptible to replay attacks especially if the password or pin is transmitted over the air in an insecure manner. *Challenge-response* identification or *strong identification* schemes are usually employed in wireless networks. Here, Alice proves her identity to Bob by demonstrating knowledge of a secret, rather than presenting the secret itself. For this purpose, a quantity called the "nonce" is used. A nonce is a value employed no more than once for the same purpose and eliminates replay attacks. Random numbers, time stamps, sequence numbers etc. are used as nonce in practice. One example of a challenge-response protocol is as follows:

1. Alice is registered with Bob via a password and user name.
2. Bob sends Alice a random number (challenge).

3. Alice replies with an encrypted value of the random number where the encryption is done by using her password as the key (response).
4. Bob verifies that Alice indeed possesses the key (the password).

An eavesdropper Oscar cannot replay the response because the challenge is different if he tries to contact Bob. Oscar also cannot determine the password because the encryption scheme is sufficiently strong and the password is *never* revealed.

Message authentication is a security service that provides two functions: sender authentication and message integrity. By sender authentication, what we mean is that the receiver can be assured that the message has been originated from the person who claims to have sent the message. Message integrity assures a receiver that no one has modified the message in transit. Both these functions can be accomplished by adding a *keyed message digest* (MD), *message authentication code* (MAC), or *message integrity checks* (MICs) to a message. The MAC here should not be confused with the medium access control layer discussed in Chapter 4. Block ciphers and hash functions can be used to create MACs. These are checksums on data created using block ciphers or hash functions with a shared secret key between the communicating parties. MACs or MICs provide message authentication and integrity. If Oscar were to fabricate a message or modify a legitimate message, the checksum would always fail alerting the receiver of a problem with the received data. The Cipher Block Chaining MAC (CBC-MAC) that uses block ciphers and HMAC that employs hash functions are popular standard implementations of MACs.

The way a MAC is used to provide message authentication is as follows. It creates a fixed length sequence of bits that depend on the message itself and a secret key shared between the communicating parties. Irrespective of whether the message is a few kilobytes long or hundreds of megabytes, the MAC creates a sequence of bits of fixed length that directly depends on the message and the key. This sequence of bits is appended to the message and then the result is transmitted over the insecure channel. Note that the message could be sent in plaintext form if confidentiality is not an issue. It is computationally infeasible to create a replica of the MAC without the message and key. If the message is modified in transit, the receiver can discover this fact by creating a MAC from the received message and comparing it with the transmitted MAC. Since the secret key is shared only between the communicating parties, it also assures the receiver of the origin of the sender.

A keyed message digest operates in a slightly different manner. The message digest depends only on the message and not the key. Hash functions are used to create message digests. The message is appended with the MD and the result is encrypted using a session key shared between the communicating parties. This way, both the message and the MD, which verifies it, are kept secure. The MD has to be sufficiently long to prevent what is known as the "birthday attack". Given a message digest of length b bits, with a good probability, a fake message with the same MD can be generated in $2^{b/2}$ trials. This result is due to the fact that good probabilities of finding two people with the same date of birth exist in a group with roughly the square root of the number of days in a year. That is, in a group of 20 people, it is quite likely that any two would have the same birthday.

Figure A7.5 shows a schematic of message authentication with hash functions. On the left hand side, Alice concatenates the message x and its hash value $h(x)$ together before encryption the result with the secret key k. The ciphertext $y = e_k[x\|h(x)]$ is transmitted over an insecure channel. Bob decrypts the ciphertext y and expects to find a message and its hash value concatenated together. He separates the message x from the hash value, computes a new hash value and compares the two together. If the ciphertext is modified or replaced in between, Bob is able discover this fact easily. No one can impersonate Alice since it is computationally impossible to create a ciphertext that decrypts into a message and its hash value without knowledge of the key k. Thus both sender authentication and message integrity is assured. The interested reader is referred to [Sta98] for

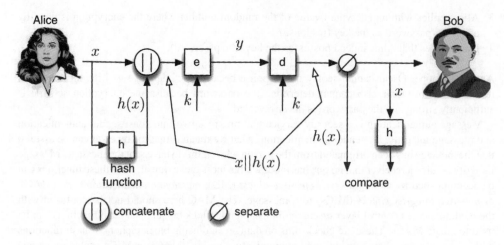

Figure A7.5   Message authentication with hash functions.

other schemes for message authentication. Using the hash function is generally preferred because of its speed.

Digital signatures are like physical signature. They attest some information and are bound to that information. Typically this involves encrypting the hash value of some information with the private key of a public-key/private-key pair. Suppose Alice generated some data and created a digital signature of the data. Anyone can verify the signature because decrypting the signature requires the public key, which is available to everyone. No one except Alice can generate the signature because she is the only one in possession of the private key. Recall that knowledge of the public key does not help Oscar or others deduce the private key.

Example A7.6: Non-repudiation and digital signatures

We considered sender authentication and message integrity in this chapter. This does not however assure non-repudiation. For instance, let us suppose that Alice is a consumer and Bob an e-commerce service provider. Bob claims that Alice placed an order with him for purchasing books worth $350 and Alice denies the transaction. Alice claims that she had requested books worth only $100. Both of them are able to produce ciphertexts and messages purportedly used in the transaction. Since both parties know the shared session key, it is impossible to verify who is being truthful and who is not. Public-key algorithms and digital signatures can be employed to resolve such situations.

We know that *only* the owner of the key knows the private key part of a public key algorithm. Consequently, this information can be used to bind the owner to a message transmitted by him. Popular public key algorithms operate such that it is possible to encrypt a message using a private-key as well. We can compare this to the following scenario. Only the owner of a mailbox can slip a message through the slot because only he has access to the private key that opens the mailbox. No one other than Alice can encrypt the message using her private key (or produce a meaningful ciphertext that can be decrypted with her public key). The problem with this encryption is that *anyone* will be able to decrypt the message because the public key dual is available to everyone. But this is exactly the concept of a signature. If Alice were to sign a document, this means that

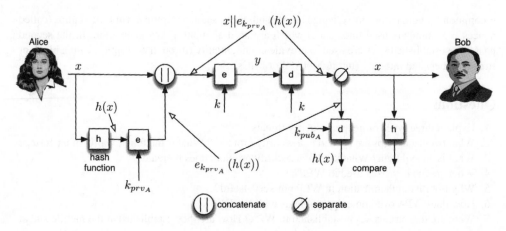

Figure A7.6   Digital signatures.

anyone should be able to verify her signature. However no one should be able to forge her signature. This is indeed the case here when a message is encrypted using the private key.

Digital signatures take the concept a step further. The entire document *need* not be encrypted. As already discussed, this process would be extremely slow. Instead a message digest of the message is "signed" or encrypted with the private key. The encrypted "signature" is appended to the message. Once again, since it is computationally impossible to derive a message from the hash, the signature and the message are bound together. If the document needs to be confidential, the usual encryption procedures can be employed after the signature is applied. Figure A7.6 shows a schematic of how digital signatures are applied. Here, k_{AB} is a session key that is used to keep the document confidential.

Cryptographic Protocols: The cryptographic primitives discussed above are used in cryptographic protocols, which are designed with specific security objectives in mind. Cryptographic protocols are famously known to be hard to design since they will likely have pitfalls that are hard to detect [Kau02]. A good example of a cryptographic protocol that fails to meet most of its security objectives is the *Wired Equivalent Privacy* (WEP) protocol used in legacy IEEE 802.11 wireless local area networks [Edn04]. Moreover, cryptographic primitives make use of keys shared between communicating parties. Establishing secret keys between legitimate parties interested in communicating, such that Oscar does not obtain any knowledge of the keys is not trivial and itself requires cryptographic protocols. Key establishment is usually based upon master keys established with trusted third parties or public key cryptography.

Most well designed cryptographic protocols have two phases (see Figure 7.2). In the first phase, the communicating entities *identify* or *authenticate* themselves to one another. In some cases the entity authentication is unilateral (i.e., Alice authenticates herself to Bob, but not vice versa). Entity authentication makes use of passwords, pins, pass phrases, biometrics, security tokens, and the like. Challenge-response protocols that do not require an entity to reveal the password, but only demonstrate knowledge of the password are commonly used for entity authentication. Also as part of the first phase, the communicating entities also establish keys for security services to be provided next. Establishment of keys can be in two ways – key transport or distribution where one party generates the keys (or a master key) and transports it securely to the other party or key agreement where both parties exchange information used in the secure creation of the same key at both ends. It is common for both parties to exchange random numbers, sequence numbers or time stamps (called nonces – or numbers used once once) which are used as input in key generation. In the second phase, the established keys are used to provide confidentiality (through encryption with a block or stream cipher) and integrity (through MACs or MICs).

Questions

1. Explain three types of general security threats.
2. What two usual steps are needed in a security protocol? What is the functionality of each step?
3. What is an evil twin? What kind of a security problem does it create?
4. What are the key sizes used in WEP?
5. Why is entity authentication in WEP not very useful?
6. How does WPA overcome the problems with WEP?
7. What are the various keys established in WPA? How are they established at the mobile and at the access point?
8. Describe the AES-CCM protocol.
9. What is pairing in the case of Bluetooth?
10. Name the threats in sensor networks. Explain the threat to data aggregation.
11. What are the privacy and authentication requirements of cellular networks?
12. How are public-key and secret key algorithms different?
13. Explain the importance of key sizes in the security of an encryption algorithm.
14. What is a challenge–response scheme? How does it work in North American 2G TDMA systems?
15. Explain why the TMSI cannot be used to track a mobile even though it is transmitted in plaintext form by a mobile when it registers with the network.
16. In what ways are the confidentiality and message integrity checks of packets in WLANs and frames in UMTS similar? How are they different?

Problems

Problem 7.1

A not so rich hacker uses an old computer and brute force to break into some wireless systems. It takes him 1 ms on average to test a key to see if it is the right one for an encryption independent of the algorithm employed. How long will it take him to break into an IEEE 802.11 system in the worst case? How long will it take him to break into a North American TDMA cellular system on average?

Problem 7.2

In Problem 7.1, the hacker realizes that the last six bits of the keys used in a private 802.11 LAN are always zeros. In what time can he break into the system in the worst case?

Problem 7.3

In Problem 7.1, the hacker manages to:

a. buy a second old computer that can test a key in 1.5 ms. With the two computers, in what time can he break into the system in the worst case?

b. upgrade his computer so that he can test a key in 1 ms. What time can he break into the system in the worst case?

Projects

Project 7.1

Use your laptop for this project. Use the free tool Netstumbler (http://www.netstumbler.com/) or Inssider (http://www.metageek.net/products/inssider) on Windows or the equivalent istumbler (http://www.istumbler.net/) on the MAC to map all WiFi access points near your house. Enumerate the ESSID, the MAC address, and channel numbers that the access points are using. How many networks can you detect from inside your house? Does the number of networks change if you try scanning for them from outside your house? How many interfering networks are near your house (i.e., those networks using the same channel number)?

Project 7.2

Note that the networks detected in Project 7.1 are only those networks that choose to broadcast their SSIDs. There are WiFi networks that do not broadcast their IDs that may also be in the area. Use the Kismet tool (http://www.kismetwireless.net/) to see if there are any hidden networks in your area.

Project 7.3

Use your laptop and smartphone for this project. Walk around your home or in the campus of the university and enumerate the various SSIDs of wireless LANs that your mobile station can see. Try to collect as many different SSIDs as possible. Determine what fraction of these WiFi networks is protected. If possible, also determine what fraction is using WEP and what fraction is using WPA. Do you see any differences between the number and types of WiFi networks seen by the laptop and the smartphone?

Part III

Wireless Local Access

8

Wireless LANs

8.1 Introduction

In the previous chapters of this book we provided an overview of the evolution of the wireless information networking industry and then explored the principles of air–interface design and networking operations. In Parts III and IV of this book we discuss examples of successful systems to provide the reader with a deeper understanding of the details of how a variety of wireless networks have evolved. This description is divided into two groups, systems for wide area networks and for local and personal area networks. In this part, our focus is on wireless local networks and we start in this chapter with an examination of IEEE 802.11. IEEE 802.3 or Ethernet is the primary wired technology that is used as the access network to the Internet. Ethernet also started as a local area networking technology that connected hosts that belonged to the same organization. In a similar manner, IEEE 802.11 or WiFi is the primary wireless local area networking (WLAN) technology, sometimes called the *wireless Ethernet*. The goal of this chapter is to provide an overview of this technology. WiFi is now a mature technology with widespread deployment in organizations, campuses, hotspots (in coffee shops, airports, and hotels) and residences. In this section, we provide a history of wireless local area networking. The next section discusses the IEEE 802.11 standard from a top-down view.

During the past two decades, as the vision of the WLAN industry evolved, WLANs were implemented based on a variety of innovative technologies and raised a lot of hopes for development of a sizable market several times. Today, the major differentiation of WLANs from wide area cellular services is the method of delivery of data to users, data rate limitations, and frequency band regulation. Cellular data services are delivered by operating companies as services while the WLAN users belong to the organization that owns the network. At a time when the 3G cellular industries are striving for tens of Mbps packet data services, WLAN standards are working on more than 1000 Mbps services. Another differentiation with other radio networks is that today almost all WLANs operate in the unlicensed bands where frequency regulations are loose and there is no charge or waiting time to obtain the band.

Further, as explained in Chapter 6, architectural and operational characteristics in WLANs are much simpler than the respective counterparts in cellular networks that are discussed in the next part of this book. For instance, there are no separate channels such as pilot, synch, random-access and traffic channels that are common in cellular networks. Instead there are MAC layer frames that carry control or management messages, or have payloads comprised of IP and other higher layer

---

Principles of Wireless Access and Localization, First Edition. Kaveh Pahlavan and Prashant Krishnamurthy.
© 2013 John Wiley & Sons, Ltd. Published 2013 by John Wiley & Sons, Ltd.

packets. Because of the high speed of the medium, lower propagation delay and fixed overhead per packet, longer packets with variable length are used as opposed to short packets with fixed length used in cellular networks. The medium access mechanism is distributed and contention-based that is better suited for low-coverage and lower traffic load environments as opposed to the cellular networks with limited bandwidth, large propagation delay and large numbers of users, where a centrally controlled medium access mechanism is preferable. The emphasis, from the beginning for WLANs, has been on packet-data applications that need the network to support higher data rates. This makes the design of wideband modems the center piece of the evolution of this technology. Finally, the variable packet sizes are carried on the backbone by routers rather than circuit or packet-switches designed for constant rates or fast fixed packet lengths.

To obtain a deeper understanding of all these issues it is very educational to go over the history of the WLAN industry to see how all of these unique issues evolved which we undertake next. The interested reader is referred to [Per08] for an in depth treatment of the latest WiFi standard, namely IEEE 802.11n.

8.1.1   Early Experiences

Gfeller at the IBM Rüschlikon Laboratories in Switzerland first introduced the idea of a wireless LAN in the late 1970s [Gfe80]. The number of terminals in manufacturing floors was growing and wiring them in that environment was difficult. In offices, wires are normally snaked under the suspended ceilings and through the interior partitions and walls, but these options are not available in manufacturing floors. In office environments, in extreme cases, it is possible to install wiring under the floor, using conduits, or even simply left over the floor with some cover. In manufacturing floors, the environment is rugged, under floor wiring is more expensive and simply leaving wires on the floor is can be dangerous because heavy machinery may roll over them. The diffused IR technology was selected at IBM labs for the implementation of a wireless LAN to avoid interference with the electromagnetic signals radiating from machinery and to avoid dealing with long lasting administrative procedures with frequency administration agencies. The principal researcher of this project abandoned the project because the goal of 1 Mbps with reasonable coverage did not materialize.

Ferrert at HP's Pal Alto Research Laboratories in California performed the second project on wireless LANs around the same time [Fer80]. In this project a 100 kbps direct sequence spread spectrum (DS-SS) WLAN operating around 900 MHz was developed for office areas that used CSMA as the method of access. This project was conducted under an experimental license agreement from the FCC. The principal of this project failed to obtain the necessary frequency bands from the FCC and discouraged with the administrative complexity, he also abandoned his project. A couple of years later Codex, Motorola attempted to implement a WLAN at 1.73 GHz and that project was also abandoned after negotiations with the FCC.

Although all the pioneering WLAN projects were abandoned, WLANs continued to attract attention and negotiations continued with the FCC to secure frequency bands for this purpose [Pah85]. These projects revealed several important challenges facing the WLAN industry that remain to this day:

1. Complexity and cost: The alternatives for implementing WLANs such as IR, spread spectrum or traditional radios are far more complex and diversified than the wired LANs.
2. Bandwidth: Data rate limitations of the wireless medium are more serious than those of wired media.

3. Coverage: The coverage of a WLAN operating within a building is less than that of a single cable (bus or ring) or even TP-based LANs.
4. Interference: WLANs are subject to interference from other overlaid WLANs or other users operating in the same frequency bands.
5. Frequency administration: Radio based WLANs are subject to expensive and untimely frequency regulations.

8.1.2   Emergence of Unlicensed Bands

Wireless LANs need a bandwidth of at least several tens of MHz while they have not yet shown a market compatible in strength with the cellular voice industry that originally started with two pieces of 25 MHz bands that produced a huge market. Comparable sizes of bands for PCS applications were auctioned in the United States for tens of billions of dollars while the market for WLANs has not yet passed a billion dollars per year. The dilemma for the frequency administration agencies was to justify a frequency allocation for a product with a weak market.

In the mid 1980s, the FCC found two solutions for this problem. The first and the simplest solution was to avoid the 1–2 GHz bands used for the cellular telephone and PCS applications and approve higher frequencies at several tens of GHz where plenty of unused bands were available. This solution was first negotiated between Motorola and the FCC and resulted in Motorola's Altair, the first wireless LAN product operating in licensed 18–19 GHz bands. Motorola had actually established a headquarters to facilitate user negotiation with the FCC for the usage of WLANs in different areas. A user who changed the location of operation of his/her WLAN substantially (from a town to another) contacted Motorola and they would manage the necessary frequency administration issues with the FCC.

The second and more innovative approach was resorting to unlicensed frequency bands as the solution. In response to the applications for bands for WLAN projects mentioned in the previous section and motivated by studies for various implementations of wireless LANs [Pah85], Mike Marcus of the FCC initiated the release of the unlicensed ISM bands in the May of 1985 [Mar85]. The ISM bands were the first unlicensed bands for consumer product development and played a major role in the development of the WLAN industry. In simple words licensed and unlicensed bands can be compared to private backyards and public gardens. If one can afford it, he or she can own a private backyard (licensed band) and arrange a barbeque dinner (a wireless product). If one cannot afford to buy a house with backyard, he or she simply moves the barbeque party to the public park (unlicensed band) where he/she should observe certain rules or *etiquette* that allows others to share the public resource as well. The rules enforced on ISM bands restricted the transmit power to 1 W and enforced the modems radiating more than 1 mW to employ spread spectrum technology. It was believed that spread spectrum communications would restrict interference and allow co-existence of several wireless applications in the same band. Table 8.1 provides a summary of the important features of the ISM bands.

Table 8.1   Summary of the properties of the ISM bands

Frequencies of operation	902–928 MHz;
	2.4–2.4835 GHz;
	5.725–5.875 GHz
Transmit power limitation	1 W for DSSS and FHSS
Low power with any modulation	

8.1.3   Products, Bands, and Standards

Encouraged by the FCC ruling and some visionary publications in wireless office information networks summarizing previous works and addressing the future directions in this field, [Pah85, Pah88a; Kav87] a number of WLAN product development projects mushroomed almost exclusively over the North American continent. By the late 1980s the first generation of WLAN products using three different technologies, licensed bands at 18–19 GHz, spread spectrum in the ISM bands around 900 MHz, and IR appeared in the market. At around the same time a standardization activity for WLANs under IEEE 802.4L was initiated that was soon converted into an independent unit – IEEE 802.11 that was finalized in 1997! The first generation products consisted of shoe box sized access points and receiver boxes or PC installed cards that could connect workstations to LANs wherever wiring difficulties for the LANs justified using a more expensive WLAN connection. Today, we call this application LAN-extension [Pah94, Pah05]. Market predictions at that time were estimating a shift of around 15% of the LAN market to WLANs that would generate a few billion dollars of sale per year by first few years in the 1990's. In May of 1991, to create a scientific forum for the exchange of knowledge on WLANs, the first IEEE sponsored WLAN workshop was organized concurrent to the 802.11 meeting, in Worcester, Massachusetts [Wor91].

In 1992, as a follow up to the initial momentum for WLAN developments, lead by Apple, an industrial alliance called WINForum was formed aiming at obtaining more unlicensed bands from FCC for the so called Data-PCS activities. WINForum finally succeeded in securing 20 MHz of bandwidth in the PCS bands that was divided into two 10 MHz bands – one for Isochronous (voice like) and one for Asynchronous (data type) applications. The original aim of WINForum was to secure 40 MHz for asynchronous applications. WINForum also defined a set of rules or etiquettes for these bands that would allow the co-existence. Figure 8.1 shows the Unlicensed-PCS bands and the spectrum etiquette associated with them. The WINForum etiquette is based on CSMA rather that CDMA and spread spectrum communications used in ISM bands. This was a better choice because implementation of CDMA needed power control and larger bandwidth that was not feasible in an un-coordinated, multi-user, multi-vendor WLANs and spread spectrum without CDMA offers a less bandwidth efficient solution.

Another standardization activity started in 1992 was the HIgh PERformance LAN (HIPERLAN). This ETSI based standard aimed at high performance LANs with data rates of up to 23 Mbps that was an order of magnitude higher than the original 802.11 data rates of 2 Mbps. To support these data rates, the HIPERLAN community was able to secure two 200 MHz bands: 5.15–5.35 GHz and

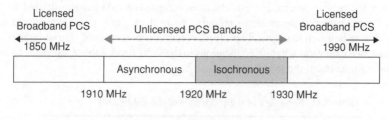

Three basic rules
1. Listen before talk (or transmit) LBT Protocol
2. Low transmitter power
3. Restricted duration of transmissions

Figure 8.1   Unlicensed PCS bands and their spectrum etiquette.

Table 8.2 Properties of the U-NII bands

Band of operation (GHz)	Maximum Tx power (mW)	Max. power with antenna gain of 6 dBi (mW)	Maximum PSD (mW/MHz)	Applications: suggested and/ or mandated	Other remarks
5.15–5.25	50	200	2.5	Restricted to indoor applications	Antenna must be an integral part of the device
5.25–5.35	250	1000	12.5	Campus LANs	Compatible with HIPERLAN
5.725–5.825	1000	4000	50	Community networks	Longer range in low-interference (rural) environs

17.1–17.3 GHz for WLAN operation. This encouraged the FCC to release the so-called Unlicensed National Information Infrastructure (U-NII) bands in 1997 when the original HIPERLAN standard (later called HIPERLAN-1) was completed. Table 8.2 summarizes the U-NII bands and their restrictions. The WINForum etiquette was evaluated for the U-NII bands but it was found not be suitable because research activities around that time favored wireless ATM that could not operate on a listen-before-talk etiquette. At the time of this writing, U-NII bands are extended and used by IEEE 802.11a/h/j/n and HIPERLAN-2 project is discontinued. However, the most popular IEEE 802.11g supporting 54 Mbps OFDM-based WLANs at 2.4 GHz was first adopted jointly by 802.11.a and HIPERLAN-2 for operation at 5 GHz. In the early 2000s, the Federal Communications Commission allocated several GHz of bandwidth around 60 GHz for wireless applications and the IEEE 802.11 and 802.15 working groups are considering extremely high data rate WLAN and WPAN related standards in these frequency bands especially in light of the plentiful bandwidth available. We discuss WLAN standards in more detail in Section 8.2.1.

8.1.4   Shift in Marketing Strategy

In the first half of the 1990s, WLAN products were expecting a sizable market of around a few billions of dollars per year for shoe box sized products used for LAN-extension in indoor areas and this did not materialize. Under this situation, two new directions for product development emerged. The first and the most simple approach was to take the existing shoe-box type WLANs, boost up their transmitted power to the maximum allowed under regulations, and equip them with directional antennas for outdoor inter-building LAN interconnects. These technically simple solutions would allow coverage of up to a few tens of kilometers with suitable rooftop antennas. The new inter-LAN wireless bridges could connect corporate LANs that were within range. The cost of the inter-LAN wireless solution was much cheaper than the wired alternative, T1-carrier lines, leased from the PSTN service providers. The second alternative was to reduce the size of the design to a PCMCIA WLAN card to be used with laptops that were enjoying a sizable growth and demanded mobility for LAN connectivity. However, this approach was not available for all existing products and it was more suitable for the spread spectrum products operating in lower frequencies. Figure 8.2 illustrates these three applications for the WLANs. Recently, there are new low cost products for

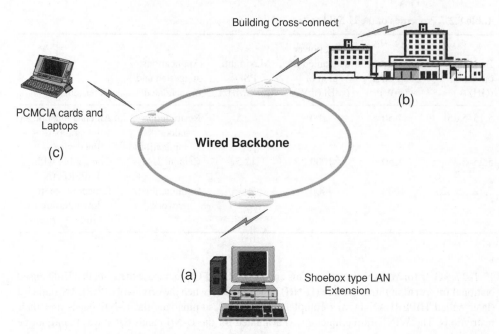

Building Cross-connect

(b)

PCMCIA cards and
Laptops

(c)

Wired Backbone

(a)

Shoebox type LAN
Extension

Figure 8.2   Different forms of WLAN products. (a) LAN extension. (b) Inter-LAN bridge. (c) PCMCIA cards for laptops.

LAN extension that can convert a serial port or Ethernet connector to a WLAN interface for desktop PCs and workstations that operate at 11 Mbps.

The original marketing strategy for a LAN-extension application was indeed a horizontal one aiming at selling individual WLAN components directly to the customers. Another major shift in the marketing strategy of a few successful companies in the mid 1990s was the move toward vertical markets where a wireless network was sold as a complete solution to an application. The major vertical markets approached by the WLAN industry were *"barcode" Industries* providing wireless inventory check and tracking in warehouses and manufacturing floors, *financial services* providing for wireless financial updates in large stock-exchanges, *healthcare* networks providing wireless mobile services inside the hospitals and *wireless campus area networks (WCANs)* providing for wireless classrooms and offices. All these efforts boosted the market for the WLAN to above half a billion dollars per year over the last few years of the 1990s.

Example 8.1: A WCAN in WPI

Figure 8.3 illustrates the schematic of an experimental NSF sponsored WCAN that was design as a testbed for performance monitoring of WLAN products at CWINS, WPI in 1996. The testbed connects five buildings with inter-LAN bridges using different technologies. Inside each building access points provide coverage to the laptops that are carried by the students. The professor broadcasts his image and writing on the electronic board to allow students to participate in the wireless classroom from different buildings in the campus. The entire wireless network is connected to the backbone through a router to isolate the traffic for traffic monitoring experimentations.

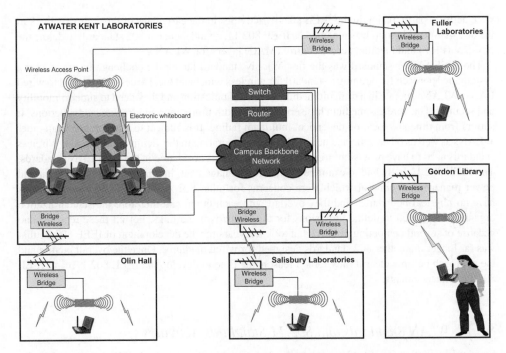

Figure 8.3 The experimental NSF sponsored Wireless Campus Area Network (W-CAN) at WPI.

Today, the horizontal market for the WLAN industry is mainly focused on WLANs as an alternative to wiring additional LAN segments wherever the cost of the WLAN is justifiable. One example of this situation is installation with frequent relocations where the additional cost of the WLAN solution is justified by the relocation costs of the wired solution. Temporary networking situations such as registration sites in conferences or fairs (jobs, food, etc.) is another example where a wireless solution is preferred to the expensive but more reliable wired alternative. Buildings with difficult or impossible-to-wire situations, such as marble buildings or historical monuments where drilling for wiring is not favored, provides another example of situations where WLANs are justified. The most prominent incentive for WLANs is the general use in the laptops in the home and offices.

8.2 Wireless Local Area Networks and Standards

The IEEE 802.11 is the first WLAN standard and so far, the only one that has secured a market. Within the IEEE, there are several standards activities carried on by different groups. The IEEE 802 LAN/MAN Standards Committee is responsible for Local Area Network (LAN) standards and Metropolitan Area Network (MAN) standards. Individual working groups are in charge of a variety of LAN/MAN standards of which the 802.11 working group is responsible for *wireless local area networking* (WLAN) *standards*. The IEEE 802.11 standardization activity originally started in 1987 as a part of the IEEE 802.4 Token Bus standard under the group number IEEE 802.4L. The IEEE 802.4 is a counterpart of the IEEE 802.3 and 802.5, which pays special attention in supporting factory environments. One of the early motives for using WLANs was in factories for control of and communication between equipment. For this reason, car manufacturers such as GM were actively participating in the IEEE 802.4L activities in the early days of this industry. In 1990 the 802.4L WLAN group was renamed as IEEE 802.11, an independent 802 standard, to define the physical (PHY) and medium access control (MAC) layers for WLANs.

The IEEE 802.11 standard was the first WLAN standard facing the challenge of organizing a systematic approach for defining a standard for wireless wideband local access. Compared to wired LANs, WLANs operate in a difficult medium for communication and they need to support mobility and security. The wireless medium has serious bandwidth limitations and frequency regulations. It suffers from time and location dependent multipath fading. It is subject to interference from other WLANs as well as other radio and non-radio devices operating in the vicinity of a WLAN. Wireless standards need to have provisions to *support mobility* that is not shared in the other LAN standards. The IEEE 802.11 body had to examine connection management, link reliability management and power management none of which were concerns for other 802 standards. In addition, WLANs have no physical boundaries and they overlap with each other, and therefore the standardization organization needed to define provisions for the *security* of the links. For all these reasons, and because of several competing proposals, it took ten years for the development of IEEE 802.11 that was far longer than other 802 standards designed for wired mediums. Once the overall picture and the approach became clear it only took a reasonable time to develop the IEEE 802.11b and IEEE 802.11a enhancements.

8.2.1   WLAN Standards and 802.11 Standards Activities

Standards activities for WLANs have evolved geographically with the ETSI's Broadband Radio Access Network (BRAN) working on High Performance Radio LAN or HIPERLAN standards in Europe and the IEEE on the 802.11 series in the US. In Japan, the Multimedia Mobile Access Communications Promotion Council (MMAC-PC) working group under the Association of Radio Industries and Businesses (ARIB) works on WLAN standards. The spectrum used by WLANs is mostly unlicensed spectrum although some systems use licensed spectrum as well (e.g., HIPER-LAN in some licensed bands). The only commercially successful WLAN standard has been the IEEE 802.11. HIPERLAN/1 was standardized in the mid-1990s and supported complex multi-hop ad hoc networking. The medium access mechanism in HIPERLAN/1 [Wil95b] was based on a form of carrier sense multiple access called Elimination-Yield Non-Preemptive Multiple Access (EY-NPMA). HIPERLAN/2 has adopted a physical layer that is very similar to IEEE 802.11a and a medium access mechanism based on reservation and TDMA. However both these standards have not been adopted successfully in commercial products. The MMAC activities include a variety of wireless access networks (some compatible with the IEEE 802.11 standards) ranging from wireless personal area networks to outdoor fixed public networks. Table 8.3 summarizes some of the 802.11, HIPERLAN and MMAC standards.

As mentioned earlier, the competing standard for wireless LANs in Europe is the HIPERLAN/2 standard that is specified for the U-NII bands. HIPERLAN/2 uses the same physical layer as IEEE 802.11a although it accommodates a few different data rates. In this standard, there are mechanisms suggested for power measurement and control and radio resources management. Previously there existed a task group H that was enhancing the current 802.11 MAC and 802.11a PHY with network management and control extensions for spectrum and transmit power management in the U-NII bands with the possibility of dynamic channel selection capabilities. In this case, APs would be able to dynamically select channels based on information they can obtain about neighboring APs that may transmitting on the same channel. This way, a laborious network planning process could be simplified. The 802.11h standard completed in October 2003 considers such transmit

Table 8.3 Summary of some WLAN standards

Standard	Standard body	Spectrum	Data rate	Primary medium access	Primary region
IEEE 802.11,b, g	IEEE	2.4 GHz ISM bands	1, 2, 5.5, 11, up to Mbps	CSMA/CA[1]	North America
IEEE 802.11a	IEEE	5 GHz U-NII and ISM bands	Up to 54 Mbps	CSMA/CA	North America
IEEE 802.11n	IEEE	2.4 GHz ISM bands	>100 Mbps	CSMA/CA	North America
HIPERLAN/1	ETSI	5 GHz bands	23 Mbps	EY/NPMA[2]	Europe
HIPERLAN/2	ETSI	5 GHz bands	Up to 54 Mbps	TDMA reservation	Europe
Wireless Access and WLAN	MMAC	3-60 GHz bands	20-25 Mbps	Various	Japan

[1]CSMA/CA – Carrier sense multiple access with collision avoidance
[2]EY-NPMA – Elimination yield non-preemptive multiple access

power control (TPC) and dynamic frequency selection (DFS) to satisfy the regulatory aspects in Europe.

Outside of the standards that have been specified already in this chapter, there are several ongoing activities in the IEEE 802.11 working group. There are several task groups that are engaged in enhancing aspects of the IEEE 802.11 standard. Some of these are as follows. *Task Group J* is considering enhancements to the current standard to provide operations in the frequency band between 4.9 GHz and 5 GHz for use in Japan. The reason for this task group is to make changes in the 802.11 standard to accommodate the regulatory demands in this spectrum that exist in Japan. There will be expected changes to both the MAC and PHY layers to meet these regulations. *Task Group K* is looking at enhanced radio resource management outside the purview of 802.11h. The primary goal of this group is to provide mechanisms to higher layers that enable radio and network measurements as necessary. Once power measurements and reporting are possible in a standardized manner, they can be exploited to make better use of the spectrum, reduce interference and so on. Future road transportation is expected to evolve into an *intelligent transportation system* or (ITS). The goal of *Task Group P* (called Wireless Access for the Vehicular Environment – WAVE) is to define enhancements to 802.11 that may be required to support ITS applications. Such definitions will include data exchange between high-speed vehicles and between these vehicles and the roadside infrastructure in the licensed ITS band (5.9 GHz). A newly formed Task Group *Y* is looking at operation in the 3.5 GHz bands that were recently opened up by the FCC for WLAN operation. One of the objectives of this protocol is to developed a fair contention protocol for access to the medium.

While not directly changing or adding to the standards, there are task groups that are involved in maintenance and other issues related to 802.11. Task group M is performing the maintenance of the 802.11 standard. A task group D is looking at regulatory domain updates. There are also some new proposed activities that are pending approval at various levels as of the time of the writing – an

802.11t that aims to recommend practices for wireless performance prediction, an 802.11u that is considering interworking with external networks, an 802.11v for network management of mobile stations and an 802.11w that provides protection (data integrity and authentication) of management frames. In addition, there are several study groups looking at harmonizing the 802.11 and the European ETSI standards and some investigating the possibility of improvements to the 802.11 standard to provide higher throughput.

8.2.2   Ethernet and IEEE 802.11

The wired local area networking (LAN) technology that evolved among many other options is the constantly evolving IEEE 802.3 standard, the Ethernet. Originally, the legacy Ethernet was developed for operation in a LAN with a bus topology and coaxial cable. Today, the Ethernet protocol and medium access mechanisms run over a variety of wired physical media and the star topology has over taken the bus topology. The IEEE 802.11 or WiFi, was originally designed as a "wireless Ethernet" and its packet format and MAC build on those in IEEE 802.3, the wired Ethernet, while the physical layer has gone through complete re-design to fit the substantially more difficult wireless medium. Today, the Ethernet packet is the most popular packet format used for data communications at the lower layers with the TCP/IP for higher layer communications.

The medium access in Ethernet is based on carrier-sense multiple access with collision detection (CSMA/CD) which was briefly described in Chapter 4. If two hosts on the same LAN transmit packets such that they may be on the medium at the same time, a collision occurs. In the legacy Ethernet for wired medium, the network interface cards detected collisions by an increase in voltage of the baseband signal above the level of transmission expected from a single terminal transmission. Unlike the wired medium, where collisions can be simply detected by observing voltage inceases above a threshold, due to extensive power variations caused by fading and antenna feedback, it is extremely challenging to detect collisions on the wireless medium. As a result, in the IEEE 802.11 standard, rather than CSMA/CD, a modified version of this protocol, carrier sensing with collision avoidance or CSMA/CA is implemented. Overall, it is preferable from a performance and cost stand-point to use collision avoidance, rather than detect collisions on the air. This is one of the primary differences between the MAC of Ethernet and that of WiFi. In the IEEE 802.11 several approaches for avoiding collisions are used as we will see later in this chapter.

The packet format of the WiFi also builds on the Ethernet packet format. The Ethernet packet comprises of a header consisting of a preamble and a starting delimiter. The preamble is used to synchronize the receiver with the timing of the transmitted packet and the starting delimiter is a special sequence denoting the start of the actual information contents of the MAC packet. The information content in the MAC packet starts with destination address, followed by the source address, followed by a field for the actual payload and a CRC error detecting code to examine the sanity of the received packets and to determine if a retransmission is necessary. In IEEE 802.11, the header is separated from the MAC information and it is called the Physical Layer Convergence Protocol (PLCP). In addition to the preamble and the starting delimeter, the PLCP carries more information such as the data rates which can vary depending on the channel and protocol type that we will explain later. The MAC packet of IEEE 802.11 has four address fields because each terminal connects to an access point. So we need two MAC address points to address a wireless terminal. Additional fields to support control packets as well as facilitating other features needed for wireless environment will be discussed later.

The physical layer in WiFi is far more complex than the simple baseband signals employed in Ethernet. We discussed some of these physical layer transmission and coding schemes in Chapter 3. We consider the details of these physical layer schemes in this chapter.

8.2.3   Overview of IEEE 802.11

A voice oriented connection-based standard like many (cellular systems) begins with the first step of specifying the services to be provided. Then the reference system architecture and its interfaces are defined and at last the detailed layered interfaces are specified to accommodate all the services. The situation in connectionless data oriented networks, such as IEEE 802.11 is quite different. The IEEE 802.11 standard provides for a general PHY and MAC layer specification that can accommodate any connectionless applications whose transport and network layers accommodate the IEEE 802.11 MAC layer. Today, TCP/IP is the dominant transport/network layer protocol hosting all popular connectionless applications such as web access, e-mail, FTP, or telnet, and it works over all MAC layers of the LANs including IEEE 802.11. Therefore, the IEEE 802.11 standard does not need to specify the services. However, IEEE 802.11 provides for local and privately owned WLANs with a number of competing solutions. In this situation, the first step in the standardization is to group all the solutions into one set of requirements with a reasonable number of options. The next step, in a manner similar to connection-based standards, is to define a reference system model and its associated detailed interface specifications.

The number of participants in the IEEE 802.11 standards soon exceeded a hundred with a number of solutions. The finalized set of requirements, that did not came about easily, are:

- Single MAC to support multiple PHY layers
- Should allow multiple overlapping networks in the same area
- Handle the interference from other ISM band radios and microwave ovens
- Mechanism to handle "hidden terminals"
- Options for time bounded services
- Provisions on privacy and access control.

In addition it was decided that the standard would not be concerned with licensed band operations. These requirements set the overall direction of the standard in adopting different alternatives. However, as it often happens in these types of standards, the actual adoptions were based on successful products that were already available in the market.

The 802.11 standard, like most LAN standards is concerned only with the lower two layers of the OSI stack, namely the physical (PHY) and medium access control (MAC) layers. The MAC and PHY layers operate under the IEEE 802.2 logical link control (LLC) layer that supports many other LAN protocols. In the case of wired LAN standards such as 802.3, there are several physical layers that correspond to the same MAC specifications. A good example is IEEE 802.3, which was originally designed for thick coaxial cable, but was subsequently revised to include thin coaxial cable, a variety of twisted pair cables, and even fiber optic links. In the same way, the IEEE 802.11 standard specifies a common MAC protocol that is used over many different PHY standards. The PHY standards are the "base" IEEE 802.11 standard, the 802.11b and g standards and the 802.11a standard. A new 802.11n physical layer is under consideration by the 802.11 working group. The MAC protocol is based on carrier-sense multiple-access with collision avoidance (CSMA/CA). An optional polling mechanism called point coordination function (PCF) is also specified. In addition to the MAC and PHY layers, the IEEE 802.11 standard also specifies a management plane that

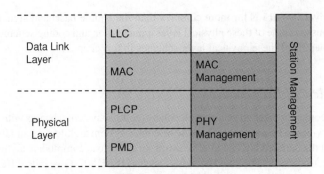

PLCP: Physical Layer Convergence Protocol
PMD: Physical Medium Dependent

Figure 8.4   Protocol stack of IEEE 802.11.

transmits management messages over the medium and can be used by an administrator to tune the MAC and PHY layers. The MAC Layer Management Entity (MLME) deals with management issues such as roaming and power conservation. The PHY Layer Management Entity (PLME) assists in channel selection and interacts with the MLME. A Station Management Entity (SME) handles the interaction between these management layers. Figure 8.4 shows the protocol stack associated with IEEE 802.11.

The base IEEE 802.11 standard specifies three different PHY layers – two using radio-frequency (RF) and one using infra-red (IR) communications. The RF PHY layers are based on spread spectrum (SS) – either direct sequence (DS) or frequency hopping (FH) while the IR PHY layer is based on pulse position modulation (PPM). Two different data rates are specified – 1 and 2 Mbps for each of the three PHY layers. The RF physical layers are specified in the 2.4 GHz industrial, scientific, and medical (ISM) unlicensed frequency bands.

Example 8.2: Origins of PHY layer solutions

The DSSS solution for IEEE 802.11 is based on WaveLAN that was designed at NCR, Netherlands [Tuc91]. The FHSS solution was highly affected by RangeLAN designed by Proxim, CA and products from Photonics, CA and Spectrix, IL, affected the DFIR standard.

The IEEE 802.11b standard specifies the physical layer at 2.4 GHz for higher data rates – 5.5 and 11 Mbps. The PHY layer makes use of a modulation scheme called complementary code keying (CCK). The transmission rate depends on the quality of the signal and it is backwards compatible with the DSSS based base-802.11 standard. Depending on the signal quality, the transmission rates could fall back to lower values. The 802.11g standard further increases the data rates to up to 54 Mbps in the 2.4 GHz ISM bands using orthogonal frequency division multiplexing (OFDM). The IEEE 802.11a standard [Kap02] deals with the PHY layer in the 5 GHz unlicensed national information infrastructure (U-NII) bands. Once again data rates up to 54 Mbps are specified in these bands with OFDM as the modulation technique. Depending on the PHY layer alternative, the frequency band is divided into several channels. Each channel supports the maximum data rate allowed by that PHY layer alternative. The proposal for a very high rate PHY layer (> 100 Mbps) called 802.11n employs multiple input and output antennas at the transceivers. The technology is popularly called multiple input multiple output (MIMO) and also uses OFDM as the modulation scheme.

In the next few sections, the IEEE 802.11 standard is discussed in a top-down manner. First, the different topologies possible in IEEE 802.11 are considered with the focus on understanding some of the management functions. Then detailed discussions of the MAC layer of 802.11 and different PHY layer alternatives are presented. Once the basic operation of the 802.11 WLAN has been considered, security issues in IEEE 802.11 will be discussed as also the recent ongoing activities to extend the standard further.

8.3 IEEE 802.11 WLAN Operations

In this section, we describe the operational aspects of IEEE 802.11, the MAC layer, the physical layer and briefly overview security aspects. Some of the architectural differences between WLANs and WWANs were discussed in Chapter 6, while more details and comparisons of the security aspects were examined in Chapter 7.

8.3.1 Topology and Architecture

The topology of an IEEE 802.11 WLAN can be one of two types – infrastructure or ad hoc (see Figure 8.5). In the infrastructure topology, an access point (AP) covers a particular area called

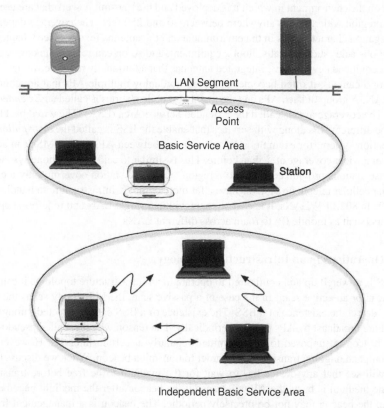

Figure 8.5 Topologies in IEEE 802.11.

the basic service area (BSA) and mobile stations (MSs) communicate with each other or with the Internet through the AP [Cro97]. The AP is connected to a LAN segment and forms the *point of access* to the network. All communications go through the AP. So an MS that wants to communicate with another MS first sends the message to the AP. The AP looks at the destination address and sends it to the second MS. The AP along with all the MSs associated with it is called a basic service set (BSS). In the ad hoc topology (also called independent BSS or IBSS), MSs that are in range of each other can communicate directly with one another without a wired infrastructure. However, it is not possible for a MS to forward packets meant for another MS not in the range of the source MS. Figure 8.5 shows schematics of both topologies. MSs and APs are identified by a 48-bit MAC address that is similar to other MAC addresses at the link layer. In an infrastructure topology, the MAC address of the AP also forms the BSSID, a unique identifier of the BSS.

If we assume that the range of communication of any WLAN device, be it a MS or an AP is a region of radius R, we can look at the comparative advantages of the two topologies. A MS can communicate with another MS that is up to $2R$ away using an AP provided both MSs are within a distance R of the AP. The cost here is the additional transmission from the AP to the destination. In the ad hoc topology, a destination MS cannot be more than a distance R from the source MS. The advantage is that the information can be received in one hop.

Extending the Coverage in Infrastructure Topology

Depending on the environment in which it is deployed and the transmit powers that are used, an AP can cover a region with a radius anywhere between 30 and 250 feet. The coverage depends upon radio propagation characteristics in the environment and the antenna features (see Chapter 2). The presence of obstacles such as walls, floors, equipment, and so on can reduce the coverage. Many new 802.11 equipped devices have integrated antennas that additionally reduce coverage. To cover a building or a campus, it often becomes necessary to deploy multiple APs that are connected to the same LAN. A group of such APs and the member mobile stations is called an *Extended Service Set* (ESS). The coverage area is called the Extended Service Area (ESA). The wired backbone that connects the different APs along with services that enable the ESS is called the *distribution system*. The distribution system, for example, supports roaming between APs so that MSs can access the network over a wider coverage area than before. This is similar to cellular telephone systems where multiple base stations provide coverage to a region, each base station covering only a cell. Note however that cellular telephone systems have a far more complex infrastructure to handle roaming and handoff. In 802.11 WLANs, it is easy to roam within a single LAN and requires support from higher layers (such as mobile IP) to roam across different LANs.

Network Operations in an Infrastructure Topology

When a MS is powered up and configured to operate in an infrastructure topology, it can perform a passive scan or an active scan. In the case of a passive scan, the MS simply scans the different channels to detect the existence of a BSS. The existence of a BSS can be detected through *beacon* frames that are broadcast by APs pseudo-periodically. The reason why it is called pseudo-periodic is that the beacon is supposed to be transmitted regularly at certain intervals. However, the AP cannot preempt an ongoing transmission in order to transmit a beacon. When we discuss the MAC layer, we will see that any device has to wait for the medium to be free before transmitting a frame. If the medium is busy, the AP will transmit the beacon after the medium becomes free in which case, the beacon may not be precisely periodic. The beacon is a management frame that announces the existence of a network. It contains information about the network – the BSSID and the capabilities of the network (the PHY alternatives it supports, if security is mandatory, whether the MAC layer supports polling, the interval at which beacons are transmitted, timing parameters, etc.). The beacon is similar to certain control channels in cellular telephone systems (for instance the Broadcast Control Channel – BCCH in GSM). The MS also performs signal strength measurements on the beacon frame. In the case of an active scan, the MS already knows the ID of the network that it wants to connect to. In this case, the MS sends a *probe request* frame on each channel. APs that hear the probe request respond with a *probe response* frame that is similar in nature to the beacon. In either case, the MS can create a scan report that provides it with information about the available BSSs, their capabilities, their channels, timing parameters, and other information. The MS makes use of this information to determine a compatible network with which it can associate itself.

In order to associate itself with an AP, the MS must authenticate itself if this is part of the capability of the network and we will look at this in a later section. Otherwise, as long as the MS satisfies the announced capabilities of the network, it can send an *association request* frame to the AP. The association request informs the AP of the intention of the MS to join the network and it also provides additional information about the MS such as its MAC address, how often it will listen to the beacon (called the *listen interval*), the supported data rates and so on. If the AP is satisfied with the capabilities of the MS, it will reply with an *association response* frame. In this message, the MS is given an association ID and this frame confirms that the MS is now able to access the network. During this association phase, the MS can be authenticated by the network and vice versa. Unlike the ad hoc mode of operation, administrators can control access to the network in the infrastructure mode of operation.

If a MS moves across BSSs or if it moves out of coverage and returns to the BSA of an AP, it will have to re-associate itself with the AP. For this purpose, it will use a *reassociation request* frame similar in form to the association request frame, except that the MAC address of the old AP will be included in the frame. The AP will respond with a *reassociation response* frame. There are three mobility types in IEEE 802.11. The "No Transition" type implies that the MS is static or moving within a BSA. A "BSS Transition" indicates that the MS moves from one BSS to another within the same ESS. The most general form of mobility is "ESS Transition" when the MS moves from one BSS to another BSS that is part of a new ESS. In this case upper layer connections may break (it will need Mobile IP for continuous connection). A MS moving from one BSS to another will have to detect the drop in signal strength from the old AP and detect the beacon of the new AP before the re-association request. It could also use a probe request message instead of detecting the beacon from the new AP. This simple handoff between two APs is MS initiated. In cellular telephone systems, the MS is instructed by entities in the network (such as base station controllers) to perform the handoff from one base station to another. They may use information supplied by the MS such as the signal strength from different base stations. The handoff procedures in a WLAN are as shown in Figure 8.6.

One of the important issues in wireless networks is roaming between different points of access to the wired network. This is possible only if equipment from different vendors supports the same set of protocols and is interoperable. Previously, there existed a Task group F that had proposed an inter-access point protocol (IAPP) to achieve multi-vendor interoperability. For example, when a MS moves from one AP to another and sends a reassociation request, the new AP must be able to converse with the old AP over the distribution system to inform it of the handoff and to free the resources in the old AP. This is achieved using the IAPP now standardized as 802.11f in July 2003. The 802.11f standard specifies the information and format of the information to be exchanged between access points and includes the recommended practice for multi-vendor access point interoperability via the IAPP across distribution systems.

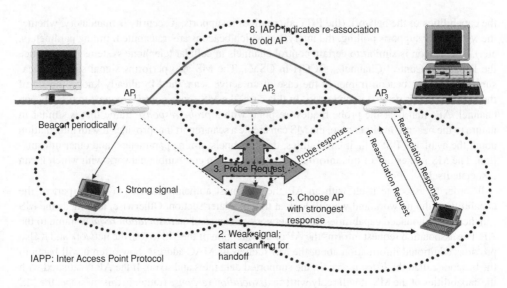

Figure 8.6   Handoff in a WLAN.

Power management is an important component of network operations in an IEEE 802.11 WLAN. The idle receive state dominates the LAN adaptor power consumption. The challenge is how we can power off during the idle periods and maintain the session. The IEEE 802.11 solution is to put the MS in sleeping mode, buffer the data at AP, and send the data when the MS is awakened. Compared to the continuous power control in cellular telephones this is a solution tailored for bursty data applications. When MSs have no frames to send, they can enter a sleep mode to conserve power. If a MS is sleeping when frames arrive at an AP for it, the AP will buffer such frames. A sleeping MS wakes up periodically and listens to the beacon frames. How often it wakes up is specified by the listen interval mentioned earlier (Figure 8.7). The beacon frame also contains a field called the traffic indication map (TIM). This field contains information about whether or not packets are buffered in the AP for a given MS. If a MS detects that it has some frames waiting for it, it can wake up from the sleep mode and receive those frames before going back to sleep. The MS uses a *power-save poll* frame to indicate to the AP that it is ready to receive buffered frames. The AP sends the buffered data when the station is in active mode.

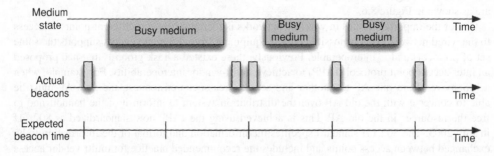

Figure 8.7   Power management in IEEE 802.11.

If the MS chooses to leave the network or shutdown, it will send a *dissociation* frame to the AP. This frame will terminate the association between the MS and the network enabling the network to free resources that were previously reserved for the MS (such as the association ID, buffer space, etc.).

Network Operations in an Ad Hoc Topology

In an ad hoc topology, there is no fixed AP to coordinate transmissions and define the BSS. A MS that operates in ad hoc mode will power up and scan the channels to detect beacons from other MSs that may be in the vicinity and that may have set up an IBSS. If it does not detect any beacons, it may declare its own network. If it does detect a beacon, then the MS can join the IBSS in a manner similar to the process in the infrastructure topology. MSs in an IBSS may choose to rotate the responsibility of transmitting a beacon. Power management works in a similar way, except that the source MS itself has to send an announcement traffic indication map (ATIM) frame to the recipient MS.

Network Operations in Mesh Topology

Recently, wireless mesh networking has received a lot of attention as it enables the deployment of wireless networks over a large area without the need for an extensive fixed (wired) infrastructure. A wireless mesh network consists of entities that connect to each other over an air interface and relay packets in the network thus created. This eliminates the need for a wired backbone to relay packets. Some of these entities may act like APs creating infrastructure WLANs and becoming points of access to the mesh network. Other entities will connect to the Internet and enable any device in the mesh network to access the Internet. Wireless mesh networks can use a variety of technologies such as IEEE 802.16 or WiMAX based devices (see Chapter 13) or IEEE 802.11 based devices. In 2004, a task group "S" of 802.11 was set up to investigate mesh networking with 802.11 and propose a standard for using a *wireless distribution system* unlike the wired distribution system in the ESS. Some elements of operations in a mesh network have been proposed in [Lee06]. In a mesh network, APs (or MSs) are required to be capable of relaying packets to one another using the air interface so that packets may be delivered from a source MS to a destination MS through multiple wireless hops. Entities with relay capabilities are called mesh points. A mechanism for determining a path from one mesh point to another is also necessary and it is expected to be implemented at the MAC layer. Mesh portals enable connectivity to other mesh networks, LANs, or the Internet. Multicast and broadcast capability at the MAC layer is also another aspect that the IEEE 802.11s standard is expected to address. This task group is also supposed to develop enhancements to the current IEEE 802.11 standard to provide a method to configure the distribution system using the four MAC addresses thereby enabling some form of mesh networking between access points. This could be a wired or wireless mesh network that allows for automatic topology learning and dynamic path configuration over self-configuring multi-hop topologies. There are already proprietary protocols performing this task to extend coverage within homes, but the standard will allow for different scenarios with different requirements (e.g., quick set up and tear down, maximizing throughput, etc.).

8.3.2   The IEEE 802.11 MAC Layer

MSs in an IEEE 802.11 network have to share the transmission medium, which is air. If two MSs transmit at the same time and the transmissions are both in range of the destination, they may collide resulting in the frames being lost. The MAC layer is responsible for controlling access to the medium and ensuring that MSs can access the medium in a fair manner with minimal collisions. The medium access mechanism is based on carrier sense multiple access but there is no collision detection unlike the wired equivalent LAN standard (IEEE 802.3) as previously discussed. In IEEE 802.3 sensing the channel is very simple. The receiver reads the peak voltage on the wire of cable and compares that against a threshold. Collisions are extremely hard to detect in RF because of the dynamic nature of the channel. Detecting collisions also incurs difficulties in hardware implementation because a MS has to be transmitting and receiving at the same time. Instead, the strategy adopted is to avoid collisions to the extent possible. In IEEE 802.11, there are two types of carrier sensing – physical sensing of energy in the medium and virtual sensing. Physical sensing is through Clear Channel Assessment (CCA) signal produced by PLCP in the physical layer of the IEEE 802.11. The CCA is generated based on "real" sensing of the air interface either by sensing the detected bits in the air or by checking the RSS of the carrier against a threshold. Decisions based on the detected bits are made slightly slower but they are more reliable. Decisions based on the RSS may create false alarms caused by high interference levels. Best designs take advantage of both carrier sensing and detected data sensing. In addition to physical sensing the IEEE 802.11 also provides for the virtual carrier sensing. Virtual sensing is implemented by decoding a duration field in the 802.11 frame that allows a MS to know the time for which a frame will last. A "length" field in the MAC layer is used to specify the amount of time that must elapse before the medium can be freed. This time is stored in a *network allocation vector* (NAV) that counts down to zero to indicate when the medium is free again. To illustrate the IEEE 802.11 MAC Layer, we will use the ad hoc topology as an example. However, the procedures are identical in an infrastructure topology as well.

The Distributed Coordination Function

We will first describe the basic medium access process in IEEE 802.11 called the distributed coordination function or DCF. Consider Figure 8.8 that shows the basic method for accessing the medium in IEEE 802.11. A MS will initially sense the channel before transmission. If the medium is free, the MS will continuously monitor the medium for a period of time called the *Distributed*

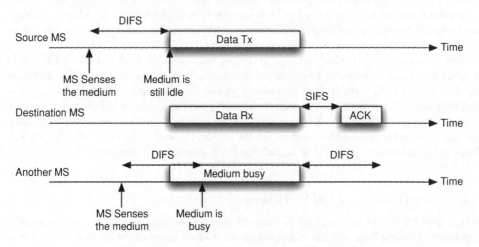

Figure 8.8   Basic medium access in IEEE 802.11.

Coordination Function (DCF) *Inter Frame Space* (IFS) or DIFS. If the medium is still idle after DIFS, then the MS can transmit its frame without waiting. Otherwise, the MS will enter a back off process. The rationale is that if another MS senses the medium after the first MS, it will also wait for DIFS. However, before a time DIFS expires, the first MS would have started its transmission. Upon hearing the transmission, the second MS will have to back off. The wireless medium is harsh and unreliable and hence all transmissions are acknowledged. The destination of the frame will send an acknowledgement (ACK) back to the source if the frame is successfully received as follows. It will wait for a time called the *Short Inter Frame Space* (SIFS) and transmits the ACK. The SIFS value is smaller than the DIFS value. All IFS values depend on the physical layer alternative. Thus, any other MS that senses the channel as idle after the original frame was transmitted will still be waiting and ACK frames have priority over their transmissions. In order to maintain fairness and avoid collisions, the MS that senses the medium as free for a time DIFS and transmits a frame will have to enter the backoff process if it wants to transmit another frame immediately. The exception is when it is transmitting one frame in many fragments. In such a case, the MS can indicate the number of fragments in the first frame to be transmitted and occupy the channel till the frame is completely transmitted.

The backoff process works as follows. Once a MS enters the backoff process, it picks a value called the *backoff interval* (BI) that is a random value uniformly distributed between zero and a number called the *contention window* (CW). The MS will then monitor the medium. When the medium is free for at least a time DIFS, the MS will start counting down from the BI value as long as the medium is free. The counter is decremented every so often (called a slot). If the medium is sensed as occupied before the counter goes down to zero, the MS will freeze the counter and continue to monitor the medium. As soon the counter becomes zero, the MS can transmit its frame. This process is shown in Figure 8.9.

The IEEE 802.11 MAC supports *binary exponential backoff* like IEEE 802.3. Initially, the CW is maintained at a value called CW_{min} which is typically $2^5 - 1 = 31$ slots. So the BI will be uniformly distributed between 0 and 31 slots. The slot time varies depending on the physical layer alternative. For example, it is 20 μs in the IEEE 802.11b standard and 9 μs is the 802.11a standard. If a packet is not successfully transmitted (this could be due to collisions or a channel error), the value of CW is essentially doubled. The MS will now pick a BI value that is uniformly distributed between 0 and $2^6 - 1 = 63$ slots. This process can be continued till CW reaches a value that is CW_{max} (usually 1023 slots). The rationale behind this approach is as follows. If there are many MSs contending for the medium, it is likely that one or more MSs may pick the same BI value. Their transmissions will then collide. By increasing the value of CW, it is likely that this probability will go down thereby reducing collisions.

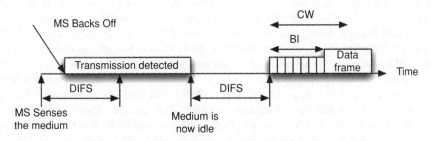

Figure 8.9 Backoff process in IEEE 802.11.

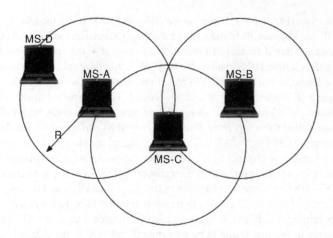

Figure 8.10 Illustrating the hidden and exposed terminal problems.

Frames may be lost due to channel errors or collisions. A positive ACK from the destination is necessary to ensure that the frame has been successfully received. In IEEE 802.11, each MS maintains retry counters that are incremented if no ACKs are received. After a retry threshold is reached, the frame is discarded as being undeliverable.

The Hidden Terminal Problem and Optional Mechanism

In wireless networks that use carrier sensing, there is a unique problem called the hidden terminal problem (see also Chapter 4). Suppose all MSs are identical and have a transmission and reception range of R as shown in Figure 8.10.

The transmission from MS-A can be heard by MS-C but not MS-B. So when MS-A is transmitting a frame to MS-C, MS-B will not sense the channel as busy and MS-A is *hidden* from MS-B. If both MS-A and MS-B transmit frames to MS-C at the same time, the frames will collide. This problem is called the hidden terminal problem. There is a dual problem called the exposed terminal problem. In this case, MS-A is transmitting a frame to MS-D. This transmission is heard by MS-C, which then backs off. However, MS-C could have transmitted a frame to MS-B and the two transmissions would not interfere or collide. In this case, MS-A is called an *exposed terminal*. Both hidden and exposed terminals cause a loss of throughput.

To reduce the possibility of collisions due to the hidden terminal problem, the IEEE 802.11 MAC has an optional mechanism at the MAC layer as shown in Figure 8.11. Suppose MS-A wants to transmit a frame to MS-C. It will first transmit a short frame called the *Request-to-Send* (RTS) frame. The RTS frame is heard in the transmission range of MS-A and includes MS-C and MS-D, but not MS-B. Both MS-C and MS-D are alerted to the fact that MS-A intends to transmit a frame and they will not attempt to simultaneously use the medium. This is achieved by the virtual carrier sensing process that sets the NAV to a value equal to the time it will take to successfully complete the exchange of frames. In response to the RTS frame, MS-C will send a *Clear-to-Send* (CTS) frame that will be heard by all MSs in its transmission range. This includes MS-B and MS-A but not MS-D. The CTS frame lets MS-A know that MS-C is ready to receive the data frame. It also alerts MS-B to the fact that there will be a transmission from some MS to MS-C. Consequently MS-B will defer any frames that it wishes to transmit in anticipation of the completion of the communication to

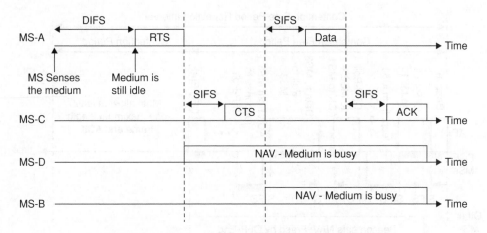

Figure 8.11 Operation of the RTS-CTS mechanism.

MS-C. This way, even though MS-B is outside the transmission range of MS-A, the CTS message can be used to *extend* the carrier sensing range thereby reducing the hidden terminal problem. Of course, it is quite possible that the RTS frame itself collided with a transmission from MS-B. In such a case, both MS-A and MS-B will have to enter the backoff process and retransmit their frames.

The RTS-CTS mechanism can be controlled in IEEE 802.11 by using an RTS Threshold. All unicast and management frames larger than this threshold will always be transmitted using RTS-CTS. By setting this value to 0 bytes, all frames will use RTS-CTS. The default value is 2347 bytes that disables RTS-CTS for all packets. When RTS-CTS signals are used, the CTS frame is transmitted by the destination MS after waiting simply for a time equal to SIFS. This way, the CTS frame has priority compared to all other transmissions that have to wait for at least a time DIFS and perhaps an additional waiting time in backoff. Using the RTS-CTS signals reduces the throughput of a WLAN, but it may be essential to use this in dense environments.

The Point Coordination Function

One consequence of using CSMA/CA as described above with DCF is that it is impossible to have any bounds on the delay or jitter suffered by frames. Depending on the traffic load and the BI values that are picked, a frame may be transmitted instantaneously or it may have to be buffered till the medium becomes free. For real-time applications such as voice or multimedia, this can result in performance degradation especially when strict delay bounds are necessary. To provide some bounds on the delay, an optional MAC mechanism called the *Point Coordination Function* or PCF is part of the IEEE 802.11 standard [Cro97]. PCF provides contention free access to frames using a polling mechanism described below.

The process starts when the AP captures the medium by sending a beacon frame after it is idle for a time called the *PCF Inter Frame Space* (PIFS). The PIFS is smaller than the DIFS and larger than SIFS. In the beacon frame, the AP, also called the point coordinator, announces a *Contention Free Period* (CFP) where the usual DCF operation will be preempted. All MSs that use only DCF will set a NAV to indicate that the medium will be busy for the duration of the CFP. The AP maintains a list of MSs that need to be polled during the CFP. MSs get onto the polling list when they first associate with the AP using the association request. The AP then polls each MS on the list for

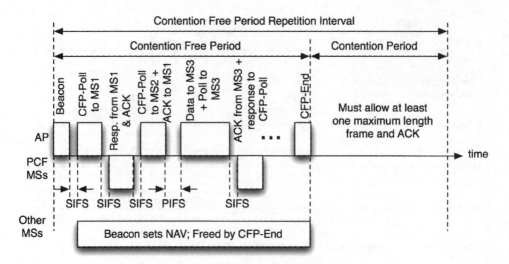

Figure 8.12   Operation of the PCF.

data. The polls are sent after a time SIFS and the ACKS to the poll and any associated data will be transmitted by the corresponding MS also after a time SIFS. If there is no response from a MS to a poll, the AP waits for a time PIFS before it sends the next poll frame or data. The AP can also send management frames whenever it chooses within the CFP. An example of PCF operations is shown in Figure 8.12.

The AP indicates the culmination of the CFP via a message called CFP-End. This is a broadcast frame to all MSs and frees the NAV in MSs that are only DCF based. Following the CFP, a contention period starts. In this period, it must be possible for a MS to transmit at least one maximum length frame using DCF and receive an ACK. The CFP can be resumed after completion of the contention period. The PCF mechanism is optional in IEEE 802.11. Most commercial systems deployed today do not support PCF and real-time services do not have very good support in WLANs today. Note that polling has a lot of overhead especially if MSs do not have frames to send when they are polled.

MAC Frame Formats

While this article will not define all of the different frame formats of an IEEE 802.11 MAC frame and discuss the fields in great detail, it will consider some examples to illustrate the MAC frame formats. Figure 8.13 shows the general format of a MAC frame. The most significant bit is last

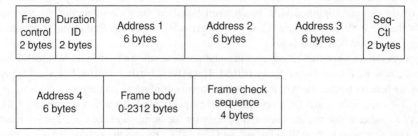

Figure 8.13   General format of a MAC frame.

(right most) and the bits are transmitted from left to right. The *Frame Control* field has two bytes and comprises of many fields. It carries information such as the protocol version, the type of frame (management – probe request, association, authentication, and so on, control – RTS, CTS, etc., or data – pure data, CFP poll and data, null,etc.), the number of retries, and whether the frame is encrypted (discussed later). The duration field is important to set the NAV during virtual carrier sensing.

There can be up to four address fields in the frame [Gas02]. The addresses can be different depending on the type of frame. Common addresses used are the source and destination addresses, the receiver address if the destination is different from the receiver (e.g., the receiver is the AP, but the destination is a wired node on the LAN segment), the transmitter address (once again if the transmitter is different from the source – it is the AP), and the BSSID. The sequence control field is used in case there is fragmentation of frames. The frame body carries the payload from upper layers and the frame check sequence is a 32 bit cyclic redundancy check used to verify the integrity of the frame at the receiver. The frame format in Figure 8.8 is used in an infrastructure topology. In an IBSS, only three address fields are used.

The RTS and CTS frames are very short frames – 20 and 14 bytes respectively and are shown in Figure 8.14. The ACK frames are very similar to the CTS frame.

Compared to Ethernet, IEEE 802.11 is a wireless network that needs to have control and management signaling to handle registration process, mobility management, power management and security. To implement these feature, the frame format of the 802.11 should accommodate a number of instructing packets, similar to those we described in wide area networks. The capability of implementing these instructions is embedded in the control field of the MAC frames. Figure 8.15 shows the overall format of the control field in the 802.11 MAC frame with description of all fields except type and subtype. These two fields are very important because they specify various instructions for using the packet. The two-bit *Type* field specifies four options for the frame type:

- Management Frame (00)
- Control Frame (01)
- Data Frame (10)
- Unspecified (11).

The 4-bit **Subtype** provides an opportunity to define up to 16 instructions for each type of frame. Table 8.4 shows all used six-bits for the Type and Subtypes in the frame control field. Combinations that are not used provide an opportunity to incorporate new features in the future.

Figure 8.16 illustrates the frame body of the beacon frame. The time stamp allows MSs to synchronize to a BSS. The beacon interval says how often the beacon can be expected to be heard. It is typically 100 milliseconds, but could be changed by an administrator. The capability

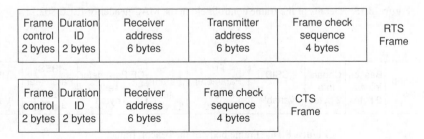

Figure 8.14 RTS and CTS frame formats.

Table 8.4   Type and subtype fields and their associated instructions

Management type (00)	Association request/response (0000/0001) Reassociation request/response (0010/0011) Probe-request/response (0100/0101) Beacon (1000) ATIM: Announcement traffic indication map (1001) Dissociation (1010) Authentication/deauthentication (1011/1100)
Control type (01)	Power save poll (1010) RTS/CTS (1011/1100) ACK (1101) CF end/CF end with ACK (1110/1111)
Data type (10)	Data/data with CF ACK/no data (0000/0001) Data poll with CF/data poll with CF and ACK (0010/0011) No data/CF ACK (0100/0101) CF poll/CF poll ACK (0101/0110)

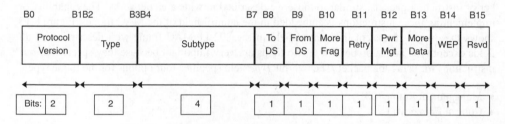

B0	B1B2	B3B4		B7 B8	B9	B10	B11	B12	B13	B14	B15
Protocol Version	Type	Subtype	To DS	From DS	More Frag	Retry	Pwr Mgt	More Data	WEP	Rsvd
Bits: 2	2	4	1	1	1	1	1	1	1	1

Protocol Version: currently 00, other options reserved for future

To DS/from DS: "1" for communication between two APs

More Fragmentation: "1" if another section of a fragment follows

Retry: "1" if acket is retransmitted

Power Management: "1" if station is in sleep mode

Wave Data: "1" more packets to the terminal in power-save mode

Wired Equivalent Privacy: "1" data bits are encrypted

Figure 8.15   Details of the frame control field in the MAC header of IEEE 802.11.

Tiime stamp 8 bytes	Beacon interval 2 bytes	Capab. Info. 2 bytes	SSID Variable	FH Parameter set 7 bytes	DS Param set 2 bytes	CF Parameter set 8 bytes	IBSS Parameter set 4 bytes	TIM Var.

Figure 8.16   Frame body of the beacon frame.

information (two bytes) provides information about the topology (whether infrastructure or ad hoc), whether encryption is mandatory, and whether additional features are supported. One such feature is *channel agility* where the AP hops to different channels after a predetermined amount of time.

We have not discussed PHY layer alternatives yet. The parameter sets in the beacon provide information about the PHY layer parameters that are necessary to join the network. For instance, if frequency hopping (FH) is used, the FH parameter set will specify the hopping pattern. The traffic indication map (TIM) field is used to support MSs that may be sleeping as described earlier.

8.3.3   The PHY Layer

The IEEE 802.11 standards body has standardized several different PHY layer alternatives. When it was first standardized in 1997, there were three PHY layer options. We will call these options as the "base" IEEE 802.11 PHY layer alternatives. The IEEE 802.11b supports up to 11 Mbps in the 2.4 GHz ISM bands, the IEEE 802.11g standard supports up to 54 Mbps in the 2.4 GHz ISM bands and the IEEE 802.11a standard supports up to 54 Mbps in the 5 GHz U-NII bands. Before we discuss these alternatives, let us look at the PHY layer in IEEE 802.11.

The PHY layer in IEEE 802.11 is broken up into two sub-layers – the *Physical Layer Convergence Protocol* (PLCP) and the *Physical Medium Dependent* (PMD) layers. The PLCP includes a function that adapts the underlying medium dependent capabilities to the MAC level requirements. The PLCP would, for instance add some additional fields to the frame to enable synchronization at the physical layer. The PMD actually determines how information bits are transmitted over the medium. When the MAC protocol data units (MPDU) arrive to the PLCP layer a header is attached that is designed specifically for the PMD of the choice for transmission. The PLCP packet is then transmitted by PMD according to the specification of the signaling techniques.

The Base IEEE 802.11 Standard

The base IEEE 802.11 standard specifies three different PHY layer alternatives. Two of these use radio frequency (RF) transmissions in the 2.4 GHz ISM bands and one uses diffused infra-red (DFIR).

The FH Option

The first option for transmission in the 2.4 GHz ISM bands makes use of frequency hopping spread spectrum (FHSS). The entire band is divided into 1 MHz wide channels and the specification makes it important to confine 99% of the energy to one such channel during transmission to reduce interference to the other channels. These restrictions are also due to the rules imposed by the Federal Communications Commission (FCC) in the United States. The standard specifies 95 such 1 MHz wide channels and they are numbered accordingly. In the United States, only 79 of these channels are allowed. Devices that use the FH option hop between these channels when transmitting frames. The dwell time in each channel is approximately 0.4s or the minimum hop rate of the IEEE 802.11 FHSS system is 2.5 hops per second that is rather slow. The hop sequences (the channel hopping pattern) depends on mathematical functions. An example hopping pattern is {3, 26, 65, 11, 46, 19, 74, ... }. In the United States, each set of hopping patterns can have at most 26 different channels. This means that it is possible to create three orthogonal hopping sets (since there are 79 channels in the United States). If three APs use these three orthogonal hopping sets, there will be no interference

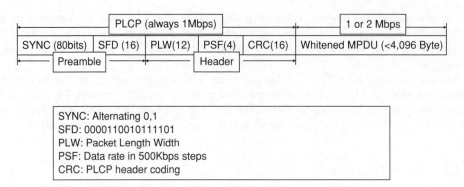

Figure 8.17   PLCP frame for the FH option in IEEE 802.11.

between these networks. The modulation scheme used with FHSS is called Gaussian Frequency Shift Keying (GFSK). This modulation scheme makes use of the frequency information to encode data. It is possible to use either two frequencies within the channel or four frequencies within the channel. In the former case, the data rate will be 1 Mbps and in the latter case, the data rate will be 2 Mbps. The advantage of the FHSS system is that the receivers are less complex to implement. The PLCP for the FHSS PMD introduces an 80-bit field for synchronization, a frame delimiter and some fields to indicate the data rate. Depending on this field, the data rate can be modified in steps of 500 kbps from 1.0 to 4.5 Mbps. However the standard only supports 1 and 2 Mbps. The values of SIFS and the slot for backoff in this option are 28 and 50 μs respectively.

Figure 8.17 shows the details of the PLCP header, which is added to the whitened MAC PDU to prepare it for transmission using FHSS physical layer specifications of the IEEE 802.11. The PLCP additional bits consist of a preamble and a header. The *Preamble* is a sequence of alternating 0 and 1 symbol for the 80 bits that are used to extract the received clock for carrier and bit synchronization. The start of the frame delimiter (*SFD*) is a specific pattern of 16-bits, shown in the figure, indicating start of the frame. The next part of the PLCP is the header that has three fields. The 12-bit packet length width (*PLW*) field identifies the length of the packet that could be up to 4Kbytes. The 4-bits of the packet-signaling field (*PSF*) identifies the data rate in 0.5 Mbps steps starting with 1 Mbps.

Example 8.3: Specification of data rate on the physical layer

The existing 1 Mbps is represented by 0000 as the first step. The 2 Mbps by 0010 that is 2×0.5 Mbps $+ 1$ Mbps $= 2$ Mbps. The maximum 3-bit number represented by this system is 0111 that is associated with $7 \times 0.5 + 1 = 4.5$ Mbps. If all four bits are used we have $15 \times 0.5 + 1 = 8.5$ Mbps. These limitations imply that data rates cannot even reach 10 Mbps.

The rest of the rates are reserved for the future. The 16-bit CRC code is added to protect the PLCP bits. It can recover from errors of up to 2-bit and otherwise identify whether the PLCP bits are corrupted or not. The total overhead of the PCLP is 16 bytes (128-bits) that is less than 0.4% of the maximum MPDU load justifying the low impact of running the PCLP at lower data rates. The received MPDU is passed through a scrambler to be randomized. Randomization of the transmitted bits, that is also called whitening because the spectrum of a random signal is flat, eliminated the dc bias of the received signal. A scrambler is a simple shift register finite state machine with special feedback that is used both for scrambling and de-scrambling of the transmitted bits.

The DS Option

The direct sequence spread spectrum (DSSS) modulation technique has been the most popular commercial implementation of IEEE 802.11. DSSS has some inherent advantages in multipath channels and can increase the coverage of an AP for this reason[Tuc91]. We will briefly discuss the features of this PMD layer. Details of DSSS will be discussed in Chapter 12.

In a DSSS system, the data stream is "chipped" into several narrower pulses (chips) thereby increasing the occupied spectrum of the transmitted signal. One common way of doing this is to multiply the data stream (typically a series of positive and negative rectangular pulses) by a spreading signal (typically another series of positive and negative rectangular pulses, but with much narrower pulses than the data stream). While the data stream is random and depends on what needs to be transmitted, the spreading signal is deterministic. Figure 8.18 (a) shows an example where the data stream $d(t)$ is multiplied by a spreading signal $a(t)$ to produce a signal $s(t)$ that is then modulated over an RF carrier. In this figure, eleven narrow pulses are contained within one broad data pulse. The pulses could have a positive $(+)$ or negative $(-)$ amplitude. This results in the bandwidth expanding by a factor of eleven and this is also called as the *processing gain* (see Chapters 3 and 4 also for a general discussion of processing gain). A specific pattern of pulses in the spreading signal is used. The pattern used in the IEEE 802.11 standard is a Barker Sequence. The interesting property of the Barker Sequence is that its autocorrelation has a very sharp peak and very narrow sidelobes as shown in Figure 8.18b. Because of this property, it is possible for a receiver to reject interference from multipath signals and recover information robustly in a harsh wireless environment. The Barker Sequence with differential binary phase shift keying (DBPSK) is used for data rates of 1 Mbps and the Barker sequence with differential quadrature phase shift keying (DQPSK) is used for data rates of 2 Mbps. In either case, the chip rate is 11 Mcps (mega chips per second).

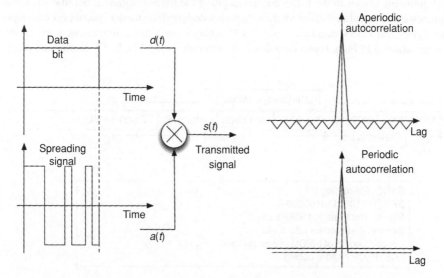

Figure 8.18   (a) Direct sequence spread spectrum. (b) Autocorrelation of the Barker pulse.

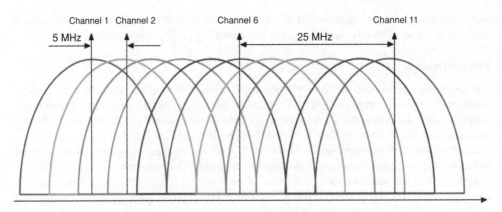

Figure 8.19   Channelization for the IEEE 802.11 DS option.

Unlike FHSS, a signal carrying 2 Mbps, now occupies a bandwidth that is as large as 25 MHz. In the IEEE 802.11 standard, 14 channels are specified for the DSSS PMD. Channel 1 is at 2.412 GHz, Channel 2 at 2.417 GHz and so on (see Figure 8.19). Only the first 11 channels are available for use in the United States. Figure 8.12 shows the channelization in the United States. Since each channel occupies roughly 25 MHz bandwidth and the channel separation is only 5 MHz, there is significant overlap between channels. If two WLANs in the same vicinity were to use adjacent channels, there would be severe interference and throughput degradation. There are three *orthogonal* channels (Channels 1, 6, 11) in the United States that can be deployed without interference. The FH option is easier in terms of implementation because the sampling rate is on the order of the symbol rate of 1 Msps. The DS implementation requires sampling rates on the order of 11 Mcps. However, because of the wider bandwidth, DSSS provides a better coverage and a more stable signal.

Figure 8.20 shows the details of the PLCP frame for the DSSS version of the IEEE 802.11. The overall format is similar to the FHSS but the length of the fields is different because transmission techniques are different and different manufacturers designed the model product for development of the FHSS and DSSS standards. The PLCP sublayer once again introduces some fields for synchronization (128 bits), frame delimiting and error checking. The PLCP header and preamble

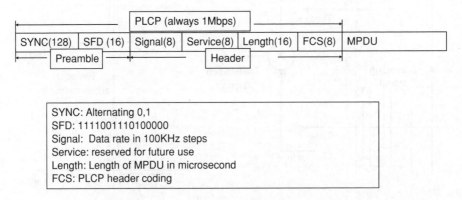

Figure 8.20   PLCP frame for the DSSS of the IEEE 802.11.

are always transmitted at 1 Mbps using DBPSK. The rest of the packet is transmitted using either DBPSK or DQPSK depending on the data rate. The values of SIFS and the slot for backoff in this option are 10 and 20 μs respectively. The MPDU from the MAC layer is transmitted either at 1 or 2 Mbps, however, analogous to the FHSS version of the standard, the PLCP of the DSSS version also uses the simpler BPSK modulation at 1 Mbps all the time. The MPDU for the DSSS does not need to be scrambled for whitening because each bit is transmitted as a set of random chips that is a whitened transmitted signal. The length of the SYNC in the DSSS is 128 bits that is longer than FHSS because DSSS needs a longer time to synchronize. The format of the SFD of the DSSS is identical to that of the FHSS but the value of the code, shown in Figure 8.20 is different. The PSF field of the FHSS is called *Signal* field and it uses 8 bits to identify data rates in steps of 100 Kbps (five times more precision than FHSS).

Example 8.4: Frame formats for various data rates in IEEE 802.11

Using the above encoding, we represent 1 Mbps for DSSS by 00001010 (10 × 100 Kbps) and the 2 Mbps by 00010100 (20 × 100 Kbps), and 11 Mbps (used in IEEE 802.11b) by 001101110(55 × 100 kbps) and 01101110 (110 × 100 kbps). The maximum number in this system is 11111111 that represents 255 × 100 Kbps = 25.5 Mbps.

The *Service* field in the DSSS is reserved for future use and it does not exist in the FHSS version. The *Length* field of the DSSS is analogous to the PLW in the FHSS, however length field specifies the length of the MPDU in microseconds. The frame correction sequence (*FCS*) field of the DSSS is identical to the CRC field of the FHSS.

The DFIR Option

The third option in IEEE 802.11 is to use infrared (IR) for transmission (Valadas, 1998). The spectrum occupied by the IR transmission is at wavelengths between 850 and 950 nm. The technique used for transmission is diffused infrared – that is communications is omnidirectional. The range specified is around 20 m, but the transmissions cannot penetrate physical obstacles. The modulation scheme used is pulse position modulation (PPM). A data rate of 1 Mbps is supported using 16-PPM and a data rate of 2 Mbps is supported using 4-PPM. This is in comparison with the IrDA (infrared data association) standard, which primarily allows communications at a few hundred kbps to a few Mbps between two devices (like a laptop and a personal digital assistant) that are within a few feet of one another.

The PMD of DFIR operates based on transmission of 250 ns pulses that are generated by switching the transmitter LEDs on and off for the duration of the pulse. Figure 8.21 illustrates the 16-PPM and 4-PPM modulation techniques recommended by the IEEE 802.11 for 1 and 2 Mbps respectively. In the 16-PPM blocks of 4-bits of the information are coded to occupy one of the 16-slots of a 16-bit length sequence according to their value. In this format each 16 × 250 ns= 4000 ns carries 4-bits of information that supports a 4bits/4000ns = 1 Mbps transmission rate. For the 2 Mbps version every 2-bits are PPM modulated into 4-slot of duration 4 × 250 ns = 1000 ns that generates data at 2 bits/1000 ns = 2 Mbps. The peak transmitted optical power is specified at 2W with an average of 125 or 250 mW.

The PLCP packet format for the DFIR is shown in Figure 8.22. The PLCP signals are shown in the unit of slots of 250 ns for one basic pulse. The SYNC and SDF fields are shorter because non-coherent detection using photosensitive diode detectors do not need carrier recovery or elaborate

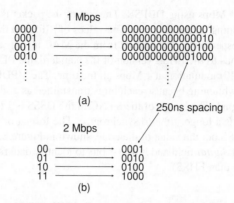

Figure 8.21   PPM using 250-ns pulses in the DFIR version of the IEEE 802.11: (a) 16-PPM for 1 Mbps, (b) 4-PAM for 2 Mbps.

random code synchronizations. The 3-slot data rate indication system starts by 000 for 1 Mbps and 001 for 2 Mbps. The Length and FCS are identical to the DSSS. The only new field is the DC level adjustment (*DCLA*) that sends a sequence of 32-slots allowing receiver to set its level of the received signal to set threshold for deciding between received "0"s and "1"s. The MPDU length is restricted to 2500 bytes.

IEEE 802.11b and 802.11g

The DS option for IEEE 802.11, although successful, consumed a lot of bandwidth for the given data rate. The chip rate is 11 Mcps, but the maximum data rate is 2 Mbps. That is, one Barker Sequence of 11 chips, transmitted every microsecond, can at most carry two bits of information. To increase the data rate, the IEEE 802.11b standard adopted a slightly different method. Instead of transmitting one 11-chip sequence every microsecond, with IEEE 802.11b, the device transmits one 8-chip codeword every 0.727 µs. Each 8-chip codeword can carry up to eight bits of information

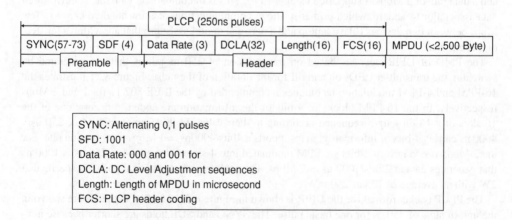

Figure 8.22   PLCP frame for the DFIR of the IEEE 802.11.

Table 8.5 Mapping for CCK

Dibit	Phase parameter	Dibit (d_{i+1}, d_i)	Phase
(d1, d0)	φ_1	(0,0)	0
(d3, d2)	φ_2	(0,1)	π
(d5, d4)	φ_3	(1,0)	$\pi/2$
(d7, d6)	φ_4	(1,1)	$-\pi/2$

for a maximum data rate of $8/(0.727 \times 10^{-6}) = 11$ Mbps. If the codeword carries only four bits of information, the data rate will be 5.5 Mbps. The codewords are derived from a technique called *complementary code keying* (CCK) [Hal99].

CCK works as follows for the case when eight bits are mapped into an 8-chip codeword. The incoming data stream is broken up into units of eight bits. Suppose the least significant bit is labeled d0 and the most significant bit is labeled d7. Then, four phases are defined to correspond to the four possible values of a pair of bits as shown in the first two columns of Table 8.5. Depending on what the bits are, the phases then take on a value as shown in the third and fourth columns in Table 8.5. For example, if d5 = 0 and d4 = 1, then the phase $\varphi_3 = \pi$. Once the phases are determined, the 8-chip codeword is given by the vector:

$$C = \{e^{j(\phi_1+\phi_2+\phi_3+\phi_4)}, e^{j(\phi_1+\phi_3+\phi_4)}, e^{j(\phi_1+\phi_2+\phi_4)}, -e^{j(\phi_1+\phi_4)},$$
$$e^{j(\phi_1+\phi_2+\phi_3)}, e^{j(\phi_1+\phi_3)}, -e^{j(\phi_1+\phi_2)}, e^{j(\phi_1)}, \}$$

This vector has elements that belong to the set $\{+1, -1, +j, -j\}$ where j is the square root of -1. These four elements can be mapped in RF to the phase of the carrier and the receiver can decode this phase information to recover the data bits. CCK can be though of either as a modulation scheme or as a coding scheme. All the terms of the above equation share the first phase, if we factor that out we have:

$$\mathbf{c} = \{e^{j(\varphi_2+\varphi_3+\varphi_4)}, e^{j(\varphi_3+\varphi_4)}, e^{j(\varphi_2+\varphi_4)}, -e^{j(\varphi_4)}, e^{j(\varphi_2+\varphi_3)}, e^{j(\varphi_3)}, -e^{j(\varphi_2)}, 1\}e^{j(\varphi_1)}$$

This coding suggests that our 256 transformation matrix can be decomposed into two transformations, one a unity transformation that maps 2-bits (one complex phase) directly and the other one that maps 6-bits (3-phases) into an 8-element complex vector with 64 possibilities determined by the inner function of the above equation. The above decomposition leads to a simplified implementation of the CCK system that is shown in Figure 8.23. At the transmitter the serial data in multiplied into 8-bit addresses. Six of the eight bits are used to select one of the 64 orthogonal codes produced as one of the 8-complex code and two bits are directly modulated over all elements of the code that are transmitted sequentially. The receiver is actually comprising of two parts: one the standard IEEE 802.11 DSSS decoder using Barker codes and one a decoder with 64 correlators for the orthogonal codes and an ordinary demodulator for IEEE 802.11b. By checking the PLCP data rate field receiver know which decoder should be employed for the received packets. This scheme provides an environment for implementation of a WLAN that accommodates both 802.11 and 802.11b devices.

The advantage of CCK is that it maintains the channelization of IEEE 802.11 while increasing the data rate by a factor of five. CCK is also fairly robust to the degradations caused by multipath in the wireless environment. The values of SIFS and the slot for backoff in this option are 10 and

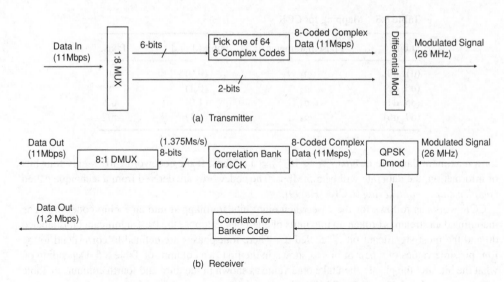

Figure 8.23 Simplified implementation of the CCK for IEEE 802.11b.

20 μs, respectively. IEEE 802.11b also has an optional modulation method called packet binary convolutional coding (PBCC) that is not widely implemented. The advantage of PBCC over CCK is the use of powerful convolutional coding for forward error correction.

The IEEE 802.11g standard [Vas05] maintains backwards compatibility with IEEE 802.11b and IEEE 802.11 DS options by adopting minimal PHY layer frame changes and by including some mandatory and optional physical layer components. In the PLCP layer, 802.11g allows for the use of short preambles to reduce packet overhead. The four physical layers specified in this standard are prefixed by the term ERP that stands for *extended rate physical*. The standard specifies *Orthogonal Frequency Division Multiplexing* (OFDM) and CCK as the mandatory modulation schemes with a data rate of 24 Mbps as the maximum mandatory data rate. With OFDM, IEEE 802.11g also provides for optional higher data rates of 36, 48, and 54 Mbps. OFDM is the same modulation scheme that is used in IEEE 802.11a and we discuss it below. PBCC is an optional modulation scheme in 802.11g that allows for raw data rates of 22 and 33 Mbps. We discuss OFDM in the context of 802.11a below, but the discussion could also apply to 802.11g with some modifications. More details of OFDM are considered in Chapter 13.

The IEEE 802.11a Standard

One of the primary problems for huge data rates in wireless channels is what is called the coherence bandwidth of the wireless channel caused by multipath dispersion. The coherence bandwidth limits the maximum data rate of the channel to that which can be supported within this bandwidth (for example, if the coherence bandwidth is B Hz and the channel bandwidth is $W \gg B$ Hz, a transmission bandwidth of W Hz will result in irrecoverable errors unless equalization or spread spectrum is used). In order to overcome this limitation, we can send data in several sub-channels each on the order of the coherence bandwidth or less, so that many of them will get through correctly. Using several sub-channels and reducing the data rate on each channel increases the symbol duration in each channel. If the symbol duration in each channel is larger than the multipath dispersion, errors

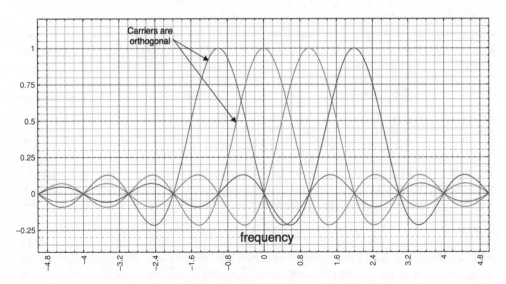

Figure 8.24 Orthogonal carriers in OFDM.

will be smaller and it will be possible to support larger data rates. This principle can be exploited while maintaining bandwidth efficiency using a fairly old technique called OFDM. OFDM has been used in digital subscriber lines (DSL) as well to overcome the variations in attenuation with frequency over copper lines. OFDM enables spacing carriers (sub-channels) as closely as possible and implementing the system completely in digital eliminating analog components to the extent possible. OFDM [Kap02] is used as the physical layer in IEEE 802.11a, HIPERLAN/2, and IEEE 802.11g.

IEEE 802.11a specifies eight 20 MHz channels [Nee99]. As shown in Figure 8.24, several sub-channels are created in OFDM using orthogonal carriers in each channel: 52 sub-channels are specified for each channel with a bandwidth of approximately 300 kHz each; 48 sub-channels are used for data transmission and four are used as pilot channels for synchronization. One OFDM symbol (consisting of the sum of the symbols on all carriers) lasts for 4 μs and carries anywhere between 48 and 288 coded bits. For example, at 54 Mbps, the OFDM symbol has 216 data symbols. With a code rate of 3/4, the number of coded bits/symbol will be $4 \times 216/3 = 288$. This is possible by using different modulation schemes – ranging from binary phase shift keying where we have one bit per sub-channel to more complex modulation schemes like Quadrature Amplitude Modulation (QAM). Error control coding also plays an important role in determining the data rate. Table 8.6 summarizes some features of the different supported data rates.

The PLCP in the case of 802.11a is a bit different in that there is no synchronization field. A rate field with 4 bits indicates the data rate that is being transmitted. This field is shown in Table 8.6 for different data rates. The preamble and header are always modulated using BPSK (lower data rates). The values of SIFS and the slot for backoff in this option are 16 and 9 μs respectively.

The IEEE 802.11n Standard

The IEEE 802.11a and 802.11g MAC and PHY layers constrain the raw data rate to 54 Mbps and the throughput to a fraction of that depending on traffic load, channel conditions and so on. Thus, a

Table 8.6   Data rates and associated parameters in IEEE 802.11a

Data rate (Mbps)	Modulation	Code rate	Data bits/ symbol	Coded bits/ sub-channel	PLCP rate field
6	BPSK	1/2	24	1	1101
9	BPSK	3/4	36	1	1111
12	QPSK	1/2	48	2	0101
18	QPSK	3/4	72	2	0111
24	16-QAM	1/2	96	4	1001
36	16-QAM	3/4	144	4	1011
48	64-QAM	2/3	192	6	0001
54	64-QAM	3/4	216	6	0011

new task group started work on an IEEE 802.11n standard that would look at both MAC and PHY enhancements to improve the throughput to more than 100 Mbps (to up to 600 Mbps). Note that this throughput was not simply the raw data rate on the air, but the actual throughput of the network. For sometime, there were two competing proposals in Task Group N for the PHY and MAC layers – WWiSE (World-Wide Spectrum Efficiency) and TGnSync with many vendors in each group. Both proposals used MIMO at the physical layer. Products conforming to parts of these proposals were also available in the market. Neither of the proposals was successful in obtaining 75% of the vote. In January 2006, these two proposals merged with a third proposal that was finally approved.

Some of the ideas that were floated to improve throughput and made it to the standard were to use directional antennas or beamforming, *channel bonding* of two 20 MHz channels for a wider 40 MHz channel, multiple-input multiple-output (MIMO) with OFDM and throughput enhancements at the MAC layer. As we have mentioned in Chapter 3, MIMO enables the spectral efficiency of links to go well above the 1 bps/Hz that is usually the order in traditional systems to tens of bps/Hz. This increase in spectral efficiency is possible through the use of *space-time techniques* (see Chapter 3 and Chapter 13) such as space-time coding, beamforming, and spatial multiplexing. These techniques either increase the reliability of the link through diversity, increase capacity by canceling interference, or from the simultaneous transmission of multiple data streams from multiple antennas.

The primary MAC enhancement to improve throughput in 802.11n is the use of frame aggregation to reduce overhead [Xia05]. At very high data rates, the overhead of waiting times, backoff and frame headers can reduce throughput significantly. One method of reducing this overhead is to aggregate frames – either at the MAC or PHY layer. Similarly, acknowledgments can also be delayed and aggregated. Frame aggregation can be used for single destinations, multiple destinations, and use multiple rates for multiple destinations.

The IEEE 802.11ac Standard

The IEEE 802.11n standard provided throughput improvements. However, technology has to evolve with the applications that emerge over time. Over the last few years, streaming of high quality video has become a very important application. More people now watch streaming video on demand, not only on their televisions, but on other devices like laptops and tablet computers. Simultaneous video streams and other real-time bandwidth-intensive applications like gaming necessitated the move towards higher throughputs with WLANs.

Table 8.7   Summary of PHY alternatives in IEEE 802.11

Standard	Spectrum – US (GHz)	Data rates (Mbps)	Transmission scheme
Base IEEE 802.11	2.402–2.479	1, 2	GFSK, FHSS
	2.402–2.479	1, 2	B/QPSK, DSSS
	850–950 nm	1, 2	PPM, IR
802.11a	5.15–5.35, 5.725–5.825	6–54	OFDM
802.11b	2.402–2.479	1, 2, 5.5, 11	CCK
802.11g	2.402–2.479	1–54	OFDM, CCK
802.11n	2.4 and 5.0	Up to 600	MIMO/OFDM
802.11ac	2.4 and 5.0	> 1 Gbps	MIMO/OFDM and multi-user MIMO

The result of this need has been the emergence of the IEEE 802.11ac standard [Nee11] that is being rolled out at the time of this writing. This standard incorporates channel bonding that exceeds that of IEEE 802.11n. Instead of bonding two 20 MHz channels, IEEE 802.11ac allows the bonding of up to four channels for a total bandwidth of 80 MHz! Further, two 80 MHz channels can be used simultaneously by a single device to increase the throughput further. Of course this is possible only in the 5 GHz bands where the bandwidth exists and not in the 2.4 GHz bands. Two other changes to the physical layer are expected to help the throughput increase. The first is the use of 256-QAM in addition to the 64-QAM modulation scheme that has been used with IEEE 802.11a/g/n. With the highest code rate of 5/6 and two spatial streams with MIMO, a link in IEEE 802.11ac can have a raw data rate of over 800 Mbps in an 80 MHz bonded channel. The second is the use of *multi-user* MIMO where the access point can transmit data over multiple spatial streams to multiple mobile stations *simultaneously*. This increases the overall throughput of the network since mobile stations need not wait for channel access and the access point can transmit at up to four times the 800+ Mbps rate.

Summary of PHY Layer Alternatives

Table 8.7 summarizes the different PHY layer alternatives in IEEE 802.11.

8.3.4   Capacity of Infrastructure WLANs

As a mobile user moves inside the coverage area of a cell, the received SNR changes and can be modeled as a random variable. Figure 8.25, for example, illustrates the statistical behavior of the SNR as a mobile station is located in different locations of a shopping mall. For the traditional circuit-switched voice applications, each user has a single data rate independent of its received SNR and we calculate the capacity in terms of the number of available voice channels in a cell or in the entire network, which is a function of total available bandwidth (see Chapter 5). All of the modern wireless data applications are multi-rate systems for which the data rate is adjusted depending on the received SNR and packet losses. Examples of multi-rate WLAN and mobile data systems are IEEE 802.11a/g and HDR services. In multi-rate systems each MS can operate at one of the multiple choices of the data rates according to the value of its received SNR. In a single user

Data rate (Mbps)	Coverage distance (m)	Area of coverage (m^2)	$p_n = \dfrac{A_i}{\pi D_4^2}$
$R_1=11$	$D_1=50$	$A_1=7850$	0.19
$R_2=5.5$	$D_2=70$	$A_2=7536$	0.18
$R_3=2$	$D_3=90$	$A_3=10048$	0.24
$R_4=1$	$D_4=115$	$A_4=16092$	0.39

(b)

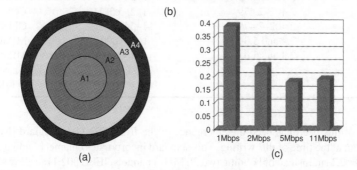

(a)   (c)

Figure 8.25 Data rates and coverage areas for the IEEE 802.11b in a semi open indoor area (a) coverage area for different data rates, (b) calculation of probabilities for data rates, (c) probability density function of the data rates.

environment the data rate of a MS is the average of all data rates it operates at while moving in the area. In a multiple user environment, the average data rate per MS is also a function of the medium access control technique and the number of MSs in the area. In the rest of this section we provide a framework for understanding the capacity of a wireless data network regardless of its medium access control.

In wireless data applications, at short distances from the access point or base station, a multirate MS has its highest data rate. As the distance increases, the received signal strength and consequently SNR reduces until a point where the necessary SNR for the highest data rate is not available and the modem has to be switched to the next lower data rate. As the distance continues to increase the data rate continues to fall to lower rates until the signal strength falls below the coverage of the access point or the base station at the lowest allowed rate. In an infrastructure WLAN with multiple APs providing a comprehensive coverage we expect that when the signal falls below one of the lower thresholds, there is another access point or base station to connect to. Therefore, if we consider an area that is covered by a wireless data service, the data rate available to the user has a spatial distribution in which associated with any location is one of the available multiple rates of the system. In other words, the data rate in a random location in the area of the coverage forms a *discrete random variable*. One way to define a capacity for this multirate system with a statistical data rate is to define the *spatial capacity* as the *average* of the data rates that a user randomly located in the area of the coverage observes. With this definition the spatial capacity will be given by:

$$R_{av} = \sum_{n=1}^{N} p_n R_n \tag{8.1}$$

where R_{av} is the average spatial data rate, R_n is one of the available multirates, and p_n is the probability of occurrence of that data rate, which is the ratio of the areas in which we have that specific data rate to overall area of the coverage of the access point or the base station.

Example 8.5: Spatial capacity of IEEE 802.11b

IEEE 802.11b supports four data rates namely 11, 5.5, 2, and 1 Mbps. In a semi-open indoor area these data rates can be used up to distances of 50, 70, 90, and 115 m, respectively. Figure 8.25a shows the area of coverage and circles around an AP that provides different data rates. If a terminal in located randomly in the area of coverage, the probability of being in each of the areas is given by the ratio of the area for the specific data rate to the total coverage area. That is, $p_n = A_i / \pi r^2$ where r is the radius of the largest circle (115 m). Figure 8.25b shows the data rate, distance coverage, annular area, and the probability of having a certain data rate. Figure 8.25c shows the probability density function of the data rates calculated from the ratio of the coverage area for a data rate and the overall coverage area. If we substitute the data rates and their probabilities from the density function in the Figure 8.25 into Equation 8.1, the average data rate or the spatial capacity of the AP is 2.584 Mbps, which is well below the expected 11 Mbps.

The above scenario provides a worst-case condition for an infrastructure WLAN. In actual deployment, network designers try to have overlapping cells and overlaps are at the low data rates reducing their contribution to the average data rate. The upper bound for spatial capacity in a cellular deployment is achieved when the adjacent APs are close enough that the user always observes the highest data rate (11 Mbps in Example 8.6). For the popular grid installation, as shown in Figure 8.26, the minimum distance between the access points (size of the grid) to support the maximum spatial

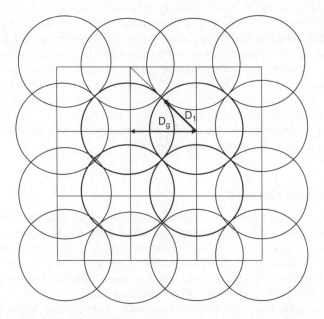

Figure 8.26   Grid deployment with optimal spatial capacity.

diversity is $r_g = \sqrt{2}r_1$ where r_1 is the radius of the circle where the maximum data rate (50 m in example 8.6) is available. For this situation the probability density function of the data rates for the system is only an impulse at 11 Mbps with a probability of one. As the grid distance between the access points increases beyond r_g the probability density function of the data rates starts to include lower data rates resulting in decreases in the spatially averaged data rate. As mentioned earlier, if we have no coverage gaps, the minimum spatial throughput cannot be less than the spatial throughput of a single access point (2.584 Mbps for the 802.11b in Example 8.6). In our discussions in this chapter we addressed single floor indoor environments. For some discussions on the multi-floor environment the reader can refer to [Hil01].

8.3.5   Security Issues and Implementation in IEEE 802.11

Security in wireless networks is an important problem especially because it is extremely difficult to contain radio signals within a protected perimeter [Edn04]. Anyone can listen to radio signals and anyone can also potentially inject signals into the network. Typically, in any network wireless or wired, it is common to deploy security features or services like confidentiality, entity authentication, data authentication and integrity and so on to protect against security threats [Sti02]. The IEEE 802.11 standard has some mechanisms to provide confidentiality, integrity and authentication at the link level (see Chapter 7 for more details). All data that leaves the 802.11 link will not be protected. For instance, a MS communicating with an AP can have all its IEEE 802.11 frames that are on the air be protected. Once the AP receives the frame, all protection is removed before it is transmitted on the distribution system. So additional security at the higher layers (such as IPSec or the Secure Sockets Layer – SSL) may be required for some applications if the payload needs to be secure.

The original mechanism for providing confidentiality and authentication in IEEE 802.11 is called *Wired Equivalent Privacy* (WEP) [Gas02; Edn04]. Over the last few years, several techniques for compromising WEP have been published in the literature. Tools such as AirCrack, Kismet and WEPcrack are freely available that can be used to extract the secret key used in WEP encryption. WEP makes use of the RC4 stream cipher with 40 bit keys (although there are options to use 128 bit keys in most commercial products today). Both the implementation of WEP and the RC4 algorithm itself have vulnerabilities that have rendered WEP not secure for today's applications. WEP was initially proposed in the standard as a self-synchronizing, exportable, efficient option. While it does satisfy these three properties, its security has left much to be desired.

In what follows we briefly discuss the original implementation of security in IEEE 802.11 and discuss enhancements that have been recently proposed.

Entity Authentication with WEP

The mandatory entity authentication mechanism in IEEE 802.11 is called *Open-System Authentication*. In this case, there is no real authentication. If one IEEE 802.11 device sends a frame to another, it is implicitly accepted. For example, a MS may simply send a frame to the AP choosing "open system" as the authentication algorithm (authentication algorithm = 0). The AP will simply accept it if open system access is allowed and send a response. From this transaction, the AP will obtain the MAC address of the MS for communication purposes.

A better authentication procedure is called *Shared-Key Authentication* where WEP is implemented [Gas02]. If the network is using WEP, shared key authentication is mandatory. The assumption is that all devices in the network share a secret key. A MS will send a frame for authentication with sequence number 0, the authentication algorithm set to 1 (to indicate shared key authentication). The AP will then send a challenge message (128 bits) in clear text to the MS along with its response. The MS will respond with an encrypted version of the challenge text. If the AP is able to verify the integrity of the reply, the MS is authenticated and it has the shared WEP key configured in it. Sometimes a MS will authenticate itself with several APs before associating itself with one of them. This process is called preauthentication. This authentication scheme is still not very secure however and also creates weaknesses in the protocol due to the way in which it is employed with a stream cipher (see Chapter 7 for details).

Several commercial products also implement address filtering where only certain MAC addresses are allowed access to the network. This is not part of the standard and it is also possible for malicious users to spoof MAC addresses easily. However, address filtering is an additional security measure that is available for IEEE 802.11 networks.

Confidentiality and Integrity with WEP

Confidentiality is simply provided in IEEE 802.11 by encrypting all packets using the RC4 stream cipher. Stream ciphers operate as follows. Using a secret key, a pseudo-random sequence of bits (called the key stream) is generated. If this sequence has a very long period and the algorithm is strong, it will be computationally impossible for someone to generate the sequence without knowing the secret key. A pseudo random generator is used along with the 40-bit secret key to create a key sequence that is simply XOR-ed with the plaintext message. The pseudorandom sequence thus generated will be XOR-ed with the MAC frame to make the contents of the frame secure from interception. RC-4 is one algorithm to generate the pseudo-random key stream. This algorithm makes use of a secret key and in the case of WEP an initialization vector (IV) that is 24 bits long. Since the key is constant for all transactions, the same pseudorandom key stream is generated if the IV is not changed. An attacker could capture two streams of encrypted frames, XOR them together and eliminate the key stream. He would then have an XOR of two data frames. If by some chance he knows the contents of one data frame, he can get the other as well. Since the IV is only 24 bits long, it is possible for an attacker to break the encryption scheme. One well-publicized attack is the Fluhrer, Mantin, Shamir (FMS) attack on RC-4. In addition, there are several weak keys that could make the encryption scheme easier to break. It is also possible for an attacker to replay packets depending on the sequence numbers that are being used. In order to ensure that an attacker has not modified a message, the WEP protocol uses the in-built CRC to verify the integrity of the message. Checking the integrity of the message using the CRC has vulnerabilities that have been publicized in recent years.

Key Distribution in WEP

The IEEE 802.11 standard does not specify how the shared keys must be distributed to devices (AP and MSs). It is usually a manual installation of keys where a user will type the key in the device driver software. This process is unfortunately not scalable and also has several human vulnerabilities. Users may write down the key on a piece of paper when they buy a new device and lose this paper. Some vendors have automated methods of key distribution. Cisco's Light Extensible Authentication Protocol makes use of the challenge-response mechanism to generate a key at the AP and an identical matching key locally in the MS that could then be used in a successful encrypted communication.

Security Features in 802.11i

The Task Group I of the IEEE 802.11 working group has prepared an enhanced security framework for IEEE 802.11 called 802.11i that was approved as a standard in June 2004. Several vendors have already implemented elements of this standard. This framework includes what is called a *robust security network* (RSN) that is similar to WEP, but has several new capabilities in devices [Edn04]. It is possible for both WEP and RSN devices to coexist in a *transitional security network* (TSN).

A consortium of major WLAN manufacturers called the Wi-Fi Alliance considered options to improve security in legacy devices while 802.11i was being standardized. The proposal from this alliance is called Wi-Fi Protected Access (WPA) that introduces enhancements to WEP called *Temporal Key Integrity Protocol* (TKIP). In this protocol, RC-4 is still used as the encryption algorithm. However, this protocol adds some features to overcome weakness of WEP. A message integrity code is used instead of the CRC check. It changes the way in which IVs are generated. It changes the encryption key for every frame, increases the size of the IV and also adds a mechanism to manage keys.

In 802.11i, RC-4 is replaced by the Advanced Encryption Standard (AES). In particular, the key stream and message integrity check will be generated by a Counter- mode Cipher-clock-chaining MAC Protocol (CCMP). AES is a block cipher – it operates on fixed blocks of data unlike a stream cipher that generates a key stream. However, any block cipher can operate in different *modes* and cipher-block-chaining (CBC) is one such mode of operation. The counter mode is another mode of operation. It is expected that the counter mode will be used to generate the key stream and the CBC will be used to generate the message integrity check. Both these modes have been used in other systems with good security.

Both TKIP and AES-CCMP provide confidentiality and message integrity. In order to perform entity authentication, the IEEE 802.11 system has to still rely on challenge response protocols. Over the years, there have been several protocols developed for dial-up entity authentication and for port security in wired LANs. These include 802.1X, the extensible authentication protocol (EAP) and Remote Authentication Dial-In User Service (RADIUS). Note that all these protocols are not equivalent – for instance, both 802.1X and RADIUS could use EAP for entity authentication and key distribution. EAP itself would use some challenge-response protocol like Challenge Handshake Authentication Protocol (CHAP) or SSL to authenticate the devices. Both WPA and RSN mandate 802.1X and EAP as part of the access control mechanism for 802.11 networks. Note that access control is increasingly becoming an important problem with the emergence of hot spot networks in airports, cafes, and so on.

Questions

1. Name three categories of unlicensed bands used in the United States and compare them in terms of size of the available band and coverage.
2. How does the current state of the art data rate of the wired (Ethernet) and wireless LANs (802.11) compare with one another?
3. Why are unlicensed bands essential for the WLAN and WPAN industries?
4. Explain the difference between the wireless Inter-LAN bridges and wireless LANs.
5. What three topologies can IEEE 802.11 WLANs operate in? What are the differences?
6. Name four major transmission techniques considered for WLAN standards and give the standard activity associated with each of them.
7. Compare OFDM and spread spectrum technology as PHY alternatives for WLANs.

8. Give the physical specification summary of the DSSS and FHSS used by the legacy IEEE 802.11.
9. What are the MAC services of IEEE 802.11 that are not provided in the traditional LANs such as 802.3.
10. Why does the MAC layer of 802.11 have four address fields compared to 802.3 that has two?
11. What is the PCF in 802.11, what services does it provide, and how is it implemented?
12. Explain the difference between a hidden terminal and an exposed terminal.
13. Explain why an AP in 802.11 also acts as a bridge?
14. What are the responsibilities of the MAC Management sublayer in 802.11?
15. What is the purpose of PIFS, DIFS, and SIFS time intervals and how they are used in the IEEE 802.11?
16. What is the difference between a Probe and a Beacon signal in 802.11?
17. Explain the operation of the timing of the beacon signal in 802.11.
18. How is authentication and integrity provided in IEEE 802.11?
19. What modulation scheme and code rates are used in IEE 802.11a/g to provide a raw data rate of 36 Mbps? Explain how this rate is achieved in terms of number of symbols and bits per second.
20. How is IEEE 802.11n different from IEEE 802.11a or IEEE802.11g?
21. How is IEEE 802.11ac different from IEEE 802.11n?
22. What are the differences between the different methods of deployment of infrastructure WLANs?

Problems

Problem 8.1

You want to transmit the information sequence 00111100 using CCK as in 802.11b. What is the CCK codeword in vector form? Show all steps. Assume that bit d0 is the left most bit.

Problem 8.2

a. Use the equation for generation of CCK to generate the complex transmitted codes associated with the data sequence $\{0,1,0,0,1,0,1,1\}$
b. Repeat (a) for the sequence $\{1,1,0,0,1,1,0,0\}$
c. Show that the two generated codes are orthogonal.

Problem 8.3

The original WaveLAN, the basis for the IEEE 802.11 uses an 11-bit Barker code of $[1,-1,1,1,-1,1,1,1,-1,-1,-1]$ for DSSS.

a. Sketch the aperiodic autocorrelation of the code (see Chapter 12 for details).
b. If we use the system using random codes with the same chip length in a CDMA environment, how many simultaneous data users we can support with an omni-directional antenna and one access point?

Problem 8.4

a. If in the PPM-IR PHY layer used for the IEEE 802.11 instead of PPM we were using baseband Manchester coding, what would be the transmission data rate? Your reasoning must be given.
b. What is the symbol transmission rate in the IEEE 802.11b? How many complex QPSK symbols are used in one coded symbol? How many bits are mapped into one transmitted symbol? What is the redundancy of the coded symbols (the ratio of the coded symbols to total number of choices)?
c. What is the symbol transmission rate of the coded symbols per channel in the IEEE802.11g ? How does this symbol rate relate to the data rates (6, 9, 12, 18, 27, 36, 54 Mbps) and convolutional coding rates (1/2, 3/4, 9/16)?

Problem 8.5

Redraw the timing diagram of Figure 8.8, assuming that all MSs use RTS/CTS mechanism to send packets.

Problem 8.6

A voice over IP application layer software generates a 64Kbps coded voice packet every 20 ms. This software is installed in two laptops with WLAN PCMCIA cards communicating with an AP connected to a Fast Ethernet (100 Mbps).

a. What is the length of the voice packets in milliseconds, if the PCMCIA cards were DSSS IEEE 802.11?
b. If the two terminals start to send voice packets almost at the same time. Give the timing diagram to show how the first packets are delivered though the wireless medium to the AP using CSMA/CA mechanism.
c. Repeat (b) and (c) if 802.11b at 11 Mbps was used instead of DSSS 802.11. How would this change if 5.5 Mbps was the data rate?

Problem 8.7

Figure P8.1 shows the layout of an office building. If the distance between the AP and the MSs 1, 2, and 3 are 50, 65, and 25 m, respectively, determine the path loss for between the AP and MSs:

a. Using path loss per wall model and free space loss (assume that the wall loss is 3 dB per wall).
b. Using the 802.11 path loss model from Chapter 2.

Using the 802.11 transmitted and received power specifications, determine whether a single AP can cover the entire building.

Problem 8.8

You are designing a wireless LAN for an office building. You are not able to perform measurements or site surveys and have to rely on statistical models and certain other information. There are also

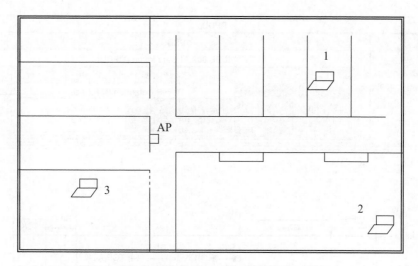

Figure P8.1 Layout of an office building.

certain constraints on where you can actually place the access point(s). You have the following information available to you:

Maximum number of walls between an access point and a mobile terminal $= 4$.

Maximum number of floors between an access point and the mobile terminal $= 2$.

Transmit power possibilities $= 250$ mW and 100 mW.

Sensitivity of receiver is -90 dBm.

Maximum distance from access point to building edge $= 30$ m.

Building has office walls, brick walls, and metallic doors.

Shadow fading margin $= 8$ dB.

What would be a conservative estimate of the number of access points required for the WLAN set-up? Why? State your assumptions, models, and provide reasons for all your assumptions and calculations. *Hint: Use path loss models from Chapter 2 that are applicable to indoor areas.*

Problem 8.9

Suppose the coverage areas where 11, 5.5, 2, and 1 Mbps data rates are reliably available have radii of 20, 30, 40, and 50 m instead of the values in Example 9.6. What is the spatial capacity of the access point? How would the spatial capacity change if the 1 Mbps data rate was available up to 75 m? Plot the spatial capacity versus the range of the 1 Mbps coverage area as it varies from a radius of 50–100 m, assuming all other values remain the same.

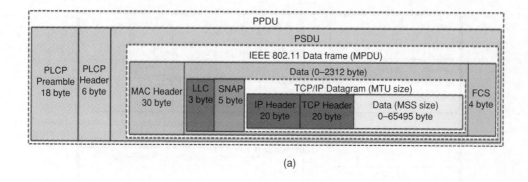

(a)

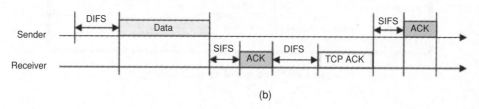

(b)

Figure P8.2 Packet transmission in the IEEE 802.11 b. (a) Overheads for the formation of a packet. (b) Overheads for successful transmission of a TCP packet.

Problem 8.10

Figure P8.2a shows the overhead for packet formation and applications using TCP packets. Each TCP packet can have a length of up to 65 495 bytes that should be fragmented to fit the maximum MAC packet of 2312 byte. The TCP/IP header is 40 bytes, the 802.2 LLC/SNAP header is 8 bytes, and the 802.11 MAC and PLCP headers and synchronization preamble are 34 and 24 bytes respectively. The TCP ACK is a TCP header with no application data and the MAC ACK is shown in Figure P8.2b. Assuming SIFS and DIFS intervals of 10 and 50 μsec, respectively, determine the application throughput of the 802.11b for data rates of 11, 5.5, 2, and 1 Mbps for data packets of length 100 and 1000 bytes.

Problem 8.11

In reality the throughput of a WLAN is a function of the channel characteristics and it fluctuates in time. Figure P8.3 shows a typical application throughput of an 802.11b terminal in an observation time of 1 min. Due to the channel fading and other imperfections this throughput varies in time as we measure it in a certain distance between the transmitter and the receiver. As the distance between the transmitter and the receiver increases and the RSS reduces this average throughput also reduces. The throughput (Mbps) versus distance (m) relation of an IEEE 802.11b in an office building is empirically determined to follow the following approximate equation:

$$S_u(r) = -0.2r + 5.5 \qquad \text{(P8.1)}$$

a. Determine the maximum throughput (in Mbps) and maximum coverage (in m)?

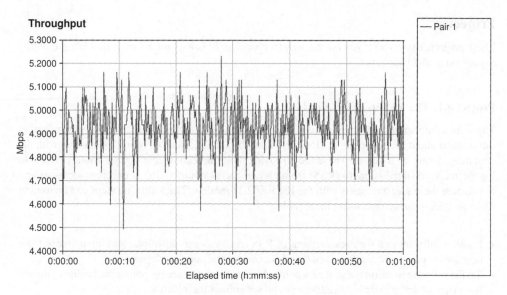

Figure P8.3   Throughput variations in one location for 802.11b.

b. Show that we can find the average throughput of a user randomly walking in the coverage area of an access point by:

$$\bar{S}_u = \frac{2 \int_0^{R_L} r \, S_u(r) dr}{R_L^2} \quad [Mbps] \tag{P8.2}$$

In which R_L is the distance for which the thoughput of the WLAN approaches to zero and the WLAN no longer has coverage.

c. Use Equations P8.1 and P8.2 to calculate the average throughput of a user randomly walking in the coverage area of the WLAN.

d. Compare your results with the minimum and maximum nominal data rates of IEEE 802.11b. Explain the difference between your results.

Problem 8.13

The average throughput versus distance relationship of an IEEE 802.11g in a typical office building is measured to fit the following function:

$$S_u(r) = \begin{cases} 22; & 0 < r < 1 \\ -22 \log 10r + 25; & r > 1 \end{cases} \tag{P8.3}$$

a. Use Equations P8.1 and P8.2 to calculate the average throughput of a user randomly walking in the coverage area of the WLAN.

b. Compare your results with the minimum and maximum nominal data rates of the IEEE 802.11g. Explain the difference between your results.

Projects

These projects assume that you have a wireless network at home and a laptop that has the ability to connect to WiFi networks.

Project 8.1: The RSS in IEEE 802.11

There are a number of software tools (e.g., WirelessMon by PassMark), which can be used to gather information about access points in close proximity of a mobile station. Many tools come with the operating system on a laptop. These tools provide multiple features but we are going to use it to log the received signal strength (RSS) from chosen access points (APs) at different locations and to compare these measurements with the IEEE 802.11 models. The following steps can be used to make an RSS measurement using these tools.

- Install a software tool for measurement of RSS (e.g., you can download wirelessmon.exe from http://www.passmark.com/products/wirelessmonitor.htm) on your laptop.
- Set the software to monitor the access point of your choice, the access points can be distinguished from each other by their MAC addresses and sometimes their SSIDs.
- Modify the logging options of the software for recording the characteristics of an AP.
- Record the RSS readings from a specific AP.

a. Do "war driving" in a specific floor of your building, for which you have a floorplan schematic available, to find the exact location of the AP in that floor. Show the locations in the schematics of the building.
b. Select five different locations on the floor of your choice which are approximately 1, 5, 10, 20, and 30 m away from your AP of choice. Spread the points over the entire floor and mark them on your schematic floor plan. Determine the distance from the selected points to each of the AP locations in that floor.
c. Measure the RSS at each location for at least 1 min. Calculate the *average* RSS received from each AP in each location and record them in a table, which relates the distance to the RSS from your target AP.
d. Use the table to generate a scatterplot of the average RSS (in dBm) versus the distance (in logarithmic scale) for all APs in your target floor.
e. Find the best fit 802.11 model for your data.
f. Use www.speakeasy.net/speedtest to record the measured data rate in each of the five locations.
g. Explain the correlation among the throughput from speakeasy and the power and distance at each location.

Project 8.2: Coverage and Data Rate Performance of IEEE 802.11 WLANs

I. Modeling of the RSS

To develop a model for the coverage of the IEEE 802.11b/g WLANs, a group of undergraduate students at WPI measured the received signal strength (RSS) in six locations in the third floor of the Atwater Kent Laboratory (AKL) at WPI, shown in Figure P8.4. After subtracting the RSS from the transmitted power recommended by the manufacturer, they calculated the path-loss for all the points that are shown in Table P8.1.

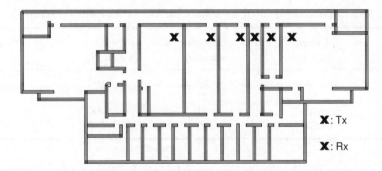

Figure P8.4 Location of the transmitter and first five locations of the receiver used for calculation of the RSS and path-loss.

To develop a model for the coverage of the WLANs they used the simple distance–power gradient model:

$$L_p = L_o + 10\alpha \log d$$

where d is the distance between the transmitter and the receiver, L_p is the path-loss between the transmitter and the receiver, L_o is the path-loss at the first meter, and α is the distance–power gradient. One way to determine L_o and a from the results of measurements is to plot the measured L_p versus $\text{Log} d$ and find the best fit line to the results of measurements.

a. Use the results of measurements by the students to determine the distance–power gradient, α and path loss at the first meter from the transmitter, L_o. In your report provide the MATLAB® code and the plot of the results and the best-fit curve.
b. Manufacturers often provide similar measurement tables for typical indoor environments. Table P8.2 shows the RSS at different distances for open areas (an area without wall), semi-open areas (typical office areas), and closed areas (harsher indoor environments) provided by PROXIM, one of the manufacturers of WLAN products. Use the results of measurements from the manufacturer and repeat part (a) for the three areas used by the manufacturer. Which of the measurements areas used by the manufacturer resembles the third floor of the Atwater Kent laboratory where the students took measurements? Assume that the transmitted power used for

Table P8.1 The distance between the transmitter and the receiver and the associated path-loss for the experiment

Distance (m)	Number of walls	L_p (dB)
3.0	1	62.7
6.6	2	70.0
9.5	3	72.75
15.0	4	82.75
22.5	5	90.0
28.8	6	93.0

Table P8.2   Data rate, distance in different areas, and the RSS for IEEE 802.11b (Proxim)

Data rate (Mbps)	Closed area (m)	Semi-open area (m)	Open area (m)	Signal level (dBm)
11	25	50	160	−82
5.5	35	70	270	−87
2	40	90	400	−91
1	50	115	550	−94

these measurements was 20 dBm. In your report include the plots used for calculations of the distance–power gradient at different locations.

II. Coverage study

IEEE 802.11b/g WLANs support multiple data rates. As the distance between the transmitter and the receiver increases the WLAN reduces its data rate to expand its coverage. The IEEE 802.11b/g standards recommend a set of data rates for the WLAN. The first column of Table P8.3 shows the four data rates supported by the IEEE 802.11b standard and the last column represents the require RSS to support these data rates. Table P8.3 shows the data rates and the RSS for the IEEE 802.11g provided by Cisco.

a. Plot the data rate versus coverage (staircase functions) for IEEE 802.11b WLANs for closed, open, and semi-open areas using Table P8.2 provided by Proxim. Discuss the coverage versus data rate performance in different areas and relate them to the value of α of different areas, calculated in part I of the project.

b. Using α and L_0 found for the third floor of the Atwater Kent Labs, plot the data rate versus coverage (staircase functions) for IEEE 802.11b and g WLANs operating in that area. Discuss the differences in data rate versus. coverage performance of 802.11b and g in the third floor of the Atwater Kent Laboratories.

Table P8.3   Data rates and the RSS for the IEEE 802.11g (Cisco)

Data rate (Mbps)	RSS (dBm)
54	−72
48	−72
36	−73
24	−77
18	−80
12	−82
9	−84
6	−90

9

Low Power Sensor Networks

9.1 Introduction

As we described in Chapter 1, the air–interface of the wireless networks evolved around two distinctive paths, one lead by the cellular telephone to connect handsets to the PSTN, focused on circuit switched telephone network access to mobile devices, and the other around WLANs enabling wireless data connection to the Internet for laptops and other devices. This evolution started in the early 1980s, when the PSTN was the most popular medium for communication, and by far the largest sector of the information networking industry. Throughout this evolution, the habits of the world population for communicating with one another gradually shifted towards the data applications supported by the Internet. In parallel to that, the WLAN industry gained increasing importance. The air–interfaces designed for WLAN data applications influenced the migration of the cellular air–interface from voice centric TDMA and CDMA to data centric OFDM designs with centralized scheduling (see Chapter 13). Besides, the wireless personal area networking (WPAN) industry emerged to complement WLANs and extend the applications to new domains. WPANs complement WLANs in two distinct ways. First by reducing the power consumption of devices used by lower speed ad hoc sensor networking applications, where the battery life needs to be extended far beyond the existing battery life of devices connecting to WLANs. Second in gigabit wireless local networking to increase the data rate of WLANs beyond the existing products and standards. These applications prompted the IEEE 802.15 standardization activities and resulted in a number of standards that evolved in that committee since the late 1990s. Figure 9.1 illustrates the relationship between the data rate and a number of popular applications and how WPAN technologies are perceived to complement the WLAN technologies in supporting these applications.

The very first WPAN to appear in the literature was the BodyLAN that emerged from a DARPA project in the mid-1990s [Den96]. The BodyLAN was a low power, small size, inexpensive, WPAN with modest data rates that could connect personal devices with a range of around 5 feet (1.75 m) on and around a human being. Motivated by the BodyLAN project, a WPAN group was originally started in June 1997 as a part of the IEEE 802.11 standardization activity. In January 1998, the WPAN group published the original functionality requirements for WPANs. In May 1998, the development of Bluetooth was announced and a Bluetooth special group was formed within the WPAN group [Sie00]. In March 1999, the IEEE 802.15 working group was approved as a separate group in the 802 community to handle WPAN standardization. At the time of writing, the IEEE

Principles of Wireless Access and Localization, First Edition. Kaveh Pahlavan and Prashant Krishnamurthy.
© 2013 John Wiley & Sons, Ltd. Published 2013 by John Wiley & Sons, Ltd.

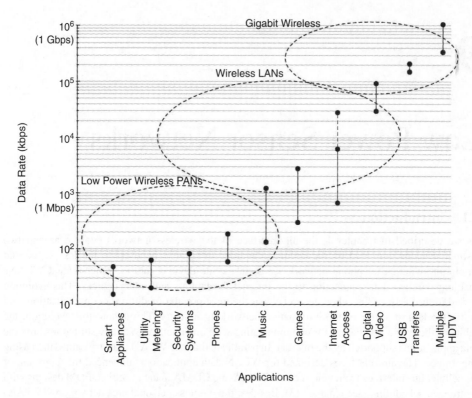

Figure 9.1   Applications, bandwidth requirements, and WLAN/WPAN.

802.15 WPAN group is a major standardization committee with a number of subcommittees for gigabit wireless and low power sensor networking.

In the standardization for low power sensor networks, the IEEE 802.15 community has completed the very successful IEEE 802.15.1 Bluetooth and IEEE 802.15.4 standards, which specifies the PHY and MAC layers and works with ZigBee applications. The IEEE 802.15.6 group is working on the emerging body area networks (BANs). In this chapter we address the technical aspects of the two most popular low power and lower data rate WPAN technologies, Bluetooth and IEEE 802.15.4 (we will call this ZigBee) and we provide an introduction to BANs the emerging BAN technology. Bluetooth and ZigBee both operate in the 2.4 GHz ISM unlicensed bands where WLAN technologies also operate. In describing the technical aspects of these technologies, we only provide the details of the architecture and communication layers. In contrast, BAN technology is emerging for operation in several bands and its focus of applications is in inside and around the human body. This is a new and unexplored medium for radio propagation and so, in describing this technology we put more emphasis on radio channel modeling.

9.2   Bluetooth

Bluetooth and ZigBee technologies draw upon the original spread spectrum based air–interface first implemented with the legacy IEEE 802.11 standard. Although the air–interface for IEEE

802.11 WLANs has since evolved to use OFDM and MIMO technology, legacy spread spectrum transmission has become the technology of choice for low power WPANs in the IEEE 802.15 standards. This is not incidental, because as we discussed in Chapter 3, spread spectrum technology is the technology of choice for low power and low data rate wireless air–interface design. At the MAC layer, Bluetooth was originally designed with the support of companies designing equipment for cellular telephony. It filled the void for mostly voice-oriented WPAN applications. ZigBee followed the IEEE 802.11 MAC as well as the PHY layer and emerged as the low-power WPAN technology for sensor data applications. In this chapter we explain these two technologies and how they relate to the legacy IEEE 802.11 technology. This discussion provides a good overview of the centralized versus random access for voice and data applications in the personal area networks designed for covering only a few meters. The examples in the discussion also include the details of FHSS and DSSS transmission techniques in a WPAN environment.

Bluetooth is an open specification for short-range wireless voice and data communications. It was originally developed as a cable replacement in personal area networks to operate all over the world. In 1994, the initial study for development of Bluetooth started at Ericsson in Sweden. In 1998, companies such as Ericsson, Nokia, IBM, Toshiba, and Intel formed a special interest group to expand the concept and develop a standard under the IEEE 802.15 banner. In 1999, the first specification, v1.0b, was released and then accepted as the IEEE 802.15 WPAN standard for 1 Mbps networks. Today, Bluetooth has penetrated the huge smart phone market as well as numerous consumer devices.

The story of the origin of the name Bluetooth is interesting and worth mentioning. "Bluetooth" was the nickname of Harald Blaatand, 940 – 981 A.D., King of Denmark and Norway. When Bluetooth was introduced to the public, a stone carving, shown in Figure 9.2, claimed to be erected from Harald Blaatand's capital city Jelling, was also presented [Blu00]. This strange carving was interpreted as Bluetooth connecting a cellular phone and a wireless notepad in his hands. This

Figure 9.2 Picture of Bluetooth on the stone.

picture was used to symbolize the vision in using "Bluetooth" to connect personal computing and communication devices. Bluetooth, the king, was also known as a peacemaker and a person who brought Christianity to Scandinavians to harmonize their beliefs with the rest of the Europe. That fact was used to symbolize the need for harmony among manufacturers of WPANs around the world and to support the growth of the WPAN industry.

Bluetooth was the first popular technology for short-range ad hoc networking that was designed for integrated voice and data applications. Unlike WLANs, Bluetooth does not strive for very high data rates. It maintains effective data rates < 1 Mbps, but it has an embedded architectural design that is suitable to support voice applications with transmission of centrally controlled shorter packets. Bluetooth 3.0 allows the use of co-located WiFi signals to increase data rates (when needed). Since 2011, Bluetooth 4.0 specifies support for devices such as pedometers and heart-rate monitors with extremely low energy consumption expanding the potential applications where it can be employed. The batteries in such devices are expected to last for months without replacement or recharging.

The Bluetooth group originally considered three basic application scenarios that are shown in Figure 9.3 [Blu00]. The first application scenario, shown in Figure 9.3a, was for "wire replacement", to connect a personal computer or laptop to its keyboard, mouse, microphone, and notepad. As the name of the scenario indicates, it avoids multiple short-range wiring surrounding today's personal computing devices. The second application scenario, shown in Figure 9.3b, was for ad hoc networking of several different devices in very short range of each other, such as in a conference room. As we saw in Chapter 7, WLAN standards and products also commonly consider this scenario. The third scenario, shown in Figure 9.3c is to use Bluetooth as an access point to wide area voice and data services provided by the cellular networks, wired connections, or satellite links. The IEEE 802.11 community also considers this overall concept of the access point. However, the Bluetooth access point is used in an integrated manner to connect to both voice and data backbone infrastructures. Today, Bluetooth is mostly used for the first scenario and increasingly for connecting so-called "smart" devices such as fitness monitors to smart phones. It is also used to stream music to speakers or headphones from devices such as computers or MP3 players.

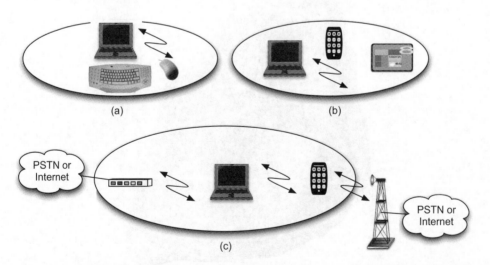

Figure 9.3 Bluetooth application scenarios: (a) cable replacement, (b) ad hoc personal network, (c) integrated access point.

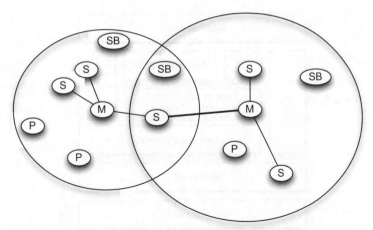

Bluetooth States: M-Master, S-Slave, P-Parked and SB-Stand-by

Figure 9.4   Bluetooth's scattered ad hoc topology and the concept of the *piconet*.

9.2.1   Overall Architecture

The topology of the Bluetooth is referred to as a *scattered ad hoc topology* that is illustrated in Figure 9.4. In a scattered ad hoc environment, a number of small networks each supporting a few terminals co-exist or possibly interoperate with one another. To implement such a network we need a plug and play environment. The network should be self-configurable, providing an easy mechanism to form a new small network, as well as participation in an existing small network. To implement that environment the system should be capable of providing different states for connecting to the network. The terminals should have options to associate with multiple networks at the same time. The access method should allow formation of small independent ad hoc connections as well as the possibility of interacting with large voice and data networks considered by Bluetooth.

To accommodate these features, the Bluetooth specification defines a small cell (similar to a basic service area in IEEE 802.11) as a *piconet* and identifies four states: master "M", slave "S", stand-by "SB" and parked/held, or parked "P" for a Bluetooth enabled terminal. The mobile terminal that initiates a connection is a Master device in that piconet. The devices connecting to the Master are called Slaves. As shown in Figure 9.4, the Bluetooth topology, however, allows "Slave" terminals to participate in more than one piconet. A Master terminal in the Bluetooth can handle seven simultaneous and up to 200 active slaves in a piconet. If access is not available, a terminal can enter the SB mode waiting to join the piconet later. A radio can also be in a "Parked" or a low-power connection. In the parked mode the terminal releases its MAC address while in the SB state it keeps its MAC address. Up to 10 piconets can operate in one area [Blu00]. Also note that in Figure 9.4, the coverage of the two master devices are different since Bluetooth allows for different classes of devices with different transmit powers.

Bluetooth specifications have selected the unlicensed ISM bands at 2.4 GHz for operation. The advantage is the worldwide availability of the bands and the disadvantage is the existence of other users, in particular IEEE 802.11 products in the same band. In the early 2000s, a sub-committee of IEEE 802.15 worked on the interference issues related to the Bluetooth and IEEE 802.11 and 11b that we considered in Chapter 5.

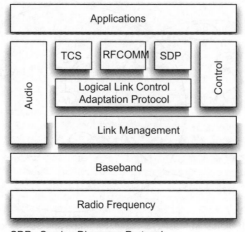

SDP : Service Discovery Protocol
TCS : Telephone Control Protocol
RFCOMM : RF Communications

Figure 9.5 Protocol stack of the Bluetooth.

9.2.2   Protocol Stack

One of the distinct features of Bluetooth is that it provides a complete protocol stack that allows different applications to communicate over a variety of devices. Other wireless local networks, such as the IEEE 802.11 usually specify only the lowest three lower layers for communications. The protocol stack for voice, data and control signaling in Bluetooth is shown in Figure 9.5 [Haa00]. The *RF layer* specifies the radio modem used for transmission and reception of the information. The *Baseband layer* specifies the link control at bit and packet level. It specifies coding and encryption for packet assembly and how the frequency hopping operation should work. The *Link Management Protocol* configures the links to other devices by providing for authentication and encryption, state of units in the piconet, power modes, traffic scheduling, and packet format. The *Logical Link Control and Adaptation Protocol* provides connection-oriented and connectionless data services to the upper layer protocols. These services include protocol multiplexing, segmentation and reassembly, and group abstractions for data packets up to 64 kilobytes in length. The audio signal is directly transferred from the application to the Baseband layer. Also the Link Management Protocol and the application exchange control messages interact to prepare the physical transport for an application. There are three other protocols above the Logical Link Control Adaptation Protocol. The *Service Discovery Protocol* finds the characteristics of the services and connects two or more Bluetooth devices to support a service such as streaming music, faxing, printing, teleconferencing, or e-commerce facilities. *Telephony Control Protocol* defines the call control signaling and mobility management for the establishment of speech for telephony application. Using these protocols legacy telecommunication applications can be developed.

Example 9.1: Telephony control protocol in Bluetooth

Figure 9.6 shows the protocol stack for implementation of the cordless telephone application. The audio signal is directly transferred to the Baseband layer while Service Discovery Protocol and

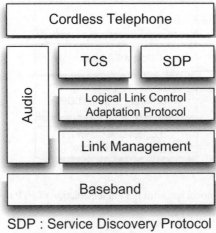

SDP : Service Discovery Protocol
TCS : Telephone Control Protocol

Figure 9.6 Protocol stack for the implementation of cordless telephone over Bluetooth.

Telephony Control Protocol operating over Logical Link Control Application Protocol and Link Management Protocol handle signaling and connection management.

The *Radio Frequency Communication (RFCOMM)* is a "cable replacement" protocol that emulates the standard short range wired serial interface conversion data signals over Bluetooth baseband. Using this interface protocol, a number of non-Bluetooth specific protocols can be implemented on the Bluetooth devices to support legacy applications that use serial interfaces.

Example 9.2: Light-weight applications in Bluetooth

Figure 9.7 shows the implementation of a vCard (digital business card) transfer application. This application protocol runs over Object Exchange Protocol that is carried by the RFCOMM protocol

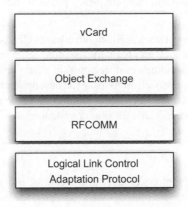

Figure 9.7 Protocol stack for the implementation of vCard exchange over Bluetooth.

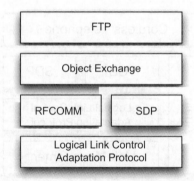

Figure 9.8   Protocol stack for the implementation of FTP over Bluetooth.

in the Bluetooth protocol stack. As shown in Figure 9.7 the resulting packets are then passed through the Logical Link Control Adaptation Protocol and then Baseband before actual radio frequency "over the air" transmission.

The protocols defined in Figure 9.5 are either developed by the Bluetooth group exclusively or are modified versions of existing protocols. Different applications may use different protocol stacks but nevertheless all of them share the same physical and data link control mechanisms. The Bluetooth special interest group has adopted a number of popular Internet and PSTN existing protocols within its specification. The Bluetooth specification is itself open and other protocols can be accommodated on top of the existing protocol stack.

Example 9.3: FTP over Bluetooth

Figure 9.8 provides a protocol stack for implementation of the File Transfer Protocol application. The Objective Exchange (OBEX) and RFCOMM protocols manage the data transfer while Service Discovery Protocol provides for the establishment of the link and similar to our last two examples the results is directed to baseband and radio frequency physical transmission protocols.

The overall structure of the protocol stack in Bluetooth does not clearly follow the traditional models (such as the seven-layer OSI model) and the corresponding acronyms. Therefore, the division of the following section may appear somehow different from other wireless local networks described in the book. However, we make every effort to make the description as close as possible to the rest of the networks to provide a coherent story that is comforting to the reader towards understanding the details and relating them to the details for similar systems.

9.2.3   Physical Layer

The traditional equivalent of the physical layer in the case of Bluetooth is embedded in the RF and Baseband layers of the Bluetooth protocol stack. The physical connection of Bluetooth uses a FHSS modem with a nominal antenna power of 0 dBm (around 10 m coverage) that has an option to operate at 20 dBm (around 100 m coverage). The low power version of the Bluetooth provides a reasonable coverage for its popular applications such as cable replacement and guaranteed modest interference with 802.11 devices, which operate in the same frequency bands.

Like the 1 Mbps option of the IEEE 802.11 FHSS standard, the Bluetooth specification uses a two level Gaussian FSK modem with a transmission rate of 1 Mbps that hops over 79 channels in the unlicensed bands starting at 2.402 GHz and stopping at 2.480 GHz. As explained in Chapter 2, the rms delay spread in indoor areas for short range coverage is under 100 ns. With that, the coherence bandwidth of the channel is around $\frac{0.1}{100 \times 10^{-9}} = 1$ MHz. Therefore, for a transmission rate at 1 MSps, the channel frequency response is relatively flat. The transmitted waveforms preserve their shape and we do not need complex signal processing techniques to equalize the channel. The two-level Gaussian FSK modem allows implementation of simple non-coherent detection using frequency demodulators. A more complex version of this modem is one that uses GMSK as the modulation scheme. GMSK is used as the transmission technique in the GSM standard (see Chapter 11). The difference between FSK and MSK is that the two frequencies used for data transmission in MSK are twice as close as the two tones separation in FSK. Therefore, GMSK is twice more bandwidth efficient than GFSK, but GFSK is implemented with simpler and more power efficient circuitry. WPANs operate in the unlicensed bands, where one can access a wider spectrum and they are designed to support low power ad hoc sensor networking. In such environments, GFSK is a better solution than GMSK, used in GSM.

Although the transmission technique of the base FHSS physical layer in the original IEEE 802.11 and Bluetooth are the same, the hopping rate and pattern and number of hops used in Bluetooth are different from those in IEEE 802.11. The Bluetooth frequency hopping rate is 1600 hops per second (625 μs dwell time) as compared to the 2.5 hops/s (400 ms dwell time) system adopted by IEEE 802.11. As we described in Chapter 2, the Doppler spread in most indoor environments is around 5 Hz. With this value of Doppler spread, the coherence time of the channel is around 200 ms. Since the slow FHSS systems used by Bluetooth and IEEE 802.11 send a packet per hop, the frame format of Bluetooth comprises of much shorter packets with respect to coherence time of the channel, which are better suited for voice oriented networks, where the network needs to avoid retransmissions for real-time streaming of voice conversations.

The Bluetooth specification assigns a specific frequency-hopping pattern for each piconet. This pattern is determined by the piconet identity and master clock phase residing in the Master terminal in the piconet. Figure 9.9 illustrates the essence of frequency hopping strategy in Bluetooth. The

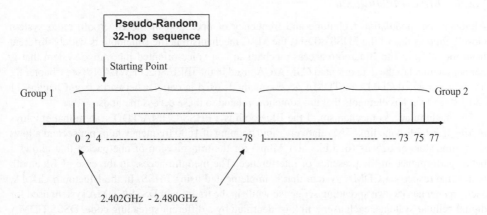

Figure 9.9 The hopping sequence mechanism in Bluetooth.

overall hopping pattern is divided into 32 hop segments. The 32-hop pseudo random hopping pattern segment is generated based on the master identity and clock phase. The 79 frequency hops at the ISM bands are arranged in odd and even classes. Each 32-hop sequence starts at some point in the spectrum and hops over the pattern that covers 64 MHz because it hops either on odd or even frequencies. After completion of each segment the sequence is altered and the segment is shifted 16 frequencies in the forward direction. The 32 hops are concatenated and the random selection of the odd or even index is changed for each new segment. This way segments slide through the carrier list to maintain the average durations for which each frequency is used close to each other (i.e., like a uniform probability distribution). A change of the clock or identity of the piconet will change the frequency hopping sequence and segment mapping. This allows different piconets to operate in the same vicinity with different sets of (pseudo) random frequency hopping sequences. These frequency-hopping sequences are not orthogonal to one another but they are randomized against each other. With 79 hops, it is difficult to find a large number of orthogonal sequences anyway [Haa00].

To protect the integrity of the transmitted data, Bluetooth uses two error-correction schemes in the baseband controllers. A forward error correcting code is always applied to the header information and if needed it is extended to the payload data for the voice packets. The optional coding of the payload in data applications reduces the number of retransmissions that is desirable to increase the throughput. Coding of the payload for voice applications increases the integrity of the real-time streaming packets that will improve the quality of the service for the users. Coding is always applied to the header because header information is short and important. The detection of an error in the header allows a fast request for re-transmission and for voice payloads, a fast decision for keeping or dropping a corrupted packet. In general, the flexibility of optionally using forward error correction for the payload provides an option to avoid overhead in favor of increased throughput when the channel is good and error free. An unnumbered automatic repeat request scheme is also applied by the baseband layer for data packets, in which the recipient acknowledges the received data. For data transmission to be acknowledged, both the header error check and the payload check, if applied, must indicate a "no error" condition. These functionalities implemented in the baseband layer of the Bluetooth protocol stack are often implemented in the data link layer of traditional protocol reference models for networks.

9.2.4 MAC Mechanism

Although the modulation technique and frequency of operation of the Bluetooth radio system closely follows that of the FHSS 802.11, the MAC mechanism in the Bluetooth is widely different from the 802.11. The Bluetooth access mechanism is a voice-oriented innovative system that is neither identical to the data oriented CSMA/CA, used in the IEEE 802.11 WLANs (see Chapter 8), nor voice oriented CDMA or TDMA access methods used in cellular networks (see Chapters 11 and 12), and yet has elements that are somehow related to these access methods.

The medium access mechanism of the Bluetooth is a FHSS/CDMA/TDD system that employs *polling* to establish the link. The relatively rapid hopping of 1600 frequency hops per second allows short time slots of 625 μs (625 bits at 1 Mbps) for the transmission of one packet that allows a better performance in the presence of interference. The medium access in the case of Bluetooth is in some respects a CDMA system that is implemented using FHSS. In the Bluetooth CDMA, each *piconet* has its own spreading sequence while in the traditional DSSS/CDMA system used for digital cellular systems, each *user's* link is identified by a different spreading code. DSSS/CDMA has not been selected for Bluetooth because DSSS/CDMA needs a centralized power control (for protection against near-far effects) that is not possible in a scattered ad hoc topology with many device classes as envisioned for Bluetooth applications. Without the need for centralized power control for CDMA operation, the FHSS/CDMA in Bluetooth allows tens of piconets to overlap in the same area providing an effective throughput that is much larger than 1 Mbps.

As we discussed in Chapter 7, The FHSS version of IEEE 802.11 operates over the same 79 hops as Bluetooth with only three sets of hopping patterns. The throughput of the Bluetooth FHSS/CDMA system, however, is less than the 79 Mbps that could be achieved in a coordinated FDM or OFDM system employed in 802.11a/g. In Bluetooth, the FHSS/CDMA is selected over simple FDM or OFDM because ISM bands at 2.4 GHz to maintain low power consumption. The access method in each piconet of Bluetooth is based on TDMA/TDD. The TDMA format allows multiple voice and data terminals to participate in a piconet. Duplexing in time (TDD) eliminates the cross-talk between the transmitter and the receiver allowing a single chip implementation in which a radio alternates between transmitter and receiver modes. To share the medium among a larger number of terminals, in each slot, the "Master" device decides and *polls* a "Slave" device allowing it to transmit. Polling is used rather than contention access methods because contention based access creates excessive overhead for short packets (625 bits). Recall from Chapter 4 that contention based access requires waiting times and back-off periods, which may be longer than the 625 μs slot used in Bluetooth.

9.2.5 Frame Formats

The Bluetooth packet format is based on one packet per hop and a basic 1-slot packet that is 625 μs long. The packet can be extended to 3-slots (1875 μs) and 5-slots (3125 μs). This frame format and the FHSS/TDMA/TDD access mechanism allow a "Master" terminal to poll multiple "Slave" terminals at different data rates for voice and data applications in the piconet.

Example 9.4: Operation of piconets

Figure 9.10 illustrates several examples of Bluetooth frame formats for operation in a piconet. In Figure 9.10a, a Master terminal (M) is communicating with three Slave terminals (S1 to S3). The TDMA/TDD format allows simultaneous connectivity to the three terminals assigning 625 μs (equivalent to 625 bits at 1 Mbps) slots for transmission and a time gap between the two packets in each direction. Terminals may run different applications (voice or data at different rates), but transmissions should occur on one of the 1-slot detailed packet formats that are specified by the Bluetooth group standardization committee. The time gap is specified at 200 μs to allow a terminal to switch from transmitter to receiver mode for the TDD operation [Haa00]. Figure 9.10b shows an asymmetric communication in which the Master uses a higher speed 3-slot link while the Slave operates at lower rate with 1-slot packets. Figure 9.10c represents a symmetric higher speed 3-slot communication link and Figure 9.10d an asymmetric high speed 5-slot Master to Slave link with a lower speed 1-slot link from the Slave to the Master.

The overall packet structure of the Bluetooth is shown in Figure 9.11. There are 74 bits for an access code field, 54 bits for the header field and up to 2744 bits for different payloads that can be as long as five slots. In IEEE 802.11 FHSS packets, the preamble and header of the physical layer, shown in Figure 8.17, were 96 and 32 bits respectively while the payload could be as long as $4096 \times 8 = 32\,768$ bits. The size of the overhead is more or less in the same range but the maximum payload of 802.11 is at least an order of magnitude larger. Bluetooth uses more flexible shorter packets for

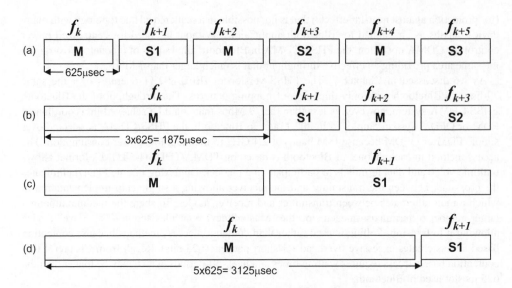

Figure 9.10 TDMA/TDD multi-slot packet formats in Bluetooth: (a) one-slot packets, (b) asymmetric three-slot, (c) symmetric three-slots (1875 μs), (d) asymmetric five-slot (3125 μs).

better performance in fading, but these gains are at the expense of a higher percentage of overhead that reduces the throughput.

The access code field consists of a 4-bit preamble and a 4-bit trailer plus a 64-bit synchronization PN sequence with a large number of codes with good autocorrelation and cross-correlation properties (see Chapter 12 for a discussion of autocorrelation and cross-correlation). The 48-bit IEEE MAC address unique to every Bluetooth device is used as the seed to derive PN-sequence

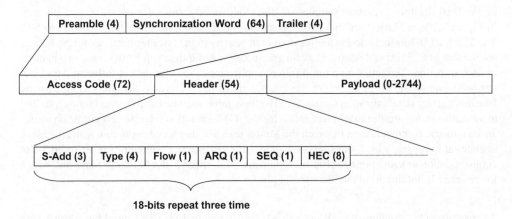

Figure 9.11 Overall frame format of the Bluetooth packets.

for hopping frequencies of the device. There are four different types of access codes. The first type identifies a Master terminal and its piconet address. The second type of access code specifies a Slave identity that is used to page a specific Slave. The third type is a fixed access code reserved for the inquiry process that will be explained later. The fourth type is the dedicated access code that is reserved to identify specific sets of devices such as printers or cellular phones.

As shown in Figure 9.10 the header field has 18 bits that are repeated three times to increase the reliability. The 18-bit starts with a 3-bits Slave address identifier, a 4-bits packet type, 3-bits for status reports, and an 8-bit error check parity for the header. The 3-bit Slave address allows addressing the seven possible active Masters in a piconet. The 4-bit packet type allows 16 choices for different grade voice services, data services at different rates, and for control packets. The 3-bit status reports are used to flag overflow of the terminal with information, acknowledgement of successful transmission of a packet, and sequencing to differentiate the sent and resent packets.

The Bluetooth special interest group specifies different payloads and associated packet type codes that allow implementation of a number of voice and data services. Different master-slave pairs in a piconet can use different packet types, and the packet type may change arbitrarily during a communication session. The 4-bit packet type identifies 16 different packet formats for the payloads of the Bluetooth packets. Six of these payload formats are primarily used for packet data communications. Three of the payload formats are primarily used for voice communications. One is an integrated voice and data packet and four are control packets common for both voice and data links.

The three voice packets, shown in Figure 9.12, are packets with different grades of protection, numbered as 1, 2, 3, to designate the level of quality. These voice packets are all single-slot packets, the length of the payload being fixed at 240 bits. They do not use the status report bits, because voice packets are sensitive to delay but not to a modest packet loss rate of less than 1%. So, they do not need to be re-transmitted. However, voice is a real-time application demanding a steady data rate. As a result, voice packets are transmitted over reserved periodic duplex intervals to support 64 kbps per voice conversations that are commonly used in the PSTN to carry digital voice. The lowest grade 1 voice packet uses all 240 bits for the user voice samples, grade 2 uses 160 bits for user voice samples and 80 bits of parity for a 1/3 forward error correction code, and grade 3 uses

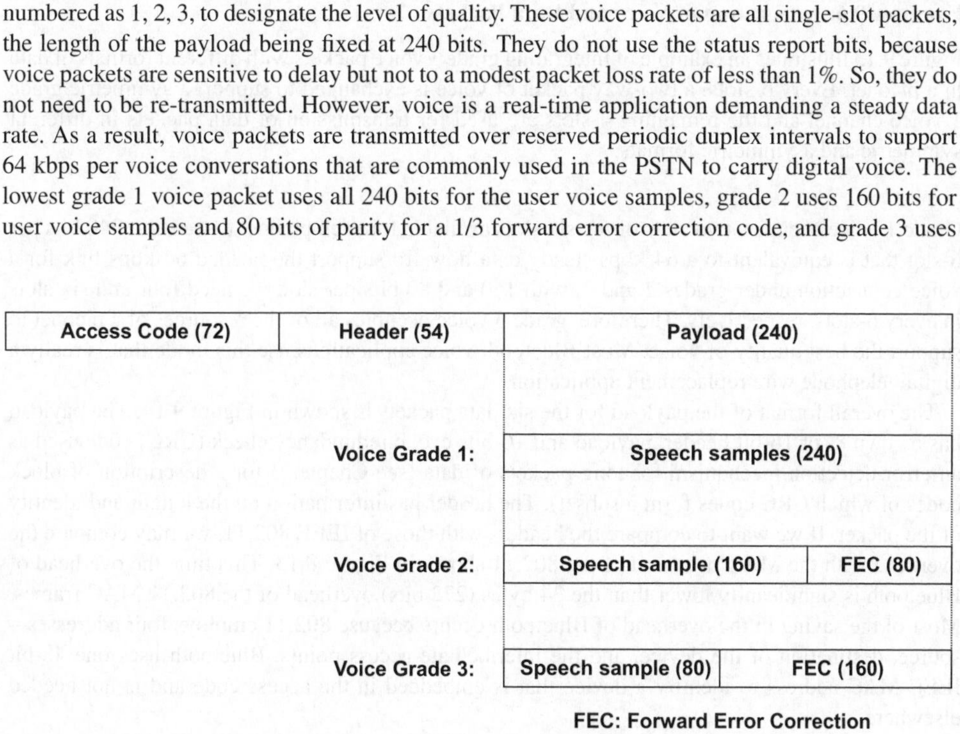

Access Code (72)	Header (54)	Payload (240)

	Voice Grade 1:	Speech samples (240)

	Voice Grade 2:	Speech sample (160)	FEC (80)

	Voice Grade 3:	Speech sample (80)	FEC (160)

FEC: Forward Error Correction

Figure 9.12 Three options for 1-slot voice packet frame formats.

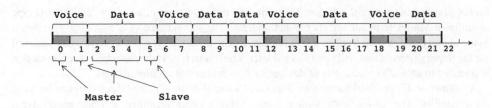

Figure 9.13   Integration of voice and data in Bluetooth medium access.

80 bits of user voice samples and 160 bits of parity for a 2/3 forward error correcting code. To keep the data rate for voice samples at 64 kbps, the grades 1, 2, and 3 packets in each direction are sent every six, four and two slots respectively.

Example 9.5: Data rate of high quality voice packets

The grade 1 voice packets are 240 bits long, and they are 1-slot packets sent every 6-slots. Slots carrying a packet are created at the rate of 1600 slots/s. Therefore, the effective data rate to support grade 1 voice is: $\frac{1600 \left(\frac{slots}{s} \right)}{6 \text{ (slots)}} \times 240$ (bits) = 64 kbps.

Example 9.6: Integration of Voice and Data Packets

Figure 9.13 illustrates an example of integrating grade 1 voice packets with different formats of data in a piconet. Every 6-slots, a two-way packet of voice is exchanged to support a symmetric grade 1 voice channel and the remaining 4-slots are used for transmission of data packets in different symmetric and asymmetric formats.

Grade 1 voice occupies two slots of the six slots available for each piconet to transmit 240 bits per 6-slot that is equivalent to a 64 kbps steady data flow. To support the needed 64 kbps link for a voice connection under grades 2 and 3, with 160 and 80 bits per slot, we need four and six slots in every 6-slots, respectively. Therefore, grade 3 voice occupies all of the resources of a piconet to support the best quality of voice. Most Bluetooth voice applications use this mode that is really a digital telephone wire replacement application.

The overall format of the payload for the six data packets is shown in Figure 9.14. The payload has its own 8- or 16-bit header, payload and 16-bits cyclic redundancy check (CRC) code used as an error detection mechanism for long packets of data (see Chapter 3 for a description of block codes of which CRC codes form a subset). The header has information on the length and identity of the packet. If we want to compare the headers with those of IEEE 802.11, we may compare the overhead with the MAC overhead of the 802.11 shown in Figure 8.13. This time the overhead of Bluetooth is significantly lower than the 34 bytes (272 bits) overhead of the 802.11 MAC frames. Most of the saving in the overhead of Bluetooth occurs because 802.11 employs four addresses – source, destination of the device, and the intermediate access points. Bluetooth uses one 48-bit IEEE MAC address to identify a device that is embedded in the access code and is not needed elsewhere.

The six data packets are divided into medium and high rate based on how much error protection they receive and there are grades for their data rate according to the number of slots: 1, 3, or

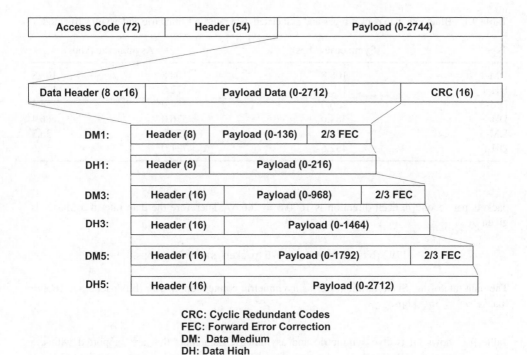

CRC: Cyclic Redundant Codes
FEC: Forward Error Correction
DM: Data Medium
DH: Data High

Figure 9.14 Frame format for data transmission in packets of 1-, 3-, and 5-slot lengths with two different levels of quality.

5 to carry the data. These differentiations allows for six classes of data packets. Figure 9.14 shows the overall frame format of all six classes of data packets. The medium rate data packets use rate 2/3 forward error correcting codes that improve the reliability of the link at the cost of lower data rates. High rate data packets do not employ coding to achieve higher data rates. Using a different number of slots for a packet data payload size, exercising the coding option, and changing the nature of the transmitted packets in each direction (symmetric or not), a number of packet data links with flexible rates can be implemented according to the Bluetooth specification.

Example 9.7: High data rate in Bluetooth

A symmetric data with high rate in 1-slot (DH1) link between a Master and a Slave terminal, shown in Figure 9.14, carries 216 bits per slot at a rate of 800 slots per second (every other slot) in each direction. The associated data rate is 216 (bits/slot) $\times$ 800 (slots/s) $=$ 172.8 kbps.

Example 9.8: Medium data rate in Bluetooth

The asymmetric data with medium rate and 5-slot length (DM5) link, shown in Figure 9.14, uses 5-slot packets carrying 1792 bits per packet. If we assign this configuration to a Master to download data, then as shown in Figure 9.10d we have to assign a 1-slot medium rate data packet (DM1) carrying 136 bits per packet by the Slave terminal. In this situation, the number of

Table 9.1   Bluetooth data packet types and associated data rates in symmetric and asymmetric modes

Type	Symmetric (kbps)	Asymmetric (kbps)	
DM-1	108.8	108.8	108.8
DH-1	172.8	172.8	172.8
DM-3	256.0	384.0	54.4
DH-3	384.0	576.0	86.4
DM-5	286.7	477.8	36.3
DH-5	432.6	721.0	57.6

packets per second in each direction is $1600/6 = 266.67$. Therefore the data rate of a Master is given by:

$$1792 \text{ (bits/packet)} \times 1600/6 \text{ (packets/s)} = 477.8 \text{ kbps.}$$

The data rate of the Slave terminal in this asymmetric connection is: 136 (bits/packet) $\times$ $1600/6$ (packets/s) $= 36.3$ kbps.

Table 9.1 shows all twelve symmetric and asymmetric data links that are supported with the frame format of the Bluetooth specification. The data rates for these links can be calculated in a manner similar to Examples 9.7 and 9.8. The maximum data rate of 723.2 kbps is available in an asymmetric channel for a single user while the reverse channel supports only 57.6 kbps. The reader should remember that data applications operate in bursts, and therefore, even if a Master node communicates with the maximum seven Slave data terminals, most of the time only one of the Slave terminals will communicate with the Master. When more than one Slave terminal simultaneously attempts to communicate with a Master terminal, the quality of service provided to the Slave terminals has to be compromised either by sharing the throughput or by providing additional delays. The decision making process to reach a compromise in the voice oriented access methods, such as the one used in Bluetooth, needs a complex algorithm to handle the quality of service as negotiated at the start of a session. Comparing this situation with CSMA/CA used in 802.11, there is no negotiation at the starting point. When more than one terminal attempts to communicate with a single access point, the medium is shared and the compromise is made automatically through the CSMA/CA access method described in Chapters 4 and 8. For distributed data only applications CSMA/CA is more appropriate. However, when voice applications become dominant, a TDMA/TDD type access methods can guarantee a certain quality of service (e.g., steady data rate) with the fast hardware at the lower layers for voice while CSMA/CA cannot do this easily and needs to implement it at software at the higher levels (e.g., using quality of service mechanisms or polling that sits on top of CSMA/CA).

The only remaining traffic packet in Bluetooth is a data-voice packet that is a mixed voice and data packet with the same access code and overall header that must be transmitted at regular intervals. The voice part carries 80 bits of voice payload without any coding and the data part is a short packet of length 0–72 bits with a 16-bit 2/3 CRC coding and an 8-bit data payload header. This packet also uses three status report bits.

The Bluetooth specification also defines four control packets. The first packet occupies only half of a slot and it carries the access code with no data or even a packet type code. This packet is used before connection establishment to only pass an address. The second and third packets have the access code and the header, and so they have packet type codes and status report bits. The second packet is used for acknowledgement signaling and there is no acknowledgement for this packet. The third packet is used for polling and its format is similar to that of the second packet but it has an acknowledgement. Master terminals use the polling packet to find the Slave terminals in their coverage area. The fourth packet carries all the information necessary to synchronize two devices in terms of access code and hopping timing. This synchronization packet is used in the inquiry and paging process that will be explained later.

9.2.6   Connection Management

The Link Management Protocol layer and Logical Link Control Application Protocol layer of the Bluetooth, shown in Figure 9.6, perform the link setup, authentication, and link configuration. An important issue in a truly ad hoc network is how to establish and maintain all the connections in a network whose elements appear and disappear in an ad hoc manner and there is no central unit transmitting signals to coordinate these terminals. In both digital cellular systems and WLANs, there is a common control signal or a beacon signal that allows a new terminal to lock to the network and exchange its identity with the networks identity. The Bluetooth specification achieves initiation of the network through a unique inquiry and page algorithm.

The overall state diagram of the Bluetooth is shown in Figure 9.15. In the beginning of the formation of a piconet, all devices are in stand-by mode. Then one of the devices starts with an

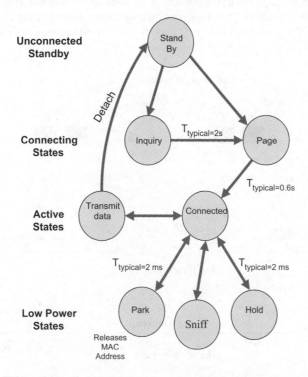

Figure 9.15   Functional overview of the Bluetooth specification.

Inquiry and becomes the Master terminal. During the Inquiry process the Master terminal registers all the stand-by terminals that then become Slave terminals. Note that it is not necessary for a more powerful device to be the master. A camera could be a master device and a laptop the slave device since the camera initiated the Inquiry process. After the inquiry process, identification and timing of all Slave terminals is sent to the Master terminal using the synchronization control packets. A connection starts with a page message with which the Master terminal sends its timing and identification to the Slave terminal. When a connection is established, the communication session takes place and at the end the terminal can be sent back to the Stand-By, Hold, Park or Sniff states. Hold, Park, and Sniff are power saving options. The Hold mode is used when connecting to several piconets or managing a low power device. In the Hold mode data transfer restarts as soon as the unit is out of this mode. In the Sniff mode, a slave device listens to the piconet at reduced and programmable intervals according to the application needs. In the Park mode a device gives up its MAC address but remain synchronized to the piconet. A Parked device does not participate in the traffic but occasionally listens to the traffic of the Master terminal to re-synchronize and check on broadcast messages.

The main innovative part of the inquiry and paging algorithms in Bluetooth is a searching mechanism for two terminals that are not synchronized but they both know a common address. The following example explains this algorithm.

Example 9.9: Search algorithm for synchronization

Two Bluetooth devices with a common 48-bit IEEE 802 address first use the common address to generate a common frequency hopping pattern of 32 hops and a common PN-sequence for the access code of all their packets. Then they start their operation as depicted in Figure 9.16. In the initial state, Terminal 1 sends two identity ID packets carrying the common access code every half

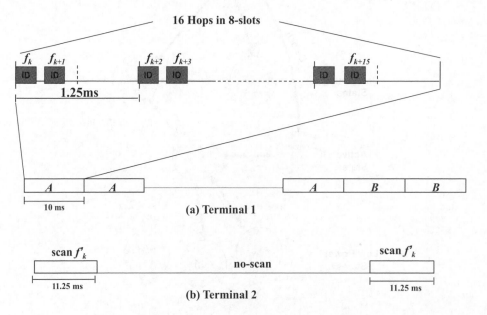

Figure 9.16 Basic search for the Paging algorithm in Bluetooth.

slot on a different hop frequency associated with the common frequency hopping pattern and listens to the response of the slave device (Terminal 2) in the next slot. If there is no response, it continues broadcasting the ID packets on the two new frequencies in the common hop pattern and repeats this procedure eight times for a period of 10 ms (eight 2-slot times). During these 10 ms, the common ID is broadcast over 16 of the total 32 different hop frequencies. If there is no response, Terminal 1 assumes that Terminal 2 is in sleep mode and repeats the same broadcast again and again until the period of transmission becomes longer than the expected sleeping time of Terminal 2. At this time, Terminal 1 assumes that Terminal 2 has scanned, but its scan frequency was not among the 16 hops, designated by A in Figure 9.16, and continues its broadcast with the second half of the 32 hop frequencies, designated by B in the figure. If Terminal 2 is in sleeping mode, it wakes up periodically for a period of 11.25 ms to scan the channel at a given frequency for its desirable access code and sleeps again. In each scan period of 11.25 ms, the sliding correlator in Terminal 2 tries to detect the desired address at 16 different frequencies. If one of these frequencies is the same as the scanning frequency, the correlator peaks and synchronization is signaled. Depending on the operation, Terminal 2 can scan the second time at the same frequency or at a new frequency for verification. In either case the objective is to maximize the probability of hitting the same frequency as the broadcast frequency.

The basic principle explained in the above example is used during the Inquiry and Paging processes. The following two examples explain these applications for the above mechanism.

Example 9.10: Paging process

As in the previous example, the Master terminal (Terminal 1) broadcasts repeating ID page trains carrying the access code of the paged terminal, two per slot, waits for the response in the next slot, repeats the page trains at new hopping frequencies of the paged terminal to cover 16 frequencies every 10 ms, and repeats this for the estimated length of the sleeping time. The Slave terminal scans for 11.25 ms with one of the 32 frequencies of its hopping pattern, sleeps and scans at the next hopping frequency. When frequencies are the same, a peak appears at the correlator output of the Slave terminal and the slave responds by sending its own ID packet as an acknowledgement for detection of frequency hopping timing. The Master terminal then stops broadcasting ID packets and sends a synchronization packet containing its own ID and timing information. The Slave terminal responds with another ID packet to correspond to the timing of the Master terminal and then connection is established and the Slave joins the piconet for information exchange. Usually, the Master terminal knows the approximate timing of the hopping pattern and the 16 most probable hops are adequate to establish the connection. In case this estimate is not correct, as in the previous example, the Master terminal resorts to the second half of the 16 hops when there is no response after the estimated sleeping time.

Example 9.11: Inquiry process

The Inquiry message is typically used for finding Bluetooth devices, including printers that have Bluetooth, fax machines and other similar devices with an unknown address. The general format of the Inquiry process is very similar to the Paging mechanism. A unique access code and frequency hopping pattern are reserved for Inquiry. In other words, the Inquiry process is universally identified with all attributes of any device. Like Paging, Inquiry starts with an "Inquirer" broadcasting an ID packet every half slot at a different hop frequency, covering 16 frequencies every 10 ms, and repeats the same process until it receives responses. The "Inquiree" scans with the sliding correlator for 11.25 ms. When frequencies are the same, the sliding correlator peaks in all devices that are scanning. To avoid collision a device detecting the Inquiry ID runs a random number generator and waits for the length of the outcome before it scans the channel again. When the peak appears the second time after random waiting time, the Inquiree terminal sends a synchronization packet, allowing the Inquirer to learn its ID and timing information. After this process is completed, the Inquirer's radio has the Device IDs and Clocks of all radios in its range of coverage. After completion of the first Inquiry, the inquired device changes its scan frequency and continues scanning for the next Inquiry and follows up with synchronization signaling.

9.2.7 Security

Bluetooth specifications provide usage protection and information confidentiality. Bluetooth has three modes of operation – non-secure, service-level, and link-level security. Devices also can be classified into trusted and distrusted. It makes use of two secret keys (128 bits for authentication and 8–128 bits long for encryption), a 128-bit long random number and the 48-bit MAC address of devices. Any pair of Bluetooth devices that wish to communicate will create a session key (called the link key) using an initialization key, the device MAC address, and a personal identity number. This protocol has been shown to have several vulnerabilities [Jak01] by which a malicious entity could obtain the personal identity numbers and keys depending on how the session initialization of the communication protocol is performed. Some other details are available in Chapter 7.

9.3 IEEE 802.15.4 and ZigBee

ZigBee is a suite of protocols capable of mesh networking between nodes that are in range, some with multi-hop routing capability, defined for operation with the IEEE 802.15.4 standard that specifies the PHY and MAC layers for low data-rate, low power, low cost wireless personal area networks. The first ZigBee specification was ratified in 2004 and latest version was released in 2007. The relation between IEEE 802.15.4 and ZigBee is also similar to the relation between IEEE 802.11 and Wi-Fi. We use the terms Zigbee and IEEE 802.15.4 interchangeably in many places.

In as much as the IEEE 802.15.1 Bluetooth standard was a low complexity, inexpensive, and low power *single-hop* ad hoc network design influenced by the IEEE 802.11 FHSS standard, the design of the IEEE 802.15.4 standard has been influenced by the IEEE 802.11 DSSS standards. In comparison with the IEEE 802.11 WiFi devices, ZigBee devices are designed for very low cost communications among scattered devices such as low-power sensors with minimal infrastructure. Many of the applications and the radio coverage of ZigBee devices are similar to those of Bluetooth. However, ZigBee intends to provide faster formation of a piconet, a larger number of active devices, a much longer battery life, but lower data rates of 20–250 kbps. The flexible data rate and faster connection time makes ZigBee more desirable for connectionless data-oriented applications to enable communication between the sensors and the Internet.

To follow the same format of presentation that we have used to present the details of other wireless networking technologies, here we start with the overall architecture and the general protocol stack followed by sections on the details of the physical layer, the medium access control layer, and how packets are formed at the these layers.

9.3.1 Overall Architecture

One of the general architectural differences of the IEEE 802.15.4 standard is that it defines two types of nodes functionality. The two types of nodes are referred to as *full-function devices* and *reduced-function devices*. A full function device serve as a *coordinator* or master of a piconet. It can also serve as a common node or a full function slave. A full-function device can communicate with any other device in the network (within its radio communication range) and it can further help in routing messages (packets or frames) throughout the Zigbee network. In contrast, reduced function devices are defined to be extremely simple with very modest resources and sparse communication capabilities, and are usually only used as slave nodes to communicate with a full function device. Consequently, such devices are often in sleep modes most of the time and communicate only very infrequently (e.g., upon detecting an event or sensing some parameter), allowing their batteries to last for months at a time. This characteristic of reduced function devices provides flexibility in the implementation of a variety of topologies that can address diversified applications. As an example, consider a typical example application of a wireless light lamp switch. The node at the lamp can be a full function device since it is connected to the main power supply and does not have power constraints, while a battery-powered light switch would be a reduced function device to conserve energy and increase the battery life. In fact, at the time of this writing, there are LED bulbs that have been introduced with Zigbee capability that communicate with a Zigbee bridge that can plug into a WiFi router. Users can control the operation of the bulbs through the web or applications on their smartphones.

In a manner similar to Bluetooth and IEEE 802.11, IEEE 802.15.4 and ZigBee support both peer to peer and star network topologies. A new additional topology for the IEEE 802.4 is the cluster tree topology. Figure 9.17 shows all three topologies for IEEE 802.15.4-based ZigBee networks. The peer-to-peer networks, shown in Figure 9.17a form arbitrary patterns of connections. The extent to which the extension of these connections is possible depends on the distance between each pair of nodes. The nodes in Figure 9.17a are all full function devices and are thus meant to serve as the basis for constructing on-the-fly networks that are capable of performing self-management and organization. Figure 9.17b shows a simple star topology with a master–slave operation that is similar to that of Bluetooth. In the ZigBee network, however, as we mentioned earlier, we have two

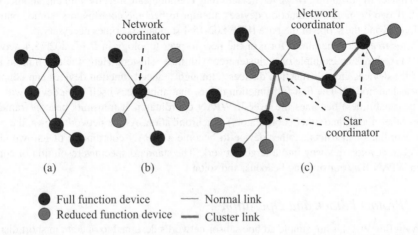

Figure 9.17 IEEE 802.4 ZigBee topologies: (a) peer to peer, (b) star for master–slave operation, (c) clustered stars.

classes of nodes which provide more flexibility in forming a network to support an application. The network coordinator must be a full function device and the other nodes in Figure 9.17b can be either full or reduced function devices depending on the application. Figure 9.17c shows an example of a more complex clustered tree topology formed by two different devices with three star coordinators and one network coordinator connecting stars together. The coordinators in this figure play the role of a router in the ad hoc network, which means the network coordinator manages the entire network. This type of topology is unique to ZigBee and it allows the formation of a hierarchical ad hoc network with simple end nodes which are in sleeping mode most of the time, allowing extremely long battery lives.

In a cluster tree topology there are three distinct classes of node operation. The first class is the reduced function or dumb end nodes devices, which are sleeping most of the time to extend the battery life. Each such end device allows up to 240 end points for separate applications sharing the same radio. For example, a three-gang light switch would have three distinct end points sharing the same radio electronics and battery. The second class of devices is the mid-level routers, which have the ability to stack up communication messages and respond to general enquiries about sleeping end devices in their vicinity. These routers are also responsible to find out the best way to pass on a message to a node that is not in the range. The third class of nodes at the top of the hierarchy is the network coordinator, which is always on and relies on connection to a good power source. In addition to being a router the network coordinator sets the rules for basic network operation such as finding an appropriate frequency channel for the network.

In the case of the IEEE 802.15.4 standard (along with the Zigbee higher layer protocols), the setting up of the network is accomplished through the use of WPAN coordinators. In the star topology, nodes communicate directly to a WPAN coordinator and all communications go through the coordinator. In many ways, this is like nodes communicating to an access point in wireless local area networks. In the peer-to-peer topology, devices can communicate directly with one another as long as they are in radio range.

In the star topology, a full function device, when deployed, automatically becomes the WPAN coordinator. This is accomplished simply by the device, by transmitting a beacon and announcing itself as the coordinator, if it does not hear any other device when it powers up. In the peer-to-peer topology, a similar mechanism works and there is a WPAN coordinator (usually the first full function device to power up). However, devices may communicate directly with one another where allowed. If two or more full function devices attempt to become coordinators, some contention resolution beyond the scope of the base IEEE 802.15.4 standard becomes necessary.

A *cluster tree* is a generalized form of the peer-to-peer topology in IEEE 802.15.4 networks. Figure 9.17c shows an example of a cluster tree with three clusters. Here, the assumption is that most of the devices are full-function devices (although reduced function devices can connect to clusters as leaf nodes). The first full-function device that announces itself (or a device with more power or capabilities) becomes the overall WPAN coordinator. It transmits beacon frames and provides other nodes (and other coordinators) synchronization. As the network grows, the overall WPAN coordinator instructs another device to become a WPAN coordinator of its own cluster. Clusters can develop this way into a large network. The standard specifies resolution of conflicts between WPAN IDs, transmitting beacons, and so on.

9.3.2 Protocol Stack and Operation

Though ZigBee WPANs are simple ad hoc sensor networks designed to operate in short distances of up to 10 m, the network has several layers designed to enable communication within the network, connection to a network of higher level, and ultimately an uplink to the Internet. Like all other

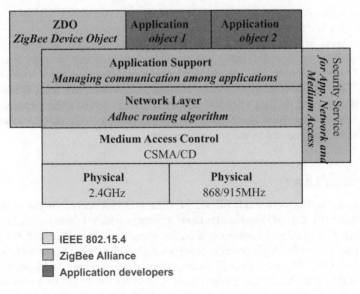

Figure 9.18   IEEE 802.15.4 ZigBee protocol stack.

communications networks IEEE 802.15.4-based ZigBee divides up the communications tasks into layers on a stack of protocols shown in Figure 9.18. The lower layers, that is, the physical and medium access control layers, are defined by the IEEE 802.15.4 standard and the higher layers are specified by the ZigBee alliance. In addition, independent application developers can define their own application layers that can then communicate with the ZigBee defined layers and protocols. The top of the stack is the application layer and the bottom is the physical radio. The middle layers are used to glue the application to the actual transmission so that nodes can communicate reliably, efficiently, and securely and with interfaces that designers can employ to develop their applications in an easier supporting environment.

As shown in Figure 9.18, on top of the ZigBee defined protocols we have a *ZigBee Device Object*, that is a special application object, available on all ZigBee nodes. The address of this application object is always "zero", and other application objects running on the ZigBee module are numbered from 1 to 240. These form the various applications that an end-device may be supporting. The Device Object is something that has its own profile, which other user application objects and other ZigBee nodes can access and it is responsible for overall device management, security keys, and policies. Each profile in a ZigBee module includes a table of other ZigBee modules on the network and the services they offer. All other applications end-points use the available information in the table to discover other devices, to manage binding, and to specify security and network setting. The *Application Support Layer* protocol in Figure 9.18 routes messages on the network to different application end points running on a ZigBee module by maintaining a "binding table" and forwarding messages to the appropriate application.

The *Network Layer* provides the routing and multi-hop capability required to turn MAC-level communications into a full star, tree or mesh network. ZigBee employs a distance-vector routing algorithm suitable for ad hoc networks called the *Ad-hoc On-demand Distance Vector* (AODV) protocol. This algorithm automatically constructs a low-speed ad hoc network of nodes by forwarding messages, discovering neighboring devices and building up a map of the routes to other nodes. In the coordinator nodes, the network layer assigns network addresses to new devices when they join the network for the first time.

The *Security Service* protocol provides for establishing and exchanging security keys, and using these keys to secure the communications link through encryption and message integrity checks (see Chapter 7 for more details). The security services works across three layers to provide security at each level. Like all other security services in wireless networks, this layer is responsible for encrypting and decrypting the data when it is generated or received and authenticating it (checking for integrity) when it is received. The ZigBee Device Objective layer dictates the security policies and configurations implemented by the security services.

9.3.3   Physical Layer

In a manner similar to Bluetooth and IEEE 802.11, IEEE 802.15.4 operates in the unlicensed radio bands. In addition to the 2.4 GHz band used in those standards, which is mostly available worldwide, the IEEE 802.15.4 standard also supports options for operation in the unlicensed 868 MHz bands in Europe and unlicensed 915 MHz bands in countries such as the United States and Australia. The physical layer of IEEE 802.15.4 defines 27 channels with three data rates of 20, 40, and 250 Kbps in these three different frequency spectrums. Figure 9.19 shows the 27 different channels and their associated bandwidth specified by the standard in the three different spectrums. The first channel with 0.6 MHz bandwidth, referred to as channel number "0", is at 868 MHz, 10 channels each with 2 MHz bandwidth are in the 915 MHz bands and the remaining 16 channels each with 5 MHz bandwidth are in the 2.4 GHz bands. The IEEE 802.15 standards committee's task group 4c has been considering standards for the 779–787 MHz bands in China and task group 4d is looking at the 950–956 MHz bands in Japan. The standard defines *channel pages* in addition to the *channel numbers*. Channel pages are used to distinguish between the different possible modulation schemes used in the frequency channels.

The basic radios use Direct Sequence Spread Spectrum (DSSS), in a manner similar to legacy IEEE 802.11, but with different chip rates and modulation techniques. The Binary Phase Shift

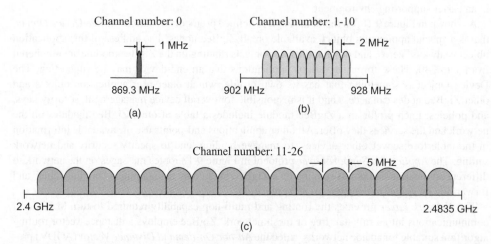

Figure 9.19   Frequency bands used by different physical layer options in the IEEE 802.15.4: (a) one channel at 868 MHz, (b) 10 channels in 915 MHz, (c) 16 channels in 2.4 GHz.

Keying (BPSK) modulation and a 15-chip Linear Feedback Shift Register (LFSR) pseudo random sequence are used for the 868 and 915 MHz bands with chip rates of 0.3 and 0.6 Mcps, respectively. This results in a data rate of:

$$\frac{0.3 \text{ (Mcps)}}{15 \text{ (cpb)}} = 20 \text{ kbps}$$

for the single channel in 868 MHz and:

$$\frac{0.6 \text{ (Mcps)}}{15 \text{ (cpb)}} = 40 \text{ kbps}$$

for each of the 10 channels in the 9.15 MHz bands. An optional amplitude shift keying modulation scheme allows the data rates to be increased by using a form of code division multiplexing (where bits are sent in parallel by spreading them using almost orthogonal sequences).

The 2.4 GHz operation uses 16-ary orthogonal coding with code words of length 32 chips. The modulation technique used for transmission of data is the off-set QPSK. This general structure is similar to the reverse channel of the IS-95 CDMA system, which also uses M-ary orthogonal coding and offset QPSK modulation. Offset QPSK provides a constant envelope, which reduces the power consumption of the last stage amplifiers in the radio. M-ary orthogonal coding reduces the error rate and adds to the integrity of the received bits. The receiver is implemented non-coherently, which does not need carrier synchronization between the transmitter and the receiver and therefore avoids excessive power consumption and complexity of the phase lock loops needed for coherent modulation. These measures are taken to satisfy the low cost and low power implementation of the radio. To form the transmitted symbols, the arriving raw data stream at a rate of 250 Kbps is used to create 4-bit blocks of data at a rate of:

$$\frac{250 \text{ (kbps)}}{4 \text{ (bpS)}} = 62.5 \text{ (kSps)}$$

Each 4-bit block is then mapped to one of 16-orthogonal symbols. The chips are then transmitted at the rate of 2 Mcps using offset QPSK modulation. Therefore, the net processing gain of this DSSS transmission scheme used in the IEEE 802.15.4 is:

$$\frac{2 \text{ (Mcps)}}{250 \text{ (}k \text{ bps)}} = 16$$

It is useful to remind the reader here that the processing gain in the basic IEEE 802.11 DSSS transmission scheme was $N = 11$, which is reasonably close to what is used in the IEEE 802.15.4 standard. It is also useful for the reader to remember that the CCK modulation scheme used in IEEE 802.11b is a different sort of M-ary orthogonal coding. These observations show the path of evolution of experimentally successful local radio networks.

The maximum transmission power of IEEE 802.15.4 radios is usually 0 dBm (1 mW) compared to a maximum transmit power of 20 dBm (100 mW) used in the IEEE 802.11 devices. This 20 dB difference, assuming free space propagation with a distance-power gradient of two, accounts for one order of magnitude higher coverage for IEEE 802.11 devices. As we explained in Chapter 2, the coverage depends on the environment. But it is customary to assume that the coverage of WPANs is roughly 10 m and that of WLANs is approximately 100 m. The lower transmission power and power conscious design of the radio is one of the major differences between the design of IEEE

802.15.4 and IEEE 802.11 devices. The receiver sensitivity in the 2.4 GHz bands is specified to be at least –85 dBm. For BPSK modulation in the 868/916 MHz bands, the receiver sensitivity should be at least –92. This allows an extra 7 dB for operation in the 868/915 MHz bands. In addition, using Equation 2.3, the path loss in the first meter at 2.4 GHz and 868 MHz we have another:

$$20 \log \left(\frac{2.4 \ (\text{GHz})}{868 \ (\text{MHz})} \right) = 8.8 \ (\text{dB}),$$

difference between the two bands that adds up to a 15.8 dB edge for operation at lower frequencies. In free space this accounts for approximately up to $10^{15.8/20} \approx 6$ times longer coverage for operation in lower frequency bands. In indoor environments, lower frequencies also penetrate the walls better, which increases the coverage at lower frequencies to even higher values.

9.3.4 MAC Layer

Medium access control for the IEEE 802.15.4 standard is based on carrier sense multipale access with collision avoidance (CSMA/CA) that is also the main medium access control technique used in the IEEE 802.11 standard. Some details of CSMA/CA were provided in Chapter 4. In general the CSMA/CA in IEEE 802.15.4 is a simpler version of the CSMA/CA used by the IEEE 802.11 standard, explained in Chapter 8, which includes different options for conserving battery and reducing power consumption. The PCF option, variety of inter-frame delays (PIFS, DIFS, and SIFS), and the RTS and CTS mechanisms which were included in the IEEE 802.11 MAC are not included in 802.15.4. The functionality that the RTS/CTS and PCF mechanisms were providing in IEEE 802.11 are here provided by simpler and more practical guaranteed time slot transmission. The medium access control layer provides reliable communications between a node and its immediate neighbors (in radio range) and manages packing data into frames prior to transmission, and then unpacks received packets to check them for errors. In addition, this layer provides beacons and synchronization to improve communications efficiency. Networks are formed either based on beacon or without a beacon. Beacons do not follow carrier sensing and they are sent on a fixed timing schedule. Acknowledgement packets are also sent without carrier sensing after the arrival of a packet or MAC frame. Two other important features of 802.15.4 MAC are: (a) the allowance for guaranteed time slots, (b) integrated MAC support for secure communications. Devices that have low latency real-time requirements can use the so-called guaranteed time slots allocated to them by a coordinator without resorting to carrier sensing. The details are as follows.

The MAC protocol in 802.15.4 operates under a superframe structure (see Figure 9.20). A superframe is defined as the period between two *beacons*, which are special management packets transmitted by the coordinator. Beacons synchronize the WPANs and provide information about the network. Within the time between two beacons, that is, the superframe, sensor nodes can have an active period and inactive period. The active period can be divided into a contention period and a contention-free period. The contention period is slotted. In each slot, nodes use carrier-sensing multiple access with collision avoidance to access the channel. The process is quite simple. Each device waits for a random period to see if the channel is idle. If it is idle, it simply transmits. Otherwise, it backs off for another random period and tries again (see bottom of Figure 9.20). This access suffers from the disadvantages mentioned previously with carrier sensing in Chapter 4 such as the hidden terminal problem. Scheduled access is possible through the use of the contention-free

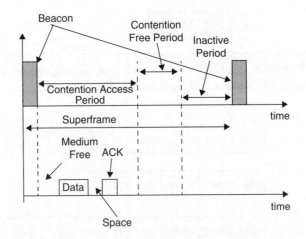

Figure 9.20 Illustration of medium access in IEEE 802.15.4.

period where the WPAN coordinator creates guaranteed time slots that nodes can use without contention from other nodes. It is possible to have a WPAN without beacons in which case non-slotted CSMA/CA is adopted by all nodes.

The standard considers "transactions" that are initiated by the low power devices, which will otherwise have the choice to be in a low-power mode. In the star topology, nodes can send data to a coordinator or receive data from a coordinator. In the former case, the node simply sends data to the coordinator using CSMA/CA and gets an acknowledgment if requested. In the latter case, the node should first request data from the coordinator, get an acknowledgment for its request followed by the data. Upon receipt of the data, it acknowledges to the coordinator. Acknowledgements do not wait for the medium to be idle as in the case of data frames. Alternatively, the beacon can have information about whether or not there is pending data for a sensor node. Peer to peer transmissions occur in a similar manner except that special steps may be necessary for synchronizing transmissions of two peer nodes. In order to allow the MAC layer to process frames, there must be some time that lapses between successive frame transmissions. The time between receipt of a frame and the transmission of an acknowledgment is the smallest while long and short inter frame spaces are used to separate long or short frames.

In general, ZigBee protocols minimize the time the radio is "on" to reduce unnecessary power consumption. In non-beacon-enabled networks, power consumption is decidedly mixed: some devices are always active, while others spend most of their time sleeping. In these networks an unslotted CSMA/CA channel access mechanism is used and routers typically have their receivers continuously active, requiring a more robust power supply. These networks allow a bipolar consumption of energy. The router consumes substantially while end nodes only transmit when an external stimulus is detected (e.g., an event such as a rise in temperature). The typical example for such operation is the light lamp operation which we discussed earlier. In networks using a beacon, nodes are only activated after a beacon is transmitted. The special ZigBee router node transmits a beacon periodically to confirm its presence to other users. Other nodes may sleep between beacons to save in energy consumption. For different options of the physical transmission the beacon interval ranges over 15–24 ms. This option is more suited for higher traffic load and it is very similar to the operations in IEEE 802.11.

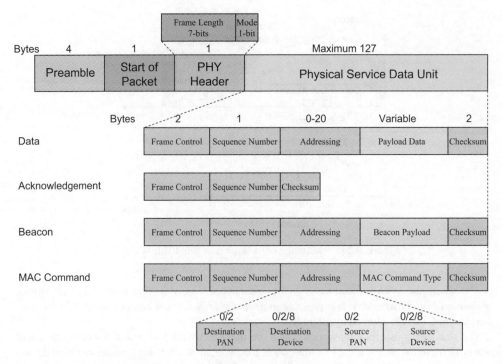

Figure 9.21   Frame format for IEEE 802.15.4 packets.

9.3.5   Frame Format

IEEE 802.15.4 uses four different types of frames for data, acknowledgment, beacon and MAC commands. Figure 9.21 shows the frame format for all four types of packets. Every frame has a 4-byte preamble, a 1-byte start of the packet delimiter, and a 1-byte start of the frame which uses 7-bits to identify the length of the packet in bytes and 1-bit to identify the addressing mode because the device address is either 2-bytes (16-bits) or 8-bytes (64-bits). In addition to the payload, the Physical Service Data Unit for all packets has a 2-byte frame control to carry control messages, 1-byte sequence number to provide for tracking the sequence of the packets during fragmentation and reassembly process (for large application packets), up to 20-bytes of addressing, and 2-bytes of frame check code for error detection. The acknowledgement packet does not have any address field but the other packets have source and destination addresses with 2-bytes for indicating the address of the coordinator node of the PAN plus 2- or 8-bytes for the device address. The payload for the data frame is the information to be delivered with a length that ensures an overall length of the physical service data unit to stay under 127 bytes. The payload for the beacon provides other devices the necessary information bits for synchronization and self-configuration. The MAC command control field carries different MAC control messages.

9.3.6   Comparison of ZigBee with Bluetooth and WiFi

Table 9.2 provides a summary comparison among WiFi (802.11g), Bluetooth and ZigBee technologies. All of these devices use the 2.4 GHz unlicensed ISM bands but ZigBee also supports

Table 9.2 Comparison of 802.11, 802.15.1, and 802.15.4

Technology/feature	802.11g (WiFi)	802.15.1 (Bluetooth)	802.15.4 (Zigbee)
Frequency	2.4 GHz	2.4 GHz	2.4 GHz to 868/915 MHz
Modulation	OFDM	FHSS/BPSK	DSSS/QPSK
MAC	CSMA/CA	TDMA/TDD	CSMA/CA
Maximum data rate	54 Mbps	1 Mbps	20/40/250 Kbps
Device types	One	One	Full/reduced function
Number of channels	11–14 (3 orthogonal)	79 (hopped)	26
Maximum number of devices	32	8	Up to 65 535
Battery life	Hours/days	Weeks	Months
Coverage	100 m	10 m	10/50 m
Topologies	Star, peer to peer	Star, peer to peer	Star and cluster-tree
Connection time	<10 s	3–5	30 ms

868/915 MHz bands which have a better coverage for the same level of transmit power. The modulation techniques in the case of Bluetooth and ZigBee both use spread spectrum technology (but different types), which is a power efficient transmission technique. Both devices also implement non-coherent modulation to reduce the complexity and power consumption in the electronic design. The MAC of WiFi and ZigBee are both based on CSMA/CA, which is a distributed MAC protocol suitable for data applications. The ZigBee implementation is a light version of the MAC protocol in WiFi because it has fewer features. Each piconet in Bluetooth uses a centralized TDMA/TDD-like technique, based on polling, which is more appropriate for connection based telephone applications. Different piconets are separated using FHSS/CDMA technique in the case of Bluetooth. The data rate of ZigBee networks is in the range of the data rates of Bluetooth which are suitable for ad hoc low-rate sensor network applications. WLANs generally provide much higher data rates (54 Mbps for 802.11g) and are more suited to modern computer networking applications for home and small office networking and pervasive access in public buildings.

ZigBee allows two different types of terminals which allow the design of simple end nodes which may sleep most of the time to stretch the life time of the battery. Although the number of channels for WiFi and ZigBee in 2.4 GHz look similar, ZigBee channels are narrower (5 MHz) allowing the implementation of 16 non-overlapping channels. The IEEE 802.11g networks have only three non-overlapping channels. Bluetooth allows up to 79 piconets which is far beyond the other two, but the number of nodes per Master has a more practically important role because usually we have a few piconets but a large number of nodes. In ad hoc sensor networking, having only seven simultaneously connected nodes is rather limited and ZigBee's higher number of nodes (each with many applications) is very desirable. All the features of ZigBee are designed to minimize power consumption which results in a longer lifetime for the battery. The coverage of WiFi is better (with a higher transmit power) and that is important for computer networking. Bluetooth and ZigBee are designed for low-rate ad hoc sensor networking and the coverage can be extended by the selected topology. The ability of ZigBee to form cluster-tree networks increases the coverage of the ad hoc network significantly making it a better choice for applications where we have a number of sensor networks spread over a large geographical area.

As a result of these differences WiFi and Bluetooth have found their specific applications. WiFi is dominating the WLAN market for home, small office, and ad hoc public building access. Bluetooth has found a number of ad hoc wire replacement applications for telephone connections inside the cars and for audio devices. It is expected that ZigBee develop a similar market for sensor networks for medical, power grid, home automation, and meter-reading sensor networking applications.

9.4   IEEE 802.15.6 Body Area Networks

The main difference between our treatment of IEEE 802.15.6 Body Area Networks and IEEE 802.15.1 Bluetooth and IEEE 802.15.4 ZigBee in this chapter is that the latter two are completed mature standards, while the first one is an ongoing standardization activity aiming at discovery of a new technology. IEEE 802.15.6 is more like early days of IEEE 802.11 where we had numerous views on what is the meaning of a wireless LAN and yet there was no horizontal market insight to justify an urgent need for any standardization work. Therefore, our treatment of this section is slightly different. We start with why the BAN industry is perceived to become a significant industry and then we describe the differences between this wireless networking technology and other existing wireless local networking technologies.

9.4.1   What is a BAN?

Engineering innovations are often motivated by metaphors described in science fiction. Later on, empirical science or engineering shapes the technology and the market to design a certain device. The wireless networking industry was motivated by Captain Kirk's communicator in the 1960s science fiction *Star Trek*. This communicator was shown in some of the presentations in the early days of wireless networking. After about half a century, modern smart phones are perhaps a close imitation of that fantasy. Smart phones are almost the same size as the communicator and they allow multimedia communications, presumably everywhere. However, we are far away from beaming Mr. Spock to safety, whenever he is in serious trouble!

The body area networking industry may be influenced by another 1960s science fiction, *Fantastic Voyage*, in which a spacecraft with its crew was shrunken to become a micro-device capable of traveling inside the human body. That spacecraft lost its communication and localization capabilities and went through an unguided dramatic travel within the human body before it exits through the eye drops of the person in whose body they were traveling. At the time of this writing, endoscopy capsules are traveling the digestive system of human bodies and people perceive other micro-robots to travel inside the blood circulation system of human beings to support health applications. We do not have the technology to shrink people. However, we have enough remote control capabilities that we can have robots do operations inside a human body without an operator inside the craft.

The main question is that "how did we arrive at this point?" In the past decade miniaturization and cost reduction of semiconductor devices has allowed the design of small low cost computing and wireless communication devices used as sensors in a variety of popular wireless networking applications and this trend is expected to continue over the next two decades. It is expected that a myriad of new applications designed around sensor technologies will emerge to stimulate the world economy for another round of industrial growth. One of the most promising areas of economic growth associated with this industry is that of body sensor networks that are also referred to as body area networks (BANs) [Yan06]. These networks are expected to connect wearable and implantable sensory nodes together and with the Internet to support numerous applications ranging from traditional externally mounted temperature meters or implanted pacemakers to emerging blood pressure sensors, eye pressure sensors for glaucoma, and smart pills for health monitoring and precision drug delivery. A number of technical challenges regarding size and cost, energy requirements, and wireless communication technology are under investigation and in the heart of these investigations is the understanding of radio propagation in and around the human body.

To support the growth of this industry, recently, the Federal Communication Commission (FCC) has allocated specific bands for Medical RadioCommunication Services (MedRadio) bands [FCC09] and the IEEE 802.15.6 standard has been formed to address standardization of these emerging technologies. The IEEE 802.15.6 standard defines the technologies and models for characteristics of the medium for wearable and implanted sensor networks [Aoy08; Hag08a, Hag08b; Kim08].

9.4.2 Overall Architecture and Applications

Figure 9.22 shows the overall architecture of a BAN connecting implant and body mounted sensor devices to a body mounted base station and an external access point connecting to the Internet. The characteristics of the implant and body mounted devices are different because they may have different sizes, communications, localization, and power consumption requirements and capabilities. The body-mounted base station is usually mounted on the belt above the hips of a person and carries larger batteries, has extended computational capabilities and on board memory. The results of the sensing and other data can be delivered to the Internet either in real-time or with opportunistic timing. As an example, in capsule endoscopy, the capsule inside the human digestive system sends the video to the body mounted base station that saves the information. Later, when the patient visits the doctor, the video is transferred to a computer for further processing and diagnosis. In a pacemaker application, a doctor may ask the patient to get close to a WiFi access point at certain times of the day to send a sample of the heart signal for remote monitoring of a patient's heart. From the wireless networking point of view, the system should be designed to accommodate all of these situations. The proper design of the network involves understanding and modeling of the channel behavior and the design of the physical and MAC layer to support envisioned applications.

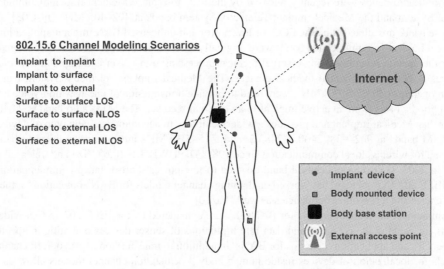

Figure 9.22   General architecture of a BAN.

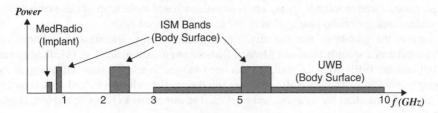

Figure 9.23   Frequency bands used for BAN.

9.4.3   Channel Measurement and Modeling

Radio propagation measurements and modeling for wireless networking applications begin with defining measurement scenarios and frequency bands. For each scenario, a measurement campaign is conducted and results of measurements are post processed and analyzed to develop statistical models for the behavior of the channel in each scenario. These models are then used for performance measurement and evaluation of alternative wireless networking solutions. Figure 9.22 illustrates a generic scenario involving implant and body-surface sensors and base stations and how they get connected to an external access point. IEEE 802.15.6 considers seven measurement scenarios for BAN channel modeling. We have four different classes of radio propagation models: Implant to Implant, Implant to body surface and external access point, body surface to body surface and body surface to external network connection. These classes are further categorized into scenarios, we have one scenario for implant to implant, two scenarios for each of the other channel model classes, which forms a total of seven scenarios, summarized in the left top part of Figure 9.22.

The unlicensed frequency bands considered by IEEE 802.15.6 for BAN applications are shown in Figure 9.23. These bands include the Medical Radiocommunication Services (MedRadio) bands [FCC09] over 401–406 MHz for measurement scenarios with implant sensors, the Industrial, Scientific and Medical (ISM) band and Ultra-Wideband (UWB) frequency bands for non-implant sensors. The MedRadio bands were recently released by the FCC to replace Medical Implant Communication System and the Medical Implant Telemetry System bands at 402–405 MHz [FCC03]. All of these bands are allocated by the FCC for low power and unlicensed BAN implant applications. Medical Implant Communication bands were originally used for data transmission to help diagnostic or therapeutic functions using external medical processing transceiver and implanted medical devices or between implanted medical devices. The Medical Implant Telemetry System bands, which operate at 403.5–403.8 MHz, were used to provide transmission of one-way non-voice data on a periodic basis from an active implant to an external receiver. The MedRadio band is 2 MHz wider than Medical Implant and provides a wider flexibility to accommodate new type of devices. The ISM bands at 902–928, 2400.0–2483.5, and 5.15–5.35 MHz are the first unlicensed bands released for wireless local communication since 1985. The UWB 3.1–10.6 GHz band, released by FCC in 2003, is also an unlicensed band devoted to low-power short-distance communications. The IEEE 802.15.6 standard has worked on defining channel models for BAN applications at these frequencies and a few other frequencies as well [Yaz10].

Communication channel models for BANs that are considered by the IEEE 802.15.6 provide a means for gauging the coverage and data rate limitations of sensor devices operating in specific frequencies and application scenarios for BANs. In addition to transmission, we also need channel models for localization of devices inside human body. Localization channel models allow us to determine limits on the accuracy and precision of the location estimates for different implant and body mounted sensors. Localization for implant and body mounted sensors is a new area of research that will likely draw on previous research on indoor geolocation [Pah98, Pah02, Pah05, Pah06; Say10]. Measurement of the radio characteristics for high-speed wireless communications for BAN enables a number of applications mentioned earlier in this section. The measurement of localization characteristics of BANs would enable another set of interesting applications for implant sensors such as localization of wireless endoscopy capsule inside body, for body surface mounted applications, and localization of surgery equipment inside operation rooms [Pah12a, b].

Challenges in Radio Propagation Analysis Inside the Human Body: At the time of this writing, channel measurement and modeling for radio propagation inside and around the human body to support waveform transmission and device localization is in its infancy. From an innovative research point of view, the measurement and modeling of radio propagation inside and around the human body offers unique challenges making this area very appealing for basic or fundamental research. These challenges are caused by several specifics of the medium and the BAN applications that are profoundly different from those that are applicable for traditional indoor radio propagation. The important difference between propagation inside a human body and the overall indoor propagation is that the medium for propagation inside the body is similar to propagation within liquids, which have substantially different conductivity than air which is the main medium for traditional indoor radio propagation. In addition, the insides of a human body comprise a continuous non-homogeneous medium with non-geometric boundaries, while traditional indoor areas are made of a discrete non-homogeneous environment with fairly geometric boundaries for radio propagation. Figure 2.1a shows a typical indoor environment and how we can employ the geometric construction and the fact that most of the propagation is through air to construct simple radio propagation models using ray optics. Figure 9.24 shows the inside of the non-geometric and continuously non-homogeneous human body that will not allow application of simple ray tracing techniques for developing radio propagation models. Radio propagation inside different organs, bones, and the muscle tissues is widely different posing a challenge for analysis and modeling.

Indoor environments are very complex propagation media as well, but we can easily measure the wideband radio channel characteristics using a network analyzer and develop empirical statistical models (discussed in Chapter 2). The second challenge for radio propagation studies for wireless communications and localization inside a human body is that we cannot simply place antennas inside a human body to collect empirical data for statistical radio propagation modeling.

In free space we use the time of flight of the signal for precise ranging to measure the distance between the transmitter and the receiver. This is done by multiplying the time of flight by the speed of radio wave propagation that is the same as speed of light in air. This is no longer possible inside the human body. Since the human body is a non-homogeneous liquid type medium the speed of radio wave propagation is different from the speed of light in air and it also differs in various organs. Therefore, the third challenge for in-body radio propagation analysis is the measurement of time of flight for localization applications.

Understanding the nature of signal propagation is key to the design of efficient and low-power, low-cost communication systems and precise localization for body area networks. Therefore, the first step in research is to start a measurement and modeling program to understand the nature of signal transmission inside human body. At the time of this writing the existing measurement results for understanding the propagation in and around a human body are fragmented and they do not pay attention to around the body localization applications such as locating surgery equipment during a knee operation [Pah12a, b] or localization of endoscopy capsules [Pah12a, b]. There is still a need for a comprehensive measurement program for wireless high-speed communication and localization applications for BANs.

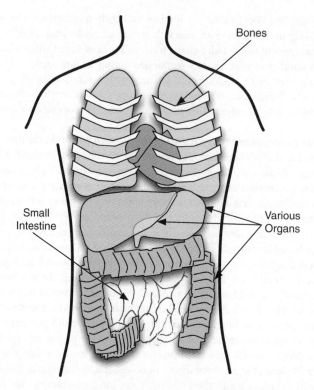

Figure 9.24 Different organs and bones inside human body.

Computational Techniques for RF Propagation Inside Human Body: The indoor environment is a very complex environment for radio propagation analysis by direct solution of Maxwell's equations using Finite Difference Time Domain (FDTD). In smaller indoor areas, building specific ray tracing and ray optics can be used for the analysis of radio propagation for wireless communications and localization [Pah05]. As a result, standardization committees have resorted to generalized statistical modeling of this environment for WLANs (discussed in Chapter 8) and low-power WPANs that we discussed in this chapter. Statistical models are based on empirical measurements of the multipath characteristics and these measurements are obtained by placing antennas in different locations inside a target building. Since measurements of signals inside a human body requires placement of antennas inside the human body, it becomes very complex, making this approach undesirable. To determine the characteristics of the radio propagation inside human body researchers resort to direct solution of Maxwell's equations, using Phantoms of a human body or taking advantage of the organs of dead animals and validate these results with limited measurements on human subject, mostly on the surface of the body.

Another reason that computational techniques such as FDTD are not used for general indoor areas is that in these technique we need to divide the area for modeling into grids that have dimensions on the order of the wavelength of the transmission frequency and then we have to solve the Maxwell equations involving calculus using numerical methods for integration and differentiation. For typical WLAN and WPAN applications, the frequency of operation is a few GHz resulting in wavelengths that are in orders of several centimeters. A 3D grid of a typical indoor area is huge and requires tremendous amount of calculations to simulate such an environment. As we consider radio propagation inside the human body, computational techniques for direct calculation of the Maxwell's equations, such as FDTD, become computationally feasible and it is indeed commercially available. These commercially available software are commonly used by researchers for modeling the path-loss and fading characteristics inside a human body [Yaz07b; Hag08a, b; Say10].

To measure the channel characteristics for wideband radio communications and localization, one needs to analyze waveform transmission inside a human body. Figure 9.25 shows a typical result of a commercially available Ansoft HFSS TM simulation software with cubic boundaries and two dipole antennas. Figure 9.25a shows the boundary conditions and location of the antennas without a body. The two horizontal black lines represent the dipoles. Figure 9.25b shows the body between the two antennas and the electric field plot inside the human body. Using this simulation we can compare the results of actual measurement with that of HFSS simulations. Figure 9.26 illustrates the measured and HFSS simulated channel impulse responses in the two scenarios shown in Figure 9.25.

Wideband simulations using Ansoft HFSS in a desktop computer takes a few days per plot. To reduce the computational time one may use commonly accepted computational platforms such as MATLAB® to speed up the process [Mak11]. This method can reduce the computational time by orders of magnitude and obtain channel impulse responses in a few minutes. Figure 9.27 shows the human body simulation on the MATLAB described in [Mak11] and the results of wideband waveform transmission inside human body studied in [Kha11]. The computational difference between HFSS and the FDTD simulation in MATLAB also resides in the selection of more computationally friendly square grids in Figure 9.27, rather than triangular grids used in HFSS in Figure 9.25.

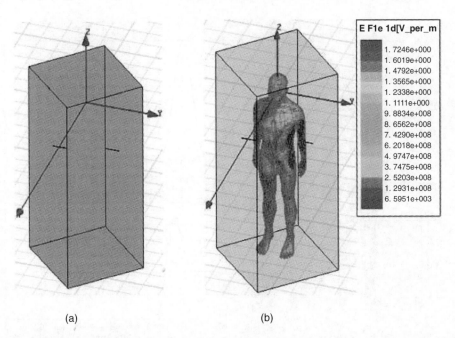

E F1e 1d[V_per_m

1. 7246e+000
1. 6019e+000
1. 4792e+000
1. 3565e+000
1. 2338e+000
1. 1111e+000
9. 8834e+008
8. 6562e+008
7. 4290e+008
6. 2018e+008
4. 9747e+008
3. 7475e+008
2. 5203e+008
1. 2931e+008
6. 5951e+003

(a)   (b)

Figure 9.25 Ansoft HFSS TM simulation setup: (a) without body, (b) with body and electric field plot. The two horizontal black lines represent dipole antennas. Reproduced with permission from [Ask11]. Copyright @ 2011, John Wiley and Sons.

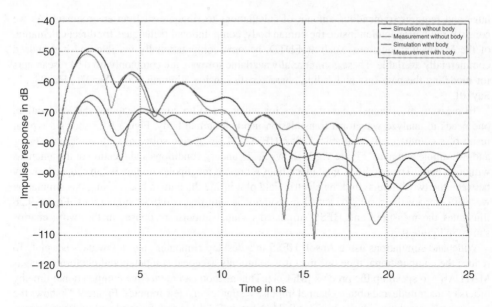

Figure 9.26 Impulse response obtained from the HFSS simulation and actual measurement in the two scenarios of Figure 9.24 [Kha11]. Reproduced with permission from [Ask11]. Copyright @ 2011, John Wiley and Sons.

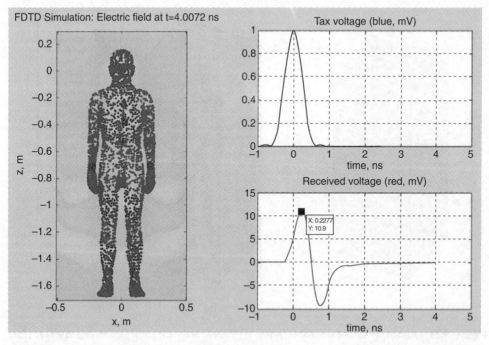

Figure 9.27 Impulse response obtained from the FDTD simulations using MATLAB [Kha11]. @ 2011 IEEE. Reprinted, with permission, from [Kha11].

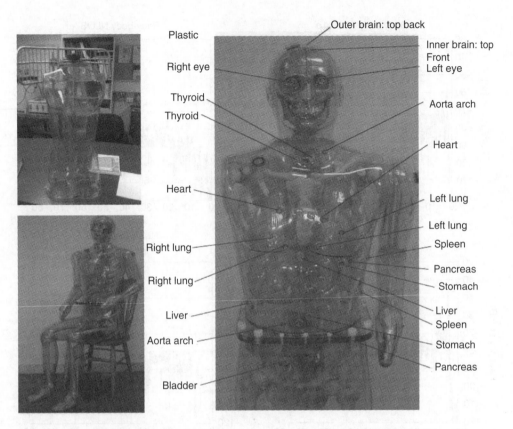

Figure 9.28 Different phantoms: (a) hollow torso phantom, (b) full-body phantom with real human skeleton and organs, (c) detail of organs.

Empirical Measurements inside Human Body: To check the accuracy of computational techniques for the analysis of radio propagation inside a human body one may use a *phantom*. Different phantom setups have been implemented to measure radio propagation inside the body and test the performance of designed antennas for body area applications. In [Joo06] researchers use fiberglass shell phantoms filled with tissue simulating liquid. The liquid is a typical sugar and a water based liquid, also commonly used in antenna propagation measurements inside the head from mobile terminals. In [Alo06] researchers have used a similar full-human hollow phantom with a plastic shell with animal organs holding an implant antenna inside, to characterize and simulate human tissues for measurements of implant radio propagation. In [Soo04] tissue-simulating material composed of TX-151, sugar, salt, and water was used to verify the results of spiral and serpentine antenna performance. In [Kim04] researchers use a human tissue-simulating liquid made from deionized water, salt, and cellulose to verify the performance of planar inverted-F antennas. In [Hig07] researchers use an implant placed inside a Perplex body (30 cm diameter cylinder), and measure the signal strength as a function of the distance.

Phantoms are designed with different degrees of details. Figure 9.28 shows a hollow torso phantom and a full body phantom with real human bones and cavities for variety of organs. Both of these phantoms are designed by Phantom Laboratory (Salem, N.Y.). For general applications

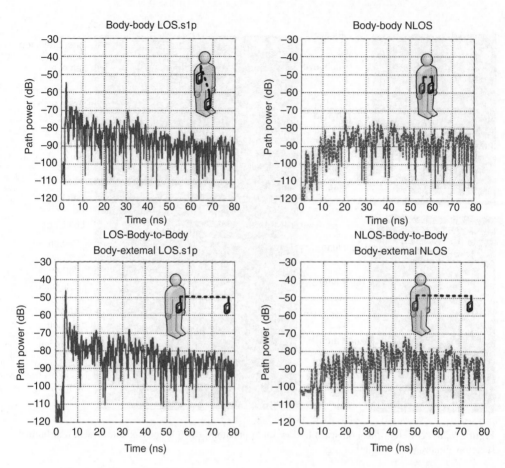

Figure 9.29   Channel impulse response in four different orientations of the body.

one fills the cavities with different mixtures of liquids. For radio propagation modeling for wireless communications and localization, we need to be able to place antennas in known locations within the body that adds to the complexity of the design. Ideally the phantom would have a large access port on the torso, which would allow items to be inserted within the phantom. The hollow phantom in Figure 9.28 costs several thousand dollars while the full phantom costs around 30 times more than that. Manufacturing of the full phantom is very complex but it more closely mimics the human body. The bones come from real human bodies and the limbs have their own access for filling a liquid. The arms and legs have simple pivot joints at the shoulders, elbows, hips and knees.

Body Surface Radio Propagation Measurement and Modeling: Several studies for RF propagation measurements around the human body for wireless high-speed BAN applications have been conducted. Authors in [Ryc04] derive a channel model for the 400, 900, and 2400 MHz bands using FDTD methods and show the impact of creeping waves for around body propagation. In [For06] and [Gou08], authors propose a UWB channel and path-loss models based on actual measurements in the 3–6 GHz bands. [Yaz07a; Min08; Saw08]. More specifically [Yaz07a] discusses the effect of human body UWB antennas and their propagation characteristics. The studies in [Min08] and [Saw08] present the channel modeling efforts for body to external signal propagation and provide results of measurements for path-loss and time-delay characteristics for these scenarios.

Channel measurement and modeling on the body: Body surface measurement campaigns have been performed with a variety of frequencies (including 902–928 MHz and 2400–24 835 MHz ISM bands and 3.1–10.6 GHz UWB frequencies) for body surface to body surface as well as body surface to external antenna [Yaz10].

Channel measurement and modeling for body surface mounted sensors is divided into Body to Body and Body to External experiments, which are further divided into line of sight (LOS) and Non-LOS (NLOS) scenarios. In LOS scenarios there is a direct unobstructed path between the transmitter and the receiver. In NLOS scenarios, the body is blocking the signal from a direct connection path between the transmitter and the receiver. Figure 9.29 shows the results of four measurement experiences for body surface to body surface and body surface to external for LOS and body obstructed NLOS conditions. In LOS experiences, we can clearly see the direct path, which is also the strongest path. In NLOS experiences, the direct path is not the strongest path and the overall channel impulse response has a clearly different pattern of behavior. The considerable differences in multipath profiles suggest the need for separating LOS and NLOS channel models for these scenarios and applications.

In Figure 9.29, if we model the path as going through the body, using empirical measurements, we can model other paths reflecting off the walls with a ray tracing algorithm to have a hybrid model for radio propagation for external BAN applications. Figure 9.30 shows the general idea behind this approach. Ray Tracing provides a ray optics approximation to propagation analysis and it can be used for larger areas where FDTD solution to propagation analysis is not practical. The advantage of Ray Optics approach is that we can find more compact solutions explaining the behavior of path arrivals [Ali02].

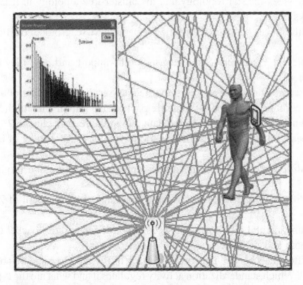

Figure 9.30 Propagation analysis using principles of geometric ray optics.

9.4.4 Physical and MAC Layer

Design of the physical and MAC layer for BANs is very much application dependent. For example, continual monitoring of the heart-beat may need a real-time low-speed link to the Internet, while transmission of a few pictures every second for endoscopy capsule diagnostics may need a reliable low power transmission using strong error control coding, but processing of the data may be done after data collection. Certainly for all BAN applications, power efficient modulation and medium access control methods are needed in principle and a number of researchers are working on these topics. The important fundamental issue is also localization of objects inside the human body and discovery of methods for localization that is suitable for in body applications. This is a new field of research that is gaining some momentum.

Questions

1. What is IEEE 802.15 and what is its relation to Bluetooth, ZigBee, and BAN technologies?
2. What are the differences between IEEE 802.15 device specifications and the device specifications of IEEE 802.11 WLAN devices?
3. Name the four states that a Bluetooth terminal can take and explain the difference among these states.
4. Name the three classes of applications that were originally considered for Bluetooth technology and identify those which are similar to 802.11 WLAN technologies.
5. What are the similarities and differences between the FHSS transmission scheme used in IEEE 802.11 and that used in Bluetooth in terms of data rate, modulation technique, available frequencies for hopping, speed of the hop, and the number and pattern of the hops?
6. What are the differences between ad-hoc networking solutions offered by 802.11 and 802.15?
7. What is the difference between a full function device and a reduced function device in IEEE 802.15.4?
8. What is the difference between a channel page and a channel number in IEEE 802.15.4?
9. Differentiate between the star topology and the cluster-tree topology in IEEE 802.15.4 based sensor networks.
10. What are the differences between the frame format and MAC protocol of Bluetooth and IEEE 802.11 FHSS?
11. How many different voice services does Bluetooth support and how they are differentiated from one another?
12. How many different symmetric and asymmetric data services does Bluetooth support?
13. What is the maximum supported asymmetric packet data rate in Bluetooth? How many slots per hop does it use? What is its associated data rate in the reverse channel?
14. Compare the header and access code of Bluetooth with the PLCP header of FHSS based IEEE 802.11.
15. What are the differences between the implementation of paging and inquiry algorithms in Bluetooth?
16. What are the differences between the frame format and MAC protocol of ZigBee and IEEE 802.11 DSSS?
17. What are the fundamental differences between supported data rates, packet format, and MAC layer of Bluetooth and ZigBee?
18. What are the differences between the protocol stack of IEEE 802.11 and that of IEEE 802.15.4?
19. What is a BAN? Which standardization activity regulates it and what are the typical applications supported by this technology?

20. What are the computational methods used for radio propagation analysis inside the human body?

Problem 9.1

Draw and explain the complete protocol stack for the implementation of e-mail application over Bluetooth.

Problem 9.2

Consider that encoded voice in Bluetooth is at 64 Kbps in each direction.

a. Using the packet format for the HV1 channels show that these packets are sent every six slots.
b. Using the packet format for the HV2 channels, calculate how often these packets are sent.
c. Repeat (b) for HV3 packets.

Problem 9.3

a. What is the hopping rate of frequencies in Bluetooth? How many bits are transmitted in each one-slot-packet transmission?
b. If each frame of an HV3 voice packet in Bluetooth carries 80 bits of the samples of speech, what is the efficiency of the packet transmission (ratio of the overhead to the overall packet length)?
c. Determine how often a HV3 packet has to be sent to support 64 kbps voice in each direction.
d. The DH5 packets carry 2712 bits for every five-slots packet. Determine the effective data rate in each direction in this case.

Problem 9.4

Repeat examples 9.7 and 9.8 for all other data rates supported by Bluetooth (see Table 9.1).

10

Gigabit Wireless

10.1 Introduction

Communication networks have evolved to provide pervasive connectivity to phones, computers, and consumer electronic devices. Wireless networking technologies have evolved to provide wireless or tetherless access to these devices. As we have discussed in Chapter 8, over the past three decades, the WLAN industry evolved to support home networking and hot-spot coverage and eventually received adoption by the smart phone industry opening the door to chip set sales on the order of billions per year. As the speed and memory of mobile computers and smart phones have increased, the need for data rate for wireless access has grown from 1–2 Mbps in the case of legacy IEEE 802.11 to hundreds of Mbps supported by 802.11n. As we have discussed in Chapter 9, over the past decade, two other wireless technologies emerged to complement the popular and ever growing WLAN industry. First were the low-power WPAN wireless networks, such as Bluetooth and ZigBee, which were designed for ad hoc sensor networking. These technologies were the focus of Chapter 9. The second class of personal area networks and devices, comprising more recent technology, has been the super high data rate complement to WLANs, which is now emerging for multimedia applications that demand ultra high-speed wireless connections. We discuss such networks in this chapter and refer to them as "Gigabit Wireless" in general.

At the time of this writing, the super high speed WPAN industry is not as established or mature as that for low power WPANs. This industry is experiencing an evolutionary path, biding its time for a sizable market to emerge in the near future. However, those engaged in research and development in this industry believe that it will serve a number of applications that include high definition video streaming, wireless Gigabit Ethernet, wireless docking stations, wireless displays, desktop point to multi-point connections, wireless back haul, and wireless ad hoc networks [Guo07; Dan10; Per10; Kum11]. The maximum data rates for 3G mobile data systems (see Chapter 12) are around 2–10 Mbps, and the WLAN industry is marketing >100 Mbps systems with the latest 802.11n technology. Wireless mobile data services provide a comprehensive coverage similar to cellular networks and WLANs are intended to cover up to 100 m for local applications. Candidate technologies for super high-speed WPAN systems aim at a minimum of hundreds of Mbps to several Gbps data rates with a coverage of up to 10 m in indoor areas.

Our objective in this chapter is to provide an overview of the evolution of this industry and provide some detailed examples of fundamentals of design and operation of these networks. The super high speed WPAN industry has evolved with two hypes. The first hype in this area started

Principles of Wireless Access and Localization, First Edition. Kaveh Pahlavan and Prashant Krishnamurthy.
© 2013 John Wiley & Sons, Ltd. Published 2013 by John Wiley & Sons, Ltd.

with the emergence of UltraWideBand (UWB) technology that resulted in detailed design of a number of alternative implementation technologies in the 3.1–10.6 GHz unlicensed bands, which we use as examples. The second hype started more recently with the availability of millimeter wave (mmWave) 60 GHz unlicensed bands, work in which is still continuing and we provide its overview.

10.1.1   UWB Networking at 3.1–10.6 GHz

In the late 1990s and early 2000s, ultra wideband (UWB) communications began to attract considerable attention as a method of transmission for short-range wireless ad hoc networking in both commercial and military applications. The first technology used for UWB communications was called *impulse radio* [Win00], which was based on direct baseband pulse transmission without using carrier modulation. One of the major advantages of impulse radio was that the implementation of periodic transmission of narrow pulses was possible with low power expenditure. Because of this low power characteristic, as discussed in Chapter 3, pulse transmission techniques are also popular in the design of remote-control devices, albeit mostly with infra-red as the transmission signal.

Pulse transmission has been used for military radar applications since the 1950s [Tay01] and the first patent for a communication application was issued in 1973 [Ros73; Gha03a, b]. The new wave of interest in UWB for military applications began with DARPA projects such as the BodyLAN project and Small Unit Operation–Situation Awareness Systems (SUO/SAS). In these projects low-power small-radio technologies were needed to support broadband multi-rate wireless communications with accurate positioning in indoor and urban areas where traditional positioning using the GPS does not perform satisfactorily.

As we described in Chapter 9, in 1997 the IEEE 802.15 WPAN standardization committee began to define specifications for BodyLANs and eventually adopted Bluetooth as its standard specification, 802.15.1b. The same standardization group considered UWB solutions within the IEEE 802.15.3a sub-committee. This committee was focused on defining standard specifications for WPANs operating in the 3.1–10.6 GHz unlicensed bands released by the FCC in 2002.

For understanding the term *ultra-wideband* let us quickly review the bandwidth occupied by various wireless technologies over time. The 1G Analog cellular systems occupied a bandwidth of 30 kHz or less and the *total allocated bandwidth* for each direction of communication was around 25 MHz, split between two service providers. The 2G digital cellular systems generally occupy the same bands as 1G systems, and additional tens of MHz of spectrum were allocated as PCS bands, which are used by network operators to support 2G technologies. The commercially successful GSM TDMA system employs 200 kHz per carrier while the 2G cdmaOne system uses 1.25 MHz per carrier. The 3G digital cellular systems employ less than 5 MHz of bandwidth per RF carrier and the total available bandwidth is on the order of 100 MHz or so. In contrast, the IEEE 802.11 WLANs occupy a maximum bandwidth of 26 MHz per carrier and the largest available *total* bandwidth is in the 5 GHz domain, around several hundreds of MHz. We can compare these bandwidths with that of UWB systems, whose specifications suggest a bandwidth on the order of GHz and the total available bandwidth is around several GHz. Figure 10.1 illustrates the relative bandwidth of cellular, PCS, WLAN, and UWB systems. Cellular and PCS systems have much higher power spectral density and they occupy relatively smaller bandwidths. WLANs have a larger allocated bandwidth and a much lower power spectral density than that of the cellular systems. The UWB systems have larger bandwidth than WLANs (by an order or more) and the lowest power spectral density.

Figure 10.2 shows the *spectral mask* mandated by the FCC for unlicensed UWB communications. This figure also identifies the frequency spectrum considered by the IEEE 802.15.3a proposal. The

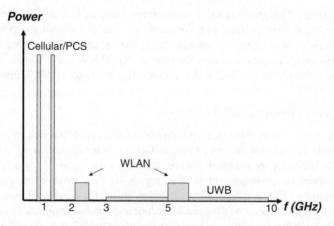

Figure 10.1 Relative power limits and the bandwidths available for cellular/PCS, WLAN, and UWB.

original petitions for the UWB band covered the entire spectrum and were intended to be used by impulse radio entirely. However, after a number of controversial debates concerning interference with other low power devices, in particular GPS receivers, the FCC decided to specify this mask, which reduces the radiation in the spectrum between 0.96 and 3.1 GHz, so as to facilitate virtually harmless co-existence with popular existing systems.

As far as standardization activities are concerned, the IEEE 802.15.3 sub-group completed a preliminary standard for 11 and 55 Mbps operation in 2003 and the objective of 802.15.3a was to use UWB technology to increase this data rate to several hundred Mbps to Gbps (that was at the time orders of magnitude higher than 802.11b operating at 11 Mbps). In the IEEE 802.15.3a group, several options were evaluated for UWB communications, among which are the historical impulse radio technology, as well as the Direct Sequence (DS-UWB) and Multi-Band OFDM

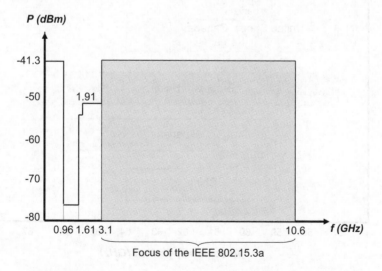

Figure 10.2 FCC spectral mask for unlicensed UWB communications.

(MB-OFDM) systems. This standard lost its momentum later and was dissolved in January 2006. However, the technical work produced by the committee has educational and pedagogical value in its consideration of alternative technologies for implementation of super high-speed WPAN wireless transmission techniques to achieve the goals of "Gigabit Wireless". More details on UWB communications are available in [Gha04; Opp04] and other references cited in them.

10.1.2   Gigabit Wireless at 60 GHz

In 2001, the FCC released unlicensed bands within 57–66 GHz, not all of which are available all over the world, but parts are available in most countries. They are generally called 60 GHz bands. Later, the release of this spectrum opened another wave of activity for Gigabit Wireless applications. The research and development community in this field argues that the mmWave has several advantages over UWB technology. The UWB regulatory constraints are not the same in different countries. The rules for using different segments of the band for indoor and outdoor applications or the way that a device should react when it interferes with another service in those bands are different in different countries. As discussed earlier, the UWB spectrum overlaps with popular applications such as WLANs at 5.2 GHz (see Figure 10.1), which can cause mutual interference among applications operating in these bands. These restrictions forces the channelization of the entire band into narrower bands for multi-carrier operation that restricts the highest achievable data rate and keeps them to values close to those achieved by 802.11n.

Figure 10.3 illustrates the 60 GHz mmWave spectra available worldwide in major industrial countries [Wig10]. The mmWave spectra provide a continuous 7 GHz of bandwidth at 60 GHz that overcomes the difficulty of channelization with UWB. In addition, increasing the frequency of operation to around 60 GHz will contain the signal inside a room and allows for the design of smaller directional antenna arrays supporting a more secure and less interfering environment that is useful for a number of applications. Motivated by these features, a number of standardization activities and alliances were formed in the past few years to discover and implement technologies

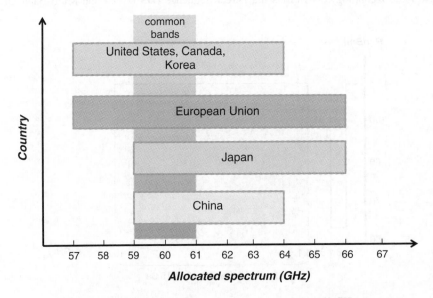

Figure 10.3   Worldwide availability of spectrum at 60 GHz.

for the 60 GHz mmWave bands. Good examples of these activities are the IEEE 802.15.3c group that emerged form the WPAN community [IEEE09] and the IEEE 802.11ad group that emerged out of the WLAN community [Per10].

The 802.15.3c was formed to specify another PHY alternative for mmWave 60 GHz to complement the 802.15.3 WPAN standard released in 2003. This group was formed in March 2005 and defined a new MAC and PHY for this channel in September 2009 [IEEE09]. This technology is expected to accommodate coexistence with other WPANs and support high definition video streaming and other streaming applications with over 2 Gbps data rates. The IEEE 802.11ad group is building on the legacy of the IEEE 802.11 commercial success and expects to complete its standard soon [Per10].

10.2 UWB Communications at 3.1–10.6 GHz

The idea of "Gigabit Wireless" first emerged with ultra-wideband (UWB) communication networks and in the IEEE 802.15.3a community. This group explored the characteristics of the UWB channel, evaluated alternative technologies for UWB communications, and interacted with the FCC to settle the rules with the unlicensed bands for these systems. In this endeavor, they defined three major transmission technologies: impulse radio, direct sequence UWB, and multi-carrier OFDM and developed UWB channel models for coverage and for characterizing the multipath arrivals for the UWB radio spectrum between 3.1 and 10.6 GHz. Here we provide an overview of these technologies, more as examples of systems that were first considered for short-range communications. In the last section of this chapter, we discuss the channel models designed for performance evaluation of these systems in the UWB and mmWave spectra.

In a manner similar to spread spectrum signals, UWB signals can be designed to coexist with other radio systems already in use. Since the bandwidth of UWB systems are much wider than those of spread spectrum systems, they can readily be overlaid on top of existing systems without causing undue interference to those systems. At the same time, since the bandwidth is ultra-wide the UWB system has a very low power spectral density and it can overlay many existing systems with different transmission requirements. As we discussed in Chapter 2, signal fading is caused by the overlap of signals received on different paths. Because of its ultra wide bandwidth, the UWB signal isolates (resolves) multipath components resulting in a stable received signal power with reduced fast fading effects. Using bandwidths on the order of GHz for UWB systems allows achieving gigabit wireless transmission and more precise localization. Both of these features were perceived to be essential in support of emerging video gaming and home entertainment industries as well as in military applications for urban fighting and first responders.

10.2.1 Impulse Radio and Time Hopping Access

As we noted earlier, the interest in UWB communications began with the idea of impulse radio transmission. With this technique, a very short-duration (on order of a few tenths of a nanosecond) and low-power (high duty cycle of several hundreds of nanoseconds) pulse is used for information transmission. The spectrum of this pulse obviously occupies a very wide band (several GHz) and this is the reason it was given the name UWB. The spectral height of the UWB signal is very low because the low transmission power is further spread over a very wide bandwidth.

Pulse Shape and Antenna: Implementation of the impulse radio UWB signal transmitter does not involve modulation and if carefully designed, the transmitter can be much simpler than those in traditional narrowband or wideband spread spectrum systems. The designers of impulse radio UWB systems chose a transmission waveform that: (1) has such a high bandwidth and signal processing gain that interference from existing systems is negligible, (2) has a spectral height comparable to the background noise so that the FCC allows it to co-exist with the systems already in place, (3) has a relatively simple implementation, and (4) has a spectrum which looks like a pass-band signal having no DC component.

Example 10.1: UWB pulse shape in Impulse Radio

Time Domain Corporation is one of the leading companies engaged in the design of impulse radio UWB systems [TDC13]. The pulse shape used in their pioneering system was a monocycle mathematically described by:

$$v(t) = 6A\sqrt{\frac{e\pi}{3}}\frac{t}{\tau}e^{-6\pi\left(\frac{t}{\tau}\right)^2} \tag{10.1}$$

where A represents the peak amplitude of the pulse, τ is a constant determining the width of the pulse, and t is the time variable. The spectrum of the monocycle pulse in frequency domain is given by:

$$v(f) = -j\frac{2f}{3f_C^2}\sqrt{\frac{e\pi}{2}}e^{\frac{\pi}{6}\left(\frac{f}{f_c}\right)^2} \tag{10.2}$$

where $f_C = 1/\tau$ is the center frequency of the pulse. Figure 10.4a shows a typical graph of the pulse for $\tau = 0.5$ ns associated with a center frequency of $f_c = 2$ GHz. The half power (3-dB) bandwidth of the pulse occupies about 2 GHz (Figure 10.4b). The pulses are transmitted periodically and the information is encoded in the location or position of the pulses, implementing a pulse position modulation system. Figure 10.5c, d illustrates the pulse repetition and its effects on the spectrum of the signal. The typical pulse width used for this implementation is between 0.2 and 1.5 ns and the time interval between consecutive pulses ranges from 25 to 100 ns [TDC13].

One of the challenges in the design of an UWB system is the choice of antennas. This is in particular more challenging for the original impulse radio systems that covered a spectrum ranging from very low frequencies up to a few GHz. The antennas designed for these bands need to have a flat response across a wide spectrum and they also have to be compact. Since these conditions are hard to maintain, the antennas used for impulse radio transmission actually change the shape of the transmitted pulse.

Example 10.2: Pulse shape and Antennas in Impulse Radio

The pulse shape introduced in Example 10.1 is in fact a received pulse shape. The actual transmitted pulse is show at the left of Figure 10.5. When this pulse is applied to a bow-tie antenna, the pulse propagates with distortion and the received signal will appear, as shown at the right of Figure 10.5. The antennas behave as filters, and even in free space, a differentiation of the pulse occurs as the wave radiates from transmitter antenna and it is absorbed by the receiver antenna [Win98].

Time-Hopping Spread Spectrum Transmission: The pulse position modulation (PPM) scheme in the case of the impulse radio is combined with time-hopping spread spectrum medium access

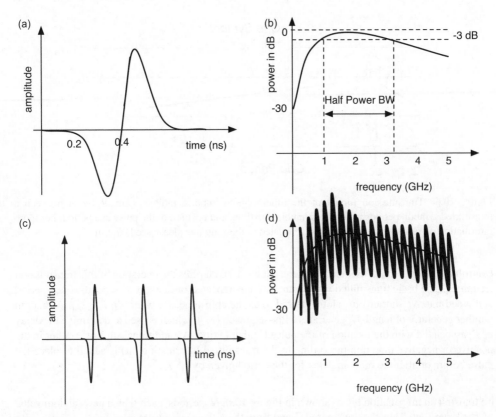

Figure 10.4 (a) Schematic of original UWB pulse used in pioneering impulse radio system designed by Time Domain Corporation. (b) Illustration of the spectrum of the pulse. (c) Repetition of the pulses. (d) Illustration of repetition of pulses resulting in spikes in the spectrum.

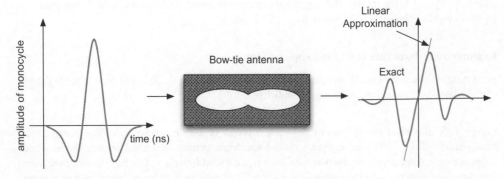

Figure 10.5 Illustration of transmission of an impulse through a bow-tie antenna pair and the resulting waveform observed at the receiver front end, see [Win98].

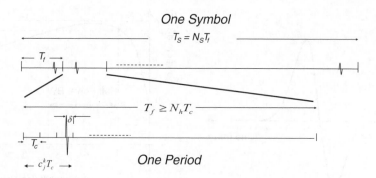

Figure 10.6 Transmission format of the time-hopping impulse radio system. Relation between a transmitted symbol, periods inside the symbol, chip time, and position of the pulse in a period. For more information on further experimental results related to these figures please see [Win00].

control to support a multi-user environment. Figure 10.6 clarifies the operation of this transmission technique. The basic transmitted waveform, $w_{tr}(t)$, is repeated once every T_f seconds. Each period is divided into N_h shorter time slots referred to as the chip-time, T_c, where $N_h T_c \le T_f$. A random number generator of length N_s generates a PN-sequence $\{c_j{}^k\}$, which is used to determine the delay $c_j{}^k T_c$ associated with the location of the pulse for that period. An additional shift of d either in the positive or negative direction (according to the transmitted information bit) is added to place the pulse before or after the random delay for the period given by $c_j{}^k T_c$.

To transmit an information bit, as shown in Figure 10.6, N_s periods, each with a new random value of $\{c_j{}^k\}$, are sent. Thus the transmitted signal from the k-th user is represented by:

$$x_{tr}^k(t^k) = \sum_{j=-\infty}^{\infty} w_{tr}(t^k - jT_f - c_j^k T_c - \delta d^k_{\lfloor j/N_s \rfloor}) \qquad (10.3)$$

where $d_j = \{0,1\}$ is the information digit that remains the same for N_s periods. Since we send one bit every $N_s T_f$ seconds, the data rate is $R_s = 1/N_s T_f$ bits/s.

Example 10.3: Data rate of a time-hopping system

For a typical values of $T_f = 100$ ns and $N_s = 500$, the data rate is $R_s = 20$ kbps and for the maximum chip duration of $T_c = 0.2$ ns we can get $\delta = 0.1$ ns.

Figure 10.7 illustrates the difference between a direct sequence spread spectrum (narrower in bandwidth) and a UWB time-hopping spread spectrum system described above. In the direct sequence spread spectrum system, the data stream is divided into chips that have smaller duration, and therefore with the same transmitted power and data rate we will have a lower power spectral density and a bandwidth expansion equal to the processing gain of the direct sequence spread spectrum system. In the case of a UWB time hopping spread spectrum impulse radio system, the transmitted pulses occupy a fraction of a chip causing another expansion in the bandwidth and consequent reduction in the power spectral density.

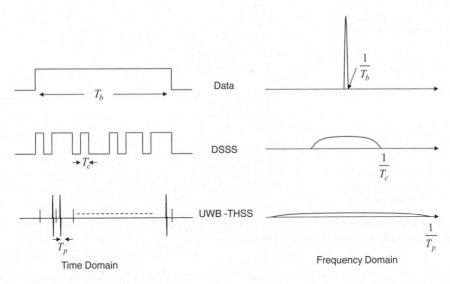

Figure 10.7 Comparison between the time- and frequency characteristics of narrowband data, DSSS, and UWB-THSS.

10.2.2 Direct Sequence UWB

In the actual operation of the impulse radio, the UWB pulses interfered with many existing systems, including cellular, WLAN, and GPS Systems [Ham02]. In particular, GPS receivers operating with signals from satellites with very low signal strength were vulnerable to UWB interference. As a result of these concerns, the FCC decided to mandate the frequency template shown in Figure 10.1. This decision in the year 2003, re-directed the attention of the IEEE 802.15.3a UWB standardization activities toward the 3.1–10.6 GHz band. Consequently impulse radio using the lower frequency bands, shown in Figure 10.5, lost their support in the standardization committee. Two leading 802.15 proposals brought forward after the 2003 FCC announcement are known as direct sequence UWB (DS-UWB) and multi-band OFDM (MB-OFDM). The DS-UWB technique is addressed in the remainder of this section and the MB-OFDM technique is described in the next section. The basic coverage cell in the WPAN industry is referred to as a *pico-cell* having a nominal coverage range of about 10 m. A network operating within that range is referred to as piconet. Different WPAN technologies support different numbers of overlapping piconets. For example Bluetooth, the first WPAN standard under IEEE 802.15.1, supports seven overlapping piconets. The UWB proposals considered multiple bands that can be combined with medium access control to support larger numbers of overlapping piconets.

Physical Layer and M-BOK: The DS-UWB system uses the direct sequence spread spectrum technique, which emerged as the physical layer of choice in 3G cellular networks in the late 1990s. In pulse transmission one can use the two polarities of the pulse as well as the absence of transmission to create a *ternary* symbol transmission technique. In the proposed DS-UWB system, as shown in Figure 10.8, the 3.1–10.6 GHz band is divided into a low band from 3.1 to 4.9 GHz and a high band from 6.2 to 9.7 GHz. The bandwidth of the high-band is twice the bandwidth of the low-band resulting in shorter time domain pulses in the high band to support higher symbol transmission rates. The 4.9–6.1 GHz band is purposely neglected to avoid interference with IEEE

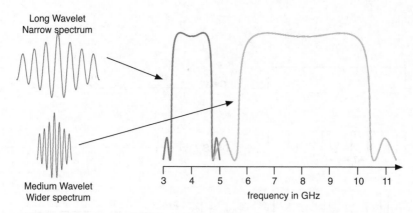

Long Wavelet
Narrow spectrum

Medium Wavelet
Wider spectrum

frequency in GHz

Figure 10.8 Illustration of the spectrum of two different pulses used in the DS-UWB proposal. For the actual frequency and time response see [Koh04].

802.11a devices operating in the 5 GHz band. Each piconet of the DS-UWB system operates in one of the two bands and piconets in the same band are separated by code division multiplexing (see Chapter 12 for a discussion of cellular CDMA) using ternary Multiple Bi-Orthogonal Keying (M-BOK) spreading codes [Mac03].

Example 10.4: Basic M-BOK codes

Table 10.1 shows the M-BOK ternary codes of length 24 used in the DS-UWB system proposed by IEEE 802.15.3a. Each chip can take three values $\{-1,0,1\}$ and hence the term ternary. For $M = 2$, the first row or its reverse, which means each chip value is flipped, are used to identify two symbols for binary transmission. Each incoming information bit is mapped to one of the two symbols. For $M = 4$ the first two rows and their reverses are used to form four symbols. The incoming information bits are grouped into two bits and one of the four symbols is selected for each information di-bit. For $M = 8$ all four rows and their inverses are used to form 8-MBOK codes. The incoming information bits are grouped into 3-bit blocks, each selecting one of the 8-BOK codes.

The M-BOK codes play two roles in the system: they act as a spreading code to differentiate the users of piconets from other piconets, and in each piconet they serve as an M-ary orthogonal coding technique to improve the performance. As compared with Walsh codes used in cdmaOne, described in Chapter 12, in that system Walsh codes are used in the forward channel to differentiate the users and in the reverse channel as M-ary orthogonal codes to improve the performance. The M-BOK system proposed by IEEE 802.15.3a serves both purposes – for coding and modulation. The M-BOK scheme was preferred to pulse-position modulation, used in the impulse radio solution,

Table 10.1 M-BOK ternary codes of length 24 for $M = 2, 4, 8$

2-BOK uses Code 1	$-1\ 1\ -1\ -1\ 1\ -1\ -1\ 1\ 1\ -1\ 0\ -1\ 0\ -1\ -1\ 1\ 1\ 1\ -1\ 1\ 1\ 1\ -1\ -1\ -1$
4-BOK uses Codes 1 & 2	$0\ -1\ -1\ 0\ 1\ -1\ -1\ 1\ 1\ -1\ -1\ 1\ 1\ 1\ 1\ -1\ -1\ 1\ 1\ -1\ 1\ 1\ -1\ 1\ 1\ 1\ 1$
8-BOK uses Codes	$-1\ -1\ -1\ -1\ 1\ 1\ -1\ 1\ 1\ -1\ 1\ 1\ -1\ -1\ 1\ 1\ -1\ -1\ 1\ 1\ -1\ -1\ 1\ 1\ 1\ 0\ -1\ 0\ 1\ 1$
1, 2, 3, & 4	$0\ -1\ 1\ 1\ 1\ 1\ -1\ -1\ -1\ -1\ -1\ -1\ -1\ -1\ 1\ 1\ -1\ 1\ 1\ -1\ 0\ 1\ -1\ 1\ 1\ 1\ -1\ -1\ 1\ 1$

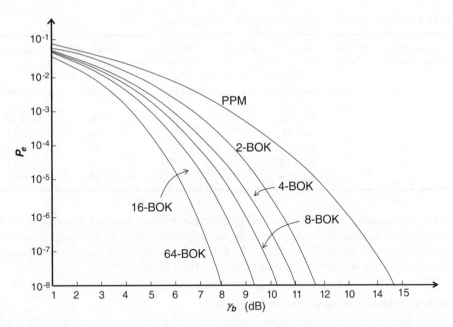

Figure 10.9 Performance of different M-BOK codes versus PPM in additive white Gaussian noise. For further experimental results related to these figures please [Koh04].

because it performs better and has more flexibilities for multi-rate and multi-channel operation. The IEEE 802.15.3a proposal used M-BOK codes of length 24 and 32 with arrays up to 64 elements to achive different data rates.

Figure 10.9 provides a comparison between the performance of Pulse Position Modulation (PPM) used in impulse radio and a variety of M-BOK coding techniques similar to those presented in the last example [Siw04]. For very low error rates, 2-BOK has a 3 dB edge over PPM and as we double the number of symbols, the requirement in terms of the signal to noise ratio per bit reduces by approximately one dB. It can be shown that as M goes to infinity, M-BOK can attain the Shannon bound. In practical situations, we have to keep the symbol transmission rate the same, and an increase in the number M will increase the data rate of the system, but we need a higher signal to noise ratio per symbol. In a manner similar to other multi-rate wireless data networks, as the distance between a transmitter and a receiver is decreased, the signal strength increases and we can use larger symbol rates to achieve higher data rates.

Variable data rates in the IEEE 802.15.3a DS-UWB proposal were achieved by changing the processing gain, and by switching between different values of M in the M-BOK coding and modulation technique. Lower data rates with higher processing gains can provide coverage over larger areas. The chip rates in the low and high bands are 1.368 and 2.736 Gcps respectively. The basic symbol transmission rate for the low-band with spreading factors of 24 and 32 are 1.368(Gcps)/24(chips) = 57 Msps and 1.368(Gcps)/32(chips) = 42.75 Msps. Similarly, high-band operation supports two basic data rates, 114 and 85.5 Msps. Different data rates are derived from these basic rates and three different forward error control coding schemes. The first forward error control coding option is a convolutional code with coding rate $r = 0.50$. The second coding option is a (55,63) Reed-Solomon code with rate $r = 0.87$. The third option is a concatenated code that has both rate $\frac{1}{2}$ convolutional and a (55,63) Reed-Solomon code resulting in an overall code-rate of

Table 10.2 Different data rates supported by DS-UWB [Koh04]

Information data rate (Mbps)	Modulation scheme	Symbol rate	Quadrature?	Code rate r
25	2-BOK	57	No	0.44
50	2-BOK	114	No	0.44
114	4-BOK	114	No	0.5
112	8-BOK	85.5	No	0.44
200	4-BOK	114	Yes	0.44
224	64-BOK	85.5	No	0.44
450	64-BOK	85.5	Yes	0.44
900	64-BOK	85.5	Yes	0.87

$r = 0.44$. The standard also allows BPSK and QPSK transmission in the band. This feature allows another degree of freedom to diversify the supportable data rates. Table 10.2 provides a summary of data rates achievable by DS-UWB solution for different combinations of coding and modulation techniques supporting data rates up to values close to 1 Gbps [Koh04].

Example 10.5: Data rate calculation in DS-UWB

For the 25 Mbps data rate in Table 10.2, we have two codes representing bits "0" and "1" and BPSK modulation (no quadrature phase). Therefore, each point in the constellation is identified by one bit and the uncoded transmission data rate is 57 Mbps, the same as the symbol transmission rate. The coded data rate is 57 Mbps $\times$ 0.44 = 25 Mbps. For the 200 Mbps rate, we have four symbols, resulting in two-bits per symbol and QPSK modulation that has another two bits per symbol. Therefore, in the overall scheme, we have four bits per symbol. With a symbol rate of 114 Msps and a coding rate of $R = 0.44$ we have a data rate of:

$$114 \text{ Msps} \times 4 \text{ bps} \times 0.44 = 200.64 \text{ Mbps}.$$

Medium Access Using Frequency, Time, and Codes: Engineers and scientists always strive to take advantage of the successful stories of the past and the evolution of technologies is based on re-inventing those past inventions. As we noted earlier, the medium access control in the case of DS-UWB combines frequency, time, and code division multiplexing techniques. This is done to take advantage of the good features of all of these techniques. The frequency–division multiplexing scheme chooses one of the two operating frequency bands to control the interference and increase the capacity, when needed. The code division multiplexing scheme uses ternary code sets in $\{\pm 1, 0\}$ with 2, 4, 8-BOK with length 24, and 64-BOK with length 32 to support multi-rate operation that is very useful for wireless data applications. The four CDMA codes within each frequency band are used to further separate the piconets by providing for implementation of logical channels. Within each piconet, time division multiplexing separates different users. The following example is provided to aid the understanding of this diversified medium-access method as an example of system engineering design.

Example 10.6: Medium access control in DS-UWB

Figure 10.10 illustrates an example scenario using frequency, time and code division multiple access in the DS-UWB scheme in a multi-room indoor area. Each piconet is separated by frequency

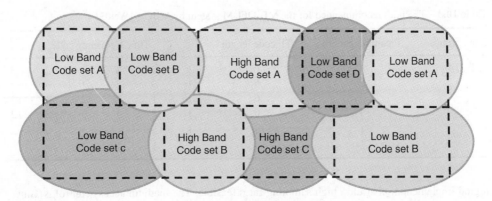

Figure 10.10 Distribution of the *piconets* in the frequency and code division medium access used in the DS-UWB proposal for the IEEE 802.15.3a. Users in each room share the channel using time division multiplexing.

between the low-band and high-band, in a manner similar to other wireless systems. Within each set using the same band, the piconets are separated by codes into channels A, B, C, and D again in a manner similar to cellular CDMA techniques. Then, within each cell, different users are separated by TDMA using a central scheduling system similar to those used in other TDMA systems such as LTE (Chapter 13).

10.2.3 Multi-Band OFDM

The second proposal from IEEE 802.15.3a suggested multi-band OFDM (MB-OFDM) technology. This approach [Sho03] adopted the OFDM technique, which first emerged as the technology of choice for IEEE 802.11 WLANs and later penetrated the cellular wireless data industry. The MB-OFDM scheme, in a manner similar to DS-UWB, was designed to operate in the 3.1–10.6 GHz unlicensed UWB bands. In this approach, shown in Figure 10.11, the spectrum is divided into 15 bands each of width 528 MHz. In each band a 128-point OFDM system using QPSK modulation is implemented to limit the required precision of mathematical operations and make

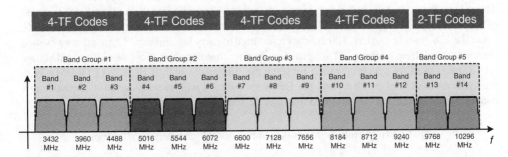

Figure 10.11 Overview of the MB-OFDM proposal for IEEE 802.25.3a at the 3.1–10.6 GHz band. The grouping of the channels and the time–frequency medium access scheme in each group of channels are illustrated.

Table 10.3   TF codes recommended for the MB-OFDM system by MBOA [Wel03]

Band groups	Preamble pattern	TF code length	Time frequency code					
1, 2, 3, 4	1	6	1	2	3	1	2	3
	2	6	1	3	2	1	3	2
	3	6	1	1	2	2	3	3
	4	6	1	1	3	3	2	2
5	1	4	1	2	1	2	–	–
	2	4	1	1	2	2	–	–

digital implementation at ultra high sampling rates feasible. The medium access control is Time-Frequency Multiple Access (TFMA), which combines the time- and frequency-diversity benefits of FHSS and DSSS into one medium access control technique.

Figure 10.11 also gives an overview of the relation between channel allocation and medium access for the MB-OFDM proposal. The 15 bands in the 3.1–10.6 GHz unlicensed UWB spectrum are divided into five groups of 528 MHz bands. Group 1 is the most desirable because Group 2 interferes with the 5 GHz band used by IEEE 802.11a/n devices, and higher groups have smaller coverage areas. Each physical piconet is implemented in a band group and several logical piconets share a band group using different TFMA codes. Groups 1–4 have four time-frequency (TF) codes and group 5 has two TF codes for logical channel separation. Since each piconet is identified by one TF code, the MB-OFDM proposal can accommodate 18 piconets in the entire UWB spectrum.

Table 10.3 provides the four patterns of TF codes used in groups 1, 2, 3, and 4 and the two patterns for group 5. Groups 1–4 have three different frequencies associated with the three bands in the group and the length of the time sequence is six. As a result, each band is used twice per symbol. Group 5 has two bands, a code length of 4, and again each band is used twice during each symbol transmission. With this technique, as shown in Table 10.3, neighboring piconets have two collisions per code length.

In each time-frequency slot the MB-OFDM system proposes transmission of one OFDM symbol. Therefore, as an example, consider one cycle of band group 1. Here, we send six OFDM symbols in the bands 1, 2, and 3 in a sequence shown in first row of Table 10.3. Like any other OFDM system, each OFDM symbol consists of three sections, one to carry the information, one for the cyclic prefix, and a time gap between the symbols. In this scheme, we have a 242.4 ns information length with a 60.6 ns prefix and a 9.5 ns guard time interval, for total symbol duration of 312.5 ns. The values of cyclic prefix and time gap are designed to accommodate the multipath delay spread of the channel (see Chapters 2 and 13) and to allow for time for switching between different bands. From the 128 tones or carriers, 100 are data tones used to carry information, 12 are pilot tones used for carrier and phase tracking, 10 are guard or dummy tones used to allow flexibility, and six are NULL tones used for synchronization (observe the similarities with IEEE 802.11 OFDM). With the symbol duration of 312.5 ns and the QPSK modulation with two bits/symbol, the basic raw data transmission rate for the proposed MB-OFDM scheme is:

$$1/312.5 \text{ (ns/symbol)} \times 2 \text{ (bits/symbol)} \times 100 \text{ (carriers)} = 640 \text{ Mbps}$$

Multirate data communication in this system is supported by adjusting the spread of the pulses in time as well as changing the coding rate. Changing the coding rate is also done in all WLAN and WPAN systems, but stretching the pulse transmission was something new presented in the

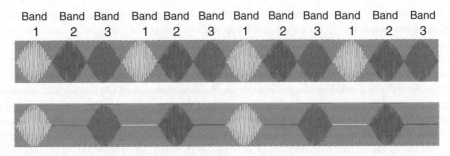

Band Band Band   Band Band Band Band Band Band Band   Band   Band
  1   2   3   1   2   3   1   2   3   1   2   3

Figure 10.12 Control of data rate in MB-OFDM using spreading time. The time spread of the stream on top is half of the time spread of the lower stream. The data rate of the top stream is twice the data rate of the lower stream.

MB-OFDM UWB proposal. To adjust the time spread, the same time sequence is spread by a factor of two or four by transmitting the symbols every other time slot or every four time slots to reduce the symbol transmission rate by factors of two or four. Figure 10.12 illustrates the basic concept behind time spreading, and consequently, the average transmitted power of the top stream in this figure is a half of the time spreading and power consumption of the lower stream. The data rate of the upper stream, however, is twice the data rate of the lower stream. Therefore, integration of time spreading allows flexibility in terms of more diversified data rates and power consumption. Table 10.4 provides the specification of data rates supported by the proposed MB-OFDM UWB system. A simple example illustrates how one can achieve different data rates shown in this table.

Example 10.7: Calculating Data Rates in the MB-OFDM UWB system

The lowest data rate in the table is 55 Mbps that is obtained by a coding rate of 11/32 with a spreading rate (repetition for variable data rate support) of four. Therefore the data rate is:

$$640 \text{ Mbps} \times 11/32 \text{ (bits/coded bits)}/4 \text{ (spreading rate)} = 55 \text{ Mbps}$$

10.2.4   Channel Models for UWB Communications

As we explained in Chapter 2, channel modeling for wireless networks is very complex and the cause of this complexity in the behavior of the wireless channels is the presence of multipath. Traditionally, standardization organizations define path-loss models for the coverage and multipath structure of the channel for devices operating according to the standards recommendation. The multipath is caused by obstacles around and in between the transmitter and the receiver and as

Table 10.4   Specification of different TF codes recommended for the MB-OFDM system by MBOA

Information data rate (Mbps)	55	80	110	160	200	320	480
Coding rate (R)	11/32	1/2	11/32	1/2	5/8	1/2	3/4
Spreading rate	4	4	2	2	2	1	1
Raw bit rate (Mbps)	640	640	640	640	640	640	640

such it is a function of the application environment. Another important factor affecting the channel behavior is the frequency of operation and bandwidth of the typical signals. As the frequency of operation is increased the penetration of the signal through the walls and other objects in the application environment changes resulting in changes in the path-loss characteristics of the channel. For example, at frequencies around a few GHz, used for Wi-Fi devices, a radio signal can easily penetrate the walls in a building. As we increase the frequency to tens of GHz, gradually the signal becomes confined in a room. With the increase in the frequency of operation, we have access to wider bandwidths. Consequently, the systems that we design may use hundreds of MHz to several GHz rather than a few tens of MHz used in Wi-Fi devices. This drastic change in the bandwidth allows transmission of much narrower pulses through the channel. These pulses resolve more multipath components and we need new models for the multipath structure as well. As we resolve more multipath components, the number of paths increases and the statistics of their amplitude fluctuations also change.

Channel models used for WLANs are valid for low power WPAN devices operating at 2.4 GHz, described in Chapter 9, but we need new channel models for Gigabit Wireless WPANs operating at 3.1–10.6 GHz and 60 GHz bands. In the remainder of this section we address path-loss and multipath modeling for 3.1–10.6 GHz UWB channels.

Path-Loss Modeling for UWB Channels: When the signal bandwidth spans several GHz, one may reasonably ask, given such a wide bandwidth, how the received signal strength should be modeled. In the previous chapters of the book we developed path loss models for the center frequency of the channel and we assumed that they are valid for the entire transmission bandwidth because the bandwidths were at most a few tens of MHz while the center frequency was around several GHz. This may not stay valid in UWB communications, where the center frequency and the bandwidth are both around several GHz. Considering the Friis' free space propagation formula (refer to Equation 2.1 in Chapter 2), the received signal strength is an inverse function of the frequency. As a result, when the bandwidth is ultra wide the received signal strength at the lower end of the spectrum can be much higher than that at the higher end of the transmission spectrum. Given this fact, one might question whether we can still apply narrowband path loss modeling techniques to UWB radio propagation.

Assuming a perfect transmitter amplifier, a perfect antenna with the same gain at all frequencies in the entire spectrum, and considering free-space propagation, using Equation 2.1, the received signal strength as a function of frequency is given by:

$$P_r(f) = P_t G_t G_r \left(\frac{\lambda}{4\pi d} \right)^2 = \frac{P_t G_t G_r c^2}{(4\pi d)^2 f^2} \qquad (10.4)$$

where P_t is the average transmit power spectral density. Then the average transmitted power of an ultra wideband system in the bandwidth W is $P_{t-ave} = P_t(f) \times W$, and the average received power around the center frequency f_c is given by:

$$P_{r-ave}^{UWB} = \int_{f_c-W/2}^{f_c+W/2} P_r(f)df = \frac{P_{t-ave}G_t G_r c^2}{W(4\pi d)^2} \left[\frac{1}{f_c - W/2} - \frac{1}{f_c + W/2} \right]$$

$$= \frac{P_{t-ave}G_r G_t c^2}{(4\pi d)^2 f_c^2} \left[\frac{1}{1 - (W/2f_c)^2} \right] \qquad (10.5)$$

The first term in Equation 10.5 is equivalent to the received signal power in a narrowband system with the same average transmitted signal power as the UWB system and the same center frequency as the UWB system. This would be the received power if we had a narrowband system operating at the center frequency of the UWB system and its power would be determined from:

$$P_{r-ave}^{NB} = \frac{P_{t-ave} G_r G_t c^2}{(4\pi d)^2 f_c^2} \tag{10.6}$$

Substituting Equation 10.6 into Equation 10.5, the received signal power of the UWB system in terms of the traditional narrowband systems is given by:

$$P_{r-ave}^{UWB} = P_{r-ave}^{NB} \left[\frac{1}{1 - (W/2f_c)^2} \right] \tag{10.7}$$

Equation 10.7 describes the relationship between the bandwidth and received signal power of an UWB system and its equivalent narrowband system. Considering the FCC regulated UWB spectrum described in Figure 10.2, the maximum allowed bandwidth of a system with uniform spectral density can occupy the frequency range 3.1–10.6 GHz for which $W = 7.5$ GHz and $f_c = 6.85$ GHz. Substituting these values into Equation 10.7 we observe that the narrowband system differs from the UWB system by only 1.5 dB. From this discussion we can conclude that the narrowband models for path loss describing the received power with one value, rather than a function of frequency, can be used to approximate the path loss for a UWB system [Sol01; Foe04]. This observation significantly reduces the complexity of empirical channel measurement and modeling techniques for UWB systems. Stated simply, we can apply all of the measurement and modeling techniques described in Chapter 2 to UWB systems.

Figure 10.13 illustrates a scatter plot of the received signal power for an UWB system for different distances, and the best one-gradient fit and partitioned two-gradient fit to the experimental data. The measurements were performed using cone antennas for the transmitter and the receiver for the 3–6 GHz spectrum in the first floor of the Atwater Kent Laboratories at Worcester Polytechnic Institute. The modeling used a theoretical path loss of 42 dB at the first meter for all cases in the figure. The single-gradient path loss model for this experiment is given by:

$$L_p = 42 + 33 \log d \tag{10.8}$$

The two-gradient partitioned model, for the same experiment, is described by:

$$L_p = 42 + \begin{cases} 2.7 \log d, d \leq 10\text{m} \\ 27 + 67 \log \frac{d}{10}, d > 10\text{m} \end{cases} \tag{10.9}$$

Similar results are also reported in [Cas02]. More extensive measurements in residential and commercial buildings are reported in [Gha03a, b].

Modeling of UWB Multipath Behavior: Figure 10.14 shows the frequency and time response of a sample UWB channel in a typical office indoor area with 3 GHz bandwidth centered at 4.5 GHz where the LOS path is obstructed. Since the bandwidth is ultra wide, the resolution is very small and many paths are resolved. The walls in indoor areas act as parallel mirrors creating an infinite number of images each reflecting a ray between the transmitter and receiver. Since these walls are a few meters apart, these paths are associated with rays with lengths that are on the order of meters

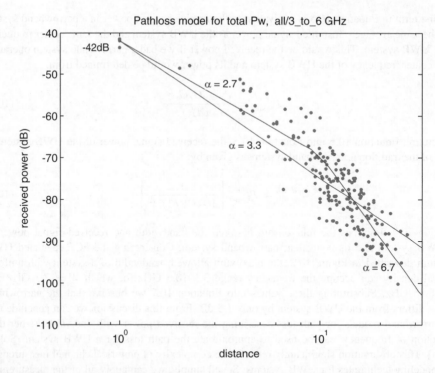

Figure 10.13   Scattered plot of the received power versus distance and the best first single slope and two-segment partitioned path loss models.

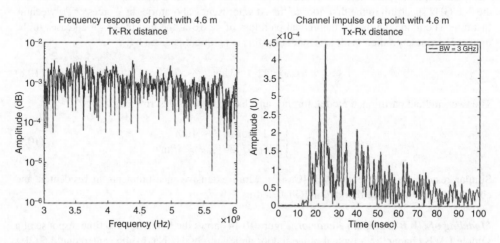

Figure 10.14   Frequency and time response of an UWB system with 3 GHz bandwidth in an obstructed LOS situation.

apart from one another. Since each meter is associated with a 3 ns delay, a system with a bandwidth of 3 GHz will resolve most of the paths. If we consider the 84 MHz of bandwidth at 2.4 GHz or 125 MHz of bandwidth at 5.2 GHz that are used for Wi-Fi devices, we see that the number of paths for UWB is an order of magnitude more (note that WiFi uses channelization which further reduces the bandwidth of the actual signal).

An increase in the number of paths does not change the total average received power and RMS delay spread of the channel but has an impact on the arrival rate and the statistics of fluctuations of the paths. In general, one can assume that for ultra wide bandwidth all paths are isolated and the amplitude of each path is relatively fixed. As the terminal is moved, or people move about in the space between the transmitter and receiver, the amplitudes of the paths change due to shadow fading, which as we discussed in Chapter 2, exhibits a lognormal distribution. As the bandwidth becomes narrower, the neighboring paths combine their amplitudes and phases, causing multipath fading and rapid fluctuation of the received signal strength following a Rayleigh or a Rician distribution. Therefore, one of the main physical characteristics of UWB transmission is the stability of the received signal from different paths.

Another issue that was considered by those designing the first channel models for the UWB channel was the arrival time of the paths. Traditional models that we have introduced earlier in this book assume that paths arrive at fixed time intervals for ease of characterization. In reality, however, paths arrive at random and several models have been developed to reflect their behavior [Sal87; Gan91]. The assumption of fixed intervals was useful because it could simplify the hardware simulation of the channel and the results obtained for performance evaluation of communication systems was not significantly affected by this assumption. The inter-arrival of the paths, however, have impact on the performance of localization systems using the time of flight of the paths that was also considered as one of the major applications of the UWB technology. These observations led those involved in channel modeling for UWB applications to resort to more complex models with closer appearance to the empirical wideband measurements. In the following few paragraphs, we describe one of these models that was considered by the IEEE 802.15 WPAN community. This is based on measurements performed at Intel [Foe04] and was sometimes referred to as *Intel's model*.

The Intel Model for UWB Multipath Modeling: Similar to the Saleh–Valenzuela model [Sal87] for traditional RF signals, assumes that paths arrive in clusters of rays and that arrivals form a Poisson distribution (i.e., inter-arrivals are exponentially distributed) and in a manner similar to [Gan91], further assumes that fluctuations of the amplitudes form a lognormal distribution. Figure 10.15 illustrates the basic concept of clusters and ray arrivals in the clusters.

The overall impulse response is now represented by:

$$h(\tau) = \sum_{l=0}^{\infty} \sum_{k=1}^{\infty} \beta_{kl} \delta(\tau - T_l - \tau_{kl}) e^{j\varphi_{kl}} \qquad (10.10a)$$

where the sum over l represents the clusters, the sum over k represents the ray arrivals within each cluster, and T_l is the delay of the l-th cluster. The strength of the paths within the clusters is determined from:

$$\overline{|\beta_{kl}|^2} = \overline{|\beta_{00}|^2} e^{-T_l/\Gamma} e^{-\tau_{kl}/\gamma} \qquad (10.10b)$$

where $\overline{|\beta_{00}|^2}$ is the average power of the first path in the first cluster and Γ and γ are the decay rates associated with the clusters and the rays within the clusters, respectively. Since the arrivals

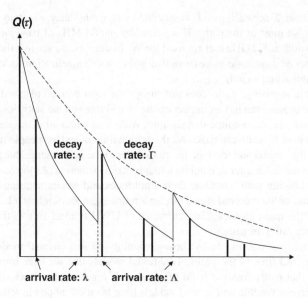

Figure 10.15   The basic concept for the clusters and rays for multipath arrival.

of the clusters and the rays within the clusters form Poisson processes, the inter-arrival rate of the clusters and rays form exponential distributions are given by:

$$p(T_l/T_{l-1}) = \Lambda e^{-\Lambda(T_l - T_{l-1})}$$

$$p[\tau_{kl}/\tau_{(k-1)l}] = \lambda e^{-\lambda[\tau_{kl} - \tau_{(k-1)l}]} \tag{10.10c}$$

where Λ and λ represent the cluster and ray arrival rates.

The Intel model begins with this framework, assuming a double-cluster arrival of the paths. The amplitude of the signal, β_{kl}, is assumed to be lognormal and the phase, φ_{kl}, is modeled as a binary number taking only two values of 0 and 180° with equal probability. To form the lognormal distribution, it is assumed that $20\log_{10}(\beta_{kl})$ has a normal distribution function with mean and variance of μ_{kl}, σ^2 respectively. The mean of the lognormal distribution is given by:

$$\mu_{kl} = \frac{10\ln(\overline{\beta^2(0,0)}) - 10T_l/\Gamma - 10\tau_{k,l}/\gamma}{\ln(10)} - \frac{\sigma^2\ln(10)}{20} \tag{10.11}$$

where $\overline{\beta^2(0,0)}$ is the average received power in the first path, $\tau_{0l} = T_l$ is the arrival delay of the *l-th* cluster, τ_{kl} is the delay of the *k-th* path in the *l-th* cluster, Γ and γ are the cluster and ray decay factors, respectively. Table 10.5 shows three sample obstructed-LOS (OLOS) and two sample LOS model parameters considered in the above model.

Figure 10.16 shows two sample simulated channel impulse responses using the IEEE 802.15 WPAN committee's recommended channel models. In this figure each path is represented by an impulse. To make it look more like an actual measurement, one needs to convolve this impulse response with the impulse response of a pulse (e.g., a raised cosine pulse) having the bandwidth of the actual UWB measurement or communication system.

Table 10.5 Five set of parameters for simulation of UWB channel behavior

Channel characteristics	OLOS	OLOS	OLOS	LOS	LOS
Mean excess delay (τ_m; ns)	17	22	27	3	4
RMS delay (τ_{rms}; ns)	15	20	25	5	9
Model parameters					
Λ (1/ns) cluster arrival rate	1/11	1/14	1/15	1/22	1/60
λ (1/ns) ray/path arrival rate	1/0.35	1/0.33	1/0.32	1/0.94	1/0.5
Γ cluster decay factor	16	22	30	7.6	16
γ ray decay factor	8.5	10	10	0.94	1.6
σ (dB) standard deviation of lognormal fading	4.8	4.8	4.8	4.8	4.8

10.3 Gigabit Wireless at 60 GHz

Wireless data networking technologies have evolved to support emerging application scenarios that are made possible by the growth of the computer industry. The wireless LAN industry emerged to support wireless local area Internet access in homes, offices and public areas. This evolution took two steps, first to discover the implementation of the technology in an appropriate frequency band and second to discover real applications and market for the technology. During the 1990s, the WLAN industry was discovering the technology, and during the 2000s it discovered successful applications for sizable markets. During the discovery of the WLAN technology, we perceived certain applications and during the actual discovery of the market, we found many other applications to emerge, which mirrored the parallel evolution of the computer industry. Similarly, Gigabit Wireless technology was discovered in the 2000s during the examination of the UWB spectrum and the current activities at 60 GHz is motivated by the emergence of popular applications with a sizeable market.

At the time of this writing, several groups are engaged in defining the future 60 GHz mmWave wireless technology and in the discovery of the markets for this technology. In this section we draw on application scenarios and architectures *perceived* by the 60 GHz mmWave Gigabit Wireless community [Wig10] with the details of implementation of the system proposed in [ECM08].

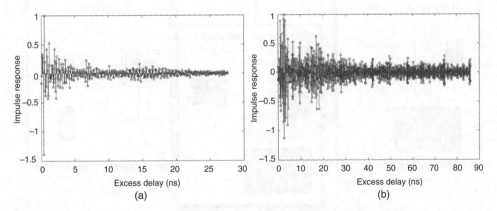

Figure 10.16 Two sample channel simulations using Intel's UWB channel models for (a) LOS and (b) OLOS conditions.

Other examples of emerging technologies at 60 GHz are available in [Per10, IEEE09]. These technologies widely draw on the experimental designs in UWB spectrum that were detailed in the previous section.

10.3.1 Architecture and Application Scenarios

Figure 10.17 (see [Wig10] for more details) provides a particular classification for Gigabit Wireless applications that are perceived to emerge. The first set of applications is for distances around one meter, typically for peer to peer file transfer using Gigabit Wireless (mostly as a cable replacement). These could be used for personal graphical information exchange and multimedia exchange between different devices such as laptops, cameras, and mobile multimedia storage systems. The second class of applications is referred to as "wireless display" and they involve multimedia applications projecting the content of a multimedia system on a large electronic display. These multimedia Gigabit Wireless applications may need to cover a range of up to 10 m. The third class of applications is to connect computer peripherals to a mobile computing device. The fourth class is to increase the throughput of traditional WiFi applications when a laptop is in close proximity of an access point. These connections have different costs, sizes and battery requirements. For example, a *wireless Universal Serial Bus (USB)* device, which may be used for file transfer, must be very inexpensive, very small and consume extremely low power, while the wireless communication device installed on a monitor may have no serious restriction on power or size and a user may be willing to pay much more to make it happen. As a result, most standard organizations working in this area work on several options for the devices to support a more flexible environment for implementation of a larger number of applications in heterogeneous network. Figure 10.18 illustrates the differences between the homogeneous and heterogeneous networks. In a homogenous network the physical

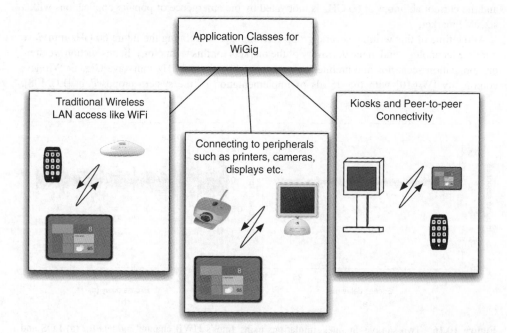

Figure 10.17 Different classes of applications perceived for Gigabit Wireless at 60 GHz.

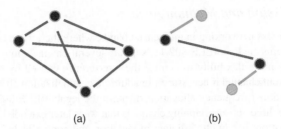

(a)   (b)

Figure 10.18 (a) A homogeneous network, where all devices have the same physical layer. (b) A heterogeneous network, where devices have different physical layers.

layer of all terminals are the same. In a heterogeneous network we have different physical layers demanding interoperability and coexistence between the terminals so that they can communicate with one another, while they do not interfere with each other's operation [ECM08]. At the time of this writing, the IEEE 802.11ad group has a standard to support 60 GHz mmWave networks with an industry alliance called WiGig allied with these efforts. The relationship between IEEE 802.11ad and WiGig may be thought of as being similar to that between IEEE 802.11 and WiFi (see also the IEEE 802.11ac standard briefly described in Chapter 8).

The proposed 60 GHz wireless system in [ECM08] has three different types of devices. Device type A is considered as the "high end" device designed to provide multimedia services over LOS and NLOS channel conditions, using trainable antennas (beamforming) and complex transmission and signal processing techniques including MIMO. Device type B, considered as an "economy" device, is expected to support multimedia services over LOS channels with non-trainable antennas and minimum signal processing. Device type C, designed for data over very short distances of less than one meter with a cheap physical layer implementation, is considered as a "bottom end" device. Figure 10.19 shows a scenario with a heterogeneous and a homogenous network in coexistence and cooperation. There are three types of physical layers used to connect devices involved in these networks: a 6.4 Gbps physical layer with a coverage of 10 m, a 3.2 Gbps physical layer with a coverage of 3 m, and a 3.2 GBps physical layer with a coverage of 1 m.

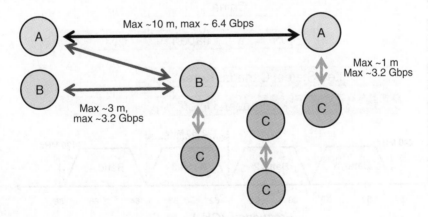

Figure 10.19 A heterogeneous and a homogeneous network coexisting and cooperating to connect three classes of terminals with different physical layers.

10.3.2   Transmission and Medium Access

To accommodate wireless networking in unlicensed bands, where there is no control over interference, traditional systems such as IEEE 802.11, support several channels in each band. This way a wireless device has more flexibility to control the interference caused by neighboring devices (e.g., by using a different channel if necessary). In addition, channelization allows for adjusting the differences in international frequency allocation rules. In the legacy IEEE 802.11 DSSS scheme, for example, there are three non-overlapping channels that were later extended to many more overlapping channels. As more bandwidth efficient modulation and coding techniques such as M-ary orthogonal coding and multicarrier OFDM techniques found their way in the wireless local data communications world, data rates on the orders of the transmission bandwidth were attained. Considering this argument, to achieve Gigabit Wireless networking, we need a spectrum that allows several channels, each with a bandwidth on the order of a GHz.

The 60 GHz mmWave unlicensed bands with a total of 7 GHz width of available spectrum allows the possibility of accommodating Gigabit Wireless with multiple channels. Figure 10.20 shows a four-channel division option at 60 GHz suggested in [ECM08], each channel having a bandwidth that is slightly more than 2 GHz. As shown in this figure, this channelization approach also accommodates for differences in channel allocation in different countries. Another advantage of the 60 GHz bands is that the path-loss characteristics for the four channels at 60 GHz are more uniform than the characteristics for UWB channels that are operating at lower frequencies with similar bandwidths. If we consider $W/2f_c$ in Equation 10.5 as a measure for uniformity of the received power, this parameter is around 0.02 for all four channels at 60 GHz. If we had similar channelization for the 2–10 GHz spectrum, in the first band, this parameter would be around 0.3 and for the fourth band it would be around 0.1. As a result, due to interference from 802.11a devices and high ratio of the bandwidth to center frequency, as we explained in the previous section, the designers of wireless networks at UWB frequency bands could not accommodate that many "close to identical" channels with such a wide bandwidth.

In terms of the physical layer, the systems operating in the 60 GHz mmWave bands have selected: (a) a single carrier, and (b) a multi-carrier OFDM option. Neither of these techniques compromises

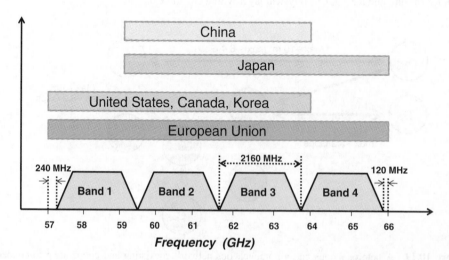

Figure 10.20   Frequency bands and worldwide availability of spectrum for Gigabit Wireless at 60 GHz.

on the data rate, as the spread spectrum techniques did with the legacy IEEE 802.11. However, as expected, the single carrier scheme has less tolerance to multipath while the more complex OFDM scheme can handle issues with multipath much more effectively. As a result, the simpler single carrier transmission is used for low power and shorter distance applications and the more robust OFDM scheme is used to support higher data rates and longer distances. For each physical layer option, in a manner similar to other multi-rate wireless data networks, 60 GHz mmWave systems support a wide variety of data rates and coverage distances by using different multi-symbol transmission modulation and coding techniques. For example, the proposed system in [ECM08] supports data rates from 0.4 to 6.4 Gbps and up to 10 m of coverage for the type A terminals using single and multi-carrier OFDM transmissions. Type B and C terminals use only the simpler single carrier modems, and they only support data rates up to 3.2 Gbps for very small distances up to 3 and 1 m.

The new physical layer feature explored further by the 60 GHz mmWave systems is the use of *phased array antennas* that allow beamforming and directional transmissions. Since the length of the array antennas is proportional to the wavelength of the transmitted signal, implementation of a small antenna array at 60 GHz is much more practical than in the 3–10 GHz range. An antenna element of length $\lambda/4$ at 60 GHz is 1.25 mm while at 2.4 GHz it increases to around 5 cm. Thus, the design of an array antenna at 60 GHz can be done on the chip surface. The simplicity and practicality of integration of array antennas for the design of 60 GHz mmWave systems has attracted considerable attention in analyzing the effects of directional antennas in all aspects of design at 60 GHz.

At the physical layer using directional antennas reduces the transmit power requirement due to the potential antenna gain and this reduction is very much needed because an increase in the frequency of operation will increase the transmission power requirement. As described by Equation 10.4, the received power is inversely proportional to the frequency of operation due to the first meter loss and other losses. Therefore, as we move from the UWB spectrum that is centered around 6 GHz to the 60 GHz mmWave systems, we need approximately 100 times or 20 dB more transmit power to maintain the same received power levels at the same distance from a transmitter. With the same transmitted power and in a LOS channel, where the distance-power gradient is around two, the 20 dB loss results in one order of magnitude reduction in coverage. This implies that a WPAN that covers 10 m at 6 GHz covers 1 m at 60 GHz. Using directional antennas compensates for some of that power loss by restricting the radiation of the signal to a narrow beam rather than in all directions uniformly. The restriction of the radiation in certain directions also reduces the interference to other communicating devices (see Chapter 5 where sectored antennas are discussed) providing for prospective reuse in space and additional spatial throughput because of this reuse. Figure 10.21 [ECM08] explains the idea of spatial reuse with directional antennas. In this figure, the use of directional antennas allows three different networks to operate in the same space with minimal interference with one another.

The traditional issues related to medium access control (see Chapter 4) remain the same for 60 GHz mmWave systems. To support a diversified set of applications, the 60 GHz mmWave networks need to accommodate both the flexibility provided by CSMA/CA for general multi-user data applications such as Internet browsing or file transfer and the deterministic control offered by TDMA for support of quality of service (e.g., latency) needed for transferring streaming real-time multimedia content [Per10]. Before any real market emerges, it is very difficult to make a solid decision on the orientation of the medium access toward CSMA/CA or TDMA. Accommodating both options, as suggested in [Per10] may be an option. The other option is to design two separate MACs and, like Bluetooth and ZigBee, let the customers pick the one that fits their application.

The new fundamental technical challenge in the design of medium access control in the 60 GHz mmWave networks is the handling of directional antennas with the allowance of overlaying a number

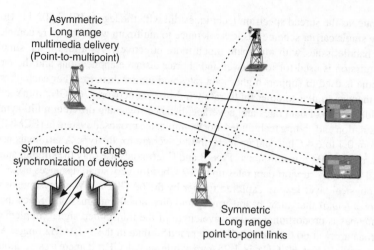

Figure 10.21   Directional antennas allow for spatial reuse using directional antennas.

of networks to take advantage of spatial reuse. We need mechanisms to support interoperability and coexistence among these networks. This problem surfaces when we start the overlaying of networks and during the operation, the orientation of the terminal and formation of the networks changes, impacting the directional transmissions. In the design of the medium access control for mmWaves using directional antennas, special attention will have to be paid for the discovery of the devices in the network, carrier sensing itself, the hidden terminal problem, and operation of devices that may be blocked without causing interference to others (exposed terminal problem) [Dan10].

10.3.3   Channel Models for 60 GHz mmWave Networks

The novelty of emerging wireless technologies mostly relies on new applications and radio propagation modeling. The transmission techniques and medium access control methods remain mostly similar among different technologies, especially in recent times, when the effectiveness of variety of alternative technologies has already been explored. Radio propagation at 60 GHz for mmWave Gigabit Wireless networks is significantly different compared to radio propagation in traditional WLANs and WPANs operating in the 2–10 GHz spectrum. At 60 GHz, the wavelength of the signal is so small that it cannot penetrate through walls or other objects. As a result, the signal is mostly confined to the room where the transmitter is located, and it is carried from a transmitter to a receiver primarily through a LOS path. As LOS is broken (e.g., by a person or an object placed between the transmitter and the receiver), the LOS path attenuates tremendously and then a few reflecting paths can carry the signal to the receiver. In addition, as we discussed earlier, to compensate for excessive path loss, mmWave systems resort to directional antennas that change the requirements for channel modeling at these frequencies. In traditional WLANs and WPAN systems, omni-directional antennas (in 2-D) are commonly used. This situation has directed standardization efforts in channel modeling at 60 GHz to include the effects of polarization and beamforming as well as the effects of moving objects into the multipath structure of the channel.

Having only a few paths, and being confined to a room makes geometric ray tracing techniques (see section 2.3) as the desirable technique for modeling the channel. Figure 10.22a [Mal09] shows

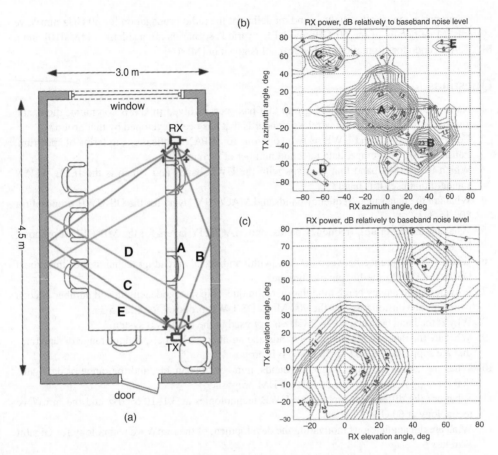

Figure 10.22 (a) An office setting scenario (b) SNR for azimuth angles (c) SNR for elevation angles: Printed with permission © IEEE [Mal09].

a typical application scenario in a conference room. This scenario is used for an extensive measurement campaign at 60 GHz in a 3m by 4.5m conference room. The results of these measurements were compared with the results of ray tracing for five paths shown in the figure: the LOS path A; two first order reflections, B and C; and two second order reflections D and E. These paths may carry the signal according to the relative azimuth angle of the transmitter and receiver antennas. Figure 10.22b shows the signal to noise ratio (SNR) for the most powerful paths corresponding to the LOS, first order and second order reflections as a function of different transmitter and receiver azimuth angles for the antennas. With a large number of combinations for the angles we have the first order reflections approximately 10 dB lower in signal strength and second order reflections around 20 dB lower in signal strength compared to the signal strength of the LOS path. For most practical considerations, an analysis that uses only the first order reflections, in a manner similar to Example 2.11 of section 2.3, is adequate. Figure 10.22c shows the total received signal strength obtained by changing the elevation angle of the transmitter and the receiver antennas. There are two clusters of elevation angles (around zero and 45 degrees) that belong to a LOS path and the first reflected path from the ceiling. These results support the intuitive notion that geometric ray tracing is an appropriate tool for the analysis and modeling of the radio propagation for 60 GHz mmWave systems. More details of the 60 GHz model for various scenarios are available in [Mal10] and a MATLAB® code for simulation of this channel is given in [Mas09].

Questions

1. Name the IEEE standardization group that has been involved in UWB communications and also name the other popular wireless standards that has been developed by that group?
2. Compare Bluetooth and UWB as two solutions for WPAN applications in terms of spectrum, transmit power, coverage, data rate, and number of piconets.
3. Which is the ISM band that overlaps with the UWB bands and which is the IEEE WLAN standard that uses this band?
4. What are the data rates, frequency bands,and MAC/PHY layers for the DS-UWB proposal for 802.15.3a?
5. What are the data rates, frequency bands, and MAC/PHY layers for the MB-OFDM proposal for 802.15.3a?
6. What is the difference between the bandwidth and signal spreading techniques of traditional DSSS and UWB time-hopping technique?
7. Provide the frequency band, modulation technique, chip rate, code length, and number of code words in the constellation for the 900 Mbps DS-UWB proposal for 802.15.3a.
8. Why is the band between 4.9–6.2 GHz is not used in the DS-UWB system?
9. What are the range of the frequency bands, the number of channels, and the range of data rates that are supported by 802.15.3a DS-UWB proposal?
10. What are the range of the frequency bands, number of channels, and the range of data rates that are supported by 802.15.3a MB-OFDM proposal?
11. What are the differences among the UWB technologies at 3.1–10.6 GHz and the mmWave technology at 60 GHz?
12. What are the applications motivating the development of the mmWave technology for Gigabit Wireless?
13. Compare the propagation environment for the UWB and the mmWave technologies.
14. What are the differences among the heterogeneous and homogeneous mmWave WPAN networks?
15. What are the practical advantages of using directional antennas in mmWave technology? What are the challenges?

Problems

Problem 10.1

Assume an UWB device with a bandwidth of 1 GHz operates in a band that includes the entire IEEE 802.11a bands. If both devices attempt to communicate with an access point supporting both technologies and they are located at the same close distance from the access point for which the distance-power gradient is two.

a. What is the signal to interference ratio for the received IEEE 802.11a signal at the access point?
b. What is the signal to interference ratio for the received UWB signal at the access point?

Problem 10.2

a. Enumerate all 8-BOK codes of length 24. Show that the first code is orthogonal to the second and third code.
b. Calculate the two basic data rates of DS-UWB for spreading factors of 24 and 32 in the high-band.

Problem 10.3

a. Show that with all time-frequency hopping patterns for MB-OFDM, shown in Table 10.3, we have two collision per code.
b. Repeat Example 10.7 for all data rates given in Table 10.4.

Problem 10.4

a. Assuming that the starting frequency of a band is given by f_{st}, use Equation 10.7 to derive the received UWB power to the narrowband power ratio as a function of bandwidth W and starting frequency.
b. Using your derived equation in part (a), plot the power ratio in dB versus $0 < f_{st} < 5$ GHz when $W = 5$ GHz.
c. Repeat (b) for $W = 10$ GHz.
d. Using the results of parts (b) and (c) discuss the sensitivity of the flatness of the spectrum to the starting frequency and bandwidth of the system.

Problem 10.5

a. The sensitivity of electronic devices is a function of bandwidth of the transmitted signal and the background noise. If the sensitivity of WLAN device with a bandwidth of 20 MHz operating at center frequency of 2.45 GHz is –94 dBm, what is the sensitivity of an UWB WPAN device with a bandwidth of 2 GHz operating at the center frequency of 5 GHz? Assume a linear relationship for simplicity.
b. Compare the path-loss at the first meter of the WLAN and WPAN devices in part (a).
c. If the transmitted power of the WLAN is 100 mW and the WPAN transmits at 1 mW, determine the maximum path-loss allowed for each of the devices.

Problem 10.6

Using the UWB path loss models given by Equations 10.8 and 10.9 determine the coverage of 1 mW UWB WPAN device described in Problem 10.5. Compare the coverage of the WPAN device with that of the WLAN device described in that problem. Use the IEEE 802.11 model C from Chapter 2 for calculation of the WLAN coverage.

Problem 10.7

Using Equation 10.10 for modeling of the multipath arrival of an UWB channel, determine the RMS delay spread of the model in terms of the given amplitude of the paths. Assume we have two clusters.

Part IV

Wide Area Wireless Access

11

TDMA Cellular Systems

11.1 Introduction

In the previous parts of this book we first provided an overview of the evolution of wireless access and localization industry followed by explanation of the principles of air–interface design and principles of network infrastructure design in Parts I and II, respectively. In Part III of the book, we discussed local and personal area wireless access technologies, which were mainly designed for local wireless access to the Internet using unlicensed frequency bands. In Part IV, starting with this chapter, we discuss wide-area wireless access technologies that are traditionally divided into different generations. First generation (1G) *analog* cellular networks evolved for wireless access to the PSTN using licensed frequency bands. This technology transformed into second generation (2G) *digital* cellular networks that employed TDMA and CDMA technologies. The 2G systems were originally deployed in the early 1990s for cellular telephone application with some attention to integrate the lower-speed wireless data applications, on the orders of several tens of Kbps, similar to voice-band modems. As the popularity of the Internet started in the mid-1990s, third generation (3G) cellular networks emerged to provide higher data rates on the orders of several Mbps using CDMA technology. More recently, with the popularity of smart phone such as iPhone, the demand for multimedia wireless Internet application has been the force behind fourth generation (4G) cellular networks supporting data rates on the orders of hundreds of Mbps using OFDM and MIMO technologies, originally evolved for WLAN applications. In this part Part VI of the book we examine these wireless cellular technologies in some detail and explain how they evolved.

This chapter is devoted to 2G TDMA systems using GSM technology as the main example system. A number of TDMA systems emerged in the 1980s during the evolution of 2G systems for cellular and personal communication systems (PCS) [Pah85]. In the 1990s, with the advancements in battery technology for handheld terminals and the overwhelming popularity of cellular telephones, the differentiation between digital cellular and PCS systems disappeared. The increasing demand for capacity diverted the attention of service providers to the availability of spectrum rather than technology. With the emergence of new PCS bands at around 2 GHz, most service providers expanded their cellular services by upgrading the frequency of operation of their digital cellular systems from around 1 GHz to around 2 GHz, without using the so-called PCS standards. Today, GSM is by far the most popular TDMA standard in the world that is used both in cellular and PCS bands, with several billions of users. The structure of the system is also very clear and useful for pedagogical purposes. Therefore, we use GSM as our example for TDMA systems.

Principles of Wireless Access and Localization, First Edition. Kaveh Pahlavan and Prashant Krishnamurthy.
© 2013 John Wiley & Sons, Ltd. Published 2013 by John Wiley & Sons, Ltd.

Wireless networks are complex multi-disciplinary systems and the description of their standards is often very long and convoluted. Our objective in explaining standards is to provide the reader with adequate understanding of the overall objectives of a system, a view of the architecture of the hardware and software elements of the network, some details of protocols and algorithms to understand how information transfer works, and the path of their evolution. This path began with emphasis on high quality wireless access to the PSTN and shifted towards wireless access to the Internet. In this chapter we first define the objectives and architecture of a general cellular network. Then we discuss mechanisms that are designed to support mobility of a wireless device and finally, we describe protocols used for communication among the elements of the infrastructure (in the case of GSM) as our example for TDMA networks. To minimize the difficulties in understanding the details of all standards, we make a conscious effort to minimize the description of all aspects of standards and present the material in a similar format. Following a similar format will help the reader grasp a better overall picture of various technologies and help the reader be more comfortable in understanding the details as needed. However, this effort will not completely eliminate all difficulties, because each standard uses its own reference model and a number of acronyms that are different from one standard to another. To further help the reader, we have also provided a number of examples that describe certain features of a standard in more depth. This way we have preserved the flow of the depth of the text, while some important features are treated in more detail.

11.2   What is TDMA Cellular?

In late 1980s, the popularity of 1G analog cellular networks was beyond the expectations of service providers. Two different issues motivated the emergence of 2G digital cellular networks in Europe and the United States. In Europe, a number of standards had emerged in different countries and the size of the countries was smaller. So, roaming for mobile users was a major issue. To approach a unified standard and solve the roaming problem, Europeans resorted to TDMA digital cellular using GSM as the standard. In the United States, the number of subscribers in major cities such as New York or Los Angeles had reached such a level that, at peak hours, mobile users were experiencing unacceptable delays or call blocking and there was a need for a more bandwidth-efficient system. Both TDMA and CDMA digital cellular technologies were adopted for 2G. GSM technology was the first 2G digital cellular system and it emerged at a time that all other countries were adopting cellular technology. As a result, it received a worldwide acceptance soon and it eventually penetrated the United States market by becoming the single most successful TDMA technology adopted worldwide.

In 1982, European countries allocated frequency bands of 890–915 MHz and 935–960 MHz for the Pan-European 2G cellular standard and the GSM standardization group was formed. The main charter of the group was to develop a 2G standard to resolve the roaming problem in the six existing different 1G analog cellular systems in Europe. After evaluating several options, the committee decided to go for a unified new digital standard as it would facilitate roaming and at the same time provide for large-volume production. In 1991, the specification of the standard was completed, and in 1992, the first deployment started. By the year 1993, 32 operators in 22 countries adopted the GSM standard; at the turn of this century close to 150 countries had adopted GSM for cellular operation; and at the time of writing it is the technology that is adopted in the largest number of countries.

Although the original goals of GSM could be met only by defining a new air–interface, the group went beyond just the air–interface and defined a system that complied with (at that time) emerging integrated voice and data services and other emerging fixed network features. To this end, the committee also defined a number of other interfaces between the hardware and software elements of the network making GSM a complete digital cellular standard that is very suitable for pedagogical purposes. One of the interesting ironies of this evolution is that GSM, and later all other 2G digital cellular systems, brought the integrated voice and data services to mobile digital services to all users while the original wired integrated system lost its popularity and never found a massive acceptance with users. This reflects the real multi-disciplinary nature of the telecommunication industry in which the behavior of the market is not always as predictable as more focused industries such as component design.

11.2.1   Original Services and Shortcomings

The first step in understanding a multi-purpose system is to identify the services that are provided by that network because the entire network is designed to support these services. Analog cellular systems were developed for a single application – voice – and in a manner similar to analog access to the telephone network other data services such as fax and voice-band modems were defined as overlay services on top of the analog voice service. GSM was an integrated voice-data service that provided a number of services beyond the analog cellular telephone. These services were divided into three categories: teleservices, bearer services, and supplementary services.

Teleservices provided communication between two end user applications according to a standard protocol. These services included: telephony, emergency speech calls, facsimile, teletex, short messaging services (SMS), and videotext. The upper most layer of the protocol stack of the standard had to be specified so that it could communicate with the protocols used in these applications.

Bearer services provided capabilities to transmit information among user-network-interfaces or access points. Traditional bearer services included a variety of circuit switched access to the telephone network and packet switched public data networks. To implement bearer services, the lower layers and frame format of the standard had to specify how these transmissions would be implemented over the air–interface.

Supplementary services are not stand-alone services but they are services that supplement a bearer- or tele-service. Supplementary services in GSM were call forwarding and call barring. They were applied to both bearer and teleservices. Other supplementary services included call waiting and calling number identification. These services were usually implemented in the wired infrastructure of the cellular network.

Reviewing these services we observe that some of these services such as telephony and SMS became popular and some like teletex, videotext, facsimile, or all the data services were rarely used in a popular application in the case of GSM. Complex multi-disciplinary technologies are designed based on a perception of emerging services, and as the emerging applications evolve in their own path of evolution, some of these services survive and some disappear. Further, the evolution creates new applications that have not been perceived at the time of the standardization and the need to support such new applications calls for the enhancement of the standards, and if needed, the emergence of new standards to replace them. The main application in the case of GSM was the cellular telephone, and SMS was an unexpected application that became popular a decade after the completion of the standard.

The wireless data applications supported by GSM were mostly circuit switched data at rates up to 9600 bps that were compatible with the voice-band modems of that time. In the meanwhile, in the mid-1990s Internet access emerged as the main route for data communications, demanding changes in the GSM standard and the emergence of other technologies to follow. Wireless access

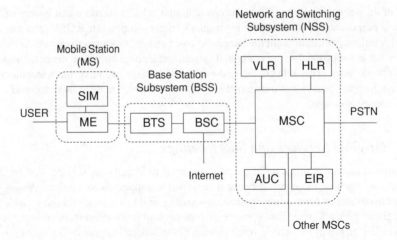

Figure 11.1   Reference model for architectural elements of a typical cellular network.

to the Internet requires higher data rates for multimedia applications and to support these, GSM needed to be modified. Increasing the data rate and access to the Internet required changes in the infrastructure and the air–interface of the standard, as explained later.

In the rest of this chapter we first describe the original GSM system and then we continue with explaining how its air–interface was modified to support high data rate packet switched access to the Internet.

11.2.2   Reference Architecture for a Cellular Network

The description of a circuit switched wireless cellular network is quite intricate and involves detailed specifications of the terminal and the fixed hardware backbone and software databases that are needed to support operations (see Chapter 6). To describe such a complex system, a reference model or overall architecture is needed to provide a comprehensive understanding of the network elements and operations by dividing the system into sub-systems. The infrastructure of most cellular networks is similar – it is the air–interface and wireless access technique that varies in different systems. Our presentation of the overall architecture of a cellular network is organized into three major segments as shown in Figure 11.1. These segments are: the Mobile Station (MS), the Base Station Sub-system (BSS), and the Network and Switching Sub-system (NSS). Figure 11.2 provides a more physical representation of the architectural elements of such a typical cellular network and the relationship among these elements. This division of the architectural elements was adopted from [Hau94] and we will follow that for the description of the system's elements in the following sections. Standardization organizations define such reference models and specify the details of protocols across the interfaces between the elements of the reference model. This allows for easy multi-vendor operation where each vendor designs a hardware or software element that complies with the standard interface. Multi-vendor operation facilitates the growth of the industry and controls the cost through competition.

Mobile Station: The MS communicates the information with the user and modifies it to meet the transmission protocols of the air–interface that allow it to communicate with the Base Station

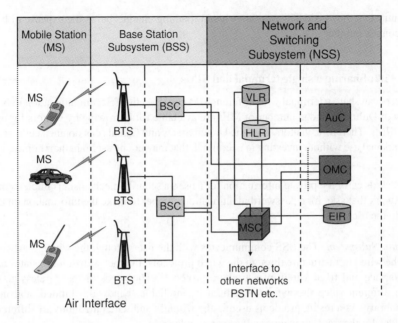

Figure 11.2 A different view of the reference architecture for a typical cellular network.

Sub-system (BSS). The user interacts with the MS through a microphone and speaker for speech, keypad and display for short messaging, touchscreen interfaces in the case of smartphones, and sometimes a cable or Bluetooth connection for other data terminals. The Mobile station has two elements. The first element is the *Mobile Equipment (ME)* that is a piece of hardware that the customer purchases from equipment manufacturer or their dealers. The mobile equipment is also called *User Equipment* in other standards. This hardware piece contains all the components needed for the implementation of the protocols to interface with the user and with the air–interface to the BSS. The components usually include a speaker, microphone, keypad, and the radio modem. Therefore, the ME is an expensive piece of hardware. To encourage more users to subscribe to wireless services, a number of service providers in the early days of the cellular industry, and even today subsidize the price of the MEs.

The second element of the MS in a typical cellular network is the *Subscriber Identity Module (SIM)* that is a smart card issued at the time of subscription, identifying the specifications of a user such as the user's address and type of service. The calls in the system are directed to the SIM rather than the terminal. Short messages are also stored in the SIM card. Using SIM cards was not a possibility with 1G analog cellular systems, and the North American 2G digital cellular standards have not implemented this option. Although implementing a SIM is a fairly simple concept, it has a significant impact on the way that a user transacts with a service provider. A SIM card carries each user's personal information that enables a number of useful applications.

Example 11.1: Roaming and SIMs

People visiting different countries, but not keen on making calls using their home number because of expensive roaming charges, can always carry their own terminal and purchase a SIM card in every country that they visit. This way they avoid roaming charges but at the expense of having a different contact number

Example 11.2: Sharing a single terminal and SIMs

Several users can share a terminal with different SIM cards. At the Telecommunication Laboratory, University of Oulu, Finland, a number of MEs are available for borrowing by visitors for use with their own SIMs. Therefore, visitors from the United States and Canada can obtain a cellular service for their personal use without investing in a terminal, that may not be useful in their home countries.

Since SIM cards carry the private information for a user, a security mechanism is implemented in a typical network that asks for a password identification number to make the information on the card available to the user.

Base Station Subsystem: The BSS communicates with the user through the wireless air–interface and with the wired infrastructure through the wired protocols. In other words, it translates between the air–interface and fixed wired infrastructure protocols. The needs of the wireless and wired media are different since the wireless medium is unreliable, bandwidth limited, and needs to support mobility. As a result, protocols used in the wireless and wired mediums are different. The BSS provides for the translation among these protocols.

Example 11.3: Speech conversion

In all cellular networks, the user's speech signal is converted into a format at a rate around 10 kbps (this is digitized voice with a speech coder) and communicated over the air–interface. This is in order to ensure a bandwidth efficient format that does not stress the scarce radio resources. The backbone wired network typically uses a 64-kbps digitized voice format in the PSTN hierarchy. Conversion from analog to around 10-kbps voice takes place at the MS and the change from the 10- to 64-kbps voice coding takes place at the BSS.

Example 11.4: Signaling in a typical cellular network

The signaling format to establish a connection in wired networks is a multi-tone frequency scheme used in traditional wired telephones. Digital cellular on the other hand, establishes the call through the exchange of a number of packets. The translation of this communication into a dialing signal is made in the BSS.

As with speech coding and dialing, explained in the above examples, data transmission protocols over the air–interface are different from that of the wired infrastructure. All these translations are performed at the BSS. As we will see in the description of high speed data services for TDMA networks in Section 11.7, to implement packet data services on the same air–interface, the BSS also separates packet switching data for the Internet from the PSTN voice traffic.

There are two architectural elements in the BSS. The *Base Transceiver System (BTS)* is the counterpart of the MS for physical communication over the air–interface. The BTS components include a transmitter, a receiver, and signaling equipment to operate over the air–interface and it is physically located in the center of the cells where the BSS antenna is installed. One BSS may have from one up to several hundred BTSs under its control.

The second architectural element of the BSS is the *Base Station Controller (BSC)* that is a small switch inside the BSS in charge of frequency administration and handoff among the BTSs inside a BSS. The hardware of the BSC in a single BTS site is located at the antenna and in the multi-BTS systems, usually in a switching center where the other hardware elements of the network and switching subsystem is located.

Network and Switching Subsystem: The NSS is responsible for network operation. It provides for communications with other wired and wireless networks as well as support for registration and maintenance of the connection with the MSs. The NSS could be interpreted as a wireless specific switch that communicates with other switches in the PSTN and at the same time supports functionalities that are needed for a cellular mobile environment. The NSS is the most elaborate element of a typical digital cellular network and it has one hardware, the mobile switching center, and four major software elements: the visitor location register, the home location register, the equipment identity register, and the authentication center. The operation and maintenance control (OMC) oversees the operation of this unit.

The *Mobile Switching Center (MSC)* is the hardware part of the wireless switch that can communicate with PSTN switches using the traditional protocol commonly used in the PSTN as well as other MSCs in the coverage area of a service provider. The MSC also provides to the network, the specific information on the status and location of the mobile terminals.

The *Home Location Register (HLR)* is database software that handles the management of the mobile subscriber account. It stores the subscribers' address, service type, current location, forwarding address, authentication/ciphering keys, and billing information. In addition to the telephone number for the terminal, the SIM card is identified by an International Mobile Subscriber Identity (IMSI) number, which is totally different from the actual telephone number. The IMSI is used primarily for internal networking applications.

Example 11.5: Numbering schemes in a typical cellular network

The telephone number of a subscriber in Finland could be 358-40-770-5246. The first three digits are the country code; the next two are the digits for the specific MSC and the rest are the telephone number. The IMSI of the same user can be 244-91 followed by a 10-digit number that is totally different from the telephone number. The first three digits of the IMSI identify the country, that is, Finland, and the next two digits, the billing company (the service provider Telia).

The *Visitor Location Registration (VLR)* is temporary database software similar to the HLR for identifying the subscribers visiting inside the coverage area of an MSC. The VLR assigns a Temporary Mobile Subscriber Identity (TMSI) that is used to avoid using the IMSI on the air. Maintenance of two databases at the home and visiting site allows a mechanism to support call routing and dialing in a roaming situation where the MS is visiting the coverage area of a different MSC. As discussed in Chapter 6 and as we will see in the later chapters, the mechanism of holding two databases to support mobility is used almost in all mobile networks.

The *Authentication Center (AUC)* holds different algorithms that are used for authentication and encryption of the subscribers. Different classes of SIM cards have their own algorithms and AUC collects all of these algorithms to allow the NSS to operate with different terminals from different geographical areas.

The *Equipment Identification Register (EIR)* is another database managing the identification of the mobile equipment against faults and theft. This database keeps the International Mobile *Equipment* Identity (IMEI) that reveals the manufacturer, country of production, and terminal type. Such information can be used to report stolen phones or check if the phone is operating according to the specification of its type. The implementation of the EIR is left optional to the service provider.

11.3   Mechanisms to Support a Mobile Environment

Now that we have described all the hardware and software elements of a typical digital cellular network for wireless access to the PSTN, we can describe how different functionalities of the network are implemented with these elements. Four mechanisms are embedded in all wide area wireless access technologies designed primarily to connect to the PSTN that allow a mobile telephone to establish and maintain a connection with the network. These mechanisms are registration, call establishment, handoff (hand-off or handover), and security. Registration takes place as soon as one turns the mobile unit on, call establishment occurs when the user initiates or receives a call, handoff helps the MS to change its connection point to the network during a call, and security protects the user from fraud and eavesdropping. General descriptions of these mechanisms were included in Chapter 6, but in this section, we describe the details of their implementation in a typical TDMA architecture for wide area wireless access to the PSTN in the licensed frequency bands. To illustrate the complexity of wireless access to these networks, when we discuss registration and call establishment in a typical cellular network, we compare these mechanisms with their counterparts in traditional wired access to the PSTN.

11.3.1   Registration

When we subscribe to a traditional wired telephone service, the telephone company brings a pair of wires that are connected to a port of a switch in a PSTN end office to our homes. Then the residential telephone number is registered in a database in the network and the registration is fixed. Therefore, the connection and registration process for a wired access to the network is a one-shot operation, and after that, the connection is active and the registration is valid as long as a subscription to service is valid. With wireless access to a cellular network, each time we turn the mobile station on, we need to establish a new connection and possibly establish a new registration with the network. We may actually connect to the network at different locations through a BTS that may not be owned by our service provider. Therefore, a wireless network needs a registration process that is far more complex than the registration in wired networks.

Technically speaking, as we turn on a MS it passively synchronizes to the frequency; bit and frame timings of the closest BS to get ready for information exchange with the BS. After this preliminary set up, the MS reads the system and cell identity to determine its location in the network. This whole process is often called cell search. If the current location is not the same as before, the MS initiates a *Registration* procedure. During the registration procedure, the network provides the MS with a channel for preliminary signaling. The MS provides its identity in exchange for the identity of the network, and finally the network authenticates the mobile station. The simplest connection takes place if the MS is turned on in the previous area and the most complex registration process occurs when the mobile is turned on in a new MSC area which needs changes in the entries of the VLR and HLR. The following example illustrates the complexity of the registration process of a typical wireless access to the PSTN when a mobile is turned on in a new MSC.

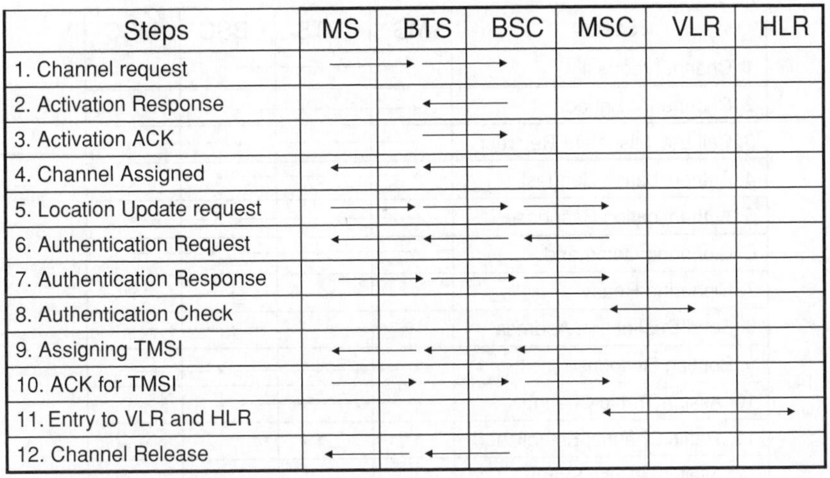

Steps	MS	BTS	BSC	MSC	VLR	HLR
1. Channel request	→—	—→				
2. Activation Response		←—				
3. Activation ACK		—→				
4. Channel Assigned	←—	←—				
5. Location Update request	→—	—→	—→			
6. Authentication Request	←—	←—	←—			
7. Authentication Response	→—	—→	—→			
8. Authentication Check				←—	—→	
9. Assigning TMSI	←—	←—	←—			
10. ACK for TMSI	→—	—→	—→			
11. Entry to VLR and HLR					←—	—→
12. Channel Release	←—	←—				

Figure 11.3 Registration procedure in a typical cellular telephone network.

Example 11.6: The registration procedure

Figure 11.3 shows the twelve-step registration process in a typical cellular telephone network that takes place when a MS is turned on in a new MSCs area. In the first four steps, a radio channel is established between the MS and BSS to process the registration. In the next four steps, the NSS authenticates the MS. In the following three steps, a TMSI is assigned and adjustments are made to the entries in the VLR and HLR. In the final step, the temporary radio channel for communication is released and transmission starts over a traffic channel.

11.3.2 Call Establishment

Call establishment in traditional wired access to the PSTN starts with a dialing process that transfers the number to the nearest PSTN switch where a routing algorithm finds the best connection through intermediate switches to the destination. After establishment of the link, the last switch (end office) at the destination sends a signal back to the source to announce whether the destination is available or busy that is signaled to the user at the source. When the destination terminal is off-hook, another signal is sent to the source end-office to stop the waiting tone and establish the traffic line. In the mobile environment we have two separate call establishment procedures for mobile to fixed and fixed to mobile calls. Mobile to mobile calls are a combination of the two. The following two examples provide the detailed procedure in a typical cellular telephone network for both types of call establishment.

Example 11.7: Mobile originated call

The five-step-procedure in a traditional wired access to the PSTN for call set-up changes to a 15-step mobile originated call establishment procedure in a typical digital cellular network. As shown in Figure 11.4, the first five steps are similar to the registration process in a typical digital cellular, except that these are done to prepare for call establishment. The next two steps start ciphering (encryption) to provide a protection against eavesdropping. The rest of the steps are similar to

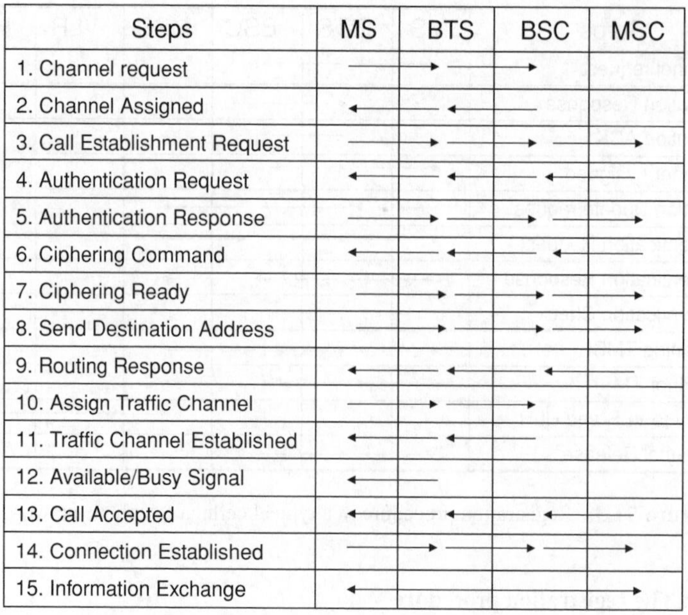

Steps	MS	BTS	BSC	MSC
1. Channel request	———→	———→		
2. Channel Assigned	←———	←———		
3. Call Establishment Request	———→	———→	———→	
4. Authentication Request	←———	←———	←———	
5. Authentication Response	———→	———→	———→	
6. Ciphering Command	←———	←———	←———	
7. Ciphering Ready	———→	———→	———→	
8. Send Destination Address	———→	———→	———→	
9. Routing Response	←———	←———	←———	
10. Assign Traffic Channel	———→	———→		
11. Traffic Channel Established	←———	←———		
12. Available/Busy Signal	←———			
13. Call Accepted	←———	←———	←———	
14. Connection Established	———→	———→	———→	
15. Information Exchange	←———			———→

Figure 11.4 Mobile-initiated call establishment procedure in a typical digital cellular telephone network.

those in wired access to the PSTN except that we have an additional traffic channel assignment procedure.

Example 11.8: Mobile terminated call

The most complicated call establishment for wireless access to the PSTN is for the situation where a fixed telephone dials a mobile visiting another MSC. As shown in Figure 11.5, after dialing, the PSTN directs the call to the MSC identified by the destination address. The MSC requests routing information from the HLR. Since, in this case, the mobile is roaming in the area of a different MSC, the address of the new MSC is given to MSC and it contacts the new MSC. At the destination MSC, the VLR initiates a paging procedure in all BSSs under the control of the MSC holding the registration. After a reply from the MS, the VLR sends the necessary parameters to the MSC to establish the link to the MS.

11.3.3 Handoff

The procedures for handoff in wireless access to the PSTN broadly follow the procedures described in Chapter 6 that dealt with mobility management in general. In this section we provide a detailed example of a mobile assisted handoff procedure used in a typical TDMA wireless access to the PSTN. There are two types of handoff – internal and external. Internal handoff is between BTSs that belong to the same BSS and external handoffs are between two different BSSs belonging to the same MSC. Sometimes there are handoffs between BSSs that are controlled by two different MSCs. In such a case, the old MSC continues to handle call management. Roaming between two MSCs in two different countries is prohibited and the call simply drops.

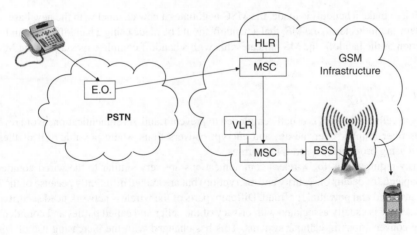

Figure 11.5 Mobile-terminated call in a visiting network.

Handoff is initiated because of a variety of reasons. Signal strength deterioration is the most common cause for handoff at the edge of a cell. Other reasons include traffic balancing where the handoff is intiated by the network to ease traffic congestion by moving calls in a highly congested cell to a lightly loaded cell. The handoff could be synchronous where the two cells involved are synchronized or it may be asynchronous. Since the MS does not have to resynchronize itself in the former scenario, the handoff delay is much smaller (100 vs 200 ms in the asynchronous case).

Figure 11.6 shows the handoff procedure between two BSSs that are controlled by one MSC. The BTS provides the MS a list of available channels in neighboring cells. The MS monitors the RSS from the neighboring cells and reports these values to the MSC. This is called *mobile-assisted* handoff as it was discussed in Chapter 6. The BTS also monitors the RSS from the MS to make a handoff decision. Proprietary algorithms are used to decide when a handoff should be initiated. If

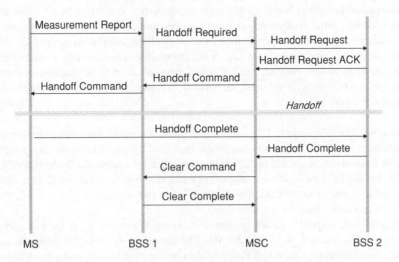

Figure 11.6 Handoff involving a single MSC but two BSSs.

a decision to make a handoff is made, the MSC negotiates a new channel with the new base station subsystem and indicates to the MS, that a handoff should be made using a handoff command. Upon completion of the handoff, the MS indicates this with a handoff complete message to the MSC.

11.3.4 Security

Security in cellular systems is usually employed to prevent fraud via authentication, avoid revealing the subscriber number over the air, and encrypt conversations where possible. All of these are achieved using proprietary (secret) algorithms.

Security requirements for wireless communications are very similar to the wired counterparts (see Chapter 6 for details of security and encryption) but are treated differently because of the applications involved and potential for fraud. Different parts of the wireless network need security. Over the air security is usually associated with privacy of the caller and called parties and confidentiality of voice conversations in cellular systems. This has changed with the increasing use of wireless data services. Message authentication, identification, authorization, and so on also become issues with cellular wireless access to the PSTN. Wireless networks are inherently insecure compared to their wired counterparts. The broadcast nature of the channel makes it easier to be tapped. Analog cellular telephones are extremely easy to tap and conversations can be eavesdropped using an RF scanner. Digital wireless network such as TDMA and CDMA are much harder to tap and RF scanners cannot do the trick anymore, but since the circuitry and chips are freely available, it may not be hard for someone to break into a system that does not employ security.

Privacy Requirements of Wireless Access to PSTN: A variety of control information is transmitted over the air in addition to the actual voice or data. These include call set-up information, user location, user ID (or telephone number) of both parties, and so on. These should all be kept secure since there is potential for misusing such information. Calling patterns (traffic analysis) can yield valuable information under certain circumstances. Voice and data traffic should be encrypted for confidentiality.

Authentication and Integrity: While privacy and confidentiality continue to be the important issue in wireless networks, other security requirements are becoming significant in recent times. There has been widespread fraud and impersonation of analog cellular telephones in the past. Although this is more difficult with digital systems, it is not impossible. There is thus a need to correctly *identify* and *authenticate* a mobile terminal. Control messages need to be checked for integrity to ensure that spoofed messages do not cause the network to behave abnormally leading to widespread disruption of communications.

Implementation: The SIM cards in 2G mobile devices have a microprocessor chip that can perform the computations required for security purposes. Figures 11.7 and 11.8 show the principles of operation for the authentication and ciphering in a typical cellular network. We have used GSM as an example, but similar procedures are adopted in other systems. A secret key K_i is stored on the SIM card and it is unique to the card. This key is used in two algorithms, A1 and A2, for entity authentication and confidentiality respectively.

For authentication purposes, shown in Figure 11.7, the secret key K_i is used in a challenge response protocol between the BSS and the MS. The secret key K_i is used to generate a privacy key K_c that is used to encrypt messages (voice or data) as the case may be using the A2 algorithm. The control channel signals are encrypted using a third encryption algorithm. The size of a typical

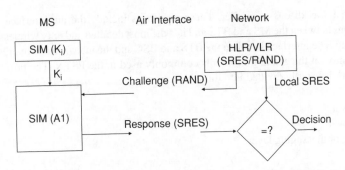

Figure 11.7 Basic principles of authentication in a cellular network.

secret key K_i and the response to the challenge play an important role in the robustness of the security. Figure 11.8 illustrates the basic principles of ciphering in a typical cellular network. The challenge random string and the secret key K_i are used with algorithm A2 to generate a new ciphering key K_c which is used with a third algorithm, A3, to encrypt the data for confidentiality and protection against eavesdropping. Another aspect of security in cellular networks is that the secret key information is not shared between systems. Instead a data structure consisting of the random number used in the challenge, the response to the challenge, and the data encryption key K_c is exchanged between the VLR and the HLR. The VLR verifies if the response generated by the MS is the same. The algorithms A2 and A3 are secret and usually not shared between different systems.

In 3G systems, message authentication (see Chapter 7) is used to ensure the integrity and authenticity of control messages.

11.4 Communication Protocols

In the previous sections of this chapter we introduced a typical circuit switched cellular network services and architectural elements as well as an overview of the mechanisms that allows this architecture to support mobile operation. In this section we will provide the description of how these elements and mechanisms are integrated with one another to implement the services. Elements of a network communicate with each other through a protocol stack that is specified by the standard committee. The standardization committees specify the interfaces among all the elements of the

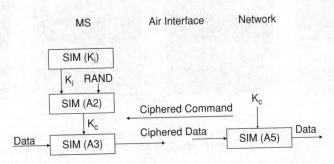

Figure 11.8 Basic principles of ciphering in a typical cellular network.

architecture that was discussed earlier. These interfaces include the air–interface that specifies communications between the MS and BTS and it is the most detailed and only wireless transmission related interface. The interface between the BTS and BSC and the interface between BSC and MSC draw significantly on the existing protocols commonly used in the PSTN. The protocol stack in a cellular network is usually divided into three layers:

1. Physical layer
2. Data link layer
3. Networking or messaging layer.

Figure 11.9, as an example, shows the protocol architecture for communication between the main hardware elements and the associated interfaces in the case of GSM. Messages between the BTS and BSC flow through wires or fiber. The support on this interface is for voice traffic at 64 kbps and data/signaling traffic at 16 kbps. Both types of traffic are carried over Link Access Protocol type D (LAPD, which is a data link layer protocol used in PSTN). The message transfer between different BSCs to the MSC are also wired and based on existing protocols in PSTNs. The physical layer is a 2 Mbps standard connection for PSTN that employs signaling protocols for communication as well. The Message Transport Protocol (MTP) and the Signaling Connection Control Part (SCCP) of PSTN are used for error free transport and logical connection respectively. These protocols are there to establish and maintain the connection to the wired PSTN.

The general structure of the reference model and protocol stack, for all cellular networks designed to provide access to the circuit switched PSTN telephone services, are the same. The main difference is in the implementation of the air–interface physical layer that can employ TDMA or CDMA technology. In the following three sections we cover more details of the three layers with specific examples to provide the readers with an understanding of how a cellular network operates to support different services.

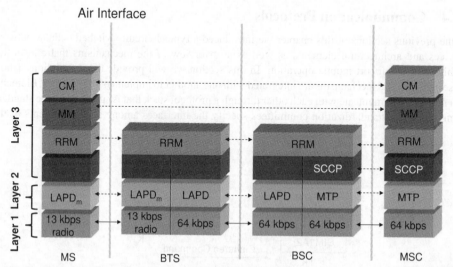

CM: Connection Management; MM: Mobility Management; SCCP: Signal Connection Control Part
RRM: Radio Resource Management; MTP: Message Transfer Part; LAPD: Link Access Protocol-D

Figure 11.9 The GSM protocol architecture.

11.4.1   Layer I: Physical Layer

The physical layer of wired interfaces follow the PSTN standards, in which the digital voice is transferred throughout the network with 64 kbps/user. The new physical layer defined in the GSM specifications is for the air–interface. This layer specifies how the information from different voice and data services are formatted into packets and sent through the radio channel. It specifies the radio modem details, structure of traffic and control packets in the air, and the packaging of variety of services into the bits of a packet. This layer specifies the modulation and coding techniques, power control methodology, and time synchronization approaches that enables establishment and maintenance of the channels.

Description of a typical physical layer for the air–interface of a cellular network is the major part of the work by an standardization committee. This involves details of how packets should be formed and transmitted to carry the traffic and signaling and control messages as well as description of the channel and signal specifications for transmission of information bits in the air. In the remainder of this section we describe how packets are formed and transmitted in the three-layer format used in cellular networks. The channel model and design of the modem for transmission of information bits will be discussed separately in Section 11.5 that follow the current section.

Power and Power Control: As discussed in Chapter 6, power management is an important issue in wireless networks in general. Power management in cellular telephone networks helps the service provider to control the interference among the users and minimize the power consumption at the terminal. Therefore, power management has direct impact on the quality of service and the life of the batteries that are extremely important to the users.

There are two major classes of mobile stations defined in GSM namely, vehicle mounted and handheld terminals. Vehicle-mounted terminals use the car battery and handheld terminals employ smaller rechargeable batteries. The antenna for the vehicle mounted mobile is mounted outside the car that is away from the user's body, while the antenna in the handheld terminals is next to the ear and brain of the user that raises health concerns at high radiated powers. In a typical cellular network, cells have radii ranging from around 100 m to 35 km depending on the base station (see Chapter 5). The size of the cells also plays a role in the required transmitted power for the BTS and the MS. To allow manufacturers and service providers to accommodate the varied requirements for different MS and BSS subsystems, a number of radiated power classes are identified by the different cellular networking standards. For example, in the case of GSM, there are five power classes for the mobile terminal from 29 dBm (0.8 W) up to 44 dBm (20 W) with a 4 dB separation between consecutive mobile classes. There are eight classes for the BTS power ranging from 34 dBm (2.5 W) up to 55 dBm (320 W) in 3-dB steps.

The transmitted radio frequency power in a given MS is always controlled to its minimum required value (that can be as small as 20 mW) to minimize the co-channel interference among different cells and maximize the life of the battery. The BSS calculates the power level for individual MSs by monitoring the interference and received signal strength and sends this information through control signaling packets to the MS.

TDMA Physical Packet Bursts: In a TDMA connection based wireless access network, the traffic and signaling control information are transmitted in bursts. In this section we use GSM as an example to describes how these bursts are formed by the transmission system. GSM in Europe uses the 890–915 MHz bands for the uplink (reverse) and 935–960 MHz bands for the downlink (forward) channels. As shown in Figure 11.10, the 25 MHz band for each direction is divided into 124 channels each occupying 200 kHz with a 100 kHz guard band at two edges of the spectrum.

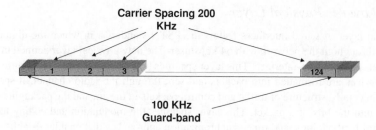

Figure 11.10 Frequency bands in GSM.

Each carrier supports eight time slots in a frame for TDMA operation. The data rate on each carrier is 270.833 kbps through a waveform that is created with a Gaussian Minimum Shift Keying (GMSK) modem, described later in Section 11.5, with a normalized bandwidth expansion factor of 0.3. With this data rate, the "duration" of each bit is 3.69 μs. The user transmission packet "burst" is fixed at 577 μs, that accommodates both information bits and a time gap between the packets (partly to avoid interference) for a total duration equivalent to 156.25 times the bit duration of 3.69 μs.

GSM supports four types of bursts for traffic and control signaling. Figure 11.11 shows all the four bursts types. The *Normal Burst,* shown in Figure 11.11a, consists of three tail bits at the beginning and at the end of the packet, equivalent to 8.25 bits of gap period, two sets of 58 encrypted bits (for a total of 116 bits), and a training sequence of 26 bits. The tail bits are three zero bits, providing a gap time for the digital radio circuitry to cover the uncertainty period to ramp on and off for the radiated power and to initiate (reset) the convolutional decoding of encoded data. The

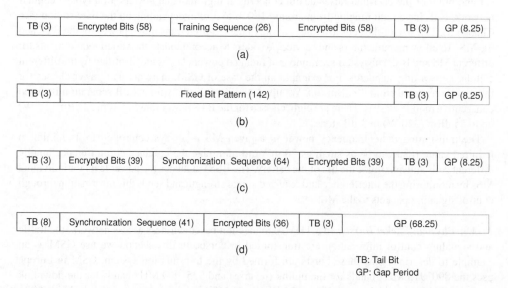

Figure 11.11 The four burst types in GSM: (a) normal burst, (b) frequency correction burst, (c) synchronization burst, (d) random access burst.

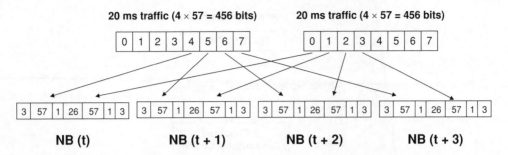

Figure 11.12 Interleaving traffic frames onto the TDMA GSM frame in the air.

26 bits training sequence is used to train an adaptive equalizer at the receiver. Since the channel behavior is constantly changing during the transmission of the packet, the most effective place for the training of the equalizer is in the middle of the burst. The 116 encrypted data bits includes 114 bits of data and two flag bits at the end of each part of the data that indicates whether the data is user traffic or signaling and control bits. The user traffic data arrives in frames of length 456 bits as shown in Figure 11.12 they are *interleaved* into the transmitted normal bursts in blocks of 57 bits plus one flag bit. The purpose of interleaving is to improve the performance of coding schemes and the quality seen by users by distributing the effects of fade hits among several blocks. The 456 bits are produced every 20 ms. Therefore, the equivalent of 20 ms of arriving information is mapped into 456 bits. The standard specifies the method that maps the 20 ms of the traffic into the 456 bits.

Example 11.9: Packetization of voice traffic

Figure 11.13 shows how the 456-bit packets are formed from the speech signal. Each 20 ms of the coded speech at 13 kbps forms a 260-bit packet. The first 50 most significant bits receives a 3-bit Cyclic Redundant Code (CRC) protection and then they are added to the second group of 132-bits

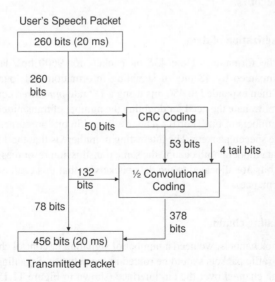

Figure 11.13 Coded speech packets in GSM.

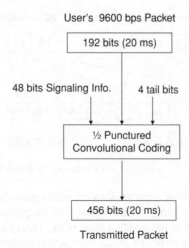

User's 9600 bps Packet

192 bits (20 ms)

48 bits Signaling Info.   4 tail bits

½ Punctured
Convolutional Coding

456 bits (20 ms)

Transmitted Packet

Figure 11.14   Coded data packets in GSM.

with lower importance and a 4-bit tail that is all zeros. The resulting $132+53+4 = 189$ bits are then encoded with a convolutional encoder with code rate 1/2 that doubles the number of bits to 378. The convolutional code provides for error correction capabilities. The 378 coded bits are added to the 78 least important speech-coded bits to form a 456-bit packet every 20 ms. The 456-bit packets are used to form transmission bursts shown in Figure 11.11. In this encoding scheme we have three classes of speech bits: the first class of 50 bits receives both CRC error detection and protection with the rate 1/2 convolutional error correcting coding. The second class of 132 bits receives only the convolutional encoding protection and the last 78 bits receives no protection. Therefore, the speech coder can protect the more important bits representing larger values of voltages by assigning them into different categories.

Example 11.10: Packetization of data

Figure 11.14 shows the formation of the 456 bit packets for 9600 bps data. The 192 bits of information are accompanied by 48 bits of signaling information and four tail bits to form a 244 bits packet that is then expanded to 456 bits using a 1/2 rate *punctured* convolutional encoder. Punctured coding can eliminate the need for doubling the number of transmitted bits by eliminating (puncturing) certain numbers of bits [Por01]. The resulting 456 bits are turned to normal bursts in a manner similar to the speech packets. The interesting point here is that the 13 kbps speech coded signal and 9600 bps data modem both occupy the same transmission resources on the air–interface. More channel coding bits are allocated to the data packets so that this can be expected to provide better error rate performance.

Example 11.11: Signaling channel

In addition to the traffic channels, we need a number of signaling or control channels that are used to determine how the traffic packets should be routed in the network. Signaling channels using the normal burst NB as the channel over the air–interface (shown in Figure 11.15) use 184 signaling bits to convey the signaling message. These bits are first block coded with 40 additional parity check bits and four tail bits to form a 228-bit block. The 228-bit block is then coded with a 1/2 rate

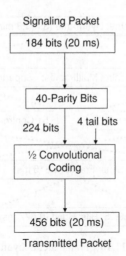

Signaling Packet

Figure 11.15   Signaling channel in GSM.

convolutional encoder to form a 456-bit packet occupying a 20 ms slot that is turned to a burst for transmission as shown in Figure 11.12.

The other three types of bursts are simpler and designed for specific tasks. The simplest of all the remaining bursts is the *Frequency-Correction Burst*, shown in Figure 11.11b. It has three tail bits at the start and the end of the burst. The rest of the packet contains all "0"s that allows simple transmission of the carrier frequency without any modulated information. An equivalent of 8.25 bits duration is used as the *gap period* between this burst and others. The BS broadcasts the frequency-correction burst and the MSs use it to synchronize with the master clock in the system. The *Synchronization Burst*, shown in Figure 11.11c, is very similar to the normal burst except that the training sequence is longer and the coded data are used for the specific task of identifying the network. The BTS broadcasts the synchronization burst and the MSs use it for initial training of the equalizer as well as initial learning of the network identity and to synchronize the time slots.

The *Random Access Burst* is used by the MS to access the BS as it registers with the network. The overall structure is similar to NB except that a longer start up and synchronization sequence is used to initiate the equalizer. Another major difference is the length of the much longer gap period that allows a rough calculation of the distance of the MS from the BTS. This calculation is possible by determining the arrival time of the random access burst. A gap period of 68.25 bits translates to 252 μs. The signal transmitted from a MS should travel more than 75.5 km (at the speed of 300 000 km/s) before arriving at the BTS to exceed this gap period.

TDMA Frame Hierarchy: When a number of different slots carry the user traffic and a variety of control signals, a hierarchy is needed to identify the *location* of certain bursts among the large stream of bursts that are directed toward different terminals. Each terminal needs a number of counters to track the related packets at different levels of the hierarchy.

Example 11.12: TDMA frame hierarchy in GSM

The GSM radio–interface standard provides a variety of traffic channels and control channels defined in a hierarchy built upon the basic 8-slot TDMA transmission format. The frame hierarchy,

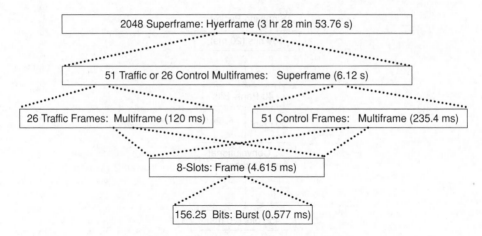

Figure 11.16   Frame hierarchy in GSM.

depicted in Figure 11.16 shows the TDMA hierarchy of the GSM network from a burst of 0.577 ms interval to a hyperframe of length of around 3.5 h. The basic building block of the frame hierarchy is a 4.615-ms frame. Each frame comprises eight bursts or time slots. The time-slot interval is equivalent to the transmission time for about 156.25 bits, for which, as we saw in Figure 11.11, durations equivalent to 8.25 (68.25 for a random access burst) bit times are used as guard times during which no signal is transmitted. The next level in the hierarchy is a GSM multiframe, shown in Figure 11.16. Each 120-ms multiframe is composed of 26 frames, each containing eight time slots. In each multiframe, 24 frames carry user information, while two frames carry system control information related to individual users. The data rate per voice user is calculated by considering that for each 120 ms, 24 voice-bursts each carrying $2 \times 57 = 114$ bits of information are transmitted. Therefore, the data rate per user is $24 \times 114/0.120 = 22{,}800$ bits/s. The speech coder has a data rate of 13 kbps and the addition of error-detection and error-correction coding brings the transmission rate up to 22.8 kbps.

Figure 11.16 shows that the eight-slot frames may be also organized into control multiframes rather than traffic multiframes. Control multiframes, are used to establish several types of signaling and control channels used for system access, call set-up, synchronization, and other system control functions. Either traffic or control-multiframes are grouped into superframes, which are in turn grouped into hyperframes. Counters at the terminals need to track the packet numbers at hyperframe, superframe and mutiframe levels to communicate with the network.

The counter for multiframes in the mobile terminal needs to keep track of the traffic channel for the terminal. Another counter needs to track the traffic superframe to identify where the location of the two control frames is. A variety of control signaling information embedded in the control superframe are extracted from their appropriate location using the counter for those frames.

Logical Channels: In the last sections, we described how traffic and control packets are inserted in a hierarchy and how terminals use counters to identify the location of specific packet bursts in the overall structure of the frames. Communication between the terminal and the base station involves both user information (speech and data) traffic as well as signaling and control. The entire communication system can be thought as a distributed real-time computer that uses a number of instructions to transfer information packets from one location to another. We have several major tasks to make such a system work. We need initial signaling for registration and call establishment, we need to maintain the synchronization among the terminals, we need to manage mobility, and we need to transfer the data traffic. In a manner similar to computers, we need a set of instructions and ports to instruct different elements of the network to perform their specified duties. In telecommunication systems, these ports are referred to as logical channels. Logical channels use a physical TDMA slot or a portion of a physical slot to specify an operation in the network.

Logical channels are usually divided into two principal categories: ***traffic channels*** and ***control channels***. Traffic channels are two-way channels carrying the voice and data traffic between the MS and BTS. Traffic logical channels are implemented over the NB physical bursts shown in Figure 11.11a. In the GSM system, these channels carry the 13 kbps speech-coding data and the 9600, 4800, and 2400 bps data. Figures 11.13 and 11.14 show the procedures to create the frames for 13 kbps speech and 9600 bps data, respectively. As we saw earlier, when we include signaling overhead each channel has a gross bit rate of 22.8 kbps for the network.

There are three classes of control channels: ***Broadcast Channels***, ***Common Control Channels***, and ***Dedicated Control Channels***. The broadcast channels are broadcast from the BTS to MSs in the coverage area of the BTS. These broadcast channels include ***Frequency Control Channel*** used by the BTS to broadcasts carrier synchronization signals. An MS in the coverage area of a BTS uses this broadcast channel to synchronize its carrier frequency and bit timing. This channel is implemented on the physical frequency correction burst, shown in Fig. 11.11b. Another broadcast control channel is ***Synchronization Channel*** used by the BTS to broadcast frame synchronization signals to all MSs. Using this channel, MSs will synchronize their counters to specify the location of arriving packets in the TDMA hierarchy and physical synchronization burst, shown in Figure 11.9c, are used to implement it. A third important logical broadcast channel is ***Broadcast Control Channel*** used by BTS to broadcast synchronization parameters, available services, ***cell ID,*** and signal strength measurements for handoff. Once the carrier bit and frame synchronization between the BTS and MS are established, this channel, implemented on a normal burst, informs the MS about the environment parameters associated with the BTS covering that area.

The common control channels are also one-way channels used for call establishment. One example of these channels are ***Paging Channel***, used by the BTS to page the MS for incoming call is a broadcast channel implemented on a normal burst. Another example is the ***Random Access Channel,*** implemented on the short random access bursts shown in Figure 11.11d. This channel is used by the MS to access the BTS for call establishment using a slotted-ALOHA protocol that contends for one of the available slots in the GSM traffic frames.

The *dedicated* control channels are two-way channels supporting signaling and control for individual users. These channels are assigned to each terminal to transfer network control information for call establishment, mobility management, and exchange necessary parameters between the BTS and the MS to maintain the link. More detailed description of the logical channels and GSM operation is available in [Goo97, Red95]. In order to implement the required functionality, for example mobile call establishment presented in Figure 11.4, each of the 15 action steps is mapped to a logical channel and the sequence of these logical channels implements the operation. For example, the channel request message is mapped to a logical random access channel, security is implemented through the dedicated stand-alone logical channel and so on.

11.4.2   Layer II: Data Link Layer

Any connection-based network can be considered to be two networks, one used for traffic and the other for signaling and control. The signaling and control may be through the same physical

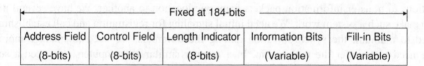

Figure 11.17 Frame format of Layer II in the air–interface of GSM.

channels or through separate physical channels. In traffic channels for GSM, as we saw in Figures 11.13 and 11.14, the information bits are encoded with strong error detection and correction codes to form packets of length 456 that are then sent with four normal bursts. Signaling and control data are conveyed through Layer II and Layer III messages. The overall purpose of Layer II is to check the flow of packets for Layer III and allow multiple service access points with one physical layer. In GSM Layer II checks the address and sequence number for Layer III and manages acknowledgements for transmission of the packets. In addition, Layer II allows two service access points for parallel signaling and short messages (SMS). Unlike other GSM data services that are carried through traffic channels, the SMS traffic channel in the GSM is not communicated through traffic channels. In GSM, the SMS is transmitted through a fake signaling packet that carries user information over signaling channels. The Layer II in GSM provides this mechanism for multiplexing the SMS data into signaling streams.

Figure 11.17 shows a typical format for the Layer II of the air–interface in the GSM. The address field is optional and it identifies the service access point, protocol revision type, and nature of the message. The control field is optional and it holds the type of the frame (command or response) and the transmitted and received sequence numbers. The length indicator identifies the length of the information field. The information field carries the Layer III payload. Fill-in bits are all "1" bits to extend the length to the desired 184 bits. In peer to peer Layer II communications, such as acknowledgements, there is no Layer III payload and fill-in bits cover this field.

11.4.3   Layer III: Networking Layer

As we discussed in Section 11.3 there are a number of mechanisms needed to establish, maintain and terminate a mobile communication session. The networking or signaling layer implements the protocols needed to support these mechanisms. The networking layer in cellular network is also responsible for control functions for supplementary and SMS services. The traffic channels, as we saw earlier, are mapped into the related logical channels and carried by normal bursts in different formats associated with different speech or data services. The signaling information uses other bursts and more complicated Layer II packaging. A signaling procedure or mechanism or protocol, such as the registration process shown in Figure 11.4, is composed of a sequence of communication events or *messages* between elements of the systems that are implemented on the logical channels encapsulated in the Layer II frames. Layer III defines the details of implementation of messages on the logical channels encapsulated in Layer II frames. Among all messages communicated between two elements of the network only a few, such as Layer II acknowledgement, do not carry Layer III information.

Figure 11.18 shows the typical format of Layer II and Layer III messages in a procedure between two elements of the network. They start with simple pure Layer II messages without Layer III information bits to initiate a procedure. Then a number of layer II messages with Layer III information follow to complete the necessary operation for the procedure. At the end, a couple of pure Layer II messages disconnect the session between the two elements. Information bits of the Layer II packets, shown in Figure 11.17, specify the operation of a Layer III message.

```
┌─────────────────────────────────────────────────┐
│ Layer II Messages                               │
│ Set the mode of communicatione                  │
│ Unnumbered Acknowledgement                      │
├─────────────────────────────────────────────────┤
│ Layer III RRM, MM, and CM Messages Started      │
│ ...................... ...........................│
│ ...................... ...........................│
│ ....................................................│
│ (Layer III Messages Ended)                      │
├─────────────────────────────────────────────────┤
│ Layer II Messages                               │
│ Disconnect                                      │
│ Unnumbered Acknowledgement                      │
└─────────────────────────────────────────────────┘
```

Figure 11.18 Typical format of the messages in a procedure used for implementation of a network operation mechanism.

The number of Layer III messages is much larger than the number of pure Layer II messages. To further simplify the description of the Layer III messages, as we showed in Figure 11.9, a typical cellular network divides these messages into three sub-category or sub-layers: Radio Resource Management (RRM), Mobility Management (MM), and Connection Management (CM) messages.

The radio resource management sublayer of Layer III manages the frequency of operation and the quality of the radio link. This sublayer does not have an equivalent in wired networks because there is no frequency assignment issue in the wired networks. The main responsibilities of the radio resource management are to assign the radio channel and hop to new channels in implementation of the slow frequency hopping option, manage handoff procedure and measurement reports from MS for handoff decision, implement power control procedure, and adapt to timing advance for synchronization.

The mobility management sublayer handles mobility issues that are not directly related to the radio. Major responsibilities of this sublayer are location update, authentication procedure, temporary MS Identity (TMSI) handling, and attachment and detachment procedures for the international MS identity (IMSI). The connection management sublayer establishes, maintains, and releases circuit-switched connection and helps in SMS. Specific procedures for the connection management sublayer are mobile-originated and -terminated call establishment, change of transmission mode during the call, control of dialing using dual-tones, and call re-establishment after MM interruption. Figure 11.19 shows the 15-step mobile initiated call establishment procedure that was discussed earlier and maps each step to a Layer III category. Note that Layer III does not handle the traffic message and therefore we have no sublayer association for that part of the procedure.

Explanation of the details of encoding of each message and a complete list of the GSM messages are beyond the scope of this book. For the complete list of the messages used in Layer III of the GSM reader can refer to [Goo97] and for further details of the operation to [Red95] or [Gar99].

11.5   Channel Models for Cellular Networks

As we explained in Chapter 2, modeling of the radio propagation is aimed at supporting tools for calculation of the coverage using a path-loss model and modeling of the effects of multipath using

Message Name	Category
1. Channel Request	RRM
2. Immediate Assignment	RRM
3. Call Establishment Request	CM
4. Authentication Request	MM
5. Authentication Response	MM
6. Ciphering Command	RRM
7. Ciphering Ready	RRM
8. Send Destination Address	CM
9. Routing Response	CM
10. Assign Traffic Channel	RRM
11. Traffic Channel Established	RRM
12. Available/Busy Signal	CM
13. Call Accepted	CM
14. Connection Established	CM
15. Information Exchange	

Figure 11.19 Mobile-initiated call establishment procedure in GSM and mapping of each step to a sublayer in Layer III.

the scattering function for performance evaluation of transmission techniques. Path-loss models are often in general given by Equation 2.7, which is reproduced below:

$$L_p = L_0 + 10\alpha \log d + X,$$

where the path-loss L_p is related to path-loss at the first meter distance from the transmitter L_0, the distance-power gradient α, distance between the transmitter and the receiver d and a zero mean Gaussian random variable X representing the shadow fading. The scattering function is given by Equation 2.22 and reproduced here as:

$$S(\tau, \lambda) = Q(\tau) \times D(\lambda)$$

that is composition of two functions, the delay power spectrum:

$$Q(\tau) = |h(\tau)|^2 = \sum_{i=1}^{L} P_i \delta(\tau - \tau_i)$$

representing the portions of the received signal power at different delays of arrival of the multipath components and the Doppler spectrum $D(\lambda)$ representing the shape of power fluctuations in each arriving multipath component. Standards organizations define these functions for different wireless networking applications in the environments, frequency of operation, bandwidth and power levels that these standards are designed for as explained in Chapter 2. Cellular networks were originally designed for urban areas and outdoor applications inside the vehicles and for frequencies under

Table 11.1 The COST-231 model for PCS applications in urban areas

General formulation: $$L_p = 46.3 + 33.9 \log f_c - 13.82 \log h_b - a(h_m) + [44.9 - 6.55 \log h_b] \log d + C_M$$ where f_c is in MHz, h_b and h_m are in m, and d is in km		
Range of values		
Center frequency f_c (MHz)		1500–2000 MHz
h_b, h_m (m)		30–200 m, 1–10 m
D		1–20 km
C_M	Large city	0 dB
	Medium city/suburban areas	3 dB

1 GHz. Advancements in micro-electronics extended the devices into the portable handheld terminals that further moved the application domain to the indoor environment. Demand for more bandwidth opened PCS bands close to 2 GHz. As a result, several organizations at different times have been involved in modeling of the channels for cellular networking applications. These organizations have divided channel models for cellular networks into a cell hierarchy consisting of macro-cells covering wide areas, micro-cells covering streets of the dense urban areas and models covering the indoor areas. In the rest of this section we provide an overview of some of the models that have evolved for path-loss and scattering function of cellular networks at various frequencies and for various cellular hierarchies.

11.5.1 Path Loss Models for Cellular Networks

In the cellular hierarchy, macro-cellular areas span a few kilometers to tens of kilometers, depending on the location. These are the traditional "cells" corresponding to the coverage area of a base station associated with traditional cellular telephony base stations. The frequency of operation is mostly around 900 MHz though the emergence of PCS has resulted in frequencies around 1800 and 1900 MHz for such cells. There have been extensive measurements in a number of cities and locations of the received signal strength in macro-cellular areas and have been reported in the literature. The most popular of these measurements for traditional cellular bands corresponds to those of Okumura–Hata model that we described as one of the most popular path-loss models in Table 2.4 (Section 2.2.5). This model was originally designed for cellular bands and was later extended to PCS bands. This extension is known as the COST-231 model and it is summarized in Table 11.1. The parameter $a(h_m)$ is defined in Table 2.4, the correction parameter C_M is added, and the path-loss in the first meter is adjusted in Table 11.1 to extend the model to frequencies beyond 1500 MHz up to 2 GHz (that now includes the PCS frequency bands of 1800–1900 MHz, which were not released for cellular telephony when the original models were designed). This type of patchwork extension is one of the characteristics of the channel models for cellular networks because they operate in expensive licensed bands (that were expanded) throughout the evolution of the cellular industry. For more samples of models for path-loss in macro-cellular environments, one can refer to [Pah05].

Micro-cells are cells that span hundreds of meters to a kilometer or so and are usually supported by base station antennas that are below the rooftop level and mounted on lampposts or utility poles.

Table 11.2   Path loss formulas for micro-cells

Environment	Scenario	Path loss expression
Low rise	Non-LOS	$L_p = [139.01 + 42.59 \log f_c] - [14.97 + 4.99 \log f_c]$ $\text{sgn}(\Delta h)\log(1+|\Delta h|) + [40.67 - 4.57 \, \text{sgn}(\Delta h)]$ $\log(1+|\Delta h|) \log d + 20 \log (\Delta h_m/11.8) + 10 \log$ $(20/r_h)$
High rise $h_m = 1.6$ m	Streets perpendicular to the LOS street	$L_p = 135.41 + 12.49 \log f_c - 4.99 \log h_b + [46.84 - 2.34 \log h_b] \log d$
	Streets parallel to the LOS streets	$L_p = 143.21 + 29.74 \log f_c - 0.99 \log h_b + [411.23 + 3.72 \log h_b] \log d$
Low rise + high rise	LOS	$L_p = 81.14 + 39.40 \log f_c - 0.09 \log h_b + [15.80 - 5.73 \log h_b] \log d$, for $d < d_{bk}$
		$L_p = [48.38 - 32.1 \log d] + 45.7 \log f_c - (25.34 - 13.9 \log d) \log h_b + [32.10 + 13.90 \log h_b] \log d + 20 \log (1.6/h_m)$, for $d > d_{bk}$

The shapes of micro-cells are also no longer circular (or close to circular) since they are deployed in streets in urban areas where tall buildings create *urban canyons*. There is little or no propagation of signals through buildings and the shape of a micro-cell is more like a cross or a rectangle depending on the placement of base station antennas at the intersection of streets or in between intersections. The propagation characteristics are quite complex with propagation of signals affected by reflection from buildings and the ground, scattering from nearby vehicles, and for obstructed paths, diffraction around building corners and rooftops. Many individual scenarios should be considered unlike radio propagation in macro-cells.

Bertoni and others [Har99] have developed empirical path loss models based on signal strength measurements in the San Francisco Bay Area that are similar to the Okumura–Hata's models for a variety of situations. The corresponding path loss models are summarized in Table 11.2.

The various quantities in Table 11.2 are explained below. As usual, d is the distance between the mobile terminal and the transmitter in kilometers, h_b is the height of the base station in meters, h_m is the height of the mobile terminal antenna from the ground in meters, and f_c is now the center frequency of the carrier in GHz that can range between 0.9 and 2.0 GHz. In addition, the following parameters are defined. The distance of the mobile terminal from the last rooftop (in meters) is denoted by r_h. A rooftop acts as a diffracting screen (see Figure 11.20) and the distance from the closest such rooftop (around 250 m in many cases) becomes important in non-LOS situations and introduces a correction factor. The height of the nearest building above the height of the receiver antenna is denoted by Δh_m and introduces a correction factor similar to r_h. The average building height in the environment is an important parameter in micro-cellular environments. The relative height of the base station transmitter compared with the average height of buildings is denoted by Δh. Usually Δh ranges between –6 and 8 m. In LOS situations, it is observed that there are two distinct slopes of the path loss curves, one in the near-end region and one in the far-out segment. A *breakpoint* distance d_{bk} is used to separate the two piecewise linear fits to the measured path loss (see Section 2.2.5 for a breakpoint model for WLANs). The breakpoint distance is dependent on the heights of the base station and mobile antennas as well as the wavelength λ of the carrier (all in m) and is given by $d_{bk} = 4 \, h_b h_m/1000\lambda$.

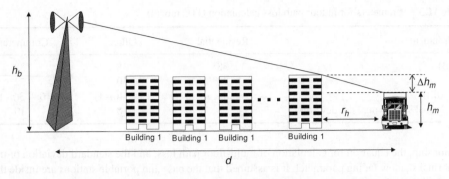

Figure 11.20   Geometry in a micro-cell; definitions of r_h and Δh_m.

Example 11.13: Path Loss Calculation in a Micro-cell

Determine the path loss between the BS and MS of a 1.8 GHz PCS system operating in a high-rise urban area. The MS is located in a perpendicular street to the location of the BS. The distances of the BS and MS to the corner of the street are 20 and 30 m, respectively. The base station height is 20 m.

Solution: The distance of the mobile from the base station is $(20^2 + 30^2)^{1/2} = 36.05$ m. Using the appropriate equation from Table 11.2, we can write the path loss as:

$$L_p = 135.41 + 12.49 \log f_c - 4.99 \log h_b + [46.84 - 2.34 \log h_b] \log d = 68.89 \text{ dB}$$

In addition to the empirical models presented above, there are theoretical models [Ber94] that predict the path loss in micro-cellular environments that have been adopted by a variety of standards bodies. Another model available for the micro-cellular environments is the JTC model that is explained in [Pah94]. This model provides for PCS microcells in a manner similar to the COST model.

In the models for the outdoor areas the relationship between the path loss and the height of the antenna is linear. As a result, the path-loss in dB, used in equations for calculation of the path-loss for macro- and micro-cells, is related to base station antenna height in logarithmic form. Results of measurements in [Ber94] do not agree with this assumption, when we are considering indoor propagation in a multi-floor setting. In one case, there is a theoretical explanation indicating that diffraction out of windows becomes significant as the number of intervening floors increases.

An improvement to the linear relation between path-loss and distance is to include a non-linear function of the number of floors in the path loss model as follows:

$$L_p = A + L_f(n) + B \log d + X \tag{11.1}$$

Here $L_f(n)$ represents the function relating the power loss with the number of floors n, and X is a lognormal-distributed random variable representing the shadow fading. This type of setting is commonly used in multi-floor building where a service provider installs an antenna on the roof-top or one of the middle floors to cover several floor of a building. Table 11.3 gives a set of suggested parameters in dB for the path loss calculation using Equation 11.1 at carrier frequencies of 1.8 GHz. The rows of the table provide the path loss in the first meter, the gradient of the distance-power

Table 11.3 Parameters for indoor path loss calculation (JTC model)

Environment	Residential	Office	Commercial
A (dB)	38	38	38
B	28	30	22
$L_f(n)$ (dB)	$4n$	$15 + 4(n-1)$	$6 + 3(n-1)$
Log normal shadowing (std. dev.; dB)	8	10	10

relationship, the equation for calculation of multi-floor path loss, and the standard deviation of the lognormal shadow fading parameter. It is assumed that the base and portable stations are inside the same building. The parameters are provided for three classes of indoor areas, residential, offices, and commercial buildings. This table is taken from an RF channel modeling for PCS applications described in [JTC94].

11.5.2 Models for Scattering Function of Cellular Networks

One of the earliest models for scattering function of the cellular networks was defined for the macro-cellular operation by the GSM group. This group defined a set of channel profiles with discrete delay power spectrums of different lengths for rural areas, urban areas, and hilly terrains [GSM91]. The Doppler spectrum choices for each path or tap of the model were either Rician or the classical Rayleigh. In a manner similar to the simulation of narrowband signals, the Doppler power spectrum for the classical Rayleigh model is:

$$D(\lambda) = \frac{1}{2\pi f_m} \times \left[1 - \left(\frac{\lambda}{f_m}\right)^2\right]^{-\frac{1}{2}} \quad \text{for} - f_m \leq \lambda \leq f_m \tag{11.2}$$

Here, f_m is the maximum Doppler frequency possible and is related to the velocity of the mobile terminal via the expression $f_m = v_m \times c/f$ where v_m is the mobile velocity and f is the center wavelength of the radio signal. Note here that we use λ as the Doppler frequency rather than the wavelength.

The Rician spectrum is the sum of the classical Doppler spectrum and one direct path, weighted so that the total multipath power is equal to that of a direct path alone:

$$D(\lambda) = \frac{0.41}{2\pi f_m} \left[1 - (\lambda/f_m)^2\right]^{-1/2} + 0.91\delta(\lambda - 0.7 f_m), \quad -f_m < \lambda < f_m, \tag{11.3}$$

To simulate the channel, the absolute power at each location is determined from the Okumura–Hata path loss model. An example of the delay power spectrum $Q(\tau)$ recommended by this group is given in Figure 2.22. A slightly different version of this model is recommended by COST-207 committee for simulation of the GSM channels [Cos86].

A more elaborate and comprehensive model is recommended by the JTC for simulation of radio propagation in different areas for 1900 MHz PCS bands [JTC94]. This recommendation includes parameters for both indoor and outdoor channels. The path loss model for this recommendation was discussed in Section 11.5.1. Here we discuss the multipath profile structures defined in [JTC94]. The general structure of the JTC model is the same as the GSM model, but the JTC model is more comprehensive. The JTC model divides the environments into one indoor and two outdoor classes.

Table 11.4 JTC parameters of the wideband multipath channel for indoor commercial buildings

	Channel A		Channel B		Channel C		
Tap	Rel[1] delay (ns)	Average power (dB)	Rel[1] delay (ns)	Average power (dB)	Rel[1] delay (ns)	Average power (dB)	Doppler spectrum $D(\lambda)$
1	0	0	0	0	0	0	Flat
2	100	−5.9	100	−0.2	200	−4.9	Flat
3	200	−14.6	200	−5.4	500	−3.8	Flat
4			400	−6.9	700	−1.8	Flat
5			500	−24.5	2100	−21.7	Flat
6			700	−29.7	2700	−11.5	Flat

The indoor areas are in turn divided into residential, office, and commercial areas. The outdoor areas include urban high-rise, urban/suburban low rise, and outdoor residential areas. Each class of outdoor areas is divided into other classes specified by the transmitter-antenna height with respect to the tops of buildings. Each tap is simulated in the same way as we described for GSM mdoel. The model defines two types of Doppler spectra – classical Rayleigh – and flat – for each tap of the discrete-time model. The classical spectra is similar to the Rayleigh spectra used in GSM model. The flat spectrum is used for simulation of the Doppler spectrum in indoor areas and is defined by:

$$D(f) = \frac{1}{2\pi f_m}, \quad -f_m < f < f_m \tag{11.4}$$

Because the multipath characteristics can be quite different in the same area from one radio link to another, this model suggests three different types of channel profiles for each environment, providing a wide variety of rms multipath delay spreads for each class of area. Tables 11.4–11.6 provide a list of the JTC wideband multipath channel models in indoor areas. Each table has three channel models A, B and C associated with good, medium, and bad conditions. In reference [Pah05] we have more details of this model and the complete description of all of JTC models are documented in [JTC94; Pah05].

Table 11.5 JTC parameters of the wideband multipath channel for indoor office buildings

	Channel A		Channel B		Channel C		
Tap	Rel[1] delay (ns)	Average power (dB)	Rel[1] delay (ns)	Average power (dB)	Rel[1] delay (ns)	Average power (dB)	Doppler spectrum $D(\lambda)$
1	0	0	0	0	0	0	Flat
2	100	−8.5	100	−3.6	200	−1.4	Flat
3			200	−11.2	500	−2.4	Flat
4			300	−10.8	700	−4.8	Flat
5			500	−18.0	1100	−1.0	Flat
6			700	−25.2	2400	−16.3	Flat

Table 11.6   JTC parameters of the wideband multipath channel for indoor residential buildings

	Channel A		Channel B		Channel C		
Tap	Rel[1] delay (ns)	Average power (dB)	Rel[1] delay (ns)	Average power (dB)	Rel[1] delay (ns)	Average power (dB)	Doppler spectrum $D(\lambda)$
1	0	0	0	0	0	0	Flat
2	100	−13.8	100	−6.0	100	−0.2	Flat
3			200	−11.9	200	−5.4	Flat
4			300	−111.9	400	−6.9	Flat
5					500	−24.5	Flat
6					600	−29.7	Flat

11.6   Transmission Techniques in TDMA Cellular

There are a number of considerations that enter into the choice of a modulation technique for use in a wireless application. These requirements are more restrictive in TDMA systems, where a single carrier without spreading the spectrum is used for transmission. In its broadest form, carrier modulation techniques can be divided into three categories of amplitude-, frequency-, and phase-modulation techniques. The TDMA digital cellular transmission techniques are designed with certain classes of power efficient amplifiers, which provide the highest power efficiency among the common types of power amplifiers. However, these amplifiers are highly nonlinear, and therefore it is necessary that the transmitted signal has an envelope that is close to constant. In addition for operation using multiple carriers they have to cope with extensive amplitude fluctuations caused by fading. As a result, amplitude modulation techniques are not desirable and frequency and phase digital modulation techniques have emerged as the traditional techniques in this industry. Radio modems designed for TDMA networks also need low side lobes to minimize interference to neighboring carrier frequencies. The adjacent-channel interference which is the interference that a transmitting radio presents to the user channels immediately above and below the transmitting user's channel is a major parameter in the design of TDMA cellular systems. The out of band interference of the mobile closer to the antenna is a serious source of interference for the mobile located in a farther distance. Therefore, when designing transmission systems for TDMA cellular networks special attention is paid to keep the side lobes of the transmitted signal in significantly low values. Another important issue in the design of a TDMA radio modem is sensitivity to multipath. Various modulation techniques have different degrees of resistance to multipath. This was a major issue in the development of the digital cellular and PCS TDMA standards, where it was necessary that each standard be written to accommodate the worst-case multipath conditions likely to be encountered by users over the entire geographical region of usage for that standard. Here we provide a description of the Gaussian Minimum Shift Keying (GMSK) modulation techniques, used in the GSM as an example of a transmission techniques used in a TDMA cellular network.

Frequency modulation is the predominant form of analog modulation used in the mobile radio industry. Digital frequency modulation is referred to as Frequency-Shift Keying (FSK) that forms a simple and popular method for wireless communications. Figure 11.21a shows the basic concept behind binary FSK modulation. The binary baseband data stream is encoded into two different frequencies before transmission in the channel. To implement this modulation in its simplest form, as shown in Figure 11.21b, one can input the binary data stream directly to a traditional analog FM transmitter and use an analog FM receiver to demodulate the signal at the receiver. In its ideal

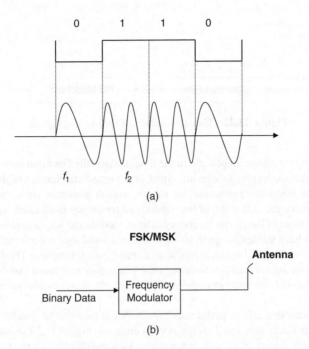

Figure 11.21 (a) Basic concept of FSK. (b) Implementation of FSK using a frequency modulated transceiver.

form an analog FM transmitter linearly maps the instantaneous amplitude of the message signal to a constant amplitude sinusoid with varying frequency at the output of the transmitter. A binary input signal takes only two levels of amplitude, so the output would be a constant envelope signal with two frequencies associated with the two different levels of signal. If the baseband input signal has multiple levels the output would be still a constant envelope signal with as many frequencies as the number of levels in the transmitted signal.

An important parameter in the design of FSK modems is the frequency spacing between the tones. This distance is representative of the occupied bandwidth of an FSK signal and to maintain optimal detection at the receiver it should take specific values that ensure orthogonality of the transmitted symbols. For non-coherent detection, when the receiver is not locked to the phase of the transmitted carrier, FSK modems use a minimum distance between the tones of $1/2T_s$ where T_s is the duration of the transmitted data symbols. For coherent demodulation, the distance between the tones can be reduced to $1/2T_s$ that is the minimum acceptable distance between the tones that ensures the orthogonality of the transmitted symbols. The FSK modulation with minimal tone distance of $1/2T_s$ is referred to as *Minimum Shift Keying* (MSK) that was a very popular transmission technique in wireless data communications.

To make an MSK signal even more attractive for radio communications, as shown in Figure 11.22, the transmitted baseband signal is filtered before frequency modulation to further reduce the side lobes. The most popular filters used for this implementation are Gaussian filters and associated modulation technique is referred to as Gaussian MSK or GMSK, which is one of the most popular modulation techniques in TDMA wireless networks. The transmitted signal at the output of the frequency modulator is still a constant envelope signal that avoids nonlinearities that

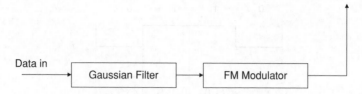

Data in

Figure 11.22   Gaussian minimum shift keying signals.

can be introduced by the power amplifiers. In the time domain, the Gaussian filter smoothens sharp transitions of the voltage levels. As a result, rather than immediate changes of the tone frequency at the output of the frequency modulator, we have a smooth transition from one tone frequency to another that reduces the side lobes of the transmitted frequency modulated signal. Since phase represents the derivative of the carrier frequency in time, modulation with a gradual transition of the frequency does not have sudden jumps in phase of the signal and they are referred to as *continuous phase modulation* techniques. Desirable modulation techniques for cellular TDMA radio channels are *constant-envelope continuous-phase-modulation* techniques that avoid non-linearity and have low side lobes. GMSK is an ideal example of a constant envelope continuous phase modulation technique.

An important factor that affects performance in GMSK is the time-bandwidth product $W \times T_s$. Here, W is the 3-dB bandwidth and T_s is the symbol duration. Figure 11.23 shows the spectrum of the GMSK signal for variety of $W \times T_s$, normalized 3-dB bandwidths, of the Gaussian filter. For $W \times T_s \to \infty$, the bandwidth of the Gaussian filter is infinity, (no filtering) and the system is indeed an MSK system. As the bandwidth of the filter becomes narrower, the power in the side lobes of the

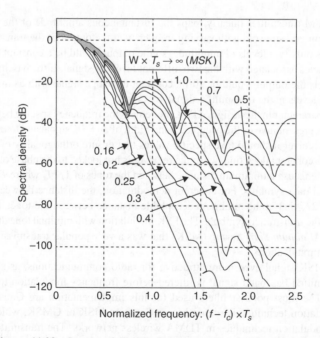

Figure 11.23   Spectra of GMSK signals for different $W \times T_s$ products.

transmitted signal and consequently adjacent channel interference reduces. In contrast, reduction of the bandwidth of the filter further smoothens the transition between levels in the time domain that increases the probability of erroneous detection at the receiver. The designer of the modem should decide on a compromise between the adjacent channel interference and detection error rate. The GSM TDMA standard designed preliminary for cellular telephone application recommends a $W \times T_s$ value of 0.25, other standards designed for mobile data applications have used higher values of up to 0.5.

For steady-signal reception in AWGN, the bit-error probability for GMSK was shown by [Mur81] to be:

$$P_b = \frac{1}{2}\mathrm{erfc}(\sqrt{\alpha \gamma_b})$$

where the parameter α is a constant related to the normalized bandwidth $W \times T_s$. Values of α were determined to be:

$$\alpha = \begin{cases} 0.68 \text{ for GMSK with } W \times T_s = 0.25 \\ 0.85 \text{ for simple MSK } (W \times T_s \to \infty) \end{cases}$$

From these values of α it can be seen that the performance of GMSK with $\alpha = 0.25$ degrades from that of MSK by about 1 dB, and thus with the $\alpha = 0.25$ of the GSM design. Figure 11.24 compares the probability of error for GMSK and QPSK modulation. Performance of GMSK is slightly worst than QPSK but it has an ideal constant envelop characteristics. Other TDMA networks have used

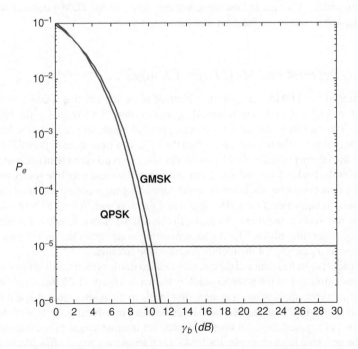

Figure 11.24 Probability of error versus signal to noise ratio per bit for GMSK ($a = 0.25$) and QPSK.

variations of QPSK with an intention to make their envelope to become closer to a constant envelope [Pah05].

Higher bandwidth efficiencies are achieved at the expense of more complex coherent implementations and lower uncoded bit error rate requirement. Data services often have more restrictions on the bit error rate than voice oriented networks. Cellular phones accept lower qualities than cordless telephones that were designed for wireline quality operation. The diversity in requirements combined with the availability of bandwidth for a particular service has created this diversity in modulation parameters for different standards.

11.7   Evolution of TDMA for Internet Access

The 2G TDMA technology that we describe in this chapter was originally designed in mid-1980s to replace analog cellular telephone and the focal point of the design was to provide a more reliable cellular telephone application with some supplementary services. In the mid-1990s, the Internet and data applications emerged as important methods of communication and wireless data communications started to attract attention resulting in emergence of a number of technologies in that area. These technologies were divided into technologies using their own infrastructure, technologies using cellular infrastructure but their own air–interface and the technologies that were integrated in the existing cellular infrastructure and air–interface [Pah94]. The latter systems finally evolved and were the first mobile data networks that gained wide popularity. These systems were building based on an existing digital cellular networks and they were capable of supporting packet switched data aiming at higher data rates for the wireless connection. Implementation of these networks needed changes in the air–interface as well as the wired infrastructure of the cellular network. In this section we explain how these systems emerged for TDMA cellular networks and what the fundamental change in evolution of these systems was.

11.7.1   Architectural and MAC Layer Changes

Networks designed for TDMA access to the Internet reuse the existing TDMA infrastructures, similar to Figures 11.1 and 11.2, which were designed to connect a terminal to the PSTN through BSS and BSC. The packets of data using the same physical bursts will arrive to the BSS and BSC but they have to redirect to the Internet rather than the MSC and from there to the PSTN. To redirect the packets to the internet, usually TDMA networks use two hardware elements, a sniffer to take away and deliver the packet switched data from and to the BSC and a mobile router that direct the packets to the Internet and from the Internet, while supporting the mobile operation (Figure 11.25). The mobile router is equivalent to the MSC that was a gateway switch for PSTN that could handle mobility of the terminals connected to its ports. In the cellular network industry sometimes these nodes are called supporting nodes. The sniffer supporting node serves the packet data service and the mobile router is a gateway for the mobile access to the Internet.

Mixing the packet switched data and the circuit switched traffic is then handled through a protocol layer. The overall structure of the protocol stack is shown in Figure 11.26 that shows how packets flow from higher layers, applications and signaling levels to the convergence, logical link control, radio link control or MAC, and the physical layer actually transferring the bits of data. A more detailed structure of a typical protocol stack to implement Internet access over a traditional cellular network can be provided in an example. Each MS has a temporary logical link identity to identify itself in the logical link control header.

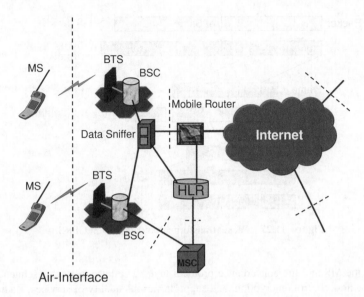

Figure 11.25 System architecture for Internet access using a circuit-switched cellular telephone infrastructure.

Example 11.14: GPRS and TDMA Internet Access

The packet transformation data flow in the MS for a General Packet Radio System (GPRS) system designed for Internet access through GSM infrastructure is shown in Figure 11.27. The mobile terminal uses the same physical radio channels as GSM for GPRS that supports IP packets at the network layer to be used by end-to-end applications. The convergence protocol supports a variety of network protocols that include IP packets used in the Internet. This layer multiplexes and de-multiplexes the network layer payload. The logical link control layer forms a logical

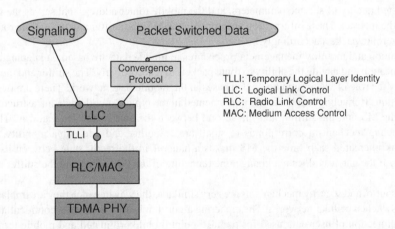

Figure 11.26 Packet switching protocol structure for a circuit switched cellular networks.

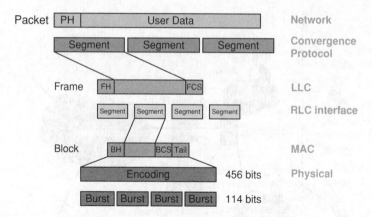

Figure 11.27   Packet transformation data flow in GPRS.

link between the MS and the convergence protocol layer. The layer performs sequence control, error recovery, flow control, encryption, and supports various quality of services. Radio link and medium access control prepares the data for the TDMA hierarchy. The 456-bit physical layer packets are finally encoded into 114-bits burst that are transmitted in the air the same way as a GSM burst.

In a manner similar to circuit-switched cellular telephone networks connected to the PSTN, there are mechanisms in packet switched mobile data networks to support mobility. These mechanisms provide for attachment procedure, location update and handoff, and power control and security of the connection. Before accessing mobile data services, the MS must register with the network and become "known" to the packet switched data network. The MS performs an attachment procedure with a mobile router (called a serving GPRS support node or SGSN) that includes authentication. The MS is allocated a temporary logical link identity by the mobile router and a packet data protocol context is created for the MS. This protocol is a set of parameters created for each session and contains the version of IP connection (IP-v4 or IP-v6), the mobile address assigned to the MS, the requested quality of service parameters, and the mobile router address that serves the point of access to the Internet. These information are stored in the MS, the sniffer and the mobile router and are used to route packets accordingly.

The location and mobility management procedures in mobile data are based on keeping track of the MSs location and having the ability to route packets to it accordingly. The sniffer and the mobile router play the role of visitor and home databases in the mobile data network. There are two levels of connections (tunneling mechanisms) implemented in the typical mobile data infrastructure, one between the MS and the sniffer, and the second between the sniffer and the gateway. The two-level tunneling mechanism corresponds to small area mobility and wide area mobility. A new logical link is created each time the MS makes a handoff in the ready state between itself and the sniffer. If the gateway does not change, the tunneling of the packet beyond the sniffer remains the same.

Power control and security mechanisms are very similar to the way in which they are implemented in circuit switched cellular networks. The ciphering algorithm is used to provide confidentiality and integrity protection of user data used for point-to-point mobile originated and mobile terminated data transmission and point to multipoint group mobile terminated data transmission. The algorithm is restricted to the MS and sniffer connection encryption. Data applications are not real time voice and often it is useful to put the terminal in sleep until data is available for the terminal. This approach saves the transmission power.

11.7.2   Data Rate in TDMA Packet Switched Networks

In general telephone application requires a real-time balanced two-way communications with a relatively low data rate. Designers of the TDMA wireless cellular networks have designed their network focused on this application, trying to maximize the number of voice users. In the GSM TDMA system, for example, we have eight users sharing a 200 KHz bandwidth and a link with a "raw" data rate of around 270 kbps. For every 114 bits of data transmitted over the air with a normal burst, as shown in Figure 11.11, we have 26 bits of training sequence, six tail bits and 8.25 bits for the time gap. In addition a portion of the packet burst are used for control and signaling. Effectively, if we consider these overheads for each of the eight users we have 456 bits every 20 ms that form a 22.8 kbps maximum rate for a user operating in one of the TDMA slots. However, parts of these 456 bits are usually used for coding that reduces this data rate to 13 kbps for voice or up to 9600 bps data rates for circuit switched data. For the bursts of packet switched data applications, the most important feature is the data rate. The higher the data rate the faster the transfer of information and the more desirable the transmission technique. To increase the data rate as we discussed earlier, one may assign additional slots to a single user. If we assign all eight slots to a user the maximum data rate for the GPRS would be $8 \times 22.8 = 184.4$ kbps. However, parts of this data should be used for coding and other purposes and we can never achieve this rate. Table 11.7 shows a set of data rates obtained for GPRS packet switched data for different number of slots and different coding schemes. Note that the maximum data rate possible is roughly 172 kbps when there is minimal coding and all the slots are allocated to one user, and this is 12.4kbps smaller than the value of 184.4 kbps. We have avoided details of these coding techniques and what is the number and type of bits transmitted in a slot, but the reader can imagine how they happen by considering other coding techniques that we have used previously to support different data rates.

To further increase the data rate one may think of increasing the data rate in the same band by increasing the number of bits per symbol in more complex modulation techniques techniques. For example, in the GSM and GPRS the transmission techniques is GMSK that sends binary data with one bit per symbol if we modify this modulation to 8-PSK we can send three bits per symbol to increase our bound for achieving data rate from 184.4 to 553.2 kbps. In these situations we can have the maximum data rate when the user is closer to the base station and receives higher signal strength. As we go further away from the base station we may fall back to GMSK and its associated data rates. Implementation of this concept on the GSM TDMA system is called the *Enhanced Data for Global Evolution* (EDGE). Table 11.8 shows sample formats for a typical EDGE implementation achieved by different number of slots and modulation techniques. The details of the coding for the GPRS

Table 11.7   Samples of GPRS data rates (in kbps) for different channel-coding schemes

Approximate channel coding rate	One slot	Four slots	Eight slots
1/2	9.2	36.8	73.6
2/3	13.55	54.2	108.4
3/4	15.75	63	126
1	21.55	86.2	172.4

Table 11.8   Sample EDGE data rates (kbps) using different modulation and channel coding schemes

Approximate channel coding rate	Modulation scheme	One slot	Four slots	Eight slots
1/2	GMSK	8.8	35.2	70.4
1	GMSK	17.6	70.4	140.8
1/3	8-PSK	22.4	89.6	179.2
1	8-PSK	59.2	236.8	473.6

and EDGE systems are slightly different, therefore the data rates with the same modulation and the same number of slots does not match exactly. The details of different data rates and performance of EDGE is available in [Yal02].

The data rates achieved by TDMA networks are bounded by the bandwidth of the carrier. Modems used for TDMA do not spread the spectrum and consequently they have narrower transmission bandwidths as they are compared with other modems designed with spreading their bandwidth used in networks with CDMA and OFDM transmission technologies. Higher bandwidth per carrier allows implementation of higher data rates. Increasing the highest achievable data rate for packet switched data applications and Internet access was the motivation for moving to 3G CDMA and 4G OFDM transmission techniques for cellular networks.

Questions

1. What are the differences between a mobile digital telephone and the public switched telephone system?
2. Name the three subsystems in the GSM architecture.
3. Name the three types of services provided by GSM.
4. What is the importance of the hierarchical framing structure in GSM?
5. What are the data services provided by GSM?
6. What are the incentives for power control in a TDMA network? Name the elements of the GSM system that are involved in handling power control.
7. What are VLR and HLR, where are they physically located and why do we need them?
8. What is the difference between registration and call establishment?
9. What are the reasons to perform handoff?
10. What is the difference between network controlled and mobile assisted handoffs?
11. What is the difference between a logical and physical channel?
12. Name five most important logical channels in GSM?
13. How does GSM convert 456 bits of the speech, data, or control signal into a normal burst of 156.25 bits?
14. How are higher data rates provided to a user in GPRS? How are they provided in EDGE? What are the differences?
15. What additional elements are needed as part of a GPRS or EDGE network in comparison to GSM?

Problem 11.1

a. Using the bit and time durations in Figure 11.13, show that speech coding rate for GSM is 13 kbps and the effective transmission rate to support one 13 kbps coded voice channel is 22.8 kbps.

b. What is the required transmission bandwidth for eight slots of the GSM system?
c. Give the overall overhead rate of the system that is the difference between the required transmission rate for the traffic and the actual transmission rate of the GSM.
d. Determine the efficiency of the system that is the ratio of the overhead over raw transmission
   rate.

Problem 11.2

a. Consider the multiframe transmission in GSM depicted in Figure 11.16. Use the overall structure
   of the multiframe, frame, and slot to show that the transmission rate of the GSM is indeed
   270.833 kbps.
b. In each GSM multiframe 24 frames are used for traffic and two for associated control signaling.
   Considering the detailed burst frame and multiframe infrastructure show that the effective
   transmission rate for each GSM voice traffic is 22.8 kpbs.
c. The slow association control channel uses 114 bits of one slot of each 26 slot traffic multiframe.
   What is the transmission rate for this channel in bps?

Problem 11.3

The stand-alone dedicated control channel (SDCCH) uses four time slots in each 51-control multiframe shown in Figure 11.16. Use the superframe timing to determine the effective data rate of
this logical channel.

Problem 11.4

Considering Figure 11.14, calculate the net data rate (data plus signaling) and the effective transmission rate of a 9600 bps GSM data service.

Problem 11.5

a. Considering the frequency allocation strategy of Figure 11.10 for the GSM systems, give the total
   number of traffic channels per 50 MHz of bandwidth used for two-way GSM communications.
b. Give the total number of GSM channels per MHz of bandwidth.
c. Give the number of channels per cell for frequency reuse factors of $N = 4$ and $N = 3$.

Problem 11.6

Repeat Problem 11.5 for the North American TDMA system assuming that this system replaces an
AMPS system with 395 traffic channels and a frequency reuse factor of $N = 7$.

Problem 11.7

a. What is the allowable power ramping time for GSM receivers? *Hint*: the time gaps of normal,
   frequency correction and synchronization bursts, shown in Figure 11.11, are designed to allow
   power ramping.

b. The time gap of the random access burst, shown in Figure 11.11, is designed to assure this packet does not collide with the normal bursts. What is the maximum coverage, the distance between the BS and MS, of a GSM base station? Assume that this gap is reserved for two-way travel and radio waves travel at 300 000 km/s.

c. The length of the synchronization sequence in a synchronization burst is designed to allow time advance for two-way bit synchronization. Use this parameter to calculate the maximum coverage of the GSM. Compare your results with that of part (b).

12

CDMA Cellular Systems

12.1 Introduction

In the cellular telephone industry, *code division multiple access* (CDMA) primarily refers to an air–interface and access technique that is based on direct sequence spread spectrum based transmission techniques described in Chapter 3[1]. Error control coding and spreading of the spectrum, soft handoffs, and strict power control play a very important role in the design and operation of CDMA based systems. While the air–interface is significantly different in the case of CDMA, compared with TDMA techniques, the core (fixed) network infrastructure that supports the wireless interface is very similar to the architecture of the Global System of Mobile Communications (GSM) core network. Although separate standards were employed in the 2G CDMA systems in the United States, the functionalities of the network elements were similar to those of GSM.

In the previous chapter, we discussed TDMA based cellular telephone standards in some detail with the primary example being GSM. The GSM standard came about in Europe as a result of initiatives by the European Telecommunications Standards Institute towards a unified digital cellular system. Although the original goals of GSM could have been met simply by defining a new air–interface, the group went beyond just the air–interface and defined a complete system that complied with wired ISDN-like services and other emerging fixed network features. To this end, the committee also defined a number of other interfaces between the hardware and software elements of the network making GSM a complete digital cellular standard.

Unlike GSM in the European Union, standards activities in the United States are based on developed or mature technologies and have considerable input from the industry. The Telecommunications Industry Association (TIA), or the T1P1 Committee of the Alliance for Telecommunications Industry Solutions (ATIS), develops North American wireless standards. The so-called *Interim Standards* (IS) developed by the TIA form the basis for deploying cellular systems till they are formally specified as TIA or ITU standards. The Advanced Mobile Phone Service (AMPS) was the predominant analog cellular service in the 1980s in the United States. While AMPS specified the air–interface, very little standardization was available in the backbone infrastructure leading to proprietary implementations and lack of interoperability. Roaming across system boundaries was very complicated requiring subscriber intervention. To solve these problems, the Telecommunications Industry Association worked on the IS-41 standard that specifies an open communications interface

---

[1] Bluetooth technology uses CDMA with frequency hopping spread spectrum and UWB technology uses CDMA with pulse position modulation.

Principles of Wireless Access and Localization, First Edition. Kaveh Pahlavan and Prashant Krishnamurthy.
© 2013 John Wiley & Sons, Ltd. Published 2013 by John Wiley & Sons, Ltd.

between two AMPS systems. Digital 2G cellular services evolved in two different directions for the air–interface – time division multiple access (TDMA) and the IS-136 standard and code division multiple access (CDMA) and the IS-95 standard. Interoperability between these standards was impossible over the air–interface except via dual mode telephones that actually implements two separate radio systems and through coordinating the mobile terminal with the available wireless service in an area. The backbone infrastructure specified by IS-41 standard, however, evolved to support both the IS-136 and IS-95 standards. The IS-136 standard has been supplanted by GSM almost everywhere. We do not provide a treatment of the North American network reference model because of its close similarity to that of GSM, described in the previous chapter. For more details of the fixed infrastructure for CDMA and other North American systems the reader is referred to [Gar00; Goo97].

IS-95 is the North American 2G digital cellular standard that employs CDMA as the access method as well as the air–interface. This technology was developed by Qualcomm around 1990 and is also called by the brand name cdmaOne. We will refer to IS-95 and its variations as cdmaOne in what follows. In 1989, Qualcomm first demonstrated its technology and developed the common air–interface specifications in 1991. In 1993, the TIA published Qualcomms's common air–interface specifications as the interim standard IS-95. The cdmaOne standard has since evolved into the 3G cellular standard called cdma2000.

In the early 2000s, the Third Generation Partnership Project (3GPP) specified a 3G cellular standard also based on CDMA. This standard is commonly referred to as the Universal Mobile Telecommunication System or UMTS. The air–interface is popularly called *Wideband* CDMA or W-CDMA. The core network architecture for UMTS is again backwards compatible with that of GSM and includes the modified entities (i.e., the GPRS support nodes) that handle data traffic. The objective of UMTS is to support higher data rates that can support multimedia applications, provide a high spectral efficiency, and make as many of the interfaces standard as possible. While voice traffic continues to be the main source of revenue, packet data for Internet access, advanced messaging services like multimedia e-mail, and real-time multimedia for applications such as telemedicine and remote security were envisaged. The requirements include improved voice quality (wireline quality), data rates up to 384 kbps everywhere and 2 Mbps indoor, support for packet and circuit switched data services, seamless incorporation of existing 2G and satellite systems, seamless international roaming, and support for several simultaneous multimedia connections.

The main emphasis in this chapter is on the description of the CDMA air–interface specifications used in the 2G (based on cdmaOne) and 3G (based on UMTS) cellular standards. There are several similarities but some differences between these systems in terms of the CDMA air–interface. The CDMA air–interface is also the enabler of higher layer protocols since many of the functionalities of the control signaling is reduced to air–interface changes. This simplifies the *logical channels* employed in CDMA based systems, and also makes them quite different from TDMA based cellular systems. The term "CDMA" now implies more than simply the air–interface. In includes features such as soft handoff and strict power control that are essential for the operation of the network. The reader is referred to several books on CDMA systems (eg., [Gar00; Hol00]) for further reading.

12.2 Why CDMA?

Code division multiple access (CDMA) is both an access method and also an air–interface. The reasons for employing CDMA in cellular telephone systems span from the advantages it provides at the system level to the advantageous properties at the physical layer that make it robust under interference and fading. All of the CDMA systems employ some form of direct sequence spread spectrum and powerful error control codes. This means that the *bandwidth* of a carrier in CDMA systems is several times (often two orders of magnitude) larger than the data rate supported per user (or logical channel) on the carrier. The significance of CDMA cellular systems is that it is possible *to reuse frequencies in all cells* unlike traditional cellular telephony described in Chapter 5. In other words, user data such as voice from all users in all cells, and in most implementations, the control channel and signaling information are transmitted *at the same frequency at the same time*. This is possible by employing a variety of physical layer schemes, such as spread spectrum with processing gain and inherent resistance to multipath dispersion, RAKE receiver based diversity gain, powerful error correcting codes, variable rate voice coders that provide considerable gain due to reduction in interference from pauses in natural conversation, a relatively fast power control mechanism to minimize interference, and soft handoffs. All of these reduce the required energy per bit to interference/noise ratio (E_b/I) for proper operation. As the required E_b/I for proper operation drops, it becomes possible to employ the same frequencies in every cell without the problem of co-channel interference observed in TDMA and FDMA cellular systems.

The advantages of CDMA became evident in the first successful CDMA implementation, namely the cdmaOne standard. Upon deployment, this standard demonstrated an increase in system capacity compared to analog and TDMA systems and improved quality of voice by using a better voice coder. The quality of voice calls was further improved through the use of soft handoffs (described later in this chapter). The natural resistance to multipath and fading, low E_b/I requirement, and implementation of strict power control resulted in less power consumption (6–7 mW on average) that was about 10% of analog or TDMA phones. Moreover, as mentioned earlier, cdmaOne implementations did not require cumbersome frequency planning since all cells employed the same frequency at the same time (although some tuning for proper deployment was necessary). This resulted in CDMA also becoming the popular choice for 3G systems.

While CDMA does provide an innate flexibility for voice traffic, its disadvantages also lie in the large bandwidth requirements, the necessity for strict power control, and implementation complexity. It has fallen out of favor in 4G systems where data traffic is predominant. In such systems, orthogonal frequency division multiplexing, which offers flexibility in bandwidth allocation, is the transmission scheme of choice and this is discussed in Chapter 13.

12.3 CDMA Based Cellular Systems

Services in CDMA based cellular systems are similar in nature to those discussed in Chapter 11. The network and system architecture are also similar in nature to TMDA systems such as GSM. *Protocols* for radio resource management, mobility management, and security of the CDMA systems are implemented in a manner that is similar to TDMA systems. Yet, differences arise as a result of using CDMA as the air–interface as we will see later in this chapter. There are differences in terms of cell search, control signaling, handling power control, and especially employing soft handoffs. In this chapter we mostly address these differences and concentrate on the implementation of the air–interface. There are two CDMA systems that we use as examples – the IS-95 based 2G digital cellular standard (that is also the basis for cdma2000, which is a 3G standard) and the WCDMA based 3G standard that is employed in UMTS. These systems have different carrier bandwidths and different implementations of CDMA.

In UMTS, four traffic classes are defined based on the quality of service needs of the associated applications. These classes are conversational (e.g., real-time voice calls), streaming (e.g., recorded video), interactive (e.g., web browsing or chat), and background (e.g., downloading files or e-mail). The delay, error rate, and bandwidth requirements of these classes vary. For example, voice calls are able to tolerate greater error rates than file downloads, but need stringent delay limits. We do not describe these services in great detail in this chapter because of their similarities to those described in Chapter 11. The major difference is the recognition that traffic classes such as web browsing and file downloads are as important as traditional voice calls.

As described in Chapter 6, wireless network operations need mechanisms such as radio resource management, mobility management and security that support user terminals that may be moving. Implementation of these mechanisms in a TDMA cellular system was described in Chapter 11. From a functional standpoint, similar mechanisms are necessary in CDMA cellular systems. For example, it is necessary to page a mobile station when a call arrives. It is necessary also to hand a call off from one base station to another when the signal quality starts deteriorating.

In CDMA systems, the differences arise in how these mechanisms are implemented because direct sequence spread spectrum is employed as the physical layer transmission scheme. Cell search makes use of the autocorrelation properties of spreading sequences. Handoff is more complicated because of the near-far effect and results in a scheme called *soft handoff*. Power control is critical again because of the near-far effect. Frame sizes and messages are often tailored to match the timing embedded in the spreading sequences.

As we will see in the next sections, implementation of some of these mechanisms will be very different and will exploit the inherent features of direct sequence spread spectrum. We thus start with the physical layer first in this chapter and explain later how various mechanisms for supporting mobility are implemented in CDMA cellular systems.

12.4   Direct Sequence Spread Spectrum

The air–interface in CDMA systems is by far the most complex of all systems and it is not typically symmetrical on the forward and reverse channels unlike TDMA systems. The ways in which spreading the spectrum and error control coding are employed on the forward and reverse channels are often different. In the forward channel, transmissions originate at a single transmitter (the base station) and transmissions for all users are synchronized. It is thus possible to employ orthogonal spreading codes to minimize the interference between users. On the reverse channel, mobile terminals transmit whenever they have to and propagation delays are different. As the transmissions are not synchronized, the way spreading is employed may be different. One example is the use of the same orthogonal codes, but for *orthogonal modulation* to reduce the error rate. In the following sections, we look at the specifics of the forward and reverse channels.

In Chapter 3, we discussed the basics of spread spectrum and error control coding techniques. In this chapter we discuss these with the details of implementation in CDMA systems as examples. Depending on the system, the actual implementation of CDMA may be different. In direct sequence spread spectrum, the transmitted symbol is sliced into many *chips*, and the pattern of chips (e.g., whether they are positive or negative in a binary symbol) is called the spreading sequence. In the binary case, the spreading sequence can be written as a vector of $+1$s and -1s representing the polarity of the chips. The spreading sequences are chosen to have "good" autocorrelation and cross-correlation properties. We elaborate on what "good" means in Section 12.4.1.

Consider a simple case shown in Figure 12.1 where originally, the baseband pulse is a wide rectangular pulse of duration T seconds. A positive rectangular pulse represents '0' and a negative rectangular pulse represents '1'. The bandwidth of such a signal comprising of random bits is pro-portional to $1/T$. With direct sequence spread spectrum, instead of transmitting the wide rectangular pulse to represent a bit, a sequence of smaller (in time) rectangular pulses called chips, each of duration T_c is transmitted. The bandwidth of this signal is proportional to $1/T_c$. Some of these pulses

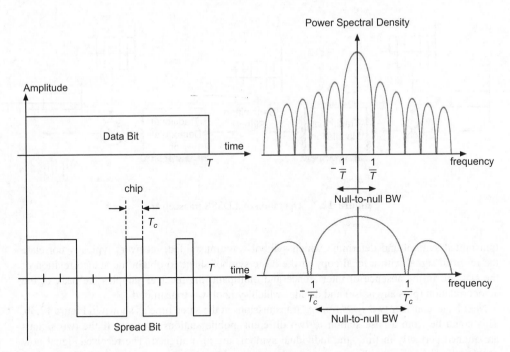

Figure 12.1 Wide rectangular baseband signal ands its PSD (top) and a direct sequence spread spectrum signal with 11 chips to the bit and its PSD – not to scale (bottom).

are of positive polarity and some are of negative polarity. In Figure 12.1, there are 11 chips in one bit and the sequence of smaller rectangular chips can be represented by the vector [1 1 1 –1 –1 –1 1 –1 –1 1 1 –1]. Thus, to transmit a '0', the sequence of pulses corresponding to [1 1 1 –1 –1 –1 1 –1 –1 1 1 –1] would be transmitted. To transmit a '1', the negative sequence [–1 –1 –1 1 1 1 –1 1 1 –1 1] would be transmitted. This particular sequence of 11 chips corresponds to the Barker sequence or Barker code that is employed in IEEE 802.11 for the 1 and 2 Mbps data rates (see Chapter 8).

The receiver needs to locally generate a copy of the spreading sequence and synchronize chips prior to decoding the data bits. When the chips are synchronized and the receiver multiplies the received signal with the locally generated spreading sequence, the wide rectangular pulse is regenerated. This is because a negative chip multiplied by a negative chip results in a positive value ($-1 \times -1 = 1$). This process can be accomplished using a correlator or matched filter receiver (Figure 12.2).

12.4.1   Receiver Processing with Direct Sequence Spread Spectrum

The autocorrelation and cross-correlation properties of spread spectrum sequences are important for their use in CDMA systems. We alluded to "good" autocorrelation and cross-correlation properties previously. We look at what "good" means in some more detail here. The objective here is not to be very mathematically rigorous, but to provide a conceptual idea as to why the autocorrelation and cross-correlation properties are important.

First let us consider a single transmission from a transmitter to a receiver as shown in Figure 12.3a. There is no multipath or interference. Symbols are transmitted every T seconds. To detect the

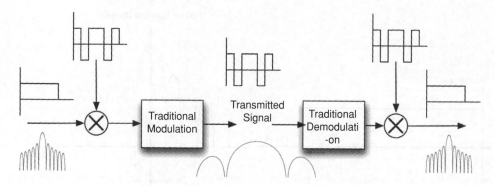

Figure 12.2   Operation of a DSSS transceiver.

transmitted symbol and determine which symbol was transmitted, receivers typically correlate the received signal with a local copy of the basic symbol shape to obtain the autocorrelation of the symbol. In the absence of interference and multipath, the receiver samples the peak of the autocorrelation (plus any noise) and decides which symbol was transmitted.

Next let us consider two *synchronized* transmissions at the same time as shown in Figure 12.3b. This could be from a base station to two different mobile stations or users. If the two signals are aligned perfectly in time, the individual symbols are also aligned. The received signal now comprises of the *sum* of the two signals. Let us consider the receiver of user 1. The receiver correlates this sum with the symbol shape. If the two signals are employing different spreading sequences, the result of the correlation is the *sum* of the *autocorrelation* of the sequence used by

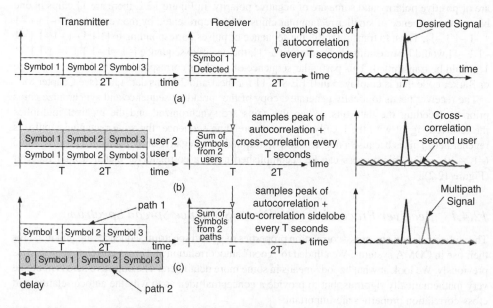

Figure 12.3   Transmission of symbols: (a) single transmission, (b) two synchronized transmissions, (c) multipath reception of a single transmission.

user 1 and the *cross-correlation* between user 1's sequence and user 2's sequence *when they are perfectly synchronized*. This cross-correlation is interference and ideally, we would like this to be zero. Unfortunately, it is not zero, but by employing orthogonal sequences for spreading and PN sequences for scrambling, the cross-correlation can be made very small.

Finally, consider multipath reception of the signal as shown in Figure 12.3c. Here, only two significant paths are shown. There is a noticeable 'delay' between the two significant paths that is not trivial compared to the symbol duration T. This corresponds to time dispersion of the channel discussed in Chapter 2. If the receiver tries to detect Symbol 1 arriving on the first path, there is interference from the signal that is arriving along the second path. The interference comprises of part of Symbol 0, shown simply as '0' in Figure 12.3c and part of Symbol 1. Since these symbols use the same spreading sequences, the interference from the second path manifests itself as the *autocorrelation sidelobe* of the spreading sequence of the user. Note from Figure 12.3c that the signal arriving along the second path manifests itself through an autocorrelation peak that is delayed compared to the peak created by the signal arriving along the first path. A RAKE receiver can exploit this second peak and this provides diversity as discussed later.

In summary, in CDMA systems, both multi-user interference and multi-path interference exist. Thus it is important to employ spreading sequences that: (a) have an almost zero cross-correlation between different sequences, (b) have an impulse-like autocorrelation with a narrow main-lobe and very low sidelobes. We explore these issues next.

12.4.2   Channelization using Orthogonal Sequences

Consider once again the situation shown in Figure 12.2b when there are multiple synchronized transmissions using direct sequence spread spectrum. If the pair-wise cross-correlation of the spreading sequences used by the multiple transmissions is zero, in essence, this is like separating the various transmissions and creating exclusive channels for each transmission. This is similar to the situation where are separated in frequency (as in FDMA) or time (as in TDMA). The difference here is that the separation is achieved using *spread-spectrum sequences or codes* and hence the term CDMA. When the cross-correlation between two synchronized codes is zero, the codes are said to be orthogonal. Orthogonal codes are also called channelization codes for this reason. The cross-correlation, when two signals $s(t)$ and $u(t)$ of duration T seconds are synchronized is given by:

$$R_{su} = \int\limits_0^T s(t)u(t)dt$$

When we consider the discrete form of the cross-correlation, we use *sequences* instead. Let us suppose that $\mathbf{S} = \{S_i\}$ and $\mathbf{U} = \{U_i\}$ are two sequences of length N whose i-th elements are written as S_i and U_i where. The cross-correlation of these sequences at zero lag (the signals are synchronized) is given by:

$$R_{su} = \sum_{i=0}^{N-1} S_i U_i$$

For orthogonal signals, built from orthogonal codes, the cross-correlation R_{su}, is zero. The cross-correlation for lags other than zero is given by:

$$R_{su}(k) = \sum_{i=1}^{N} S_i U_{i-k}$$

The index $[i–k]$ is calculated modulo N when the periodic correlation is computed. In the case of the aperiodic correlation, for indices outsider the range $[1, N]$, the elements are assumed to be zero.

Walsh Codes: The orthogonal codes that are commonly employed in CDMA systems are the Walsh codes, previously described in Chapter 3. The rows of the Hadamard matrix form Walsh codes such that each row in the matrix is orthogonal to every other row. It is possible to generate a Hadamard matrix recursively as explained in the example below.

Example 12.1: Recursive generation of Hadamard matrices and Walsh codes

The Hadamard matrix of order 1 is $\mathbf{H_1} = [0]$. Higher order Hadamard matrices can be obtained via the recursion $\mathbf{H_{2N}} = \begin{bmatrix} H_N & H_N \\ H_N & \overline{H_N} \end{bmatrix}$. Here the matrix $\overline{H_N}$ is the matrix H_N with all zeros and ones interchanged. It is easy to see that the Hadamard matrix of order 2 is $\mathbf{H_2} = \begin{bmatrix} 0 & 0 \\ 0 & 1 \end{bmatrix}$. Proceeding in this fashion, it is easy to generate H_{64} that is employed in the 2G CDMA systems. Each *row* of the Hadamard matrix corresponds to a Walsh code. Consider

$$H_8 = \begin{bmatrix} 0 & 0 & 0 & 0 & 0 & 0 & 0 & 0 \\ 0 & 1 & 0 & 1 & 0 & 1 & 0 & 1 \\ 0 & 0 & 1 & 1 & 0 & 0 & 1 & 1 \\ 0 & 1 & 1 & 0 & 0 & 1 & 1 & 0 \\ 0 & 0 & 0 & 0 & 1 & 1 & 1 & 1 \\ 0 & 1 & 0 & 1 & 1 & 0 & 1 & 0 \\ 0 & 0 & 1 & 1 & 1 & 1 & 0 & 0 \\ 0 & 1 & 1 & 0 & 1 & 0 & 0 & 1 \end{bmatrix}.$$

The first Walsh code from this matrix is [0 0 0 0 0 0 0 0], that is the all zero code. The last Walsh code is [0 1 1 0 1 0 0 1]. Note that all pairs of Walsh codes are orthogonal. In other words, with the above matrix, if the individual elements of any two rows are XOR-ed and then the resulting elements are all XOR-ed, the result is zero. The matrix can also be written in terms of –1s and 1s based on the polarity of the rectangular pulse representing the zero or one. In such a case, instead of an element-wise XOR operation to compute the cross-correlation, element-wise multiplication is performed followed by summing of the individual products as described previously.

Figure 12.4 shows two orthogonal signals derived from Walsh codes described in Example 12.1. The two signals $s(t)$ and $u(t)$ shown in this figure correspond to the Walsh codes in row 2 and row 4 in the matrix $\mathbf{H_{12}}$. The "0" is represented by a positive rectangular pulse and the "1" by a negative rectangular pulse. The product of the two signals comprises of two positive areas and two negative areas that cancel out when the integration is performed resulting in a zero cross-correlation. In discrete form, replacing the zeros by 1s and ones by –1s we get $\mathbf{S} = [1 –1 1 –1 1 –1 1 –1]$ and $\mathbf{U} = [1 –1 –1 1 –1 1 1 –1]$. Multiplying $\mathbf{S}$ and $\mathbf{U}$ element wise, we get $\mathbf{S.U} = [1 1 –1 –1 1 –1 –1 1 1]$. When we sum the elements in $\mathbf{S.U}$, we get zero, which is the cross-correlation, indicating that the two sequences are orthogonal. In a mathematical form:

$$R_{su} = \sum_{i=0}^{7} S_i U_i = 1 + 1 - 1 - 1 + 1 - 1 - 1 + 1 + 1 = 0$$

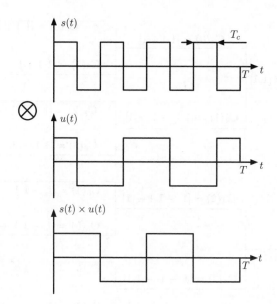

Figure 12.4 Zero cross-correlation between orthogonal signals.

Note that the data rate is fixed for a given length of the Walsh code and a fixed chip rate. As an example, consider a raw data rate $R_b = 19.2$ kbps and a Walsh code of length 64. The coded chip rate is $R_c = 64 \times 19.2 \times 10^3 = 1.2288 \times 10^6 = 1.2288$ Mcps. When the chip rate is 1.2288 Mcps and Walsh codes of length 64 are used, the data rate is always 19.2 kbps. Lower data rates are possible through *repetition* of the data symbols. If a mobile station needs higher data rates, it has to be allocated more Walsh codes. Each code-separated signal can carry 19.2 kbps and the overall data rate can be higher. This method of supporting different data rates is called the *multi-code* approach.

An alternative approach to increasing data rates is to *modify* the spreading factor or length of the spreading code. For instance, if the Walsh code had a length $N = 32$ chips, the data rate would be $1.2288 \times 10^6/32 = 312.4$ kbps. This approach where the spreading factor can be changed to support different data rates is called the *variable spreading factor* approach. However, Walsh codes with lengths smaller than 64 need not always be orthogonal to Walsh codes of longer lengths. To maintain orthogonality between codes of different lengths, a modified approach to selecting orthogonal codes has to be employed. This is discussed next.

OVSF Codes: The other orthogonal spreading code employed in CDMA systems is the Orthogonal Variable Spreading Factor or OVSF code. The OVSF codes are identical to the Walsh codes except that their usage is different. OVSF codes are created recursively using the rows of a lower dimensional matrix rather than the entire matrix itself. An example of how this is done for OVSF codes of up to length 8 is shown in Figure 12.5. For example, instead of using a 2×2 matrix to generate a 4×4 matrix as we did in Example 12.1, the row $C_2[0] = [1\ 1]$ is used to generate two codes of length 4, namely $C_4[0] = (C_2[0]\ C_2[0]) = [1\ 1\ 1\ 1]$ and $C_4[1] = (C_2[0]\ -C_2[0]) = [1\ 1\ -1\ -1]$. Similarly, $C_2[1] = [1\ -1]$ can be used to generate $C_4[2] = (C_2[1]\ C_2[1]) = [1\ -1\ 1\ -1]$ and $C_4[3] = (C_2[1]\ -C_2[1]) = [1\ -1\ -1\ 1]$. Note that $C_2[0]$, when it is repeated, is orthogonal to both $C_4[2]$ and $C_4[3]$, but it is *not orthogonal* to $C_4[0]$ and $C_4[1]$ which were its children.

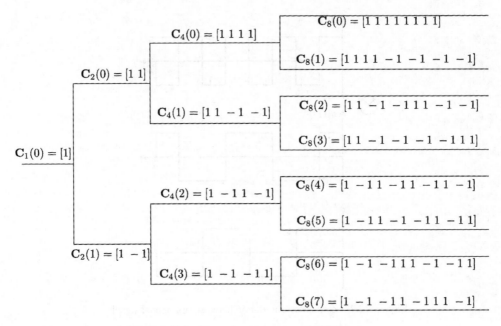

Figure 12.5   Recursive generation of OVSF codes.

The way OVSF codes are generated enables separation of codes that have non-zero partial correlations. When a variable spreading factor or processing gain is used (some signals have a higher/lower processing gain or there are more/less chips per data symbol compared to the normal case), it is still possible to maintain orthogonality of signals. In other words, OVSF codes maintain orthogonality between spreading codes of different lengths. Although every OVSF code cannot be always used, the tree shown in Figure 12.5 allows the system to assign codes to channels based on their data rates in a manner that ensures orthogonality between the channels or *block* the assignment of higher rate channels if it is not possible.

While orthogonal codes are very useful for channelization, when signals are not synchronized (as in the uplink) or in the presence of multipath delay (time dispersion), the cross-correlation between different signals can be excessive. Figure 12.6 shows the absolute value of the cross-correlation of two 64-chip Walsh codes. Notice that the cross-correlation is zero when the lag is 64 (the signals are synchronized). But a minor shift by one chip causes the cross-correlation to be as high as 44. Compared to the autocorrelation peak, which is 64, the cross-correlation value is very high. This could cause a significant increase in the bit error rate at the receiver.

To overcome this problem, after spreading a signal using an orthogonal code, the chips are further scrambled with what are called pseudo-noise or PN codes that are similar in nature to the Barker code used in IEEE 802.12. We discuss PN codes and their properties next.

12.4.3   Multipath Diversity with PN Sequences

PN sequences are typically employed in CDMA systems to reduce correlation sidelobes and also to exploit multipath diversity. The autocorrelation properties of PN sequences make them attractive

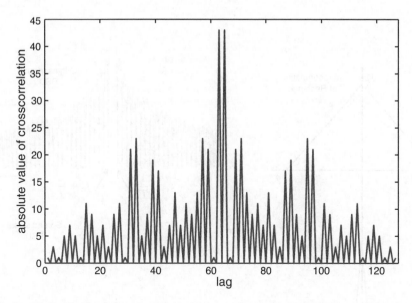

Figure 12.6   Non-zero cross-correlation between two 64-chip Walsh codes at various lags.

for mitigating the effects of multipath delay spread and time dispersion and actually exploiting it for diversity. The mathematical nature of autocorrelation is examined in Examples 12.2 and 12.3 below.

The autocorrelation of a rectangular pulse has a triangular shape as shown in Figure 12.7. This is reasonable for channels without multipath dispersion and inter-symbol interference where the receiver is able to sample the autocorrelation close to its peak value to determine whether a '0' was transmitted or a '1' was transmitted. In a multipath dispersive channel (see Chapter 2), the fat autocorrelation results in severe inter-symbol interference and irreducible error rates. Instead, by using direct sequence spread spectrum that makes use of spreading sequences with 'good' autocorrelation properties, multipath dispersion can be exploited.

As shown in Figure 12.7, the Barker sequence has an aperiodic autocorrelation that looks more like an impulse function. It has a narrow triangular peak of width $2T_c$ and small sidelobes, which cause some, but a much-reduced inter-symbol interference compared to the wider rectangular pulse. Since the basic transmission unit is a chip of duration T_c seconds, the bandwidth of the resulting signal is now spread and it is proportional to $1/T_c$ which is $N = T/T_c$ times larger than the original signal (see Figure 12.1). The value N is called the *processing gain* of the spread spectrum signal. Notice that the ratio of the autocorrelation peak to the sidelobes for the Barker sequence is also approximately equal to the processing gain N. Maximal-length sequences or M sequences (employed in 2G CDMA systems) also have good periodic autocorrelation properties as discussed in Example 12.3 and described later below.

Each resolvable significant multipath component in a time dispersive or frequency selective multipath channel produces a peak in the output at the receiver (but with a delay) as shown in Figure 12.3c. The resolvability of the multipath component depends on the chip duration T_c and components that have delays of at least T_c between them can be resolved. As described in Chapter 3, a RAKE receiver can be used to exploit the diversity in these multipath components. Figure 12.8 shows a block diagram of a RAKE receiver with three 'fingers'. This means that the receiver can

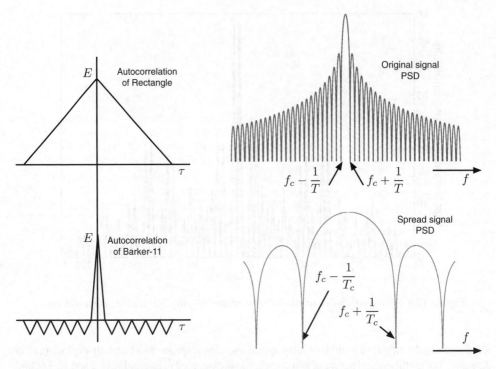

Figure 12.7   Autocorrelation and power spectral density of original baseband and spread signals.

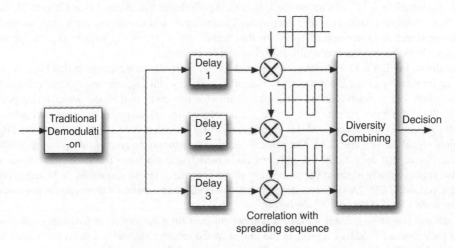

Figure 12.8   RAKE receiver with three fingers.

exploit up to three significant resolvable multipath components. If one of the peaks is suffering a deep fade, it is likely that another peak is not thereby improving the reliability of communications in a fading channel. Further, these peaks can be combined with appropriate diversity combining schemes for additional gains.

Maximal Length or M Sequences: PN sequences are typically generated using a linear feedback shift register as briefly described in Chapter 3. As shown in Figure 12.9, a linear feedback shift register consists of m shift registers with feedback connections. Feedback connections are labeled with their feedback coefficients c_i that are either zero or one. If $c_i = 0$, there is no feedback. An m-stage linear feedback shift register is represented through its characteristic polynomial, which identifies the feedback coefficients. For example, if $c(x) = 1 + x + x^4$, it is a four-stage linear feedback shift register with $c_0 = 1$, $c_1 = 1$, $c_2 = 0$, and $c_3 = 0$. Note that the degree of the characteristic polynomial is m. For certain specific feedback connections, the sequence or code generated by an m-stage linear feedback shift register has a maximum period of $2^m - 1$. Such sequences are called *maximal length* or *M sequences*.

M sequences are used for scrambling in 2G CDMA systems. M sequences can be generated using linear feedback shift registers that correspond to *primitive polynomials*, which are polynomials that play the role of prime numbers. There is an exhaustive literature on primitive polynomials of various degrees allowing the generation of M sequences of specific periodicity. M sequences are important because of their interesting autocorrelation properties.

Consider the linear feedback shift register with $m = 3$ stages shown in Figure 12.10. The characteristic polynomial for this linear feedback shift register is $1 + x + x^3$. Let the initial contents of the shift registers labeled A, B, and C be 1, 0, and 0 respectively as shown in the table in Figure 12.10. When the linear feedback shift register is clocked, the register contents are shifted forwards i.e., the bit in C moves into B, the bit in B moves into A and the bit in register A is shifted out (as the output). At the same time, the contents of registers A and B are XOR-ed and moved into register C. Thus the new *state* of the linear feedback shift register will be A $= 0$, B $= 0$, C $= 1$, as shown in the table. As the circuit gets clocked further, the state rotes through $2^3 - 1 = 7$ possibilities before returning to the original state of 0, 0, and 1 (reading the rows of the table from right to left). The sequence generated has a period of seven and

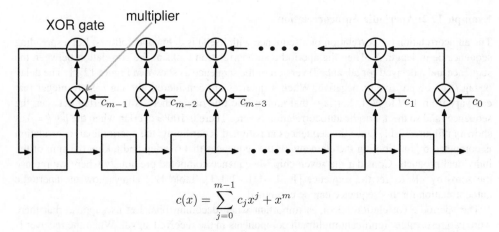

$$c(x) = \sum_{j=0}^{m-1} c_j x^j + x^m$$

Figure 12.9 An m-stage linear feedback shift register and its polynomial representation.

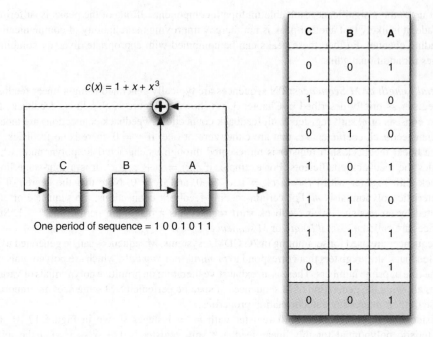

C	B	A
0	0	1
1	0	0
0	1	0
1	0	1
1	1	0
1	1	1
0	1	1
0	0	1

$c(x) = 1 + x + x^3$

One period of sequence = 1 0 0 1 0 1 1

Figure 12.10 M Sequence of length $N = 7$ generated from an $m =$ three-stage linear feedback shift register.

corresponds to 1 0 0 1 0 1 1. The repeating sequence looks like 1 0 0 1 0 1 1 1 0 0 1 0 1 1.... Depending on the initial state, the sequence may look different, but there will always be a run of three 1s, two 0s, and a single 0 and a single 1. This *run-length distribution* is one property of M sequences that results in good autocorrelation properties. We explore this in Examples 12.2 and 12.3 below.

Example 12.2: Aperiodic Autocorrelation

The autocorrelation (or correlation of a sequence with itself) is defined as follows. Let a spreading sequence be of length N. Then the aperiodic autocorrelation looks at the correlation between the sequence and a delayed (or advanced) version of the sequence as shown in Figure 12.11. The delay or lag l can be positive or negative. When sequences are considered, the lag l is an integer that can vary from $1-N$ to $N-1$. For lags that are outside this range, there is no overlap between the sequences and so the aperiodic autocorrelation is zero. There is 100% overlap when the lag $l = 0$ as shown in Figure 12.11. Once the sequences are aligned accordingly, the aperiodic autocorrelation is computed by multiplying the elements of the sequences that overlap and taking the sum of the individual products. Consider the seven chip M sequence computed previously where we replace the zeros by -1s to get the sequence $[1 \ -1 \ -1 \ 1 \ -1 \ 1 \ 1]$. Table 12.1 shows how the aperiodic autocorrelation for this sequence can be computed.

The aperiodic correlation becomes important when decoding bits. Let us suppose that there were two resolvable significant multipath components in the received signal. When the receiver is synchronized to the first path, the signal from the second path is interference.

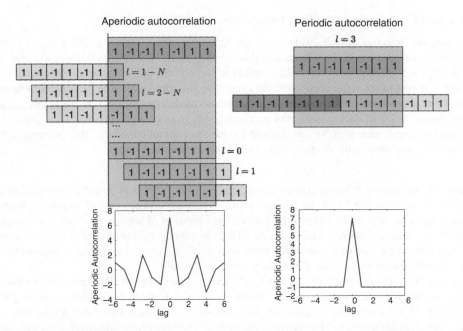

Figure 12.11 Aperiodic and periodic autocorrelation of an M sequence of length $N = 7$.

This interference can be split into two parts. The first part comes from the *previous bit* and second part comes from the *current bit* that is being decoded. Note that these two bits can have the same polarity or be different. Let the relative delay between the two paths be two chips or $2T_c$. The first part of the interference corresponds to the lag of $l = 2–N$ and the second part of the interference corresponds to the lag of $l = 2$. In Figure 12.7, these would be 0 and –1 respectively, for a total interference of –1, which adds to the desired signal value of 7 reducing it to 6.

Table 12.1 Computation of the aperiodic autocorrelation for the sequence [1 –1 –1 1 –1 1 1]

Lag	Overlapping sequences	Autocorrelation
$l = 1–N = 1–7 = –6$	[1 –1 –1 1 –1 1 1] and [1 0 0 0 0 0 0]	1
$l = 2–N = –5$	[1 –1 –1 1 –1 1 1] and [1 1 0 0 0 0 0]	0
$l = 3–N = –4$	[1 –1 –1 1 –1 1 1] and [–1 1 1 0 0 0 0]	–3
$l = 4–N = –3$	[1 –1 –1 1 –1 1 1] and [1 –1 1 1 1 0 0 0]	2
$l = 5–N = –2$	[1 –1 –1 1 –1 1 1] and [–1 1 1 –1 1 1 0 0]	–1
$l = 6–N = –1$	[1 –1 –1 1 –1 1 1] and [–1 –1 1 –1 1 1 0]	–2
$l = 0$	[1 –1 –1 1 –1 1 1] and [1 –1 –1 1 –1 1 1]	7
$l = N–6 = 1$	[1 –1 –1 1 –1 1 1] and [0 1 –1 –1 1 1 –1 1]	–2
$l= N–5 = 2$	[1 –1 –1 1 –1 1 1] and [0 0 1 –1 –1 1 –1]	–1
$l = N–4 = 3$	[1 –1 –1 1 –1 1 1] and [0 0 0 1 –1 –1 1]	2
$l = N–3 = 4$	[1 –1 –1 1 –1 1 1] and [0 0 0 0 1 –1 –1]	–3
$l = N–2 = 5$	[1 –1 –1 1 –1 1 1] and [0 0 0 0 0 1 –1]	0
$l = N–1 = 6$	[1 –1 –1 1 –1 1 1] and [0 0 0 0 0 0 1]	1

Example 12.3: Periodic or Cyclic Autocorrelation

The periodic autocorrelation considers the correlation between a sequence and a delayed and wrapped around version of the sequence as shown in Figure 12.11. In this case, there will always be N chips in both sequences when the correlation is computed unlike the aperiodic correlation where only part of the sequence is used for computing the correlation. The periodic correlation comes into play when the spread spectrum signal has no modulating data bits, as is the case in the pilot channel in 2G CDMA systems. For maximal length sequences, the periodic autocorrelation always has a peak value of N for zero lag and a value of -1 for all other lags. This property of M sequences is exploited in the pilot channel of cdmaOne as described later.

Other PN Sequences: While maximal length sequences have excellent periodic autocorrelation properties (see Example 12.3), the periodic *cross-correlation* between two M sequences of the same length may be poor for certain lags. Further, the number of M sequences of any given length is not large. In fact, as discussed below, in 2G CDMA systems, the same M sequences are used by all base stations but with offsets to differentiate between their signals. This however requires base stations to be synchronized using the global positioning system.

In UMTS, where base stations are not synchronized, other types of PN sequences are used. Gold sequences and Kasami Sequences, which are derived from M sequences are employed. There are $2^m + 1$ distinct Gold sequences with period $2^m - 1$ greatly *increasing* the number of such sequences compared to M sequences. All pairs of Gold sequences also have good periodic cross-correlation properties unlike pairs of M sequences, which may or may not have good cross-correlation properties. The periodic autocorrelation of a Gold sequence is not as good as that of an M sequence, but the other advantages make this a suitable spreading code for use in UMTS. Gold sequences are also used in the global positioning system (GPS). In GPS, Gold sequences of length 1023 ($m = 10$) are used. In UMTS, truncated Gold sequences of length 38 400 are employed, as described later in this chapter.

12.5   Communication Channels and Protocols in Example CDMA Systems

In this section, we consider the examples of a 2G CDMA system (IS-95 or cdmaOne) and a 3G CDMA system (UMTS) to examine how direct sequence spread spectrum and coding is employed in real cellular systems.

12.5.1   The 2G CDMA System

The Forward Channel: The forward channel is between the base station and the mobile station. The forward channel in cdmaOne occupies the same frequency spectrum as the 1G AMPS FDMA and 2G North American TDMA standards. Each carrier of cdmaOne occupies a 1.25 MHz of bandwidth, while carriers of AMPS and IS-136 each occupy 30 KHz of bandwidth. The cdmaOne forward channel consists of four types of logical channels – *pilot channel, synchronization channel, paging channel, and traffic channels*. As shown in Figure 12.12 each carrier contains a pilot, a synchronization, up to seven paging, and a number of traffic channels. These channels are separated from one another using different spreading codes. The modulation scheme employed for transmission of spread signal in the forward channel is QPSK.

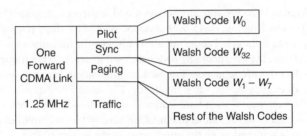

Figure 12.12 Forward channel in cdmaOne.

The fundamental format of the spreading procedure for all channels is shown in Figure 12.13. Any information contained in the form of *symbols* (after coding, interleaving, etc.) is spread (modulated) by 64-chip *Walsh Codes* that are obtained from the *Hadamard Matrices* discussed earlier and in Chapter 3. Each Walsh code identifies one of the possible 64 forward channels. Various Walsh codes are used for spreading various logical channels in cdmaOne. The Pilot channel employs the all zero Walsh code W_0. The synchronization channel is assigned the Walsh code W_{32} and so on. The assignment of some Walsh codes is shown in Figure 12.12.

After the channel symbols are spread using the orthogonal codes to create *channels*, they are further *scrambled* in the in-phase and quadrature phase lines by what are called the *short PN-spreading codes*. The same channel symbols are passed to both the in-phase and quadrature phase lines. This modulation approach is called *dual-BPSK*. This is different from true QPSK where half the channel symbols are passed to the in-phase line and half to the quadrature line. The PN spreading codes are not orthogonal as described previously, but possess excellent autocorrelation and cross-correlation properties to minimize interference across different cells or cell sectors and to

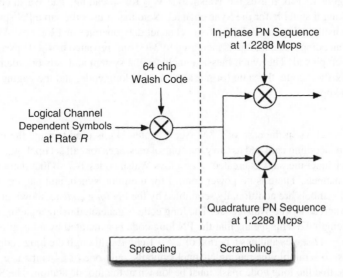

Figure 12.13 Basic spreading procedure on the Forward channel in cdmaOne.

exploit multipath diversity. The PN-spreading codes are M sequences generated by linear feedback shift registers (LFSRs) of length $m = 15$ with a period of 32 767 chips plus an extra zero for a total of 32 768 chips. The orthogonal codes are used to isolate the transmissions between different channels *within* a cell and the PN spreading codes are used to separate the transmissions between different cells. In effect the PN sequences are used to differentiate between several base stations in the area that are all employing the same frequency. The same PN sequence is used in all base stations, but The PN sequence of each base station is *offset* from those of other base stations by some value. For this reason, base stations in cdmaOne have to be synchronized on the downlink. Such synchronization is achieved using the global positioning system (GPS).

The period of the PN sequences are employed for timing and synchronization. The chip duration is 0.8138 μs. One period of the PN sequences lasts for $0.8138 \times 10^{-6} \times 32\ 786 = 26.67$ ms. There are 37.5 repetitions of the PN sequence per second or 75 repetitions in 2 s. The first bit of any frame typically starts at an even second so that it is possible to synchronize to a frame rapidly using the periods of the PN sequence. We consider the creation of the Pilot and Sync channels in Example 12.4 below.

Example 12.4: The Pilot and Sync Channels

The way the pilot channel is created is shown in Figure 12.14a. The pilot channel is intended to provide a reference signal for all mobile stations within a cell that provides the phase reference for coherent demodulation. It is about 4–6 dB stronger than all other channels. The pilot channel is used to lock onto all the other logical channels. It is also used for signal strength comparison. It uses the all zero Walsh code and contains no information except the RF carrier. It is also spread using the PN-spreading code to identify the base station. The way to identify the base station is to *offset* the PN sequence by some number of chips. In IS-95, the PN sequences are used with offsets of 64 chips that provide 512 possible spreading code offsets also providing for unique BS identification in dense microcellular areas.

The Sync channel is used to acquire initial time synchronization and the way in which it is formed is shown in Figure 12.14b. It uses the Walsh code W_{32} for spreading. The Walsh code 32 has 32 transitions making it suitable for the Sync channel. Note that it uses the same PN spreading codes for scrambling as the pilot channel. The Sync channel data operates at 1200 bps. After a rate 1/2 convolutional encoding, the data rate is increased to 2400 bps, repeated to 4800 bps and then block interleaving is employed. The synch message includes the system and network identification, the offset of the PN short code, the state (or mask) of the PN long code, and the paging channel data rate (4.8 or 9.6 kbps).

The paging channel, as in the case of GSM (see Chapter 11), is used to page the mobile station when there is an incoming call, and to carry the control messages for call set up. Figure 12.15 shows how a paging channel message is created. It employs Walsh codes 1–7 so that there may be up to seven paging channels. There is no power control for the pilot, synch, and paging channels. The paging channel symbols are additionally scrambled by the PN *long code* as shown in Figure 12.15 before they are spread by the Walsh code. The long code is generated using a paging channel long code mask of length 42. This means that the PN long code is generated by a linear feedback shift register with $m = 42$ stages and has a period of $2^{42} - 1$ chips. Although the long code is generated at 1.2288 Mcps, to scramble the paging channel symbols, only one of 64 chips is used through a decimator. Note that the long code mask must be known at the mobile station. The Sync channel, which would have been previously decoded by the mobile station, carries this information.

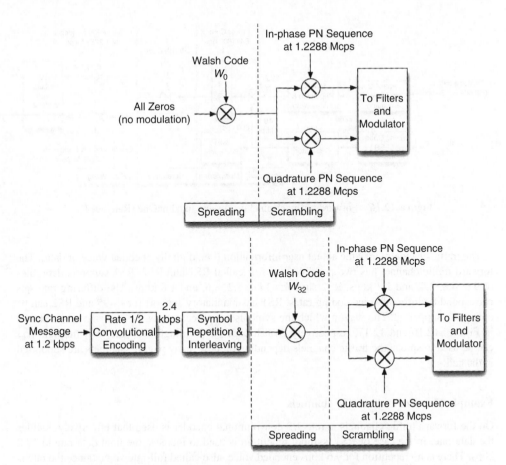

Figure 12.14 (a) Pilot and (b) Sync channel processing in cdmaOne.

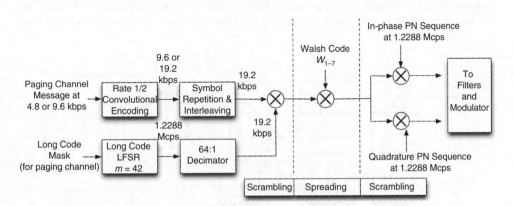

Figure 12.15 Paging channel processing in IS-95.

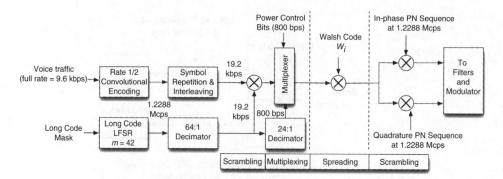

Figure 12.16 Forward traffic channel processing in cdmaOne (Rate Set 1).

The traffic channels carry the actual user information (i.e., digitally encoded voice or data). The forward traffic channel has two possible *rate sets* called RS1 and RS2. RS1 supports data rates of 9.6, 4.8, 2.4, and 1.2 kbps. RS2 supports 14.4, 7.2, 3.6, and 1.8 kbps. The differing rate sets correspond to different voice coding rates. RS1 has mandatory support for IS-95 and RS2 can be optionally supported. The way in which the symbols are processed for the two rate sets is shown in Figures 12.16 and 12.17, respectively. Walsh codes W_2 through W_{31} and W_{33} through W_{63} can be used to spread the traffic channels depending on how many paging channels are supported in the cell.

Example 12.5: Forward traffic channels

On the forward traffic channels, a rate 1/2 convolutional encoder is used that effectively doubles the data rate. In the case of RS1, symbol repetition is used to increase the final data rate to 19.2 kbps. There is no repetition for 9.6 kbps encoded voice, also called full-rate voice (since the rate is already 19.2 kbps after rate 1/2 convolutional encoding). There is repetition of four times for voice at 2.4 kbps. Lower rate voice traffic, such as that at 2.4 kbps, is created when the user is reducing his speaking or has no activity. In the case of RS2, the full-rate voice traffic is at 14.4 kbps. The final rate at the output of the symbol repeater is 212.8 kbps that is *punctured* by selecting only four

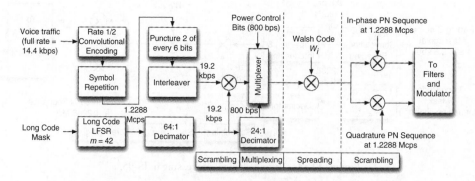

Figure 12.17 Forward traffic channel processing in cdmaOne (Rate Set 2).

out of every six bits (two bits are dropped). This reduces the data rate to 19.2 kbps at the input of the block interleaver. The forward traffic channels are multiplexed with power control information for the reverse link as shown in Figures 12.16 and 12.17. Power control bits are multiplexed with the scrambled voice bits at 800 bps. The locations at which the power control bits are inserted is determined by the long code. Note that the voice traffic is scrambled by the PN long code before it is spread by a Walsh code and the spread signal is further scrambled by the PN short codes for differentiating sectors/cells, for diversity, and to reduce interference among channels.

The Reverse Channel: The 2G CDMA reverse channel is fundamentally different from the forward channel. It employs offset QPSK rather than dual BPSK used in the forward channel. Offset QPSK is closer to a constant envelope modulation scheme with smaller phase discontinuities than plain QPSK. If the envelope of the modulation scheme is not constant and there are significant phase discontinuities, non-linear amplification tends to restore spectral sidelobes (previously suppressed through pulse shaping filters) that can cause adjacent channel interference. Thus, offset QPSK provides for a more power efficient implementation of the transmitter at the MS. The dual BPSK modulation is easier for demodulation at the MS and so it is employed on the forward channel. The overall structure of the logical reverse channels in cdmaOne is shown in Figure 12.18. There are only two types of channels: access channels and traffic channels.

Compared to the forward channel, there is no spreading of the data symbols using orthogonal codes in the reverse channel. Instead, the orthogonal codes are used for *waveform encoding*. This means that the reverse link employs an orthogonal modulation scheme that consumes more bandwidth than traditional modulation schemes like PSK or QAM, but reduces the error rate improving the performance of the system.

Example 12.6: Waveform encoding in 2G CDMA reverse link

As a simple example of waveform encoding, consider the example of the Hadamard matrix H_{12}. There are eight orthogonal Walsh codes. We can perform a mapping between inputs of three bits to one of eight waveforms as shown in Figure 12.19. This mapping is done arbitrarily in Figure 12.19.

A different mapping scheme is employed in IS-95. Consider the Walsh codes of length 64. There are 64 such codes and they are orthogonal to one another. If these codes are used as waveforms to represent a group of information bits, we can *encode* $\log_2 64 = 6$ bits using a Walsh code. For example, an input data stream 0 0 0 0 0 0 can be transmitted using the all zero Walsh code W_0.

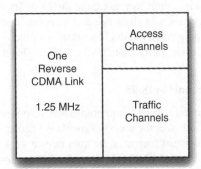

Figure 12.18 Reverse channel in cdmaOne.

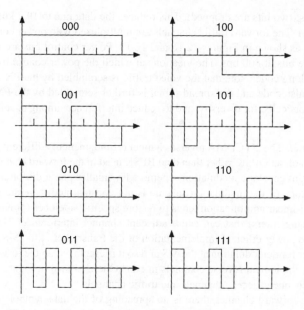

Figure 12.19 Mapping data bits to Walsh-encoded symbols.

This is something like a 64-ary modulation scheme where there are 64 symbols or alphabets for transmission. The receiver employs correlation to detect the alphabets. In IS-95, the Walsh code that is used for encoding is determined by the equation:

$$i = c_0 + 2c_1 + 4c_2 + 8c_3 + 16c_4 + 32c_5$$

where c_5 is the most recent bit. For instance, if the input six bits are (1 1 1 0 1 0), the Walsh code that will be selected is $i = 1 + 2 \times 1 + 4 \times 1 + 8 \times 0 + 16 \times 1 + 32 \times 0 = 23$; that is, W_{23} is transmitted.

As mentioned earlier, there are only two types of reverse channels in IS-95 – the access channels and the reverse traffic channels. The creation of the access channel is considered in Example 12.7. It carries the registration message, short messages, the dialed digits of a phone number to place a call, security related messages such as challenge and response, response to a paging message, and so on (see Chapter 6 for more details of the functionality of these messages). Mobile stations contend on the access channel using an Aloha-like scheme (see Chapter 4) to set up calls.

Example 12.7: The access channel in IS-95

The MS transmits control information such as call origination, response to a page, etc. to the BS via the access channels. The data rate over the access channels is fixed at 4800 bps. It is sent through a rate 1/3 convolutional encoder that increases the data rate to 14.4 kbps. Symbol repetition is employed to increase the data rate to 212.8 kbps. Every six bits is now mapped into 64 bits using the 64-ary orthogonal modulator. This process can be thought of as "spreading" the signal to a chip rate of $(64/6) \times 212.8 = 307.2$ kcps. The long PN code is used to distinguish between different

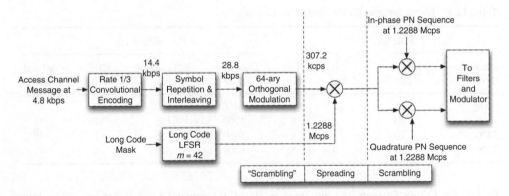

Figure 12.20 Access channel processing in cdmaOne.

access channels. It spreads each of the chips at the output of the 64-ary orthogonal modulator by a factor of four that yields a final chip rate of 1.288 Mcps. Details are shown in Figure 12.20.

The reverse traffic channel is shown in Figure 12.21. The data burst after coding and interleaving, but just before the 64-ary orthogonal modulation is at a rate of 212.8 kbps. The output of the 64-ary orthogonal modulator is again at $212.8 \times 64/6 = 307.2$ kcps. After spreading by the long PN code by a factor of four, the final chip rate is $307.2 \times 4 = 1.2288$ Mcps. A data randomizer is used in the fundamental code channel to mask out redundant data in case of symbol repetition. This process deletes so-called "power control groups" to reduce the transmit power and thus the interference on the reverse link. More about this masking out of redundant data to reduce the interference is discussed in the section on power control. The reverse traffic channel sends information related to the signal strength of the pilot and frame error rate statistics to the BS. It is also used to transmit control information to the BS such as a handoff completion message and a parameter response message. These control messages are sent in frames that may partially contain voice traffic (dim and burst) or no voice traffic (blank and burst).

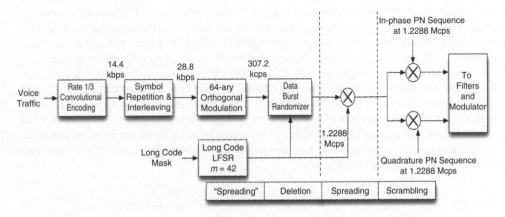

Figure 12.21 Reverse traffic channel processing in cdmaOne.

Table 12.2   Frame contents for forward traffic channels

Rate set 1				Rate set 2			
Data rate (bps)	Information bits	CRC bits	Tail bits	Data rate (bps)	Information bits	CRC bits	Tail and reserved
9600	172	12	8	14 400	267	12	9
4800	80	8	8	7200	125	10	9
2400	40	0	8	3600	55	8	9
1200	16	0	8	1800	21	6	9

Packet and Frame Formats: As discussed in above, the forward logical channels are of four types – the pilot, the sync, the paging, and the traffic channels. The reverse channels are either access channels or traffic channels. The forward traffic channel carries user data (either data bits or encoded voice) at 9600, 4800, 2400, or 1200 bps in RS1 and 14 400, 7200, 3600, or 1800 bps in RS2. The forward traffic channel frame is 20 ms long. Each 20 ms frame is further divided into 16 power control groups, each lasting for 1.25 ms. The granularity of power control depends on the size of these power control groups. As we have discussed previously, power control in 2G CDMA is performed at the rate of $1/1.25 \times 10^{-3} = 800$ times/s. Table 12.2 shows the number of information bits, frame error control check bits, and tail bits in each case.

The sync channel provides the MS information about the system identification (SID), the network ID (NID), PN short sequence offset, the PN long code state, and the system time among other things. Such *messages* can be long and are fragmented into *sync channel frames* of 32 bits shown in Figure 12.22a. Three of the sync channel frames are combined into a synch channel superframe of 96 bits. The sync channel frame lasts for 26.67 ms, which is exactly one period of the PN sequence. Thus, the sync channel frame is aligned to this period of the PN sequence. The superframe lasts for 80 ms. This embedded timing is necessary to set the states of the linear feedback shift registers in the mobile station. The "start of message" (SOM) bit is one for the first synch channel frame and zero for subsequent ones that belong to the same message. The message itself (shown in the top part) consists of the message length, the data, an error checking code, and some padding. Padding with zeros is used to ensure that every new message starts in a new superframe (Figure 12.22).

The paging channel, shown in Figure 12.22b announces a number of parameters to the MS that includes the traffic channel information, the temporary mobile subscriber identity, response to access requests, and list of neighboring base stations and their parameters. Paging can be slotted or unslotted. In the former case, which enables the MS to save on battery power, the channel is divided into 80 ms slots. The paging channel message is similar in structure to the sync channel message (it has a message length, data, CRC, etc.). Since it is too long for transmission in one slot, it is fragmented into 47 or 95 bits (data rate of 4800 or 9600 bps) and transmitted over a paging channel *half-frame* (10 ms long). The half-frame has one bit called the synchronization capsule indicator (SCI) that has functionality similar to the SOM bit. In this case however, a message can start anywhere (not necessarily in a half-frame) and a zero value for the SCI could indicate that one paging message ends and another starts within the same half-frame. Eight paging half-frames are combined into one paging slot of 80 ms.

Example 12.8: Number of bits in the paging channel half-frame and slot

How many bits are there in a paging channel half frame and slot?

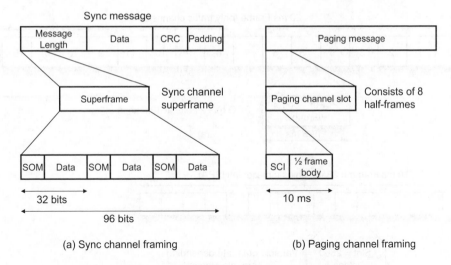

(a) Sync channel framing   (b) Paging channel framing

Figure 12.22   Framing in some cdmaOne forward channels.

Solution: The number of bits depends upon the data rate. If the data rate is 9600 bps, a 10 ms half-frame will carry 96 bits (one bit is the SCI) and 48 bits if the data rate is 4800 bps. Consequently, a paging slot, that has eight half-frames together, will contain $96 \times 8 = 768$ bits at 9600 bps and $48 \times 8 = 784$ bits at 4800 bps.

The access channel data rate is 4800 bps and each access channel message (very similar in structure to a sync message) is composed of several access channel frames lasting 20 ms. Thus an access channel frame is 96 bits long. An *access channel preamble* always precedes an access channel message and it consists of several 96-bit frames with all bits in the frame equal to zero. The actual message itself is fragmented into 96-bit frames that have 88 bits of data and 8 tail bits set to zero.

The reverse traffic channel is once again broken into 20 ms traffic channel frames. The frame is further divided into 1.25 ms power control groups (see Figure 12.23a). There are again 16 power control groups in one frame. A data burst randomizer randomly masks out (deletes) individual power control groups depending on the data rate to reduce the interference on the reverse channel. For instance, at 4.8 kbps (half the data rate), eight PCGs are masked (deleted) in each frame. In addition to voice traffic, the traffic channel can also be used to transfer signaling or secondary data. In the *blank and burst* case, the entire frame carries data. In the *dim and burst* case, part of the frame carries voice and part of it data. The frame structures for the reverse traffic channel are very similar to that of the forward traffic channel.

12.5.2   The 3G UMTS System

In UMTS, the base stations are called Node Bs and the mobile stations are called UEs (user equipment). The base station controller in GSM is replaced by a radio network controller or RNC (which behaves in a manner similar to the base station controller). The rest of the UMTS network resembles the GSM network described in Chapter 11 with the enhancements made with GPRS. The GPRS support nodes are still used for handling data traffic in UMTS. A simplified diagram of the UMTS architecture is shown in Figure 12.24.

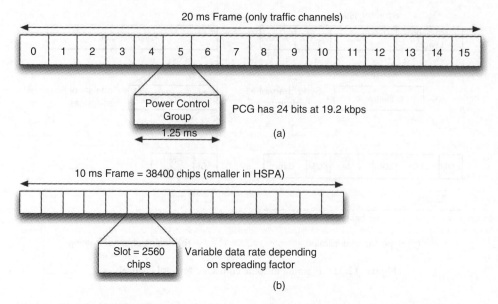

20 ms Frame (only traffic channels)

| 0 | 1 | 2 | 3 | 4 | 5 | 6 | 7 | 8 | 9 | 10 | 11 | 12 | 13 | 14 | 15 |

Power Control Group

PCG has 24 bits at 19.2 kbps

1.25 ms

(a)

10 ms Frame = 38400 chips (smaller in HSPA)

Slot = 2560 chips

Variable data rate depending on spreading factor

(b)

Figure 12.23   Power control groups (a) in cdmaOne and slots (b) in UMTS.

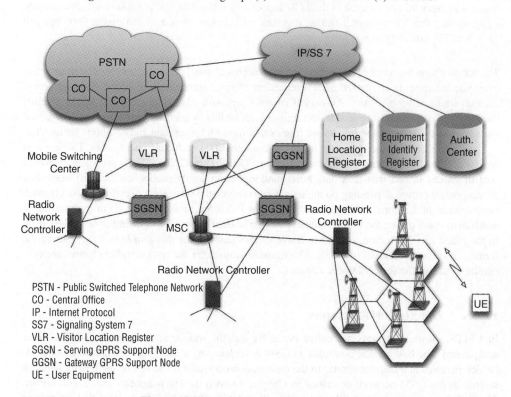

PSTN

IP/SS 7

CO

CO

CO

Home Location Register

Equipment Identify Register

Auth. Center

Mobile Switching Center

VLR

VLR

GGSN

Radio Network Controller

SGSN

MSC

SGSN

Radio Network Controller

Radio Network Controller

UE

PSTN - Public Switched Telephone Network
CO - Central Office
IP - Internet Protocol
SS7 - Signaling System 7
VLR - Visitor Location Register
SGSN - Serving GPRS Support Node
GGSN - Gateway GPRS Support Node
UE - User Equipment

Figure 12.24   Simplified architecture of UMTS.

The logical channels in UMTS are called *transport channels*. Dedicated transport channels are used to carry user data (e.g., voice calls) and call related signaling data (e.g., measurement reports and handoff commands) while common transport channels are used to carry general signaling and control information (e.g., paging messages and access messages). These logical channels are mapped to *physical channels* that are real signals generated by either the base station or the mobile station. A physical channel could carry exactly one transport channel and in some cases two transport channels. Some transport channels could also be split and carried over two physical channels. Thus there is additional complexity in the creation of channels in UMTS compared to cdmaOne, which we do not consider in detail in this book. However, the functionality of some of the transport channels is similar (e.g., there is a paging channel and an access channel which are used, respectively, to page the mobile and to set up calls).

The primary requirements of 3G systems are that they should be able to support a variety of application data rates (from 384 kbps in outdoor areas to 2 Mbps in indoor areas) and operation environments. This means that there must be support for quality of service and operation from megacells to picocells. It would be difficult to have all of these types of base stations to be synchronized as in the case of IS-95. In WCDMA, the base stations can operate in an asynchronous fashion that obviates the need of GPS availability to synchronize base stations. Consequently, PN sequences are employed differently in the case of UMTS.

The chip rate in UMTS is 3.84 Mcps that is much larger than the chip rate of 1.2288 Mcps in cdmaOne. The frequency carrier bandwidth is correspondingly larger and it is approximately 5 MHz. The chip duration is smaller enabling better resolution of multipath components in time dispersive channels. The orthogonal codes separate transmissions to different mobile stations on the downlink and different flows from the same mobile station on the uplink and actually spread the spectrum as in cdmaOne. PN codes only scramble the chips (do not increase the bandwidth) like cdmaOne and are employed to separate transmissions from different base stations or cell-sectors on the downlink and separate the signals from multiple mobile stations on the reverse link. True QPSK is the modulation scheme after direct sequence spreading. Different data symbols are carried on the in-phase and quadrature-phase lines unlike cdmaOne where the same data is carried on both lines. Thus the physical layer is more complicated than that in cdmaOne.

The frame sizes in UMTS are 10 ms long compared to the traffic frame size of 20 ms in cdmaOne. With a chip rate of 3.84 Mcps, the number of chips in 10 ms is 38 400 chips, which corresponds to one period of the Gold sequences used in UMTS. Recall that the Sync channel frames in IS-95 are 26.67 ms long, which corresponds to one period of the short PN scrambling sequence there. Thus, the periodicity of the PN scrambling sequences contains inherent timing information that can be exploited by the receiver. This is different from TDMA systems like GSM where an explicit framing structure and framing hierarchy becomes essential. The frames in UMTS are further divided into 15 slots as shown in Figure 12.23b. Each slot has 2560 chips and the number of data bits depends on the spreading factor. These slots are similar to the power control groups in cdmaOne in that power control is performed at the granularity of slots. Each slot lasts for $10/15 = 0.67$ ms and power control is performed $1/0.67 \times 10^{-3} = 1500$ times/s.

The Forward Channel: The forward channels in UMTS include transport channels mapped to physical channels and also some physical channels that are employed only for signaling without any higher layer data (i.e., no corresponding transport channels). The scrambling codes are used differently as are the behaviors of the Sync and Common Pilot Channels (see Section 12.6 for how cell search is different in UMTS compared to IS-95). A broadcast channel is used to carry information related to a cell. These channels have no power control. The forward access channel carries control information or packet data (e.g., short messages) to a mobile station. The paging channel is used to page the mobile stations when they have incoming calls. The forward access channel and the paging channel share a physical channel and do not have power control. A dedicated physical channel (corresponding to exactly one mobile station) carries both user data and signaling information with some pilot symbols. The user data and control signaling data are *time-multiplexed*. The dedicated channel has fast power control at 1500 Hz as described later. The standard also specifies a downlink shared channel for data traffic which has evolved into channels used for high speed packet access described in Section 12.7.

The Reverse Channel: Support for variable data rates and operation in a variety of environments once again governs the implementation of the reverse link for 3G systems. The dedicated channel comprises of a dedicated *data channel* and a dedicated *control channel* from each mobile station to the base station. These channels are *multiplexed in code* unlike the forward link where they are multiplexed in time. To separate these channels, orthogonal codes (OVSF codes) are employed in *each* mobile station. This is very different from the reverse channel in cdmaOne where orthogonal codes are used for waveform coding rather than separating flows originating from the same mobile station. Note that orthogonal codes are not used for separating transmissions from different mobile stations. To separate the transmissions of different mobile stations in the cell, typically, truncated Gold codes of length 38 400 chips are used. These codes also provide good diversity with a RAKE receiver.

12.6   Cell Search, Mobility, and Radio Resource Management in CDMA

Of all the second-generation cellular systems, the IS-95 standard is the most complex because of the use of spread spectrum that brings with it a set of advantages not available to TDMA based systems. These include a frequency reuse of one, robust performance in the presence of interference and multipath, and the ability to increase capacity. The pilot channel that employs M sequences is used for synchronization and demodulation, cell search, diversity, handoff, and in power control. Operation with a RAKE receiver is an important characteristic of CDMA. It provides inherent diversity in the presence of fading, thereby improving voice quality. The fingers of a RAKE receiver can select either a multipath signal or a signal from another base station if it is within range of the MS. This ability is employed in cdmaOne to perform what are known as *soft handoffs* that improve voice quality during handoff. Mobility management outside of soft-handoff is based on the general mobility management procedures discussed in Chapter 6. In the case of CDMA, specific messages are additionally included.

Using spread spectrum has a disadvantage in that the near–far effect becomes predominant and in order to prevent the signal from one user overwhelming that of another user, strict power control has to be implemented. The advantage of implementing strict power control is that the MS can operate at the minimum *required E_b/N_0* for adequate performance. This increases battery life and reduces the size and weight of the mobile terminal.

The use of pilot signals, soft handoffs, and power control also extends to third-generation cellular systems. UMTS employs soft handoffs and power control, but in ways that are somewhat different from IS-95. We describe these operations below.

12.6.1   Cell Search

Cell search refers to how a mobile station discovers who is providing service in a geographic area and the necessary system parameters, timing, and so on when it powers up. As discussed in

Chapter 6, each wireless system needs a *beacon* signal that a mobile station can latch onto, which it can then use to determine information about the networks that are available and the services it may obtain. In CDMA systems, it is typically the *pilot channel* that is used as a beacon for this purpose through its spreading sequence although there are differences between 2G and 3G systems.

Cell search in 2G: In systems based on cdmaOne, M sequences with a periodicity of 32 768 chips are used to scramble the spread signals as described previously. The pilot channel in cdmaOne has no data (all symbols are zeros) and it is further spread by the Walsh code W_0 which has *no zero transitions* (since all the chips in W_0 are 0s). If we assume a positive rectangular pulse represents a zero, after the data symbols are spread by the Walsh code in Figure 12.14a, there is no change in the baseband signal. In essence the, the pilot channel is being *spread* by the M sequences in the I and Q channels. At the mobile station, when the receiver correlates the signal with locally generated M sequences, if only one pilot signal exists, a single peak is observed in the (periodic) autocorrelation as shown in Figure 12.11. In a real cellular system, there will always be multiple pilot channels that exist because all base stations employ the same carrier frequencies and the same M sequences (although the M sequences have different offsets).

When multiple pilots exist (from different base stations), many peaks are observed as shown in Figure 12.25a. The locations of the peaks depend on two factors: (1) the propagation delay between the base station transmitting the pilot and the mobile station, and (2) the offset of the PN sequences. Offsets are multiples of 64 chips. Each chip lasts for $1/1.2288 \times 10^6 = 0.813$ μs. The duration of 64 chips is 52 μs and at the speed of light, a signal can travel 15 km in this time. Thus, the offset is

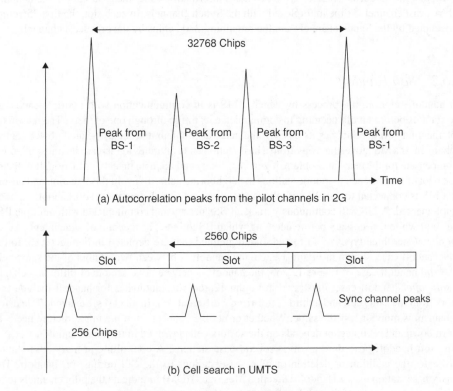

(a) Autocorrelation peaks from the pilot channels in 2G

(b) Cell search in UMTS

Figure 12.25   Cell search in CDMA systems.

really the dominating factor for the location of the peaks. Once various peaks are detected, the MS picks the strongest pilot from the set of detected peaks. This pilot channel already provides timing (26.67 ms for one period) as the peaks are periodic and can be used to detect the Sync channel, which provides additional information about the base station or cell sector.

Cell search in UMTS: Unlike systems based on cdmaOne, in UMTS, different cells use different scrambling PN sequences. The mobile station cannot search through all the different possible scrambling sequences to find the one that a base station or cell sector is using. Instead of trying to find the different unmodulated pilot channels that it can hear, the mobile station first tries to find a primary Synch channel that uses the same 256-chip scrambling code in all cells (see Figure 12.25b). The primary Synch channel is on for only the first 256 chips of every slot and so the peaks corresponding to the primary Synch channel also provide timing information about the slot boundary. A secondary Synch channel, aligned with the primary Synch channel, also operates in each cell, but each cell uses a different 256-chip code for the secondary Synch channel. There are 64 different codes that may be used here. The mobile discovers the strongest secondary Synch channel and decodes its spreading code. To ensure that the code has been correctly detected, the correlation process is performed in each of the 15 slots in a frame. This also provides timing information about the frame boundary. Once the code used in the secondary Synch channel is determined, the primary scrambling code for the cell or sector can be determined. There are eight primary scrambling codes associated with each of the 64 codes used in the secondary Synch channel. The primary scrambling code is determined through the unmodulated common pilot channel, which is then used for decoding other channels, in particular the broadcast channel, which contains information about the base station. Incidentally, the broadcast channel is time multiplexed with the Synch channels. In each slot, the first 256 chips are occupied by the Synch channels and the remaining 2304 chips by the broadcast channel.

12.6.2   Soft Handoff

Soft handoff refers to the process by which a MS is in communication with multiple candidate base stations before finally deciding to communicate its traffic through one of them. The reason for implementing soft handoff has its basis in the near-far problem and the associated power control mechanism. If a MS moves far away from a base station and continues to increase its transmit power to compensate for the near far problem, it will very likely end up in an unstable situation. It will also cause a lot of interference to mobile stations in neighboring cells. To avoid this situation and ensure that a MS is connected to the BS with the largest received signal strength, a soft handoff strategy is implemented. A MS will continuously track all BSs nearby and communicate with multiple BSs for a short while if necessary before deciding which BS to select as its point of attachment.

In cdmaOne, three types of soft handoffs are defined that are depicted in Figure 12.26. In the *softer* handoff case, shown in Figure 12.26a, the handoff is between two sectors of the same cell. In the *soft* handoff case of Figure 12.26b, the handoff is between two sectors of different cells. In the *soft-softer* handoff case, illustrated in Figure 12.26c, the candidates for handoff include two sectors from the same cell and a third sector from a different cell. In all cases, the handoff decision mechanism is more or less the same. Whether or not the connection in the infrastructure needs to be torn down and set up again depends on the sectors/cells involved in the final handoff.

The soft handoff procedure involves several base stations. A controlling primary base station coordinates the addition or deletion of other base stations to the call during soft handoff. The primary base station uses a Handoff Direction Message (HDM) to indicate the pilot channels to be used or removed as part of the soft handoff process. At some point of time, the primary base station

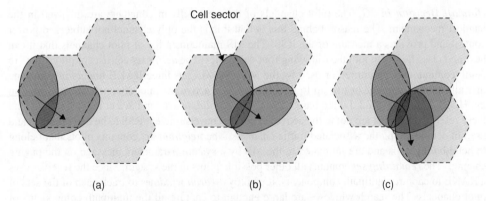

Figure 12.26   (a) Softer, (b) soft, and (c) soft–softer handoff.

is also changed after handoff. The signals from multiple base stations are combined in the BSC or MSC and processed as a single call. This process is achieved using a *frame selector join* message. Figure 12.27 shows an example of the setup and ending of handoff in a two-way soft handoff. The MS detects a pilot signal from a new base station and informs the primary base station. After a traffic channel is set up with the new base station, the frame selector join message is used to select signal from both base stations at the BSC/MSC. After a while, the pilot signal from the old base station starts falling and the MS will request its removal, which is achieved via a *frame selector remove* message.

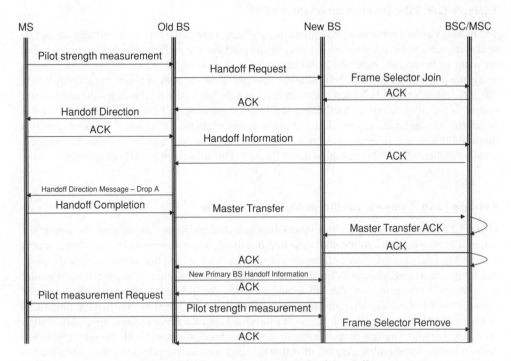

Figure 12.27   Setup and ending of soft handoff.

Handoff Decision in 2G: The pilot channels of different cells in cdmaOne are involved in the handoff mechanism. The reason behind this is that this is the only channel not subject to power control and provides a measure of the RSS. The MS maintains a list of pilot channels that it can hear and classifies them into the following four categories. The *active set* consists of pilots that are being continuously monitored or used by the MS. The MS has three RAKE fingers in cdmaOne that allows it to monitor or use up to three pilots. The active set pilot channels are indicated in the HDM on the downlink by the base station. The *candidate set* can have at most six pilots and these refer to pilots that are not in the active set but that have sufficient RSS to be demodulated and used in demodulating the associated traffic channels. The *neighbor set* contains pilots that belong to neighboring cells and are intimated to the MS by a system parameters message on the paging channel. The *remaining set* contains all other possible pilots in the system. Since the receiver uses a RAKE to capture multipath components, it employs *search windows* to track each of the sets of pilot channels. The search windows are large enough to capture all the multipath components of the pilot from a base station but small enough to minimize searching time. The multipath delays are a function of the distance between the MS and the BS and consequently, the search windows are also affected.

Several thresholds are used in the soft handoff procedure. These are similar to the RSS thresholds discussed in Chapter 6. More details of these thresholds are available in [Gar00]. Whenever the strength in a pilot falls below a threshold, the MS starts a dwell timer. Unless the pilot strength goes back above the threshold before the timer expires, the MS will drop it from a given set. There is a tradeoff in setting high or low values for these thresholds and timers (e.g., the ping-pong effect) as discussed in Chapter 6.

Example 12.9: Pilot detection threshold in IS-95

The mobile station maintains a list of pilots that are being used in the active set. Initially the mobile station is connected to one base station and only its pilot and the multipath components of the pilot are in the active set (and indicated by the handoff direction message). As the mobile station moves away, the pilot of the adjacent cell becomes stronger. If its strength is above the *pilot detection threshold* this pilot must be **added** to the active set and the MS enters what is called the soft handoff region. If the pilot detection threshold is too small, there may be false alarms caused by noise or interfering signals leading to adding of pilot channels into the active set. If the pilot detection threshold is too large, useful pilots are not added to the active set and the call may be dropped. Thus a suitable value of this threshold has to be used. This may often be cell-site specific.

Example 12.10: Using various thresholds in soft handoffs

Figure 12.28 shows an example (from [Gar00]) of how the handoff thresholds work. As soon as the strength of the pilot exceeds the pilot detection threshold, it is transferred to the candidate set (1) and the MS sends the pilot strength measurement to the base station that is transmitting the pilot. The base station sends a handoff direction message to the mobile station (2) at which time this pilot is transferred to the active set. The MS acquires a traffic channel and sends a handoff completion message (3). After the pilot strength drops below a certain "drop" threshold, the handoff drop timer is started (4). If the strength is still below this threshold after the timer expires, the mobile station sends another pilot strength measurement to the base station associated with the pilot (5). When it receives the corresponding handoff direction message without the pilot in it, the mobile station moves the pilot to the neighbor set (6) and sends a handoff completion message (7). At some point,

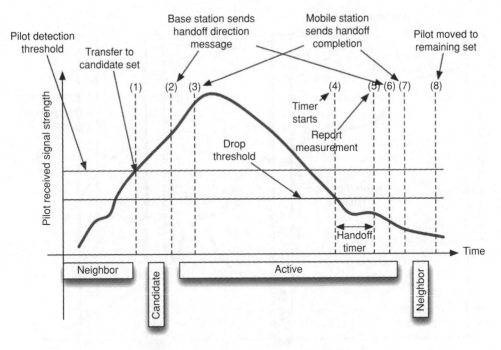

Figure 12.28   Handoff thresholds in IS-95.

the active base station may send it a neighbor update list message that no longer contains this pilot and it is moved into the remaining set (8).

All signal strength measurements are based on the pilot channel. The handoff is also a mobile assisted handoff since the mobile station reports the signal strength measurements to the network. The handoff thresholds may be adjusted dynamically to provide improvement in system performance.

Handoff Decision in 3G UMTS: In UMTS, the handoff procedure is somewhat different. Once again, different sets of pilots are maintained by a mobile station and the active set corresponds to the pilot channels being used for completing the call. Relative threshold values and hysteresis margins are employed instead of absolute values as in cdmaOne (i.e., the pilot strengths are compared with each other instead of thresholds for pilot detection and adding and dropping that need tuning depending on the environment in cdmaOne). The working of the algorithm is illustrated by the following example from [Hol00].

Example 12.11: Soft handoff in WCDMA

Figure 12.29 shows an example of soft handoff in UMTS. The events indicated along the abscissa (*x*-axis) correspond to adding a pilot to the active set if the active set is not full or removing a pilot from the active set. At the first event (1A), a pilot is added to the active set because its strength is greater than the strength of the best pilot minus a reporting range plus a hysteresis margin for more

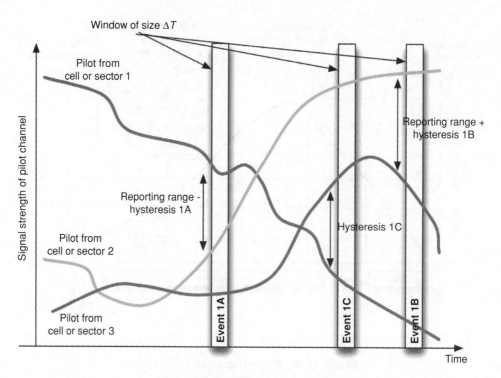

Figure 12.29   Soft handoff in UMTS.

than a time ΔT. A pilot is removed from the active set (event 1B) if its strength is below that of the best pilot by the sum of a reporting range and a hysteresis margin for a time greater than ΔT. Event 1C corresponds to a combined addition and deletion of pilots. This happens when the active set is full and the worst pilot in the set is smaller than the best pilot minus a hysteresis margin for a time ΔT. In this case, the worst pilot is deleted and the best candidate pilot is added to the active set. The reporting range is a threshold for soft handoff.

It should be noted that when the signal strength comparisons are made, averaged values are used and not instantaneous samples.

12.6.3   Power Control

Like all cellular telephony systems, CDMA is also interference limited. However, co-channel and adjacent channel interference are not the major problems here. Instead the interference is from other users transmitting in the same frequency band at the same time. In order to avoid the near–far effect, it is important to implement good power control. Also, in order to maintain a good subjective voice quality, effects such as fading and shadowing need to be countered by increasing the transmit power. In the case of CDMA, an important factor is that the signal strength may be reasonable, but frames are still received in error because of interference. There is also a non-linear relationship between the received signal strength and the frame error rates. Consequently, using the frame error rate for power control decisions is preferred over using the received signal strength. It is usually assumed that a frame error rate of 1% with maximum error bursts of two frames is optimum for maintaining voice quality and a range of 0.2–3.0% is allowed, with error bursts of up to four frames. Note that this approach is very different from the design criteria used in analog and TDMA based cellular systems that need a certain signal to co-channel interference ratio for good voice quality.

In cdmaOne, power control is very important especially on the reverse link where non-coherent detection is employed. Two types of power control are implemented – an open loop and a closed loop as discussed in Chapter 6. A slow mobile assisted power control is employed on the forward link.

Example 12.12: Open loop reverse link power control in 2G CDMA

Before a traffic channel is assigned, there is no closed loop power control in CDMA because the closed loop power control involves feedback from the base station that is delivered on the traffic channel 800 times/s. For this reason and in order to prevent sudden fall of signal strength, an open loop power control scheme is implemented. The rule here is to use a transmit power that is inversely proportional to the received signal strength of pilots from all BSs. On the access channel, the MS sends a request using a weak signal if the pilot is strong. An acknowledgement may not be received because of collisions or because the transmit power was low. If no acknowledgement is received, a stronger access probe is transmitted. This is continued a few times and then the attempt is stopped after a maximum power level is reached. Then the process is repeated after a back-off delay. Up to 15 attempts can be made to obtain a traffic channel. The disadvantages of the open loop power control are the assumption that the forward and reverse link characteristics are identical, slow response times (30 ms) and using the total power received from all base stations in calculating the required transmit power.

Example 12.13: Closed loop reverse link power control in 2G CDMA

On the downlink traffic channel, a power control bit is transmitted every 1.25 ms (800 times/s; see Figure 12.30). A zero bit indicates that the MS should increase its transmit power and a one that the MS should decrease its transmit power. Every 1.25 ms, in the BS, the receiver determines the received E_b/I_t (the signal to interference ratio) by sampling it sixteen times and if it is above a preset target, the MS is instructed to reduce its power by 1 dB. If it is not above the target, the MS is instructed to increase its power by 1 dB. This is called the *inner-loop power control* as it enables changing the transmit power value in the MS. The target value in the base station controls the long-term frame error rate. The FER is not linearly dependent on the E_b/I_t, but is also a function of the velocity, fading, environment, and so on. The target E_b/I_t is also varied over time to reflect accurate values. It is reduced by a value of x dB every 20 ms if the FER is small enough. Typically, the value of $100x$ is 3 dB. The target value may be rapidly increased if the FER starts to increase. This mechanism to change the target E_b/I_t is called *the outer loop power control*.

Example 12.14: Forward link power control in 2G CDMA

Power control on the forward link is employed to reduce inter-cell interference. Within a cell, multiple users employ orthogonal sequences and the primary source of interference is from users of other cells or from multipath. A *mobile assisted* power control is used. The MS periodically reports the FER on the forward link to the base station, which will then adjust its transmit power

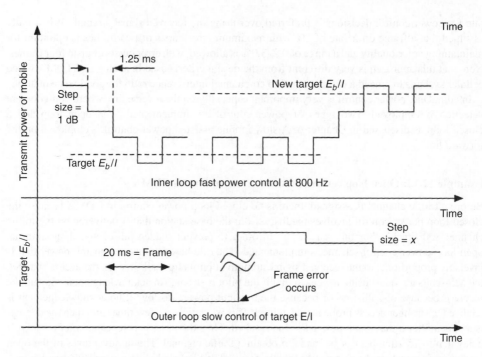

Figure 12.30   Inner- and outer-loop closed loop power control on the reverse link in cdmaOne.

accordingly. A maximum and minimum transmit power values are preset to prevent excessive interference and to avoid allowing voice quality to drop respectively.

In UMTS, closed loop power control is implemented in a manner similar to 2G CDMA with the power control bits transmitted 1500 times/s. This allows a very fast control of power and provides significant power gains in UMTS, especially at pedestrian speeds. Both inner and outer loop power control mechanisms are employed in a manner similar to cdmaOne. The difference in the case of UMTS is that fast power control is implemented for both the forward and reverse links (for specific physical channels – recall that not all channels have power control). The target E_b/I_t varies between 5.5 and 7.0 dB, depending on the vehicular speed and channel conditions.

12.7   High Speed Packet Access

Phone calls and voice were the primary applications and revenue generators for cellular wireless networks for 30 years since their inception in the early 1980s. The emergence of the Internet in the 1990s and smart phones and tablets in the mid-2000s resulted in significant shifts in the way people communicate. Short messaging or texting exploded in the late 1990s and the early 2000s. This was followed by e-mail, microblogging, social networking, voice over IP, and other methods of communication that made use of IP and were ill-suited for circuit switched connections that dominated the standards and the data networking standards that were built on top of the voice-oriented standards.

Some examples of the deficiencies of the data networking standard are as follows. In the wired part of the network, radio link-level "connections" were set up between the mobile station and the base station controller or radio network controller (see Figure 12.24 for an example architecture comprising of serving and gateway GPRS support nodes – SGSNs and GGSNs). There were two hops between the mobile station and the radio network controller resulting in delays when resources had to be freed or allocated. There were also packet tunnels that were created between other entities in the network. For example, a logical link identifier at the link level was used to identify the connection that carried the IP payload from a mobile station to an SGSN. A GPRS tunneling protocol was employed to carry traffic from an SGSN to a GGSN where the IP packets would eventually enter the Internet. This two-level tunneling introduced additional latency. At the radio level on the air–interface, the physical layer was suited for voice applications (using BPSK or QPSK and having fairly high frame error rates of 1%). Strict power control of traffic channels for example made it more difficult for lower frame error rates to be achieved.

High Speed Packet Access or HSPA was designed to overcome some of these deficiencies. It was also designed to allow service providers to enhance their data services while radical changes to the network architecture and the physical layer were slowly implemented through *Long Term Evolution* or LTE [Rao09]. In this section, we highlight some of the differences in implementation between HSPA and 3G UMTS systems originally designed for voice traffic with data traffic overlays.

Architectural Changes: The radio link for voice traffic (which is circuit switched in 2G and 3G systems) is set up between the mobile station and the base station or radio network controller. Allocation of radio resources (channels, time slots, spreading codes, and transmit power) is controlled by the base station or radio network controller. In the case of HSPA, to enable fast scheduling of resources, the base station or Node B is allowed to locally handle its radio resources and allocate slots and spreading codes on the downlink to mobile stations. As explained below, the allocation can change very rapidly. The two-tunnel protocol architecture is simplified as follows. The control signaling still happens with the SGSN. But the RNC can directly send user data packets to the GGSN instead of a link layer tunnel to the SGSN and a second tunnel from the SGSN to the GGSN as shown in Figure 12.31. As we will see in Chapter 13, this flat architecture without a hierarchy is similar to the architecture in 4G systems.

Medium Access Changes: To allocate and release resources rapidly, the 10 ms frames that were used in UMTS (see Figure 12.23b) are changed in HSPA to shorter 2 ms frames. Thus, a base station or Node B can allocate resources to individual mobile stations every 2 ms on the downlink. The downlink uses a shared channel between all active mobile stations.

On the uplink, the 10 ms frame is divided into five sub-frames to have a similar effect. Each mobile station has its own dedicated channel which is code multiplexed with the dedicated channels of other mobile stations. In UMTS, a mobile station is connected as long as it continuously transmits control information (which is code multiplexed with user information as described above). The mobile station moves to a different state when there is no data to be sent in order to reduce interference on the uplink and conserve the battery life. Unfortunately, to return to a state where data can be transmitted, the mobile station will have to go through a random access procedure that can be as long as 700 ms detracting from the idea of quick allocation and release of radio resources. To overcome this problem, HSPA uses what is called "continuous packet connectivity". Here the control information is masked or deleted when no data needs to be sent and it is sent periodically to maintain connection with the network.

Resources are allocated to mobile stations based on *channel quality information* that is reported by the mobile stations to the base station. This channel quality information and the usage of

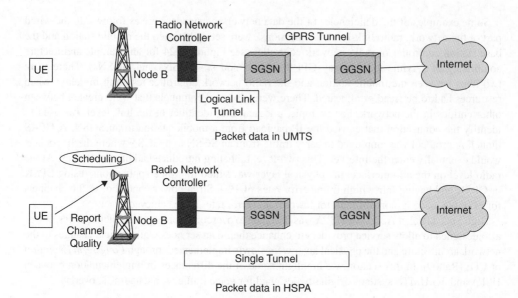

Figure 12.31 Architectural change in HSPA that tunnels user data from the RNC to the GGSN directly.

bandwidth by mobile stations is employed by the base station to decide how much of bandwidth is allocated to a mobile station in a given 2 ms frame. The medium access also uses retransmissions in a novel way. The retransmission scheme is called Hybrid-ARQ or H-ARQ. Instead of discarding packets that are received in error, the receiver (the base station on the uplink and the mobile station on the downlink) keeps the erroneous packet and requests retransmission. Turbocoding with puncturing is used for error correction. This implies that all of the redundant coding bits are not transmitted initially. When a retransmission request is received, the transmitter can send only the additional bits needed for improving error correction or transmit the entire packet again, but with a different set of coding bits. This improves the probability of successfully receiving the packet with a retransmission significantly.

Physical Layer Changes: The physical layer can be modified for every frame in HSPA based on the channel quality information. The highlights are as follows:

1. The spreading factor used on the downlink is always 16 unlike regular UMTS, where OVSF codes can be used to change the spreading factor. Thus, up to 16 different orthogonal codes can be used in each frame. All of these 16 orthogonal codes can be allocated to a single mobile station to increase its data rate or different spreading codes can be allocated to different mobile stations. The allocation of spreading codes depends on the scheduling algorithm used by the base station or Node B.
2. HSPA can employ higher-level modulation schemes. On the downlink, 64-QAM and 16-QAM can be used to increase data rates. On the uplink, 16-QAM can be used if the frame size is 2 ms.
3. HSPA enabled networks and devices can use MIMO to increase capacity or link quality with two antennas at both the transmitter and receiver. Each antenna can transmit a different data stream with a different modulation and coding scheme applied.

Questions

1. Give four reasons why CDMA was a good choice for 3G cellular systems.
2. Explain the nature and reasons for the two types of spread spectrum codes used in CDMA systems.
3. What is the difference between Walsh codes and OVSF codes?
4. What is a RAKE receiver? How is it used in cdmaOne?
5. What are the bandwidths and chip rates used in UMTS and how do they compare with those used in cdmaOne?
6. Compare the spread spectrum codes used in cdmaOne and UMTS for channelization and for scrambling.
7. How many physical channels are available in each cdmaOne carrier and what type of coding separates these channels from one another?
8. Name the forward and reverse channels used in cdmaOne.
9. How are Walsh codes employed in the cdmaOne forward and reverse channels? Explain the difference.
10. Draw a linear feedback shift register that is represented by the characteristic polynomial: $c(x) = 1 + x + x^4$.
11. Explain the basic spreading procedure in cdmaOne with a diagram.
12. What are the differences between the long PN code and the short PN codes used in cdmaOne in terms of periodicity and how they are used by the system?
13. What two approaches can be used for enabling multi-rate transmissions in CDMA systems without changing the modulation or error control coding rates? Explain briefly.
14. What does PN offset mean in cdmaOne?
15. Why is power control important in CDMA?
16. How is power control different in the forward and reverse links in cdmaOne?
17. At what rate is power control implemented in cdmaOne? How is this different from the implementation in UMTS?
18. Handoff decision in wireless networks is performed using received signal strength measurements. Name the forward channel in IS-95 that is used for this purpose.
19. What is cell search? How is it different in UMTS compared to cdmaOne?
20. What is soft handoff? Explain using the various pilot sets in cdmaOne.
21. Why are several pilot channels monitored in IS-95? When does a pilot channel from a base station move from an active set to a candidate set?

Problems

Problem 12.1

The diagram in Figure 12.17 shows the forward traffic channel for the IS-95 CDMA systems. Answer the following questions related to this diagram:

a. What is the purpose of the convolutional encoder, symbol repeater and block codes?
b. What is the purpose of the Walsh code?
c. What is type of code used as long code and what does it serve in the system?
d. What is objective of short code used in Pilot PN sequence and how does it relate to the identity of a base station?

Problem 12.2

a. Sketch all four of the four-bit Walsh functions.
b. Sketch the aperiodic autocorrelation function of all four functions.
c. Sketch the aperiodic cross-correlation function of the first and the second functions.

Problem 12.3

Repeat Problem 12.1 for 16-bit Walsh codes.

Problem 12.4

a. Give all the eight-bit Walsh codes.
b. Take the first and the fourth codes and show that they are orthogonal.
c. If we use these codes for M-ary orthogonal coding on a link with data transmission rate of 10 Mbps, what is the user data rate?

Problem 12.5

The M sequence is a class of spread spectrum sequences that is used in 2G CDMA systems. Compute the aperiodic autocorrelation of an M sequence, which is given by the following vector. Assume that the chip duration is T_c and the duration of the M sequence "pulse" is T. The M sequence is: $[1 -1 -1 1 1 1 -1 1 1 1 -1 -1 -1 -1 1]$. You can use the xcorr function in MATLAB® to verify your results.

Problem 12.6

Consider the LFSR shown in Figure 12P.1

a. Determine one period of the output stream. Assume the initial values to be 0, 0, 0, 1 from right to left.
b. What is the period of the sequence? Is it a maximal length sequence?
c. What is the polynomial representation of this LFSR?
d. Repeat (a) and (b) for the LFSR in Question 10.

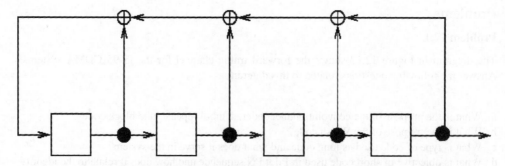

Figure 12P.1   A state machine for generating the LFSR code in Problem 12.7.

Problem 12.7

The maximum chip rate in a CDMA system is 1.28 Mcps. There are six users in the system. Three of them require 160 kbps, two require 320 kbps, and one requires 640 kbps. Use the OVSF code tree and assign codes to each of these six users. Say why it works (or does not work).

Problem 12.8

Show by mathematical induction that:

$$\mathbf{H_N H_N}^T = N\mathbf{I_N}$$

where $\mathbf{I_N}$ is the identity matrix of size N and ^{T} refers to the transpose operation. Clearly show all steps.

Problem 12.9

Consider the primitive polynomials of degree 5 given below. Write a MATLAB script to generate one period of the sequences generated by the LFSRs corresponding to these two polynomials. Using MATLAB, plot the aperiodic and periodic autocorrelations of these two sequences. Comment on your results.

$$F(x) = 1 + x^2 + x^5 \text{ and } G(x) = 1 + x + x^2 + x^4 + x^5.$$

Problem 12.10

Consider the two primitive polynomials associated with a four-stage LFSR, namely:

$$p_1(x) = x^4 + x^3 + 1 \text{ and } p_2(x) = x^4 + x + 1$$

Determine one period of the M sequences generated by the corresponding LFSRs. Compute the periodic cross-correlation of these two sequences and plot the result.

Problem 12.11

a. Using Table 3.1 (BER in AWGN), calculate required γ_b for the BER of 10^{-3} assuming a QPSK modulation scheme. (Hint: you can use the MATLAB function erfc for calculation of the complementary error function).

b. Use Equation 4.4 with γ_b (the same as S_r) determined in part (a) to calculate the number of simultaneous users, M, in a cell operating in one sector of a three-sector antenna with one 2G CDMA carrier. Assume a data transmission rate of $R = 9600$ bps and a performance improvement factor of $K = 4$ (6 dB).

c. Repeat parts (a) and (b) for different values of BER between 10^{-2} and 10^{-12} to produce a computer plot of BER in logarithmic scale versus number of users, M, in linear scale. Using this curve explain the effects of error rate requirement on the capacity of a CDMA system.

d. Repeat the plot in part (c) for normalized number of users per MHz of band. If we change the system to UMTS, does this plot or the plot in part (c) change? Explain why.

Problem 12.12

a. Assume we want to support a 19.2 kbps data service with minimum required error rate of 10^{-3} over a W-CDMA system. What is the minimum chip rate and bandwidth needed to support 100 simultaneous users with one carrier of this WCDMA system? Assume a performance improvement factor of $K = 4$ (6dB) and use Table 3.1 and Equation 4.4.

b. What would be the bandwidth requirement in (a) if the number of users were increased two times?

c. What would be the bandwidth requirement in (a) if the data rate requirement were increased to 192 kbps?

d. What would be the bandwidth requirement in (a) if the error rate requirement were increased to 10^{-4}?

Problem 12.13

The CDMA network was originally designed for telephone voice applications, where many users need a constant stream of information at a low rate during the telephone conversation. Later, data applications demanding high data rates to transmit short packets of information became demanding and network designers had to make modifications to their design so that each user can use multiple codes, and if the terminal is close to the base station, change the modem to a higher bit per symbol transmission. This problem gives a quantitative example for that transformation that has been applied as the design principle for many manufacturers of high data rate cellular wireless data networks.

a. In a CDMA network we have 50 different voice users in the reverse channel, each with a data rate of 9600 bps, sharing the medium simultaneously. What is the signal to interference ratio for each user. Include the effects of antenna sectorization (2.75), voice activity (2.0), and extra CDMA interference (1.67). Use 1.25 MHz for the carrier bandwidth.

b. What is the bit error rate (BER) for each user if the system uses BPSK modulation?

c. If we assign all 50 codes in the forward channel to one user to send a short packet of data, what would be the effective data rate of that user during the transmission of the packet?

d. If the user in part (c) uses 16-QAM rather than QPSK, what would be the effective data rate and the BER for the short packet?

13

OFDM and MIMO Cellular Systems

13.1 Introduction

Cellular networks, when they were first envisaged in the late 1970s, were primarily designed for two-way voice communications, in particular telephone calls that would be untethered. Over the last decade, the way in which human beings communicate has morphed significantly. While voice calls still constitute a large fraction of traffic on cellular networks, short messaging or texting, instant messaging, e-mail, video calling using software applications like Skype, broadcast or multicast microblogs such as Twitter, and social networking sites such as Facebook and Google+ are increasingly becoming the means by which communication occurs between people. The traffic characteristics of such communications no longer follow traditional voice conversations. The traffic is bursty, delay tolerant in many cases, and often not strictly in real time like traditional voice calls. However, transmissions have to be mostly error free since data is presented in various formats to communicating entities (e.g., postings on a web page).

As we described in Chapters 11 and 12, 2G and 3G cellular networks were designed, built, and deployed with voice calls as the primary application and source of revenue for operators. Frame error rates of up to 1% were allowed for voice calls in the case of CDMA networks. With the emergence of large volumes of data traffic, the *throughput*, *data rates*, and *latency*, rather than the number and quality of supported voice calls (see Chapter 4 for a discussion of Erlangs) became an important metric of interest. The network architecture and protocols were modified to account for the new reality of the importance of data traffic. Significant changes to the network architecture can be seen in 4G networks with the network *less hierarchical* than the network in 2G and 3G systems, and with all of the entities supporting *packet switching* using IP as the networking protocol. The reduced hierarchy results in a lower latency for user data packets. 4G systems have also been designed to exploit *a variety of carrier bandwidths*, ranging from 1.4 to 20 MHz. We recall here that the carrier bandwidth in TDMA based GSM is fixed at 200 kHz, in CDMA based IS-95 it is 1.25 MHz, and in CDMA based UMTS it is 5 MHz. The flexibility requirements in terms of bandwidth made it challenging to use direct-sequence spread spectrum as the transmission technique. Further, to support a raw data rate of say 2 Mbps using direct sequence spread spectrum, with a processing gain of 128 and a modulation scheme with low spectrum efficiency (like QPSK), a bandwidth of around 100 MHz would be needed per channel. With data rates increasing further, using direct sequence spread spectrum as the transmission scheme of choice was not considered to be the best option. At the same time, improvements in technology made the use of multi-carrier modulation possible. Thus, the fourth generation or 4G cellular systems adopted *orthogonal frequency division multiplexing* (OFDM) as the base transmission scheme of choice to support packet data traffic. Further, advances in physical layer techniques such as MIMO (see Chapter 3) allowed the increase of spectrum efficiencies beyond what was possible with 3G systems.

In the previous chapters, we have discussed 2G and 3G cellular networks that have become commercially successful. There were two types of air–interfaces in 2G cellular networks – those based on TDMA and those based on CDMA. The most popular TDMA based 2G cellular network is GSM that was used as the example in Chapter 11. The 2G CDMA systems were commercially called cdmaOne and this was used as one of the examples in Chapter 12. Also in Chapter 12, UMTS was used as an example for 3G CDMA cellular systems. UMTS employs CDMA, but it is different from cdmaOne in several ways as described in Chapter 12.

The third generation partnership project – 3GPP has been responsible for the development of UMTS standards. Towards the end of Chapter 12, we described High Speed Packet Access (HSPA), a technology that is built on UMTS foundations, but was modified to accommodate and support data traffic. Many of the architectural changes seen in 4G systems are somewhat implemented in HSPA. The physical layer in HSPA is still based on CDMA and direct sequence spread spectrum, which is not efficient in its support for data traffic. The standardization process in 3GPP considered the so-called *long-term evolution* of UMTS in the mid-2000s to support packet data traffic exclusively with OFDM. The acronym used here was LTE to denote this long-term evolution. LTE has now become a standard itself and is considered to be the 4G cellular system of choice by most service providers in the world. Around the mid-2000s, another standard for wide area packet data services using OFDM emerged, which is commercially called WiMax. WiMax was standardized by the IEEE as part of the IEEE 802.16 standardization group. WiMax was the first 4G technology to be commercially deployed, but it appears to have been supplanted by LTE in recent years. For more details of LTE, the readers are referred to [Rao09; Gho11] and for details of WiMax, readers are referred to [Das06; And07].

In this chapter, we look at 4G technology with OFDM as the underlying physical layer. We will use LTE as the primary standard as the example, although we will refer to WiMax occasionally. Our focus will be on the architectural and physical layer connections in OFDMA based cellular technology. We also briefly describe the core (wired or fixed part) network that is based entirely on IP.

13.2   Why OFDM?

As described in Chapter 3, OFDM is not a new technology. It was invented in the 1960s, but did not become a commercial success till technology advances made this possible in the 1990s. OFDM has been used in wired networks, on the access links of digital subscriber lines or DSL under the name *discrete multi-tone* or DMT. In the 2000s, the ability of OFDM to provide resilience against multipath dispersion or frequency selective fading made it an attractive option for wireless local area networks. The IEEE 802.11a standard and subsequently, the IEEE 802.11g standard adopted OFDM as the physical layer transmission mechanism. The IEEE 802.11n standard continues to employ OFDM at the physical layer.

WiMax in its earliest incarnations picked single carrier transmission to support data traffic instead of direct sequence spread spectrum, which was the choice for 3G networks described in Chapter 12. Single carrier transmission made sense for point-to-point links with directional antennas where the impact of fading and multipath dispersion could be minimized. As the end devices became mobile with small form factors, multipath dispersion made single carrier transmissions unattractive and eventually OFDM was also selected by WiMax as the physical layer of choice. The standards bodies developing LTE picked OFDM as the physical layer of choice as well making it the only option for 4G cellular systems.

13.2.1 Robustness in Multipath Dispersion

As explained in Chapter 3 and in the discussion of WLANs in Chapter 8, OFDM as a transmission technique essentially converts the wideband radio channel into a large set of narrowband radio channels. This implies that subcarriers in OFDM face flat fading rather than frequency selective fading. This makes it easier to combat the degradation caused by the radio channel. MIMO schemes for improving reliability and capacity, that often work best in flat fading radio channels but not necessarily in frequency selective channels, can significantly benefit from the use of OFDM. Further, OFDM can be implemented using inexpensive chips that perform the Fast Fourier Transform (FFT) with little additional complexity compared to traditional techniques that use complex adaptive equalization for ameliorating the effects of multipath dispersion.

Combatting Multipath Dispersion: Figure 13.1 explains the reason why OFDM is attractive for data applications. The coherence bandwidth B_c of the channel determines the transmission bandwidth

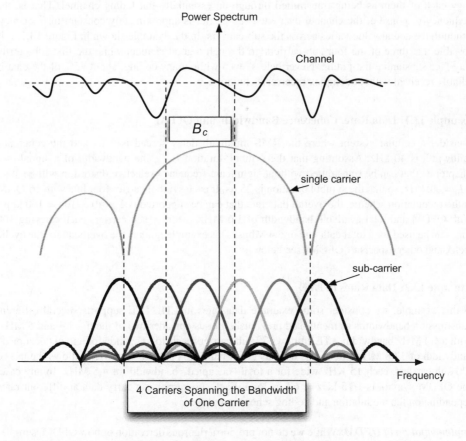

Figure 13.1   The use of OFDM to combat frequency selectivity while supporting high data rates.

that can be used such that the received signal is not very distorted with a high probability. If the transmission bandwidth is smaller than B_c, then the signal is received with relatively no inter-symbol interference. Even if there is some distortion, fairly simple equalizers could be used to eliminate the resulting inter-symbol interference. If the transmission bandwidth is much larger than B_c, then, the signal will be received with significant distortion that results in irreducible errors unless complex adaptive equalizers are used at the receiver to invert this distortion.

The reason why this impacts high data rates is as follows. The transmission bandwidth W is related to the symbol duration T_s as $W \approx 1/T_s$. As data rates increase, the symbol duration has to reduce which implies that the transmission bandwidth increases. For example, assuming a data rate of 96 Mbps and six bits per symbol (for example, 64-QAM), the symbol rate is 16 Msps or the symbol duration is $T_s = 62.5$ ns. The corresponding transmission bandwidth is on the order of 16 MHz. The RMS multipath delay spread in urban areas can be as high as a few μs. Assuming that the RMS multipath delay spread is 5 μs, one estimate of the coherence bandwidth is $B_c = 1/(5 \times 5 \, \mu s) = 40$ kHz. Clearly, a high data rate transmission scheme cannot employ single carrier modulation without facing severe inter-symbol interference.

Figure 13.1 shows how a single carrier modulation scheme can be replaced by an OFDM transmission scheme. In this figure, four OFDM sub-carriers replace a single carrier. The channel's frequency response is also shown in the figure. If the subcarriers have small bandwidth, we can view each of them as being transmitted through an essentially flat fading channel. That is, the frequency response of the channel over each sub-carrier is approximately constant in frequency, although the actual value varies across the sub-carriers. In the example shown in Figure 13.1, it is possible that three of the four sub-carriers go through a good channel while the first sub-carrier may face substantial flat fading. Nevertheless, even without any coding, about 75% of the data is reliably received without complex receiver techniques.

Example 13.1: Data Rate, Coherence Bandwidth, and OFDM

Consider a cellular system where the RMS multipath delay spread is 5 μs and the coherence bandwidth is 40 kHz. Assuming that the symbol duration is T_s, the bandwidth of a signal (see Chapter 3) that can be transmitted without significant frequency selective distortion will be $W \approx 1/T_s = 40$ kHz, so that the symbol duration is 25 μs. If each symbol carries four bits with 16-QAM as the modulation scheme, the overall data rate that can be supported is $4 \times 40 \times 10^3 = 160$ kbps. With OFDM and total available bandwidth of 1.6 MHz, 40 forty sub-carriers each carrying 160 kbps can be used for a total data rate of 6.4 Mbps. This example ignores the overhead for the cyclic prefix and other aspects of OFDM (see below).

Example 13.2: Data Rates in LTE

In this example, we consider some example data rates in LTE. LTE supports several different transmission bandwidths as mentioned previously. Let us consider two of them – 1.4 and 5 MHz. With a 1.4 MHz bandwidth, LTE employs 72 sub-carriers each 15 kHz wide for a total "occupied" bandwidth of 1.08 MHz (the rest is used as guard bands). A 5 MHz bandwidth allows the use of 300 sub-carriers, each 15 KHz wide for a total "occupied" bandwidth of 4.5 MHz. In any case, one OFDM symbol is 1/15 kHz = 67 μs long. Each sub-carrier can carry data at different rates depending on the modulation and coding scheme employed.

Implementation of OFDM: While we do not provide a rigorous derivation of how OFDM is implemented, an intuitive view is useful for understanding OFDM implementation. Figure 13.2 shows

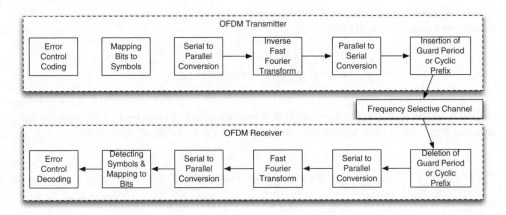

Figure 13.2 Block diagram of OFDM as a transmission scheme.

the block diagram of how OFDM as a transmission scheme is typically implemented. Before we discuss this block diagram, let us briefly consider a simplified view of how OFDM works in analog form. Let us suppose that the center carrier frequency is f_c and that there are four sub-carriers, at the frequencies f_1, f_2, f_3, and f_4 such that the four frequencies are orthogonal. Figure 13.3 shows an example with three orthogonal sub-carriers for clarity. Typically, the separation between the frequencies is $1/T_s$ where T_s is the OFDM symbol duration. The composite signal will be of the form:

$$s(t) = \cos(2\pi f_1 t) + \cos(2\pi f_2 t) + \cos(2\pi f_3 t) + \cos(2\pi f_4 t) = \sum_{i=1}^{4} \cos(2\pi f_i t)$$

When the signal $s(t)$ is considered for the duration of T_s, it becomes one OFDM symbol. Note that we can write this symbol also as $s(t) = \sum_{i=1}^{4} \text{Re}\{\exp(2\pi f_i t)\}, 0 \leq t \leq T_s$. In the general case when we have N sub-carriers, we are using quadrature modulation (both sine and cosine), and the

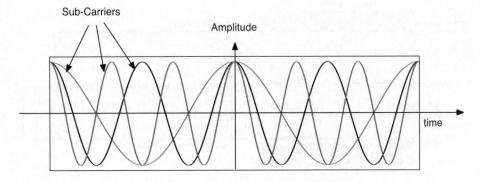

Figure 13.3 Illustration of three orthogonal subcarriers in OFDM.

complex symbol a_i represents the information in the symbol on the i-th sub-carrier, it is possible to write the *composite OFDM symbol* as:

$$s(t) = K \sum_{i=1}^{N} a_i \exp(2\pi f_i t), \, 0 \le t \le T_s$$

Here K is simply a constant. If we assume that both frequency (f_i) and time (t) are sampled and there are N samples in time, we can write the samples of $s(t)$ in the above equation as:

$$s[n] = K \sum_{i=1}^{N} a_i \exp(2\pi i n/N)$$

The above equation, which is the discrete form of OFDM, is very similar to the *inverse discrete Fourier transform* that can be efficiently implemented in hardware using the FFT algorithm. This makes the implementation of OFDM simple and efficient.

Now we can look at the details of the block diagram in Figure 13.2. As described in Chapter 3, the source data bits are encoded using an error control coding scheme (such as block coding, convolutional coding, or turbo-coding). The encoded bit stream is mapped into symbols based on the modulation scheme that is employed. For example, two bits can be encoded into one symbol if QPSK is used. The symbols arrive in a serial manner and need to be buffered till N symbols are available. This is represented by the serial to parallel conversion block in Figure 13.2. After this serial to parallel conversion, the inverse Fast Fourier Transform (IFFT) block operates to create the samples of $s(t)$, which are then transmitted serially over the frequency selective fading channel.

Before one OFDM symbol is transmitted, a guard period or cyclic prefix is concatenated to the symbol. This cyclic prefix or guard period helps when the multipath delay spread is longer than the OFDM symbol thereby causing *inter-carrier interference*. The idea of the cyclic prefix is shown in Figure 13.4 where the signal arrives at the receiver along two paths that have a relative delay

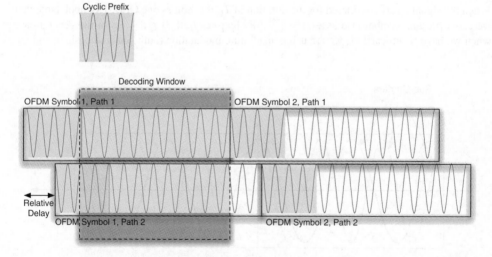

Figure 13.4   Reducing inter-carrier interference using the cyclic prefix.

somewhat larger than the OFDM symbol duration. Only one sub-carrier is shown to avoid clutter in this figure. The cyclic prefix is essentially a piece of the OFDM symbol that is copied from the end of the symbol and attached at the beginning of the symbol. Concatenating the cyclic prefix to every OFDM symbol allows the various carriers to maintain orthogonality despite the presence of an excess multipath delay. As Figure 13.4 shows, the carriers appear to be cosines within the decoding window because of the presence of the cyclic prefix. While the cyclic prefix reduces the errors caused by multipath delays, it does add to the overhead of OFDM transmission. A couple of examples will clarify the overhead.

Example 13.3: Length of Cyclic Prefix and Overhead

Suppose the largest excess multipath delay expected in a multipath channel where OFDM is employed is 1 μs. If the OFDM symbol is 20 μs long, adding a cyclic prefix that can overcome the excess delay of 1 μs implies that the overhead will be $1/(20+1)= 4.76\%$. The overhead incurred is in terms of bandwidth, because there is transmission in time that is not useful for carrying information. There is also an overhead incurred in terms of power for the same reason.

Example 13.4: Cyclic Prefix in LTE

One OFDM symbol in LTE is 1/15 kHz $\approx$ 67 μs long. The length of the cyclic prefix varies depending on where the symbol is located in a time-slot. As we will see later, there are usually seven OFDM symbols in a 0.5 ms slot in LTE. The actual duration of the slot occupied by an OFDM symbol is $7 \times 67 = 0.46$ ms long. The rest of the time is occupied by the cyclic prefix. The first OFDM symbol has a cyclic prefix that is 5.2 μs long, while the remaining six OFDM symbols have a cyclic prefix that is 4.7 μs long for a total of $4.7 \times 6 + 5.2 = 33.4$ μs. The overhead in every 0.5 ms time slot is $33.4 \times 10^{-6}/0.5 \times 10^{-3} = 6.7\%$. Note that the cyclic prefix is long enough to handle typical multipath delay spreads in outdoor environments.

In comparison, in IEEE 802.11a, the OFDM symbol is 3.2 μs long and the cyclic prefix is 0.8 μs long. Thus the overhead in every symbol is as large as $0.8/4 = 20\%$.

13.2.2   Flexible Allocation of Resources

The use of OFDM as the transmission technique allows an innovative view of radio resources that has subsequently been exploited for medium access in the form of *Orthogonal Frequency Division Multiple Access* or OFDMA. OFDMA (1) allows the flexible allocation of radio resources (bandwidth/power) for various applications, especially over the downlink, and (2) provides diversity across users in the system as described below.

Radio Resource Allocation: In cellular systems that were deployed before 4G systems based on OFDM, the carrier bandwidth was fixed. For example in the North American TDMA systems, the carrier bandwidth was fixed at 30 kHz, while in GSM, the carrier bandwidth was fixed at 200 kHz. In 2G CDMA, the carrier bandwidth was 1.25 MHz, while in 3G UMTS, the carrier bandwidth was 5 MHz. In all of these cases, once a carrier was assigned to a mobile station for any given duration, the bandwidth allocated for that duration was fixed, irrespective of whether it was necessary or not.

One way of changing the bandwidth allocation was to reallocate the bandwidth in time. This was the option selected in HSPA+, as described in Chapter 12. In HSPA+, the amount of bandwidth allocated to a mobile station could be varied every transmission time interval or frame, which

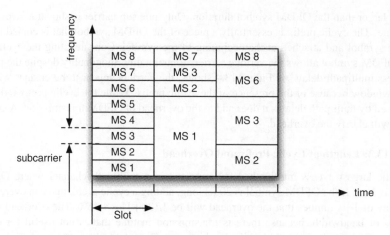

Figure 13.5   Flexible resource allocation in OFDMA.

varied between 2 and 10 ms depending on the implementation. Further, in HSPA, the base station (or Node-B) had the ability to allocate bandwidth to mobile stations without the direct involvement of the base station controller, thereby reducing the latency in making resource allocation decisions. In HSPA+, it is also possible to allocate fewer or more orthogonal codes to a mobile station (up to 16 orthogonal codes could be allocated to one mobile station in one frame). The carrier bandwidth is still 5 MHz and this cannot be easily modified to accommodate various application requirements. For example, voice traffic may require constant allocation of resources over time, but the required bandwidth may be low. It is not easy to make such flexible allocations in systems employing fixed bandwidth per mobile station.

In OFDMA based systems, on the downlink, the medium access technique allows for extremely flexible allocation of resources to individual mobile stations at a granularity that is not possible in legacy systems. Figure 13.5 shows how resources in OFDMA can be allocated on a per-slot basis. For simplicity, in this figure, we are assuming that *every* subcarrier can be flexibly allocated to any mobile station in each time slot. Usually, cellular systems such as LTE define what is called as a *physical resource block* or *PRB* comprising of a set of subcarriers and a transmission time interval, which can be flexibly allocated between different mobile stations. The physical resource block in LTE comprises of 12 sub-carriers, each with a bandwidth of 15 kHz (total bandwidth is 120 kHz) for a time slot of duration 0.5 ms (more details are available in Section 13.4).

Multi-user Frequency Diversity: The use of OFDMA, which comes with the flexibility in allocating subcarriers on the downlink, allows 4G systems to exploit what is called multi-user diversity, but in both the frequency domain and time domain. The general idea behind multi-user diversity is that channel conditions can be very different for different users or mobile stations due to their locations within the cell. By scheduling transmissions of mobile stations when they have good channel quality, the aggregate throughput of the network can be increased. In 3G systems, since the channel is constrained by the frequency carrier, the only variations observed in channel conditions are over time. So a mobile station with "good channel conditions" can be scheduled, but only in time. In OFDMA systems, such a scheduling can be done across frequency sub-carriers and in time.

Figure 13.6 shows an example of multi-user diversity as it applies to OFDMA systems. As the figure shows, different mobile stations at different locations may see different channel conditions

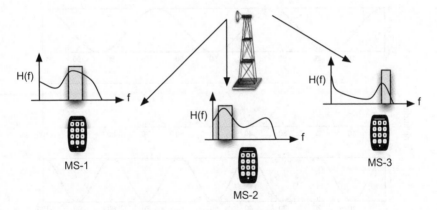

Figure 13.6   Multi-user diversity with OFDMA.

as a function of frequency. Thus, by allocating the best sub-carriers to mobile stations in each time slot, the diversity in the channel can be exploited.

However, the benefits of multi-user diversity diminishes with the number of users, the amount of coding and interleaving, and other forms of reliability employed in the transmission scheme. For example, if transmit diversity with MIMO is employed, it provides resilience against narrowband small-scale fading making the quality of the channels of various mobile stations approximately the same. Thus, multi-user diversity, while useful, provides benefits only in certain scenarios.

13.2.3   Challenges with OFDM

The use of OFDM as the transmission scheme comes its own challenges. While the implementation of OFDM using FFTs has made the hardware simpler, there are RF challenges that arise due to the multi-carrier nature of the transmission scheme. In this section, we consider the issue of the *peak to average power ratio* or PAPR problem with OFDM and the use of *single-carrier* FDMA on the uplink as the solution adopted in LTE. Other challenges in OFDM implementation arise because of the strict synchronization needs in OFDM. If the frequencies of the sub-carriers are not aligned, they may no longer be orthogonal causing inter-carrier interference. Similarly, if the sub-carriers are not aligned in time, they may not be orthogonal any more causing degradation in performance.

Peak to Average Power Ratio (PAPR) Problem: One of the major issues with multi-carrier modulation is that the peak power in the composite signal (comprising of all of the subcarriers) is substantially larger than the average power in the signal. This occurs because of the alignment of amplitudes of the various sub-carriers. At certain points in time, the sub-carriers are aligned such that their amplitudes add up to be a large value and at other points in time, the sum of the amplitudes can be small. This is illustrated in Figure 13.7 with only four sub-carriers, each using binary modulation (BPSK) and carrying a "0". At the beginning and end of the composite OFDM symbol, which is the sum of the four sub-carriers, the amplitude is large. It is in fact, four times the individual sub-carrier amplitudes. However, in the middle of the OFDM symbol, the amplitude is small, because the sub-carriers add destructively. In some ways, this is like small-scale fading within the OFDM symbol, but it is caused deterministically by the sub-carriers. As the number

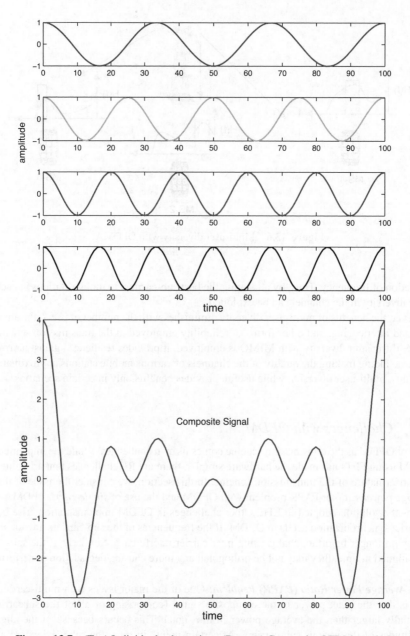

Figure 13.7 (Top) Individual sub-carriers. (Bottom) Composite OFDM symbol.

of sub-carriers increases, and the type of modulation scheme changes to 64-QAM, the difference between the peak and the average value becomes larger.

This problem is important due to the hardware considerations for implementation of an OFDM based network. The RF amplifiers used in radio transmission, either in the mobile station or at the base station will need to have *linear* characteristics over a large power range if the peak to average power ratio is large. This makes the amplifiers very expensive and inefficient. If the amplifiers are operated in the non-linear range, their efficiency is higher and costs are lower. But two problems occur. First, using the amplifiers in the non-linear region distorts the signal creates *sidelobes* in the spectrum of the resulting signal. These sidelobes cause adjacent channel interference. A solution to this problem is to reduce the overall average power of the signal before it is amplified, so that the amplification is linear. But this results in a lower signal to noise ratio at the receiver. Alternatively, the signal can be clipped between certain amplitude levels before amplification, but this also creates sidelobes and increases the bit error rate.

To address the PAPR problem, LTE employs what is called *single-carrier* FDMA (SC-FDMA) on the uplink. On the downlink, the base stations can be made complex and can employ expensive techniques to handle the PAPR problem, but the expense may be too high for mobile devices that also operate using battery power. The SC-FDMA technique essentially retains the advantages of OFDM, but in effect uses a *single carrier* instead of multiple carriers for transmitting data in a clever manner. This reduces the peak-to-average power ratio problem. We will look at how SC-FDMA is implemented next.

Single Carrier Frequency Division Multiple Access: Single Carrier Frequency Division Multiple Access is used on the uplink in LTE. SC-FDMA still maps data to only a few of the sub-carriers out of the entire block of sub-carriers in the allocated bandwidth. But the way each of the sub-carriers carries the symbol is different. The use of SC-FDMA implies that it appears as if a wideband signal is being transmitted for a shorter duration in time although the signals are not generated this way.

We describe this in an informal way here (see Figure 13.8). Let us suppose that an OFDM symbol lasts for T_s seconds and comprises of N sub-carriers. Each sub-carrier carries itself carries one data symbol (which depends on the modulation scheme used on that sub-carrier). In other words, the symbol and modulation scheme is constant for T_s seconds. This is implemented by taking N symbols in parallel, computing their IFFT and then transmitting the IFFT samples in serial. The SC-FDMA "symbol" also lasts for T_s seconds. However, on any one sub-carrier, the data symbol lasts for roughly $1/N$ of the time. In other words, each sub-carrier has many data symbols over the duration of the SC-FDMA symbol. However, the data symbol *across* a group of sub-carriers will be the same. Such sub-carriers may be contiguous or distributed in the allocated spectrum. If the sub-carriers are contiguous, the amount of frequency diversity seen by one mobile station may be limited, but multi-user frequency diversity is exploited. If the sub-carriers are distributed, each mobile station

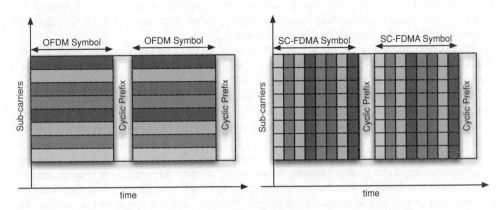

Figure 13.8 Illustration of OFDM and SC-FDMA symbols.

may benefit from frequency diversity. Note that the SC-FDMA symbol still has a concatenated cyclic prefix for protection against excess multipath delay.

Achieving all of these requires the N data symbols to be pre-coded before the IFFT is taken. This pre-coding is performed using an FFT operation in SC-FDMA along with frequency shifting so that the signal occupies the right piece of the spectrum on the uplink. We recall here that multiple mobile stations will be transmitting at the same time on the uplink and they have to be separated appropriately in frequency before transmission. At the base station receiver, separate IFFTs have to be performed after the usual FFT operation to distinguish the signals that belong to different mobile stations. Although it appears from Figure 13.8 that the data symbols have been broken up into very small pieces, we emphasize here that the rectangular block of sub-carriers over time comprises the actual transmitted "symbol", whether it is OFDM or SC-FDMA. Thus, the block is a unit in itself although it is easier to conceptualize the behavior of OFDM and SC-FDMA by referring to the sub-carriers and the data symbols that they carry over time.

13.3   Multiple Input Multiple Output

Multiple Input Multiple Output (MIMO) schemes where both a transmitter and its corresponding receiver have multiple antennas with their own RF circuitry can markedly improve the capacity and/or reliability of wireless links as described briefly in Chapter 3. An important characteristic of 4G wireless networks such as LTE and WiMax is that they all support MIMO in some form. In this section, we review some aspects of MIMO. Our treatment of MIMO here is fairly informal and limited as the subject is intricate and needs detailed knowledge of linear algebra and digital communications.

Consider Figure 13.9, where the transmit and receive antennas are shown separately. In the general case, there are M transmit antennas and N receive antennas. Such a MIMO link is sometimes called an $M \times N$ link. It is possible to imagine a radio channel as existing between *each pair of antennas* in Figure 13.9. That is, there is a channel between antenna A and antenna P, another between antenna A and antenna Q and so on. In Figure 13.9, we have represented the channel between antenna I and antenna J by a term $\tilde{h}_{ij}$. This is the low-pass equivalent channel impulse response (similar to the $h(t)$ discussed in Chapter 2). Depending on the modulation scheme employed, the symbol duration, and the nature of the environment, each of these channels may be subject to flat fading, frequency selective fading, or have LOS conditions, all in addition to path loss and shadow fading. This makes the MIMO channel complex to characterize. Many of the MIMO schemes work best when there is little or no frequency selective multipath distortion, which is the case when OFDM is employed. In such cases, it is possible to represent $\tilde{h}_{ij}$ as a Rayleigh distributed random variable, which changes in time according to the Doppler spectrum (see Chapter 2).

In 4G systems such as LTE, the use of OFDM with MIMO enables many of the advantages that MIMO brings to wireless communications. As far as a single mobile station is concerned, the traditional benefits of MIMO are exploited in 4G systems. Transmit and receive diversity are employed to improve the reliability of the links between the base station and the mobile station. With OFDM, it is possible to employ *space-frequency block codes,* which operate just like space–time block codes for transmit diversity. The difference is that, instead of employing symbol durations in time for coding, adjacent sub-carriers can be used on different antennas in the base station for transmitting the coded symbols. If the mobile station has multiple antennas, it can also benefit from additional receiver diversity. Beamforming with multiple antennas to increase the gain for specific mobile stations is possible. Finally, spatial multiplexing of different data streams over different antennas to a mobile station is possible increasing the data rates to a single mobile station. We describe these briefly below.

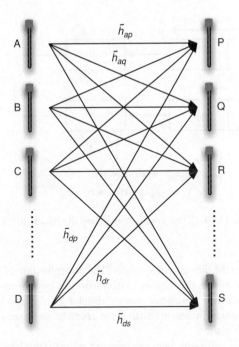

Figure 13.9 Illustration of MIMO.

13.3.1 Diversity

Diversity, discussed in Chapter 3, is the first benefit with MIMO systems. As mentioned in Chapter 3, diversity occurs when there are multiple uncorrelated or independent copies of a signal that a receiver can exploit. Even if one copy of the signal is in fade, other copies will likely not be in fade and thus, it is possible to recover the transmitted data reliably. To quickly summarize, diversity moves the bit error rate curve towards the left, thereby requiring a smaller average signal-to-noise ratio per bit to achieve the same average bit error rate. This translates into a lower transmit power or increased coverage for the same transmit power (if no other coding/interleaving is used).

In older cellular systems and also in near-recent systems, mobile devices such as phones were not equipped with multiple antennas, while it was common for the base stations to have multiple receive antennas for diversity. In the late 1990s, the question arose as to whether it was possible to exploit these multiple receive antennas to provide diversity to a mobile device with one (or more antennas). Space-time block codes (STBCs) were suggested by Alamouti [Ala98] and others [Tar98] that enable this *transmit diversity* even for mobile devices with a single antenna. We briefly describe how transmit diversity works next.

Alamouti Code: Let us suppose that the base station has two transmitting antennas (labeled 0 and 1) and the mobile station has a single receive antenna. This is called a 2×1 MIMO system. Assuming that the channels between the two transmit antennas and the receive antenna are independent, we can achieve two orders of diversity in a flat Rayleigh fading channel as follows (see Figure 13.10). Other assumptions here are that the fading coefficient of the channel can be determined at the receiver by sending some pilot symbols from the transmitter and that these coefficients change

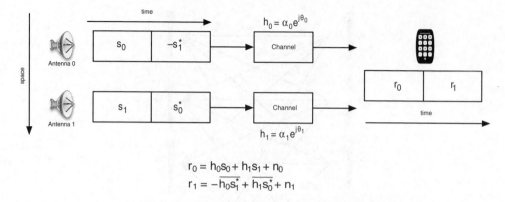

Figure 13.10   Transmit diversity using the Alamouti STBC.

slowly in time compared to the duration of symbols. The determination of the channel coefficients by the receiver, but not requiring this channel information to be sent to the transmitter is often referred to as *open-loop* schemes which are usually simpler than *closed-loop* schemes where the transmitter obtains the channel information from the receiver and precodes the transmission to exploit the channel characteristics.

Let the complex channel coefficients between transmit antenna 0 and the receiver be $h_0 = \alpha_0 e^{j\theta_0}$ and between antenna 1 and the receiver be $h_1 = \alpha_1 e^{j\theta_1}$ where the αs are normally and the θs are uniformly distributed. If a complex symbol s is transmitted by antenna $i, I = 0,1$, the receiver gets $r = \alpha_i e^{j\theta_i} s$. When both transmitters send a signal, the receiver sees the sum of the two received signals.

The cleverness of the Alamouti STBC lies in the way symbols are transmitted by the two antennas in time. As shown in Figure 13.10, antenna 0 transmits the symbol s_0 in the first symbol duration and the negative complex conjugate of symbol s_1 (i.e., $-s_1*$) in the second symbol duration. Antenna 1, on the other hand transmits the symbol s_1 in the first symbol duration and the complex conjugate of symbol s_0 (i.e., s_0*) in the second symbol duration. Assuming that the propagation times from both antennas are close, during the first symbol duration, the received signal will be:

$$r_0 = h_0 s_0 + h_1 s_1 + n_0$$

where n_0 is additive white Gaussian noise. During the second symbol duration, the received signal will be:

$$r_1 = -h_0 s_1^* + h_1 s_0^* + n_1$$

The receiver performs a linear combination of the two received signals as follows. It computes estimates for s_0, namely $\widehat{s_0} = h_0^* r_0 + h_1 r_1^*$ and for s_1, namely $\widehat{s_1} = h_1^* r_0 - h_0 r_1^*$. Substituting the expressions for h_0 and h_1 into these two equations, we can see that:

$$\widehat{s_0} = h_0^* r_0 + h_1 r_1^* = h_0^* \left[h_0 s_0 + h_1 s_1 + n_0 \right] + h_1 \left[-h_0 s_1^* + h_1 s_0^* + n_1 \right]^*$$
$$= \left(|h_0|^2 + |h_1|^2 \right) s_0 + h_0^* n_0 + h_1 n_1$$

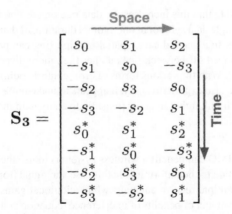

Figure 13.11 A rate 1/2 STBC with $M = 3$ transmit antennas, $m = 8$ symbol durations, and $k = 4$ symbols.

and similarly, $\widehat{s_1} = \left(|h_0|^2 + |h_1|^2\right) s_1 + h_1^* n_0 - h_0 n_1$. Note that the estimates for s_i is of the form $\widehat{s_i} = \left(|h_0|^2 + |h_1|^2\right) s_i +$ noise. Thus, if h_0 corresponds to a deep fade (i.e., it is close to zero), the estimate has a good chance of being good because h_1 may not be in a deep fade and vice versa. In other words, this is similar to having two copies of the symbol s_i with noise or having a diversity of order 2.

Note that the transmitted symbols in the Alamouti code are complex – we can use quadrature modulation. Also, two different symbols are transmitted every two symbol durations (all together, we have four symbols transmitted over two antennas in this time, but two of them are repeated in complex conjugate form). The *rate* of the Alamouti code is $2/2 = 1$. That is, there is no loss in bandwidth efficiency by using the Alamouti code. Further, it is possible to have receive diversity in addition to transmit diversity if the mobile device has multiple receive antennas.

Other STBCs and SFBCs: In the general case, if a transmitter has M antennas, and it transmits a total of k (repeated as negative or complex conjugate) symbols over m symbol durations, the *rate* of a space-time block code is k/m. Usually, the diversity gain is of order M. The only full rate complex STBC is the Alamouti code based on the technique called orthogonal design [Tar98]. For such codes, a matrix **S** of dimensions $m \times M$ that indicates how symbols are transmitted in time and across the antennas (space) defines the code. The property of such codes is that $\mathbf{S}^\dagger \mathbf{S} = \sum_{j=0}^{k-1} |s_j|^2 \, \mathbf{I}$, where **I** is an identity matrix. A rate 1/2 STBC is shown in Figure 13.11. Here there are $M = 3$ transmit antennas, $m = 8$ symbol durations, and $k = 4$ symbols that are repeated in the matrix. This complex STBC provides diversity of order 3.

Instead of using symbol durations in time, it is possible to use symbols in different frequency carriers to achieve the same objective and such codes are called space-frequency block codes (SFBCs) and they are employed in 4G systems with OFDM for achieving diversity with MIMO.

13.3.2 Spatial Multiplexing

Spatial multiplexing is the way in which MIMO enables increasing the capacity of a wireless link between two devices. In Figure 13.9, each of the M transmitting antennas could potentially transmit different data symbols at the *same time* in parallel. Given appropriate channel characteristics, the N receiver antennas can decode all the data (despite these signals adding up when they arrive at the receiver) if $N \geq M$. Note that this increases the data rate on the link by a factor of M because instead of transmitting a single data symbol in one symbol duration, M data symbols are transmitted (and received) in this time. In a general case, spatial multiplexing can be achieved with diversity gain and the two can be traded off between one another (i.e., more diversity gain and less spatial multiplexing or vice versa). With the various forms of error control coding and OFDMA employed in 4G wireless systems that also exploit diversity, it appears to make more sense to use the antennas to increase capacity of each link rather than for transmit diversity gain in most scenarios.

13.3.3   Beamforming

The final way in which MIMO can benefit a wireless system is to use the multiple antennas in the base station and mobile station for beamforming. In this case, the signal from interfering transmitters can be nulled using the multiple receive antennas while additional gains can be provided for the desired transmission. The amount of benefits of interference reduction and beamforming depend on the types of schemes employed, details of which are beyond the scope of this chapter. Beamforming by the base station in LTE is possible to increase the antenna gains towards specific mobile stations. Beamforming is also used in IEEE 802.11n and more recent WiFi devices.

13.4   WiMax

We keep the discussion on WiMax brief with more details on LTE instead. In 1998, the idea of a standard for *point to multipoint* outdoor applications, also referred to as a wireless Metropolitan Area Network (WMAN), was initiated at a National Institute of Standard (NIST) meeting and it was welcomed by the IEEE resulting in initiation of the IEEE 802.16 standardization committee in 1999. The term point-to-multipoint refers to an architecture where an entity such as a base station (point) serves multiple user stations (multiple points) as shown in Figure 13.12. The charter of this committee was to define an air–interface standard for *fixed* and *mobile* wireless systems, capable of supporting very high data rates (broadband) using a point-to-multipoint design and/or a mesh networking technology. Mesh networking is similar to ad hoc networking, where the physical entities act as routers that relay packets while also serving as wireless transmitters and receivers. However, the mesh routers are typically fixed and not mobile. The mesh technology portion of the standard was later moved to the IEEE 802.20 working group that was considering Mobile Broadband Wireless Access (MBWA) as a separate standard.

The initial IEEE 802.16 standard for point to multipoint wireless networking was developed for operation in frequency bands in the range of 10–66 GHz, which were much higher than the frequency bands used for cellular telephony, and it was approved in 2001. This early standard supported raw data rates of up to 134 Mbps using plain QPSK, 16-QAM, and 64-QAM single-carrier modulation schemes to cover an area that had a radius of 1–3 miles (1.6–4.8 km) in line of sight (LOS) conditions. As we discussed in Section 13.2, single carrier modulation does not work well in frequency selective channels unless adaptive equalization is employed. Thus, this approach was more suitable for LOS fixed terminal applications where the delay spread of the channel can be expected to be small and a single path dominates (see Chapter 2 for additional details).

Figure 13.13 shows a schematic illustrating the approximate relation between the area of coverage and the complexity of the modulation techniques for the application of point to multipoint networking systems. As expected, a higher level modulation scheme can be used closer to the base station while end systems that are farther away can only employ lower-level modulation schemes that provide lower data rates. In any case, the original IEEE 802.16 standard, also called the Local Multipoint Distribution System (LMDS), did not turn into a commercially successful product. However, it did attract considerable attention from cellular network equipment manufacturers.

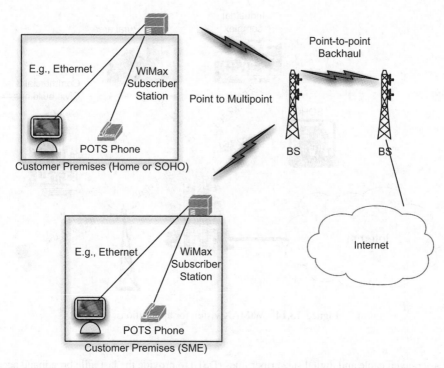

Figure 13.12   View of broadband wireless access to the home during the evolution of WiMax.

Around 2001, the WiMAX forum industrial group was formed to improve upon the IEEE 802.16 early standard as well as to promote/approve compliance with standards and ensure interoperability of equipment from various vendors and manufacturers. Today, in the same way that the popular IEEE 802.11 standard is also called Wi-Fi, the IEEE 802.16 standrad is called WiMAX. The WiMAX forum presents this technology as a competitor to wired broadband access technologies

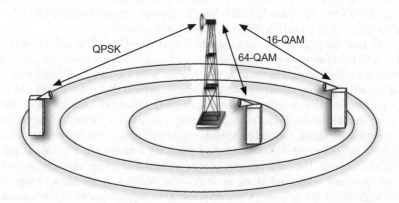

Figure 13.13   Early WiMax PHY layer transmission schemes and coverage.

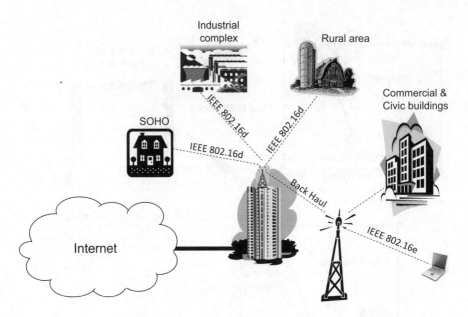

Figure 13.14   WiMAX vision for applications.

such as co-axial cable and digital subscriber lines (DSL) to provide the last mile broadband access to the Internet. Although this has not been successful in this respect, the movement of this standard in such a direction resulted in the revival of the IEEE 802.16 standard.

As part of this revival, the IEEE 802.16a standard was formalized in 2003, as an modification to the early IEEE 802.16 standard to enable operations in the 2–11 GHz bands, using not single carrier transmission, but OFDM technology and thus be able to extend the coverage to non-LOS (NLOS) conditions and environments. The IEEE 802.16b standard expanded this standard to include opeartions in the 5–6 GHz bands and also added measures to support quality of service. The next modification to the IEEE 802.16 standard was the IEEE 802.16c standard, which delivered a system profile for the 10–66 GHz 802.16 standard. The 802.16d completed in 2004, was a revision project aligning with ETSI's pan-European HIPERMAN (the outdoor version of the HIPERLAN standard for WLANs) superseding the earlier 802.16 a, b, and c amendments.

Around this time the peak of interest in WiMAX started and the IEEE 802.16e was formally approved in 2005 to add mobility to the standard and make it more like a cellular network. The IEEE 802.16e standard is sometimes referred to as the "Mobile WiMAX" standard, and it uses OFDMA as the medium access scheme and a more detailed specification of the QoS. Figure 13.14 illustrates the WiMAX view of the services after completion of the 802.16d and 802.16e standards. This vision suggests a comprehensive wireless network connecting remote farms, factories, small offices/home offices, commercial buildings, and mobile terminals with a number of wireless connections suitable for deployment in remote areas and developing countries with limited existing wiring for the backbone of the network. Thus, it is possible to view WiMax technology as something in between WLANs and traditional cellular telephone networks. The WiMAX forum defines a more complex architecture than that of WLANs (see Chapters 6 and 8) to support quality of service and later to support mobility using outdoor antennas, which makes it closer to cellular networks (see discussion below). However, the transmission technology is based on OFDM and it has been impacted by

HIPERLAN and other WLAN standards. Next, we consider the general architecture of WiMax followed by a brief description of the physical and MAC layers of WiMax.

13.4.1   General Architecture of WiMax

In WiMax [Das06], the move towards a flat architecture (as is the norm in LTE discussed later) has been considered. The IEEE 802.16 standard only specifies the physical and medium access control layers and does not specify the network architecture, but this is part of the responsibilities of the WiMax Forum. The general types of architectures of WiMAX are shown in Figure 13.15 with a pared down schematic version of the 3G architecture (see Chapter 11) for comparison. In WiMax, there are three sub-systems as shown in Figure 13.16: the Mobile Subscriber Station (MSS), the Access Service Network (ASN), and the Connectivity Service Network (CSN).

In the hierarchical case, the WiMax network resembles a 3G network to a large extent. The mobile subscriber station is like the user equipment in 3G and LTE networks. The access services network comprises of *base stations* and an *access services network gateway*. Base stations are connected to each other and the gateway (that may handle multiple base stations) connects to the connectivity services network. The gateway also houses a foreign agent to handle mobility (see Chapter 6 for a discussion of mobile IP). The connectivity services network includes a home agent function to handle mobility as well as *accounting*, *authentication*, and *authorization* functions for security and billing along with user profiles. The connectivity services network is attached to the Internet.

In the flat architecture, gateway functionality is included in the base station itself. It is possible to view the flat architectrure as having a gateway for each base station. The base station/gateway

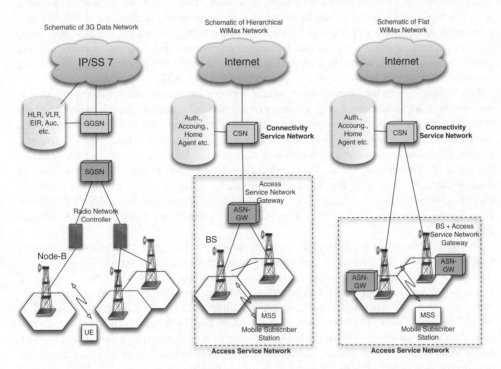

Figure 13.15   Hierarchical and flat architectures of WiMax.

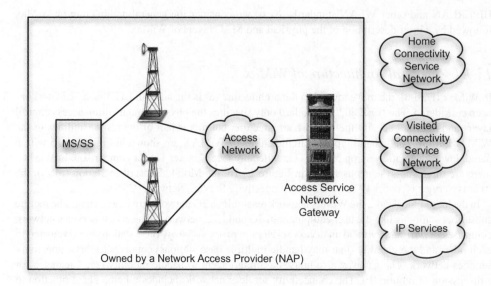

Figure 13.16   General architecture of WiMax.

now directly connects to the connectivity services network. This reduces the number of interfaces (e.g., gateway–gateway interface and base station–gateway interface), reduces the latency through a reduction in the number of hops, and allows flexibility to the base station in scheduling transmissions, as radio resources are handled locally rather than in a remote access services network gateway. This approach, where the base station handles the radio resources and scheduling locally is similar to the approach taken in HSPA, as described in Chapter 12.

The WiMax architecture further differentiates between the access and network providers. A *Network Access Provider* consists of a number of access service networks and a *Network Service Provider* owns the connectivity service network that supports the access service networks. This allows for the possibility of different types of service providers in WiMax. WiMAX defines a number of interconnections between the entities in Figure 13.16 to support multi-vendor operation that we do not discuss in this book. The architecture is flexible and can adapt to different hardware configurations allowing adaptation to variety of fixed and mobile terminals and different types of BSs (those handling larger coverages for example). Mobility is managed using Mobile IP with the home agent in the connectivity services network and the foreign agent usually in the access services network.

Figure 13.17 shows an overview of the protocol stack of the IEEE 802.16 standard. In a manner similar to all other 802 standards, this standard also defines only two layers. The MAC layer sandwiches the common MAC sublayer with convergence and security sublayers. A variety of *Convergence Sublayers* services for differentiated treatment of data are defined to support QoS and to describe how wireline technologies such as Ethernet, ATM, and IP are encapsulated on the 802.16 air–interface. The MAC layer of the standard also describes a secure communications procedure using secure key exchange during authentication, and encryption during data transfer. These features facilitate application development as well as faster and more reliable security and they do not exist in 802.11 MAC protocol stack. In addition, the MAC layer of WiMAX specifies power saving mechanisms for sleeping and idle terminals and supports handoff mechanisms. These

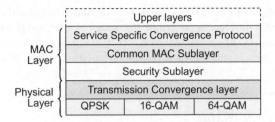

Figure 13.17   WiMax protocol stack.

features of WiMAX are very similar to those of cellular networks. The physical layer also has its own convergence protocols to support rate adaptive operation for different modulation techniques.

13.4.2   MAC Layer of WiMAX

The IEEE 802.16 standard spells out a common MAC that employs a scheduling algorithm for centrally assigned connection based access. In the original IEEE 802.16 standard, the medium access was based on TDMA and in the later versions of the standard, namely IEEE 802.16d/e, OFDMA has been used which can be interpreted as a modern implementation of FDMA/TDMA schemes.

The IEEE 802.16 standard supports both time division duplexing (TDD) and frequency division duplexing (FDD) operations in both licensed and unlicensed bands. WiMax can operate in the FDD mode where local spectrum regulatory requirements comprise of paired spectrum bands that either bar TDD or are more appropriate for FDD deployment. While FDD is suitable for wider coverage for two way interactive and delay sensitive telephony with symmetric traffic requirement, TDD enables adjustment of the downlink and uplink ratio to efficiently support asymmetric traffic characteristics for applications such as Internet access. In such cases, the downlink date rates may be substantially larger than uplink data rates. Further, with TDD, it is possible to make use of channel reciprocity for better support of channel prediction based link adaptation, channel estimation for MIMO, and other "closed loop" advanced antenna technologies where the transmitter can use feedback from the receiver to adjust the way it transmits information. TDD only requires a single frequency band for both downlink and uplink providing greater flexibility for adaptation to varied global spectrum allocation (e.g., in situations where no paired spectrum bands exist) and resulting in a less complex implementation of RF circuitry.

Compared to the contention based IEEE 802.11 MAC which employs CSMA/CA, the MAC of the IEEE 802.16 uses connection oriented, centrally assigned access which is better suited for the support of QoS and consequently a better voice quality traditional telephone connections and emerging VoIP and IP-TV applications. The IEEE 802.11 MAC (see Chapters 4 and 7) uses contention based CSMA/CA with a limited support for QoS that is provided primarily by changing the waiting times (variable interframe-spacing; IFS), using the point co-ordination function (PCF) that employs polling, and a request to send/clear to send mechanism to protect against hidden terminals. In IEEE 802.16, when a time slot with certain sub-carriers is allocated by the base station or the access point, the time slot can enlarge and contract but other mobile stations cannot use the same slots. Under high traffic loads, this approach is more stable and bandwidth efficient than the IEEE 802.11 contention access. The advantage of contention based access like CSMA/CA is simplicity – there is no signaling necessary to reserve channels or to indicate who should occupy what parts of the channel and when. This contrast between the assigned and random access MAC is one of the fundamental differences governing the design of short range wireless networks and wide area wireless networks. In general assigned access techniques are better suited for real-time streaming applications such as telephony and contention based access is more suited for bursty data applications.

13.4.3   PHY Layer of WiMax

The physical layer of the legacy 802.16 LMDS systems was defined for the 10–66 GHz bands using three single carrier modulation options, QPSK, 16-QAM and 64-QAM, with bandwidths of 20, 25, and 28 MHz, respectively. Note that the bandwidths considered here are much larger than the coherence bandwidths that may be expected in a wide area mobile environment. For example, with an RMS delay spread of 5 µs, the coherence bandwidth is 40 kHz, as discussed in Example 13.1, compared to the bandwidth of 20 MHz with single carrier transmission. The highest data rate of 134 Mbps is associated with 64-QAM and a bandwidth of 28 MHz. At these high carrier frequencies (> 10 GHz) and by employing only simple modulation techniques, the coverage of a base station would be limited to at most a few miles in line of sight (LOS) environments. Thus, the technology was suitable only for fixed last-mile access. This severely limited the practicality of the single-carrier technology. This shortcoming was considered to be one of the major reasons for the failure of the original IEEE 802.16 standard to become a commercial winner. In terms of coverage, IEEE 802.11 technology operating at 2.4 GHz already existed in the market, and at a reasonable price, and it could provide a cheaper and possibly better solution than LMDS for last-mile wireless Internet access.

The revised IEEE 802.16d standard resorted to OFDM transmission technology with 256 sub-carriers in the 2–11 GHz bands using QPSK, 16-QAM, and 64-QAM modulations to support up to 75 Mbps in 20 MHz of bandwidth. The carriers from different users are combined using Orthogonal Frequency Division Multiple Access (OFDMA) with a total of 2048 carriers for all users. OFDMA, as described previously, is a multiple-access or multiplexing scheme that provides multiplexing operation of data streams from multiple users across frequency and time in a flexible manner.

The lower frequency of operation and using the more robust OFDM transmission (which had by then emerged as the transmission scheme of choice for IEEE 802.11*a* and *g* standards in indoor applications) was expected to extend the outdoor coverage for point to multipoint 802.16d base stations to around 3–5 miles (4.8–8.0 km), even in non-LOS conditions. The IEEE 802.16e standard operates in frequency bands at 2–6 GHz with the objective of supporting *mobile* terminals. This standard uses Scalable OFDMA (S-OFDMA) to carry data supporting channel bandwidths of between 1.25 and 20 MHz with up to 2048 sub-carriers. The scalability is supported by adjusting the FFT size while fixing the sub-carrier frequency spacing. Changing the size of the FFT and subsequently the number of subcarriers adds another degree of flexibility to the choice between QPSK, 16-QPSK, and 64-QAM for individual carriers providing a more flexible environment for rate adaptive transmissions.

13.5   Long Term Evolution

Long Term Evolution (LTE) was developed by the 3GPP standards organization as a way to migrate from CDMA based 3G UMTS systems to a cellular system that has excellent support for data traffic as well as real-time applications such as Voice over IP. The entire communications in an LTE network employs IP packets and everything is packet switched as against traditional circuit-switched connections in 2G and 3G cellular networks. OFDMA is employed as the transmission/medium access scheme on the downlink while SC-FDMA is employed as the transmission/medium access scheme on the uplink. In what follows, we briefly describe the LTE network architecture, packet flow, framing in the downlink and uplink, and some operational aspects. The reader is referred to [Gho11] for more details.

13.5.1 Architecture and Protocol Stack

In this section, we describe the typical network architecture in LTE networks and also discuss how data packets are created and how they move in an LTE network.

Flattened Network Architecture: One of the main objectives of LTE was to reduce the latency for user data (so-called user plane latency), that is, the time it is transmitted by the mobile station to the time it leaves the radio access network, to something that is as small as 5 ms or less. For this reason, the network architecture in LTE was flattened as shown in Figure 13.18. As compared to 3G or 2G cellular architectures, there is little hierarchy in the network architecture, with most entities performing the needed functionalities.

The network is divided into an *evolved packet core* or EPC and an *evolved radio access network*. Notice that the radio access network consists of the user equipment or mobile station and an enhanced base station that is called *evolved Node-B* or *eNode-B*. The eNode-B in LTE performs all of the functions of the base station controller in GSM (see Chapter 11) or the radio network controller in 3G UMTS systems (see Chapter 12). eNode-Bs are interconnected as needed. For example, if there is a potential for handoff between two e-NodeBs, there is a connection between them to enable them to communicate. The functions of the serving GPRS support node – SGSN, are incorporated into the mobility management entity (MME) shown in Figure 13.18. The serving gateway (S-GW) may be physically co-located with the MME or be separate. The S-GW lies between the radio access part and the core network. All the radio transmissions end in the S-GW.

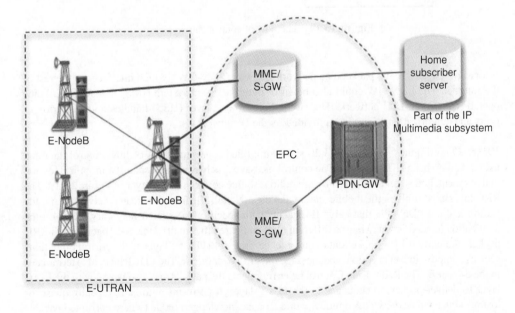

Figure 13.18 General architecture of LTE.

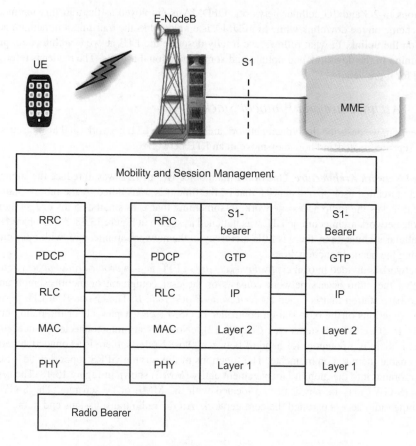

Figure 13.19   The flow of control data in LTE.

It forwards packets to the packet data network gateway (PDN-GW), which interfaces to the rest of the Internet. The PDN-GW could also be an IP multimedia subsystem (IMS) that supports Voice over IP calls in an LTE network. The Home Subscriber Server (HSS) handles authentication of users and authorization of services provided by the IP network.

Packet Flow: Figures 13.19 and 13.20 show simplified representations of how control data and user data flow in an LTE network. The control messages such as paging, security, mobility, session management, and so on are carried by the radio resource control (RRC) layer in Figure 13.19. The RRC layer exists only in the mobile station and the e-NodeB. Bearers are created between a mobile station and the PDN-GW that carry IP packets with specific QoS which may vary based on the application (e.g., voice may have a different QoS compared to e-mail). Bearer IP packets and RRC packets are carried by a packet data convergence protocol (PDCP) layer that performs functions such as compression, encryption, and message integrity over the air. The PDCP layer also terminates in the e-NodeB. The Radio Link Control layer fragments the packets and also works with the MAC layer to deliver packets in sequence to the PDCP layer. It performs automatic repeat request for lost packets in an acknowledged mode, while an unacknowledged mode (where corrupted packets

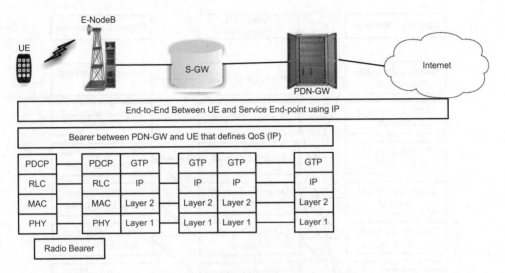

Figure 13.20 The flow of user data in LTE.

are not retransmitted) is also possible. A transparent mode is used for random access where the RLC is not really used. The MAC layer handles selection of modulation format, coding, MIMO scheme, transmit power levels, and error correction for a given packet. The physical layer transmits the packet using physical resource blocks described later. GTP is the GPRS tunneling protocol mentioned in Chapters 11 and 12. The S1-bearer carries data along the S1-interface between an e-NodeB and an MME/S-GW.

For the bearer IP packets that have the mobile station and the PDN-GW as the end points (see Figure 13.20), there may be a guaranteed bit rate with different packet error rates and latency specifications or a best effort service where the bit rates are not guaranteed. The guaranteed bit rate is used for applications such as real-time voice/video conversations, gaming, and streamed media. The best effort service is used for e-mail, file sharing, signaling for Voice over IP, and so on.

Channels that map to the Radio Level: As in UMTS, LTE specifies logical channels, transport channels, MAC layer control information, and physical channels. Logical channels are created between the RLC and the MAC layer and contain the functionality needed for network operation. Transport channels that are created between the MAC and the physical layer include the modulation, coding and other details. MAC layer control information, also carried by specific physical channels, includes scheduling information, power control commands, and so on for physical layer procedures. The physical channels actually carry the transport channel data or MAC layer control information on the air. As an example, the logical broadcast control channel that contains system information is mapped into two transport channels – a downlink shared channel and a broadcast channel. The downlink shared channel is carried by a physical downlink control channel on the air while the broadcast channel is carried by a physical broadcast channel on the air.

A number of logical, transport, and physical channels are defined in LTE, which we do not discuss in detail in this book. Figure 13.21 shows a mapping between the various logical, transport and physical channels as well as the mapping of the MAC layer control information to physical channels. The MIMO options supported for the physical channels also vary. Spatial multiplexing which increases the data rate is primarily used with the physical downlink shared channel, while the other channels mostly employ open loop transmit diversity similar to the Alamouti code discussed

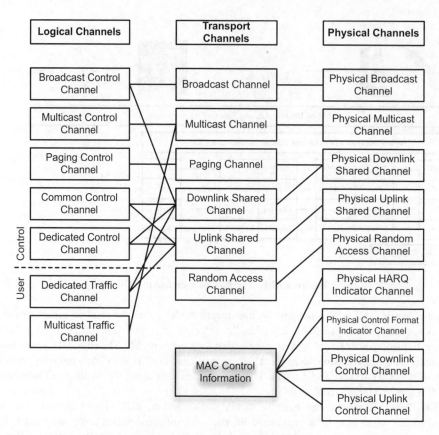

Figure 13.21 Mapping between logical, transport, and physical channels.

in Section 13.3. Not all modulation schemes can be employed on all the physical channels. For example, the physical HARQ indicator channel can only use BPSK as the modulation scheme while the physical downlink shared channel which carries user data can use 64-QAM, 16-QAM, and QPSK as the modulation schemes.

13.5.2   Downlink in LTE

As mentioned previously, LTE supports a variety of bandwidths and data rates. Obviously, smaller bandwidths can only support smaller data rates, but the smallest bandwidth supported (1.4 MHz) also has a higher overhead in terms of a guard band, which is 22.8% of the bandwidth (see Example 13.5 below). Table 13.1 shows the bandwidths supported on the downlink and some of the associated

Table 13.1   Transmission bandwidths supported on the downlink in LTE

Bandwidth (MHz)	1.4	3	5	10	15	20
FFT size	128	256	512	1024	1536	2048
Number of sub-carriers	72	180	300	600	900	1200
Number of PRBs	6	15	25	50	75	100

parameters (the FFT size and the number of sub-carriers). Recall that each sub-carrier in LTE is 15 kHz wide.

Example 13.5: Guard Band Overhead in LTE

Let us compute the overhead from the guard band in LTE for some example transmission bandwidths. With a transmission bandwidth of 1.4 MHz, the number of sub-carriers is 72. The bandwidth occupied by the sub-carriers is 72×15 kHz $= 1.08$ MHz. The rest of the bandwidth of 0.32 MHz is used as a guard band for a total overhead of $0.32/1.4 = 22.8\%$. On the other hand, with a 20 MHz transmission bandwidth, there are 1200 sub-carriers which occupy 1200×15 kHz $= 18$ MHz. The guard band occupies 2 MHz for an overhead of $2/20 = 10\%$.

The frame structure on the downlink in the case of frequency duplexing is shown in Figure 13.22. Downlink transmissions are divided into radio frames that are 10 ms in time. This is identical to the frame size in UMTS. However, the radio frame is itself made of 1 ms sub-frames with each sub-frame comprising of two time slots. The 1 ms sub-frame compares with the reduced frame size of 2 ms in HSPA that enables the base station to rapidly schedule mobile stations based on the quality of the link (and on fairness) to exploit multi-user diversity. Each time slot in a sub-frame is

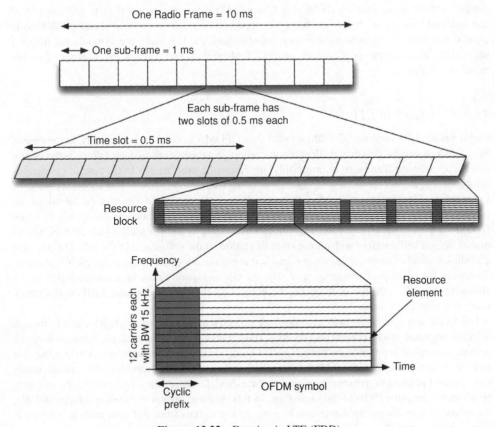

Figure 13.22 Framing in LTE (FDD).

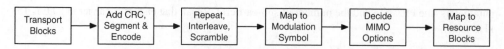

Figure 13.23   Processing a transport channel message on the downlink.

made up of seven OFDM symbols. As discussed previously in Example 13.4, in the most common case, the cyclic prefix for the first OFDM symbol is longer than the cyclic prefix of the rest of the symbols. Each OFDM symbol is made of at least 12 sub-carriers. A 0.5 ms time slot comprising of seven OFDM symbols, each with 12 sub-carriers of bandwidth 15 kHz each is called a *physical resource block* or PRB in LTE. Two PRBs is the minimum amount of radio resources allocated to any mobile on the downlink. Note that the bandwidth of a physical resource block is 12×15 kHz $= 180$ kHz. In the time division duplexed mode, the frame structure is similar, except that special sub-frames are needed to separate the uplink and downlink transmissions on the same carrier.

Figure 13.23 shows how the transport channels are processed on the downlink. A transport channel message is segmented into blocks and a cyclic redundancy check code is added to the block. The blocks are further fragmented to be suitable for encoding using convolutional coding or turbo-coding. These two coding schemes are used for error correction, while the cyclic redundancy check is used for error detection with hybrid automatic repeat request (HARQ), which is very similar to the scheme used in HSPA (see Chapter 12). Then the encoded units are interleaved and repeated if necessary before the bits are mapped into a modulation symbol and MIMO mode (spatial multiplexing or transmit diversity, antenna port, etc.). Finally, the symbols are mapped into the physical resource blocks made up of OFDM symbols and carriers and transmitted on the respective antennas.

13.5.3   Uplink in LTE

On the uplink, LTE employs SC-FDMA rather than OFDMA as described previously. In the case of SC-FDMA, it may be possible to allocate sub-carriers in a contiguous manner (which is the option that is currently used) and also to (uniformly) distribute the sub-carriers. In the case of contiguous sub-carriers, it is possible to view SC-FDMA as being similar to FDMA, in that separate chunks of spectrum are allocated to mobiles that are transmitting at the same time. Once again, the allocation is made in units of the physical resource block which is 180 kHz wide and comprises of seven OFDM symbols in the typical case. This allocation is made by the e-NodeB which ensures that no mobile station will interfere with another mobile station on the uplink within the cell. The physical downlink control channel carries the information about the allocated PRBs to the mobile stations. To exploit the benefits of frequency diversity, the channel quality may be a consideration in the allocation, as well as the use of frequency hopping (e.g., shifting the allocated PRBs in frequency every few time units or so).

Unlike the downlink, most mobile stations are expected to have only a single RF chain (although multiple antennas may be present on a device). Thus, transmission using multiple antennas is not an option, although it is possible for the e-NodeB to indicate which of the antennas provides the best quality of the signal. This is similar to selection diversity, and in this situation, the mobile station uses the best antenna to transmit the signal to the e-NodeB. Sometimes, two mobile stations may be allocated the same PRBs at the same time. In this scenario, which is called *multi-user MIMO*, the e-NodeB uses its multiple antennas to separate the signals from different mobile stations. It is possible to view this as being similar to spatial multiplexing, although the transmissions are from different transmitters. The processing of transport blocks on the uplink is similar to that of the downlink (see Figure 13.23) with the exception of the DFT precoding that is needed for SC-FDMA.

13.5.4   LTE Operational Aspects

In this sub-section, we describe some of the interesting system level operational issues in LTE such as cell search, scheduling and channel access, network planning, mobility management, and security.

Cell Search: When a mobile station first powers up, it needs to discover the services that are available in its geographical area. In Chapter 12, we discussed the cell search process in CDMA systems where the pilot channel is employed in 2G systems to discover the base stations that are providing service in the area. In OFDMA systems, things get more complicated because every cell may not be using the same frequencies. Moreover, the support for flexible bandwidth implies that the mobile station may also have to scan for multiple transmission channels of different sizes.

In LTE, the use of frames and physical resource blocks is exploited for cell search also. In a 10 ms frame, two synchronization signals are transmitted two times. They are called the primary synchronization signal (used for time slot timing and physical layer ID) and the secondary synchronization signal (used for frame timing, cell ID, and other parameters). As shown in Figure 13.22, a 10 ms frame has 10 sub-frames. Each sub-frame has two time slots, for a total of 20 slots in a frame. The primary synchronization signal is carried in the last OFDM symbol of the first and 11th slots. The secondary synchronization signal is placed just prior to the primary synchronization signal. From Table 13.1, we observe that irrespective of the transmission bandwidth, at least six PRBs are always available. The synchronization signals are transmitted in the central six PRBs in all cases, so that the mobile can detect these signals. This is illustrated in Figure 13.24.

To make the detection of the primary synchronization signal easy, LTE uses the elements of a length 63 Zadoff-Chu sequence (with one element deleted and zeros in the edges) as the data symbols on the 72 sub-carriers making up the central six PRBs. The Zadoff-Chu sequence is a complex sequence (elements of the sequence are complex numbers) that has an ideal periodic autocorrelation (a peak for zero lag and zero elsewhere) enabling the mobile device to quickly detect the primary synchronization signal. The secondary synchronization signal uses M sequences (see Chapter 12) of length 31. Once the primary synchronization signal is detected, the mobile station knows the slot boundary. It still does not know the length of the cyclic prefix (although in

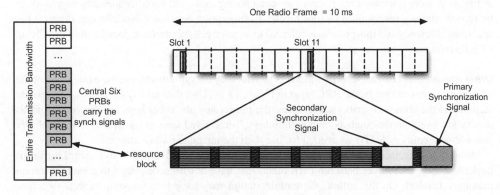

Figure 13.24   Cell search in LTE

the typical situation it is 4.7 µs long), which has to be blindly detected by finding the secondary synchronization signal. Once both the signals are correctly identified, the broadcast channel can be decoded to obtain additional cell specific information.

Scheduling and Channel Access: As mentioned previously, LTE exploits multi-user diversity in frequency and time. The e-NodeB is responsible for scheduling the resources for mobile stations both on the uplink and downlink based on channel quality and fairness. The physical downlink control channel is used to intimate the allocation of resources to mobile stations. The channel quality is reported to the e-NodeB by the mobile stations. The channel quality may be over the entire bandwidth (e.g., 20 MHz) or over a certain set of PRBs. Reporting of the channel quality incurs an extra overhead. While reporting the average quality over the band is less expensive, the benefits of diversity are better with a reporting of quality at a finer granularity on the order of a few PRBs. The channel quality may be reported by the mobile stations periodically or as needed. The modulation and coding depend on the reported channel quality. Moreover, transmissions use hybrid ARQ or HARQ where retransmissions are combined with the original transmission to improve reliability (see related discussion in the case of HSPA in Chapter 12).

As in the case of most cellular wide area systems, in the beginning, a random access procedure by mobile stations is necessary for allocation of resources. Mobile stations may by synchronized to the network or may have lost synchronization on the uplink although cell search would have allowed synchronization on the downlink. The random access procedures accommodate both cases. Mobile stations transmit a preamble on the physical random access channel to the e-NodeB. If multiple mobile stations transmit, the e-NodeB may be able to separate these transmissions and allocate some limited resources to each of the requesting mobile stations. Upon obtaining the initial resources, mobile stations can request additional resources from the e-NodeB. Zadoff Chu sequences are used to identify specific mobile stations. Where necessary, the uplink timing information is also transmitted by the e-NodeB to the mobile stations.

Network Planning: As all cellular systems, frequency reuse is necessary in LTE systems. In general, the frequency reuse problem described in Chapter 5 applies here as well. LTE employs the idea of *fractional frequency reuse,* which is similar to reuse partitioning, but at a resource block level to increase spectrum efficiency (see Chapter 5 for details). The basic idea here is that transmissions in specific frequency groups are at different transmit powers. Those at a lower transmit power can be used only by mobile stations in the center of a cell, but they do not interfere with transmissions in neighboring cells. Thus, almost universal frequency reuse is possible.However, transmissions at higher powers can cause interference in neighboring cells and such frequencies may have to be reused sparsely (every third or fourth cell). Coordination between e-NodeBs can improve the spectrum efficiency and this process is referred to as inter-cell interference coordination or ICIC in LTE literature.

Other operational Issues: Security in LTE is provided by encryption and message integrity for the control messages carried by the RRC layer in Figure 13.19. User data is only encrypted. The mobile station and the home subscriber server share the master key and other keys are derived from this master key using nonces such as random numbers. The derived keys are delivered to the mobility management entity or S-GW as needed for use with the intended mobile station.

Mobility management primarily involves eNodeBs, which have the ability to forward (or flush) packets at the PDCP layer between each other, that have not been delivered to a mobile station making a handoff. On the uplink, the mobile station may have to retransmit packets to a new e-NodeB (see also Chapter 6).

13.5.5 Miscellaneous

LTE Channel Models: The International Telecommunications Union and the WINNER project that was funded by the European Union have developed channel models suitable for the evaluation of LTE. The models are in many respects similar to the channel models discussed in Chapters 2 and 7. The path-loss models have a frequency correction factor and the shadow fading has a standard deviation that is dependent on the environment. It varies between 3 dB in LOS situations and 8 dB in NLOS situations. Multipath profiles with exponentially dropping multipath powers are available for various scenarios.

Support for Voice Conversations: Support for voice telephony in LTE is not similar to that in traditional 2G and 3G cellular networks because there are no circuit switched connections that are readily available. Although bit rates may be guaranteed for certain services, the data is still transmitted in a packet switched mode. This implies that Voice over IP is an attractive option for voice conversations. As of this writing, carriers are converging on Voice over LTE (VoLTE) as a standard for voice conversations with the IP multimedia subsystem (IMS) as the backbone support. Details of these standards are beyond the scope of this chapter.

13.6 LTE Advanced

Looking forward to the future, standards bodies are already considering advances to LTE. The set of standards corresponding to this are sometimes referred to as LTE-Advanced. Some of the primary features that are being discussed for this standard are enumerated below.

1. To support data rates on the order of 1 Gbps, it becomes necessary to have more bandwidth. Towards this, LTE-advanced is expected to support carrier aggregation, similar to IEEE 802.11n. Instead of stopping at 100 PRBs in 20 MHz, it may be possible to aggregate five carriers of 20 MHz for a total of 500 PRBs per link. This may also increase the benefits of frequency and multi-user diversity.
2. Support for small cells such as femtocells to increase reliability and capacity is being actively considered. Coordination between small cells and macro-cells is considered part of LTE-advanced.
3. LTE mostly supports two transmit and two receive antennas. LTE-advanced may support up to eight antennas on the downlink.
4. The architecture of the network may be modified to include relay nodes and heterogeneous entities with varying capabilities. Cooperative transmissions to enhance reliability or capacity are also being considered in this respect.

Questions

1. Explain the reasons why OFDMA was the choice for 4G cellular systems over CDMA.
2. Compare the channel bandwidths in 4G systems and the corresponding channel bandwidths in 3G and 2G cellular systems.
3. What are the major architectural changes in 4G systems compared to 3G and 2G systems?
4. Explain how OFDM provides robustness to frequency selective fading?
5. What is the cyclic prefix? Why is it important in OFDM?
6. What is multi-user diversity? How is it provided in OFDMA systems?
7. What is the peak to average-power ratio problem? What approach is used in LTE to address this problem?

8. What are the three types of benefits of MIMO?
9. Explain the idea of transmit diversity and space–frequency block codes.
10. Compare the PHY layer of 4G cellular systems with that of IEEE 802.11 WLANs.
11. Compare the MAC layer of 4G cellular systems with that of IEEE 802.11 WLANs.
12. How is the uplink in LTE different from the downlink in LTE?
13. What problems arise when performing cell search in LTE? How are these addressed?
14. What is fractional frequency reuse? Compare it with reuse partitioning in Chapter 5.
15. How are voice conversations carried over the air in LTE?

Problem 13.1

Consider a four carrier OFDM system. Let the frequencies of the carriers be $f_c, f_c + 1/T, f_c + 2/T,$ and $f_c + 3/T$, where $T = 100/f_c$ is the symbol duration. If BPSK is used on all carriers and $f_c = 1$ MHz, plot one OFDM symbol where the bits carried by each carrier is "0". Assume that the amplitude of each carrier is 1.

Problem 13.2

Consider the space–time block code given below where the columns represent the antennas and the rows the time slots. How many transmit antennas are required for this code? What is the rate of this code? Explain how you arrive at the answers.

$$
S = \begin{bmatrix}
s_0 & s_1 & s_2 & s_3 \\
-s_1 & s_0 & -s_3 & s_2 \\
-s_2 & s_3 & s_0 & -s_1 \\
-s_3 & -s_2 & s_1 & s_0 \\
s_0^* & s_1^* & s_2^* & s_3^* \\
-s_1^* & s_0^* & -s_3^* & s_2^* \\
-s_2^* & s_3^* & s_0^* & -s_1^* \\
-s_3^* & -s_2^* & s_1^* & s_0^*
\end{bmatrix}
$$

Problem 13.3

Show that the space–time code in Problem 13.2 satisfies the following property for $m = 3$: $\mathbf{S}^\dagger \times \mathbf{S} = \left(\sum_{j=0}^{m} |s_j|^2 \right) \mathbf{I}$, where $\mathbf{I}$ is the identity matrix of the appropriate size and $\dagger$ represents transposition followed by taking the complex conjugate of every element in a matrix. This is one of the properties of the space–time block codes derived using orthogonality design.

Project 13.1: Simulation of a simplified IEEE 802.11a/g OFDM

The OFDM has been adopted in several wireless LAN standards, including IEEE 802.11a, IEEE 802.11g, HIPERLAN/2, as well as the local multipoint distribution service (LMDS) and digital audio broadcast (DAB) systems. In this project we implement IEEE 802.11a, g, and HIPERLAN-2 OFDM modulation and demodulation techniques in the MATLAB® software. The simulation model to be implemented in this project is a simplified version of the IEEE 802.11a standards. The complete standard document can be downloaded from *IEEE Xplore* (http://ieeexplore.ieee.org).

Figure P13.1 shows a simplified version of the IEEE 802.11a/g OFDM system, which implements baseband OFDM modulation and demodulation techniques. A complete simulation needs additional

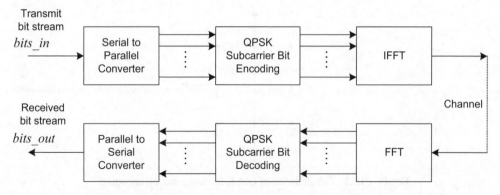

Figure P13.1   Simplified simulation model of a baseband OFDM transceiver.

details to include error control coding, interleaving, cyclic prefix, pilot subcarriers, waveform shaping, carrier modulation, and the effects of radio propagation channels that are beyond the scope of this project.

The system model to be implemented in this project is shown in Figure P13.1 and the simulation parameters are listed in Table P13.1. The serial transmit bit stream is first converted to parallel data by a Serial to Parallel Converter, which outputs two serial bits at each of its output line (52 output lines in this simulation, since N_st = 52). Then each group of two bits is encoded into a complex QPSK symbol with the constellation shown in Figure P13.2. Following that, 52 complex QPSK symbols are mapped to the input of the IFFT block according to the mapping scheme shown in Figure P13.3, where the *null* input are set to zero. The IFFT is performed to obtain a time–domain OFDM symbol (with 64 complex samples) of the duration of 3.2 μs (T_FFT = 3.2 μs).

At the receiver, the transmitted bit streams can be recovered through inverse process, as shown in the lower part of Figure P13.1.

In this project we implement the baseband OFDM modulation and demodulation system, shown in Figure P13.1, in MATLAB. The following are detailed requirements on the implementation and deliverables:

a. Implement all the blocks shown in Figure P13.1, except IFFT and FFT, for which you may use the MATLAB function *ifft*() and *fft*().
b. Generate a vector of 104 random binary bits, *bits_in*, and use this vector as the input of the system shown in Figure P13.1 that you implemented. Receive the bit vector at the output of

Table P13.1   Simulation parameters for OFDM project

Parameter	Value
N_st: number of subcarriers	52
N_FFT: number of IFFT/FFT inputs	64
BW: bandwidth	20 MHz
delta_F: subcarrier frequency spacing	0.3125 MHz (= BW/N_FFT)
T_FFT: IFFT/FFT period	3.2 μs (= 1/delta_F)

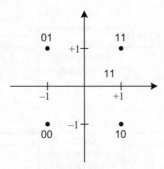

Figure P13.2 Constellation of QPSK for subcarrier bit encoding.

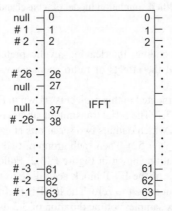

Figure P13.3 Inputs and outputs of IFFT.

the receiver, *bits_out*. Note that the bit vectos *bits_in* and *bits_out* are both of the same length of 104. Verify in your MATLAB program that the received bit stream is exactly the same as the transmitted bits by observing the output of the following code segment, which should be included in your MATLAB program:

```
if sum(abs(bits_in - bits_out)) == 0
    disp('Transmitted bits are successfully received !');
else
    disp('Transmitted bits are received in ERROR !');
end;
```

c. Plot the real part and the image part of the time–domain OFDM symbol, that is, the in-phase and quadrature signals, that you generated in your simulation in separate figures with proper time axis. Since the duration for an OFDM symbol is 3.2 μs according to Table P13.1, the time axis should be from 0 to 3.2 μs.

d. Plot the spectrum of the OFDM symbol that you generated in this simulation. Since the bandwidth is 20 MHz, the frequency axis of the plot should be from −10 to 10 MHz.

The required submission for this project includes the MATLAB source code for the simulations and the generated plots.

Part V
Wireless Localization

14

Geolocation Systems

14.1 Introduction

As discussed in Chapter 1, the association of information with *time* (temporal) and *location* (spatial) is extremely important for human beings and consequently, measuring time and location have become an essential part of information processing. Today, cellphones are replacing traditional watches as the instruments to look up time. At the same time, they are also becoming capable of locating themselves and associating their positions with maps through global or local coordinates.

Geolocation, position location, localization, and radiolocation are terms that are widely used today to indicate the ability to determine the location of a mobile station in different environments. Location usually implies the coordinates of the mobile station that may be in two or three dimensions and usually includes information such as the latitude and longitude where the mobile terminal is located. In indoor areas and within buildings, alternative coordinates and visualization techniques may be employed to indicate the location of a mobile station. In general, the essence of positioning or localization is a map that uses such coordinates, some landmarks that we can identify on the map, and a method to measure the distance from those landmarks. Historically engineers and scientists have designed a variety of maps for navigation and were used by travelers from one city to another and then through the city and ultimately inside particular buildings where specific information can be used. More recently, the advances have lead to ideas related to localization and navigation inside the human body.

Geolocation technologies are gaining prominence in the wireless market for several reasons, primarily the United States Federal Communications Commission (FCC) mandate requiring all wireless cellular carriers to be able to provide the location of emergency 911 callers to a *public safety answering point* (PSAP). However, geolocation technology has proved to be significant for both military and commercial applications in general beyond emergency location. The use of wireless devices such as cell phones, PDAs, and laptops has become the enabler of viable location based services and applications that need position location information [Bar03; Dru01; War03]. Examples of commercial location based services include locating patients and equipment in a timely fashion in hospitals, locating children and pets for personal and residential applications, and concierge and location aware services (e.g., locating the nearest coffee shop or providing information about exhibits in museums based on the customer's location). The monetary benefits of such location based services for service providers is expected to increase in the next few years. In the military and public sector, enabling soldiers, policemen, and fire fighters with knowledge

Principles of Wireless Access and Localization, First Edition. Kaveh Pahlavan and Prashant Krishnamurthy.
© 2013 John Wiley & Sons, Ltd. Published 2013 by John Wiley & Sons, Ltd.

of their location and the location of other personnel, victims, exits, dangers, the enemy, and so on proves to be invaluable.

The breakthrough in localization industry started in the past few decades through the use of RF signals for measuring the *distance* (*ranging*) between a landmark used as a reference point and a mobile electronic device. The *global positioning system* (GPS) has been the most successful positioning technique in outdoor areas and we now see the GPS receiver as an inexpensive commonplace gadget. Today, GPS technology is widely available in the civilian market for personal navigation applications. A full description of GPS system is beyond the scope of this book, but the interested reader can find much information in the literature [Kap96]. While GPS has been hugely successful, it also has several drawbacks for use in the several applications that we described above, especially in indoor areas. GPS was designed for satellite to mobile links where a direct line of sight often exists between the transmitter and the receiver. The receivers measure the time of arrival (TOA) of the received signal to measure the distance to the satellite landmarks and from that locating itself on the earth within a few meters. However the accuracy of GPS positioning is significantly impaired in urban and indoor areas, where received signals can suffer from extensive multipath effects. But it is in such urban and indoor areas that there exist a number of applications that can benefit from location information. In the mid-1990s GPS alternatives using existing RF infrastructures emerged as a new industry. Localization has now started the use of cell towers and WiFi localization for indoor areas and more recently RF signals for localization inside the human body.

In this chapter, we discuss position location issues in today's wireless networks, alternative technologies that are being investigated and standardized for position location in outdoor and indoor areas, and trends in this area. The fundamentals of the behavior of sensors used for positioning and performance bounds are discussed in Chapter 15. Practical issues related to positioning are discussed in Chapter 16. It is possible to view this chapter as a broad overview of RF based geolocation. Some topics are revisited in Chapters 15 and 16 as needed.

In the next section we consider example applications and regulatory issues related to position location. In Section 14.3, the sensing process and the location algorithms are discussed. These influence: (a) where the position location is determined (at the MS or in the wireless network), (b) the number and nature of reference locations necessary, and (c) the accuracy and precision of the estimated location. This section will also consider the standard specifications for cellular telephone systems that make use of some of these algorithms. Section 14.4 describes the location services architecture specified for use in cellular networks. Section 14.5 provides a brief overview of localization in ad hoc and sensor networks.

14.2   What is Wireless Geolocation?

Geolocation can simply be defined as the ability to locate a device or human being's position in terms of well known global or local coordinates. The coordinates can be the latitude and longitude on the earth's surface with a height above the earth. The coordinates can also be relative and local – for example, the floor of a building and the room on that floor. The position location information is usually specified with certain accuracy and precision. Accuracy refers to the error in distance from the determined position and the actual position. Precision refers to the fraction of time that the error is smaller than the above number. Any system that locates a MS needs to *sense* some characteristics related to the MS to determine its location. The position is determined using some algorithms that use these sensed characteristics as input. In addition, there are protocols that are required to transport the sensed information to entities that determine the location or provide other services. In this chapter, we will describe these aspects related to locating the position of a mobile station.

The term "location-based service" is used to denote services provided to mobile users based on their geographic location, position, or known presence. These are primarily based on a geolocation infrastructure and system put in place to obtain location information of users. As mentioned in the introduction, positioning systems have found a variety of applications both in the civilian and military environments. There are numerous such applications that are already available today such as mapping services (that provide driving directions or locations of businesses in an area), information services (that provide local news, weather, traffic, etc.), and concierge services (for making dinner reservations, movie tickets, directory services, etc.). Commercially, content, advertising and personalization services that are location dependent are being deployed today. Each of these applications requires accuracy and precision of position that is specific to its needs. In indoor areas, for applications such as inventory and asset management (such as locating wheelchairs in hospitals), the accuracy has to be within a few meters on a given floor of the building. For E-911 emergency response, the required accuracy is much lower as described later. We discuss example indoor and outdoor applications below that are becoming increasingly important.

Indoor geolocation applications traditionally have been directed towards locating people and assets within buildings. Finding mentally impaired patients in hospitals and portable equipment such as projectors, wheelchairs, and so on that are often moved and never returned to a traceable location are two common examples. The so-called *personal locator services* (PLS) [Kos00] that could also operate outdoors, employ a *locator device* that reside with a person whose location is to be determined. There are two possible scenarios – in the first case, someone requests a service to provide them with the location of the individual and appropriate steps are taken to determine the person's location. In the second case, the person is lost or in some other dire straits and can employ a panic button to request help. Here, the locator service will determine the location and provide the requested assistance. For locating equipment, only the former scenario applies.

Existing communications and computing environments, both in residential areas and offices typically have been statically configured, making the task of reconfiguration extremely complex and cumbersome requiring manual intervention. To overcome this inconvenience, smart spaces and smart office environments are being considered for deployment that can automatically change their functionality depending on the context [War03]. Such *context-aware* networks are based on awareness of who or what is present around them. With location awareness, computing devices ranging from small PDAs to desktops and Internet appliances could personalize and adapt themselves to their current set of users, each requiring their own services from the smart environment. For this purpose, not only should the smart space be aware of who is present, it should also be aware of where the user is located and whether there are other mobile devices in the vicinity. For instance, a handheld computer should be able to automatically determine the closest printer to print a document in an office environment. Such non-traditional applications also demand geolocation services.

There are several outdoor geolocation applications, the most common of which is simply the application of locating ones own self using GPS while traveling on the road. Information technology has increased the number of applications far beyond this simple self-location application. The term *telematics* used to imply the convergence of telecommunications and information processing and it has since then evolved to refer to automobile systems that combine GPS location mechanism combined with wireless communications for services such as automatic roadside assistance, remote diagnostics, and content delivery (information and entertainment) to the automobile. A good example of such a system is General Motors' OnStar system [OnStar]. *Intelligent Transportation Systems* (ITS) refer to the ability to autonomously navigate vehicles while making use of the latest traffic information, road conditions, travel duration, etc. This includes fleet management as well as automatic steering of vehicles. In order to obtain relevant information from service providers or servers across a network or the Internet, the vehicles should be able to provide their location and destination information. Alternatively, the service provider should be able to determine the vehicle's location.

14.2.1   Wireless Emergency Services

In this section, we will provide an introduction to the requirements of location based services and E-911 in wireless environments. The requirements are primarily in terms of performance metrics of interest such as accuracy and precision. We will also describe some of the regulatory aspects related to the provisioning of location information and performance metrics. We provide a brief summary of different cellular standards that we refer to later in thechapter. Cellular systems are considered in more detail in Chapter 7 of this book.

Wireless enhanced-911 or E-911 services, by far, have provided to be the biggest catalyst for investment and development of geolocation technology suitable for cellular communications. A caller on a wired telephone to an E-911 service is immediately located because the location of the fixed telephone is known with an accuracy of within a couple of rooms in a building. If the same caller is on a mobile telephone, the technology that can obtain the location is more complex. In the simplest case, the only knowledge that is available is that the caller was connected to a particular base station.

Cellular service providers may use the position information for network related issues such as handoff or location management [3GPPa; Dra98]. However, much of the accuracy and precision levels for mobile phones are being driven by the values mandated by the FCC for E-911 public safety applications in the United States. The FCC is responsible for regulatory issues related to telecommunications services in the United States. The E-911 service provides emergency assistance to callers through a public safety answering point. FCC has specified accuracy and precision values for E-911 calls. The service provider must be able to locate the caller within an accuracy of 100 m at least 67% of the time and within 300 m at least 95% of the time and provide this information to the public safety answering points. The reader is referred to [Mey96; FCC03; Ree98] for more details. In Europe, emergency services are referred to as E-112 services and regulatory authorities are in the process of specifying accuracy and precision levels for E-114.

Cellular systems use many different physical layer and networking standards making position location different in different systems. The first generation cellular systems (also called 1G systems) use analog modulation schemes and do not support any position location services. Second generation or 2G systems use digital modulation and two primary multiple access technologies – time division multiple access (TDMA) in the Global System for Mobile Communications (GSM) based systems and code division multiple access (CDMA) in IS-95 or cdmaOne systems. Third generation or 3G systems are based entirely on CDMA, but again have two primary standards: cdma2000 and UMTS. UMTS is the successor of GSM and cdma2000 is the successor of IS-95. All of the 2G and 3G standards are expected to satisfy the FCC mandate. While the positioning schemes in all of them have some common features, the standards and terms are somewhat different as discussed in Section 14.3.3.

There are no mandated positioning requirements for wireless local area networks (WLANs). Most deployed WLAN equipment follows the IEEE 802.11 standard or its enhancements like 802.11a, b, or g. There are proprietary solutions for positioning with 802.11 WLANs, which primarily use RF signatures or variations thereof described below. Accuracy with such systems ranges from a few meters to tens of meters with varying levels of precision depending on the environment (building material, architecture, campus area, etc.).

Table 14.1 Comparison of performance measures for telecommunications and geolocation systems

Telecommunications systems	Geolocation systems
Quality of service • Signal to interference ratio • Packet error rate • Bit error rate	Accuracy of service • Percentage of location requests located to within an accuracy of δ meters • Distribution of distance error at a geolocation receiver
Grade of service • Call blocking probability • Availability of resources • Unacceptable quality	Location availability • Percentage of location requests not fulfilled • Unacceptable uncertainty in location
Coverage • Where communications is possible	Coverage • Where location information is available
Capacity • Subscriber density that can be handled	Capacity • Location requests/frequency that can be handled
Miscellaneous • Delay in call set up • Reliability • Database lookup time • Management and complexity	Miscellaneous • Delay in location computation • Reliability • Database lookup time • Management and complexity

14.2.2   Performance Measures for Geolocation Systems

In this section, we consider performance benchmarking of geolocation systems primarily employing the work and discussion in [Tek98]. Wireless systems have traditionally focused on telecommunications performance issues such as quality of service, grade of service, bit error rates, capacity, reliability, and coverage. For geolocation systems, some of these performance issues are still valid although new performance benchmarks are necessary as described below. Table 14.1 compares performance measures for telecommunications and geolocation systems based on [Tek98].

One of the most important performance measures of a geolocation system is the accuracy with which the location is determined. This is similar to the bit error rate or packet error rate requirements in telecommunications systems. As in the case of BER, the actual benchmark values may be different depending on the application in question. For example, voice packets can tolerate a BER of 1%, but data packets need a BER of at least 10^{-6}. In the same way, outdoor position location applications demand a lower accuracy compared to indoor applications.

Location system accuracy is often defined as the area of uncertainty around the exact location where a percentage of repeated location measurements are reported. For example, 67% of the measurements of the location of a MS lie within 50 m of the actual location or 95% of the measurements lie within 100 m of the actual location. This accuracy heavily depends on the radio propagation environment, receiver design, noise and interference characteristics, number of redundant measurements available for the same location, and the complexity of signal processing performed. In [Kri98; Pah98], the distribution of the distance error in indoor areas that ultimately affects the area of uncertainty is discussed in detail based on measurements and simulations.

The grade of service for telecommunication systems is usually the call-blocking rate during the peak hour. In a similar manner, the probability that a location request will not be fulfilled is a measure of the grade of service for a geolocation system. The location request will not be fulfilled if location sensing measurements are not available in sufficient numbers or the measurements lead to unacceptable location accuracy.

Coverage in telecommunications systems is related to the service area where at a bare minimum, access to the wireless network is possible. For geolocation systems, coverage corresponds to the availability of a sufficient numbers of measurements of a sensed characteristic to perform a location computation.

Finally, several other issues [Tek98] are also important in geolocation systems in manners similar to telecommunications systems – delay in triggering a location measurement, location algorithm calculation time, network transmission delay, database lookup time, end to end delay between the time a location request is made and the location information is received and so on. Reliability – the mean time between failures and the mean time to repair, network management, and complexity are also important issues.

14.3   RF Location Sensing and Positioning Methodologies

In this section, we will describe *basic* positioning methodologies. In the following section, we consider a generic geolocation system architecture and define different positioning approaches. Section 14.3.2 describes RF location sensing approaches and algorithms.

14.3.1   Generic Architecture

A functional architecture of a geolocation system is shown in Figure 14.1a [Cha99]. The two essential functional ingredients for position location are the location estimation of the mobile station and sharing this information with appropriate attributes with some entity in the wireless network.

Geolocation systems *measure* or *sense* RF parameters of radio signals that travel from a mobile to a fixed set of receivers OR from a fixed set of transmitters to a mobile receiver. There are thus two ways in which the actual estimate of the location of the mobile station can be obtained. In a *self-positioning system*, the mobile station locates its own position using measurements of its distance or direction from known locations of transmitters (e.g., GPS receivers). In some cases, *dead reckoning*, a predictive method of estimating the position of the mobile by applying the course and distance traveled by a mobile station since a location was previously determined, could be employed. Self-positioning systems are often referred to as mobile-based or terminal-centric

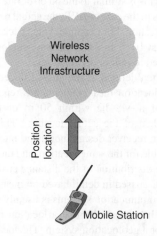

Figure 14.1   Functional architecture of indoor geolocation system.

[Mey96] positioning systems. In *remote positioning systems*, receivers at known locations on a network together compute the location of a mobile transmitter using the measurements of the distance or direction of this mobile from each of the receivers [Dra98]. Remote positioning systems are also called network-based or network-centric [Mey96] positioning systems. Network-based positioning systems have the advantage that the mobile station can be implemented as a simple transceiver with small size and low power consumption for easy carrying or attachment to assets that need tracking as a simple and inexpensive tag. In addition, it is possible to have *indirect* remote or self positioning systems where the mobile may transmit information about its location to a location control center or the location control center transmits the location of the mobile to itself through an appropriate communications channel.

Example 14.1: Indirect remote positioning

In indirect remote positioning, an E-911 public safety answering point requires the location information of a caller. If a mobile-based positioning system is used, the MS determines its own position either using GPS or signals from multiple base stations. This information has to be transmitted to the location control center by the mobile terminal through one of the base stations.

An example of a geolocation system architecture [Kos00] is shown in Figure 14.2. A geolocation service provider provides location information and location aware services to subscribers. Upon a request from a subscriber for location information about a mobile station, the service provider will contact a location control center querying it for the co-ordinates of the mobile station. This

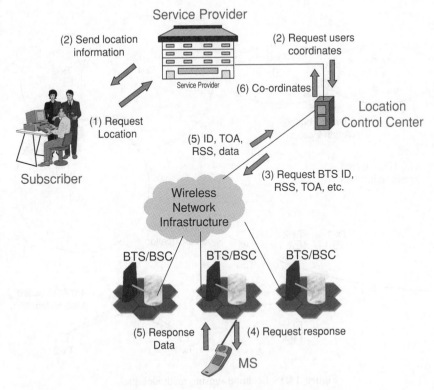

Figure 14.2   General architecture of a geolocation system.

subscriber could be a commercial subscriber desiring to track a mobile device or a public safety answering point trying to answer an E-911 call. The location control center will gather information required to compute the mobile station's location. This information could be parameters such as received signal strength, base station ID, time of arrival of signals, and so on that we will discuss later. Depending on past information about the mobile station, a set of base stations could be used to page the mobile station, and directly or indirectly obtain the location parameters. These are sometimes called *geolocation base* stations (GBSs). Once this information is collected, the location control center can determine the location of the mobile with certain accuracy and convey this information to the service provider. The service provider will then use this information to visually display the MS's location to the subscriber. Sometimes the subscriber could be the mobile station itself, in which case the messaging and architecture will be simplified especially if the application involves self-positioning.

14.3.2   Positioning Algorithms

Positioning processes use *closeness* to the point of association (POA) of a MS with the network, *measures of distance* such as the time of arrival (TOA), time difference of arrival (TDOA), phase difference or received signal strength (RSS) of signals, *measures of direction* such as the angle or direction of arrival (A/DOA) of signals, a fingerprint or signature of signal characteristics at a location, or a combination of these to estimate the location of a MS [3GPPa]. Figure 14.3 shows how some of these positioning methodologies operate. In the following discussion, we explain

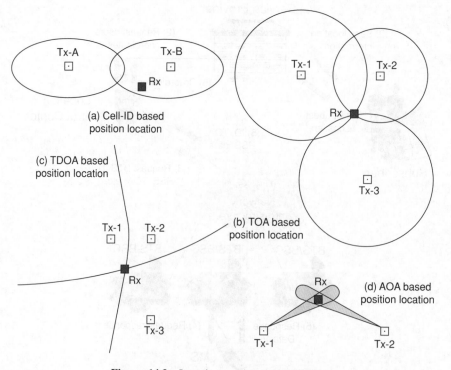

Figure 14.3   Location sensing methodologies.

positioning approaches based on self-positioning or remote positioning. However, the alternative approach (or even indirect positioning) is also possible in similar ways.

The location of a mobile terminal can be determined as follows. Let us consider for example a remote-positioning system where the geolocation base stations are together determining the MS's position (a similar approach is applicable for self-positioning systems). It is possible to exploit characteristics of radio signals transmitted by a MS to fixed receivers of known location to determine the location of the MS. The GBSs measure certain signal characteristics and make an estimate of the location of the MS based on the knowledge of their own location. The general problem can be stated as follows:

The locations of N receivers (GBSs) are known via their co-ordinates (x_i, y_i) for $i = 1, 2, 3, \ldots, N$. We need to determine the location of the MS (x_m, y_m) using characteristics of the signals received by these transmitters.

In order to determine (x_m, y_m), traditionally, the distance or direction (or both) of the MS must be estimated by several of the GBSs from their received signals. Distances can be determined using properties of the received signal such as the signal strength, the signal phase, or the time of arrival. The direction of the MS can be determined from the angle of arrival of the received signal. Alternative approaches such as closeness of match of RF signatures or closeness to a point of association have been used in recent times.

Closeness to the Point of Association

Most wireless networks use a fixed point of access to the network. These points of access could be base stations (BSs) in cellular networks and access points (APs) in WLANs. BSs or APs have radio transceivers that provide service over a specific geographical area called a "cell" (see Chapters 5 and 6). In the closeness to the point of access (POA) approach, the position of the MS is known to be somewhere within the cell. For example, in Figure 14.3a, the receiver (Rx) is known to be in the coverage area (cell) of transmitter (Tx) B. Unfortunately, the size of a cell may be as small as 100 m in microcellular areas and as large as 15 km in macrocellular areas. Thus the accuracy and precision can vary significantly depending on the density of deployment of BSs or APs and their coverage areas. In this case, the MS location (x_m, y_m) is associated with the location of the BS with coordinates (x_i, y_i).

Distance Based Techniques

If the distance to a mobile station is known from at least three distinct transmitters (whose locations are known), it is possible for the mobile receiver to construct three overlapping circles as shown in Figure 14.3b. The point where the three circles intersect is the position of the receiver. Distances between transmitters and receivers are estimated using received signal strength, the time of arrival or the time difference of arrival of a transmitted signal.

Three measurements are required to estimate the position of the mobile in two dimensions and four measurements are required for estimating the position in three dimensions. In Figure 14.3b, the need for three measurements for estimating the position in two dimensions is illustrated. If the distance between the receiver and the mobile is estimated to be d, it is obvious that the mobile could be located on a circle of radius d centered on the receiver. A second measurement reduces the position ambiguity to the end points of the chord that is common to the two circles. The third measurement provides a fix on the location of the mobile.

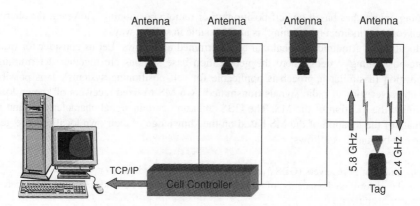

Figure 14.4   Local indoor positioning system.

Time/Time Difference of Arrival: A transmitted signal travels 0.3 m/ns in air or free space and this property can be exploited to determine the distance between the transmitter and receiver. This is the time of arrival (TOA) technique that is employed with some modifications in current GPS receivers [Kap96] as well as certain E-911 location systems. When a geolocation base station detects a signal, its absolute time of arrival is determined. If the time at which the MS transmitted the signal is known, the difference in the two times will give an estimate of the time taken by the signal to arrive at the geolocation base station from the MS. Errors in determining the time of arrival often make the intersection of the circles used to determine the location a region rather than a point. In such a case, the position location estimation algorithm will pick some point within this region as the estimated position. Wireless systems that employ the TOA (or TDOA) technique employ pulse transmission, phase information or spread spectrum techniques to form time estimates. For instance, the time difference between two signals received for either self-positioning or remote positioning, can be estimated from their cross-correlation.

Example 14.2: Commercial indoor location systems based on TOA

Recently, a few commercial product for indoor geolocation appeared in the market [Wer98].

The overall system architecture of these systems is shown in Figure 14.4. These systems use simple-structured *tags* that can be attached to valuable assets or personnel badges. Indoor areas are divided into cells with each cell being served by a *cell controller*. The cell controller is connected to a number of antennas (16 in [Wer98]) located at known positions. To locate the tag position, a cell controller transmits 2.4 GHz spread spectrum signal in the unlicensed ISM bands through different antennas in time division multiplexed mode. Upon receiving signals from the cell controller antenna network, tags simply change the frequency of the received signal to another portion of the available unlicensed bands, in either 2.4 or 5.8 GHz, and transmit the signal back to the cell controller with tag ID information phase-modulated onto the signal. The distance between tag and antenna is determined by measuring round trip time of flight. With the measured distances from tag to antennas, the tag position can be obtained using the TOA method. A host computer is connected to each cell controller, through a TCP/IP network or other means, to manage the location information of the tags. Since the cell controller generates the signal and measures round trip time of flight, there is no need to synchronize the clocks of tags and antennas.

The multipath effect is one of the limiting factors for indoor geolocation (see Chapter 2 for a discussion). Without multipath signal components, the time of arrival (TOA) can be easily determined from the autocorrelation function of the spread spectrum signal. The autocorrelation is two chips wide and the time to rise from the noise floor to the peak is one chip. If the chipping rate were 1 MHz, it would take 1000 ns to rise from the noise floor to the peak, providing a "ruler" with a thousand 30-cm increments. In this manner, a 40-MHz chipping rate, chosen for PinPoint system, provides a ruler of 25 ns that provides real-world increments of about 3.8 m. Because of regulatory restrictions in the 2.4 and 5.8 GHz unlicensed bands, faster chipping rates are not easy to achieve, and signal-processing techniques must be used to further improve the accuracy. If different frequency bands are used for uplink and downlink communications the interference between the channels can be further isolated.

If the absolute times of arrival are unknown, the *time difference of arrival* (TDOA) technique, which uses time differences between pairs of transmitters, is preferred. In this case, the measured times from two transmitters are subtracted and the result is the intersection of two hyperbolas as shown in Figure 14.3c. In GPS, the TDOA technique is used where the differences in the times of arrival of signals from satellites are used to locate the mobile. The TDOA technique defines hyperbolas (rather than circles) on which the transmitter must be located with foci at the receivers. Three or more TDOA measurements provide a position fix at the intersection of hyperbolas. Exact solutions and Taylor series approximations are available [Com98] for solving these equations. Compared to TOA method, the main advantage of the TDOA method is that it does not require the knowledge of the transmit time from the transmitter. As a result, strict time synchronization between the MS and the GBSs is not required. However, the TDOA method requires time synchronization among all the receivers used for geolocation.

   While geometric interpretation can be used to calculate the intersection of circles or hyperbolas, when there are errors, estimates have to be used. Multipath propagation of signals (see Chapter 2) impact errors in the time (difference) of arrival of signals as signals may take more time to reach the receiver because of multiple reflections. A recursive least squares estimate (see project at the end of this chapter) is used when there are errors in the distance measurements. Let the distance d_i from the i-th GBS be determined from the absolute TOA of the signal it receives as $d_i = c \times \tau$ where c is the velocity of light and τ is the time taken by the signal to reach the GBS. If the location of the i-th GBS is (x_i, y_i) and the location of the mobile is (x, y), we have N equations of the form:

$$f_i(x, y) = (x - x_i)^2 + (y - y_i)^2 - d_i^2 = 0 \qquad (14.1)$$

for $i = 1, 2, \ldots, N$. As a geolocation problem, extensive research has been done to improve upon the accuracy of algorithms that are used to estimate the position of a mobile. Especially when N is more than three or four (thus providing redundancy in the measurements), information in the redundant measurements can be used to reduce errors that are introduced by noise, environment, multipath, and so on [Kap96]. Figure 14.5 shows an example of using the recursive least squares technique to arrive at the location of the MS.

Received Signal Strength: If the transmitted power at the MS is known, measuring the received signal strength (RSS) at the GBS can provide an estimate of the distance between the transmitter and the receiver using known mathematical models for radio signal path loss that depend on distances (see Chapter 2). As with the TOA method, the measured distance will determine a circle, centered on the receiver, on which the mobile transmitter must lie. This technique results in a low complexity receiver for a self-positioning system. This method is however very unreliable because of the wide variety in path loss models and the large standard deviations in the errors associated with these models due to shadow fading effects. Receivers do not distinguish between signal strength in the LOS path and in reflected paths [Mey96]. Especially indoors, the power distance gradients can

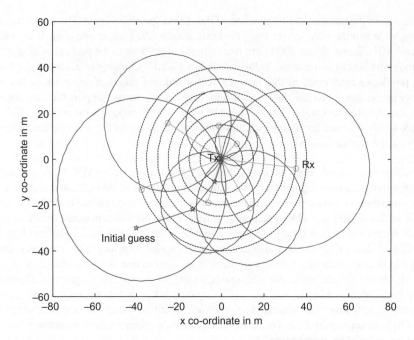

Figure 14.5   Recursive least squares to determine the MS location using measurements at seven GBSs.

vary from anywhere between 15–20 dB/decade to as high as 70 dB/decade. Also, these gradients and other parameters employed in path loss models are site-specific. As a result, this technique cannot be employed in situations where the required accuracy is a few meters. The accuracy of this method can be improved by utilizing pre-measured received signal strength contours centered at the receiver [Fig69] and multiple measurements at several base stations [Kos00]. A fuzzy logic algorithm was shown in [Son94] to be able to significantly improve the location accuracy.

Example 14.3: A commercial geolocation system based on RSS

The infrastructure of Paltrack indoor geolocation system [Pal13], developed by Sovereign Technologies Corp., consists of tags, antennas, cell controllers and an administrative software server system. The PalTrack system utilizes a network structure that resides on an RS-485 node platform. A network of transceivers is located at known positions within the serving area while the transmitter tags are attached to assets. The tag transmitters transmit unique identification code at 418 MHz frequency band to a network of transceivers when on motion or at predefined time intervals. Transceivers estimate the tag location by measuring received signal strength (RSS) and utilizing a robust RSS-based algorithm patented by Sovereign Technologies Corp. The master transceiver collects measured information from the transceivers and relays it to a PC-based server system. The accuracy for PalTrack is 0.6–2.4 m. The key component of the PalTrack system is the RSS-based geolocation algorithm.

Received signal phase: The received signal phase is another possible geolocation metric. It is well known that with the aid of reference receivers to measure the carrier phase, differential GPS (DGPS) can improve the location accuracy from about 20 m to within 1 m compared to standard

GPS, which only uses pseudorange measurements [Kap96]. One problem associated with the phase measurements lies in the ambiguity resulting from the periodic property (with period 2π) of the signal phase while the standard pseudorange measurements are unambiguous. Consequently, in DGPS, the ambiguous carrier phase measurement is used for fine-tuning the pseudorange measurement. A complementary Kalman filter is used to combine the low noise ambiguous carrier phase measurements and the unambiguous but noisier pseudorange measurements [Kap96]. For indoor geolocation systems, it is possible to use the signal phase method together with the TOA/TDOA or RSS method to fine-tune the location estimate. However unlike DGPS, where the LOS signal path is always observed, the serious multipath condition of the indoor geolocation environment causes more errors in the phase measurements.

Direction Based Techniques

If the positions of two fixed transmitters are known, a receiver can compute its own position by determining the angles at which these transmitters are located with respect to itself as shown in Figure 14.3d. This is called the direction or angle of arrival technique for position location. If the accuracy of the direction measurement (roughly the beam width of the antenna array) is $\pm\theta_s$, the AOA measurement at the receiver will restrict the transmitter position around the line-of-sight (LOS) signal path with an angular spread of $2\theta_s$. Two such AOA measurements will provide a position fix as illustrated in Figure 14.3d.

Consequently, one of the problems with this technique is the wide angular ranges for transmissions of most signals in wireless systems. Antennas in most 2G cellular systems are either omnidirectional or transmit over angles as wide as $120°$. D/AOA techniques are thus not specified in any of the positioning standards for cellular systems. Moreover, the accuracy of the position estimation depends on where the transmitter is located with respect to the receivers. If the transmitter lies in between the two receivers along a straight line, AOA measurements will not be able to provide a position fix. As a result, more than two receivers are usually needed to improve the location accuracy. Also, radio signals propagate through reflections and diffraction rendering the direction from which a signal arrives potentially random. For macro-cellular environments, where the primary scatterers are located around the transmitter and far away from the receivers (the geolocation base stations), the AOA method can provide acceptable location accuracy [Caf98]. But dramatically large location errors occur if the LOS signal path is blocked and the AOA of a reflected or a scattered signal component is used for estimating the direction. In indoor environments, surrounding objects or walls mostly block the LOS signal path. Thus the AOA technique is not suitable for indoor geolocation systems. In addition, this requires placing expensive array antennas at the receivers to track the direction of arrival of the signal. While this is a feasible option in next generation cellular systems where smart and narrow-beam antennas may be deployed to increase capacity, it is in general not a good solution for low cost indoor applications.

Signature Based Techniques

Problems in TOA or AOA schemes arising due to multipath propagation can be overcome by exploiting multipath propagation for position location. The idea behind this technique is that specific positions in a given environment have specific or unique *radio signatures* or *fingerprints* [Kos00]. The received signal can be extremely site-specific because of its dependence on the terrain and intervening obstacles. So the multipath structure of the channel is unique to every location and can be considered as a fingerprint or signature of the location if the same RF signal is transmitted from that location. This property has been exploited in proprietary systems to develop a "signature or fingerprint database" of a location grid in specific service areas. The received signal is measured as a vehicle moves along this grid and recorded in the signature database. When another vehicle moves in the same area, the signal received from it is compared to the entry in the database and thus its location is determined. Such a scheme may also be useful for indoor applications where the multipath structure in an area can be exploited. By creating a database of signatures and the associated positions, it will be possible to estimate the position of a MS if it can measure the radio signature at its own position and compare it with entries in the database.

Example 14.4: Commercial geolocation systems based on fingerprinting

A company called U.S. Wireless Corporation has designed and implemented a system based on this scheme called RadioCamera in downtown Oakland in California. The mobile transmits RF signals, which scatter because of multipath conditions. RadioCameraTM takes measurements of the RF signals and collects all the multipath rays. A Location Pattern Signature is developed using the multipath rays. The Location Signature is compared to a learned database and a location is determined. Continuous measurements of the location pattern signature provide tracking. Ekahau is a vendor of a geolocation system for WLANs that uses RSS fingerprints for estimating the location of a mobile station.

In WLANs, the received signal strength from multiple different APs at a given position can be used as a location specific signature [Bah00]. Since WLANs are widely deployed and MSs that can connect to WLANs can scan for multiple acces points and measure the RSS, it is possible to deploy a fingerprint based geolocation system without additional infrastructure or need for special hardware. In this case, typically, the RSS from N APs forms an N-dimensional vector. The mean vector $\mathbf{R}_{(i)} = [r_{1(i)}\ r_{2(i)}\ r_{3(i)} \cdots r_{N(i)}]$ is stored in the database for locations $i = 1,2,3, \ldots L$, typically on a square or rectangular grid in the physical service area. Thus the database has L entries, each corresponding to the mean RSS vector at a location. When a mobile station wants to determine its location, it measures a *sample* RSS vector $\mathbf{S} = [s_1\ s_2\ s_3\ \cdots\ s_N]$. The Euclidean distance between $\mathbf{S}$ and $\mathbf{R}_{(i)}$ is given by:

$$D_{(i)} = \|\mathbf{S} - \mathbf{R}_{(i)}\| = \sum_{n=1}^{N} (s_n - r_{n(i)})^2 \qquad (14.2)$$

The Euclidean distances between the sample vector and all entries in the database are computed (i.e., $D_{(i)}$ is computed for $i = 1,2,3, \ldots L$). The location associated with the entry in the database that has the smallest Euclidean distance to the measured sample is chosen as the estimate of the location of the mobile. This process, illustrated in Figure 14.6, is similar to matching a received signal to one in the signal constellation as described in Chapter 3. The primary difference is that the "noise" in this case need not be Gaussian (although this assumption is commonly made). Further, the entries in the database do not necessarily form a structured constellation. More details of matching samples to database entries and the associated errors are described in [Kae04; Swa08].

Unfortunately, for several reasons, signature based techniques are not free of disadvantages. Radio signatures change with time due to changes in the environment. Moreover, the size of the database needs to be reasonable and may not capture all possible signatures. It is very laborious to collect RSS data to populate the fingerprint database. Finally, it is unlikely that N access points are visible at a location all the time. It is possible to distinguish between locations based on the visibility of access points, but the accuracy is likely to be lower.

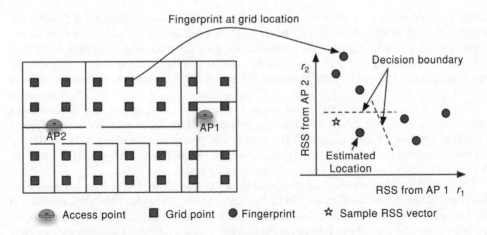

Figure 14.6   Illustration of estimating locations using RSS fingerprints.

Hybrid Positioning Schemes

Recently, attempts have been made to use a combination of sensing methods and technologies to improve the location estimates. A good example is the ability of PDAs to use both WLANs and cellular signals to estimate the location of the MS. A popular commercial enterprise involved in using WLANs for location estimates is Skyhook. Recent research has also considered the use of both Bluetooth and WLANs to improve the accuracy of location estimates within buildings.

14.3.3   Positioning Standards for Cellular Telephone Systems

In this section, we will describe techniques specified by cellular telephone standards for RF location sensing. These standards employ the techniques described in the previous section. The options for E-911 services when they were first mandated included traditional GPS or a network centric approach based on TDOA techniques. GPS, especially after the elimination of *selective availability* of the signal by the United States government provides sufficient accuracy for E-911 systems. The one disadvantage of GPS is that the time to first fix can be very long depending on what satellite constellation a MS may be able to see. There is also the problem of using GPS in urban canyons. However, compared to stand-alone GPS, network-centric approaches can provide a faster time to first fix but are unreliable and inaccurate. A variety of approaches are now part of cellular standards for geolocation.

Techniques based on Closeness to the Point of Access: The standard specified for locating MSs using POA is called the Cell-ID technique [Tre04; 3GPPa]. In most cellular standards, the BS serving a cell broadcasts information about itself. In GSM, the broadcast control channel carries this information. In IS-95 or cdma2000 systems, the pilot channel and sync channels together provide information about the BS. In 3G UMTS, the Cell-ID technique can use paging, location or routing area updates, or cell update messages to get information about the serving BS [3GPPa]. When a MS associates itself with the BS, it is aware of the cell (or cell sector) in which it is located. One may assume that this will be the BS closest to the MS. This is true only if the received signal strength of the broadcast control channel or pilot channels from the nearest cell are the strongest. In most cases, MSs latch on to the BS with the strongest signal. Due to radio propagation effects, a MS may sometimes associate itself with a BS that is farther away. As reported in [Tre04], this could be as high as 43% of the cases. The coverage of a cell is irregular and needs to be known a priori to know in what region a MS is located. The accuracy reported in [Tre04] was 800 m in the New York area and 500 m in Italy using the Cell-ID technique. The accuracy of the Cell-ID technique can be improved by using the round-trip time in CDMA systems or timing advancement information used in the framing structure in TDMA systems.

Techniques based on TOA/TDOA: Most standard cellular positioning schemes use TOA or TDOA for estimating the position of the MS. The standards include Enhanced Observed Time Difference (E-OTD), Observed Time Difference of Arrival – Idle Period on DownLink (OTDOA-IPDL), Uplink Time Difference of Arrival (UTDOA), Assisted GPS (AGPS), and Advanced Forward Link Trilateration (A-FLT). We discuss these standards briefly below.

The E-OTD standard was the earliest standard for positioning of MSs and it is suggested for GSM and EDGE (Enhanced Data for Global Evolution) systems. The idea here is that the MS will determine the TDOA from multiple BSs and determine its position using standard TDOA techniques or using improved algorithms. It is called "enhanced" because additional *location measurement units* are required to compute the timing difference between the clocks in different GSM base stations (that are not synchronized). The accuracy and precision with this method does not usually meet the FCC mandate because of clock accuracy and timing accuracy issues with E-OTD. The accuracy of E-OTD has been reported to be between 50 and 400 m with availability varying between 70 and 95% depending on whether the area covered is urban or rural [Mar02]. Moreover, the time to obtain the position estimate can be as large as 5 s and software changes are necessary in the handset to enable it to work with E-OTD.

The A-FLT standard is similar to E-OTD except that it uses the pilot signals in IS-95 based cellular systems. Pilot signals from the serving base station and a neighboring base station are used to compute the TDOA hyperbolas. The advantage here is that base stations in IS-95 are already time synchronized using GPS. The chip duration in the spread spectrum signals used in IS-95 is 0.813 μs, which results in better accuracy. In A-FLT, the resolution used for reporting is actually 1/8 of the chip duration. Measurement reports in [Nis00] indicate that the accuracy is 48 m for 67% of the time and 130 m 90% of the time. The Enhanced Forward Link Trilateration (E-FLT) is similar to A-FLT, but it uses a different signaling protocol with the network and it also could use additional information from location measurement units or radio signatures to improve accuracy. Because of cdma2000's similarity to IS-95, A-FLT and E-FLT are also used with that standard.

The OTDOA standard also uses TDOA measurements, but in the UMTS standard. The MS measures the time difference between frames transmitted by multiple base stations. Location measurement units are used in a manner similar to E-OTD to account for asynchronous transmissions from base stations. Since UMTS uses CDMA as the access scheme, a MS close to a BS may not be able to hear signals from other BSs because they are swamped by the high power of signals from the closer BS. To allow the MS to make measurements from other BSs, each BS ceases its transmission for short periods of time (called idle periods). This technique is called OTDOA-IPDL for this reason. Either the MS can compute its own position or it can report the measurements to the network where a stand-alone mobile location center computes the MS's position. Simulation results in [Por01] indicate that the OTDOA method can provide good accuracy and precision. In rural areas, the accuracy is 17 m 67% of the time and 27 m 95% of the time. In urban areas, the accuracy is between 68 and 86 m 67% of the time and is 156–193 m for 95% of the time, depending on the type of urban environment. In the UTDOA standard, different location measurements units compute the position of a MS by using the TDOA of a signal transmitted by the MS that they all receive. This method does not need the MS to do anything, but location measurement units have to be deployed by the service provider.

Network assisted GPS is a method that significantly improves the accuracy of position estimates in a cellular system [Dju01]. It is possible to install a GPS receiver in each MS so that a MS can determine its own position. GPS by itself in a MS is not a viable solution for many reasons. The time to first fix of a cold receiver can be several minutes. The MS needs a clear view of the skies to observe at least 4 satellites and so this approach does not operate well in urban canyons. If the MS has to scan for satellites, the signals from satellites and determine its own position, this could consume the MS battery power significantly. In the case of assisted GPS, a partial (or low-complexity) GPS receiver is built into the MS. Additionally, AGPS servers are placed in the network as appropriate. By predicting what signal a MS may see and sending that information to the MS, the network entity can enable a faster time to first fix, shortening it from minutes to a second or less [Dju01]. The wireless network signals information about the reference time, the visible satellite list, the satellite signal spread spectrum code phase, and search windows appropriate for these signals to the MS reducing the burden on the MS. This improves the time to first fix and also the accuracy of position estimates. Assisted GPS also enables the network entity to detect signals with weaker signal strength than a MS and send a *sensitivity assistance message* to the MS. AGPS can be used in GSM, IS-95, cdma2000, and UMTS systems, and in conjunction with other techniques such as Cell-ID, E-OTD, A-FLT, or OTDOA when GPS signals are not available. Accuracy estimates for AGPS range between 5 and 30 m.

Techniques based on Location Signatures: While there is no standard specified for using RF location signatures for estimating the position of a MS, some companies [Pol13] and research work have demonstrated the feasibility of this approach. In [Aho03], the multipath intensity profile at a given location is used as the RF signature. A database correlation mechanism is used where the measured multipath profile is correlated with profiles stored in a database. The profile in the database with the highest correlation is chosen as the estimated position of the MS. Simulation results in [Aho03] show that 67% of the position estimates are within 25 m of the exact position and 95% are within 140 m.

Table 14.2 provides a summary of these different geolocation approaches based on some of the performance measures discussed previously. The table is self-explanatory.

14.4   Location Services Architecture for Cellular Systems

The focus of this section is to provide an overview of the network architecture and protocols that are required to be in place in cellular wireless systems to provide location based services once the location sensing is completed. The complexity and large number of protocols, the different types of cellular networks, and the numerous standards that apply under different circumstances make it difficult to completely document such information in a short section. Consequently, we present a simplified overview with a classification that closely follows how someone outside the field would perceive the network and its services. Readers should refer to the detailed standards produced by the third generation partnership project such as [3GPPa] and its equivalents in the United States for a complete understanding of the subject. We would also like to point out that the messaging considered in this article is primarily related to control messages (i.e., signaling required to enable the actual communication or transaction).

As described in Section 14.3, there are several techniques such as Cell ID, observed time difference of arrival, advanced forward link trilateration, etc. that are used to determine the position of the MS in cellular networks. In cellular wireless systems, the normal infrastructure for transporting voice and data traffic is leveraged in a *location services* (LCS) *architecture* for transporting the location sensing information and query management for location-based

Table 14.2 Comparison of geolocation techniques in cellular systems

Geolocation technique	Coverage	Accuracy	Delay	Complexity/cost/other
Cell-ID	Almost everywhere	Poor – as high as 800 m	Low; information is included in signaling messages	Low
E-OTD	70–95% depending on type of area	Medium – because of clock accuracy (50–400 m)	Can be as high as 5 s	Needs additional measurement units to account for lack of synchronization
A-FLT or E-FLT	Good	Reasonable (48–130 m)	Medium	Makes use of the small chip duration in CDMA systems No changes to handsets Privacy is network controlled
OTDOA	Good	Good; dependent on type of area	Medium	Needs base stations to stop transmissions periodically to avoid near–far effect
Assisted GPS	Good; indoor coverage is suspect	Superior	Short time to first fix but needs a lot of signaling between network and mobile	Computation is offloaded to the network so that there is minimum impact on handset battery life
Radio fingerprinting	Good	Good	Computational effort to match fingerprints	Collecting a fingerprint database may be labor-intensive

services. For simplicity, we break up the signaling and communication of location information into three parts:

1. *Over the air* transport (or access network communications) that requires communication between the mobile station (MS) and the rest of the network. For example, the IS-801 standard handles this part of the communication for CDMA based cellular systems such as IS-95 and cdma2000.
2. *Signaling within the fixed part of the cellular system* (core network communications) that is necessary to enable position determination and account for mobility issues. For example, the JSTD-036 standards specified by the EIA/TIA for wireless emergency services and location services extend the ability of the signaling network to transport location and emergency related information in standardized formats.
3. *Application protocols that make use of the location services architecture*. We consider the *Mobile Location Protocol*, developed by the Open Mobile Alliance (OMA) that defines a set of constructs and services to enable location based services for applications.

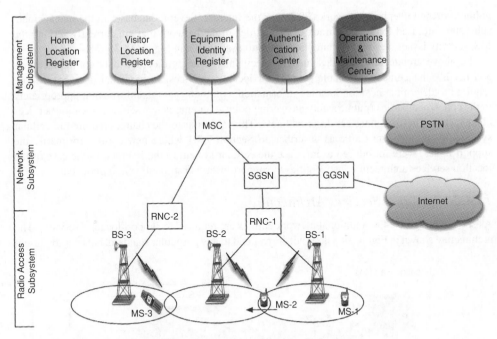

Figure 14.7   Generic cellular network architecture.

This section summarizes the three types of communications required for location based services in cellular networks. The goal is not to cover all standards and architectures, but to provide an overview and summary of some of the standards and messaging that happens in cellular networks for location-based services.

14.4.1   Cellular Network Architecture

Figure 14.7 shows a schematic of the architecture of a cellular system. While this schematic is not particular to a specific standard, it provides an idea of the different components in the network. The reader is referred to Chapter 6 and 11 for more details of architectures of cellular systems. In the radio access subsystem, the mobile station, sometimes called user equipment is the device whose position is to be determined. Base stations, as before, are fixed transmitters that are points of access to the rest of the network. A MS communicates with a BS during idle periods (signaling), cellular phone calls (voice) or other data transmission. Base stations are controlled by radio network controllers (RNCs) that also manage the radio resources of each BS and MS (frequency channels, time slots, spread spectrum codes, transmit powers, and so on).

The network subsystem carries voice and data traffic and also handles routing of calls and data packets. The mobile switching center (MSC) and the serving and gateway GPRS[1] support nodes (SGSN and GGSNs) are responsible for handling voice and data respectively. These network entities perform the task of mobility management, where they keep track of the cell or group of cells where a MS is located and handle routing of calls or packets when a MS performs a handoff, that is, it moves from one cell to another (e.g., see MS-2 in Figure 14.7). They connect to the

---

[1]GPRS stands for General Packet Radio Service

public switched telephone network (PSTN) or the Internet. Several databases in the management subsystem are used for keeping track of the entities in the network that are currently serving the MS, security issues, accounting and other operations as shown in Figure 14.7.

The above architecture was designed to specifically handle voice and data communications and needs enhancements to enable support for location services. In particular, new entities are required to determine position location information, communicate this information appropriately to the concerned parties (public safety answering point – for emergency services, location services clients, etc.). These changes are described next. This is similar to the changes required in cellular systems to support data traffic as described in Section 7.7.1 where new entities (primarily the support nodes) were introduced to interface the cellular system to the Internet. In the case of the location services architecture, new entities to support location estimation are introduced.

14.4.2   Location Services Architecture

As shown in Figure 14.8, additional network entities are required to support location services. The architecture shown in Figure 14.8 does not correspond to any particular standard, but tries to present

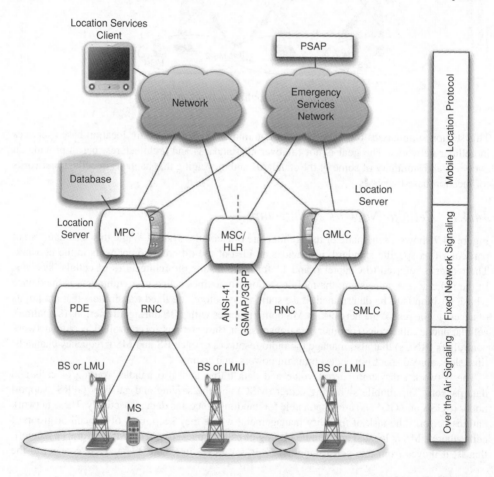

Figure 14.8   Architecture for location services in cellular networks.

some of the important network entities that are part of different standards such as the J-STD-036 and 3GPP TS 25.305. The goal here is to provide a general discussion of these entities rather than describe each standard individually. Also, some of these entities may be co-located although they are shown separately in Figure 14.8.

The location measurement unit (LMU) is a device that assists the MS in determining its position or uses signals from the MS to determine the position of the MS. It is used with assisted GPS to help the MS determine its position. With other positioning techniques such as uplink time difference of arrival, it makes measurements of radio signals and communicates this information to network entities such as the RNC. A location measurement unit may be associated with a BS, in which case it communicates with the RNC over a wired link. Alternatively, it may be a stand alone location measurement unit which uses the air interface to communicate with the RNC.

The Mobile Positioning Center (MPC) is the entity that handles position information in cellular networks that use ANSI-41 for signaling – typically North American cellular systems [TR-45-02]. It uses a Position Determining Entity (PDE) to determine the MS's position using a variety of technologies such as assisted GPS or some observed time difference of arrival. The position determining entity can determine a MS's position while the MS is in call or when it starts a call (using information from the MS or location measurement units). There may be multiple position determining entities that are used by one mobile positioning center. The mobile switching center is associated with a mobile positioning center. The same mobile positioning center may be associated with multiple mobile switching centers. The mobile positioning center and mobile switching center communicate with the emergency services network as described later. The mobile positioning center also handles access restrictions to the position information.

In 3GPP-based networks [3GPPa] such as the Universal Mobile Telecommunications System (UMTS) or the Global System of Mobile Communications (GSM), the Gateway Mobile Location Center (GMLC) and the Serving Mobile Location Center (SMLC) take up the responsibilities for positioning similar to the mobile positioning center and position determining entity. The gateway mobile location center is the first point of contact when the position of a MS is required. The serving mobile location center coordinates the resources necessary to determine the MS's position, sometimes calculating the position and accuracy itself (using information from the MS or location measurement unit).

An emergency call is ultimately answered by a public safety answering point (PSAP). The public safety answering point connects to the fixed infrastructure in the cellular system through the emergency services network (ESN), which interfaces with the mobile positioning center or gateway mobile location center and the mobile switching center. Two types of calls are considered by the emergency services network – the first where the position information is pushed to the ESN along with the signaling that occurs with the emergency call and the second where the ESN has to pull the position information. In the former case (called Call Associated Signaling or CAS), an emergency services network entity (ESNE) communicates with the mobile switching center serving the emergency call and obtains the position information. In the latter case (called Non-Call Associated Signaling or NCAS), an emergency services message entity (ESME) interfaces with the mobile positioning center or gateway mobile location center to pull the position information. The database shown attached to the mobile positioning center in Figure 14.8 translates the position of the MS into a number specifying the emergency service zone where the MS is located. It is the emergency service zone that is assigned to a public safety answering point and emergency services such as police, fire, or ambulance.

Finally at a higher layer we can consider location services clients and location servers (often co-located with the gateway mobile location center or mobile positioning center). These are often independent of the underlying network technology. The location services client may be requesting position information either for emergency services or for some other purposes (e.g., concierge services). In any case, it has to communicate over some network (usually using the Internet protocol – IP) with the location server to obtain the MS's position.

14.4.3 Over the Air (Access Network) Communications for Location Services

Certain communications have to take place over the air interface in the process for locating a MS, which are specified by standards that are different for IS-95/cdma2000 and 3GPP systems. Although such signaling happens over the air, it is also important from the point of view of functionality – it enables querying MSs for measurements related to positioning such as signal strength, timing, round trip times, GPS information, and so on.

The IS-801 standard [IS-801-99] defines the signaling messages between a MS and a BS to support position determination in IS-95 or cdma2000 networks. The standard specifies the formats of messages and procedures to be adopted by the MS and BS when messages are received. Most of the messages are in the form of requests and responses. Some of the request messages sent by the MS include those that ask for BS capabilities, GPS assistance, GPS almanac, and GPS ephemeris. Some of the response messages sent by the MS include MS information, pilot phase, time offset measurements, and pseudorange measurements. The BS makes requests for the response messages from the MS and provides responses to the MS requests.

In 3GPP, the signaling messages to support position determination are carried by the radio resource control (RRC) messages [3GPPb]. The form of messaging is similar to the request-response scheme in IS-801. The serving radio network controller generates radio resource control measurement control messages that are sent to the MS through the BS. They include data about the serving mobile location center, assistance data related to GPS, or instructions to the MS to perform measurements. In response, the MS sends a radio resource control measurement report that contains the position of the MS or other measurements that will help the serving mobile location center to determine the position. Several control and report messages may be necessary before success of failure of the position determination.

14.4.4 Signaling in the Fixed Infrastructure (Core Network) for Location Services

When a request for the position of the MS arrives, entities within the fixed infrastructure in Figure 14.8 have to communicate information between them to support the determination of the position and delivering this information to the appropriate destination. The signaling for this is specified in the JSTD-036 and 3GPP TS 25.305 V. 7.2.0 standards. These standards consider a variety of scenarios – such as CAS and NCAS calls, automatic detection of emergency calls, handling position determination and delivery in the case of calls that are in handoff, and so on. We provide a brief summary of some of the signaling in the fixed infrastructure.

When a MS initiates an emergency services call, the delivery of position information to the public safety answering point is called Emergency Location Information Delivery or ELID. The mobile switching center serving the MS that makes the emergency call, contacts the mobile positioning center or gateway mobile location center for position information. A position determining entity or serving mobile location center may have automatically detected the invocation of an emergency call and started procedures for computing the MS's position. Alternatively, the mobile positioning center or gateway mobile location center may contact the position determining entity or serving mobile location center to start such procedures and obtain the information. Once the mobile positioning center or gateway mobile location center has the position information, it will send such information to the mobile switching center. In the case of a CAS call, the mobile switching center sends the position information to the ESNE. The information sent includes information such as the calling party number and the position of the MS. In the case of an NCAS call, the mobile switching center serving the MS making the emergency call will send the Emergency Service Routing Digits (essentially information about the BS or cell sector serving the MS) to the ESNE. Then, the ESME associated with the ESNE autonomously requests the position information from the appropriate mobile positioning center or gateway mobile location center. Two or more mobile switching centers may be involved in delivering position information if the MS is in handoff during the call.

Several standard formats are used for protocols between different networking entities involved in ELID. In ANSI-41 based systems (such as cdma2000 and IS-95), ANSI-41 is used for communication between mobile switching centers, and also between mobile switching centers and other entities. Two special protocols are also used in such systems. The *Location Services Protocol* (LSP) is used between the PDE and mobile positioning center or between the routing database and mobile positioning center. The *Emergency Services Protocol* (ESP) is employed between the mobile switching center and the ESME. Protocols developed for integrated services digital network (ISDN) are used for communication between the mobile switching center and ESNE. The GSM Mobile Applications Part (GSMAP) has been extended in 3GPP systems to enable similar communications.

14.4.5   Mobile Location Protocol

The mobile location protocol (MLP) is an example of an application level mechanism used by a location services client to obtain position information about a MS from a location server (such as the mobile positioning center or gateway mobile location center). This protocol can also be used for emergency services. The MLP was initially developed by the location interoperability forum (LIF) [LIF02]. The goal of the MLP specification was to develop standard methods (using XML) for Internet applications to obtain position information from cellular network entities. The work of the LIF was later rolled over into the activities of the Open Mobile Alliance (OMA) [Bre05], an industry form consisting of hundreds of telecommunications and related companies for generating market driven specifications for mobile services to ensure interoperability between these services.

Some protocol/service layers associated with the MLP are shown in Figure 14.9. At the top, three types of MLP services are defined. The Basic MLP service corresponds to emergency services and

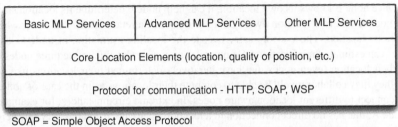

SOAP = Simple Object Access Protocol
WSP = Wireless Session Protocol

Figure 14.9   Part of the mobile location protocol stack.

ELID as it is defined by 3GPP. Advanced and other services can be developed to follow the MLP specifications as required. Some services that have been already defined include standard and emergency location services (further classified into *immediate* for delay sensitive single location response and *reporting* for an LCS client requesting position information). A triggered location reporting service is specified for the instance when a MS's location needs to be reported when an event occurs or a certain time elapses. The core location elements are in the form of Document Type Definitions (DTDs) that form the building blocks for the extensible markup language (XML). The transport of these XML messages is specified in the lowest layer. Mapping to standard web/web services protocols like HTTP and Simple Object Access Protocol (SOAP) are specified here, as well as mappings to the wireless session protocol, which is part of the wireless application protocol developed by the OMA.

The way the MLP operates is fairly simple. A location services client requests position information from a location service using an MLP request, which may, for example, be transported using XML in HTTP and SSL (see Chapter 11 for a brief discussion of SSL). The location request is an XML document that could include the MS's identification in either the North American mobile identification number format or the GSM mobile subscriber identifier format, the age of the position information, the response time, and accuracy. The response from the location server is also delivered to the LCS client using MLP. The location response is also an XML document, which provides information such as the accuracy, response time, and so on.

14.5   Positioning in Ad Hoc and Sensor Networks

In infrastructure-free ad hoc networks or sensor networks, geolocation is referred to as "localization", where each node (mobile or stationary) in the network needs to determine its location. In sensor networks, the location of a sensor is important because the physical sensed quantity (e.g., temperature in a reactor) is often correlated with the location (e.g., where in the reactor is the temperature too high?). The resource constraints of nodes makes it infeasible to embed GPS chips in them. The lack of fixed infrastructure in many cases makes geolocation a challenging problem.

The problem of localization in ad hoc and sensor networks can be explained with reference to Figure 14.10. The connectivity of nodes in this network is indicated by dashed lines (only nodes with lines between them can communicate directly; to reach other nodes, multi-hop communications is necessary). The assumption in these networks is that a certain fraction of nodes in the network are location aware (either because they have a GPS chip embedded or because the network administrator has manually included the location information). In general, the problem of localization in an ad hoc network then becomes one of determining the location of the remaining nodes in the network. For example, in Figure 14.10, node B can directly see three location-aware nodes. By measuring the distance to these three nodes (using signal strength or TOA), node B can estimate its own location. Once node B knows its own location, it can help node F determine its location since node F can see node B and two location aware nodes. However, there are errors that occur when node B estimates its location. These errors carry over to the location estimation of node F. This way, most nodes can estimate their locations. Some nodes may never be able to see three nodes that are aware of their location (even nodes that have iteratively determined their locations like node F). However, they may collaboratively be able to compute their locations as in the case of nodes C and G, which connect to different location aware nodes. In extreme circumstances, for example in the case of node A, that has no direct connection to a location-aware node and has very few neighbors, the location error may be significant.

Most work in this area (e.g., [Sav01a, b]) employs either the signal strength or time of arrival measurements to determine the distance. Connectivity has also been suggested as a metric to

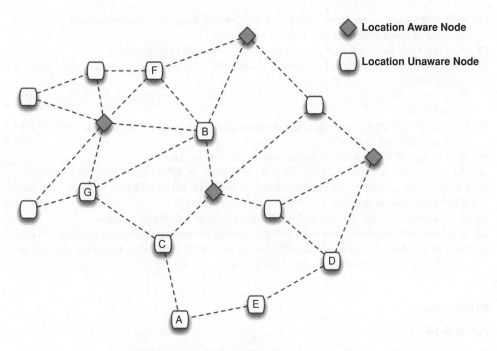

Figure 14.10 Localization in ad hoc networks.

estimate the range to a set of location-aware nodes (e.g., [Bul00]). The use of multiple ultrasound devices to estimate the direction of arrival is proposed in [Nic03] with ranging to estimate the positions of nodes in infrastructure-free networks.

Questions

1. Why are cellular service providers interested in location based services? Give some examples of location based services.
2. Explain the differences between GPS, wireless cellular assisted GPS, and indoor geolocation systems.
3. What are E-911 services and who has mandated these services?
4. Explain three differences between performance metrics for telecommunications and geolocation services.
5. Compare mobile centric and network centric geolocation techniques in terms of complexity and accuracy.
6. What are the basic elements of a wireless geolocation system?
7. Differentiate between remote and self-positioning systems.
8. What is the advantage and what is the disadvantage of using closeness to the point of association as the estimate of a mobile station's location?
9. Name the three major distance based techniques used for location finding and explain how they are implemented in a system
10. How is TOA different from TDOA for geolocation?

11. Why is RSS not a very good measure of the distance between a transmitter and a receiver? How can distance estimates with RSS be improved?

12. Why are AOA techniques not popular in indoor geolocation applications?

13. What are the advantages and disadvantages of RSS fingerprint based location techniques in indoor areas?

14. Why is it not certain that a MS is associated to the base station or access point that is physically closest to it?

15. In using TOA/TDOA in cellular systems, what two reasons make the accuracy better in CDMA systems compared to GSM?

16. How is the OTDOA scheme different from the UTDOA scheme?

17. Why do cellular networks employ assisted GPS instead of a full GPS receiver in each MS?

18. Compare the accuracy performance of the different TOA/TDOA based location estimation schemes in cellular systems.

19. Describe the functionality of a mobile positioning center in cellular systems.

20. What new entities are required in the cellular network architecture to support location services?

21. How can nodes in an ad hoc or sensor network determine their locations even if they are not directly connected to nodes that are location aware?

Problems

Problem 14.1

Two base stations located at (x, y) coordinates (500, 150) and (200, 200) are measuring the angle of arrival of the signal from a mobile terminal with respect to the x-axis. The first base station measures this angle as $45°$ and the second as $75°$. What are the co-ordinates of the mobile terminal?

Problem 14.2

In Problem 14.1, what happens if the first base station incorrectly measures the AOA from the mobile terminal as $50°$? $30°$?

Problem 14.3

Base stations A, B, and C located at (50, 50), (300, 0), and (0, 134) are found to be at distances 90, 200, and 100 m from a mobile terminal. Draw circles corresponding to these values and try to determine the location of the mobile terminal.

Problem 14.4

In Problem 14.3, what happens if the mobile incorrectly measures the distance from base station B as 100 m? 300 m?

Problem 14.5

Consider a fingerprint database with five fingerprints given by [−40 −65 −70], [−75 −33 −57], [−67 −55 −71], [−38 −59 −59], and [−55 −55 −55] (all values are in dBm) associated respectively with five rooms labeled A, B, C, D, and E respectively. The fingerprint elements are mean RSS values at various locations measured from three access points. A MS samples the received signal strength from the same three access points in the area. The sample RSS vector is $[-41\ -60\ -66]$ dBm. Compute the Euclidean distance between the sample RSS vector and the five fingerprints. Sort the locations according to the ascending order of Euclidean distances. Where is the MS most likely located? Is there is chance that this estimate is erroneous? Why?

Projects

1. In this project, you will develop a crude positioning system that makes use of the observed SSIDs of WLANs in the area. Use any of the tools mentioned in the projects section in Chapter 9 like Netstumbler, iStumbler, or Inssider to discover the *visible* SSIDs of networks at different locations along your street. Create a table that associates a set of visible SSIDs with a particular location. The sets of visible SSIDs must be different, so that you can distinguish between locations. Then try to predict where you are by looking at the list of visible SSIDs at random locations along your street. Describe how accurate this method is according to your experiments. Suggest alternatives to improve the accuracy.

2. Let us suppose that there are N estimates d_i of the distance of a MS from N known locations with coordinates (x_i, y_i) for $i = 1, 2, 3, \ldots, N$. We then have N equations of the form:

$$f_i(x, y) = (x_i - x)^2 + (y_i - y)^2 - d_i^2 = 0$$

where (x, y) is the unknown location of the MS. The *least squares* technique provides a method of estimating x and y when there are errors in the estimates d_i. The technique works as follows. Let $\mathbf{F} = [f_1(x, y) f_2(x, y) \ldots f_N(x, y)]^T$. Construct the *Jacobian* matrix given by:

$$\mathbf{J} = \begin{bmatrix} \dfrac{\partial f_1(x, y)}{\partial x} & \dfrac{\partial f_1(x, y)}{\partial y} \\ \cdots & \cdots \\ \dfrac{\partial f_N(x, y)}{\partial x} & \dfrac{\partial f_N(x, y)}{\partial y} \end{bmatrix}$$

Pick an estimate of the solution $\mathbf{U} = [x^*\ y^*]$. Determine the error in the solution as $\mathbf{E} = -(\mathbf{J}^T\mathbf{J})^{-1}\mathbf{J}^T\ \mathbf{F}$ evaluated at $\mathbf{U}$. The new solution is $\mathbf{U} + \mathbf{E}$. Iteratively, the error in the solution is reduced by computing a new error that is added to the previous solution to obtain a new solution till a point is reached where the solution does not change. Write a MATLAB® program that takes as input the N known locations, the N range estimates, a guess of the solution, and provides as output the final solution for the location of the MS. Also, add additional code to plot the locus of the intermediate solutions starting from the initial guess.

15

Fundamentals of RF Localization

15.1 Introduction

In Chapter 14, we provided a broad overview of wireless localization along with some system level descriptions of the architecture and protocols in cellular networks. As described there, the motivator for positioning with cell phones was the E-911 service. In 1996 the FCC introduced regulations requiring wireless service providers to be able to locate mobile callers in emergency situations with specified accuracy (100 m in 67% of the time). In the late-1990's, the military and commercial need for indoor localization was also recognized and this led to a number of research activities that have continued over the last decade for understanding and quantifying the performance of positioning systems. In this chapter, we describe some of the approaches for modeling and analyzing the performance of various sensors for localization. In particular, we discuss the performance bounds on the ranging used for localization.

Figure 15.1 illustrates the functional block diagram of a wireless geolocation system, which zooms into the functionality for positioning in the larger system shown in Figure 14.1. The main elements of the system are as follows: (i) a number of location-sensing devices that provide metrics related to the relative position of a mobile station (MS) with respect to a known landmark, called reference point (RP), (ii) a positioning algorithm that processes metrics reported by location sensing elements to estimate the location coordinates of MS, and (iii) a display system that illustrates the location of the MS on a map. The location metrics may indicate the approximated received signal strength (RSS), direction of arrival (DOA) of the signal or its time of arrival (TOA). The positioning algorithm processes the received metrics to determine the coordinates of the MS. As the measurement of metrics become less reliable or exact, the complexity of the position algorithm increases. The display system pins the estimated location coordinates to a map of an area. This display system could be a service or an application residing in a server or a mobile locating unit, a locally accessible software in a local area network, or a universally accessible service on the web such as Google maps. Obviously, as the horizon of the accessibility of the information increases, the design of the display system becomes more complex. In navigation applications, when we have some information regarding the movement of the MS, we combine the RF position estimates with the information on the pattern of movements to refine the location estimates.

There are two basic approaches that have been used to design an RF localization system. The first approach is to develop a signaling system and a network infrastructure of location sensors focused primarily for geolocation applications. The second approach is to use an existing wireless network

Principles of Wireless Access and Localization, First Edition. Kaveh Pahlavan and Prashant Krishnamurthy.
© 2013 John Wiley & Sons, Ltd. Published 2013 by John Wiley & Sons, Ltd.

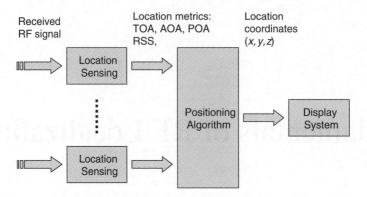

Figure 15.1 Functional block diagram of a wireless geolocation system.

infrastructure to locate a MS. The advantage of the first approach is that the physical specification, and consequently the quality of the location sensing results, is under the control of the system designer. The MS can be designed as a very small wearable tag or a sticker and the density of sensors in the infrastructure can be adjusted to meet the required accuracy of the location finding application. The advantage of the second approach is that it avoids expensive and time-consuming design and deployment of a special infrastructure. These systems, however, need to use more intelligent algorithms to compensate for the lack of granularity of the measured metrics.

As we transform a localization application from one environment to another, the behavior of the metrics changes and we need to study RF propagation in the area to understand this behavior. As a result, RF propagation analysis remains essential for comparative performance evaluation of different approaches to localization.

In principle, the behavior of the RSS and TOA are adequate for understanding of RF based technologies even when we use DOA techniques. Both these approaches try to estimate the distance between a mobile device whose location is unknown and a set of RPs whose locations are known. The behavior of the RSS for communication and localization is the same and therefore models developed for the RSS are also useful for performance analysis of RSS-based localization. However, the wideband models designed for multipath characterization of communication links may not be as useful for analysis of the TOA based localization systems because these models are designed to describe the multipath delay spread of the channel to analyze the intersymbol interference in the received signal, while in TOA-based localization we are interested in the power and *arrival time* of the first path and other multipath components arriving in its vicinity.

15.2 Modeling of the Behavior of RF Sensors

As the number of applications for RF location sensing increases, and in response to that a number of different technologies for localization are developed, we need a framework for comparative performance evaluation of these technologies. Most performance evaluation techniques require statistical models for the behavior of the sensors that describes the deviation of the measured metrics from the expected value (if the sensor was operating under ideal conditions). These models are needed to relate the performance of the system to the density and deployment strategy of the sensors. The behaviors of different sensors using TOA, RSS, DOA, or location signature of the delay-power profile, are quite different; therefore we need to develop separate models for different metrics.

15.2.1 Behavior of RSS Sensors

Models to characterize the extensive multipath in indoor and urban areas represent the overall channel impulse response for a given location of the transmitter and receiver by:

$$h(\theta, \tau, t, d) = \sum_{i=1}^{L} \beta_i^d(t)e^{j\varphi_i^d(t)}\delta[t - \tau_i^d(t), \theta - \theta_i(t)] \qquad (15.1)$$

where β, φ, θ, and τ represent the amplitude, phase, angle, and delay of arrival of each path respectively, and d is the distance between the transmitter and the receiver. The received signal strength based RF sensors process the received signal to determine the average RSS and use it to estimate $\hat{d}$, the distance between the target object and the location sensor. The average RSS, P_d, in dB at a given distance is:

$$P_d = 10 \, \log[RSS_d] = 10\log\left[\sum_{i=1}^{L} \overline{|\beta_i^d(t)|^2}\right] \qquad (15.2)$$

where β_i is the amplitude of the arriving paths defined in Equation 15.1. The measurement of the average RSS is independent of the bandwidth of the measurement device and therefore, the measured distance using RSS is independent of the bandwidth. In the case of wideband measurements, the effects of multipath fading is averaged over the spectrum of the signal and this average is computed by measuring the strength of each arriving path and using it in Equation 15.2. For narrowband systems, where we have only one arriving pulse with fluctuating amplitude according to the multipath fading characteristics, we need to average the signal over a longer period of time to make sure that the multipath fading is averaged out.

To calculate the distance between the transmitter and the receiver, we use the measured average RSS and a *distance-power relationship* to determine $\hat{d}$, the distance between the target object and the location sensor. If we define the distance measurement error as the difference between the measured and actual value of distance, $\varepsilon_d = \hat{d} - d$, this error in RSS systems is also independent of the bandwidth of the system. The measurement of the RSS is relatively simple and accurate, but the relation between the measured RSS and the distance is complex and diversified. Therefore, the accuracy of RSS-based techniques depends on the accuracy of the model used for the estimation of the distance using the RSS.

A number of statistical models for relating RSS to the distance between the transmitter and the receiver in indoor areas, developed for wireless communication applications, are presented in Chapter 2. As we discussed in Chapter 2, ray tracing algorithms provide a much more reliable estimation of the received power by using the layout of the building. Therefore, they can be used to improve the performance of sensors using RSS. Ray tracing algorithms are however computationally intensive and an alternative is to use geometrical statistical models. The advantage of geometrical statistical modeling and its ability to model site specificity while eliminating the complexity of ray tracing computations are described in [Has02]. One of the pioneering applications of ray tracing for indoor positioning and intruder detection is reported in [Hat04].

15.2.2 Behavior of TOA Sensors

In TOA systems, the measurement of the TOA requires more complex receivers and the accuracy of the measurements depends on the bandwidth of the system and multipath conditions. A TOA sensor estimates the distance between the transmitter and receiver from the relation $\hat{d}_w = c\hat{\tau}_{1,w}$ where c is the speed of light and $\hat{\tau}_{1,w}$, given in Equation 15.1, is the estimate of the *TOA of the direct straight-line path between the transmitter and the receiver*. The estimate of the TOA is obtained by detecting the *first peak* of the received signal and this value is a function of the signal bandwidth and the environment's multipath conditions. For example, in the case of a free-space channel that can be used for GPS applications when the receiver is in an open area, the multipath effects are negligible. We can thus measure the distances from the satellites that are tens of thousands of kilometers away to an accuracy that is within a value smaller than a few meters using a signal with a bandwidth of 1 MHz. However, in multipath rich environments, such as indoor areas, attaining similar accuracies even by using the TOA with RPs around the building and UWB devices with several GHz of bandwidth faces serious challenges.

Figure 15.2 [Ala06a] illustrates the effects of multipath on the estimation of the TOA. The multipath components in the channel model described by Equation 15.1 are represented by impulses. The lower parts of Figure 15.2 show these impulses obtained by ray tracing in a typical indoor area. In a practical implementation of a localization system, we have a finite bandwidth and each impulse is replaced by a waveform. To measure the TOA, we detect the timing of the peak of the first arriving path and use it as the TOA of the direct path between the transmitter and the receiver. In line of sight conditions, the first path in the profile is the strongest path and also the representative of the direct path between the transmitter and the receiver and the timing of the peak of this path is measured to determine the TOA. As shown in Figure 15.2a, in multipath conditions, the peak of this pulse will shift from its expected value because of the effects of other multipath components that are close to the first path. The shift in the peak causes an error in estimating the TOA and consequently the estimated distance between the transmitter and the receiver. This error is a function of the width of the pulse and consequently the bandwidth of the system that is inversely proportional to the width of the pulse.

In obstructed line-of-sight conditions, sometimes the direct path is blocked by objects (e.g., a metallic elevator) and if the strength of this path falls below the detection threshold of the receiver, as shown in Figure 15.2b we have an undetected direct path (UDP) condition that causes a large error in the measurement of the TOA [Pah98]. In principle this error will occur no matter how large the bandwidth of the system is. To understand the behavior of the TOA systems in multipath

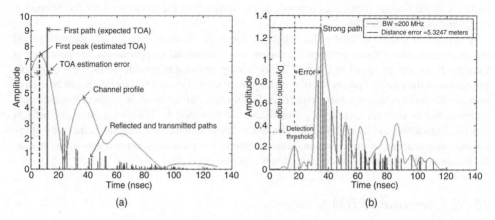

Figure 15.2 Effects of multipath on TOA estimation: (a) multipath components close to the arrival of the direct path, (b) undetectable direct path.

rich areas we need to model the relation between the TOA estimation error and the multipath conditions in an environment. Since this relationship is very complex, in a manner similar to other statistical models for RF propagation, we need to resort to statistical and empirical modeling. We proceed with the description of one of these models relating the distance measurement error to the bandwidth and the distance described in [Ala06a].

We define the distance measurement error for a system with a bandwidth of w as:

$$\varepsilon_{d,w} = \hat{d}_w - d \qquad (15.3)$$

where d is the actual distance between the RP and the mobile device and $\hat{d}_w = c\hat{\tau}_{1,w}$ is the estimate of the distance obtained from measurement of the timing of the first peak of the received channel profile for a given bandwidth. In [Ala06a] the distance measurement error is divided into two components, one caused by multipath arrivals close to the first detected peak, $\varepsilon_{m,w}$, and the other component the UDP error that is added to the multipath error whenever an object significantly blocks the direct path and a UDP condition occurs, $\varepsilon_{U,w}$:

$$\varepsilon_{d,w} = \varepsilon_{m,w} + \xi(d)\,\varepsilon_{U,w} = \gamma_w \cdot \log(1+d) + \xi(d) \cdot \varepsilon_{U,w} \qquad (15.4)$$

The multipath error has two components, a $\log(1+d)$ scaling factor that adjusts the amount of error with distance using a logarithmic scale starting with a minimum of zero and logarithmic growth after that, and a Gaussian random variable γ_w with the mean and variance of m_w and σ_w^2 with the probability density function:

$$f_{\gamma_w} = \frac{1}{\sigma_w\sqrt{2}} e^{-\frac{(x-m_w)^2}{2\sigma_w^2}}.$$

The statistics of this random variable adjusts the error to the bandwidth of the system. This approach isolates the effects of distance and bandwidth on the distance measurement error. The UDP error is multiplied by a binary random variable $\xi(d)$ that reflects the probability of occurrence of a UDP condition as a function of distance d and another Guassian random variable $\varepsilon_{U,w}$ with the mean and variance of $m_{U,w}$ and $\sigma_{U,w}^2$, and a probability density function:

$$f_{\varepsilon_{U,w}} = \frac{1}{\sigma_{U,w}\sqrt{2}} e^{-\frac{(z-m_{U,w})^2}{2\sigma_{U,w}^2}}$$

As the distance increases, the probability of occurrence of the UDP condition increases. The variance of the error decreases with the increase in bandwidth of the system. Table 15.1 provides the parameters for this model collected through a measurement campaign at the Atwater Kent Laboratories, Worcester Polytechnic Institute, using 405 UWB measurements in the same floor of the laboratory. Figure 15.3 compares the CDF of the results of empirical measurements versus the model for bandwidths of 200 MHz and 1 GHz.

15.2.3 Models of the Behavior of DOA

RF sensors using DOA for positioning measure and estimate the *angle of arrival* of the signal along the direct path, $\hat{\theta}_1(t)$ in Equation 15.1, from or to each reference point to calculate the location fix.

Table 15.1   Parameters used for the modeling of the ranging error using Equation 15.4 [Ala06a].
© 2006 IEEE. Reprinted, with permission, from [Ala06a]

W (MHz)	m_W (m)	σ_W (cm)	$P_{closeU,W}$	$P_{farU,W}$	$m_{U,W}$(m)	$\sigma_{U,M}$ (cm)
20	3.66	515	0	0.005	−12.83	0
50	1.57	205	0	0.009	24.48	21.1
100	0.87	115	0	0.091	5.96	358.5
200	0.47	59	0.006	0.164	3.94	289.0
500	0.21	26.9	0.064	0.332	1.62	80.9
1000	0.09	13.6	0.064	0.620	0.96	60.4
2000	0.02	5.2	0.070	0.740	0.76	71.5
3000	0.004	4.5	0.117	0.774	0.88	152.2

The only model addressing the DOA is the Spencer's model. This model assumed that the paths arrive in clusters and the DOA is determined by:

$$\theta_i(t) = \Theta_l - \omega_{kl}$$

in which Θ_l, ω_{kl} are the angle of arrival of the cluster and the ray within the cluster, respectively. The distribution of the arrival angle of a cluster Θ_l is assumed to be uniform between 0 and 2π and the distribution of the angle of arrival of the rays within the clusters is assumed to have a Laplacian distribution given by:

$$p(\omega_{kl}) = \frac{1}{\sqrt{2\pi}}e^{-|\sqrt{2}\omega_{kl}/\sigma|}$$

where σ is the variance of the arrivals. A typical value for the variance of the arrival angles within a cluster is around $22°$ [Spe00]. A variation of Spencer's model is considered for IEEE 802.11n wideband channel model [Pah05]. In the IEEE 802.11n model arrival of the clusters are modeled

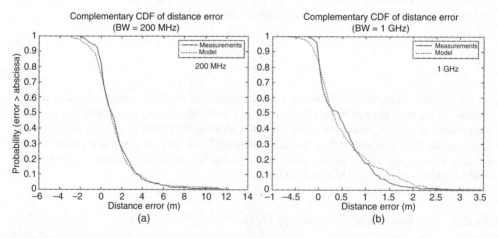

Figure 15.3   CDF of the empirical data versus the model for distance measurement error of the TOA based systems for bandwidths of: (a) 200 MHz, (b) 1 GHz.

as Poisson process and the angle of arrival for rays within the clusters is modeled with Laplacian distribution. A novel measurement of the DOA and limited empirical data of the distribution of the DOA related to an indoor area is available in [Tin01].

As we mentioned previously, since DOA in urban and indoor geolocation systems have not yet proven to be reasonably cost effective compared with more popular RSS- and TOA-based systems, they have not attracted considerable amount of research work that needs a model for the behavior of the channel. Further research in this area would be helpful to determine the limitations on the performance of directional antennas for indoor and urban area location finding applications. Existing models such as those in [Spe00] and [Tin01] are developed and verified for communication applications. Further research is needed to verify the accuracy of these models for the analysis of the behavior and design of algorithms for DOA positioning systems. One can expect that problems in detecting the direction of arrival of the direct path are likely when it is obstructed as in the case of TOA measurements. Further, the direction of arrival of the nearby multipath components may be significantly different due to reflections or other radio propagation vagaries.

15.3 Performance Bounds for Ranging

In the design of an RF localization system, we need to compare the performance of different alternatives for localization. In the same way that we were using the Shannon–Hartley Bound to compare modulation schemes for achieving a maximum data rate given a signal to noise ratio, in localization, it is common to use the Cramer–Rao Lower Bound (CRLB) on the variance, which is a measure of the *spread* of the error associated with a location estimate, for comparing the precision of location estimations by alternative approaches for localization. The smaller the variance, the smaller is the chance that the error in location estimate is large. In the same way that different information transmission applications have different error rate requirements, different localization applications have different precision requirements. For a conceptual system design, a positioning engineer may compare the CRLB for different metrics used for localization to select the appropriate technology or decide on the density for installation of the infrastructure to meet certain precision.

15.3.1 Fundamentals of Estimation Theory and CRLB

To explain the application of the CRLB for localization, we start with a simple example, shown in Figure 15.4.

Let us assume we have a parameter α that we want to measure, for example, the distance between a mobile device and a reference point. Let us suppose that we measure that parameter as, O, from a metric such as the TOA of the received signal. The measured observation is not the same as the parameter and if we measure the metric multiple times, each time we may observe a different value. If the probability distribution function of the observation, given the actual value of the measurement,

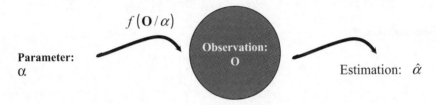

Figure 15.4 Basics of the estimation process and the CRLB for a single parameter.

is given by $f(O|\alpha)$, the *smallest* variance of the estimate of the parameter can be determined by the CRLB. This is given by what is called the inverse of the Fisher information matrix [Van68]:

$$\mathbf{F} = E\left[\frac{\partial \ln f(O/\alpha)}{\partial \alpha}\right]^2 = -E\left[\frac{\partial^2 \ln f(O/\alpha)}{\partial \alpha^2}\right]$$

In other words, the CRLB is given by:

$$CRLB = Var\left[\hat{\alpha}(O) - \alpha\right] \geq \mathbf{F}^{-1}$$

Example 15.1: CRLB

For a single observation that is corrupted by zero mean Gaussian noise η with variance σ^2 given by:

$$O = \alpha + \eta$$

where the conditional probability density function of O is given by: $f(O|\alpha) = \frac{1}{\sqrt{2\pi}\sigma} \exp\left(-\frac{(O-\alpha)^2}{2\sigma^2}\right)$. The Fisher matrix can be computed as:

$$\mathbf{F} = -E\left[\frac{\partial^2 \ln f(O/\alpha)}{\partial \alpha^2}\right] = \frac{1}{\sigma^2} \Rightarrow CRLB = \mathbf{F}^{-1} = \sigma^2$$

In this case, this result implies that the variance of our estimated value, based on one measured sample, is the same as variance of the measurement noise, which is intuitively correct.

To have another simple example related to the calculation of the CRLB, let us consider multiple measurements (observations) of a parameter and the CRLB on the estimation of the parameter.

Example 15.2: CRLB for N observations

For N observations that are each corrupted by independent zero mean Gaussian noise samples η with variance σ^2, the observations are given by:

$$O_i = \alpha + \eta_i, \text{ for } i = 1, 2, 3, \ldots, N$$

Then the joint probability density function of the observations is given by:

$$f(O|\alpha) = \prod_{i=1}^{N} \frac{1}{\sqrt{(2\pi)^N}\sigma_i} \exp\left(-\frac{(O_i-\alpha)^2}{2\sigma_i^2}\right) = \frac{1}{\sqrt{(2\pi)^N}\prod_{i=1}^{N}\sigma_i} \exp\left(-\sum_{i=1}^{N}\frac{(O_i-\alpha)^2}{2\sigma_i^2}\right)$$

The Fisher matrix is:

$$\mathbf{F} = -E\left[\frac{\partial^2 \ln f(O|\alpha)}{\partial \alpha^2}\right] = \frac{N}{\sigma^2}$$

and the CRLB is given by $F^{-1} = \frac{\sigma^2}{N}$.

The result from Example 15.2 indicates that using N sample observations can help us reduce the estimate by N times. A simple algorithm to achieve this bound is to simply average all the observations and use that as the estimate of the distance. Having an intuitive understanding of the meaning of the CRLB will help us in the next section to quantitatively compare RSS-, TOA-, and DOA-based localization techniques.

15.3.2   RSS-based Localization

In an RSS-based localization system, we use the measured (observed) received power to determine the distance between a mobile device and a reference point. As we described in Section 2.2, the RSS approximately drops linearly with the logarithm of the distance. Equation 2.7 describes the general relation between the path-loss and distance in a given environment, and it is a function of the distance power gradient α and the shadow-fading random component X that is modeled as a Guassian distributed random variable with standard deviation of σ. If we rewrite this equation in terms of the received power, the observed power at the receiver is:

$$O = P_r = P_0 - 10\alpha \log d + X$$

Since we want to use this observation to estimate the distance d, the probability distribution function of the observation is:

$$f(O/d) = \frac{1}{\sqrt{2\pi}\sigma} e^{-\frac{(O-P(d))^2}{2\sigma^2}} = \frac{1}{\sqrt{2\pi}\sigma} e^{-\frac{(P_r - P_0 + 10a \log d)^2}{2\sigma^2}}$$

Then,

$$\mathbf{F} = -E\left[\frac{\partial^2 \ln f(O/d)}{\partial d^2}\right] = E\left[\frac{\partial \ln f(O/d)}{\partial d}\right]^2 = \frac{(10)^2 \alpha^2}{(\ln 10)^2 \sigma^2 d^2}$$

and the CRLB will be:

$$CRLB = \mathbf{F}^{-1} = \frac{(\ln 10)^2}{100} \frac{\sigma^2}{\alpha^2} d^2$$

Since the CRLB is the variance of the estimate the standard deviation of error is the square root of this value:

$$\sigma_P \geq \frac{\ln 10}{10} \frac{\sigma}{\alpha} d.$$

That means the spread of the error around its mean value is on the order of the distance.

Example 15.3: Example

For an indoor environment with a distance power gradient of $\alpha = 3.5$ and variance of shadow fading of $\sigma^2 = 8$ dB for a distance of $d = 10$ m between the transmitter and the receiver, we will have a CRLB given by $\sigma_P \geq 1.86$ m. That means any ranging estimation algorithm and system using RSS with 10 m distance between the target and the reference point can not exceed an accuracy of more than 1.86 m.

As we observe in the following section these errors are substantially higher than the TOA-based measurement of the distances. Direct measurement of the distance from the RSS is unreliable because of the wide variety in statistical path loss models and the large standard deviations in the errors associated with these models due to shadow fading effects. To make RSS-based localization systems more reliable we need to build a certain intelligence in the system to recognize the radio propagation characteristics of the geographical area by offline calibration measurements, use intricate building-specific models using ray tracing, or using complex pattern recognition algorithms for location finding [Pah02]. Therefore, the complexity of RSS based systems is in the processing of the unreliable RSS reports.

15.3.3   TOA-based Localization

The TOA based systems measure the distance based on an estimate of signal propagation delay. The system is designed so that the received signal waveform in free space has a sharp peak and the variations of the TOA of the peak is measured to determine the distance between a transmitter and a receiver. For a TOA-based localization system operating in free space, if the transmitted pulse is $s(t)$ the observed signal at the receiver is given by

$$O(t) = s(t - \tau) + \eta(t)$$

where τ is the time of flight of the signal between the transmitter and the receiver and $\eta(t)$ is the additive white Gaussian noise component, with a spectral height of $N_0/2$ observed at the receiver. To form the probability density function of the observation given the value of the parameter τ we should note that we are observing the entire pulse in a Gaussian noise with variance of σ^2, as if we were observing K points on the signal when K grows to infinity. In other words, the probability density function of the observation would be:

$$f(O|\tau) = \frac{1}{(\sqrt{2\pi}\sigma)^K} \exp\left\{-\frac{1}{2\sigma^2} \sum_{k=1}^{K} [O_k - s_k(\tau)]^2\right\}\Bigg|_{k\to\infty} \propto \exp\frac{1}{N_0} \int_{T_0} [O(t) - s(t - \tau)]^2 dt$$

The Fisher matrix is now calculated from the second derivative of natural log of this function:

$$\ln[f(O/\tau)] = \frac{1}{N_0} \int_{T_0} [O(t) - s(t - \tau)]^2 dt = \frac{1}{N_0} \int_{T_0} [O^2(t) - 2O(t)s(t - \tau) + s^2(t - \tau)]^2 dt$$

Since

$$\frac{d^2}{d\tau^2} \int_{T_0} E[O^2(t)]dt = \frac{d^2}{d\tau^2} \int_{T_0} E[s^2(t - \tau)]dt = 0,$$

the Fisher matrix for the TOA estimation is given by

$$F_\tau = E\left[\frac{d^2}{d\tau^2}\{\ln[f(O/\tau)]\}\right] = \frac{2}{N_0} \int_{T_0} \frac{d^2}{d\tau^2} E[O(t)s(t - \tau)]dt$$

$$= \frac{2}{N_0} \int_{T_0} \frac{d^2}{d\tau^2} s^2(t - \tau)dt = -\frac{1}{\pi N_0} \int_{-\infty}^{+\infty} \omega^2 |S(\omega)|^2 d\omega$$

Therefore the CRLB representing the variance of the estimate is given by:

$$CRLB = F^{-1} = \frac{\pi N_0}{\int\limits_{-\infty}^{+\infty} \omega^2 |S(\omega)|^2 \, d\omega}$$

Since the energy per symbol is defined as

$$E_s = \int\limits_{-\infty}^{+\infty} s^2(t) dt = \frac{1}{2\pi} \int\limits_{-\infty}^{+\infty} |S(\omega)|^2 \, d\omega$$

and the signal to noise ratio as:

$$\rho^2 = \frac{2E_s}{N_0}$$

If we define the normalized bandwidth of the pulse as

$$\beta^2 = \frac{\int\limits_{-\infty}^{+\infty} \omega^2 |S(\omega)|^2 \, d\omega}{\int\limits_{-\infty}^{+\infty} |S(\omega)|^2 \, d\omega}$$

The CRLB will for TOA-based ranging will be:

$$CRLB = \frac{1}{\rho^2 \beta^2}$$

which is a function of the inverse of the signal-to-noise ratio and the normalized bandwidth of the transmitted waveform used for TOA measurements.

Example 15.4: CRLB for TOA with a flat spectrum

If we use the flat spectrum shown in Figure 15.5 for the calculation of the CRLB we have:

$$\beta^2 = \frac{\int\limits_{-\infty}^{+\infty} \omega^2 |S(\omega)|^2 \, d\omega}{\int\limits_{-\infty}^{+\infty} |S(\omega)|^2 \, d\omega} = \frac{2 \int\limits_{f_0 - \frac{W}{2}}^{f_0 + \frac{W}{2}} (2\pi f)^2 \frac{S_0}{2} 2\pi \, df}{2 \int\limits_{f_0 - \frac{W}{2}}^{f_0 + \frac{W}{2}} \frac{S_0}{2} 2\pi \, df}$$

$$= \frac{4\pi^2}{3} \frac{\left(f_0 + \frac{W}{2}\right)^3 - \left(f_0 - \frac{W}{2}\right)^3}{W} = 4\pi^2 \left(f_0^2 + \frac{W}{12}\right)$$

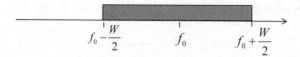

Figure 15.5 Flat spectrum used for calculation of CRLB for the TOA-based systems in Example 15.4.

and

$$\rho^2 = \frac{2E_s}{N_0} = \frac{2P_sT_0}{\sigma^2/W} = 2 \times SNR \times WT_0$$

Then the CRLB is given by

$$\sigma_\tau^2 \geq \frac{1}{\rho^2\beta^2} = \frac{1}{8\pi^2} \frac{1}{SNR} \frac{1}{WT_0} \frac{1}{f_0^2 + \frac{W}{12}}$$

Example 15.5: CRLB in the 2.4 GHz ISM Bands

Some commercial products use the ISM bands at 2.4 GHz to design TOA-based localization systems. As we have seen in Chapter 3 with approximately 10dB signal to noise ratio we can establish a reliable digital communication link and the maximum bandwidth in the 2.4 GHz ISM bands is 84 MHz. If we assume $WT_0 = 1$, using a TOA-based ranging technique we can achieve a spread of error on the order of:

$$\sigma_d \geq c \sqrt{\frac{1}{8\pi^2} \frac{1}{SNR} \frac{1}{WT_0} \frac{1}{f_0^2 + \frac{W}{12}}} \approx 3 \times 10^8 \sqrt{\frac{1}{8\pi^2} \frac{1}{10} \times \frac{1}{2.4 \times 10^9}} = 4.5 \times 10^{-3}$$

Comparing this with the values of a few meters in a typical indoor area, we notice that with the TOA systems, achieving centimeter precision is possible. However, we shall consider that every nanosecond error in measuring the peak of the signal would result in a 30 cm error in accuracy.

15.3.4   DOA-based Localization

In a DOA system, the location sensor measures the direction of arrival of the received signals (i.e., angle of arrival) from the target transmitter using directional antennas or antenna arrays. Figure 15.6 shows a very simple and fundamental technique that uses the DOA for navigating a ship. A sailor with a compass identifies the angle of two landmarks, a lighthouse and a radio tower, and by matching the angles with lines passing through the landmarks identifies the location of the boat. This simple example illustrates the practicality of using the DOA for localization with the help of a compass and a map.

In localization using RF devices, as shown in Figure 15.7a, if we have a reference point, RP, equipped by a directional antenna with a beamwidth of θ_s, and another metric such as the mean RSS or TOA to measure the distance, we can locate an object (or tag, T_g) with an accuracy proportional to $d \tan \theta_s$. As shown in Figure 15.7b, with two RPs with directional antennas, we can achieve the same level of accuracy as well. Therefore, the DOA metric can be used with a simple single antenna and

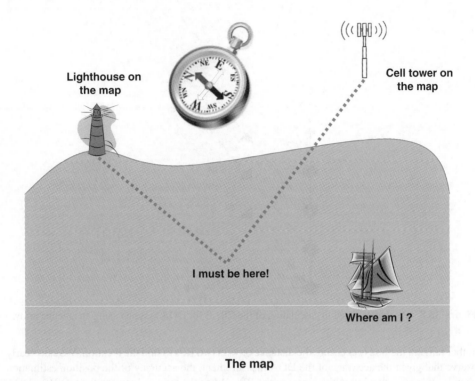

Figure 15.6 Basic concept revealing the importance of DOA in localization using a simple map with landmarks and a compass.

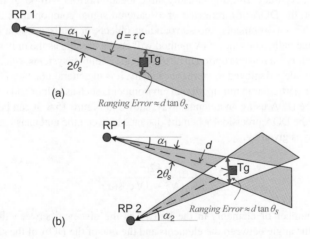

Figure 15.7 Basic concept behind using DOA for location estimation with: (a) one and (b) two antennas.

Figure 15.8 Parameters used in calculation of the CRLB for DOA systems using an antenna array.

another ranging method to measure the distance or with two RPs to find the location. We can clearly observe that given the accuracy of the DOA measurement, the accuracy of the position estimation depends on the tag position with respect to the RPs. When the tag lies between the two RPs, DOA measurements will not be able to provide a position fix. As a result, more than two RPs are normally needed to improve the location accuracy. In macro-cellular environments where the primary scatters are located around the transmitter and far away from the receivers, the DOA method can provide acceptable location accuracy. But dramatically large location errors will occur if the LOS signal path is blocked and the DOA of a reflected or a scattered signal component is used for location estimation. In indoor environments, the surrounding objects or walls inside a building usually block the LOS signal path. Thus the DOA method will not be useable as the only metric for indoor geolocation. While this is a feasible option in next generation cellular systems, where smart antennas are expected to be widely deployed to increase capacity, it is in general not a good solution for low cost applications, in particular in multipath rich environments such as indoor and dense urban areas.

If we measure the DOA using an antenna array, shown in Figure 15.8, it can be shown [Mal07] that the CRLB for the DOA precision, when the distance between the antennas is much larger than the size of the array is given by:

$$\sigma_D^2 \geq \frac{12c^2}{N_a(N_a^2 - 1)l \cos(\alpha)}\sigma_\tau$$

where N_a is the number of elements in the array, l is the distance between the antenna array elements and α is the angle between the elements and the line of the DOA of the signal. Therefore, the maximum percision is gained when the signal from the target arrives perpendicular to the arrays and if the signal is in parallel to the array, the measurement of the DOA is unpredictable. Increasing the number of antennas and the distance among them will increase the accuracy of measurements. The accuracy would then be comparable with that of the TOA-based systems.

15.4 Wireless Positioning Algorithms

In the last section, we analyzed localization approaches and bounds on their percision when we use the RSS, TOA, or DOA as a metric to determine the distance between a landmark and a mobile terminal. To find the position of a device, as shown in Figure 15.1, we need a positioning algorithm to combine the metrics read from different RPs (landmarks) to locate the device. Algorithms with well defined properties are available for satellite based GPS systems. There are least-squares algorithms and maximum-likelihood algorithms; there are algorithms based on a single snapshot of the measurements, and those using measurement and movement history. There are various kinds of sequential filters, which adaptively estimate some unknown parameters of the noise processes [Kap96; Mis10].

GPS, in particular, has focused a great deal of attention on positioning algorithms based on TOA with considerable success. GPS can provide positioning accuracy ranging from tens of meters to centimeters in close to real time depending upon the user's resources [Mis10]. In essence, these techniques are readily applicable to indoor location sensing systems. However, indoor location sensing involves quasi-stationary applications and a number of unreliable reference points for which existing GPS algorithms, designed for mobile systems with a few reliable reference points, does not provide the best solution. The unreliability of localization in indoor and urban areas is caused due to the fact that more precise TOA and DOA techniques, as we explained in Section 15.2, become unreliable in multipath environments. As a result, today, RSS-based localization that has less accuracy but performs more consistently in multipath conditions is the most popular approach used in popular commercial applications such as localization in smart devices [Wor08]. The RSS-based localization techniques use the existing WLAN infrastructure in buildings for localization in commercial applications. In public safety and military first responder applications, however, hybrid localization techniques using variety of RF and mechanical location sensors have been under investigation for the past decade or more [Moa11].

As we discussed previously, TOA and RSS metrics are the most popular in wireless positioning systems for urban and indoor areas. TOA metrics provide a more accurate measure of distance but may need additional infrastructure. The RSS is an easier metric to measure and integrates well with the existing communications infrastructure, but it is less reliable (widely varying) and often needs more complex algorithms and additional calibration procedures. Our discussion of the algorithms will emphasize TOA and RSS based localization systems.

15.4.1 Relation between Ranging and Positioning

Based on the discussion of the bounds for ranging in the previous section, one may next think of the relation between ranging and localization error. To examine this issue, we provide the analysis of the positioning estimation error for RSS-based localization systems, following [Che02]. Consider a case with N reference points used as landmarks to locate a device using RSS. The received signal power and the distance from the reference points are given by:

$$P(r_i) = P_0 - 10\alpha_i \log r_i + X, i = 1, 2, 3, \dots, N$$

and:

$$r_i = \sqrt{(x - x_i)^2 + (y - y_i)^2}$$

where X represents the shadow fading and (x, y) and (x_i, y_i) are the location of the device and RPs, respectively. If the model for calculation of the received power is accurate, the error in localization is caused by the effects of shadow fading that can be modeled as a Gaussian random variable with variance σ^2. Shadow fading causes variations in the received power, $d\mathbf{P}$, and the variation in this power causes a variation in the range estimate, $d\mathbf{r}$. Since the variation in power from a given reference point is given by:

$$dP_i(x, y) = -\frac{10\alpha_i}{\ln 10}\left(\frac{x - x_i}{r_i^2}dx + \frac{y - y_i}{r_i^2}dy\right); \quad i = 1, ...N,$$

in vector form, the relation between $d\mathbf{P}$ and $d\mathbf{r}$ would be:

$$d\mathbf{P} = \mathbf{H}d\mathbf{r} \quad \Rightarrow \quad d\mathbf{r} = \left(\mathbf{H}^T\mathbf{H}\right)^{-1}\mathbf{H}^T d\mathbf{P}$$

where:

$$d\mathbf{P} = \begin{bmatrix} dP_1 \\ dP_2 \\ . \\ dP_N \end{bmatrix}; \quad d\mathbf{r} = \begin{bmatrix} dx \\ dy \end{bmatrix}; \quad \mathbf{H} = -\frac{10}{\ln 10}[\alpha_1........\alpha_N] \begin{bmatrix} \dfrac{x - x_1}{r_i^2} & \dfrac{y - y_1}{r_i^2} \\ . & . \\ . & . \\ \dfrac{y - y_N}{r_N^2} & \dfrac{y - y_i}{r_N^2} \end{bmatrix}$$

Since the shadow fading is a zero mean Guassian random variable:

$$\text{cov}\left(dP_i, \ dP_j\right) = \begin{cases} \sigma^2, & i = j \\ 0, & i \neq j \end{cases}; \quad i, j = 1, 2, ...N$$

Therefore, the covariance of the location estimate is:

$$\text{cov}\left(d\mathbf{r}\right) = \sigma^2\left(\mathbf{H}^T\mathbf{H}\right)^{-1} = \begin{bmatrix} \sigma_x^2 & \sigma_{xy}^2 \\ \sigma_{xy}^2 & \sigma_y^2 \end{bmatrix}$$

The standard deviation of location error caused by shadow fading is given by:

$$\sigma_r = \sqrt{\sigma_x^2 + \sigma_y^2}$$

Example 15.6: Localization error in a room

Figure 15.9 shows the results of the above analysis to determine the contour of RSS-based localization errors in a 30×30 m room. The distance power gradient for the path-loss model is $\alpha = 2$ and the standard deviation of the shadow fading is assumed to be $\sigma = 2.5$ dB. Figure 15.9a uses three access points (APs) as the reference landmarks and Figure 15.9b uses five APs. In both cases the variance of position error is higher along the sidelines of the area. In the central areas we have lower errors.

In general the positioning error is in the orders of the ranging error. However, the distribution of error in the area would be different and fluctuates around the ranging error. When we are in the central parts we get equally accurate ranges from all reference points and that gives us a better estimate.

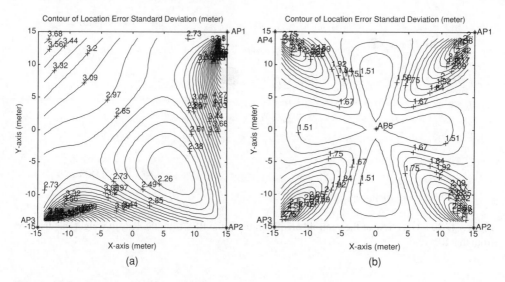

Figure 15.9 Contour of RSS-based positioning error in a 30×30 m room: (a) with 3-APs, (b) with 5-APs [Che02]. © 2002 IEEE. Reprinted, with permission, from [Che02].

15.4.2 RSS-based Pattern Recognition Algorithms

Popular commercial products using RSS-based localization for urban and indoor areas use different classes of pattern recognition algorithms. All of these algorithms rely on "war driving" in the area to create a database of the RSS measurements to create a fingerprint of the reference points (RPs), in which the RSS of landmarks such as cellular tower base stations (BSs) or Wi-Fi access points (APs) are measured. These radio map databases can be used to estimate the location of the BSs or APs as well as the location of a device that reports certain readings from the BSs or APs. In urban areas, these databases are formed by driving in the streets and tagging the RSS readings from landmarks in the RPs by GPS reported locations.

Example 15.7: The simple centroid algorithm

The most common method for localization of landmarks with unknown location such as a WiFi AP is using the simple centroid algorithm, described in Figure 15.10.

Let there be given a set of N reference points, RPs, each identified by a GPS reading

$$\{l(x_n, y_n) \; ; \; n = 1, 2, \ldots N\}$$

Also, let in a given RP location, $l(x,y)$, a device measures (observes) the RSS from M-APs that we do not know their locations:

$$\{\mathbf{O} = [p_1, \ldots\ldots, p_M]/l(x, y)\} \tag{15.5}$$

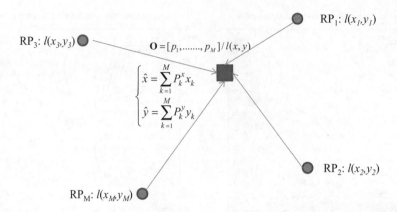

Figure 15.10 Overview of the centroid algorithm using the RSS from the RPs to weight the distance from that RP.

Further, let in *L-RP* locations we have measured the *RSS* from a given WiFi AP. Then the centroid algorithm estimates the location of that AP from the GPS readings of *L-RP* locations as:

$$\begin{cases} \hat{x} = \sum_{k=1}^{L} P_k^x x_k \\ \hat{y} = \sum_{k=1}^{L} P_k^y y_k \end{cases}$$

where P_k^x and P_k^y are the *weights* of the averaging along the *x*- and *y*-axis for 2D localization. The simplest centroid algorithm assigns the same probability to all reference points:

$$P_k^x = P_k^y = \frac{1}{L}$$

More complex centroid algorithms may weigh the power readings from each RP with a factor proportional to their power reading so that the location estimate in the center of all the RPS is shifted closer to the RPs with higher readings.

The nearest neighbor algorithm was first used as a simple approach for localization in indoor areas [Bah00]. Since GPS does not work in indoor areas, in this method we tag the location of the RPs manually on the map of an indoor area.

Example 15.8: Nearest neighbor algorithm

Figure 15.11 illustrates the general idea behind the nearest neighbor algorithm, we collect a database of RSS from *M* APs at *N* known RPs in the map:

$$\left\{ \mathbf{O}_n = [p_{n,1}, \ldots \ldots p_{n,M}]; l(x_n, y_n) \right\} ; \quad n = 1, 2, \ldots N$$

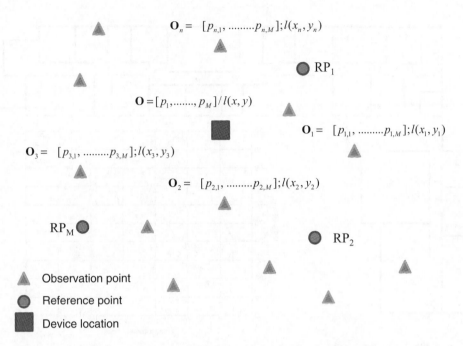

Figure 15.11 Overview of the nearest neighbor algorithm using the data base of the measured received power from the RPs in different locations to estimate the location of a device.

Given the power readings in an unknown location by Equation 15.5, we form the distance metric for each RP as:

$$d_n = \sqrt{\sum_{m=1}^{M} \left(p_n - p_{n,m}\right)^2}, \quad n = 1, 2, ...N$$

We select the location of the RP with minimum distance as the location of the device:

$$l(\hat{x}, \hat{y}) = l(x_i, y_i) \; ; \; i \text{ is the index of } d_{\min}$$

The advantage of this algorithm is that we do not need to know the location of the APs and we can build a larger database by measuring the RSS at more points to increase the accuracy of the algorithm.

A simple modification to the nearest neighbor algorithm is the K-nearest neighbor algorithm in which we select K locations with the lowest distances:

$$l(x_k, y_k); k = 1, .., K$$

and we declare the location of the device as the centroid of these lowest distance points:

$$l(\hat{x}, \hat{y}) = \frac{1}{K} \sum_{k=1}^{K} l(x_k, y_k)$$

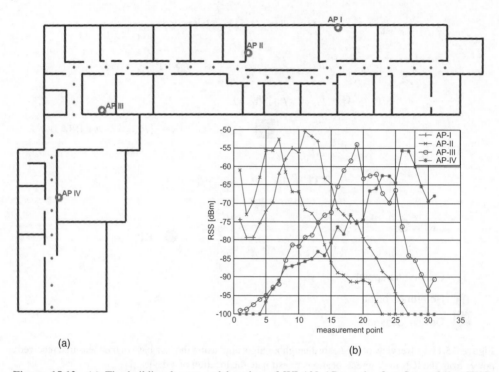

(a)   (b)

Figure 15.12 (a) The building layout and location of WLAN APs at the first floor of the CWC Laboratory at the University of Oulu, Finland. (b) RSS signature in different locations [Pah02]. © 2002 IEEE. Reprinted, with permission, from [Pah02].

Example 15.9: Performance of the nearest neighbor algorithm

Figure 15.12a [Pah02] presents a partial layout of the Telecommunications Laboratory at the Centre for Wireless Communications at University of Oulu, Finland. The locations of four 802.11b Access Points (AP) and 31 RP measurement locations along a long corridor, with about 2 m separation between adjacent RPs, are illustrated in the figure. A mobile device is carried along the corridor and the RSS values are measured at each location. Figure 15.12b shows the measured RSS at all four AP as the terminal travels from the right corner close to the AP-I to the end of the vertical corridor after AP-IV. When the nearest neighbor algorithm is applied to the measurement data, the standard deviation of the positioning error is 2.4 m, and at about 80% of locations the position error is less than 3 m.

Although the nearest neighbor or closest neighbor algorithm can be used for any measurable metrics that provides information about the distance to a landmark, such as TOA [Kan04b] or RSS [Bah00], this algorithm is more popular for RSS based systems. To form the reference data base for the nearest neighbor algorithm to associate known locations with RSS readings from the access points, it is also possible to calculate the RSS from the distance between the reference point and the mobile and the path-loss model for a given environment. This approach will avoid time consuming measurement process to form the database. Since path-loss models are not reliable, the measurement of the RSS has proven to provide a more practical solution [Bah00]. This argument lead to the point that one may think of using statistical models with the measurements to improve the performance.

Example 15.10: Statistical Kernel method

One of the pioneering work using statistical approaches for indoor positioning is reported in [Roo02a]. In this approach, referred to as the Kernel method, the measured data is collected from N locations as RPs from M APs with K measurements per location:

$$\mathbf{O}_n = \begin{bmatrix} \mathbf{o}_{n,1}, & \ldots\ldots\ldots \mathbf{o}_{n,M} \; ; \; l(x_n, y_n) \end{bmatrix} \; ; \; n = 1, 2, \ldots .N$$

$$\mathbf{o}_{nm} = (p_{nm1}, \; p_{nm2}, \; \ldots p_{nmk}); \quad n = 1, \ldots N; \; m = 1, 2, \ldots M$$

The location estimate is based on measurements given in Equation 15.5. The Kernel method defines a mass probability distribution function based on all measured signal strengths and assumes that the difference between two measured powers in the same location forms a Gaussian random variable. Then the joint probability density function of the observed data and the values in the measurement database is given by:

$$p(\mathbf{O}|l_n) = \frac{1}{M} \sum_{m=1}^{M} K(\mathbf{O}, \mathbf{O}_{nm})$$

where:

$$K(\mathbf{O}, \mathbf{O}_{nm}) = \frac{1}{(\sqrt{2\pi}\sigma)^K} e^{-12\sigma^2 \sum_{k=1}^{K} (p_m - p_{nmk})^2}$$

From Bayes theorem:

$$p(l_n|\mathbf{O}) = \frac{p(\mathbf{O}|l_n).p(l_n)}{p(\mathbf{O})} = \eta \cdot p(\mathbf{O}|l_n)$$

The estimated location in this approach is the expected value of the location given the observed valued of the power:

$$l(\widehat{x}, \widehat{y}) = E[l|\mathbf{O}] = \sum_{n=1}^{N} l(x_n, y_n)p(l_n|\mathbf{O}) = \eta \sum_{n=1}^{N} l(x_n, y_n)p(\mathbf{O}|l_n)$$

In this equation η is determined so that the total probability is normalized, that means:

$$\eta = 1/\sum_{n=1}^{N} p(\mathbf{O}|l_n)$$

Example 15.11: Performance of statistical Kernel method

Figure 15.13 provides a comparison among nearest neighbor and Kernel techniques [Roo02a]. The test area is a 16×40 m office with concrete, wood and glass structures with 10 access points

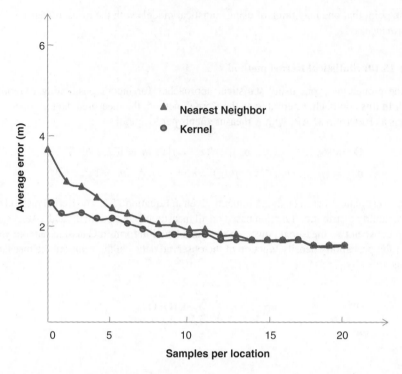

Figure 15.13 Comparison of the performance of nearest neighbor and kernel methods. © 2003 IEEE. Reprinted, with permission, from [Unb03].

(8 around the perimeter and 2 in the center). The training data is 155 points using a 2-m grid, and at each grid (*calibration*) point, 40 RSS observations are recorded. The test data is collected independently by using a similar 2-m grid, but by selecting the test points to be as far as possible from the calibration points. At each of the 120 test points, 20 observations were gathered. The vertical axis is the average distance measurement error and the horizontal axis represents the number of measurements in each location.

The algorithms presented in Examples 15.9 and 15.11 do not take advantage of the channel modeling knowledge presented in Chapter 2 of this book. In principle, if we have a reliable channel model we can avoid measurements and train the algorithms with predicted values of the RSS calculated from the channel model. We have statistical models and building specific models for the behavior of the radio propagation. The statistical models, described in Chapter 2, are less accurate and need extensive empirical measurements to calculate their parameters for a specific area. If we do extensive measurements to determine an accurate model, we might as well use these measurements directly in positioning application. In Chapter 2, however, we showed that ray tracing algorithms provide a close estimate of the RSS. Ray tracing algorithms need the electronic map of the area and in most wireless positioning applications we do have an electronic map to show the location of the terminal in the area. The following example provides some preliminary results in that area.

Example 15.12: Indoor positioning with ray tracing

In [Hat04] a 2D ray tracing software is used for positioning applications in the first floor of the Atwater Kent Laboratories (AKL) at the Worcester Polytechnic Institute. Figure 15.14 shows

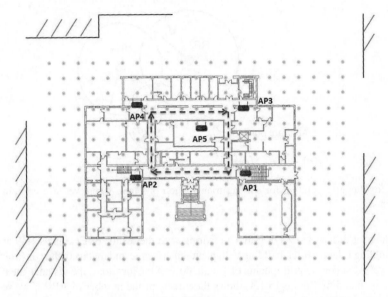

Figure 15.14 Layout of the first floor of AKL with outside walls from other buildings, location of the APs, and the grid used for ray tracing. The dashed line is used for positioning performance evaluation.

the layout of the building and the surrounding walls, the paths used for performance analysis, the locations of five access points, and the area covered by the grid used as the reference points. The ray tracing software generates a vector of five RSSs from the access points (AP) for every node of the grid. Using the nearest neighbor algorithm, the location of the mobile host is calculated. The estimated location is used to calculate the distance measurement error. Figure 15.15a shows the cumulative distribution function of the distance measurement error in the

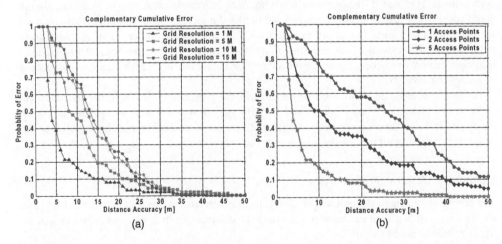

Figure 15.15 Statistics of the distance measurement errors for positioning using 2D ray tracing. (a) Effects of grid size. (b) Effects of number of APs [Hat04]. © 2004 IEEE. Reprinted, with permission, from [Hat04].

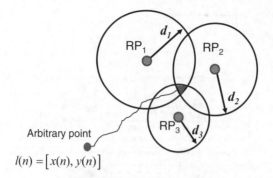

$$l(n) = \left[x(n), y(n)\right]$$

Figure 15.16 Iterative solution for solving triangulation problem, when we have error in distance measurements.

AKL building for different sizes of grid. Comparing with the results of Examples 15.9 and 15.11, the AKL building is larger brick building with plenty of concrete and metals in the construction, and we have fewer APs. For the grid spacing of 1 m, in 60% of the locations, the distance measurement errors are less than 5 m. Figure 15.15b shows the effects of the number of APs. If we reduce the number of APs from five to two, for 60% of locations, we have increases in error from 5 m to around 13 m.

In diversity channels, rather than using the total power in each location we may measure the signature of RSS is each *diversity branch* to add to the accuracy of the system. When these measurements are already available in the system, performance enhancement can be implemented with low costs. For example, in DSSS CDMA systems it is possible to use the measured time and signal strength of all fingers of the RAKE receiver in place of the total RSS to improve the positioning performance.

15.4.3   TOA-based Least Square Algorithms

With reliable TOA-based distance measurements from landmarks with known locations, simple geometrical triangulation methods can be used to find the location of a device without any need for finger printing efforts. As shown in Figure 15.16, due to estimation errors of distances from the landmark RPs caused by inaccuracies in TOA measurement, the geometrical triangulation technique can only provide a *region* for the location of the mobile device with uncertainty, instead of a single position fix.

To obtain an estimate of location coordinates in the presence of errors in the measurement of location metrics, a variety of direct and iterative statistical positioning algorithms have been developed to solve the problem by formulating it as a set of non-linear iterative equations. The simplest iterative algorithms are the least square (LS) algorithms.

Example 15.13: LS algorithm

The general statement of the problem in 2D localization (see also the project in Chapter 14) is that we have a set of N non-linear equations defining the distances from N known landmarks that we use as reference points (RPs):

$$f_i(x, y) = (x_i - x)^2 + (y_i - y)^2 - d_i^2; \quad i = 1, 2, \ldots N$$

In these N non-linear equations, d_i is the measured distance from the i-th RP, and (x, y) and (x_i, y_i) are the unknown location of the device and the known location of the landmarks used as RPs, respectively. The function $f_i(x, y)$ reflect the ranging error for the distance of the device from the i-th RP. If we define the quadratic vector function F as:

$$F = [f_1(x, y), \ f_2(x, y), \ \ldots\ldots, \ f_N(x, y)]^T$$

The Jacobian matrix of the F is defined as:

$$\mathbf{J} = \begin{bmatrix} \dfrac{\partial f_1(x, y)}{\partial x} & \dfrac{\partial f_1(x, y)}{\partial y} \\ \cdots & \cdots \\ \dfrac{\partial f_N(x, y)}{\partial x} & \dfrac{\partial f_N(x, y)}{\partial y} \end{bmatrix}$$

If we start from an arbitrary location:

$$l(n) = [x(n), y(n)]$$

then we can update the location by:

$$l(n + 1) = l(n) + E_n$$

where:

$$E_n = -\left(J^T J\right)^{-1} J^T F.$$

This algorithm can start from an arbitrary location and solve the quadric set of equations iteratively. Figure 15.16 shows a typical path that algorithm may take to update the location from an arbitrary location to a location in the region of intersection among all circles.

Variations of the LS algorithm have been used for localization in urban and indoor area. Example 15.10 considers one such variation.

Example 15.14: LS and RGWH algorithms for indoor geolocation

In [Kan04a], the performance of least square (LS) and residual weighted (RGWH) LS algorithms in a square area with four reference pointss in the corners of the room is evaluated. The LS algorithm is the simple traditional gradient algorithm described above, also used in basic GPS systems, which iteratively minimizes the square of the error in estimation of the position. For 2D applications, considered in this example, it needs a minimum of three RPs. With more RPs, the algorithm is expected to provide a more accurate positioning. The RGWH, originally proposed for cellular positioning [Che99a, b], is another version of the LS algorithm that calculates all the possible solutions for positioning, in our case all possible three reference points combinations and four reference points solution. The final estimate of the location is made by a weighted average of all the estimates. The weighting factor is the inverse of the residual error. This means that less reliable positions with higher residual errors are counted with less emphasis. Using a random number generator, a uniformly distributed location in the room is selected. Using the distance measurement error model for TOA-based ranging of [Ala06a, b], provided in Section 15.2.2, the estimated

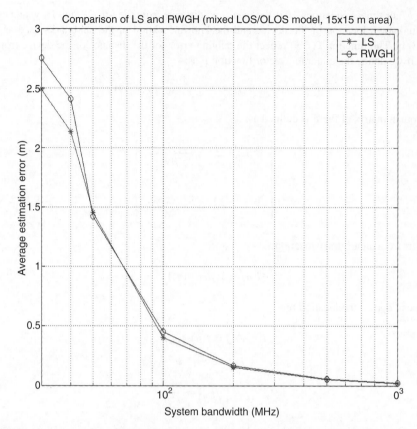

Figure 15.17   The average estimated positioning error versus bandwidth of a TOA system for LS and RWGH algorithm in a 15 × 15 m room with mixed LOS and OLOS conditions [Kan04a].

distances from the four reference points are determined for a variety of system bandwidths. These distances are then used with the LS or RGWH algorithms to determine the estimated location of the terminal in the area. The difference between the estimated location and the actual randomly selected location is used as the positioning error. By repeating this experiment the statistics of the positioning error associated with each of the algorithms is determined for different bandwidths of the system. Figure 15.17 shows the comparison of the performance of the two algorithms for a variety of system bandwidths. The RGWH algorithm provides a slightly better performance in lower bandwidths.

Questions

1. Explain the differences between GPS, wireless cellular assisted GPS, and indoor geolocation systems.
2. Why does GPS not function adequately in indoor areas?
3. Why can GPS use satellites that are tens of thousands of meters away from a mobile device and still have accuracies close to existing indoor geolocation systems on smart phones?

4. Why is the RSS not a very good measure of the distance between a transmitter and a receiver? How can distance estimates with RSS be improved?
5. Why are AOA techniques not popular in indoor geolocation applications?
6. What is CRLB?
7. Why does the CRLB improve when multiple samples or observations are made in additive Gaussian noise?
8. What is the error relationship between ranging and position location?
9. Explain the operation of the centroid algorithm.

Problems

Problem 15.1

a. Using [Ala06b] model, determine the mean and variance of TOA-based ranging error for distances of 3, 5, and 15 m if we have a 1 GHz bandwidth and a LOS condition with no UDP condition.
b. Repeat (a) for bandwidths of 100, 10, and 1 MHz at a distance of 5 m.
c. Repeat (a) and (b) when we have the UDP conditions with the same statistics as in the model.

Problem 15.2

Let us suppose that there are N estimates d_i of the distance of a MS from N known locations with coordinates (x_i, y_i) for $i = 1, 2, 3, \ldots, N$. We then have N equations of the form:

$$f_i(x, y) = (x_i - x)^2 + (y_i - y)^2 - d_i^2 = 0$$

where (x, y) is the unknown location of the MS. The *least squares* technique provides a method of estimating x and y when there are errors in the estimates d_i. The technique works as follows. Let $\mathbf{F} = [f_1(x, y) \, f_2(x, y) \, \ldots \, f_N(x, y)]^T$. First, we have to construct the *Jacobian* matrix given by:

$$\mathbf{J} = \begin{bmatrix} \dfrac{\partial f_1(x, y)}{\partial x} & \dfrac{\partial f_1(x, y)}{\partial y} \\ \cdots & \cdots \\ \dfrac{\partial f_N(x, y)}{\partial x} & \dfrac{\partial f_N(x, y)}{\partial y} \end{bmatrix}$$

Next we pick an estimate of the solution $\mathbf{U} = [x^* \, y^*]$ and we determine the error in the solution as $\mathbf{E} = - (\mathbf{J}^T \mathbf{J})^{-1} \mathbf{J}^T \, \mathbf{F}$ evaluated at the estimate $\mathbf{U}$. The new solution is $\mathbf{U} + \mathbf{E}$. Iteratively, the error in the solution is reduced by computing a new error that is added to the previous solution to obtain a new solution till a point is reached where the solution does not change. Let the known locations of reference points be (10, 10), (0, 15), and (−5, 5). The measured distances from these reference points are 15, 16, and 5 m. Use the least squares approach to determine the estimate of the location by using the location (2, 2) as the initial estimate.

Problem 15.3

Show that for RSS based ranging the CRLB is given by:

$$\sigma_D^2 \geq \frac{(\ln 10)^2}{100} \frac{\sigma_{sh}^2}{n_p} d$$

Where:

σ_{sh} : standard deviation of zero mean gaussian random variable representing log-normal shadowing

n_p : path loss factor

d : distance between two nodes

Problem 15.4

Extend the 2D CRLB for RSS-based localization to 3D scenarios. Name an application where the 3D bounds helps.

Problem 15.5

Using the parameters of the antenna array defined in Figure 15.8 show that the CRLB for DOA estimation is given by:

$$\sigma_D^2 \geq \frac{12c^2}{N_a(N_a^2 - 1)l \cos(\alpha)} \sigma_\tau$$

where N_a is the number of elements in the array, l is the distance between the antenna array elements and α is the angle between the elements line the DOA of the signal.

16

Wireless Localization in Practice

16.1 Introduction

The evolution of wireless localization techniques for indoor and urban area applications, where GPS does not work properly, has been an active area of research for commercial and public safety applications since the mid-1990s. At the time of writing, RSS-based Wi-Fi localization is dominating the commercial market, is complementing GPS chipsets in all popular smart phones, and operates by itself in many other devices such as pads, book readers, notebooks, and laptops where no GPS chipsets exist. This technology takes advantage of the random deployment of Wi-Fi devices worldwide to support indoor and urban area localization with reasonable accuracies for tens of thousands of applications on smart devices, from modern location based services for finding businesses such as Yelp, to traditional turn by turn localization [Pah10; Wor12].

Public safety and military applications demand more precise localization for first responders and they are discovering more sophisticated techniques for such precise indoor geolocation using hybrid techniques [Moa11]. RF localization techniques for these hybrid technologies are intent on more precision through TOA-based localization. In this chapter we address the practical aspects of these technologies used for geolocation in indoor and urban areas for commercial and military applications. In addition we examine challenges in localization inside the human body that is emerging as an important area of research for the future of the wireless health industry [Pah12a, b].

16.2 Emergence of Wi-Fi Localization

Since the inception of WLAN in the 1980s [Pah85], it has been one of the wonders of the wireless revolution by nurturing ground-breaking innovative technologies for popular applications. These innovations took place because Wi-Fi technology was designed for data applications, and the underlying networking always strives for higher data rates (in the order of 100 Mbps and beyond) for wireless Internet access to foster access to ever-growing multimedia applications. These Internet applications are commonly used in indoor areas, where the substantial multipath conditions necessitate sophisticated innovative transmission technologies to achieve high data rates. As a result, from what we have explained in Part I of this book, WLANs introduced the first popular commercial application of spread spectrum technology, and later OFDM and MIMO antenna systems. In contrast to the WLAN industry, the prosperous cellular networking industry with

Principles of Wireless Access and Localization, First Edition. Kaveh Pahlavan and Prashant Krishnamurthy.
© 2013 John Wiley & Sons, Ltd. Published 2013 by John Wiley & Sons, Ltd.

approximately seven billion subscribers, emerged with a focus on lower-speed (around 10 kbps) cellular telephone applications demanding comprehensive coverage to support a continual quality of service while the user moves across a large metropolitan area. These conditions for cellular networks nurtured the evolution of CDMA technology for 3G cellular networks. At the time of writing, the 4G cellular industry is using LTE technology, which borrows OFDM and MIMO technologies from successful wireless Internet access experience with Wi-Fi networks.

In the year 2000, three years after completion of the first IEEE 802.11 standard, using Wi-Fi signals for localization appeared in the literature [Bah00; Li00]. The first commercial application of this technology was in the Real Time Location Systems (RTLS) used for asset and personnel tracking in local indoor areas with an accuracy of around a few meters. In the past few years, Wi-Fi localization found its way into Wi-Fi or Wireless Positioning Systems (WPS), used in emerging smart devices, such as the iPhone, to complement GPS and cell tower localization in numerous everyday consumer applications in metropolitan areas, ranging from social networking to tagging photos or videos with location information. These applications commonly accept accuracies on the order of a few tens of meters, but they demand an immediate location fix and comprehensive coverage. This expansion in user demand poses a challenge to the industry because the GPS technology is not fast enough and it does not adequately cover indoor areas, where almost all of these applications are initiated. The cell tower and GPS assisted localization techniques are fast enough but they may not be able to provide the needed accuracy when they use the existing infrastructure. The WPS industry using Wi-Fi localization emerged to resolve these deficiencies. The incentive for this emergence was that in addition to 24 or 32 GPS satellites and hundreds of thousands of cell towers, there are hundreds of millions of IEEE 802.11 WLAN access points worldwide, which can be exploited opportunistically to locate terminals in a variety of outdoor situations and indoor environments.

WPS complements GPS for indoor coverage, reduces the time to fix (time needed for the GPS to provide a reliable estimate), improves power consumption, and offers resistance to interference. GPS complements WPS to provide comprehensive outdoor coverage and a universal coordinate reference frame. In 2008, the leading wonder of emerging smart devices, the iPhone, started to carry Wi-Fi chip sets to complement its 3G CDMA cellular chip sets for providing high-speed wireless Internet access wherever possible and in addition used Wi-Fi signals for localization and tracking to complement GPS and cell tower localization in a hybrid system. At the time of writing, Wi-Fi localization in smart phones and other devices, receives several billion hits per day. Figure 16.1

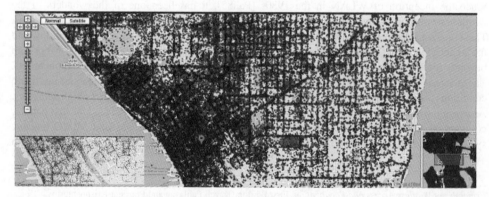

Figure 16.1   The AP database of Skyhook (Boston, Mass.) in the Seattle area. @ 2012 Skyhook, Inc.

shows the map of APs in Skyhook's (a company in Boston, Mass.) database in the Seattle area. As an example of the size of a database, at the time of writing, the Skyhook database has approximately half a billion access points worldwide.

This section describes the evolution of the Wi-Fi localization industry with particular emphasis on its application to emerging smart devices. We describe how received signal strength (RSS)-based Wi-Fi localization technology evolved out of time of arrival (TOA)-based GPS technology and how these two technologies are intertwined to address the needs of the emerging glamorous applications in smart devices.

16.2.1 Evolution of Wi-Fi Localization

Traditional GPS is not designed for indoor applications and it does not perform satisfactorily in these areas (in terms of availability or accuracy). In the second half of the 1990s, around the same time that DARPA lunched its small unit operation situation awareness system (SUO/SAS) program aiming at 1 m accuracy for indoor geolocation in military and public safety operations [Pah98], venture capitalists started funding startup companies such as PinPoint (Woburn, Mass.) [Wer98] and WhereNet (Santa Clara, Calif.) for the implementation of indoor geolocation technologies with accuracies comparable with those of SUO/SAS. The great success of TOA-based localization used in GPS pointed researchers in the military and commercial arenas to start thinking in that direction.

The idea sounded very straightforward, and according to the CRLB for TOA systems explained in Section 15.3, the variance of the ranging error for the TOA systems is given by:

$$\sigma_D \geq \sqrt{\frac{1}{8\pi^2} \frac{1}{SNR} \frac{1}{T_0 W} \frac{1}{f_0^2} \frac{1}{1 + \frac{W^2}{12 f_0^2}}} \qquad (16.1)$$

where T_0 is the observation time, SNR is the signal to noise ratio, f_0 is the center frequency of operation, and W is the bandwidth of the system. For operating frequencies around those used in GPS systems, this bound indicates that a spread of errors of around several meters is achievable if we can wait for a few minutes. If we wish to extend this technology to indoor geolocation, we have three challenges:

1. We need an increased precision to identify objects within reasonable measurement times.
2. We need to cope with an additional path loss of around 20–30 dB as the signal penetrates buildings.
3. We need algorithms to cope with the multipath conditions.

To address these challenges, in the late 1990s a number of TOA-based military and commercial indoor geolocation systems were designed; but none of them met the expectations. DARPA had to compromise on its accuracy requirements and commercial startups simply failed [Pah06].

The results of radio propagation studies for indoor geolocation applications in the SUO/SAS project revealed that the scientific source of the problem was severe multipath conditions in obstructed line of sight (OLOS) indoor environments frequently causing unexpectedly large ranging errors [Pah98]. In the following years, to remedy the situation and find a solution to the TOA-based indoor geolocation challenges caused by multipath, researchers for military and public safety applications resorted to UWB, super-resolution, multipath diversity, and cooperative localization [Pah02, Pah06]. More recently, inertial navigation systems are finding their way for more accurate and precise indoor geolocation in the research community to complement RF indoor geolocation deficiencies [Moa11].

16.2.2   Wi-Fi Localization: TOA versus RSS

During the discovery of indoor geolocation science in the late 1990s, for commercial applications, a major problem for widespread deployment was the cost of new proprietary hardware and the corresponding infrastructure needed in order to make indoor geolocation technically feasible and useful. The cost factor led the industry to examine indoor geolocation techniques using existing WLAN infrastructures, which were growing rapidly in office environments with a very reasonable cost of deployment. In the year 2000, both TOA-based [Li00] and RSS-based [Bah00] WLAN localization techniques appeared in the literature and the Wi-Fi localization industry was born. Emergence of the idea of Wi-Fi localization technology in the 2000s created a significant amount of enthusiasm in the industry. For TOA-based indoor geolocation, a number of patents were filed by different companies based on pioneering papers in that area [Li00]. The general idea of using WiFi wireless networking infrastructure for localization applications spread to other emerging networking standardization activities, such as IEEE 802.16.3 for UWB communications (see Chapter 10) and IEEE 802.16.4 for sensor networking using ZigBee technology (see Chapter 9). Although TOA-based Wi-Fi localization systems use existing infrastructure, there is a need to modify the hardware design to extract the TOA estimate from the received Wi-Fi signal. Besides, any implementation of a precise TOA-based localization system faces the same multipath challenges as military systems, demanding complex algorithms and technologies to address them. Because of these reasons, regardless of the substantial amount of research in TOA-based indoor geolocation, the commercial market is still waiting for popular products in these areas, while the burden of research is mostly carried in the military and public safety sectors.

RSS-based systems use the existing Wi-Fi hardware infrastructure without any hardware modification in the access points and the Wi-Fi device in the terminal. These systems use a localization "software patch" to read and process the RSS from different access points. As we explained in Section 15.3, the relative precision of RSS-based localization is not that sensitive to multipath and bandwidth, and consequently, the system does not need any synchronization between terminals and infrastructure. As a result, RSS-based indoor geolocation systems became an instant commercial success and a few startup companies such as Ekahau (Helsinki, Finland) and Newberry Networks (Boston, Mass.) were established immediately after the publication of the pioneering papers in that area [Bah00]. At the time of writing, companies in this domain manufacture Wi-Fi RFID tags for asset or personnel tracking and they sometime refer to their industry as the Real Time Location Systems (RTLS) industry, and since Google announced the availability of indoor maps for popular building late 2012's, a number of startup companies have emerged in that area [Wor12].

The largest development in RSS-based Wi-Fi localization was towards metropolitan area localization with the emergence of the iPhone and the popularity of smart devices in the late 2000s. This idea was examined by Intel's Place Lab [Che05] and was implemented as a commercial product by Skyhook (Boston, Mass.). The Wi-Fi localization systems used in these devices are sometimes referred to as Wi-Fi or Wireless Positioning Systems (WPS) and differ substantially from RTLS systems in their application domain, performance expectations, database collection techniques, and localization algorithms. In the rest of this section we explain how RSS-based Wi-Fi localization systems work along with the differences between RTLS and WPS.

16.2.3 How does RSS-based Wi-Fi Localization Work?

RSS is not a reliable metric for range estimation. As we explained in Chapters 2 and 15, the general statistical indoor propagation model used for calculating the received *RSS* at a distance d from a transmitter is given by:

$$P_r = P_0 - 10\alpha \log d + X \qquad (16.2)$$

where P_r is the received power, α is the distance power gradient of the environment, and X represents a zero mean Gaussian random variable describing the effects of shadow fading. Also, as we explained in Section 15.3, the CRLB of the ranging error, using Equation 16.2 to relate the distance to the power, is given by:

$$\sigma_P \geq \frac{\ln 10}{10} \frac{\sigma}{\alpha} d \qquad (16.3)$$

where σ is the standard deviation of the shadow fading.

The distance power gradient takes different values, from below two in corridors, acting as waveguides for radio propagation, up to six in buildings with significant metallic infrastructure and it is around four for most OLOS scenarios in indoor areas. The variance of the shadow fading in indoor areas is typically around 5–10 dB [Pah05]. Using these numbers the spread of the ranging error for measurement of distance using the RSS comes to values on the order of the distance between the transmitter and the receiver, which has a maximum value of around 30–50 m for a typical Wi-Fi infrastructure deployed in indoor areas. These values are not acceptable for typical RTLS commercial applications such as tracking assets or the elderly inside buildings, but they are often very reasonable for WPS applications such as turn by turn direction finding or location-based services. As a result, although RTLS and WPS follow the same principles of operation and we can call both of them "RSS-based Wi-Fi localization techniques", the technical details of their implementation are completely different and they serve two different sectors of the industry. What the two industries have in common is that we "war drive" the network coverage area, inside a building for RTLS and in a metropolitan area for WPS, to collect a database of the observed RSS measurements from different Wi-Fi access points at known or unknown locations. Later, we use the database with pattern recognition localization algorithms to find the location of unknown access points as well as location of a device that is reading certain RSS values from its surrounding Wi-Fi access points. The RSS of the Wi-Fi access point are either measured by the device passively, by reading the beacon signals periodically broadcast in the coverage area of access points, or actively, by probing the access points periodically (see Chapter 8 for a discussion of the beacon and probe messages in WLANs). Both RTLS and WPS cannot support universal coverage and they usually integrate with GPS. The details of differences between the two industries and how they integrate with GPS are described in the next section.

16.3 Comparison of Wi-Fi Localization Systems

As we explained earlier in this chapter, two industries using RSS-based Wi-Fi localization have emerged. The RTLS industry, fostering indoor geolocation for vertical applications, aims to find people or equipment in specifically surveyed buildings such as hospitals, museums, or warehouses. The second industry is the WPS industry that aims to support numerous applications on smart phones and other emerging devices such as book readers and laptops wherever they are located.

16.3.1   RTLS: Wi-Fi Localization for RFID Applications

Market and Applications: The most popular applications of RTLS have been asset tracking in warehouses, locating "in demand" equipment and personnel inside hospitals, developing map-guides for visitors to public areas such as museums, locating the elderly and patients with special needs in nursing homes, and monitoring children or pets away from visual supervision. The first generation of RTLS products were software programs running on laptops and palm-type computers equipped with Wi-Fi devices. The software operates in two modes: a *data collection mode* in which the user builds up the reference database inside a building, and the *localization* mode in which the software locates the terminal based on the RSS readings from the surrounding access points. Later, Wi-Fi chipsets were integrated in a small RFID localization tag to form an embedded system for RTLS applications. More recently, some of the manufacturers have integrated GPS chipsets within the RFID tag to provide continual tracking of the tags when the device is moved between two surveyed sites. The business model for generating income from the software solutions focused on the site licensing and the cost of site survey for collection and maintenance of the database. The introduction of Wi-Fi RFID tags added a new source of revenue out of sales of the individual tags. This market was large enough to sustain a few small companies in this field worldwide.

Performance Expectations and Database: In order for RTLS to support the corresponding applications, RTLS systems need accuracies on the order of meters in indoor areas and as we explained in the previous section, the CRLB given by Equation 16.3 shows that due to the spread of these values, such accuracies are not often attainable. To remedy this situation, RTLS systems need to resort to site survey databases taken at known locations with much smaller spatial separations than the normal separation of the access points and use these databases with a pattern matching algorithm to locate the terminal. To build up the database, the known location where the system collects the reference measurements is visually identified on the layout of a floor of a building and is typically manually entered into the database to create a *radiomap*. Since manual geo-tagging of the measurement locations is time-consuming, manufacturers of these systems usually recommend acquiring several sets of measurements in different terminal positions in a given location. It would be ideal if we could geo-tag the reference locations automatically to save time and build up the database much faster. However, there exist no other means to know the location reliably in indoor areas because GPS does not cover indoor areas and very often, it does not provide the needed accuracy of around few meters, which is the objective of RTLS applications. Figure 16.2 shows the performance of Ekahau's software for different numbers of access points and different numbers of training points in a typical laboratory building shown in Figure 16.3. Figure 16.2 shows that, with three access points and 27 training points, performance accuracies of around 1 m are achieved.

Algorithms and Coverage: RTLS products are usually sold to private enterprises that are in charge of the site survey in their own buildings. Since the enterprise also owns the APs in the buildings, the exact location of the APs is usually available. However, as we discussed earlier, RSS is not a reliable measure for distance estimation and for that reason pattern recognition algorithms, described in Section 15.4.2, which match the RSS signature of the existing APs with the database of measured RSS of all APs at known locations are used for these applications. The first of these algorithms was a simple nearest neighbor algorithm used in [Bah00], which determines the power difference between the measured RSS of the access points in unknown locations with the RSS of APs in known locations in the database and declares the estimated location as the one with the minimum RSS difference as the estimate of the unknown location (see also Examples 15.8 and 15.9).

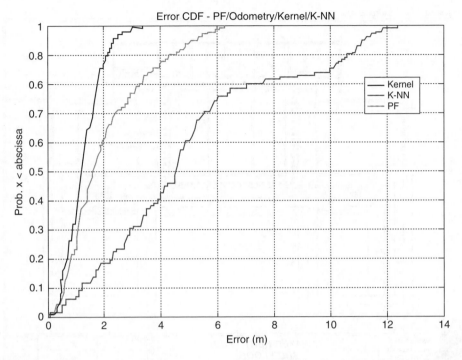

Figure 16.2   CDF of the localization error for two different Wi-Fi localization algorithms and a particle filter (PF) combining the results of Wi-Fi localization and an inertial system in the indoor route shown in Figure 16.3.

An improved performance in terms of accuracy is possible by using stochastic algorithms [Roo02a, b], which use larger numbers of measurements in a reduced number of known locations and determine the unknown location based on the relation between the weighted location to the known points in the database (see Examples 15.10 and 15.11). The advantage of the stochastic algorithms is that they need a fewer number of known locations in the database. Both nearest neighbor and stochastic algorithms do not need the actual location of the APs (which can be obtained for these applications because the owner of the localization system and the APs in the enterprise are the same). If the locations of the APs are available one may think of using radio propagation characteristics to improve the accuracy and reduce the time needed for data collection. The most popular models for radio propagation are the statistical models described by Equation 16.2, but these models do not provide an accurate enough estimate of the RSS which can help localization accuracy and this fact was established in the early days of this industry [Bah00]. Statistical models are not the only solution for RSS estimation; ray-tracing algorithms provide a more accurate building specific estimate of the RSS in indoor areas [Pah05]. Such ray-tracing algorithms use the building layout to provide a more accurate estimate of the RSS inside a building. Availability of an accurate model for radio propagation and knowledge of the location of the APs potentially eliminates the need for labor extensive measurements to collect the reference point database. This approach has been examined in [Hat06].

The RTLS are systems sold to enterprises which already have Wi-Fi networks in their buildings and the coverage of the system is the same as coverage of the Wi-Fi networks owned by the

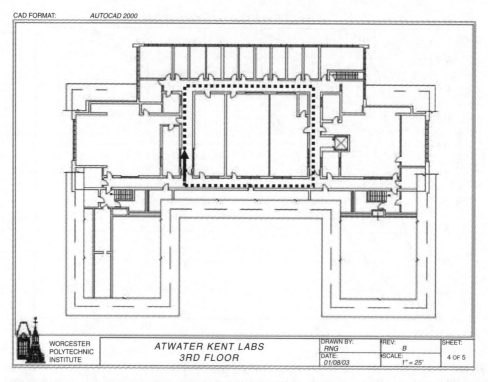

Figure 16.3 An indoor scenario for performance evaluation of the RTLS algorithms on the third floor of the Atwater Kent Laboratory, Worcester Polytechnic Institute (Worcester, Mass.).

enterprise. To increase the accuracy of the location estimation in areas with limited RSS readings, one may add additional access points without an Internet connection. These extra access points do not need full functionality of a wireless device and they can be designed with minimal cost. To extend the coverage of the system beyond the enterprise buildings, which are sometime located in geographically separated areas, the Wi-Fi RFID tags integrate GPS chipsets to provide for outdoor coverage. The algorithm for integration is quite simple: use Wi-Fi localization as default and resort to GPS readings in the absence of Wi-Fi. Another advantage of this integration is that the Wi-Fi location coordinates are defined based on local coordinates specified in the layout of a building but now they can be mapped into the global GPS location coordinates.

16.3.2 WPS: Software GPS

In WPS the database is collected through war driving in the streets of a metropolitan area, using GPS to tag the location and time of measurements and discover the location of APs (see Example 15.7). At a later time, when a user's device reads the RSS from the surrounding Wi-Fi access points, it sends a request to the database to calculate its location by comparing its RSS readings with its database and previous GPS readings using a pattern recognition algorithm. Therefore, a WPS system can be viewed as a software GPS system, which associates GPS reading in a given time to a number of RSS readings from surrounding Wi-Fi access points at that time. Since the locations of the Wi-Fi devices are fixed but the locations of the satellites providing for GPS location

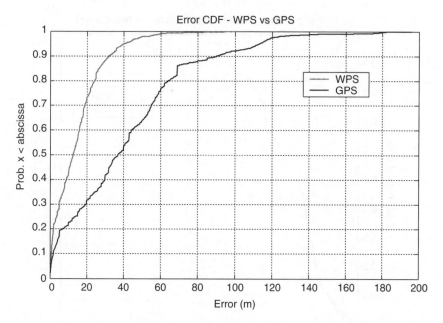

Figure 16.4 Performance of WPS versus GPS on a test route in San Francisco, as shown in Figure 16.5.

estimates are changing, we can associate several GPS readings with different levels of accuracy to the same Wi-Fi access points, or we can use pattern recognition techniques to correct GPS readings with the actual driving map. This allows a WPS system the possibility of providing a better accuracy than GPS itself. Figure 16.4 shows the performance results for Skyhook (Boston, Mass.) in the test route shown in Figure 16.5. This figure illustrates the fact that, in dense urban areas, Wi-Fi localization can have a better performance than GPS in terms of the cumulative distribution of the error. This situation reverses as we go to suburban areas, where the density and coverage of Wi-Fi signals are restricted and we can see GPS satellites directly that can now provide higher precision.

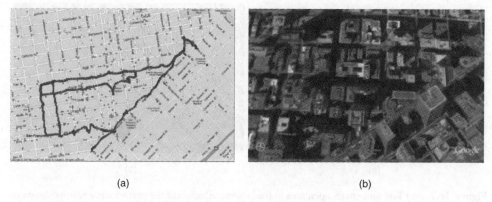

(a) (b)

Figure 16.5 (a) Test route in San Francisco for the performance results shown in Figure 16.4. (b) Satellite map of the downtown area in San Francisco. @ 2012 Skyhook, Inc.

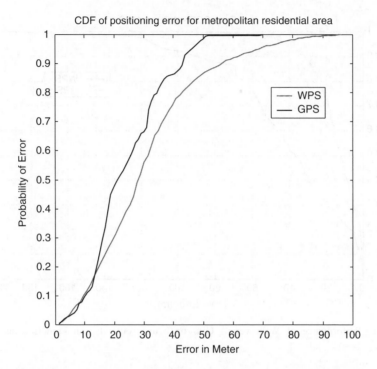

Figure 16.6 Performance of WPS versus GPS on a test route in an open area in the Boston suburbs, as shown in Figure 16.7.

Figures 16.6 and 16.7 provide performance evaluations in a route in the Boston suburban area with more open space in which GPS performs better than WPS.

Looking from this viewpoint, a WPS system can be considered as a software system for memorizing and refining GPS locations at later times without the need for the physical presence of GPS hardware. This software GPS system provides a low-cost, low-power, and fast localization technique with potential for more accurate localization. The most important feature of WPS is

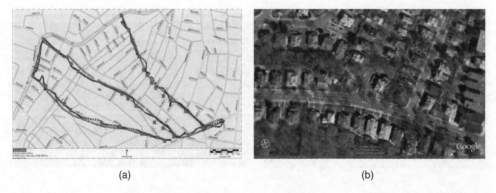

(a)   (b)

Figure 16.7 (a) Test route in an open area in the Boston suburbs for the performance results shown in Figure 16.6. (b) Satellite map of the area. @ 2012 Skyhook, Inc.

that it operates based on Wi-Fi readings, which extends the coverage of WPS to indoor areas, where GPS faces serious challenges. GPS provides a universal coverage outdoors but it does not work in a large fraction of locations in indoor and urban areas, where most location-based mobile computing applications are used. Integration of WPS software and GPS hardware provides for that comprehensive coverage, wherever a device can afford a GPS chip set.

Evolution of Hybrid Localization in Smart Devices: Around the year 2010, location technologies emerged as a core element in smart phones, tablets, and netbooks, opening a rapidly growing market for devices integrating mobile computing and navigation technologies to the point where average customers started using navigation technologies in everyday applications such as finding local services, tagging pictures and videos, finding local news, and of course receiving turn by turn directions. These applications commonly accept accuracies on the order of tens of meters but they demand an immediate location fix and comprehensive coverage everywhere. These features cannot be supported by GPS and cell tower localization alone, and so the demand arose for hybrid system integration with Wi-Fi localization techniques. This revolution in application market and user expectations for localization and tracking techniques started with the iPhone, extended to the iPad and netbooks, and then found its way into other emerging smart devices such as e-book readers, cameras, and other popular applications such as social networking.

One of the fundamental advantages of WPS is that it can be used as a stand-alone software solution for tablets, netbooks, and laptops when they are not equipped with GPS or cell phone chip sets. This solution is natural, because tablets, netbooks, and laptops use Wi-Fi chip sets to establish Internet connection and, where Wi-Fi is available, WPS also works. Smart phones have cellular network connections as well as Wi-Fi chipsets. Wi-Fi signals from hot spots, home routers, and public access and enterprise wireless networks cover most of the necessary indoor and urban areas, where Internet applications are commonly used. The less accurate cell tower localization can complement this coverage in places such as interstate highways, where Wi-Fi signals may not be available all the time. Therefore, a combination of RSS-based WPS and cell tower localization can provide fast-fix, low-power, and low-cost software localization solutions with a comprehensive coverage to these smart phones. This solution was the one used in the legacy iPhone, when it was first introduced to the market. Today, the latest editions of the iPhone as well as most other leading smart phones also carry GPS chipsets, which complement the coverage and increases the accuracy of localization in outdoor areas.

The integration of WPS and RSS-based cell tower localization is very straightforward. The more accurate WPS is always the default and cell tower localization is the backup localization engine when ever WPS is not available. Since both WPS and cell tower localization are implemented in software, there is no preference in selecting between the two governed by power consumption and battery life or the time to fix. Therefore, they can easily integrate with a simple approach: when Wi-Fi localization is available we use it because it is more accurate, and we use the cell tower localization as a backup.

Integration of WPS with GPS, however, is much more complex and technically involved because the two systems have significant complementary attributes, which can be exploited to improve the overall performance of the integrated system. WPS provides a better performance than GPS in indoor and dense urban areas, but most times, a warm GPS system on the road is preferred to WPS localization. GPS needs a few minutes to get a fix in indoor areas, where most applications are initiated, and during that period, the hardware is draining the precious battery life of the smart phones or netbooks. Therefore, in indoor areas WPS is the default and if the application is something like turn by turn direction finding, which starts in the indoor and continues outdoors, after GPS is warm and the user is out of the indoor areas the localization engine of the device can switch to GPS. How to combine these two location technologies to optimize the accuracy, power consumption and time to fix is a multi-dimensional engineering challenge demanding complex engineering solutions. This demand has stimulated a few startup companies to specialize in commercial application of WPS and its integration with GPS and cell tower localization. In military and public safety applications, further research is needed to address rapid database collection and the electromagnetic and radio signal interference effects on WPS and GPS localization techniques.

Database Collection and Algorithms: In WPS the database is collected through war driving in the streets of a metropolitan area, using GPS to tag the location and time of each measurement. The size of this database is huge compared to the database of an RTLS system and the collection procedure involves a number of drivers across many metropolitan areas.

In general the distribution of the actual Wi-Fi access points in metropolitan areas can be modeled as a stochastic process with particular spatial and temporal characteristics because the number of access points and their locations inside a metropolitan area are constantly changing and there is no way to track them precisely. In any given time interval, new access points are installed and some old access points disappear or are re-located. No single authority has control on the ownership or installation and relocation of these access points, and it is not practical to determine the actual location of all Wi-Fi access points at a given time. Therefore, the database obtained by war driving is a snapshot of this stochastic phenomenon and does not contain the entire ground truth on existence and location of all APs in a given time. As time passes, the coverage and accuracy of the database decays and we need to refresh the database periodically. This procedure is very challenging and demands a careful driving plan and re-scanning scheduling to optimize the cost of database collecting. Therefore, the collection and maintenance of a good database for WPS is an expensive and continual process; and the quality of a database collected with this approach varies substantially, depending on the depth and complexity of the method used for collecting the database.

An effective technique to reduce the cost of maintenance, expand the coverage, and increase the re-scanning intervals is to take advantage of the so called user's *organic data* to update the database. The organic data is collected by the user terminal either at a time when the user starts a localization application or with an automated program installed in the terminal, collecting data periodically. Using this organic data to update the database will expand the size of the database and reduce the refreshment intervals, which results in a significant reduction of the maintenance cost.

Integration of the organic data into the systematically collected database with coordinated war driving procedures requires *data mining algorithms* to ensure that the additional organic data does not reduce the overall accuracy of localization. In a database collected for WPS, the geo-tag carries over GPS errors, the number of measurements spread over the area depends on the speed of the drivers, and the geographical coverage of the database depends on coordination plan among drivers. We need another set of algorithms for post-processing of the collected database to minimize GPS geo-tagging errors and to make the spatial distribution of the database close to uniform. These algorithms are separate from the actual localization algorithms used for WPS.

The localization algorithms designed for WPS have to cope with database uncertainties caused by the stochastic spatial and temporal characteristics of Wi-Fi APs and the uneven distribution of the data associated with individual APs. These algorithms should process a huge database for which direct use of nearest-neighbor based algorithms may not be the optimum solution all the time [Che05]. The radio propagation environment for WPS involves a variety of complex indoor to outdoor scenarios, which are more unpredictable than indoor to indoor radio propagation in the case of RTLS. These characteristics can be sometimes exploited to improve the accuracy of the algorithms.

These complexities for the design of WPS have opened a field for innovative engineering and science for companies that engage in database collection and post-processing for Wi-Fi localization in metropolitan areas. In the same way that many companies have search engines, while Google leads the others by having more and better processed data, a better Wi-Fi localization in metropolitan areas is becoming an expertise for the companies who have the largest databases and the best algorithms for post-processing the data to locate a device with more accuracy and wider availability.

16.4   Practical TOA Measurement

Measurement of the RSS is very simple and straightforward and mostly independent of the bandwidth of the system. In contrast, TOA measurement requires time synchronization between the transmitter and receiver and is highly affected by the bandwidth of the system. As a result, the measurement of the TOA by itself has received considerable attention and has been the core of research for TOA-based localization. In free space or in air, radio signals travel at the constant speed of light. The TOA of a signal can be practically measured either by measuring the phase of the received narrowband carrier signal or by direct measurement of the arrival time of a wideband narrow pulse. The wideband pulses for measuring the TOA can be generated either by a wideband direct sequence spread spectrum (DSSS) signal or directly by ultra wideband (UWB) pulses. Therefore, techniques for the measurement of the TOA can be divided into three classes: narrowband, wideband, and UWB techniques.

16.4.1   Measurement of TOA using a Narrowband Carrier Phase

We can measure the TOA or time of flight of the signal by measuring the received signal carrier phase and comparing it with that of the phase of the transmitted carrier. Figure 16.8 explain this approach; the narrowband phase difference between the received and transmitted carrier signals is used to measure the distance between two points. The phase of a received carrier signal, ϕ, and the relative TOA of the signal, τ, are related by $\tau = \frac{\phi}{2\pi f_c}$, where f_c is the carrier frequency (see Example 2.9). It is well known that differential GPS, using the measured reference carrier phase at the receiver, can improve the location accuracy of traditional GPS from about 20 m to within 1 m [Kap96]. One general problem associated with the phase measurements lies in its ambiguity, shown in Figure 16.8a. This ambiguity results from the periodic property (with period 2π) of the signal phase, and we cant measure delays that are associated with phases more than 2. The traditional DSSS and UWB measurements using pulse transmission for calculation of time of flight are unambiguous and theoretically they can measure any value of delay. Consequently, in GPS, the ambiguous carrier phase measurement is used for refining the DSSS measurements. In GPS systems, a complementary Kalman filter is used to combine the low noise ambiguous carrier phase measurements and the unambiguous but noisier TOA measurements [Kap96].

Unlike GPS, where the direct path is always assumed to be present, the severe multipath conditions in indoor and urban geolocation environments cause substantial errors in the phase measurements. When a narrowband carrier signal is transmitted in a multipath environment, as shown in Figure 16.8b, the composite received carrier signal is the sum of a number of carriers arriving along different paths, of the same frequency but different amplitudes and phases (see Section 2.3.1). The frequency of the composite received signal remains mostly unchanged but, the phase can be substantially different from that of the direct LOS signal resulting in substantial distance measurement errors. The immediate conclusion is that the phase-based distance measurement using narrowband carrier signals cannot provide accurate estimates of the distance in heavy multipath environments.

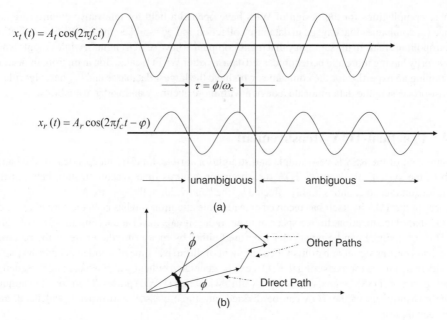

$x_t(t) = A_t \cos(2\pi f_c t)$

$\tau = \phi/\omega_c$

$x_r(t) = A_r \cos(2\pi f_c t - \varphi)$

unambiguous | ambiguous

(a)

$\hat{\phi}$

Other Paths

ϕ Direct Path

(b)

Figure 16.8 (a) Narrowband TOA measurement using phase of the received signal and the ambiguity in the measurement. (b) Phasor diagram for multipath arrival.

16.4.2   Wideband TOA Measurement and Super-resolution Algorithm

Direct sequence spread spectrum (DSSS) wideband signals have been used in GPS and other ranging systems for many years. In such a system, as we explained in Section 3.5.4, a signal modulated (spread) by a known pseudo-random (PN) sequence is sent by a transmitter. Then a receiver cross correlates the received signal with a locally generated PN sequence to obtain a narrow pulse with a sharp peak. The distance between the transmitter and the receiver is determined from the arrival time of the first correlation peak. Because of the processing gain of the correlation process at the receiver, the DSSS ranging systems perform much better than competing systems in suppressing interference from other radio systems operating in the same frequency band.

In multipath rich environments such as the indoor channel, as we explained in Section 15.2.2 and Figure 15.2, multipath causes two problems in measurement of the TOA. The paths close to the direct path (DP) between the transmitter and the receiver cause shifts in the peak of the first path, causing errors due to multipath. Also, sometimes the first path itself is blocked by metallic objects, causing substantial errors in the TOA measurements. In Figure 15.2a the DP is represented by the first path, which is also the strongest path. The location of this path in the profile is the expected value of the TOA. Other paths with a number of reflections and transmissions arrive after the DP with lower amplitudes. These paths would have been observed at the receiver as impulses, if the bandwidth of the system was infinity. In practice, however, the bandwidth is limited and each path results in a pulse shape at the receiver (see Section 2.4.2). As we observe in Figure 15.2a, the received waveform will be the superimposition of a number of transmitted waveforms whose amplitude and arrival time are the same as impulses but convolved with a pulse shape. The addition of all these pulse shapes forms the received waveform shown in the figure, which we refer to as the channel profile.

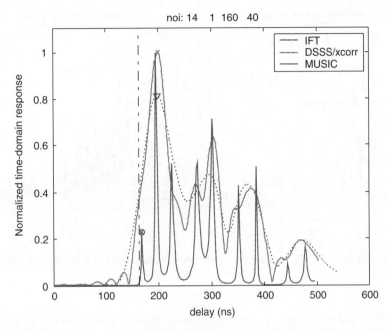

Figure 16.9   Effectiveness of super-resolution algorithm to resolve multipath components.

Due to the scarcity of the available bandwidth in practice, in some indoor geolocation applications, the DSSS ranging systems cannot provide adequate accuracy. But, it is always desirable to achieve a higher ranging accuracy using the same bandwidth. Inspired by high-resolution spectrum estimation techniques, a number of researchers have studied super-resolution techniques for this purpose. Figure 16.9 [Li04] illustrates the usefulness of the super-resolution technique for a sample measured indoor channel profile. In this figure, the MUltiple SIgnal Classification (MUSIC) algorithm is used as an example of a super-resolution technique. In the first of the other two techniques, the frequency domain channel response is directly converted to time domain using inverse Fourier transform (IFT) and then the arrival time of the direct path is detected. The second technique uses the traditional cross-correlation techniques with DSSS signals (DSSS/xcorr). The super-resolution technique can determine the TOA with a much higher resolution from the frequency channel response.

16.4.3   UWB TOA Measurement

As we mentioned previously, the signal bandwidth is one of the key factors that affects the TOA estimation accuracy in multipath propagation environments. In general, the larger the bandwidth, the higher is the ranging accuracy. UWB systems can exploit bandwidths in excess of several GHz. Naturally, these systems have attracted considerable attention as a means of measuring accurate TOA for indoor geolocation applications. Figure 16.10a illustrates a typical measurement of the channel impulse response in an office area with 500 MHz bandwidth, and Figure 16.10b provides the same profile with a 3 GHz bandwidth. The expected delay of arrival depicting the actual distance between the transmitter and the receiver is 40.5 ns and the estimated arrival with 500 MHz and 3 GHz bands are 45.5 and 40.7 ns, respectively. The 5 and 0.2 ns errors in TOA estimate

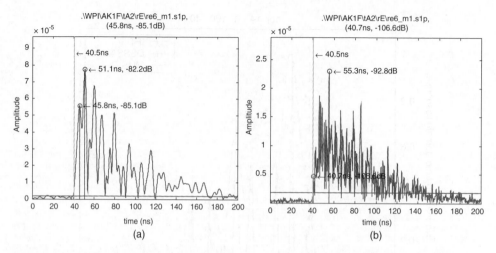

Figure 16.10 A typical UWB channel profile for: (a) a bandwidth of 500 MHz, (b) a bandwidth of 3 GHz.

result in 1.67 m and 6.7 cm errors, respectively, clearly illustrating the higher precision with the 3 GHz bandwidth. A closer look at the figure reveals that the amplitude of the first path with the 500 MHz bandwidth is larger than that with the 3 GHz bandwidth. This is due to the fact that with a 500 MHz bandwidth we resolve several times less of the multipath arrivals than a 3 GHz bandwidth and more number of paths that are combined in a pulse results in a stronger path amplitude. Therefore, the amplitude of the paths with a 500 MHz bandwidth are statistically larger than those with the 3 GHz bandwidth. In other words, one expects that the coverage of the direct path for 500 MHz system be larger than that of the 3 GHz UWB system because its paths are stronger. The natural question arises at this point is that what happens if the first path is not detected.

As a mobile terminal moves away from a base station, the strength of the DP and the total received signal power decay exponentially. In obstructed line of sight environments, when the DP goes below some threshold but the other paths are still detectable, the receiver assumes the very first detectable path in the profile to be the DP, and this mistake causes a substantial error in wideband TOA measurements. We have been referring to this situation as the undetected direct path (UDP) condition, as explained in Section 15.2.2. Figure 15.2b shows the occurrence of the UDP scenario from the results of ray tracing for a transmitted pulse with a bandwidth of 200 MHz. Since the difference between the strength of the strongest path and the first path is more than the dynamic range (the range of detectable signal level below the strongest path) of the receiver, we have a clear UDP in which the first *detectable* path is detected and declared as the DP, resulting in a 5.23 m distance measurement error. Figure 16.11 illustrates two cases of UDP with UWB measurements using 500 MHz and 3 GHz bandwidths. We have errors of 13.5 ns (3 m) and 20.4 ns (6.8 m) for 500 MHz and 3 GHz bandwidths, respectively. With a 500 MHz bandwidth several paths, otherwise resolvable with a 3 GHz bandwidth, combine and the overall profile has paths with larger path strengths. For the example in Figure 16.11, some of the early paths for the 3 GHz bandwidth (Figure 16.11b) that were under the threshold have been combined in the 500 MHz bandwidth case (Figure 16.11a) and the resulting "combined paths" have crossed the threshold, resulting in a smaller TOA measurement error.

Earlier in this chapter we stated one of the limitations in the accuracy of wideband TOA system is caused by restrictions in bandwidth. In narrower bandwidths, the paths close to the DP combine

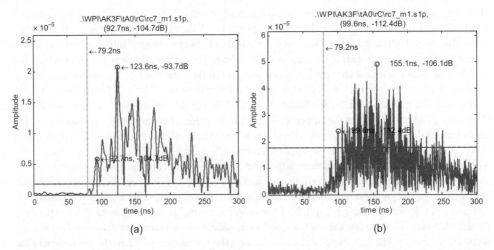

Figure 16.11   A sample UDP for UWB measurements for: (a) a bandwidth of 500 MHz, (b) a bandwidth of 3 GHz.

resulting in a shift in the location of the estimated DP (first peak of the profile). This problem is resolved when we increase the bandwidth. In this section we showed that UDP conditions cause large errors in wideband TOA measurement that at certain points increase when we increase the bandwidth. With the release of UWB bands we have adequate bandwidth for accurate distance measurement and the main challenge for implementation of accurate wideband TOA systems is to find a remedy for the UDP conditions.

16.5   Localization in the Absence of DP

In this section, we introduce a dynamic scenario of operation with defined walking routes in a typical office building to continue our discussions with more quantitative performance analysis. Figure 16.12 shows a dynamic scenario on the third floor of the Atwater Kent Laboratories (Worcester Polytechnic Institute) where a user walks along different routes in the central part of the building.

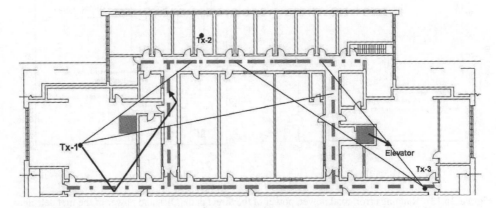

Figure 16.12   A dynamic scenario of operation on the third floor of Atwater Kent Laboratory, Worcester Polytechnic Institute, to illustrate the effects of UDP conditions in indoor geolocation.

There are two large metallic objects on the two sides of this floor plan: an elevator on the right and an RF-isolated chamber on the left. These two objects block the signal paths and in particular they cause shadowed UDP conditions when they are situated between the transmitter and the receiver. Figure 16.12 shows three walking routes in the building, the upper and lower routes in which the mobile user walks across a straight line from one end to the other end of a corridor, and a loop route in which the mobile user walks loops in the central part of the building. Transmitter 1 (Tx-1) is located in the middle of the large laboratory on the left. For this transmitter, we have substantial shadowed UDP conditions in the upper and left side corridors, but there are no shadowed UDP conditions in the lower and right side corridors. When the mobile takes the central loop route, it observes a shadowed UDP condition approximately 40% of the time. With Tx-1, as we trace the mobile along the routes, we can observe different possibilities for occurrence of UDP conditions to analyze the behavior of the large ranging errors and effectiveness of different techniques in mitigating them. In the last part of our discussions where we discuss cooperative localization techniques, in addition to Tx-1, we also use Tx-2 and Tx-3 for two-dimensional positioning needed for explaining these techniques. To carry out the analysis, we use a wideband measurement calibrated ray tracing software to generate channel impulse responses along the routes every 13 cm (the resolution of the graphical user interface).

16.5.1   Ranging Error in the Absence of DP

As the first step in our analysis, we take the upper and lower routes in Figure 16.12 for Tx-1 to demonstrate the behavior of the TOA-based ranging error in shadowed and natural UDP conditions. The shadowed UDP is caused by large metallic objects between the DP connecting the transmitter and the receiver. Natural UDP occurs when the distance between the transmitter and receiver is large enough so that the DP is fading away but still some signal arrives along other paths. Our objective here is to relate these UDP conditions to other important propagation parameters such as total power, power of the direct path (DP) and the power of the first detected path (FDP). Figure 16.13

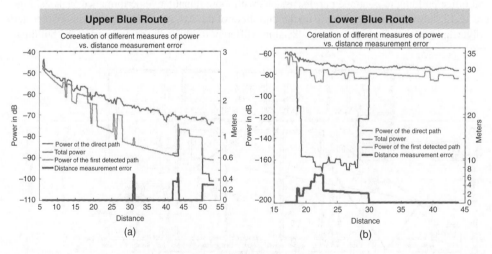

Figure 16.13   Ranging error, total power, power of the direct path (DP), and power of the first detected path (FDP) as a function of traveled distance for: (a) upper route, (b) lower route.

shows these three parameters and the resulting ranging error along the upper and lower routes. The received power of the DP and the FDP are the same, except when a UDP condition occurs. The graph on the left representing the lower route does not have any shadowing UDP conditions. There are three short bursts of natural UDP errors with distance measurement (ranging) error values of less than 0.5 m occurring at around the 30, 45, and 50 m markers between the transmitter and the receiver. In these areas, the difference between the power of the DP and the FDP is around 10 dB. The graph on the right demonstrates a clear example of shadowed UDP conditions, when the mobile is following the upper route. As the mobile moves along that route, from the distance markers at around 18–30 m, the metallic chamber creates a shadowed UDP condition causing ranging errors on the order of several meters and a difference of around several tens of dB between the DP and the FDP. This analysis demonstrates that, during shadowed UDP conditions, we have a substantial drop in the power of the DP, resulting in large ranging errors, while in natural UDP conditions we have a moderate drop in power of the DP and relatively smaller ranging errors.

16.5.2 Effects of Bandwidth

As we explained earlier, ranging error is caused either by the limitations in the bandwidth of the system or by the occurrence of UDP conditions. To demonstrate the effects of bandwidth on the overall performance, we consider our loop scenario and Tx-1, for which the occurrence of UDP conditions is around 40% of the locations. Figure 16.14 shows the CDF of the ranging error in our loop scenario for Tx-1 and a variety of bandwidths. The solid line is associated with the direct results of ray tracing in which each path is represented by an impulse with infinite bandwidth. In around 60% of the locations, on the lower parts of the loop, we detect the DP and with infinite bandwidth, we estimate the exact distance between the transmitter and the receiver with zero ranging error. For the remaining 40% of the locations the RF-isolation chamber blocks the DP causing positive values of up to 7 m for the ranging error due to erroneous detection of the FDP rather than the DP in the channel impulse response. As we gradually decrease the bandwidth to 300, 200, and 100 MHz in

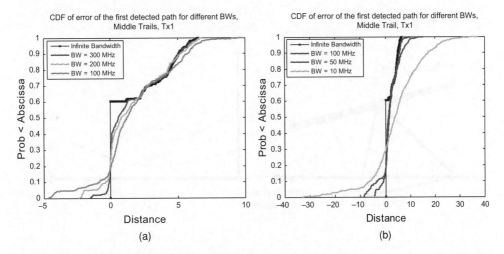

Figure 16.14 The CDF of ranging error for: (a) 300–100 MHz bandwidths, (b) lower 100–30 MHz bandwidths.

Figure 16.14a, the larger ranging error values due to the bandwidth effects appear in the CDF of the plots. Since ranging error due to the bandwidth limitations can shift the FDP of the channel profile in either direction, we can also now observe negative ranging errors. Reduction in the bandwidth spreads the range of the errors. For example, with a 100 MHz bandwidth we have errors between −5 and 10 m.

Figure 16.14b uses narrower bandwidths up to 10 MHz for which the ranging error is spread between −30 and 40 m in a loop in which the maximum distance between the transmitter and the receiver is less than 40 m. Therefore, bandwidths on the orders of 1 MHz used in GPS systems are not sufficient and we need bandwidths on the order of several hundred MHz to provide reasonable protection against the extensive multipath in indoor areas. For example, for a 200 MHz bandwidth, the spread of ranging error is in the order of −3 to +7 m, fairly comparable to UDP errors of up to 7 m observed with infinite bandwidth. To reduce the bandwidth requirements below these values, one may consider using super-resolution algorithms for post-processing, as described in Section 16.4.2. However, to reduce ranging errors below those observed in the UDP areas with an infinite bandwidth, we need fundamentally different approaches, which we examine in the following section.

Traditional radio communication techniques such as frequency diversity, time diversity, or space diversity using MIMO techniques are not effective in mitigating the large ranging errors, occurring in the absence of a DP. Two promising approaches to precise indoor localization in the absence of a DP are: (1) localization exploiting non-direct paths, and (2) cooperative localization [Pah06].

16.5.3   Localization using Multipath Diversity

Figure 16.15 illustrates the basic principle underlying the relationship between the TOA of the DP and a path reflected from a wall, for a simple two-path scenario. As the mobile receiver moves along the x-axis, the change in the distance in that direction is related to the length of the DP by $dx \cos \alpha = dl_{DP}$. As the geometry in Figure 16.15 shows, for the reflected path we also have

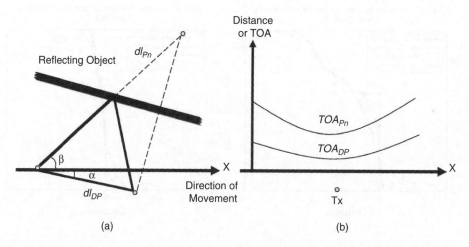

Figure 16.15   (a) Basic two path reflection environment. (b) Relation between the TOA of the paths.

$dx \cos \beta = dl_{P_n}$. Therefore, we can calculate the change in the length of the DP from the change in the reflected path, using:

$$dl_{DP} = dl_{P_n} \frac{\cos \alpha}{\cos \beta} \quad \text{or} \quad d\left(TOA_{DP}\right) = d\left(TOA_{P_n}\right) \frac{\cos \alpha}{\cos \beta} \tag{16.4}$$

In other words, knowing the angle, β, between the arriving path and direction of movement and the angle, α, between the direction of movement and the DP, we can estimate the changes in the TOA of the DP from changes in the TOA of the reflected path. This basic principle can be extended to paths reflected from many objects and also to the three-dimensional case. This general treatment is available in [Pah06].

In indoor geolocation applications, we can think of applying this principle to locating a mobile in UDP areas in the absence of a DP. Knowing the previous location of the transmitter and the direction of movements, we can always calculate α, even in the absence of the DP. If we can find a way to measure β, using values of α and β in Equation 16.4, we can track the location as the mobile receiver moves along in a UDP environment.

In order to use a path other than the DP for tracking the location, we should be able to identify that path among all other paths, and the number of reflections for that path should remain the same in the region of interest. In the simple two-path model shown in Figure 16.15, the second path consistently reflects from one wall as we move along the region and hence we can identify that path easily because it is the only path other than the DP. Since both conditions hold for the second path, the behavior of the TOA of that path, shown in Figure 16.15b, is smooth and we can use it for tracing the DP. In realistic indoor scenarios, in the absence of a DP, we have numerous other paths to use and the simplest paths to track are the first detected path (FDP) and the strongest path (SP). Both SP and the FDP have inconsistent behavior in the UDP region of interest. This inconsistent behavior is caused by changes in the path index of these paths. In other words, if we associate a path number or index with a path which is associated with a specific reflection scenario from given walls, then as we move along in a region, the path index or reflection scenario for the FDP or the SP changes. Each of these changes causes a jump in the behavior of the TOA of the path, thus impairing the smoothness needed for our estimation process [Pah06].

16.5.4   Cooperative Localization Using Spatial Diversity

In two-dimensional localization we need at least three links or connections to reference terminals with known locations. These links may have different qualities of estimate for the distance between the reference and target terminals, depending on the availability of DP in the channel. In cooperative localization using spatial diversity, in multipath-rich environments, we simply avoid ranging estimates reported from the links with UDP conditions. In other words, the redundant information provided by the additional reference points situated in a better spatial location is used to reduce the localization error. This overall situation is common in ad hoc and sensor networks where we have a fixed infrastructure of known reference points for positioning and a number of mobile users in the area. When we want to locate a mobile terminal, in addition to the distances from the respective fixed reference points, we can also use the relative distances from other mobile users. We refer to this approach as cooperative localization since the localization is conducted through a cooperative method. A similar approach is also used for general localization in sensor and ad hoc networks when we have a limited number of dispersed references and a number of ad hoc sensor terminals

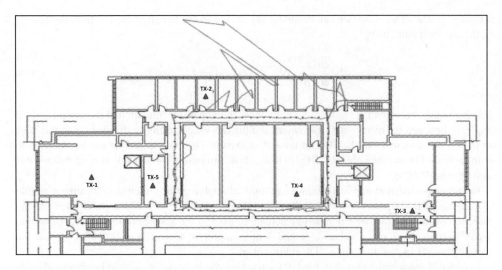

Figure 16.16   Demonstration of space diversity on the third floor of the Atwater Kent Laboratory.

with less than an adequate number of connections to reference points [Sav01a, b, Sav05]. For a general localization, we only need the whereabouts of the terminals; and the literature in that field does not address the large error caused by UDP conditions. The concept introduced here uses the redundancy of the links embedded in the sensor location and ad hoc network environments to achieve precise indoor localization. To clarify our new concept we resort again to an example using our scenario.

Figure 16.16 shows a positioning scenario with three reference transmitters in our selected office building and the loop-route scenario. The Tx-1 located in a large laboratory on the left side of the building has UDP conditions caused by the RF-isolation chamber in 40% of locations around the loop, Tx-2 located in the small office on upper parts of the building layout, covers the entire loop without any UDP location, and Tx-3 located in the lower corridor has around 50% UDP conditions around the loop caused by the elevator. The line with extensive variations around the route in Figure 16.16 shows the results of location estimation using the traditional least square algorithms for TOA-based localization with the three known reference transmitters along the loop. Whenever the DP is present, for example, in the lower and right hand routes, the ranging error is small. As we have one or two UDP conditions for our three links to the references, for example, in the upper route, the ranging error is substantially large. This observation suggests that, whenever a DP is available for all links, we can achieve a precise localization, but as soon as one of the links loses the DP we have large localization errors. In other words, if we avoid UDP conditions we can achieve precise positioning. Therefore, if we have more than the minimum number of references, assuming we can detect the UDP conditions, we can avoid links with UDP and achieve precise localization.

To demonstrate the effectiveness of this approach, we consider an example where we have two other users, Tx-4 and Tx-5, which are located in good positions, where each has three DP connections to the main reference transmitters. As shown in Figure 16.16, when we use the three main reference transmitters to estimate the location of Tx-4 and Tx-5 we have very good estimated locations for them. In an ad hoc sensor environment we can assume that our target receiver moving along the loop route can also measure its distances from Tx-4 and Tx-5. In this particular example these ranging estimates are also very accurate because they are based on the availability of the DP. The line that is very close to the route in Figure 16.16 shows the estimate of location for the mobile terminal as it moves along the loop when it uses the estimated locations of Tx-4 and Tx-5 and the actual location of Tx-2 to locate itself with the traditional least squares algorithm. As shown in Figure 16.16, our estimates are now substantially more accurate. The drastic improvement in the accuracy of localization is a result of avoiding UDP conditions and taking advantage of the redundancy of the ad hoc sensor networks to achieve precise cooperative localization.

In the above example, we showed the potential advantage of using the redundancy in sensor and ad hoc networks to achieve precise cooperative positioning. In practice, we need to develop algorithms for implementation of this concept. These algorithms need the intelligence to discover the quality of ranging estimates and possibly occurrence of the UDP conditions to use them for positioning. The algorithms for general cooperative localization first suggested in [Sav01a, b] and later discussed in the follow-up literature [Sav05] are not applicable to our approach. We need new algorithms to address specific methods to handle the behavior of the ranging errors in the absence of DP. We need to find techniques for relating a quality of estimate to each ranging and positioning estimate in order to develop precise cooperative localization algorithms for sensor and ad hoc networks. These algorithms should take advantage of redundancy to avoid unreliable reference sources and achieve robust localization using spatial diversity. A preliminary algorithm using the channel behavior to implement a practical solution for precise cooperative localization is available in [Als08]. A method for identification of UDP conditions is provided in [Hei09].

16.6 Challenges in Localization inside the Human Body

In the past decade, miniaturization and the cost reduction of semiconductor devices have allowed the design of small, low-cost computing and wireless communication devices used as sensors in a variety of popular wireless networking applications, and this trend is expected to continue in the next few decades. It is expected that a myriad of new applications designed around sensor technologies will emerge to stimulate a huge industrial growth. One of the most promising areas of industrial growth associated with this industry lies in body sensor networks (also referred to as body area networks; BAN) [Yan06]. These networks are expected to connect wearable and implantable sensory nodes together and with the Internet to support numerous applications, ranging from traditional externally mounted temperature meters or implanted pacemakers up to emerging blood pressure sensors, eye pressure sensors for glaucoma, and smart pills for health monitoring and precision drug delivery.

To support the growth of this industry, the Federal Communication Commission (FCC) has allocated specific bands for Medical Radio Communication Services (MedRadio) [FCC09] and the IEEE 802.15.6 has been formed to address the standardization aspects of these emerging technologies. The IEEE 802.15.6 models the characteristics of the radio propagation inside and around the human body and defines wireless networking technologies for wearable and implanted sensor networks. The standards recommend that the transmission power should be around 25 µW to keep the EM emissions at a healthy level [FCC09]. Certainly for all BAN applications, power-efficient modulation and medium access control methods are needed in principle and a number of researchers are working on that topic [Kim08]. The important and fundamental issue presented in this section is the localization of objects inside the human body to assist in the discovery of methods for navigating emerging micro-robots in wireless medical applications such as capsule endoscopy. This is a new field of research that has gained some momentum in recent years [Aoy09; Ban11].

Understanding the nature of signal propagation is the key to designing precise localization for any wireless network. Therefore, the first step in research is to start a measurement and modeling program to understand the nature of signal transmission inside the human body. Today, the existing literature in measurement and modeling for understanding the propagation in and around the human body is fragmented and does not pay attention to localization inside the human body [Aoy09]. The IEEE 802.15.6 is working on creating a comprehensive channel model for different scenarios and frequency bands used for communication applications [Yaz10]. There is a need for research in understanding the behavior of RF signal propagation inside the human body for localization applications. Localization techniques fundamentally work on the basis of either the RSS or TOA of the signal from the target device to the reference points. The channel model for localization and communication for RSS-based systems are the same. However, for more precise TOA-based systems we need channel models that may be different from those traditionally designed for communication applications.

From an innovative research point of view, the measurement and modeling of radio propagation inside and around the human body for RF localization applications offers unique challenges, making this area very appealing for fundamental research. These challenges are raised by several specifics of the human body medium that are in principle different from the traditional indoor radio propagation challenges. Inside the human body is a non-homogenous liquid-like environment for radio propagation, and this poses a challenge for precise localization techniques using the TOA of the signal between a transmitter and a receiver to estimate the distance. To localize a device inside the human body, the reference points are naturally mounted as sensors on the human body that constantly moves, even when we are standing still [Fu12]. Therefore unlike indoor geolocation, here the infrastructure and the environment is constantly moving causing inaccuracies in location estimates. To measure the characteristics of multipath arrivals and their effects on localization using TOA, we usually refer to statistical empirical modeling based on UWB measurements of the channel characteristics by placing antennas at different locations in the application environment. Placing antennas inside the human body is not practical and we need to resort to computational techniques [Say10], or using Phantoms, or using the dead body of an animal for empirical measurements [Pah12a, b]. The most popular computational methods for simulating radio propagations inside the human body are the Finite Element Method (FEM) [Ask11] and the Finite Difference Time Domain (FDTD) [Kha11]. To validate the results of these simulations, we need to match them with the results of empirical measurements based on body-mounted sensors. Inside the human body, distances are on the orders of centimeters and it is desirable to have simulation techniques and measurement devices that have accuracies on the order of 1 cm. That demands very fine grids for simulations using numerical techniques and extremely wide bandwidths for measurements using Phantoms or the bodies of dead animals.

16.6.1   Bounds on RSS-based Localization inside the Human Body

We begin by determining the bounds on the performance of RSS-based localization of a micro-robot inside the digestive system of the human body using the known locations of body-mounted sensors. There are a couple of path-loss models for inside the human body that can be used for this purpose. We use the models presented in [Say10] that provide the path loss gradient and variance of the shadow fading needed for calculating the Cramer Rao Lower Bounds (CRLB) for variance in RSS-based localization. For calculation of the bounds we use 3D localization techniques needed for inside the human body, described in [Ye12].

Table 16.1 Channel parameters used for performance evaluation of RSS-based cooperative localization

Implant to body surface	$L_p(d_0)$	α	σ_{dB}
Deep tissue	47.14	4.26	7.85
Near tissue	49.81	4.22	6.81

In the model presented in [Say10], similar to path loss models described in section 2.2.4, the path loss in dB at a distance d between the transmitter and receiver is modeled by:

$$L_p(d) = L_p(d_0) + 10\alpha \log \frac{d}{d_0} + X(d > d_0) \qquad (16.5)$$

where d_0 is the reference distance, set to 50 mm, and α is the path loss gradient, which is determined by the propagation at different depths inside the human body. As we already mentioned, the human body tissue strongly absorbs RF signals. Therefore, we expect values of path loss gradient that are much higher than two for free space propagation. The random variable X in Equation 16.5 is a log-normally distributed random variable representing the deviations caused by the shadowing effect of human tissue. The parameters used by the model for the implant to body surface path loss modeling are summarized in Table 16.1. In this table we have two sets of parameters for path loss from deep and near surface implants to the body surface and σ in dB is the standard deviation of shadow fading X. According to the model developed in [Say10], if the distance is less than 10 cm, we use the near surface to surface path loss model, otherwise the deep tissue to surface model is used. Using this path loss model, similar to section 15.4.1, we can calculate the positioning error for a device inside the human body with body mounted sensors as reference points for localization.

Figure 16.17 shows the location of the body-mounted sensors with respect to the grid of points and the location of stomach, small intestine, and large intestine inside the human body. The CRLB is

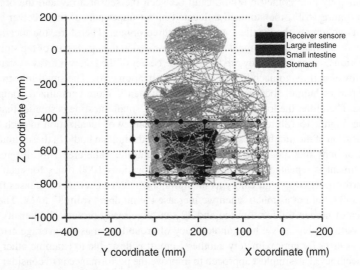

Figure 16.17 Typical simulation scenario for localization inside GI tract with 32 receiver sensors on the body surface.

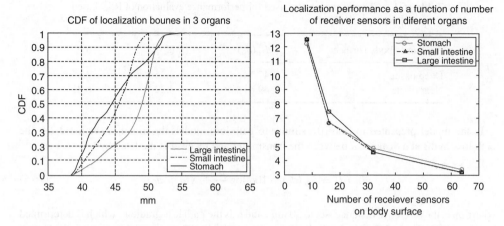

Figure 16.18 Localization bounds for RSS-based cooperative capsule endoscopy: (a) CDF of errors in different organs, (b) 90% minimum mean square error (MMSE) versus number of body-mounted sensors.

calculated in a 3D grid inside each organ. Figure 16.18 shows three sets of results for performance evaluation of the RSS-based localization using 3D-CRLB described in [Ye12]. Figure 16.18(a) shows the CDF of the localization error for the capsule in the small intestine and stomach are smaller than error inside the large intestine. The median value of error for the small intestine and the stomach is approximately 45 mm, while the median for the large intestine is approximately 50 mm. The localization error for the capsule in the stomach has the lowest average value but distributed in a wider range compared to the errors in the other two organs. These observations can be explained by the geometric relationship between the sensor array and the organs. As we can see from Figure 16.17, the stomach is located in the upper part of the receiver sensor array system, and its volume is the smallest among the three organs. Therefore, the localization error varies more in the stomach environment. The points located in the upper part of the stomach have a larger localization error value as they are far from the center of the receiver array system, while the points in the lower part of stomach have a smaller localization error. The small intestine is located in the center part of human abdomen cavity and the lumen is more centralized compared to the large intestine. Therefore, the localization error inside the small intestine is smaller than that in the large intestine. Figure 16.18b represents the average accuracy of localization in each of the three organs as a function of the number of sensors mounted on the human body. In these simulations four different sets of body mounted sensors are used, covering the same area with different densities and the experiment is repeated for each set of sensor locations 1000 times for each organ. The localization error decreases significantly as the number of receiver sensors increases from 8 to 16 and from 16 to 32, but not as much when we increase the numbers from 32 to 64. The reason for this is the area of human torso is limited and 32 sensors already provide the density needed for localization. With 32 sensors we have an accuracy of around 5 cm as the average error, using 64 sensors reduces error for approximately another 1 cm. It is desirable to examine other methods to improve the accuracy. The simplest approach to improve the performance is to consider TOA based localization, which has its own challenges as it is applied for in-body localization.

16.6.2 Challenges in TOA-based RF Localization inside the Human Body

In the previous section we showed that one may achieve accuracies around a few centimeters inside the human body by using RSS-based localization techniques. As we explained in Section 15.3.3, to achieve more accuracy in localization we need to consider the TOA-based localization techniques and these techniques are very sensitive to the multipath behavior of the medium. In traditional TOA localization applications the time of flight of a transmitted pulse with a sharp peak is measured at the receiver and distance is estimated by multiplying the time of flight with the velocity of propagation that is the same as velocity of light. Human body is similar to liquids and propagation velocity inside the liquid is a function of the relative permittivity:

$$v(\omega) = \frac{c}{\sqrt{\varepsilon_r(\omega)}}.$$

Here, velocity, v, is a function of permittivity, ε_r, and the permittivity is a function of the (radial) frequency of operation, ω. In addition, the human body is formed by various organs with complex structures, each having different characteristics of conductivity and relative permittivity. Therefore, we do not know the exact velocity of propagation and that causes error in time of flight estimations.

Considering the total distance travelled through the body is the added distances within each organ or tissue, the total distance can be expressed as:

$$d_{total} = d_1 + d_2 + \ldots + d_n,$$

where d_1 to d_n are the distances travelled in each organ or tissue. In reality, one may use the *average* permittivity of the human body to estimate the average propagation velocity inside the human body, which is:

$$\bar{v} = \frac{c}{\sqrt{\bar{\varepsilon}_r}},$$

Therefore, the estimated distance is expressed as:

$$\hat{d} = \hat{\tau}\bar{v} = (\widehat{\tau_1} + \widehat{\tau_2} + \ldots + \widehat{\tau_n})\frac{c}{\sqrt{\bar{\varepsilon}_r}} = \sum_{i=1}^{n}\frac{d_i}{v_i}\frac{c}{\sqrt{\bar{\varepsilon}_r}} = \left(\frac{d}{\frac{c}{\sqrt{\varepsilon_1}}} + \frac{d}{\frac{c}{\sqrt{\varepsilon_2}}} + \ldots + \frac{d}{\frac{c}{\sqrt{\varepsilon_n}}}\right)\frac{c}{\sqrt{\bar{\varepsilon}_r}}$$

The difference between d_{total} and $\hat{d}$ is the ranging error that is caused by the non-homogeneous nature of the human body. This error between the actual distance and the distance measured by TOA and average velocity of the propagation is caused by using a single velocity rather than multiple velocities.

To show the expected amount of ranging error in TOA-based localization due to the effects of the non-homogeneous medium inside the human body, [Ye12] conducted a 3D simulation inside the human torso. The human torso includes eight major organs, each with a different volume and conductivity, shown in Table 16.2. Figure 16.19 shows the scenario for collecting the data. Approximately 500 locations on the torso were selected randomly. The DP between pairs of locations in different sides of the body was established and each path was segmented into paths

Table 16.2   Permittivity and volume of organs (in mm^3) used to simulate the effects of the non-homogeneity of the human body

Intestine (50.7, 3936.3)	Stomach (67.8, 357)	Gall bladder (52.3, 12.4)
Lung (23.77, 4320)	Heart (65.97, 625.4)	Kidney (68.0, 325.1)
Spleen (63.1, 160.2)	Liver (51.15, 1357)	Muscle (47.8, 32 403.4)

through the different organs. Then, the above equations were used for the estimated and the actual distance calculated from the time of flight and consequently the ranging error caused by the non-homogenous medium inside the body. The average permittivity used for the time of flight estimation was calculated by weighting the permittivity of each organ according to its volume, giving an average permittivity of 46.35 in the torso environment. The permittivity and volume of different organs used for this simulation are shown in Table 16.2.

Figure 16.20 presents the results of simulating the TOA-based ranging error caused by the non-homogeneous characteristics of the human body and the best fit Gaussian distribution to the results of simulations. The standard deviation of the ranging error is 24.3 mm, while the mean value is −3.92 mm. The mean value of the ranging error is a negative value because the largest organ in the torso cavity are the lungs, with a much smaller permittivity value than the average permittivity of human tissues. In practice the value of bias in the estimation does not play an important role because it can be removed easily. The standard deviation of 2.43 mm is caused by variations in the velocity of the wave propagation across the entire torso. In applications such as the movement of an endoscopy capsule inside the small intestine, this value can be reduced due to the fact that variations in the medium are expected to be much smaller than variations in the entire torso.

To further analyze the behavior of the TOA-based systems for in-body applications, we need empirical wideband data for measurement of the TOA of the signal that passes through the human body. Measurements using antennas inside the human body are not practical and we need to resort to using phantoms or a computer simulation of the radio propagation inside the human body. If these measurements are for the purpose of TOA estimations, the problem will have its own particularities that we will discuss next.

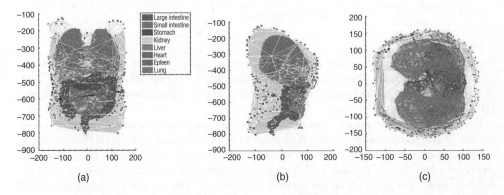

Figure 16.19   Human torso simulation scenario for 3D measurements of the ranging error due to non-homogeneity of the human body for radio propagation: (a) front view, (b) side view, (c) top view.

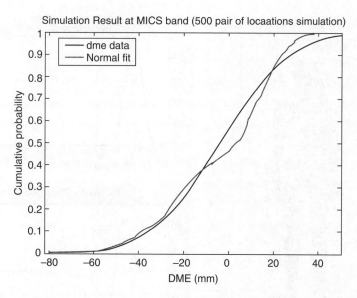

Figure 16.20 CDF of ranging error for TOA-based localization caused by non-homogeneous characteristics of the human body, and best-fit Gaussian distribution to the data.

16.6.3 Modeling of Wideband RF Propagation from inside the Human Body

The main challenge for the design of accurate algorithms for RF localization inside the human body is the lack of wideband wireless channel models for in-body localization applications. Since it is not practical to make RF measurements inside the human body, researchers resort to emulating measurements using phantoms, dead animal bodies, or computational techniques to measure the RF characteristics inside the human body. It is very difficult to emulate complex paths such as those inside the small intestine in a phantom or a dead animal body, and computational techniques may be perceived as less accurate and un-realistic. However it should be possible to use limited measurements on phantoms and the surface of human body to validate and calibrate a software simulation of RF propagation for computing Maxwell's equations inside the human body.

In the literature, there are three software simulation tools for RF propagation inside the human body: the commercially available SemCAD X used in [Aoy09; Kur09], Ansoft HFSS used in [Say10; Ask11], and a proprietary FDTD software on MATLAB® developed at CWINS/WPI [Kha11]. For TOA-based localization applications inside the human body, we need to use these software tools to analyze the wideband characteristics of a signal transmitted from inside the human body. Figure 16.21 shows a measurement scenario with multiple sensor locations and the results of a FDTD simulation of wideband radio propagations inside the human body using MATLAB [Kha11]. The body frame used for the simulation is homogeneous with a uniform dielectric constant and a uniform conductivity of 1.56 and 0.5 S/m representing average muscle tissues. The absorbing boundary conditions were used for this simulation to isolate the path passing on the surface of the body from multipath arrivals from other objects surrounding the body. Point source antennas were used to generate normalized plots of the channel impulse response independent from the antenna radiation pattern. Figure 16.21b, c show the transmitted and a sample received signal from FDTD

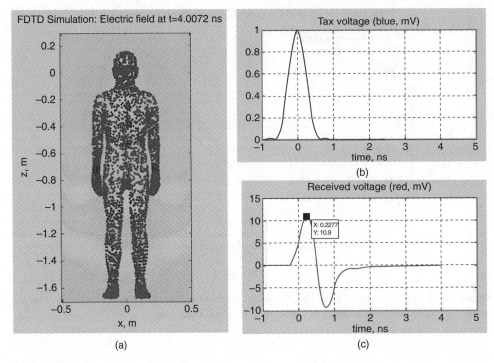

Figure 16.21 FDTD simulations on MATLAB. (a) Map of where the sensors were placed. (b) Transmitted waveform. (c) Sampled received simulated waveforms.

simulations. The antennas act as differentiators for the given 100 MHz bandwidth used for the simulation. The TOA of the "first path" is at 0.2277 ns, representing a 5 cm distance.

To analyze the accuracy of a simulation for TOA-based localization, [Kha11] carried out a number of simulations, using the scenario shown in Figure 16.21a. With the transmitter kept at a fixed position, the position of the receiver is moved to different location with different distances from the transmitter. Figure 16.22 shows the plot of measured TOA versus actual measured distances across the simulated homogeneous human body and the best fit line to relate the TOA to the distance. The difference between the actual measurements and the best fit line are the ranging error caused by overall quantization errors embedded in the computational method. The variance of the ranging error obtained from this plot came out to be about 0.72 mm. The human body model used in the FDTD simulations was homogenous to isolate the effects of non-homogeneity from the computational error and to simplify the simulation model. The error observed by computational simulations was considerably lower than the error caused by the effects of non-homogenous characteristics of human body discussed in Section 16.5.

Follow-up research on the validity of the FDTD simulations used in Figure 16.21 shows that the results of a waveform simulation with a grid of size 12.5 mm and the actual wideband measurements inside an anecdote chamber are very close for signals with a bandwidth of less than 100 MHz [Swa12]. For wider bandwidths we need shorter grids that increase the computational time exponentially.

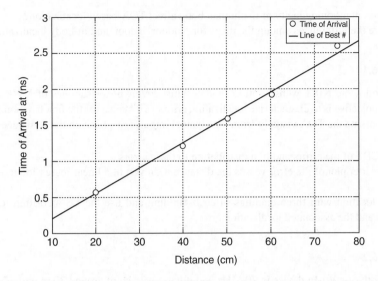

Figure 16.22   Estimated distance using TOA of FDTD simulations versus actual distance for various sensor positions.

Questions

1. What are the differences between RTLS and WPS technologies in terms of precision require-ments, database collection technique, environments where they are used, and localization algorithms?
2. How can a WPS system (which uses GPS to tag the location) provide results that are more accurate than GPS?
3. Why is WPS referred to as software GPS?
4. What are the basic differences among LS based algorithms using the distance to location of the RPs and the stochastic algorithms using training points?
5. What is the benefit of Kalman and particle filtering in localization applications? Does it help in geolocation of semi-stationary and stationary mobiles?
6. How does an inertial system help improving RF localization techniques?
7. What are the typical mechanical sensors used for localization?
8. In RSS-based localization, what are the advantages and disadvantages of using the exact location of APs and using RSS reading at arbitrary locations?
9. What is specific about RFID localization that is different from other localization schemes?
10. What are the typical localization techniques used in hybrid indoor geolocation systems and why do we need a hybrid technique?
11. How is the precision of TOA estimation using the phase of a carrier affected by multipath conditions?
12. What is the usefulness of super-resolution algorithms for TOA-based localization?
13. How is UWB technology useful for TOA-based localization in indoor areas?
14. How does multipath diversity help localization in multipath conditions?
15. How does cooperative localization differ from traditional localization techniques?
16. What are the differences among the localization environments for GPS, indoor geolocation, and inside the human body?

17. How does non-homogeneity of the human body affect TOA-based localization?
18. What are the differences among the maps for outdoor, indoor, and in-body localization?

Problem 16.1

Go to the following NIST website http://snad.ncsl.nist.gov/uwb/ for samples of UWB measurements. The objective is to characterize localization errors vs distance for the first five points in each building. Estimate the TOA visually from the channel impulse response on the web page.

a. Plot the CDF of the errors and fit a distribution to these errors.
b. Make a scatter plot of the error versus the distance. Can you fit a linear regression model to the data?
c. Select a location with three readings in each building and use the LS algorithm to find the locations and the associated localization error.

Problem 16.2

The bandwidth of a Wi-Fi device is 20 MHz and it has coverage of around 30 m with a minimum acceptable SNR of 10 dB.

a. Determine the CRLB for the accuracy of RSS-based localization at the edge of the coverage of the system for a typical indoor area using the IEEE 802.11 path loss model C from Chapter 2.
b. Repeat (a) for TOA-based localization.

Problem 16.3

Derive Equation 16.4 using Figure 16.15a.

Problem 16.4

a. In a three-path channel, the distance travelled by the arriving three paths are 3, 5, and 8 m. Determine, sketch, and label the impulse response in this multipath medium if the frequency of operation is 2.4 GHz and the transmission is only in air. Assume the received power of the signal along the first path is 0 dBm.
b. Repeat (a) if the experiment is conducted in water where the distance power gradient is 7 and the conductivity of the medium is approximately 85.

Problem 16.5

Derive the TOA-based ranging error in a non-homogeneous environment in terms of conductivity and the length of the direct path segment in each medium.

Project 16.1: Wi-Fi Localization

In this project we use software (e.g., the WirelessMon platform) to collect data using a laptop or a smart phone in a building (e.g., the second floor of AKL). We then use the database to compare the performance of some basic RSS-based localization algorithms.

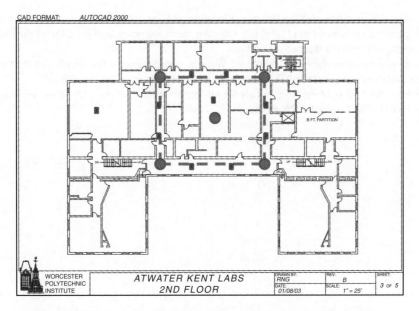

Figure P16.1   Layout of the second floor at the Atwater Kent Laboratory.

Part I: Database Collection

a. Walk on the specified route similar to Figure P16.1 of the second floor at the Atwater Kent Laboratory and observe the RSS and number of MAC addresses along the path. Go to each of the five points identified by circles in Figure P16.1 and log 20 RSS readings at each location while you stay in the same location with a normal laptop or smart phone holding posture and turning around slowly. We refer to this database as the *training* database. Attach a sample printout of your logged data showing the MAC addresses and the associated RSS readings. Enumerate the number of APs and the CDF of the RSS from different APs for your sample logs.

b. Do "war driving" to find the location of APs on the second floor of the building. Identify these locations in the layout of the floor map. Attach the log for all AP readings under each floor AP and identify the RSS of the floor APs in the log of each location. Note that you might have several APs in one physical location. Include the number of APs and the range of RSS from different APs for your sample logs.

c. Go to the 10 locations identified by small squares and create another database with one RSS reading log per location. Let us call this the *test* database.

Part II: Testing Algorithms

1. Use the AP locations and the IEEE 802.11 path loss model C to calculate the distance of each test point from the APs in the second floor. Use the distances and an iterative algorithm to determine an estimate for the locations using the algorithm introduced in Problem 15.2.

2. Use the training database in Part I (a) and the Kernel algorithm introduced in Example 15.10 to determine the location of the 10 test points.

3. Calculate the error between the estimated location and the actual location at all 10 points for the algorithms introduced in (1) and (2). Give the mean, variance, and the CDF of the distance measurement errors for the two algorithms.
4. How do you compare the two algorithms? What is your advice for the location of the training points and the location of the APs to improve the performance of the algorithms?
5. Determine the CRLB of the RSS signals for all locations and plot the CDF of its errors against the errors obtained from the algorithms.

References

[3GPPa] 3GPP TS 25.305 V. 7.2.0 (2006-03), Technical Specification: Group Radio Access Network; Stage 2 functional specification of User Equipment positioning in UTRAN (Release 7), 3rd Generation Partnership Project, 2006.

[3GPPb] 3GPP TS 25.331 V. 7.1.0 (2006-06), Technical Specification: Group Radio Access Network; Radio Resource Control (RRC); Protocol Specification (Release 7), 3rd Generation Partnership Project, 2006.

[Abr70] N. Abramson, "The ALOHA System – Another Alternative for Computer Communications," *AFIPS Conf. Proc., Fall Joint Comput. Conf.*, 37, 281–285(1970).

[Agr98] P. Agrawal, "Energy Efficient Protocols for Wireless Systems", Proceedings of PIMRC'98, pp. 564–569, September 1998.

[Aho03] S. Ahonen and H. Laitinen, "Database correlation method for UMTS location," In Proceedings of the IEEE Vehicular Technology Conference (VTC), Vol. 4, pages 2696–2700, April 2003.

[Aky98] I.F. Akyildiz et al., "Mobility management in current and future communications networks" *IEEE Network Magazine*, pp. 39–50, July/August 1998.

[Ala06a] Bardia Alavi, *Distance Measurement Error Modeling for Time-of-Arrival Based Indoor Geolocation*, Ph.D. dissertation, Worcester Polytechnic Institute, May 2006.

[Ala06b] B. Alavi and K. Pahlavan, "Modeling of the TOA based Distance Measurement Error Using UWB Indoor Radio Measurements," *IEEE Communication Letters*, Vol. 10, No. 4, pp: 275–277, April 2006.

[Ala98] S. Alamouti, "A Simple Transmit Diversity Technique for Wireless Communications," *IEEE JSAC*, Oct. 1998, pp. 1451–1458.

[Ali02] M.H. Ali and K. Pahlavan, "A New Statistical Model for Site-specific Indoor Radio Propagation Prediction based on Geometric Optics and Geometric Probability", IEEE JSAC on Wireless Networks, Jan. 2002.

[Alo06] Alomainy, A. Hao, Y. Yuan, Y. Liu, Y. "Modelling and Characterisation of Radio Propagation from Wireless Implants at Different Frequencies" Wireless Technology, 2006. The 9th European Conference on, pp. 119–122, 10–12 Sept. 2006.

[Als08] NayefAlsindi, "Indoor Cooperative Localization for Ultra Wideband Wireless Sensor Networks," *PhD dissertation,* ECE Department, WPI, May 2008. http://www.wpi.edu/Pubs/ETD/Available/etd-042308-115256/unrestricted/nalsindi.pdf.

[And07] J.G. Andrews, A. Ghosh, and R. Muahmed, *Fundamentals of WiMAX: Understanding Broadband Wireless Networking*, Prentice Hall, 2007.

[Aoy08] Takahiro Aoyagi, Jun-ichi Takada, Kenichi Takizawa, Norihiko Katayama, Takehiko Kobayashi, Kamya Yekeh Yazdandoost, Huan-bang Li and Ryuji Kohno, "Channel model for wearable and implantable WBANs," IEEE 802.15-08-0416-04-0006, November 2008.

[Aoy09] T. Aoyagi, K. Takizawa, T. Kobayashi, J. Takada, and R. Kohno, "Development of a WBAN channel model for capsule endoscopy," Proc. of 2009 International Symposium on Antennas and Propagation, Charleston, SC, U.S.A., pp. 1–4. Jun. 2009. http://ieeexplore.ieee.org/stamp/ stamp.jsp? tp=&arnumber=5172160.

[Ask11] F. Askarzadeh, Y. Ye, U. Khan, F. Akgul, K. Pahlavan and S. Makarov, "Computational Methods for Localization in Close Proximity," chapter in *Position Location – Theory, Practice and Advances: A Handbook for Engineers and Academics*, John Wiley and Sons, 2011.

[Bah00] P. Bahl and V. N. Padmanabhan. "RADAR: An In-Building RF-based User Location and Tracking System," Proc. IEEE INFOCOM'00, pp. 775–784, March 2000.

[Ban11] *The 1st Invitational Workshop on BAN Technology and Applications,* June 19–20, Worcester, MA.

[Bar03] S. Barnes, "Location-Based Services: The State of the Art," e-Service Journal 2.3, pp. 59–70, 2003.

[Ber87] D. Bertsekas and R. Gallagher, *Data Networks*, Prentice Hall, New York, 1987.

[Ber94] H. L. Bertoni, W. Honcharenko, L.R. Maciel, and H.H. Xia, "UHF propagation prediction for wireless personal communications", *Proceedings of the IEEE*, Vol. 82, No. 9, pp. 1333–1359, September 1994.

[Bla92] K.L. Blackard et al., "Path Loss And Delay Spread Models As Functions Of Antenna Height For Microcellular System Design," *Proc. 42nd IEEE Vehicular Technology Conference*, Denver, CO, 1992.

[Blu00] Bluetooth Special Interest Group, "Specifications of the Bluetooth System, vol. 1 v. 1.1, 'Core' and vol. 2 v. 1.0 B 'Profiles'," 2000.

[Bra00] R.C. Braley, I.C. Gifford, and R.F. Heile, "Wireless Personal Area Networks: An Overview of the IEEE P802.15 Working Group," *ACM SIGMOBILE Mob. Comput. Commun. Rev.*, Vol. 4, No. 1, pp. 26–33, 2000.

[Bre05] M.R. Brenner, M.L.F. Grech, M. Torabi, and M.R. Unmehopa, "The Open Mobile Alliance and Trends in Supporting the Mobile Services Industry," Bell Labs Technical Journal, 10(1), pp. 59–75, 2005.

[Bud97] K.C. Budka, H.J. Jiang, and S.E. Sommars, "Cellular Packet Data Networks", Bell Lab Technical Journal, Summer 1997.

[Bul00] N. Bulusu, J. Heidemann, D. Estrin, "GPS-less Low-Cost Outdoor Localization for Very Small Devices," IEEE Personal Communications, pp. 28–34, October 2000.

[Bur00] E. Buracchini, "The software radio concept", IEEE Communications Magazine, pp. 138–143, September 2000.

[Cac95] R. Cacares and L. Iftode, "Improving the performance of reliable transport protocols in mobile computing environments", IEEE JSAC, pp. 850–857, June 1995.

[Caf98] J. Caffery, Jr. and G.L. Stuber, "Subscriber Location in CDMA Cellular Networks", *IEEE Trans. Veh. Technol.*, Vol. 47, No. 2, May 1998.

[Cas02] D. Cassioli, M.Z. Win, and A.F. Molisch, "The Ultra-Wide Bandwidth Indoor Channel: From Statistical Model to Simulations", IEEE Journal On Selected Areas In Communications, Vol. 20, No. 6, August 2002, pp. 1247–1256.

[Cha01] M.V.S. Chandrashekhar, P. Choi, K. Maver, R. Sieber, K. Pahlavan, "Evaluation of Interference Between IEEE 802.11b and Bluetooth in a Typical Office Environment", Proc. PIMRC '01, San Diego, 2001.

[Cha99] S. Chakrabarti and A. Mishra, "A network architecture for global wireless position location services," Proc. ICC'99, pp. 1779–1783, 1999.

[Che02] Y. Chen and H. Kobayashi (2002). Signal Strength Based Indoor Geolocation. Proceedings of the IEEE International Conference on Communications. pp 436–439. 28 April – 2 May 2002. New York.

[Che05] Y.C. Cheng, Y. Chawathe, A. LaMarca, and J. Krumm, "Accuracy Characterization for Metropolitan-scale Wi-Fi Localization," *Proceedings of Mobisys 2005*, pp. 233–245, 2005.

[Che99a] P-C. Chen, *"Mobile Position Location Estimation in Cellular Systems"*, PhD thesis, WINLAB, Electrical and Computer Engineering, Rutgers University, 1999.

[Che99b] P-C. Chen, "A non-line-of-sight error mitigation algorithm in location estimation", Proceedings of the IEEE Wireless Communications and Networking Conference, 1999.

[Ches03] W.R. Cheswick, S.M. Bellovin, and A.D. Rubin, Firewalls and Internet Security, Addison–Wesley, 2003.

[Cla08] H. Claussen, L.T.W. Ho, and L.G. Samuel, "An Overview of the Femtocell Concept," *Bell Labs Technical Journal*, Vol. 13, No. 1, pp. 221–246, 2008.

[Com98] IEEE Communications Magazine on Geolocation Applications, 1998.

[Cos86] COST 207 TD(86)51-REV 3 (WG1), "Propagation on Channel Transfer Functions to be Used in GSM Tests ", Sep. 1986.

[Cox99] A. Lozano and D.C. Cox, "Integrated Dynamic Channel Assignment and Power Control in TDMA Mobile Wireless Communication Systems", IEEE JSAC, Vol. 17, No. 11., pp. 2031–2040, Nov. 1999.

[Cro97] B.P. Crow, I. Widjaja, L.G. Kim, and P.T. Sakai, "IEEE 802.11 Wireless Local Area Networks", IEEE Communications Magazine, Vol. 35, No. 9, pp. 116–126, Sept. 1997.

[Dan10] R.C. Daniels, J.N. Murdock, T.S. Rappaport, and R.W. Heath, Jr. "60 GHz Wireless: Up Close and Personal", IEEE Microwave Magazine, Dec. 2010 Supplement, pp. S44–S50.

[Das06] S. Das, T. Klein, A. Rajkumar, S. Rangarajan, M. Turner, and H. Viswanathan, "System Aspects and Handover Management for IEEE 802.16e," *Bell Labs Technical Journal*, Vol. 11, No. 1, pp. 123–142, 2006.

[Dem06] I. Demirkol et al., "MAC Protocols for Wireless Sensor Networks: A Survey," IEEE Communications Magazine, April 2006.

[Den96] L.R. Dennison, "BodyLAN: A wearable personal network", Second IEEE Workshop on WLANs, Worcester, MA, 1996.

[Dha02] N. Al-Dhahir, C. Fragouli, A. Stamoulis, W. Younis, and R. Calderbank, Space–time processing for broadband wireless access, Communications Magazine, IEEE, Vol. 40 Issue 9, Sep. 2002, pp. 136–142.

[Dju01] G.M. Djuknic and R.E. Richton, "Geolocation and Assisted GPS", IEEE Computer, February 2001.

[Dra98] C. Drane, M. Macnaughtan, and C. Scott, "Positioning GSM telephones," IEEE Communications Magazine, 36(4):46–54, 59, April 1998.

[Dru01] M-A. Dru and S. Saada, "Location-based mobile services: the essentials," Alcatel Telecommunications Review, 2001.

[ECM08] ECMA TC48 draft standard for high data rate 60 GHz WPANs, October 2008.

[Edn04] J. Edney and W.A. Arbaugh, Real 802.11 Security: Wi-Fi Protected Access and 802.11i, Pearson Education, 2004.

[Ela04a] M. Elaoud, D. Famolari, A. Ghosh, "Experimental VoIP Capacity Measurements for 802.11b WLANs," IEEE Consumer Communications and Networking Conference, 2004.

[Ela04b] M. Elaoud and P. Agrawal, "VoIP Capacity in IEEE 802.11 Networks," Proc. IEEE PIMRC, 2004.

[Eng94] P.K. Enge, "Global Positioning Systems: Signals, Measurements and Performance," Intl J. Wireless Info. Networks, vol. 1, no. 2, Apr. 1994.

[Enn98] G. Ennis, Doc. IEEE P802.11-98/319, Impact of Bluetooth on 802.11 Direct Sequence, September 15, 1998.

[Ert98] R.B. Ertel, P. Cardieri, K.W. Sowerby, T.S. Rappaport, and J.H. Reed, "Overview of Spatial Channel Models for Antenna Array Communication Systems", IEEE Personal Communications, Feb. 1998.

[Erc99] V. Erceg et al., "An empirically based path loss model for wireless channels in suburban environments," *IEEE JSAC*, Vol. 17, No. 7, pp. 1205–1211, July 1999.

[Fal96] A. Falsafi, K. Pahlavan, G. Yang, "Transmission techniques for radio LAN's – a comparative performance evaluation using ray tracing", IEEE Journal on Selected Areas in Communications, Vol. 14, No. 3, pp. 477–491, April 1996.

[FCC03] FCC E-911 webpage 2003 (http://www.fcc.gov/e911).

[FCC09] FCC Rules and Regulations, "MedRadio Band Plan", Part 95, March 2009 (http://www.cwins. wpi.edu/workshop11/ppt/business_Charles.pdf).

[Fei99] J. Feigin and K. Pahlavan, "Measurement of characteristics of voice over IP in a wireless LAN environment" *IEEE International Workshop on Mobile Multimedia Communications (MoMuC '99)*, pp. 236–240, 1999.

[Fer80] P. Ferert, "Application of Spread Spectrum Radio to Wireless Terminal Communications," *Proc. NTC '80*, Houston, TX, 244–248 (Dec. 1980).

[Fig69] W. Figel, N. Shepherd, and W. Trammell, "Vehicle location by a signal attenuation method", IEEE Trans. Vehicular Technology, vol. VT-18, pp. 105–110, Nov. 1969.

[Fis80] M.J. Fischer, "Delay Analysis of TASI with Random Fluctuations in the Number of Voice Calls," *IEEE Trans. Commun.,* **COM-28**, 1883–1889 (1980).

[Foe04] J. Foerster and Q. Li, UWB Channel Modeling Contribution from Intel, 24 June, 2002, IEEE P802.15 Working Group for Wireless Personal Area Networks (WPANs), IEEE P802.15-02/279r0-SG3a.

[For06] A. Fort, J. Ryckaert, C. Desset, P. De Doncker, P. Wambacq, and L. Van Biesen, "Ultrawideband channel model for communication around the human body, "IEEE Journal on Selected Areas in Communications, vol. 24, pp. 927–933, April 2006.

[Fri46] H.T. Friis, "A Note on a Simple Transmission Formula", Proceedings of the I.R.E. and Waves and Electrons, May 1946, pp. 254–256.

[Fu12] R. Fu, Y. Ye, N. Yang, and K. Pahlavan, "Characteristic and Modeling of Human Body Motions for Body Area Network Applications", invited paper, *Wireless Health special issue based on the IEEE PIMRC'11 best papers, International Journal of Wireless Information Networks, Springer*, Vol. 19, No. 3, August 2012, 219–228.

[Gan91] R. Ganesh and K. Pahlavan, "Modeling of the Indoor Radio Channel", IEE Proceedings-I, June 1991.

[Gar00] V.K. Garg, IS-95 and CDMA2000, Prentice Hall, Upper Saddle River, 2000.

[Gar03] S. Garg and M. Kappes, "Can I add a VoIP call?," Proc. ICC'03, pp. 779–783, 2003.

[Gar99] V.K. Garg and J.E. Wilkes, Principles and Applications of GSM, Prentice Hall, Upper Saddle River, NJ, 1999.

[Gas02] M.S. Gast, 802.11 Wireless Networks: The definitive guide, O'Reilly & Associates, 2002.

[Ger10] C.G. Gerlach, I. Karla, A. Weber, L. Ewe, H. Bakker, E. Kuehn, and A. Rao, "ICIC in DL and UL with Network Distributed and Self-Organized Resource Assignment Algorithms in LTE," *Bell Labs Technical Journal*, Vol. 15, No. 3, pp. 43–62, 2010.

[Get93] I.A. Getting, "The Global Positioning System," IEEE Spectrum, pp. 36–47, Dec. 1995.

[Gfe80] F.R. Gfeller, "Infranet: Infrared microbroadcasting network for in house data communication", IBM research report, RZ 1068 (#38619), April 27, 1981.

[Gha03a] S.S. Ghassemzadeh, L.J. Greenstein, A. Kavèiæ, T. Sveinsson, and V. Tarokh, "UWB indoor path loss model for residential and commercial buildings", *in Proc. IEEE VTC – Fall 2003*, pp. 3115–3119.

[Gha03b] S.S. Ghassemzadeh, L.J. Greenstein, A. Kavèiæ, T. Sveinsson, and V. Tarokh, "UWB indoor delay profile model for residential and commercial buildings", *in Proc. IEEE VTC – Fall 2003*, pp. 3120–3125.

[Gha04] M. Ghavami, L.B. Michael, and R. Kohno, Ultra-Wideband, Signals and Systems in Communication Engineering, John Wiley and Sons, 2004.

[Gho11] A. Ghosh, J. Zhang, J.G. Andrews, and R. Muhamed, *Fundamentals of LTE*, Prentice Hall, 2011.

[Goo89] D.J. Goodman, R.A. Valenzuela, K.T. Gayliard, and B. Ramamurthi, "Packet reservation multiple access for local wireless communications", IEEE Transactions on Communications, Vol. 37, No. 8, pp. 885–890, Aug. 1989.

[Goo91] D.J. Goodman and S.X. Wei, "Efficiency of packet reservation multiple access", IEEE Transactions on Vehicular Technology, Vol. 40, No. 1 Part: 2, pp. 170–176, Feb. 1991.

[Goo93] D.J. Goodman, J. Grandhi, and R. Vijayan, "Distributed Dynamic Channel Assignment Schemes", Proc. of the 43rd IEEE Veh. Tech. Conf., pp. 532–535, 1993.

[Goo97] D.J. Goodman, Wireless Personal Communications Systems, Addison–Wesley, 1997.

[Gou08] A.A. Goulianos, T.W.C. Brown, and S. Stavrou, "A Novel Path-Loss Model for UWB Off-Body Propagation", IEEE Vehicular Technology Conference, pp. 450–454, May 2008.

[Gra93] S.A. Grandhi, R. Vijayan, D.J. Goodman, and J. Zander, "Centralized power control in cellular radio systems", IEEE Transactions on Vehicular Technology, Vol. 42, No. 4, pp. 466–468, November 1993.

[Gra94] S.A. Grandhi, R. Vijayan, and D.J. Goodman, "Distributed Power Control in Cellular Radio Systems", IEEE Transactions on Communications, Vol. 42, pp. 226–228, February 1994.

[GSM91] GSM Recommendation 05.05, "Radio Transmission and Reception," ETSI/PT 12, Jan. 1991.

[Gue97] S. Guerin, Y.J. Guo, and S.K. Barton, "Indoor propagation measurements at 5 GHz for HIPERLAN", *10ᵗʰ Int. Conf. Ant. Prop.*, pp. 306–310, April 1997.

[Guo07] N. Guo, R.C. Qiu, S.S. Mo, and K. Takahashi, "60-GHz Millimeter-Wave Radio: Principle, Technology, and New Results", EURASIP Journal on Wireless Communications and Networking, Volume 2007.

[Haa00] J.C. Haartsen and S. Mattisson, "Bluetooth-a new low-power radio interface providing short-range connectivity", Proceedings of the IEEE, Vol. 88, No. 10, pp. 1651–1661, Oct. 2000.

[Hag08a] J. Hagedorn, J. Terrill, W. Yang, K. Sayrafian, K. Yazdandoost, and R. Kohno, "MICS Channel Characteristics; Preliminary Results", IEEE 802.15-08-0351-00-0006, September 2008.

[Hag08b] J. Hagedorn, J. Terrill, W. Yang, K. Sayrafian, K. Yazdandoost, and R. Kohno, "A Statistical Path Loss Model for MICS," IEEE 802.15-08-0519-01-0006, September 2008.

[Hal83] S.W. Halpern, "Reuse partitioning in cellular systems", Proc. Of the IEEE Vehicular Technology Conference, pp. 322–327, 1983.

[Hal96] C.J. Hall and W.A. Foose, "Practical Planning for CDMA Networks: A Design Process Overview", Proc. Southcon'96, pp. 66–71, 1996.

[Hal99] K. Halford, S. Halford, M. Webster, and C. Ander, "Complementary code keying for RAKE-based indoor wireless communication", IEEE International Symposium on Circuits and Systems, Vol. 4, pp. 427–430, Orlando, FL, 1999.

[Ham02] M. Hamalainen, et al., "On the UWB System Coexistence with GSM900, UMTS/WCDMA, and GPS," IEEE J. Sel. Areas. Comm., Vol. 20, No. 9, 2002.

[Ham86] J.L. Hammond and P.J.P. O'Reilly, Performance Analysis of Local Computer Networks, Addison–Wesley, Reading, MA (1986).

[Har06] S. Harsha, A. Kumar, and V. Sharma, "An Analytical Model for the Capacity Estimation of Combined VoIP and TCP File Transfers over EDCA in an IEEE 802.11e WLAN," Proc. IEEE IWQoS06, 2006.

[Har99] D. Har, H.H. Xia, and H.L. Bertoni, "Path-loss prediction model for micro-cells", IEEE Transactions on Vehicular Technology, Vol. 48, No. 5, pp. 1453–1462, September 1999.

[Has02] M. Hassan-Ali and K. Pahlavan, "A new statistical model for site-specific indoor radio propagation prediction based on geometric optics and geometric probability," IEEE JSAC Wireless, Vol. 1, No. 1, Jan. 2002.

[Hat04] A. Hatami and K. Pahlavan, "In-building Intruder Detection for WLAN Access", The IEEE Aerospace and Electronic Systems Society conference, PLANS, Monterey, CA, April 2004.

[Hat06] H. Hatami, Application of Channel Modeling for Indoor Localization Using TOA and RSS, Ph.D. Dissertation, Worcester Polytechnic Institute, 2006.

[Hat80] M. Hata, "Empirical formula for propagation loss in land mobile radio services", IEEE Transactions on Vehicular Technology, Vol. VT-29, No. 3, pp. 317–324, August, 1980.

[Hau94] T. Haug, "Overview of GSM: Philosophy and Results", International Journal of Wireless Information Networks, Jan 1994.

[Hay91] V. Hayes, "Standardization efforts for wireless LANS", IEEE Network, Vol. 5, No. 6, pp. 19–20, 1991.

[Hei09] M. Heidari, N.A. Alsindi, and K. Pahlavan, "Identification of the Absence of Direct Path Component in Indoor Localization Systems," IEEE Transactions on Wireless Communications, Vol. 8, Issue 7, 2009, pp. 3597–3607.

[Hei98] R. Heille, WPAN functional requirement, Doc. IEEE 802.11/98/58, Jan 22nd, 1998.

[Hig07] H. Higgins, "Body implant communications – is it a reality?" in Proceedings of the IET Seminar on Antennas and Propagation for Body-Centric Wireless Communications, pp. 33–36, London, UK, April 2007.

[Hil01] A. Hills, "Large Scale Wireless LAN Design", IEEE Communication Magazine, pp. 98–105, Nov. 2001.

[Hol00] H. Holma and A. Toskala (eds), WCDMA for UMTS: Radio Access for Third Generation Mobile Communications, John Wiley and Sons, NY, 2000.

[How90] S.J. Howard and K. Pahlavan, "Measurement and Analysis of the Indoor Radio Channel in the Frequency Domain," IEEE Trans. Instr. Meas., No. 39, pp. 751–55, 1990.

[IEE01] Proc. IEEE Workshop on Wireless LANs, Newton, MA (Sep. 2001).

[IEEE09] Part 15.3: Wireless Medium Access Control (MAC) and Physical Layer (PHY) Specifications for High Rate Wireless Personal Area Networks (WPAN). Ammendment 2: Millimeter-wave-based Alternative Physical Layer Extension, IEEE Std 802.15.3c™-2009 (Amendment to Std 802.15.3™-2003.), October 12, 2009.

[IS-801-99] IS-801, Position Determination Service Standard for Dual Mode Spread Spectrum Systems, Telecommunications Industry Association, 1999.

[Jak01] M. Jakobsson and S. Wetzel, "Security Weaknesses in Bluetooth", RSA Conference'01, April 8–12, 2001.

[Jay84] N.S. Jayant and P. Noll, *Digital Coding of Waveforms*, Prentice–Hall, Englewood Cliffs, NJ, 1984.

[Joo06] J. Krogerus, C. Icheln, and P. Vainikainen, "Experimental Investigation of Antenna Performance of GSM phones in Body-Worn and Browsing Use Positions," Proceedings of the 9th European Conference on Wireless Technology, pp. 330–333, 10–12 Sept. 2006.

[JTC94] JTC Technical Report on RF Channel Characterization and Deployment Modeling, Air Interface Standards, Sep. 1994.

[Kae04] K. Kaemarungsi and P. Krishnamurthy, "Modeling of Indoor Positioning Systems Based on Location Fingerprinting," Proc. IEEE Infocom, March 2004.

[Kan04a] M. Kanaan and K. Pahlavan, A comparison of wireless geolocation algorithms in the indoor environment, Proceedings of the IEEE WCNC, April 2004.

[Kan04b] M. Kanaan and K. Pahlavan, CN-TOA – a New Algorithm for Indoor Geolocation, IEEE PIMRC, Sep. 2004.

[Kap02] S. Kapp, "802.11a. More bandwidth without the wires", IEEE Internet Computing, Volume:6, Issue:4, July–Aug. 2002.

[Kap96] E.D. Kaplan, Understanding GPS: Principles and Applications, Artech House Publishers, 1996.

[Kat96] I. Katzela and M. Naghshineh, "Channel assignment schemes for cellular mobile telecommunication systems: a comprehensive survey", IEEE Personal Communications, pp. 10–31, June 1996.

[Kau02] C. Kaufmann, R. Perlman, and M. Speciner, Network Security: Private Communication in a Public World, Prentice Hall PTR, 2002.

[Kav87] M. Kavehrad and P.J. McLane, "Spread spectrum for indoor digital radio", *IEEE Communications Magazine*, Vol. 25, No. 6, pp. 32–40, 1987.

[Kei89] G.E. Keiser, *Local Area Networks*, McGraw–Hill, New York (1989).

[Ker00] J.P. Kermoal, L. Schumacher, P.E. Mogensen, and K.I. Pedersen, "Experimental investigation of correlation properties of MIMO radio channels for indoor picocell scenarios", 52nd IEEE Vehicular Technology Conference, 2000, pp. 14–21, 2000.

[Kha11] U. Khan, K. Pahlavan, and S. Makarov "Computational Techniques for Wireless Body Area Networks Channel Simulation Using FDTD and FEM" at the *33rd Annual International Conference of the IEEE Engineering in Medicine and Biology Society (EMBC)*, pp. 5602–5607, 2011.

[Kim04] J. Kim and Y. Rahmat-Samii, "Implanted antennas insidea human body: simulations, designs, and characterizations," *IEEE Transactions on Microwave Theory and Techniques*, vol. 52, no. 8, part 2, pp. 1934–1943, 2004.

[Kim08] J. Kim, H. Soo Lee, J.K. Pack, and T.H. Kim, "Channel modeling for medical implanted communication systems by numerical simulation and measurement," IEEE 802.15-08-0274-02-0006, May 2008.

[Kle75] L. Kleinrock and S.S. Lam, "Packet Switching in a Multiaccess Broadcast Channel: Performance Evaluation," *IEEE Trans. Commun.*, COM-23, 410–423 (1975).

[Koh04] R. Kohno, M. Welborn, and M. Mc Laughlin, DS-UWB Proposal, IEEE P802.15 Working Group for Wireless Personal Area Networks (WPANs), Document number: IEEE 802.15-04/140r2, March 2004.

[Koi04] G.M. Koien, "An introduction to access security in UMTS," *IEEE Wireless Communications*, Vol. 11, No. 1, 2004.

[Kos00] H. Koshima and J. Hoshen, "Personal Locator Services Emerge," *IEEE Spectrum*, February 2000, pp. 41–48.

[Kri98] P. Krishnamurthy, K. Pahlavan, and J. Beneat, "Radio propagation modeling for indoor geolocation applications", Proceedings of IEEE PIMRC'98, September 1998.

[Kri99a] P. Krishnamurthy and K. Pahlavan, "Analysis of the probability of detecting the DLOS path for geolocation applications in indoor areas", 49th IEEE Vehicular Technology Conference, Vol. 2, pp. 1161–1165, 1999.

[Kri99b] P. Krishnamurthy and K. Pahlavan, "Distribution of Range Error and Radio Channel Modeling for Indoor Geolocation Applications", Proc. PIMRC'99, Osaka, Japan, 1999.

[Kri99c] P. Krishnamurthy, "Analysis and Modeling of the Indoor Radio Channel for Geolocation Applications", Ph.D. Thesis, Worcester Polytechnic Institute, August 1999.

[Kue92] S.S. Kuek and W.C. Wong, "Ordered Dynamic Channel Assignment Scheme with Reassignment in Highway Microcells", IEEE Transactions on Vehicular Technology, Vol. 41, No. 3, pp. 271–276, August 1992.

[Kum11] R. Kumaralingam and G. Rahul, "The 60GHz Wireless Network Infrastructure", White Paper February 2011, http://ers.hclblogs.com/wp-content/uploads/2010/07/The-60GHz-Wireless-Network-Infrastructure.pdf.

[Kum74] K. Kummerle, "Multiplexer Performance for Integrated Line- and Packet-Switched Traffic," ICCC, Stockholm, 1974.

[Kur09] D. Kurup, W. Joseph, G. Vermeeren, and L. Martens, "Path loss model for in-body communication in homogeneous human muscle tissue," *IET Electronics Letters*, pp. 453–454, April 2009.

[Lee06] M.J. Lee and J. Zheng, "Emerging Standards for Wireless Mesh Technology," IEEE Wireless Communications, pp. 56–63, April 2006.

[Lee91] W.C.Y. Lee, "Smaller Cells for Greater Performance", IEEE Communications Magazine, pp. 19–23, November 1991.

[Leh99] P.H. Lehne and M. Pettersen, "An Overview of Smart Antenna Technology for Mobile Communications Systems", IEEE Communications Surveys, pp. 2–13, Vol. 2, Fourth Quarter, 1999.

[Li00] X. Li, K. Pahlavan, M. Latva-aho, and M. Ylianttila, "Indoor Geolocation using OFDM Signals in HIPERLAN/2 Wireless LANs", IEEE PIMRC 2000, London, Sep. 2000.

[Li04] X. Li and K. Pahlavan, "Super-resolution TOA estimation with diversity for indoor geolocation," *IEEE Trans on Wireless Comm.* Jan 2004.

[LIF02] LIF TS 101 Specification, Location Inter-operability Forum (LIF) Mobile Location Protocol, Version 3.0.0, June 2002.

[Lio94] G. Liodakis and P. Stravroulakis, "A Novel Approach in Handover Initiation for Microcellular Systems," Proc. Vehicular Tech. Conf. '94, Stockholm, Sweden, 1994.

[Lor98] J.R. Lorch and A.J. Smith, "Software strategies for portable computer energy management", IEEE Personal Communications Magazine, pp. 60–73, June 1998.

[Mac03] J. McCorkle, "DS-CDMA: The Technology of Choice For UWB," IEEE P802.15-03/277r0. July 19, 2003.

[Mac79] V.H. MacDonald, "The Cellular Concept", *The Bell System Technical Journal*, Vol. 58, No. 1, pp. 15–41, January 1979.

[Mag06] T. Magedanz and F.C. de Gouveia, "IMS – the IP Multimedia System as NGN Service Delivery Platform," *Elektrotechnik und Informationstechnik*, Vol. 123, August 2006.

[Mak11] S.N. Makarov, U.I. Khan, M.M. Islam, L. Reinhold, and K. Pahlavan, "On Accuracy of Simple FDTD Models for the Simulation of Human Body Path Loss", *the Proceedings of the IEEE Sensor Application Symposium*, San Antonio, TX, February 22–24, 2011.

[Mal07] A. Mallat, J. Louveaux, and L. Vandendrope, "UWB based positioning in multipath channels, CRBs for AOA and for hybrid TOA-AOA based methods," *in Proceedings of the IEEE International Conference on Communications* (ICC), Glasgow, Scotland, June 2007.

[Mal09] A. Maltsev, R. Maslennikov, A. Sevastyanov, A. Khoryaev, and A. Lomayev, "Experimental Investigations of 60 GHz WLAN Systems in Office Environment", *IEEE Journal on Selected Areas in Communications*, Vol. 27, No. 8, pp. 1488–1499, October 2009.

[Mal10] A. Maltsev et al., Channel Models for 60 GHz WLAN Systems, IEEE P802.11-09/0334r8, May 20, 2010, https://mentor.ieee.org/802.11/dcn/09/11-09-0296-16-00ad-evaluation-methodology.doc.

[Mar02] I. Martin-Escalona, F. Barcelo, and J. Paradells, "Delivery of non-standardized assistance data in E-OTD/GNSS hybrid location systems," Proc. IEEE PIMRC, Vol. 5, pp. 2347–2351, September 2002.

[Mar85] M.J. Marcus, "Recent US regulatory decisions on civil use of spread spectrum", Proc. IEEE Globecom, 16.6.1–16.6.3, New Orleans, December 1985.

[Mas09] R. Maslennikov and A. Lomayev, "Implementation of 60 GHz WLAN Channel Model," IEEE 802.11-09/854r0, July 2009 https://mentor.ieee.org/802.11/dcn/09/11-09-0854-00-00adimplementation-of-60ghz-wlan-channel-model.doc.

[McD98] J.T.E. McDonnell, "5 GHz indoor channel characterization: measurements and models", *IEE Coll. on Ant. and Prop. for future mobile communications*, 1998.

[Med04] K. Medepalli et al., "Voice Capacity of IEEE 802.11b, 802.11a, and 802.11g Wireless LANs," Proc. Globecom, 2004.

[Mey96] M.J. Meyer, T. Jacobson, M.E. Palamara, E.A. Kidwell, R.E. Richton, and G. Vannucci, "Wireless enhanced 9-1-1 service – making it a reality," Bell Labs Technical Journal, Vol. 1, No. 2, pp. 108–202, Autumn 1996.

[Min08] D. Miniutti, L. Hanlen, D. Smith, A. Zhang, D. Lewis, D. Rodda, and B. Gilbert, "Characterization of small-scale fading in BAN channels," IEEE 802.15-08-0716-01-0006, October 2008.

[Mis10] P. Misra and P. Enge, *Global Positioning System: Signals, Measurements, and Performance*, Revised Second Edition, Ganga–Jamuna Press, 2010.

[Moa11] N. Moayeri, J. Mapar, S. Tompkins, and K. Pahlavan (eds), Special Issue on Localization and Tracking for Emerging Wireless Systems, *IEEE Wireless Communications*, April 2011.

[Mor10] T. Morgan, An Intelligent Data Mining Technique for Emerging Location Based Applications, 2nd Invitational Workshop on Opportunistic RF Localization for Next Generation Wireless Devices, WPI, Worcester, MA June 13–14, 2010 (http://www.cwins.wpi.edu/workshop10/pres/exec_2.pdf).

[Mur81] K. Murota and K. Hirade, "GMSK Modulation for Digital Mobile Radio Telephony," *IEEE Transactions on Communications*, Vol. 29, No. 7, pp. 1044–1050, 1981.

[Nee11] R. Van Nee, "Breaking the Gigabit per Second Barrier with 802.11ac," IEEE Wireless Communications, April 2011.

[Nag98] A. Naguib et al., "A Space Time Coding Modem for High Data Rate Wireless Communications," *IEEE J. Sel. Areas. Comm.*, pp. 1459–1477, October 1998.

[Nee99] R. van Nee et al., "New high-rate wireless LAN standards", *IEEE Communications Magazine*, Vol. 37, No. 12, pp. 82–88, Dec. 1999.

[Nic03] D. Niculescu and B. Nath, "Ad hoc positioning system (APS) using AOA," Proc. IEEE Infocom, pp. 1734–1743, April 2003.

[Nis00] D.N. Nissani and I. Shperling, "Cellular CDMA (IS-95) location, A-FLT proof-of-concept interim results," The 21st IEEE Convention of Electrical and Electronic Engineers in Israel, 2000, pp. 179–182, 2000.

[OnStar] General Motors OnStar website: http://www.onstar.com.

[Opp04] I. Oppermann, M. Hamalainen, and J. Iinatti, UWB Theory and Applications, John Wiley and Sons, 2004.

[Pah00] K. Pahlavan, P. Krishnamurthy, et al., "Handoff in hybrid mobile data networks", *IEEE Personal Communications Magazine*, April 2000.

[Pah02] K. Pahlavan, X. Li, and J. Makela, "Indoor Geolocation Science and Technology", *IEEE Comm. Mag.*, Feb. 2002.

[Pah05] K. Pahlavan and A. Levesque, Wireless Information Networks, 2nd edn, John Wiley and Sons, 2005.

[Pah06] K. Pahlavan, F. Akgul, M. Heidari, A. Hatami, J. Elwell, and R. Tingley, "Indoor Geolocation in the Absence of Direct Path", in *IEEE Wireless Communications Magazine*, 2006.

[Pah09] K. Pahlavan and P. Krishnamurthy, *Networking Fundamentals, Wide, Local and Personal Area Communications*, John Wiley and Sons, 2009.

[Pah10] K. Pahlavan, F. Akgul, Y. Ye, T. Morgan, F. A. Shabdiz, M. Heidari, and C. Steger, "Taking Positioning Indoors: Wi-Fi Localization and GNSS", InsideGNSS, vol. 5, no. 3, May, 2010.

[Pah12a] K. Pahlavan, Y. Ye, R. Fu, and U. Khan, "Challenges in Channel Measurement and Modeling for RF Localization Inside the Human Body," Invited paper, *International Journal on Embedded and Real-Time Communication Systems* (IJERTCS), 3(3), 18–37, July–September, 2012.

[Pah12b] K. Pahlavan, G. Bao, Y. Ye, S. Makarov, U. Khan, P. Swar, D. Cave, A. Karellas, P. Krishnamurthy, and K. Sayrafian, "RF Localization for Wireless Capsule Endoscopy", invited paper, *special issue on localization, International Journal of Wireless Information Networks*, Springer, on line, October 14, 2012.

[Pah85] K. Pahlavan, "Wireless Communications for Office Information Networks," *IEEE Commun. Mag.*, **23**, No. 6, 19–27 (1985).

[Pah88a] K. Pahlavan, "Wireless Intra-Office Networks," *ACM Trans. Office Inf. Syst.*, 6, 277–302 (1988).

[Pah88b] K. Pahlavan and J.L. Holsinger, "Voice-Band Data Communication Modems: a Historical Review, 1919–1988," IEEE Communi Mag., 26(1) 16–27 (1988).

[Pah90] K. Pahlavan and M. Chase, "Spread Spectrum Multiple Access Performance of Orthogonal Codes for Indoor Radio Communications," IEEE Trans on Communi, COM-38, 574–577 (1990)

[Pah94] K. Pahlavan and A.H. Levesque, "Wireless data communications", Proceedings of the IEEE, Vol. 82, No. 9, pp. 1398–1430, Sept. 1994.

[Pah97] K. Pahlavan, A. Zahedi, and P. Krishnamurthy, "Wideband local access: wireless LAN and wireless ATM", *IEEE Communications Magazine*, Vol. 35, No. 11, pp. 34–40, Nov. 1997.

[Pah98] K. Pahlavan, P. Krishnamurthy, and J. Beneat, "Wideband radio propagation modeling for indoor geolocation applications", IEEE Comm. Magazine, pp. 60–65, April 1998.

[Pal13] PalTrack Tracking Systems, http://www.sovtechcorp.com/.

[Pat03] W. Pattara-atikom, P. Krishnamurthy, and S. Banerjee, "Distributed Mechanisms for Quality of Service in Wireless LANs", IEEE Wireless Communications: Special issue on "QoS in Next-generation Wireless Multimedia Communications Systems", Vol. 10, No. 3, pp. 26–34, June 2003.

[Ped00] K.I. Pedersen, J.B. Andersen, J.P. Kermoal, and P. Morgensen, "A stochastic multiple-input-multiple-output radio channel model for evaluation of space-time coding algorithms", 52nd IEEE Vehicular Technology Conference, 2000, pp. 893–897, 2000.

[Per08] E. Perahia and R. Stacey, *Next Generation Wireless LANs*, Cambridge University Press, 2008.

[Per10] E. Perahia, C. Cordeiro, M. Park, and L.L. Yang, "IEEE 802.11ad: Defining the Next Generation Multi-Gbps Wi-Fi", *7th IEEE Consumer Communications and Networking Conference*, Las Vegas, Nevada, USA, 9–12 January 2010.

[Per97] C.E. Perkins, Mobile IP: Design Principles and Practices, Addison Wesley Communications Series, 1997.

[Pol96] G.P. Pollini, "Trends in Handover Design", IEEE Communications Magazine, March 1996.

[Pol13] Polaris Wireless, http://www.polariswireless.com.

[Por01] D. Porcino, "Performance of a OTDOA-IPDL positioning receiver for 3gpp-fdd mode," in Second International Conference on 3G Mobile Communication Technologies, pp. 221–225, March 2001.

[Pra92] G.J.M. Janssen and R. Prasad, "Propagation measurements in an indoor radio environment at 2.4 GHz, 4.75 GHz, and 11.5 GHz", *Proc. of the 42nd IEEE VTC*, pp. 617–620, 1992.

[Pro08] J.G. Proakis and M. Salehi, Digital Communications, 5th edn, McGraw–Hill, 2008.

[Rao09] A.M. Rao, A. Weber, S. Gollamudi, and R. Soni, "LTE and HSPA+: Revolutionary and Evolutionary Solutions for Global Mobile Broadband," *Bell Labs Technical Journal*, Vol. 13, No. 4, pp. 7–34, 2009.

[Rap02] T.S. Rappaport, *Wireless Communications: Principles and Practice*, 2nd edn, Prentice Hall, New Jersey, 2002.

[Red95] S. Redl, M.K. Weber, M. Oliphant, and W. Mohr, An Introduction to GSM, The Artech House Mobile Communications Series, Artech House, 1995.

[Ree98] J.H. Reed, K.J. Krizman, B.D. Woerner, and T.S. Rappaport, "An overview of the challenges and progress in meeting the e-911 requirement for location service," IEEE Communications Magazine, 36(4):30–37, April 1998.

[Rez95] R. Rezaiifar, A.M. Makowski, and S. Kumar, "Optimal control of handoffs in wireless networks", Proc. IEEE VTC'95, pp. 887–891, 1995.

[RFC96] IETF RFCs, "IP Mobility Support," available at http://www.ietf.org/rfc/rfc2002.txt and "Mobile IP Network Access Identifier Extension for IPv4," available at http://www.ietf.org/rfc/rfc2794.txt.

[Roo02a] T. Roos, P. Myllymaki, H. Tirri, P. Miskangas, and J. Sievanen, "A Probabilistic Approach to WLAN User Location Estimation," *International Journal of Wireless Information Networks*, Vol. 9, No. 3, July 2002.

[Roo02b] T. Roos, P. Myllymaki, and H. Tirri, *"A Statistical Modeling Approach to Location Estimation"*, IEEE Transactions on mobile computing, Vol. 1, No. 1, Jan.–Mar. 2002.

[Ros73] G.F. Ross, Transmission and Reception System for Generating and Receiving Baseband Duration Pulse Signals without Distortion for Short Baseband Communication Systems, US Patent 3,728,632, April 1973.

[Rot08] V. Roth, W. Polak, E. Rieffel, and T. Turner, "Simple and effective defense against evil twin access points," *ACM WiSec*, 2008.

[Ryc04] J. Ryckaert, P. De Doncker, R. Meys, A. de Le Hoye, and S. Donnay, "Channel model for wireless communication around the human body", Electronics Letters, Vol. 40, No. 9, pp. 543–544, April 2004.

[Sal87] A.M. Saleh and R.A. Valenzuela, "A Statistical Model for Indoor Multipath Propagation," *IEEE J. Selected Areas Commun.*, **SAC-5**, 128–137 (1987).

[Sav01a] C. Savarese, J.M. Rabaey, and J. Beutel, "Locationing in Distributed Wireless Ad-Hoc Sensor Networks", *Proceedings of the ICASSP*, May 2001.

[Sav01b] A. Savvides, C-C. Han, and M.B. Srivastava, "Dynamic fine-grained localization in Ad-Hoc networks of sensors," Proc. Mobicom, pp. 166–179, 2001.

[Sav05] A. Savvides, W. Garber, R. Moses, and M. Srivastava, "An analysis of error inducing parameters in multihop Sensor node Localization", *IEEE Transactions on mobile computing,* November 2005.

[Saw08] H. Sawada, T. Aoyagi, J-i. Takada, K.Y. Yazdandoost, and R. Kohno, "Channel model between body surface and wireless access point for UWB band," IEEE 802.15-08-0576-00-0006, August 2008.

[Say10] K. Sayrafian-Pour, W.B. Yang, J. Hagedorn, J. Terrill, and K.Y. Yazdandoost, "Channel Models for Medical Implant Communication", *Special issue on BAN, Int. Journal of Wireless Information Networks*, Vol. 17, No. 3/4, pp. 105–112, Springer, 2010.

[Sex89] T. Sexton and K. Pahlavan, "Channel modeling and adaptive equalization of indoor radio channels", IEEE JSAC, Vol. 7, pp. 114–121, 1989.

[Sha04] S.S. Shankar et al., "Optimal Packing of VoIP Calls in an IEEE 802.11a/e WLAN in the Presence of QoS Constraints and Channel Errors," Proc. IEEE Globecom, pp. 2974–2980, 2004.

[Sha48] C.E. Shannon, "A Mathematical Theory of Communication," *Bell Syst. Tech. J.*, **27**, 379–423, 623–656 (1948). [Reprinted in book form with postscript by W. Weaver, University of Illinois Press, Urbana, IL (1949).]

[Sho03] G. Shor, TG3a-Wisair-CFP-Presentation, DS-UWB Proposal, IEEE P802.15 Working Group for Wireless Personal Area Networks (WPANs), Document number: IEEE 802.15-03/151r3, May 03.

[Sie00] T. Siep, I. Gifford, R. Braley, and R. Heile, "Paving the Way for Personal Area Network Standards: An Over View of the IEEE P802.15 Working Group for Wireless Personal Area Networks", IEEE Personal Communications, Feb. 2000.

[Sil00] R.D. Silverman, "A Cost-Based Security Analysis of Symmetric and Asymmetric Key Lengths", *RSA Bulletin Number 13,* April 2000.

[Sim85] M.K. Simon et al., *Spread Spectrum Communication*, Computer Science Press, 1985.

[Siw04] K. Siwiak and D. McKeown, Ultra-Wideband Radio Technology, John Wiley & Sons, 2004.

[Sko08] D. Skordoulis et al., "IEEE 802.11n MAC Frame Aggregation Mechanisms for Next Generation High Throughput WLANs," IEEE Wireless Communications, pp. 40–47, February 2008.

[Sol01] S. Soliman, "Report of Qualcomm Incorporated," In the matter of revision of Part 15 of the Commission's Rules Regarding Ultra-Wideband Transmissions Systems, ET Docket No. 98-153, March 5, 2001.

[Son94] H-L. Song, "Automatic Vehicle Location in Cellular Communications Systems", IEEE Trans. Vehicular Technology, vol. 43, No. 4, pp. 902–908, Nov. 1994.

[Soo04] P. Soontornpipit, C.M. Furse, and Y.C. Chung, "Design of implantable microstrip antenna for communication with medical implants," *IEEE Transactions on Microwave Theoryand Techniques*, Vol. 52, no. 8, part 2, pp. 1944–1951, 2004.

[Spe00] Q.H. Spencer, B.D. Jeffs, M.A. Jensen, and A.L. Swindlehurst, "Modeling the Statistical Time and Angle of Arrival characteristics of an Indoor Multipath Channel", IEEE JSAC, Vol. 18, No. 3, pp. 347–360, March 2000.

[Sta00] W. Stallings, Local and Metropolitan Area Networks, Sixth Edition, Prentice Hall, New Jersey, 2000.

[Sta03] W. Stallings, Network Security Essentials, Second Edition, Prentice Hall, 2003.

[Sta98] W. Stallings, Cryptography and Network Security, Prentice Hall, 1998.

[Sti02] D. Stinson, Cryptography: Theory and Practice, CRC Press, 2002.

[Swa08] N. Swangmuang and P. Krishnamurthy, "Location Fingerprint Analyses Toward Efficient Indoor Positioning," Proc. Percom, March 2008.

[Swa12] P. Swar, K. Pahlavan, and U. Khan, "Accuracy of Localization System inside Human Body using a Fast FDTD simulation Technique" 6th IEEE International Symposium on Medical Information and Communication Technology, La Jolla, CA, March, 2012.

[Tak85] H. Takagi and L. Kleinrock, "Throughput analysis for persistent CSMA systems", IEEE Trans. Comm., Vol. 33, pp. 627–638, 1985.

[Tan10] A.S. Tanenbaum and D.J. Wetherall, Computer Networks, fifth edition, Prentice Hall, 2010.

[Tar98] V. Tarokh, N. Seshadri, and A.R. Calderbank, "Space-Time Codes for High Data Rate Wireless Communications: Performance Criterion and Code Construction," IEEE Trans. Info. Theory, Mar. 1998, pp. 744–765.

[Tay01] J.D. Taylor (ed.), Ultra-Wideband Radar Technology, CRC Press, 2001.

[TDC13] Time Domain Corporation website (http://www.timedomain.com).

[Tek91] S. Tekinay and B. Jabbari, "Handover policies and channel assignment strategies in mobile cellular networks", IEEE Communications Magazine, Vol. 29, No. 11, 1991.

[Tek98] S. Tekinay, E. Chao, and R.E. Richton, "Performance benchmarking for wireless location systems," IEEE Communications Magazine, 36(4):72–76, April 1998.

[Tew09] E. Tews and M. Beck, "Practical attacks against WEP and WPA," ACM WiSec, 2009.

[Tin01] R. Tingley and K. Pahlavan, "Time–space measurement of indoor radio propagation," IEEE Trans. Instrumentation and Measurements, Vol. 50, No. 1, February 2001, pp. 22–31.

[Tob75] F.A. Tobagi and L. Kleinrock, "Packet Switching in Radio Channels, Part II: The Hidden-Terminal Problem in Carrier Sense Multiple Access and the Busy-Tone Solution," IEEE Trans. Commun., COM-23, 1417–1433 (1975).

[Tob80] F.A. Tobagi, "Multi-access protocols in packet communication systems", IEEE Trans. Comm., Vol. 28, pp. 468–488, 1980.

[TR-45-02] TR-45, Enhanced Wireless 911 Phase 2, TIA/EIA-J-STD-036-A, Revision A, March 2002.

[Tre04] E. Trevisani and A. Vitaletti, "Cell-ID Location Technique, Limits, and Benefits: An Experimental Study," Proc. 6th IEEE Workshop on Mobile Computing Systems and Applications, WMCSA'04, 2004.

[Tri97] N. Tripathi, "Generic Adaptive Handoff Algorithms using Fuzzy Logic and Neural Networks", Ph.D Thesis, Virginia Polytechnic Institute and State University, August 1997.

[Tri98] N.D. Tripathi, J.H. Reed, H.F. VanLandingham, "Handoff in Cellular Systems", IEEE Personal Communications Magazine, December 1998.

[Tuc91] B. Tuch, "An ISM Band Spread Spectrum Local Area Network: WaveLAN," Proc. IEEE Workshop on Wireless LANs, Worcester, MA, 103–111 (May 1991).

[Unb03] M. Unbehaun and M. Kamenetsky, "On the Deployment of Picocellular Wireless Infrastructure", IEEE Wireless Communications Magazine, pp. 70–80, Dec. 2003.

[Van68] H.L.V. Trees, Detection, Estimation, and Modulation Theory. New York: Wiley, 1968.

[Vas05] D. Vassis, G. Kormentzas, A. Rouskas, and I. Maglogiannis, "The IEEE 802.11g standard for high data rate WLANs", IEEE Network, pp. 21–26, May/June 2005.

[Wan05] W. Wang, S.C. Liew, and V.O.K. Li, "Solutions to Performance Problems in VoIP Over a 802.11 Wireless LAN," IEEE Trans. Vehicular. Tech., Vol. 54, No. 1, January 2005.

[War03] J. Warrior, E. McHenry, and K. McGee, "They know where you are," IEEE Spectrum, 40(7):20–25, July 2003.

[War97] A. Ward, A. Jones, and A. Hopper, "A New Location Technique for the Active Office", IEEE Personal Communications, October 1997.

[Wel03] M. Welborn, M. Mc Laughlin, P. Ceva, and R. Kohno, DS-UWB Proposal, IEEE P802.15 Working Group for Wireless Personal Area Networks (WPANs), Document number: IEEE 802.15-03/334r3, September 2003.

[Wer98] J. Werb and C. Lanzl, "Designing a positioning system for finding things and people indoors," *IEEE Spectrum*, Vol. 35, No. 9, September 1998, pp. 71–78.

[Wig10] WiGig White Paper: Defining the Future of Multi-Gigabit Wireless Communications, WiGig Alliance, July 2010.

[Wil95a] J.E. Wilkes, "Privacy and Authentication Needs of PCS", *IEEE Personal Communications*, August 1995.

[Wil95b] T.A. Wilkinson, T. Phipps, and S.K. Barton, "A Report on HIPERLAN Standardization," International Journal on Wireless Information Networks, Vol. 2, pp. 99–120, March 1995.

[Win00] M.Z. Win and R.A. Scholtz, "Ultra-wide bandwidth time-hopping spread spectrum impulse radio for wireless multiple-access communications, IEEE Transaction on Communications, Vol. 48, No. 4, pp. 679–691, April 2000.

[Win98] M. Win and R. Scholtz, "On the performance of ultra-wide bandwidth signals in dense multipath environment," *IEEE Comm. Letters*, Vol. 2, No. 2, Feb. 1998, pp. 51–53.

[Woe98] H. Woesner, J-P. Ebert, M. Schlager, and A. Wolisz, "Power saving mechanisms in emerging standards for wireless LANs: A MAC level perspective", IEEE Personal Communications Magazine, pp. 40–48, June 1998.

[Wol82] J.K. Wolf, A.M. Michelson, and A.H. Levesque, "On the probability of undetected error for linear block codes", IEEE Trans. Comm., Vol. 30, pp. 317–324, 1982.

[Won00] V.W.S. Wong and V.C.M. Leung, "Location management for next generation personal communication networks", IEEE Network Magazine, pp. 18–24, September/October 2000.

[Wor08] The First International Workshop on Opportunistic RF Localization for Next Generation Wireless Devices, WPI, Worcester, MA, June 16–17 http://www.cwins.wpi.edu/workshop08/index.html.

[Wor12] The Third Invitational Workshop on Opportunistic RF Localization for Next Generation Wireless Devices, New Orleans, LA, May 7, 2012 http://www.cwins.wpi.edu/workshop12/index.html#.

[Wor91] First IEEE Workshop on WLANS, Worcester, MA, 1991.

[Xia05] Y. Xiao, "IEEE 802.11n: Enhancements for Higher Throughputs in Wireless LANs," IEEE Wireless Communications, December 2005.

[Yal02] R. Yallapragada, V. Kripalani, and A. Kripalani, EDGE: a technology assessment, IEEE International Conference on Personal Wireless Communications, Dec. 2002.

[Yan06] G.-Z. Yang and M. Yacoub, Body sensor networks, Springer-Verlag, London, 2006.

[Yan94] G. Yang and K. Pahlavan, "Sectored Antenna and DFE Modem for High Speed Indoor Radio Communications," IEEE Trans. Vehic. Tech., Nov. 1994.

[Yaz07a] K.Y. Yazdandoost and R. Kohno, "An antenna for medical implant communications system", European Microwave Conference, pp. 968–971, Oct. 2007.

[Yaz07b] K.Y. Yazdandoost and R. Kohno, "The Effect of Human Body on UWB BAN Antennas," IEEE802.15-07-0546-00-0ban.

[Yaz10] K.Y. Yazdandoost and K. Sayrafian-Pour, "Channel Model for Body Area Network (BAN)", IEEE P802.15-08-0780-12-0006, November 2010.

[Ye12] Y. Ye, P. Swar, and K. Pahlavan "Accuracy of RSS-Based RF Localization in Multi-Capsule Endoscopy," invited paper, *Wireless Health special issue based on the IEEE PIMRC'11 best papers, International Journal of Wireless Information Networks*, Vol. 19, No 3, Springer, August 2012, pp. 229–238.

[Zah97] A. Zahedi and K. Pahlavan, "Terminal Distribution and the Impacts of Natural Hidden Terminal," *Electronic Letters*, Vol. 33, No. 9, pp. 750–751, April 1997.

[Zha89] M. Zhang and T-S.P. Yum, "Comparisons of Channel Assignment Strategies in Cellular Mobile Telephone Systems", IEEE Transactions on Vehicular Technology, Vol. 38, No. 4, pp. 211–215, November 1989.

[Zha90] K. Zhang and K. Pahlavan, "An integrated voice/data system for mobile indoor radio networks", IEEE Trans. Vehicular Technology, Vol. 39, pp. 75–82, 1990.

[Zha91] M. Zhang and T-S.P. Yum, "The Non-uniform Compact Pattern Allocation Algorithm for Cellular Mobile Systems", IEEE Transactions on Vehicular Technology, Vol. 40, No. 2, pp. 387–391, May 1991.

[Zha92] K. Zhang and K. Pahlavan, "Relation between transmission and throughput of slotted ALOHA local packet radio networks", IEEE Trans. Comm., Vol. 40, pp. 577–583, 1992.

[Zor97] M. Zorzi and R. Rao, "Energy-Constrained Error Control for Wireless Channels", IEEE Personal Communications Magazine, pp. 27–33, December 1997.

[Zor99] M. Zorzi and R. Rao, "Is TCP Energy Efficient?", Proc. of MoMuC'99, pp. 198–201, 1999.

Index

Principles of Wireless Access and Localization, First Edition. Kaveh Pahlavan and Prashant Krishnamurthy.
© 2013 John Wiley & Sons, Ltd. Published 2013 by John Wiley & Sons, Ltd.